STRUCTURE REPORTS
for 1970
Vol. 35 A

Structure Reports is prepared under the guidance of a Commission of the International Union of Crystallography. The members of the Commission sometime concerned with the preparation of this volume are listed below.

STRUCTURE REPORTS

FOR 1970

Volume 35 A

GENERAL EDITOR

W. B. PEARSON

SECTION EDITORS

L. D. CALVERT (Metals)
I. D. BROWN (Inorganic Compounds)
C. CALVO (Inorganic Compounds)

SPRINGER FACHMEDIEN WIESBADEN GMBH

First published in 1975

ISBN 978-94-017-3111-9 ISBN 978-94-017-3109-6 (eBook)
DOI 10.1007/978-94-017-3109-6

Koninklijke Drukkerij Van de Garde B.V., Zaltbommel

TABLE OF CONTENTS

SYMBOLS

The letters *a*, *b*, *c*; α, β, γ are used consistently for the edges and angles of the unit cell. Other letters used consistently are as follows.

U	Volume of unit cell
D_m	Measured density in g/cm³ or specific gravity
D_x	Density in g/cm³ calculated from cell volume and contents
Z	Number of times the formula quoted is repeated in the unit cell (Number of atoms per unit cell in alloys of simple structure)
x, y, z	Atomic coordinates as fractions of cell edge (Occasionally u, v, w or other letters are used)
X, Y, Z	Atomic coordinates in Ångström units
X', Y', Z'	Atomic coordinates in Ångström units, referred to orthogonal axes (Used only in the Organic Section)
F.W.	Formula weight
A, B, C	Types of layer in layer structures
M, A, B	Variable metal atom(s) in a sequence of related structures
X, H	Variable non-metals, usually halogen, in a sequence of related structures
R	Variable organic radical, or reliability index
s, m, w, v b	Strong, medium, weak, very, broad
TSNR	Taylor-Sinclair-Nelson-Riley (extrapolation function)

LIMITS OF ERROR

Errors are generally quoted in units in the last place. Thus 4·8754 ± 3 means 4·8754 ± 0·0003, 4·87 ± 3 means 4·87 ± 0·03, and 4·875 ± 15 means 4·875 ± 0·015. Occasionally a very doubtful last digit is given below line.

TRANSLITERATION OF RUSSIAN

а	a	и	i	р	r	ш	š
б	b	й	j	с	s	щ	šč
в	v	к	k	т	t	ы	y
г	g	л	l	у	u	ъ	”
д	d	м	m	ф	f	ь	’
е	e	н	n	х	kh	э	ė
ж	ž	о	o	ц	c	ю	ju
з	z	п	p	ч	č	я	ja

INTRODUCTION

Starting with Volume **30**, 1965, *Structure Reports* is produced in a new format by photo-offset printing from typed manuscript with unjustified lines. At the time when the decision for this change was taken, the cost of setting the manuscript in type was becoming so high as to render the cost of individual subscription prohibitive. At that time automatic typing methods giving justified lines, etc. for photo-offset reproduction did not offer any saving over type setting, but hand typing of the manuscripts could give a considerable saving in production costs. In the belief that a publication that is too expensive to buy is of little value, the format has been changed sacrificing elegance to availability.

The new format does not lead to increased length of the volumes since the information content of the typed and typeset pages is practically identical. However, the amount of work to be reported demands the eventual separation of *Structure Reports* into two volumes, A. *Metals and Inorganic* and B. *Organic*. It was convenient to introduce this change also at Volume **30**, and with Volume **31** further to restrict the publication of crystal data, so that from 1966 onwards the reports deal almost entirely with complete structure determinations only.

In the past the aim of *Structure Reports* has been to present critical reports on all work of crystallographic structural interest, whether it is derived directly from X-ray, electron or neutron diffraction, or even indirectly from other experiments. The reports were intended to be critical and not mere abstracts, except in some cases when a brief indication of the content of a paper of related interest was included in the form of an abstract. In selecting topics for reporting, the criterion "of structural interest" was freely interpreted in terms of what was topically interesting. However, the amount of literature covering matters of structural interest has become so large that this policy can no longer be followed. From Volume **28** onwards, critical reports are only given on actual structure determinations. That is to say reports are only written on papers recording the determination of the positions of the atoms in structures. Nevertheless, Volumes **28** to **30** still do record much information on lattice parameters and space groups of the structures of alloys and compounds, although in the form of tables. From Volume **31** onwards, even this information is generally omitted and only full structure determinations are reported. One of the reasons for this omission is that such data are obtainable from the continuing volumes of *Crystal Data*.† Only in this way is it possible to keep yearly volumes of *Structure Reports* to a fairly uniform and usable size in view of the ever increasing amount of published material.

Ideally, the reports have been prepared in such a way that no further structural information would be gained by consulting the original paper itself, although from Volume **21** onwards, atomic parameters are not generally reproduced for structures containing more than about 30 independent atoms. The main reason for this is

† ***Crystal Data*** **Third Edition, Ed. J. D. H. DONNAY and H. M. ONDIK; U.S. Department of Commerce and JCPDS.**

that the chance of including typographical errors in reproducing extensive tables of data is such, that anybody wishing to make detailed use of them would in any case consult the primary references.

Although the data in *Structure Reports* must be presented as briefly as possible, every effort is made to avoid jargon, so that the information is readily understandable by the non-crystallographer as well as the crystallographer. Nevertheless it is assumed that the reader has available Volume I of the *International Tables**, wherein he can obtain details of space group settings and equivalent positions of the atoms.

The arrangements in individual reports is generally: Name, Formula, Papers reported, Unit cell, Space group, Atomic positions, Interatomic and intermolecular distances, Material, Discussion, Details of analysis and References. Editorial comments are enclosed in square brackets, and it may be assumed that material not distinguished in this manner is based directly on the papers reported. The volumes are divided into three main sections: *Metals, Inorganic Compounds* and *Organic Compounds.* In the earlier volumes of *Structure Reports* the arrangement of the *Metals* section was strictly alphabetical, with crossreferences given in the text, so that the preparation and use of an index to find data on alloys was unnecessary. In order to save space this practice was discontinued in Volume **18** and subsequently. Although the arrangement of the *Metals* section remains roughly alphabetical, *it is now essential to use the Subject and Formula indexes when seeking work on elemental metals, intermetallic compounds and alloys.* The arrangement of the *Inorganic* and *Organic* sections is roughly in the order of increasing complexity of composition, related substances and related structures being kept together as far as possible. Inorganic and organic compounds should therefore also be sought in the Subject or Formula indexes. The Subject index is arranged alphabetically by the names printed as the headings of the reports, and it also includes other common names and information. The Formula index in the A volumes, *Metals and Inorganic Compounds*, is arranged in alphabetical order of the chemical symbols. In the B volumes, *Organic Compounds*, the classification is by the number of carbon atoms and secondary classification by the number of hydrogen atoms; other constituents then follow alphabetically. In the formula indexes solvents of crystallization follow at the end of the formulae, but double entries are frequently made, especially for inorganic compounds containing water, possibly or certainly as OH groups.

The scheme generally employed for the transliteration of Russian is reproduced on p. VI, and the usual abbreviations of journal titles are listed in earlier volumes. Transliteration is in accordance with draft recommendation no. 6 of the International Organization for Standardization, and the abbreviations are based on the World List of Scientific Periodicals.

W. B. Pearson

University of Waterloo
Waterloo, Ont., Canada.
25 September, 1974

* *International Tables for X-ray Crystallography*, Vol. I, *Symmetry Groups.* International Union of Crystallography, Kynoch Press, Birmingham, 1952.

STRUCTURE REPORTS

SECTION 1

METALS

EDITED BY
L. D. CALVERT

WITH THE ASSISTANCE OF THE FOLLOWING REPORTER

J. K. BYRON

ARRANGEMENT

In contrast to previous volumes of *Structure Reports,* the arrangement of the Metals section in Volume 19 and subsequent volumes is no longer strictly alphabetical with cross references given under subject headings. *It is therefore essential now to make use of the Subject and/or Formula Indexes when seeking work on metals and alloy systems.* Apart from this important difference, the arrangement of the Metals section remains generally similar to that of earlier volumes.

The names and spellings aluminum, beryllium, caesium, niobium, sulphur and wolfram should be noted.

ALUMINUM BORON COPPER

I. Darstellung und Kristallstruktur von $Cu_2Al_{2.7}B_{104}$. R. MATTES, L. MAROSI and H. NEIDHARD, 1970. *J. Less-Common Metals*, 20, 223-228.

$Al_{2.7}Cu_2B_{104}$ see also BCr this volume pp. 34-36.

Rhombohedral, β-B structure type (1,2), a = 10.21 ± 1 Å, α = 65°9' ± 7', *hexagonal* axes, a = 10.99, c = 23.98 Å, c/a = 2.18, U = 2508 Å³, D_m = 2.64.

Space group $R\bar{3}m$ (D^5_{3d})

Atomic positions (hexagonal axes)

			x	y	z	B (Å²)
3B(1)	in	3(b)	0	0	1/2	0.72±280
6B(2)	in	6(c)	0	0	0.38524±50	0.182±156
18B(3)	in	18(h)	0.05532±45	-0.05532±45	-0.05543±29	0.360±99
18B(4)	in	18(h)	0.08668±44	-0.08668±44	0.01360±27	0.253±94
18B(5)	in	18(h)	0.11079±41	-0.11079±41	-0.11243±27	0.207±86
18B(6)	in	18(h)	0.16982±45	-0.16982±45	0.02752±26	0.215±90
18B(7)	in	18(h)	0.12923±46	-0.12923±46	-0.23387±27	0.322±89
18B(8)	in	18(h)	0.10218±45	-0.10218±45	-0.30226±27	0.266±90
18B(9)	in	18(h)	0.05649±44	-0.05649±44	0.32631±28	0.276±96
18B(10)	in	18(h)	0.09027±44	-0.09027±44	0.39794±27	0.230±90
18B(11)	in	18(h)	0.05631±61	-0.05631±61	-0.44373±38	1.244±147
36B(12)	in	36(i)	0.17687±66	0.17654±64	0.17737±20	0.364±68
36B(13)	in	36(i)	0.31956±58	0.29566±58	0.12879±18	0.204±59
36B(14)	in	36(i)	0.26279±64	0.21788±60	0.41882±18	0.248±66
36B(15)	in	36(i)	0.23589±64	0.25194±65	0.34697±21	0.355±70
0.42x6A*	in	6(c)	0 ±	0	0.13404±15	0.367±84
0.25x18D†	in	18(h)	0.46298±29	-0.46298±29	0.15340±19	1.255±109

* Occupancy Cu atoms. † Occupancy Cu and Al atoms.

Interatomic distances (Å)

A - 3B(3)	2.159	D - 2B(11)	2.379
- 3B(5)	2.172	- 2B(12)	2.404
- 6B(12)	2.202	- 2B(13)	2.385
D - 1B(1)	2.488	- 2B(13)	2.489
- 1B(10)	2.257	- 2B(14)	2.380
- 1B(11)	2.175	- 2B(14)	2.456

[B-B distances range from 1.72 to 1.88 Å].

Discussion

The boron framework is identical to that of β-B; the copper and aluminum atoms partially occupy the sites A and D and the remaining Al atoms partly occupy some of the B sites. (Fig. 1).

Material and Details of analysis

Cu, Al and B were pressed into a tablet and annealed at 1300-1400°C in an Al_2O_3 crucible inside an Al_2O_3 tube with an atmosphere of argon. Intensities were obtained photographically and also with a diffractometer (651 hkl, MoK_α). A trial structure based on β-B (1,2) was refined by Fourier and difference syntheses and then by least squares.

1. R.E. HUGHES, C.H.L. KENNARD, D.B. SULLENGER, H.A. WEAKLEIM, D.E. SANDS and J.L. HOARD, 1963. *J. Amer. Chem. Soc.*, 85, 361.
2. J.L. HOARD and R.E. HUGHES, 1967. *"The Chemistry of Boron and its Compounds"*, ed. E.L. Muetterties, Wiley, New York.

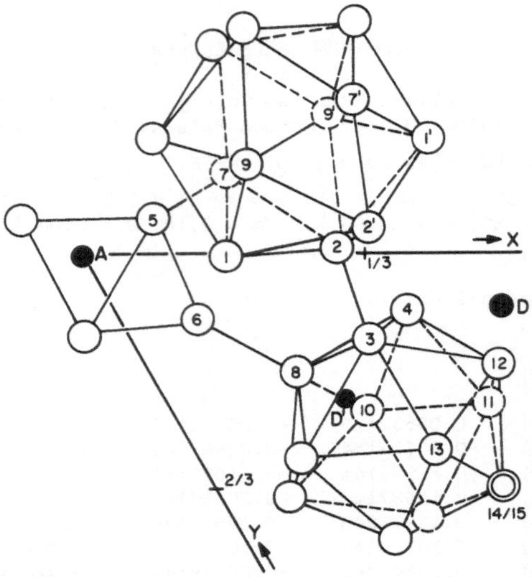

Fig. 1. A portion of the $Al_{2.7}Cu_2B_{104}$ structure. The atoms are numbered as in the coordinate table and *z* coordinates can be derived from this table. [After I.]

ALUMINUM BORON MAGNESIUM

I. Structure of $MgAlB_{14}$ and a brief critique of structural relationships in higher borides. V.I. MATKOVICH and J. ECONOMY, 1970. *Acta Cryst.*, B 26, 616-621.

$AlB_{14}Mg$ approximately

See also AlBC, p. 6.

Orthorhombic, *a* = 5.848, *b* = 10.313, *c* = 8.115 Å, [*U* = 489.42 Å^3], D_m = 2.60-2.68.

Space group *Imma* (D_{2h}^{28}) assumed from structure analysis

Atomic positions [transformed from original setting, given as *Imam**]

				x	*y*	*z*	*B* (Å^2)
∿ 3Al	in	4(*d*)	: with	0.2500	0.2500	0.7500	0.172±95
∿ 1Al ∿ 2Mg	in in	} 4(*e*)	: with	0.0000	0.2500	0.3590±90	1.205±95
16B(1)	in	16(*j*)	: with	0.1580±10	0.0630±7	0.1640±90	0.070±95
8B(2)	in	8(*h*)	: with	0.000	0.0860±7	-0.1730±90	0.103±95
16B(3)	in	16(*j*)	: with	0.2500±10	0.0800±7	-0.0450±90	0.070±95
8B(4)	in	8(*h*)	: with	0.0000	0.1670±7	0.0290±90	0.045±95
8B(5)	in	8(*h*)	: with	0.0000	0.1520±7	0.6220±90	0.063±95

* [If the original setting is taken literally a B(4) - B(4) distance of 0.47 Å results via a centre of symmetry at the origin].

Interatomic distances (Å)

Intra-icosahedral

B(1) -	B(3)	1.791±15[1.788]
	B(3)	1.837±17[1.843]
	B(1)	1.853±12[1.848]
B(2) -	B(1)	1.787±13[1.794]
	B(3)	1.798±15[1.794]
B(3) -	B(3)	1.793±16[1.804]
B(4) -	B(1)	1.788±17[1.790]
	B(3)	1.809±13[1.817]
	B(2)	1.835±15[1.839]

Extra-icosahedral

B(1) -	B(1)	1.752±15[1.762]
	Mg	2.666±17[2.660]
	Mg	2.787±7[2.785]
B(2) -	B(5)	1.787±16[1.7977]
	Al	2.324±8[2.321]
B(3) -	B(5)	1.755±17
	Al	2.422±8[2.417]
	Mg	2.736±17[2.737]
B(4) -	B(4)	1.751±14[1.712]
B(5) -	B(5)	2.040±14[2.021]
	Mg	2.363±16[2.361]
	Al	2.061±8[2.059]

Material and Details of analysis

$MgAlB_{14}$ was made by heating Mg, B and Al (1:2:14) at 900°C for six hours, then cooling and treating with HCl. A small black crystal was used in collecting 290 reflexions (Cu radiation, Ni/Co filters, stationary-counter stationary-crystal technique; Lorentz and polarization corrections). A Patterson synthesis was used to establish a partial trial structure which was refined and completed by Fourier and block-diagonal least-squares methods ($R = 0.12$).

Discussion

The structure (Fig. 2) can readily be derived from a simple modification of an hexagonal packing of icosahedra designated 4HPo (1, 2) stacked directly one above the other along the unique axis. The stacking of B_{12} icosahedra is discussed.

1. V.I. MATKOVICH, R. GIESE and J. ECONOMY, 1965. *Z. Kristallogr.*, 122, 116.
2. V.I. MATKOVICH and J. ECONOMY, 1968. *Ibid.*, 126, 182.

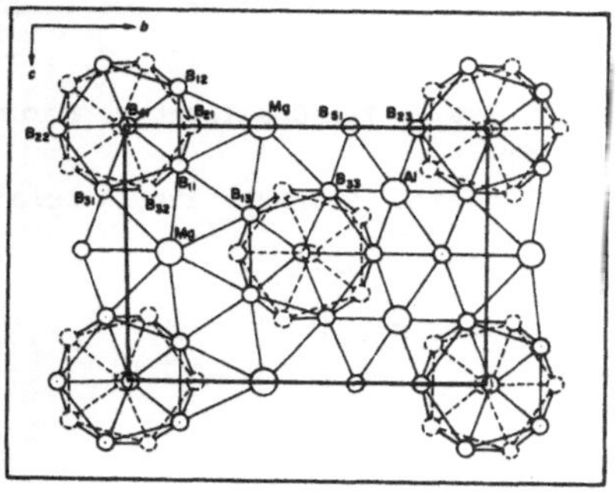

Fig. 2. A layer of icosahedra in $AlB_{14}Mg$ showing the extra-icosahedral atoms. The atoms shown lie between $x = \pm 1/4$. [Original setting].

ALUMINUM BORON CARBON

I. Zur Darstellung und Struktur eines aluminiumhaltigen Borcarbids.
H. NEIDHARD, R. MATTES and H.J. BECHER, 1970. *Acta Cryst.*, B 26, 315-317.

$\sim AlC_4B_{40}$

Rhombohedral, a = 5.253 Å, α = 64.9°, [U = 113.45 Å^3], D_m = 2.52, Z = 1, D_x = 2.48. *Hexagonal* cell, a = 5.642 ± *2*, c = 12.367 ± *4* Å, [U = 340.9 Å^3], Z = 3.

Space group $R\bar{3}m$ (D^5_{3d})

Atomic positions (hexagonal axes)

		x	*y*	*z*	*B* (Å^2)
18B (1)	in 18(*h*)	0.1715±7	-0.1715±7	0.0240±4	0.66±3
18B (2)	in 18(*h*)	0.1081±7	-0.1081±7	0.8877±4	0.66
1.82 B(*m*)	in 3(*b*)	0	0	0.5000	0.66
2.0 B } 4.0 C	in 6(*c*)	0	0	0.3808±8	0.66
0.84 Al	in 18(*h*)	0.4546±50	-0.4546±50	0.1542±23	0.66

Interatomic distances (Å)

B(1) - B(1)	* 1.78		Al - C(1)	1.79 [1.77]
- B(2)	* 1.80	* 1.83		2.02 [2.01]
B(2) - B(2)	* 1.83		- B(1)	1.95 [1.96]
C(1) - B(1)	1.61			2.16
- B(*m*)	1.47		- B(2)	1.92 [1.98]
				2.16 [2.20]

* Within one icosahedron.

Discussion

The structure is similar to those of AlC_4B_{24} and AlB_{10} (1, 2) and $C_8Al_{2.1}B_{51}$ (5), and based on that of B_4C (3)

Material and Details of analysis

The compound was prepared by melting B_4C with excess of Al (15 times) at 1550°C. Analysis gave $B_{13.34}C_{1.34}Al_{0.33}$; Al 5.2%, C 9.5%, boron content was obtained by difference. Guinier photographs were used to obtain lattice parameters. Intensities for 113 independent $h0\bar{h}\ell$ reflexions were obtained photometrically from single-crystal photographs. A trial structure based on $B_{12}C_3$ (3, 4) was refined by least squares (R = 11.7%); a difference Fourier showed no large peaks.

1. V.I. MATKOVICH, J. ECONOMY and R.F. GIESE, Jr., 1964. *J. Amer. Chem. Soc.*, 86, 2337.
2. G. WILL, 1969. *Acta Cryst.*, B 25, 1219; *Idem*, 1967. *Ibid.*, 23, 1071.
3. *Structure Reports*, 9, 154; 17, 315; 23, 270; 26, 67.
4. J.L. HOARD and R.E. HUGHES, 1967. *The Chemistry of Boron and its Compounds*. pp. 62, 88. New York, Wiley.
5. A.J. PERROTTA, W.D. TOWNES and J.A. POTENZA, 1969. *Acta Cryst.*, B 25, 1223.

ALUMINUM GOLD

I. Kristallstruktur von AuAl. K. FRANK and K. SCHUBERT, 1970. *J. Less-Common Metals*, 22, 349-354.

AlAu F.W. 223.95

Monoclinic distorted MnP (*B*31) type structure, *a* = 6.415 ± 1, *b* = 3.331 ± 1, *c* = 6.339 ± 1 Å, β = 93.04 ± 1°, [*U* = 135 Å^3, *Z* = 4, D_x = 11.00].

Space group $P2_1/m$ (C_{2h}^2)

Atomic positions

	x	*y*	*z*	*B* (Å^2)
2Au(1) in 2(*e*)	0.02_9	1/4	0.18_9	3.18
2Au(2) in 2(*e*)	0.50_0	1/4	0.31_3	3.18
2Al(1) in 2(*e*)	0.19_5	1/4	0.57_6	3.18
2Al(2) in 2(*e*)	0.69_5	1/4	0.96_4	3.18

[*Interatomic distances* (Å)]

Au(1)	- 2Au(1)	2.93	Al(1)	- 2Au(1)	2.70
	- 1Au(2)	3.08		- 1Au(1)	2.62
	- 2Al(1)	2.70		- 3Au(2)	2.64
	- 1Al(1)	2.62		- 2Al(1)	3.11
	- 1Al(2)	2.51	Al(2)	- 2Au(1)	2.65
	- 2Al(2)	2.65		- 1Au(1)	2.51
Au(2)	- 1Au(1)	3.08		- 2Au(2)	2.68
	- 2Au(2)	2.90		- 1Au(2)	2.60
	- 3Al(1)	2.64		- 2Al(2)	3.06
	- 2Al(2)	2.68			
	- 1Al(2)	2.60			

Material and Details of analysis

Guinier photographs for lattice parameters and intensities; *R* = 0.17 for observed *hkℓ*.

ALUMINUM COBALT ZIRCONIUM
ALUMINUM IRON ZIRCONIUM
ALUMINUM NICKEL ZIRCONIUM

I. [Crystal structure of the compounds Zr_6FeAl_2, Zr_6CoAl_2, and Zr_6NiAl_2.] P.I. KRIPJAKEVIČ, V.V. BURNAŠOVA and V.JA. MARKIV, 1970. *Dop. Akad. Nauk Ukr., RSR*, 32, 828-831.

Al_2CoZr_6 F.W. = 660.2

Hexagonal, β_1-K_2UF_6 structure type (1), *a* = 7.93, *c* = 3.31 Å, *c/a* = 0.417, [*U* = 180 Å^3, *Z* = 1, D_x = 6.08]. [Note the resemblance to Fe_2P]

Space group $P\bar{6}2m$ (D_{3h}^3)

Atomic positions

			x	y	z
Co	in	(*a*)	0	0	0
2Al	in	2(*d*)	1/3	2/3	1/2
3Zr(1)	in	3(*f*)	0.608	0	0
3Zr(2)	in	3(*g*)	0.247	0	1/2

Interatomic distances (Å)

Co	- 6Zr(2)	2.56	Zr(1)	- 4Al	2.98 [2.95]
	- 3Zr(1)	3.11		- 1Co	3.11
	- 2Co	3.31		- 4Zr(2)	3.20 [3.19]
Al	- 6Zr(1)	2.98 [2.95]		- 2Zr(2)	3.30 [3.31]
	- 3Zr(2)	3.07 [3.04]		- 2Zr(1)	3.31
	- 2Al	3.31	Zr(2)	- 2Co	2.56
				- 2Al	3.07
				- 4Zr(1)	3.20 [3.19]
				- 2Zr(1)	3.30 [3.31]
				- 2Zr(2)	3.31
				- 2Zr(2)	3.40 [3.39]

Details of analysis

Calculated powder intensities (Cr*K*, 57 mm. Debye-Scherrer) agreed with observed values.

1. *Structure Reports*, 11, 329.

ALUMINUM MAGNESIUM ZINC

I. An electron microscope investigation of the microstructure in an aluminum-zinc-magnesium alloy. J. GJØNNES and C.J. SIMENSEN, 1970. *Acta Met.*, 18, 881-890.

η'$MgZn_2$ F.W. = 155.05

Monoclinic, distorted Ni_2In ($B8_2$) structure (1), $a = c = 4.97$, $b = 5.54$ Å, $\beta = 120°$, [$U = 118.51$ Å^3], $Z = 2$, [$D_x = 4.34$]. This phase transforms to η-$MgZn_2$, *hexagonal*, $a = 5.21$, $c = 8.60$ Å.

Space group $P2_1/m$ (C_{2h}^2) from structure

Atomic positions [transformed from *c*-axis setting]

			x	y	z
2 Zn(1)	in	2(*a*)	0	0	0
2 Zn(2)	in	2(*e*)	0.26	1/4	0.74
2 Mg	in	2(*e*)	0.60	1/4	0.40

[*Interatomic distances* (Å)]

Mg	- 4 Zn(1)	2.97	Zn(1)	- 2 Zn(1)	2.77
	- 2 Zn(2)	3.02		- 2 Zn(2)	2.63
	- 2 Zn(2)	2.84		- 4 Mg	2.97
	- 1 Zn(2)	2.93	Zn(2)	- 2 Zn(1)	2.63
	- 2 Mg	3.26		- 2 Mg	2.84
				- 1 Mg	2.93
				- 2 Mg	3.02

Material and Details of analysis

This phase is found as an oriented precipitate in an Al(Zn,Mg) alloy (5.77 wt % Mg, 1.08 wt % Al) annealed at 485°C, quenched to 20°C and aged at 130-150°C. Single-crystal electron diffraction photographs of thin foils; observed and calculated intensities are in fair agreement.

1. *Strukturbericht*, 1, 84; *Structure Reports*, 9, 90.

ANTIMONY BARIUM LITHIUM
ANTIMONY LITHIUM STRONTIUM

I. Darstellung und Kristallstruktur von $Sr_3Li_4Sb_4$ und $Ba_3Li_4Sb_4$. O. LIEBRICH, H. SCHÄFER and A. WEISS, 1970. *Z. Naturf.*, 25B, 650-651.

$Li_4Sb_4Sr_3$ F.W. = 777.62

Orthorhombic, *a* = 15.84 ± 3, *b* = 4.88 ± 2, *c* = 7.50 ± 2 Å, *U* = 579.7 Å^3, D_m = 4.52, *Z* = 2, D_x = 4.49.

Space group *Immm* (D_{2h}^{25}) from absences and Patterson symmetry

Atomic positions

			x	*y*	*z*
2 Sr(1)	in	2(*d*)	1/2	0	1/2
4 Sr(2)	in	4(*e*)	0.631	0	0
4 Sb(1)	in	4(*i*)	0	0	0.312
4 Sb(2)	in	4(*f*)	0.212	1/2	0
8 Li	in	8(*m*)	0.180	0	0.190

B = 0.12 Å^2 for all atoms.

Material and Details of analysis

The elements were melted in an iron crucible under argon at 980°C and annealed at 600°C. Precession and Weissenberg photographs were obtained; Patterson, Fourier and difference Fourier syntheses were calculated; a cylindrical absorption correction was applied. The Li atoms were placed from spatial considerations.

Discussion

$Ba_3Li_4Sb_4$ is isomorphous, see Table I, p. 117.

Interatomic distances (Å)

Sr(1)	- 4 Sr(2)	4.29	Sb(1)	- 2 Sr(1)	3.38	Li	- 2 Sr(1)	4.01
	- 4 Sb(1)	3.38		- 4 Sr(2)	3.50		- 2 Sr(2)	3.46
	- 2 Sb(2)	3.36		- 1 Sb(1)	2.82		- 1 Sr(2)	3.32
	- 8 Li	4.01		- 2 Li	2.99		- 1 Sb(1)	2.99
Sr(2)	- 2 Sr(1)	4.29	Sb(2)	- 1 Sr(1)	3.36		- 1 Sb(2)	2.89
	- 1 Sr(2)	4.15		- 2 Sr(2)	3.48		- 2 Sb(2)	2.87
	- 4 Sb(1)	3.50		- 2 Sr(2)	3.96		- 1 Li	2.85
	- 2 Sb(2)	3.48		- 4 Li	2.87		- 2 Li	3.33
	- 2 Sb(2)	3.96		- 2 Li	2.89			[3.42]
	- 4 Li	3.46						
	- 2 Li	3.32						

ANTIMONY, BISMUTH

I. [Investigation of the crystal structure of the antimony and bismuth high pressure phases.] S.S. KABALKINA, T.N. KOLOBJANINA and L.F. VEREŠČAGIN, 1970. *Ž. Ėksp. Teor. Fiz., SSSR*, 58, 486-493 [*Soviet Physics - JETP*, 31, 259-263].

Sb III see also 1.

Monoclinic, Sb III, distorted GeS (*B*16 same as *B*29(SnS)), (2) type structure, *a* = 5.56, *b* = 4.04, *c* = 4.22 Å, β = 86°, *U* = 93.8 Å^3 [94.6 Å^3], *Z* = 4 (at ≈ 140 kb).

Bi III, Sb III type, *a* = 6.05 [also given as 6.6], *b* = 4.20, *c* = 4.65 Å, β = 85°20', *U* = 117.8 Å^3 [129.4 Å^3], *Z* = 4 (at 35.5 kb).

Space group $P2_1/m$ (C_{2h}^2)

Atomic positions

		x	y	z
2 Sb or Bi(1)	in 2(*e*)	0.77	1/4	0.08
2 Sb or Bi(2)	in 2(*e*)	0.33	1/4	0.39

Interatomic distances (Å)

Sb(1) -	2 Sb(1)	3.28	[3.29]
-	2 Sb(1)		[3.22]
-	[1] Sb(2)	2.64	[2.69]
-	[2] Sb(2)	2.93	[2.91]
-	2 Sb(2)	3.06	[3.04]
-	1 Sb(2)	3.41	[3.46]
[Sb(2) -	2 Sb(1)	2.91	
-	2 Sb(1)	3.04	
-	1 Sb(1)	2.69	
-	1 Sb(1)	3.46	
-	2 Sb(2)	2.96]	

Materials and Details of analysis

Intensities of X-ray powder patterns of Sb (130-160 kb) and Bi (35.5 kb) at pressure (filtered Mo radiation) were compared with those calculated with a model derived by distorting the GeS structure; R = 15% (F^2) for coordinates given.

1. S.S. KABALKINA, L.F. VEREŠČAGIN and V.P. MILOV, 1963. *Dokl. Akad. Nauk, SSSR*, 152, 585 [*Soviet Physics - Dokl.*, 8, 917, (1964)].
2. *Strukturbericht*, 2, 8; 3, 14; *Structure Reports*, 20, 178.

ANTIMONY CHROMIUM

ANTIMONY IRON

I. Compounds with the marcasite type crystal structure. VI. Neutron diffraction studies of $CrSb_2$ and $FeSb_2$. H. HOLSETH and A. KJEKSHUS, 1970. *Acta Chem. Scand.*, 24, 3309-3316.

$CrSb_2$ see also 1. F.W. = 295.5

Orthorhombic, FeS_2 marcasite type (2), a = 6.0275$_6$, b = 6.8738$_9$, c = 3.2715$_7$ Å at 298°K, [U = 135.5 Å^3, Z = 2, D_x = 7.24]; a = 6.0183 ± 4, b = 6.8736 ± 2, c = 3.2704 ± 2 Å at 80°K

Space group *Pnnm* (D_{2h}^{12})

Atomic positions

		x	y	z	B (Å^2)
2*M*	in 2(*a*)	0	0	0	0.20
4*B*	in 4(*b*)	*x*	*y*	0	0.20

MB_2	Temp.°K	x_B	y_B	R
$CrSb_2$	298	0.1797±2	0.3662±2	0.031
$CrSb_2$	80	0.1792±5	0.3643±4	0.048
$FeSb_2^*$ (3)	298	0.1885±2	0.3561±4	0.076
$FeAs_2^*$ (4)	298	0.1636±9	0.3608±7	0.075
$NiSb_2^*$ (4)	298	0.2196±15	0.3601±6	0.049

*Lattice parameters from 3 and 4 respectively.

[*Interatomic distances* (Å)]

Cr -	2Sb	2.74	Sb -	2Cr	2.69
-	4Sb	2.69	-	1Cr	2.74
			-	1Sb	2.84

Material and Details of analysis

Neutron diffraction patterns of monophase powder samples prepared by 1 (λ = 1.186 and 1.860 Å) were obtained between 4.2 and 298°K. Least-squares refinement of trial structure (2-4) converged for *Pnnm* but not for *Pnn*2. The magnetic cell for $CrSb_2$ at 4.2°K and 80°K has $a_m = a_{RT}$, $b_m = 2b_{RT}$, $c_m = 2c_{RT}$; the components of the moment along the *a*, *b* and *c* axes (S_1, S_2, S_3) refined to values S_1 = 0.70 ± 2, S_2 = 0 (fixed), S_3 = 0.67 ± 3.

1. H. HOLSETH and A. KJEKSHUS, 1968. *Acta Chem. Scand.*, 22, 3273.
2. *Strukturbericht*, 1, 495.
3. H. HOLSETH and A. KJEKSHUS, 1969. *Acta Chem. Scand.*, 23, 3043.
4. Idem, 1968. *J. Less-Common Metals*, 16, 472.

ANTIMONY COPPER LEAD SULPHUR

I. Verfeinerung der Kristallstruktur von Bournonit [$(SbS_3)_2/Cu_2{}^{IV}Pb^{VII}Pb^{VIII}$] und von Seligmannit [$(AsS_3)_2/Cu_2{}^{IV}Pb^{VII}Pb^{VIII}$]. A. EDENHARTER, W. NOWACKI and Y. TAKÉUCHI, 1970. *Z. Kristallogr.*, 131, 397-417.

$CuPbS_3Sb$ (bournonite) see also 1. F.W. = 488.7

$AsCuPbS_3$ (seligmannite) see also 1. F.W. = 441.8

Orthorhombic, *a* = 8.153 ± 3, *b* = 8.692 ± 3, *c* = 7.793 ± 2 Å, [*U* = 552.3 Å^3], D_m = 5.83, *Z* = 4, D_x = 5.84, for bournonite. *a* = 8.076 ± 2, *b* = 8.737 ± 5, *c* = 7.634 ± 3 Å, [*U* = 538.7 Å^3], D_m = 5.38, *Z* = 4, D_x = 5.41, for seligmannite.

Space group $Pn2_1m$ (C_{2v}^7) [Non-standard setting, compare 1]

Atomic positions

$CuPbS_3Sb$

	x	*y*	*z*	*B* (Å^2)*
2Pb(1) in 2(*a*)	0.07380±14	0.99050±17	0	1.58
2Pb(2) in 2(*a*)	0.55709±14	0.17696±18	1/2	1.53
2Sb(1) in 2(*a*)	0.07050±23	0.04091±24	1/2	1.38
2Sb(2) in 2(*a*)	0.50655±26	0.14958±30	0	1.76
4Cu in 4(*b*)	0.27597±49	0.42275±52	0.24384±45	1.91
2S (1) in 2(*a*)	0.23995±97	0.27917±95	0	1.51
2S (2) in 2(*a*)	0.23187±89	0.28860±88	1/2	1.30
4S (3) in 4(*b*)	0.10572±70	0.63870±61	0.23753±66	1.46
4S (4) in 4(*b*)	0.56333±66	0.48165±67	0.26830±73	1.66

$AsCuPbS_3$

	x	*y*	*z*	*B* (Å^2)
2Pb(1) in 2(*a*)	0.07903±8	0.00061±9†	0	1.38
2Pb(2) in 2(*a*)	0.55368±8	0.19272±10	1/2	1.30
2As(1) in 2(*a*)	0.07360±23	0.07360±22	1/2	1.05
2As(2) in 2(*a*)	0.49593±26	0.15737±22	0	1.10
4Cu in 4(*b*)	0.27430±26	0.42756±27	0.24108±27	1.82
2S (1) in 2(*a*)	0.24854±54	0.27905±44	0	1.31
2S (2) in 2(*a*)	0.22196±48	0.29237±42	1/2	1.13
4S (3) in 4(*b*)	0.10420±40	0.64340±32	0.22935±41	1.41
4S (4) in 4(*b*)	0.55870±34	0.49398±37	0.27527±42	1.38

*Isotropic equivalent; full anisotropic values are also given.
†[Misprinted as .00610 - private communication - author.]

Interatomic distances (Å)

Bournonite			Seligmannite		
Pb(1)	- 2S(3)	2.825±5	Pb(1)	- 2S(3)	2.831±3
	- 1S(1)	2.851±8		- 1S(1)	2.791±4
	- 1S(2)	3.048±7		- 1S(2)	3.036±3
	- 2S(4)	3.466±5		- 2S(4)	3.391±2
	- 2S(3)	3.584±5		- 2S(3)	3.584±3
Pb(2)	- 1S(2)	2.823±7	Pb(2)	- 1S(2)	2.816±3
	- 2S(4)	2.866±5		- 2S(4)	2.873±3
	- 2S(4)	3.205±5		- 2S(4)	3.142±3
	- 2S(3)	3.330±5		- 2S(3)	3.299±3
Sb(1)	- 2S(3)	2.492±5	As(1)	- 2S(3)	2.294±3
	- 1S(2)	2.523±7		- 1S(2)	2.301±4
Sb(2)	- 2S(4)	2.390±6	As(2)	- 2S(4)	2.275±3
	- 1S(1)	2.448±8		- 1S(1)	2.263±4
Cu	- 1S(1)	2.292±6	Cu	- 1S(1)	2.261±3
	- 1S(3)	2.335±7		- 1S(3)	2.334±3
	- 1S(2)	2.339±5		- 1S(2)	2.341±2
	- 1S(4)	2.405±6		- 1S(4)	2.383±3
S (1)	- 2Cu	2.292±6	S (1)	- 2Cu	2.261±3
	- 1Sb(2)	2.448±8		- 1As(2)	2.263±4
	- 1Pb(1)	2.851±8		- 1Pb(1)	2.791±4
S (2)	- 2Cu	2.339±5	S (2)	- 2Cu	2.341±2
	- 1Sb(1)	2.523±7		- 1As(1)	2.301±4
	- 1Pb(2)	2.823±7		- 1Pb(2)	2.816±3
	- 1Pb(1)	3.048±7		- 1Pb(1)	3.036±3
S (3)	- 1Cu	2.335±7	S (3)	- 1Cu	2.334±3
	- 1Sb(1)	2.492±5		- 1As(1)	2.294±3
	- 1Pb(1)	2.825±5		- 1Pb(1)	2.831±3
	- 1Pb(2)	3.330±5		- 1Pb(2)	3.299±3
	- 1Pb(1)	3.584±5		- 1Pb(1)	3.584±3
S (4)	- 1Sb(2)	2.390±6	S (4)	- 1As(2)	2.275±3
	- 1Cu	2.405±6		- 1Cu	2.383±3
	- 1Pb(2)	2.866±5		- 1Pb(2)	2.873±3
	- 1Pb(2)	3.205±5		- 1Pb(2)	3.142±3
	- 1Pb(1)	3.466±5		- 1Pb(1)	3.391±2

Materials and Details of analysis

Bournonite from "Kupfergruebli" in Calanda (Graubünden Canton); seligmannite from Binnatal (Wallis Canton). Spheres of both were prepared and etched with H_2SO_4 and NH_4OH; r = 0.074 mm (bournonite) and r = 0.097 mm (seligmannite). Lattice parameters were obtained from precision back-reflexion Weissenberg photographs, calibrated with Si. Intensities were obtained using a diffractometer (CuK_α) and corrected for Lorentz and polarization factors, and absorption; 590 independent reflexions were observed for bournonite and 572 for seligmannite.

The parameters of 1 were refined by block-diagonal least squares using anisotropic temperature factors and applying dispersion corrections for the Pb scattering factors; R = 4.8% (bournonite) and 2.6% (seligmannite).

1. *Structure Reports*, 20, 30.

ANTIMONY LANTHANUM

I. The crystal structure of La_2Sb. W.N. STASSEN, M. SATO and L.D. CALVERT, 1970. *Acta Cryst.*, B26, 1534-1540.

La_2Sb see also 1. F.W. = 399.60

Tetragonal, a = 4.626 ± 3, c = 18.06 Å (at 25°C), c/a = 3.904, U = 386.5 Å^3, Z = 4, D_x = 6.87.

Space group $I4/mmm$ (D_{4h}^{17})

Atomic positions

			x	*y*	*z*	*B* (Å^2)
4 La(1)	in	4(*c*)	0	1/2	0	0.73±5
4 La(2)	in	4(*e*)	0	0	0.32043±19	0.42±4
4 Sb	in	4(*e*)	0	0	0.13771±22	0.46±5

Interatomic distances (Å)

La(1)	- 4 La(1)	3.271±2	Sb	- 4 La(1)	3.395±4
	- 4 La(2)	3.983±4		- 1 La(2)	3.357±3
	- 4 Sb	3.395±4		- 4 La(2)	3.302±6
La(2)	- 4 La(1)	3.983±4			
	- 4 La(2)	4.144±4			
	- 1 Sb	3.357±3			
	- 4 Sb	3.302±6			

These distances are the weighted mean of the *h*0*ℓ* and *h*1*ℓ* refinements.

Discussion

The structure (Fig. 3) consists of two *C*38 (Cu_2Sb) structure units joined by a mirror plane perpendicular to the *c* axis. The structure of Ti_2Bi (2), which was originally described in *P*4$_2$/*mmc*, is in fact the same as that of La_2Sb.

Material and Details of analysis

Pellets of La (99.8+%) and Sb (99.99+%) filings were reacted by 3 using an electric-ignition process to initiate an exothermic reaction which melted the elements while they fell through purified argon; the melt was quenched on a copper hearth, recrushed and homogenized; crystals were obtained from an ingot annealed at 1100°C and slow cooled. Lattice parameters were derived from monochromatic Debye-Scherrer photographs ($CuK\alpha_1$, 1.5405 Å) and intensities, derived from multiple-film Weissenberg photographs (2 crystals, one with Cu/Ni and the other with Mo/Zr radiation), were corrected for absorption, Lorentz and polarization factors. A trial structure, derived from that of LaSb, was refined by block-diagonal least squares (144 observed *h*0*ℓ*, 154 observed *h*1*ℓ*); $R_{internal}$ = 0.15, R_{final} = 0.14; alternative atomic arrangements were tested and rejected.

1. *Structure Reports*, 18, 101.
2. H. AUER-WELSBACH, H. NOWOTNY and A. KOHL, 1958. *Mh. Chem.*, 89, 154.
3. M. SATO, J.B. TAYLOR and L.D. CALVERT, 1967. *J. Less-Common Metals*, 12, 419.

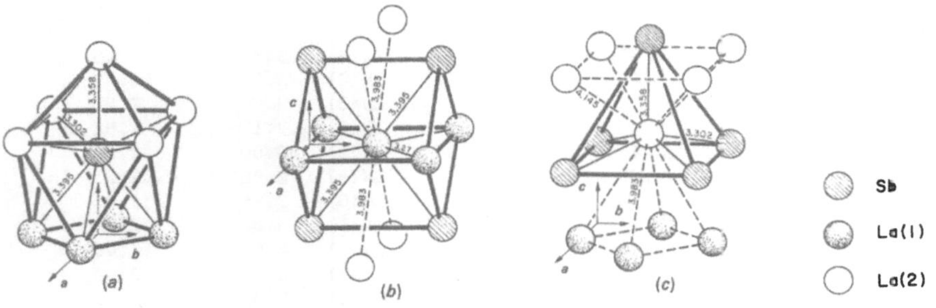

Fig. 3. Atomic coordinations in La_2Sb. Solid lines for nearest neighbours, broken lines for next-nearest neighbours. (a) Sb atom at 0,0,0. (b) La(1) at 0, 1/2,0. (c) La(2) atom at 1/2,1/2,0.180.

ANTIMONY PALLADIUM

ARSENIC NICKEL

I. Kristallstruktur von Pd_5Sb_2 und Ni_5As_2 und einigen Varianten.
M. EL-BORAGY, S. BHAN and K. SCHUBERT, 1970. *J. Less-Common Metals*, 22, 445-458.

Pd_5Sb_2 see also 1 - 5.

Hexagonal, a = 7.60$_6$, c = 13.86$_3$ Å, c/a = 1.82, U = 693.28 Å^3, Z = 6.

Space group $P6_3cm$ (C^3_{6v})

Atomic positions

			x	y	z	B (Å^2)
2Pd	in	2(*a*)	0	0	0.951	0.39
4Pd	in	4(*b*)	1/3	2/3	0.093	0.81
6Pd	in	6(*c*)	0.259	0	0.115	0.69
6Pd	in	6(*c*)	0.630	0	0.212	1.50
6Pd	in	6(*c*)	0.306	0	0.320	0.26
6Pd	in	6(*c*)	0.624	0	0.425	2.73
2Sb	in	2(*a*)	0	0	0.235	0.92
4Sb	in	4(*b*)	1/3	2/3	0.288	0.01
6Sb	in	6(*c*)	0.661	0	0.018	0.91

As_2Ni_5

Hexagonal, Pd_5Sb_2 structure type, a = 6.82$_5$, c = 12.51$_3$, c/a = 1.83, U = 503.9 Å^3, Z = 6.

Space group $P6_3cm$ (C^3_{6v})

Atomic positions

			x	y	z	B (Å^2)
2As	in	2(*a*)	0	0	0.256	0.41
4As	in	4(*b*)	1/3	2/3	0.298	0.36
6As	in	6(*c*)	0.670	0	0.023	1.15
2Ni	in	2(*a*)	0	0	0.948	1.99
4Ni	in	4(*b*)	1/3	2/3	0.100	2.82
6Ni	in	6(*c*)	0.244	0	0.122	0.94
6Ni	in	6(*c*)	0.630	0	0.215	1.23
6Ni	in	6(*c*)	0.312	0	0.320	4.47
6Ni	in	6(*c*)	0.614	0	0.440	5.90

[*Interatomic distances* (Å)]

		Pd_5Sb_2 (M_5X_2)	Ni_5As_2 (M_5X_2)
M(1) -	3M(3)	3.01	2.74
	3M(5)	2.95	2.66
	3M(6)	2.88	2.63
	1X(1)	2.99	2.40
	3X(1)	2.74	2.44
M(2) -	3M(3)	2.87	2.65
	3M(4)	2.92	2.59
	3M(6)	3.34	2.91
	1X(2)	2.70	2.48
	3X(3)	2.72	2.48
M(3) -	1M(1)	3.01	2.74
	2M(2)	2.87	2.65
	2M(3)	3.41	2.88
	2M(4)	2.84	2.51
	1M(4)	3.12	2.88
	1M(5)	2.86	2.52
	1M(6)	2.78	2.47
	1X(1)	2.58	2.36
	2X(3)	2.69	2.37
M(3) -	1X(3)	3.34	3.16
M(4) -	2M(2)	2.92	2.59
	2M(3)	2.84	2.51
	1M(3)	3.12	2.88
	2M(5)	3.00	2.69
	1M(5)	2.88	2.53
	1M(6)	2.95	2.82
	1X(1)	2.83	2.58
	2X(2)	2.63	2.40
	1X(3)	2.70	2.42
M(5) -	1M(1)	2.95	2.66
	1M(3)	2.86	2.52
	2M(4)	3.00	2.69
	1M(4)	2.88	2.54
	2M(6)	3.01	2.85
	1M(6)	2.82	2.55
	1X(1)	2.61	2.27
	2X(2)	2.68	2.37
	1X(3)	2.76	2.54

	Pd_5Sb_2 (M_5X_2)	Ni_5As_2 (M_5X_2)
M(6) - 1*M*(1)	2.88	2.63
2*M*(2)	3.34	2.91
1*M*(3)	2.78	2.47
1*M*(4)	2.95	2.82
2*M*(5)	3.00	2.85
1*M*(5)	2.82	2.55
2*X*(2)	3.05	2.76
2*X*(3)	3.02	2.67
1*X*(3)	2.52*	2.20*
X(1) - 1*M*(1)	2.99	2.40
3*M*(3)	2.58	2.36
3*M*(4)	2.83	2.58
3*M*(5)	2.61	2.27

	Pd_5Sb_2 (M_5X_2)	Ni_5As_2 (M_5X_2)
X(2) - 1*M*(2)	2.70	2.48
3*M*(4)	2.63	2.40
3*M*(5)	2.68	2.37
3*M*(6)	3.05	2.76
X(3) - 1*M*(1)	2.74	2.44
2*M*(2)	2.72	2.48
2*M*(3)	2.69	2.37
1*M*(3)	3.34	3.16
1*M*(4)	2.70	2.42
1*M*(5)	2.76	2.54
1*M*(6)	2.52*	2.20*
2*M*(6)	3.02	2.67

*[These are the shortest distances.]

Material and Details of analysis

The elements (99.9+%) were heated in evacuated SiO_2 ampoules to prepare the compounds. Guinier powder patterns (CuK_α, Si standard) were used for deriving lattice parameters and Weissenberg photographs (MoK_α) for intensities.

1. *Structure Reports*, 23, 133.
2. K. FRANK and K. SCHUBERT, 1971. *Acta Cryst.*, B 27, 916.
3. *Structure Reports*, 21, 43.
4. G.S. SAINI, L.D. CALVERT and J.B. TAYLOR, 1964. *Can. J. Chem.*, 42, 150.
5. G.S. SAINI, L.D. CALVERT, R.D. HEYDING and J.B. TAYLOR, 1964. *Ibid.*, 42, 620.

ANTIMONY SILVER SULPHUR

I. [Crystal structure of stephanite, Ag_5SbS_4.] A.A. PETRUNINA, B.A. MAKSIMOV, V.V. ILJUKHIN and N.V. BELOV, 1970. *Dokl. Akad. Nauk, SSSR*, 188, 342-344 [*Soviet Physics - Doklady*, 14, 833-835].

II. Die Kristallstruktur von Stephanit, $[SbS_3(S)Ag_5^{III}]$. B. RIBÁR and W. NOWACKI, 1970. *Acta Cryst.*, B 26, 201-207.

Ag_5SbS_4 F.W. = 789.36

See also 1.

Orthorhombic, I. a = 7.830 ± 2, b = 12.450 ± 5, c = 8.538 ± 2 Å, [U = 832.3 Å^3], D_m = 6.28-6.32, Z = 4, [D_x = 6.30]. II. a = 7.837 ± 3, b = 12.467 ± 6, c = 8.538 ± 2 Å, [U = 834.2 Å^3], D_m = 6.26, Z = 4, D_x = 6.28.

Space group $Cmc2_1$ (C_{2v}^{12}) from morphology (I), piezoelectric effect (I, II) and symmetry of Patterson synthesis (I) and absences (I, II).

*Atomic positions**

			x	*y*	*z*	
8Ag(1)	in	4(*a*)	0.50000	0.35543±62	0.16922±105	II
			0.500	0.355	0.158	I
8Ag(2)	in	8(*b*)	0.18902±72	0.06219±50	0.32683±68	II
			0.174	0.063	0.310	I
8Ag(3)	in	8(*b*)	0.31406±63	0.12312±40	0.01357±52	II
			0.312	0.122	0	I
4Sb	in	4(*a*)	0	0.33072±32	0.09900±54	II
			0	0.331	0.083	I
4S (1)	in	4(*a*)	0	0.03031±114	0.02449±174	II
			0	0.033	0	I

*Atomic positions**

	x	y	z	
8S (2) in 4(a)	0.50000	0.01522±111	0.20585±168	II
	0.500	0.014	0.200	I
8S (3) in 8(b)	0.23005±123	0.26679±76	0.27307±136	II
	0.229	0.269	0.246	I

* The coordinates of I have been transformed to conform with the values of II wherever necessary. I gives $B = 1.6$ Å^2 for all atoms, II gives anisotropic β's for all atoms.

Interatomic distances (Å)†

		II	I
Sb	- S(3)	2.47	2.39[2.40]
	- S(3)	2.47	2.39[2.40]
	- S(2)	2.48	2.49
Ag(1)	- S(1)	2.51	2.58[2.59]
	- S(1)	[3.35]	3.23[3.26]
	[2] S(3)§	2.55	2.49
Ag(2)	- S(1)	2.52	2.42[2.43]
	- S(3)	2.61	2.65[2.66]
	- S(2)	2.71	2.78[2.79]
	- S(1)	3.00	2.99[3.00]
	- Ag(3)	2.95	- [2.95]
	- Ag(2)	2.96	- [2.72]
	- Ag(3)	2.97	- [3.02]

		II	I
S(3)	- S(3)	3.61	-
	- S(2)	3.63	-
	- S(2)	3.63	-
Ag(3)	- S(3)	2.49	2.54[2.58]
	- S(2)	2.57	2.64[2.62]
	- S(1)	2.72	2.68
	- S(3)	2.92	2.88[2.86]
	- Ag(3)	2.91	- [2.94]
	- Ag(2)	2.95	- [2.95]
	- Ag(2)	2.97	- [3.02]

† Angles are also given in II.
§ Misprinted as S(2) and S(3) in I, only 1 given in II.

Material and Details of analysis

I. An iron-black, prismatic crystal of "freibergite" (0.4 x 0.3 x 0.4 mm) was used to obtain cell-parameters (diffractometer, CuK_α) and 488 non-zero reflexions (0$k\ell$ - 6$k\ell$ and hk0 - hk2, Weissenberg camera, MoK_α, $\sin\theta/\lambda < 0.8$ Å^{-1}, eye-estimation using a scale). Patterson and Fourier syntheses gave $R = 0.15$. Twinning was very prevalent, 500 fragments yielded only one untwinned crystal.

II. A cube-shaped crystal (0.075 mm) from Hodrusa (Czechoslovakia) was used to obtain the space group (Weissenberg photographs) and to obtain 443 $hk\ell$ intensities (diffractometer CuK_α (Ni filter). The intensities were corrected for absorption, Lorentz and polarization. The trial structure was obtained from a Patterson and refined by difference Fouriers and least squares ($R = 0.094$); scattering factors were corrected for dispersion.

Discussion

The structure can be described as consisting of SbS_3, trigonal prisms (Fig. 4(a)), Ag(1)-S_3, almost planar triangles (Fig. 4(b)), Ag(2)-S_4 tetrahedra (Fig. 4(c)) and Ag(3)-S_4 tetrahedra (Fig. 4(d)); Ag(2) and Ag(3) are also co-ordinated to 3 Ag atoms.

1. *Structure Reports*, 8, 170; *Strukturbericht*, 2, 348.

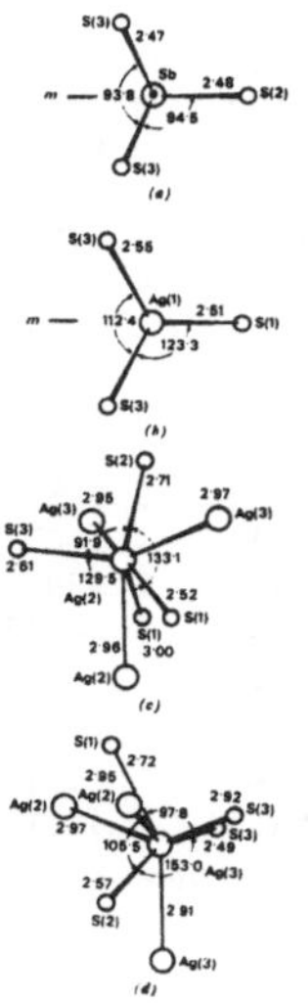

Fig. 4. The coordination around Sb, (*a*), Ag(1) (*b*), Ag(2) (*c*), and Ag(3) (*d*).

ANTIMONY YTTERBIUM

I. The crystal structure of $Yb_{11}Sb_{10}$. H.L. CLARK, H.D. SIMPSON and H. STEINFINK, 1970. *Inorg. Chem.*, 9, 1962-1964.

$Yb_{11}Sb_{10}$ F.W. = 3121

Tetragonal, $Ho_{11}Ge_{10}$ type (1, 2), a = 11.86 ± 1, c = 17.10 ± 1 Å, [U = 2405 Å^3], D_m = 8.17 ± 7, Z = 4, D_x = 8.6.

Space group $I4/mmm$ (D_{4h}^{17}) from absences and structure type

Atomic positions

			x	*y*	*z*	B_{11}*	B_{22}*	B_{33}* (Å^2)
16 Yb(1)	in	16(*n*)	0	0.2518±2	0.1886±1	38±1	22±1	6±0
8 Yb(2)	in	8(*h*)	0.1671±5	0.1671±5	0	71±4	71±4	14±1[a]
4 Yb(3)	in	4(*e*)	0	0	0.3374±2	32±4	32±4	11±1
16 Yb(4)	in	16(*n*)	0	0.3364±2	0.3974±1	26±1	36±1	4±0
8 Sb(1)	in	8(*i*)	0.3494±4	0	0	26±3	45±3	5±1
4 Sb(2)	in	4(*e*)	0	0	0.1283±3	23±5	23±5	8±2
8 Sb(3)	in	8(*h*)	0.3736±9	0.3736±9	0	104±8	104±8	19±2
4 Sb(4)	in	4(*d*)	0	0.5000	0.2500	25±4	25±4	3±2
16 Sb(5)	in	16(*m*)	0.2062±4	0.2062±4	0.3245±2	35±2	35±2	8±1[b]

* All x 10^4

a, B_{12} = 40±3 *b*, B_{12} = 6±2 Remaining B_{ij} = 0 or -1

Interatomic distances (Å)

Yb(1) to		Yb(2) to		Yb(3) to	
[2] Yb(1)	4.223	[4 Yb(1)	3.916]	[4 Yb(1)	3.923]
[2] Yb(2)	3.910[3.916]	[2] Yb(2)	3.963	[4] Yb(4)	4.119
1 Yb(3)	3.920[3.923]	[4] Yb(4)	4.319[4.321]	[Sb(2)	3.576]
[2] Yb(4)	3.819[3.820]	[2] Sb(1)	2.933	[4] Sb(3)	3.491[3.497]
1 Yb(4)	3.701[3.709]	[2] Sb(2)	3.556[3.559]	[4] Sb(5)	3.465
1 Sb(1)	3.420[3.426]	1 Sb(3)	3.464		
1 Sb(2)	3.159	[2] Sb(5)	3.672[3.677]		
1 Sb(4)	3.125				

Interatomic distances (Å)

Yb(1) to		
[2]	Sb(5)	3.413[3.417]
[2]	Sb(5)	3.527

Yb(4) to		
[1	Yb(1)	3.709
2	Yb(1)	3.820
2	Yb(2)	4.321
1	Yb(3)	4.119
1	Yb(4)	3.509]
1	Yb(4)	3.881
[2]	Sb(1)	3.165[3.168]
[2]	Sb(3)	3.393[3.395]
1	Sb(4)	3.176[3.181]
[2]	Sb(5)	3.148[3.149]

Sb(1) to		
[2	Yb(1)	3.426
2	Yb(2)	2.933
4	Yb(4)	3.168]
1	Sb(1)	3.572
[2	Sb(3)	4.440]

Sb(3) to		
[1	Yb(2)	3.464
2	Yb(3)	3.497
4	Yb(4)	3.395]
2	Sb(1)	4.440
[2]	Sb(3)	2.998
[2]	Sb(5)	3.280[3.286]

Sb(4) to		
[4	Yb(1)	3.125
4	Yb(4)	3.181]
8	Sb(5)	4.444]

Sb(2) to		
[4	Yb(1)	3.159
4	Yb(2)	3.559
1	Yb(3)	3.576]
1	Sb(2)	4.388

Sb(5) to		
[2	Yb(1)	3.417
2	Yb(1)	3.527
1	Yb(2)	3.677
1	Yb(3)	3.466
2	Yb(4)	3.149
1	Sb(3)	3.286
2	Sb(4)	4.444]
1	Sb(5)	2.936[2.941]

e.s.d.'s quoted range from 0.003 to 0.022 Å.

Discussion

Magnetic measurements show that nearly all the Yb in $Yb_{11}Sb_{10}$ is divalent, consequently the number of available valence electrons (2 + 5) is considered to be equal to that for $Ho_{11}Ge_{10}$ (3 + 4).

Material and Details of analysis

An irregular fragment (3) about 0.06 x 0.07 x 0.10 mm was used in collecting 4500 integrated intensities which, after averaging equivalent values, yielded 1370 independent reflexions of which 997 were non-zero. Crystal monochromatized MoK_α radiation was used with a Weissenberg geometry diffractometer using the ω scan technique for levels $h0\ell$ to $h12\ell$; Lorentz, polarization, dispersion (real and imaginary) and absorption corrections (transmission factors from 0.10-0.77) were applied. A trial model based on 1 was refined by full matrix least squares with isotropic (8 cycles, R = 0.13) and anisotropic (5 cycles, R = 0.080) temperature factors. The hypothesis that the Sb(3) site might be partially occupied was tested by a full-matrix least-squares refinement, the B's for Sb(3) with half occupancy were reduced to values equivalent to those of other atoms but R = 0.084 and the hypothesis was not retained. A difference Fourier synthesis had no peaks greater than 0.4 e/$Å^3$.

1. A.G. THARP, G.S. SMITH and Q. JOHNSON, 1966. *Acta Cryst.*, 20, 583.
2. G.S. SMITH, Q. JOHNSON and A.G. THARP, 1967. *Ibid.*, 23, 640.
3. R.E. BODNAR and H. STEINFINK, 1967. *J. Inorg. Chem.*, 6, 327.

ARSENIC CADMIUM

I. The crystal structure of $CdAs_2$. L. CERVINKA and A. HRUBÝ, 1970. *Acta Cryst.*, B26, 457-458.

$CdAs_2$ F.W. = 262.24

Tetragonal, a = 7.95 ± 6, c = 4.67 ± 4 Å, c/a = 0.59, [U = 295.85 $Å^3$], D_m = 5.80, Z = 4, D_x = 5.88.

Space group $I4_122$ (D_4^{10}) from absences and intensities

Atomic positions

	x	y	z
4 Cd in 4(b)	0	0	1/2
8 As in 8(f):	0.06±1	1/4	1/8

[*Interatomic distances* (Å)]

Cd - 4 As	2.69	As - 2 Cd	2.69
		- 2 As	2.43

Materials and Details of analysis

The elements (6N Cd, 5N As) were heated at 700°C in an evacuated silica tube and air cooled. Guinier photographs with CuK_α gave d_{calc} agreement with d_{obs} - no calculated intensities are given [a check showed fair overall agreement].

ARSENIC CADMIUM SILICON

I. [Structure of $CdSiAs_2$]. B.B. SARMA, A.A. VAJPOLIN and Ju.A. VALOV, 1970. Ž. *Strukt. Khim.*, 11, 781-782. [*J. Struct. Chem.*, 11, 726-727].

As_2CdSi see also 1. F.W. = 290.33

Tetragonal, chalcopyrite ($CuFeS_2$), (El_1) type structure (2), a = 5.884 ± 1, c = 10.882 ± 2 Å (from 1), [U = 376.8 Å^3, Z = 4, D_x = 5.12].

Space group $I\bar{4}2d$ (D_{2d}^{12})

Atomic positions

	x	*y*	*z*
4 Cd in 4(*a*)	0	0	0
4 Si in 4(*b*)	0	0	1/2
8 As in 8(*d*)	0.298±1	1/4	1/8

B = 0.58 Å^2, determined graphically.

Interatomic distances (Å)

As - 2Cd	2.662	Cd - 4As	2.662	Si - 4As	2.330
- 2Si	2.330				

Material and Details of analysis

Crystals up to 4 x 1 x 0.4 mm were prepared by vapour transport ($SiCl_2$); a (112) plate was used to record 24 independent *hkl* intensities (integrating camera, microphotometric measurement of densities) which were corrected for absorption. The strongest reflexions were omitted as were those affected by change of rotation. The parameter was found by systematic trial and error and gave R = 0.122.

1. A.A. VAJPOLIN, E.O. OSMANOV and D.N. TRET'JAKOV, 1967. *Izv. Akad. Nauk, SSSR, Neorg. Mater.*, 3, 260.
2. *Strukturbericht*, 2, 48.

ARSENIC COPPER SULPHUR

I. Strukturverfeinerung von Enargit, Cu_3AsS_4. G. ADIWIDJAJA and J. LÖHN, 1970. *Acta Cryst.*, B26, 1878.

$AsCu_3S_4$ see also 1, 2. F.W. = 393.8

Orthorhombic, a = 7.407 ± 1, b = 6.436 ± 1, c = 6.154 ± 1, [U = 293.4 Å^3, Z = 2], D_m = 4.40 (3).

Space group $Pmn2_1$ (C_{2v}^7) from absences

Atomic positions

			x	y	z	B (Å^2)
2As	in	2(*a*)	0	0.8268±5	0	0
2Cu(1)	in	2(*a*)	0	0.1514±7	0.4983±11	0.8125
4Cu(2)	in	2(*b*)	0.2466±4	0.3255±4	0.9866±8	0.9309
2S(1)	in	2(*a*)	0	0.8226±13	0.6454±15	0
2S(2)	in	2(*a*)	0	0.1436±13	0.1166±15	0.0793
4S(3)	in	4(*b*)	0.2598±7	0.3364±8	0.6188±10	0

Interatomic distances (Å)

As	- 1S(1)	2.182	[S(1)	- 1As	2.182
	- 1S(2)	2.162		- 1Cu(1)	2.302
	- 2S(3)	2.192		- 2Cu(2)	2.321
Cu(1)	- 1S(1)	2.302	S(2)	- 1As	2.162
	- 1S(2)	2.350		- 1Cu(1)	2.350
	- 2S(3)	2.381		- 2Cu(2)	2.312
Cu(2)	- 1S(1)	2.321	S(3)	- 1As	2.192
	- 1S(2)	2.312		- 1Cu	2.381
	- 1S(3)	2.323[2.267]		- 1Cu(2)	2.267
	- 1S(3)	2.324		- 1Cu(2)	2.324]

Material and Details of analysis

A crystal, 0.16 x 0.10 x 0.24 mm, from Butte, Montana, was used to record 321 reflexions, (diffractometer, CuK_α) which were corrected for Lorentz and polarization factors. The space group was determined from Weissenberg photographs in agreement with 4 and the parameters of 2 were refined by least squares (R = 0.08).

1. *Strukturbericht*, 2, 347.
2. *Ibid.*, 3, 96, 438.
3. A. SCHÜLLER, 1965. *Die Eigenschaft. der Miner.*, Teil I, S 62, 72. Berlin: Akad. Verlag.
4. *Strukturbericht*, 3, 439.

ARSENIC GERMANIUM

GERMANIUM PHOSPHORUS

I. Synthesis, structure and superconductivity of new high pressure phases in the systems Ge-P and Ge-As. P.C. DONOHUE and H.S. YOUNG, 1970. *J. Solid State Chem.*, 1, 143-149.

See also 1

GeP (i) F.W. = 103.56

GeAs (ii) F.W. = 147.51

The phases GeP_5 and GeP_3 are also characterized.

Tetragonal

(i) a = 3.544 ± 2, c = 5.581 ± 1 Å, [U = 70.1 Å^3], D_m = 4.73, Z = 2, D_x = 4.90.
(ii) a = 3.715 ± 1*, c = 5.832 ± 1 Å, [U = 80.36 Å^3], D_m = 6.06, Z = 2, D_x = 6.10.

[* Given also as 3.712]

Space group $I4mm$ (C_{4v}^9) from absences and structure

Atomic positions

(i)			z	B (Å^2)	(ii)	z	B (Å^2)
2 Ge	in	2(*a*)	0	1.0±7		0	1.3±6
2 P	in	2(*a*)	0.427±7	1.0±7	2 As	0.414±40	1.3±6

Interatomic distances (Å)

GeP			GeAs		
Ge - 1 P	2.384[2.383]		Ge - 1 As	2.413[2.414]	
- 4 P	2.541[2.539]		- 4 As	2.671[2.674]	
- 1 P	3.204[3.198]		- 1 As	3.423[3.418]	

Material and Details of analysis

Quenched high-pressure phases (GeP, 65kb, 800°C; GeAs, 65kb, 900°C) were made by reacting the elements. The unit cell was determined from precession photographs. The parameters were obtained by least-squares refinement of powder data (GeP, Debye-Scherrer film, densitometer measurement; GeAs, diffractometer data).

1. J. OSUGI, R. NAMIKAWA and Y. TANAKA, 1966. *J. Chem. Soc. Japan, Pure Chem. Sect.*, 87, 1169; *Idem*, 1967. *Rev. Phys. Chem. Japan*, 37, 2, 81.

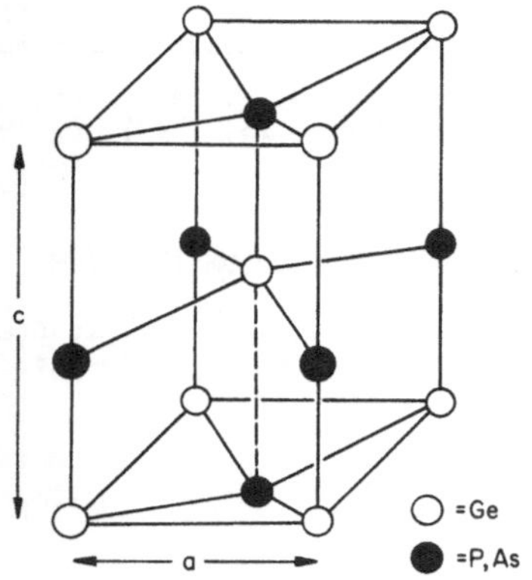

Fig. 5. The tetragonal AsGe high pressure structure. It may be considered as a distorted NaCl structure.

ARSENIC MANGANESE PHOSPHORUS

I. Crystal and magnetic structure of $MnAs_{0.92}P_{0.08}$. E.L. HALL, L.H. SCHWARTZ, G.P. FELCHER and D.H. RIDGLEY, 1970. *J. Appl. Phys.*, 41, 939-941.

See also 1, 2.

$MnAs_{0.92}P_{0.08}$

Orthorhombic, MnP structure type (?) below 440°K, a = 5.615, b = 3.560, c = 6.280 Å at 298°K, [from 1].

Space group *Pnma* (D_{2h}^{16}) from structure

Atomic positions (both atoms in 4(c))

	Mn			As,P		
T (°K)	x	y	$z^§$	x	y	$z^§$
†	0	1/4	0.250	0.250	1/4	0.9167
427	0.995±5	1/4	0.234*	0.265*	1/4	0.922±5
∿380	0.995	1/4	0.225*	0.276*	1/4	0.922
∿298	0.995	1/4	0.211*	0.291*	1/4	0.922
∿298	0.995	1/4	0.214	0.291	1/4	0.922

* values from neutron diffraction data.
§ values read from a small figure.
† for hexagonal NiAs form, temperature not specified.

Atomic positions (both atoms in 4(*c*))

	Mn			As,P		
T (°K)	*x*	*y*	$z^§$	*x*	*y*	$z^§$
∿230	0.995	1/4	0.207	0.297	1/4	0.922
∿200	0.995	1/4	0.207	0.300	1/4	0.922
∿175	0.995	1/4	0.206	0.2984	1/4	0.922
∿150	0.995	1/4	0.204	0.2985	1/4	0.922
∿120	0.995	1/4	0.204	0.2988	1/4	0.922
∿ 85	0.995	1/4	0.203	0.2988	1/4	0.922
∿ 30	0.995	1/4	0.100	0.3028	1/4	0.922

§ values read from a small figure.

Material and Details of analysis

Powders were prepared by reaction of Mn with As and P in sealed SiO_2 tubes, then ground and annealed at 400°C (1). X-ray (30-209°K) and neutron (298-427°K) powder diffraction data - no details of analysis methods.

1. D.H. RIDGLEY and J.H. GEISMAN, 1968. *J. Appl. Phys.*, 39, 592.
2. H. IDO, 1968. *J. Phys. Soc. Japan*, 25, 1543; H. IDO and T. SUZUKI, 1968. *Ibid.*, 24, 964.

ARSENIC PALLADIUM THALLIUM

I. Über eine verzerrte dichteste Kugelpackung mit Leerstellen. M. EL-BORAGY and K. SCHUBERT, 1970. *Z. Metallk.*, 61, 579-584.

$AsPd_5Tl$

Tetragonal, a = 4.00_5, *c* = 7.04_2 Å, Z = 1.

Space group *P4/mmm* (D^1_{4h})

Atomic positions

			x	*y*	*z*
1 Pd(1)	in	1(*c*)	1/2	1/2	0
4 Pd(2)	in	4(*i*)	0	1/2	0.300
1 Tl	in	1(*a*)	0	0	0
1 As	in	1(*d*)	1/2	1/2	1/2

[*Interatomic distances* (Å)]

Pd(1)	- 4 Tl	2.83	Tl	- 4 Pd(1)	2.83
	- 8 Pd(2)	2.91		- 8 Pd(2)	2.91
Pd(2)	- 2 As	2.45	As	- 8 Pd(2)	2.45
	- 4 Pd(2)	2.83			
	- 1 Pd(2)	2.82			
	- 2 Pd(1)	2.91			
	- 2 Tl	2.91			

Discussion

The structure is closely related to that of Cu_2Sb; $PMgPt_5$ also has this structure.

Material and Details of analysis

The elements (99.9+%) were melted under argon in sealed SiO_2 ampoules; qualitative agreement between observed and calculated powder intensities for a Guinier pattern (CuK_α).

BARIUM LEAD THALLIUM

I. The crystal structure of $Ba(Pb_{0.8}Tl_{0.2})_3$. E.E. HAVINGA and J.H.N. VAN VUCHT, 1970. *Acta Cryst.*, B 26, 653-655.

$Ba(Pb_{0.8}Tl_{0.2})_3$ F.W. = 757.22

Hexagonal, a = 7.342, c = 39.45 Å, [U = 1842 Å^3, Z = 14, D_x = 9.55].

Space group $P6_3/mmc$ (D_{6h}^4)

Atomic positions

			x	*y*	*z*
2 Ba(1)	in	2(*d*)	0.333	0.666	0.750
4 Ba(2)	in	4(*e*)	0	0	0.319*
4 Ba(3)	in	4(*f*)	0.333	0.666	0.034*
4 Ba(4)	in	4(*f*)	0.333	0.666	0.607
6($Pb_{0.8}Tl_{0.2}$)(1)	in	6(*h*)	0.182	0.363	0.250
12(Pb, Tl)(2)	in	12(*k*)	0.485	0.970	0.321
12(Pb, Tl)(3)	in	12(*k*)	0.167	0.333	0.393
12(Pb, Tl)(4)	in	12(*k*)	0.848	0.697	0.464

* These values give Ba-3(Pb,Tl) = 3.59 Å to layers above and below; values 0.002 larger could not be ruled out with certainty.

[*Interatomic distances* (Å)]

Ba(1)	- 6(Pb,Tl)(1)	3.68
	- 6(Pb,Tl)(2)	3.64
Ba(2)	- 3(Pb,Tl)(1)	3.58
	- 6(Pb,Tl)(2)	3.68
	- 3(Pb,Tl)(3)	3.59
Ba(3)	- 3(Pb,Tl)(3)	3.59
	- 3(Pb,Tl)(4)	3.58
	- 6(Pb,Tl)(4)	3.68
Ba(4)	- 3(Pb,Tl)(2)	3.64
	- 6(Pb,Tl)(3)	3.67
	- 3(Pb,Tl)(4)	3.64

(Pb,Tl)(1)	- 2Ba(1)	3.68
	- 2Ba(2)	3.58
	- 2(Pb,Tl)(1)	3.34
	- 4(Pb,Tl)(2)	3.41
(Pb,Tl)(2)	- 1Ba(1)	3.64
	- 2Ba(2)	3.68
	- 1Ba(4)	3.64
	- 2(Pb,Tl)(1)	3.41
	- 2(Pb,Tl)(2)	3.34
	- 2(Pb,Tl)(3)	3.47
(Pb,Tl)(3)	- 1Ba(2)	3.59
	- 1Ba(3)	3.59
	- 2Ba(4)	3.67
	- 2(Pb,Tl)(2)	3.47
	- 4(Pb,Tl)(3)	3.67
	- 2(Pb,Tl)(4)	3.47
(Pb,Tl)(4)	- 1Ba(3)	3.58
	- 2Ba(3)	3.68
	- 1Ba(4)	3.64
	- 2(Pb,Tl)(3)	3.47
	- 2(Pb,Tl)(4)	3.41
	- 2(Pb,Tl)(4)	3.34

Material and Details of analysis

The elements were melted in a sealed Fe container enclosed in a sealed SiO_2 ampoule; sample preparation in a N_2 atmosphere dry-box; measurements carried out in an N_2 atmosphere. Powder intensities from a diffractometer (monochromatized CuK_α); random orientation was achieved by sprinkling very fine powder onto a greased plate. Parameters were refined by systematic variation; R_I = 0.18.

Discussion

This structure, with stacking sequence *hhhchhc*, was predicted by 1.

1. E.E. HAVINGA, J.H.N. VAN VUCHT and K.H.J. BUSCHKOW, 1969. Paper presented at the 3rd Bolton Landing Conference; see also *Philips Res. Repts.*, 24, 407.

BARIUM MAGNESIUM SILICON
BARIUM MAGNESIUM GERMANIUM

I. Neue Vertreter des $ThCr_2Si_2$-Typs und dessen Verwandtschaft zum Anti-PbFCl-Gitter. B. EISENMANN, N. MAY, W. MÜLLER, H. SCHÄFER, A. WEISS, J. WINTER and G. ZIEGLEDER, 1970. *Z. Naturf.*, 25B, 1350-1352.
II. Intermetallische Phasen im Anti-PbFCl-Typ: BaMgSi, BaMgGe. B. EISENMANN, H. SCHÄFER and A. WEISS, 1970. *Ibid.*, 25B, 651-652.

I. *Tetragonal*, Cr_2Si_2Th type, (1), Z = 2.

Compound	*a* (Å)	*c* (Å)	*c/a*	*z*
$BaMg_2Si_2$	4.65	11.09	2.39	0.388
$BaMg_2Ge_2$	4.67	11.33	2.43	0.386
$CaCu_2Si_2$	4.04	10.00	2.48	0.384
$SrCu_2Si_2$	4.20	10.00	2.48	0.379
$CaCu_2Ge_2$	4.139	10.232	2.48	0.379
$CaAg_2Ge_2$	4.33	10.93	2.52	0.389
$SrAg_2Ge_2$	4.45	10.85	2.44	0.386
$BaAg_2Ge_2$	4.58	10.69	2.36	0.380
$CaGe_2Zn_2$	4.21	10.85	2.58	0.386

Space group *I4/mmm* (D_{4h}^{17})

Atomic positions

			x	*y*	*z*
2*A*	in	2(*a*)	0	0	0
4*B*	in	4(*d*)	0	1/2	1/4
4*C*	in	4(*e*)	0	0	*z*

The values of *z* for the individual compounds are given in the table above.

Interatomic distances (Å)

AB_2C_2	*A*-8*B*	*A*-8*C*	*A*-2*C*	*B*-4*B*	*C*-4*B*	*C*-1*C*
$BaMg_2Si_2$	3.62	3.52	4.30	3.29	2.78	2.48
$BaMg_2Ge_2$	3.67	3.55	4.36	3.30	2.80	2.58
$CaCu_2Si_2$	3.21	3.08	3.84	2.86	2.42	2.32
$SrCu_2Si_2$	3.27	3.21	3.79	2.97	2.47	2.42
$CaCu_2Ge_2$	3.29	3.18	3.88	2.93	2.45	2.48
$CaAg_2Ge_2$	3.49	3.29	4.25	3.06	2.65	2.43
$SrAg_2Ge_2$	3.51	3.38	4.18	3.11	2.67	2.48
$BaAg_2Ge_2$	3.52	3.48	4.06	3.28	2.68	2.57
$CaGe_2Zn_2$	3.43	3.22	4.19	2.98	2.57	2.47

Material and Details of analysis

Compounds were made by melting stoichiometric mixtures of the elements in Al_2O_3 crucibles under argon.

II. BaMgSi F.W. = 189.74

BaGeMg F.W. = 234.24

Tetragonal, anti-PbFCl type (2), [note also AlGeMn, (3)], *Z* = 2.

	a (Å)	*c* (Å)	*c/a*	*U* (Å^3)	D_m	D_x
BaMgSi	4.61±1	7.87±1	1.71	167.2	3.81	3.76
BaMgGe	4.64±1	7.92±2	1.71	170.5	4.54	4.56

Space group *P4/nmm* (D_{4h}^{7})

Atomic positions (origin at centre)

	BaMgSi			BaMgGe		
	x	*y*	*z*	*x*	*y*	*z*
2Ba in 2(*c*)	1/4	1/4	0.161	1/4	1/4	0.161
2Mg in 2(*b*)	3/4	1/4	1/2	3/4	1/4	1/2
2Si in 2(*c*)	1/4	1/4	0.706	1/4	1/4	0.705

Interatomic distances (Å)

BaMgSi		BaMgGe	
Ba - 4Ba	4.61	- Ba	4.64
- 4Ba	4.13	- Ba	4.16
- 4Mg	3.53	- Mg	3.55
- 1Si	3.58	- Ge	3.61
- 4Si	3.42	- Ge	3.45
Mg - 4Ba	3.53	- Ba	3.55
- 4Mg	3.26	- Mg	3.28
- 4Si	2.82	- Ge	2.83
Ge - 1Ba	3.58	- Ba	3.61
(Si)- 4Ba	3.42	- Ba	3.45
- 4Mg	2.82	- Mg	2.83

Material and Details of analysis

The elements were melted under argon in Al_2O_3 crucibles; both compounds were readily attacked in moist air. [No details of analysis given, presumably similar to Antimony-Lithium-Strontium, p. 9.]

1. Z. BAN and M. SIKIRIKA, 1965. *Acta Cryst.*, 18, 594.
2. *Strukturbericht*, 2, 45.
3. *Structure Reports*, 26, 9.

BARIUM NICKEL SULPHUR

I. Crystal structure and properties of barium nickel sulfide, a square-pyramidal nickel (II) compound. I.E. GREY and H. STEINFINK, 1970. *J. Amer. Chem. Soc.*, 92, 5093-5095.

$BaNiS_2$ F.W. = 260.2

Tetragonal, a = 4.430±1, c = 8.893±2 Å, [U = 174.5 Å^3], D_m = 4.94, Z = 2, D_x = 4.93.

Space group $P4/nmm$ (D^7_{4h})

Atomic positions

	x	*y*	*z*	*B* (Å^2)
2 S(1) in 2(b)	3/4	1/4	1/2	0.57±7
2 S(2) in 2(c)	1/4	1/4	0.1528±6	0.97±9
2 Ba in 2(c)	1/4	1/4	0.8048±1	0.84±5
2 Ni in 2(c)	1/4	1/4	0.4135±3	0.52±6

Interatomic distances (Å)

Standard deviations range from 0.001 to 0.006 Å.

S(1) - 4 Ni	2.345	Ba - 1 Ni	3.496[3.480]
- 4 Ba	3.500	- [4 Ni	3.685]
- 4 S(1)	3.132	- [4 Ba	4.430]
- 4 S(2)	3.799	- 1 S(2)	3.097[3.095]
- 4 S(1)	4.430	- 4 S(2)	3.155
		- 4 S(1)	3.500

Interatomic distances (Å)

Standard deviations range from 0.001 to 0.006 Å.

S(2) -	1 Ni	2.316[2.318]	Ni -	4 S(1)	2.345
	1 Ba	3.097[3.095]		1 S(2)	2.318
	4 Ba	3.155		4 Ba	3.685
	4 S(1)	3.799		1 Ba	3.480
	[4 S(2)	4.147]		4 Ni	3.490
	4 S(2)	4.430		4 Ni	4.430

Material and Details of analysis

Ni, S and BaS were heated at 820°C, yielding black crystals. Data were collected from a crystal (0.01 x 0.017 x 0.004 cm) for 334 *hkl* (diffractometer, MoK_α, balanced filters, Zr and Y, $2\theta < 60°$) of which 184 were independent. Absorption, polarization and Lorentz corrections were made. A trial structure based on packing considerations was refined by least squares; final R = 0.035.

Discussion

The structure (Fig. 6) contains 5-coordinated nickel atoms, each nickel being coordinated to 5 S at the apices of an almost regular square pyramid.

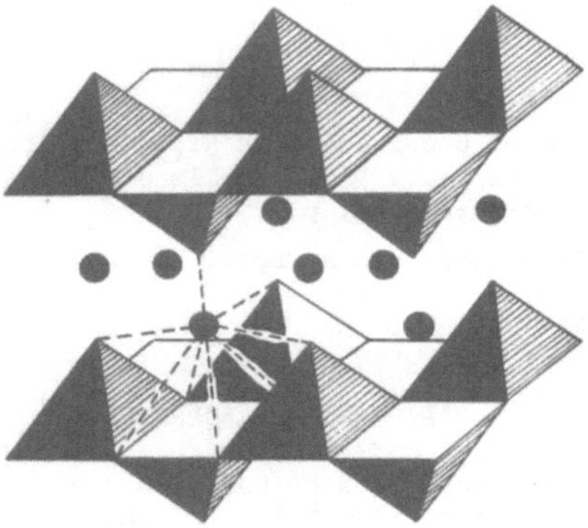

Fig. 6. The tetragonal $BaNiS_2$ structure showing the packing of the Ba atoms (●) between the NiS_5 pyramids.

BARIUM NITROGEN

I. Verfeinerung der Kristallstruktur von Bariumazid $Ba(N_3)_2$. E.M. WALITZI and H. KRISCHNER, 1970. *Z. Kristallogr.*, 132, 19-26.

$Ba(N_3)_2$ see also 1, 2, 4.

Monoclinic, $PbCl_2$ type (3) structure, a = 9.63, b = 4.41, c = 5.43 Å, β = 99°34', U = 227.4 $Å^3$, Z = 2.

Space group $P2_1/m$ (C_{2h}^2)

Atomic positions

			x	y	z
2 Ba	in	2(*e*)	0.2181±2	1/4	0.1715±3
2 N(1)	in	2(*e*)	0.093±3	3/4	0.860±5
2 N(2)	in	2(*e*)	0.108±2	3/4	0.646±6
2 N(3)	in	2(*e*)	0.112±2	3/4	0.437±4
2 N(4)	in	2(*e*)	0.584±2	1/4	0.779±8
2 N(5)	in	2(*e*)	0.460±3	1/4	0.750±5
2 N(6)	in	2(*e*)	0.333±3	1/4	0.702±5

Interatomic distances (Å)

Ba	- 2 N(4)	2.895[2.898]	N(3)	- 2 Ba	2.910
	- 2 N(3)	2.910		- 1 N(1)	2.334
	- 2 N(1)	2.914		- 1 N(2)	1.143
	- 1 N(6)	2.912[2.909]	N(4)	-[2 Ba	2.898]
	- 1 N(6)	2.943		- 1 N(5)	1.182[1.178]
[N(1)	- 1 Ba	2.972		-[N(6)	2.384]
	- 2 Ba	2.915]	N(5)	- 1 N(4)	1.182[1.178]
	- 1 N(2)	1.196		- 1 N(6)	1.203[1.207]
	-[1 N(3)	2.334]	[N(6)	- 1 Ba	2.910
N(2)	- N(1)	1.196		- 1 Ba	2.943
	- 1 N(3)	1.143		- 1 N(4)	2.384
				- 1 N(5)	1.207]

Material and Details of analysis

A crystal (0.17 mm) was used to obtain integrated Weissenberg photographs (MoK_{α}, $\sin\theta/\lambda \leq 0.9$ Å^{-1}); intensities were measured by microdensitometer (weak ones visually) and corrected for Lorentz, polarization and absorption factors. The trial structure of 2 was refined by Fourier and difference syntheses and block-diagonal least squares using anisotropic temperature factors; strong reflexions were corrected for extinction; R = 8.2% for 320 observed *hkl*.

Discussion

The structure can be derived from the $BaCl_2$ type.

1. K. TORKAR, H. KRISCHNER and H. RADL, 1965. *Mh. Chem.*, 96, 932.
2. E.M. WALITZI and H. KRISCHNER, 1969. *Z. Kristallogr.*, 129, 153.
3. H. BRAEKKEN, 1932. *Z. Kristallogr.*, 83, 222.
4. C.S. CHOI, 1969. *Acta Cryst.*, B 25, 2638.

BARIUM PALLADIUM

I. [X-ray investigation of platinum-barium and palladium-barium alloys in the region of Pt_5Ba and Pd_5Ba compositions.] N.N. ŽURAVLEV, N.P. ESAULOV, and I.V. RALL', 1970. *Kristallografija, SSSR*, 15, 374-376 [*Soviet Physics - Crystallography*, 15, 315-316].

$BaPd_5$ F.W. = 669.34

$BaPt_5$ F.W. = 1112.8

Hexagonal, $CaCu_5$ (1) type structure, $BaPd_5$, a = 5.54 ± 1, c = 4.33 ± 2 Å, [U = 115 Å^3, Z = 1, D_x = 9.66]. $BaPt_5$, a = 5.505 ±6, c = 4.337 Å, [U = 114 Å^3, Z = 1, D_x = 16.21], stable over a wide range of temperature, contrary to 2.

Space group *P6/mmm* (D_{6h}^1) from 1

Atomic positions

			x	*y*	*z*
1 Ba	in	1(*a*)	0	0	0
2 Pd(1) or Pt(1)	in	2(*c*)	1/3	2/3	0
3 Pd(2) or Pt(2)	in	3(*g*)	1/2	0	1/2

[*Interatomic distances* (Å)]

Ba	- 6Pd(1)	3.20	Pd(1)	- 3Ba	3.20	Pd(2)	- 4Ba	3.52
	- 12Pd(2)	3.52		- 3Pd(1)	3.20		- 4Pd(1)	2.69
				- 6Pd(2)	2.69		- 4Pd(2)	2.77
Ba	- 6Pt(1)	3.18	Pt(1)	- 3Ba	3.18	Pt(2)	- 4Ba	3.50
	- 12Pt(2)	3.50		- 3Pt(1)	3.18		- 4Pt(1)	2.69
				- 6Pt(2)	2.69		- 4Pt(2)	2.75

Material and Details of analysis

Elements (99.9+%) were arc-melted under argon and homogenized at 1200-1350°C and then slowly cooled. Observed powder diffraction intensities agreed well with calculated values.

1. *Structure Reports*, 11, 59.
2. *Ibid.*, 13, 31.

BARIUM PHOSPHORUS

I. Struktur und Bindungscharakter von Ba_3P_2. K.E. MAAS, 1970. *Z. anorg. Chem.*, 374, 11-18.

$Ba_4P_{2.67}$ F.W. = 632.05

Cubic, anti-Ce_2S_3 structure (defect Th_3P_4 type), a = 9.775 ± 5 Å, [U = 934.01 Å^3], D_m = 4.1, Z = 4, D_x = 4.52.

Space group $I\bar{4}3d$ (T_d^6)

Atomic positions

16	Ba	in	16(*c*):	with x = 0.0665 ± 10
10 2/3	P	in	12(*a*)	

Interatomic distances (Å)

Ba	- 3 P	3.22[3.21]	P	- 4 Ba	3.22[3.21]
	- 3 P	3.57		- 4 Ba	3.57
	- 3 Ba	3.81[3.82]		- 8 P	4.57
	- 3 Ba	4.23			

Details of analysis

A powder photograph with CuK_α (Ni filter) was used to obtain the lattice parameter by extrapolation against 1/2 ($\cos^2\theta/\sin\theta + \cos^2\theta/\theta$); the x parameter was obtained by systematic variation of x and comparison of observed and calculated ratios for groups of 12 selected lines.

BARIUM SULPHUR URANIUM

I. Préparation et structure cristalline du sulfure d'uranium et de baryum: $BaUS_3$. R. BROCHU, J. PADIOU and D. GRANDJEAN, 1970. *C.R. Acad. Sci. Paris*, C 271, 642-643.

BaS_3U F.W. = 471.6

Orthorhombic, perovskite, $GdFeO_3$, $CaTiO_3$ type, (1), a = 7.44 ± 2, b = 10.38 ± 2, c = 7.24 ± 2 Å, [U = 559.1 Å^3], D_m = 5.57, Z = 4, D_x = 5.60.

Space group $Pnma$ (D_{2h}^{16}) from absences

Atomic positions

			x	*y*	*z*
4 U	in	4(*b*)	0	0	1/2
4 Ba	in	4(*c*)	0.055	1/4	0
4 S(1)	in	4(*c*)	0.48	1/4	0.04
8 S(2)	in	8(*d*)	0.27	0.03	-0.27

Interatomic distances (Å)

Ba	- 1 S(1)	3.18	U	- 2 S(1)	2.62
	- 1 S(1)	3.38		- 2 S(2)	2.62
	- 1 S(1)	3.95		- [2 S(2)	2.63]
	- 2 S(2)	3.53	[S(1)	- 2 U	2.62
	- [2 S(2)	2.59]		- 1 Ba	3.18
	- 2 S(2)	3.40	S(2)	- 1 U	2.62
				- 1 U	2.63]

Material and Details of analysis

BaS and βUS_2 were ground and pelletized and then heated at 1100°C in evacuated, sealed SiO_2 tubes. Powder diffractometer intensities were corrected for Lorentz-polarization factors and a trial structure, based on the coordinates of 1, was refined only on Ba positions to R = 0.086 for 80 *hkl* (*I*'s).

1. *Structure Reports*, 11, 454; 21, 317.

BERYLLIUM RHODIUM

I. The crystal structure of $RhBe_{6.6}$. Q. JOHNSON, G.S. SMITH, O.H. KRIKORIAN and D.E. SANDS, 1970. *Acta Cryst.*, B 26, 109-113.

$RhBe_{6.6}$

Hexagonal, related to $CaZn_5$ ($D2_d$) type structure, a = 4.191 ± 1, c = 10.886 ± 3 Å, c/a = 2.597, [U = 165.59 Å^3], D_x = 3.792.

Space group $P\bar{6}m2$ (D^1_{3h}) probably, based on structure analysis

Atomic positions

			x	y	z	B (Å^2)
2 Rh(1)	in	2(*g*): with	0	0	0.1931±2	0.43±4
Rh(2)†	in	1(*d*): with	1/3	2/3	1/2	(0.5)
2 Be(1)	in	2(*h*): with	1/3	2/3	0.1233±39	0.47±66
2 Be(2)	in	2(*i*): with	2/3	1/3	0.1592±62	1.31±67
2 Be(3)	in	2(*g*): with	0	0	0.4018±24	0.64±40
6 Be(4)	in	6(*n*): with	0.4998±63	-0.4998	0.3170±14	1.16±30
3 Be(5)	in	3(*j*): with	0.8378±65	0.6757	0	1.16±42
Be(6)§	in	1(*f*): with	2/3	1/3	1/2	(1.0)

† 0.36±3 atoms.
§ 0.34±48 atoms, presence uncertain.

Interatomic distances (Å)

	Ligancy	Distance
Rh(1)	1 Be(3)	2.274
	3 Be(5)	2.411
	3 Be(2)	2.448
	6 Be(4)	2.493
	3 Be(1)	2.537
Rh(2)	6 Be(4)	2.332
	3 x 1/3 Be(6)	2.420
	6 Be(3)	2.646
Be(1)	3 Be(4)	2.432
	3 Be(2)	2.452
	6 Be(5)	2.490
	3 Rh(1)	2.537
	1 Be(1)	2.688
Be(2)	3 Be(4)	2.103
	3 Be(5)	2.134
	3 Rh(1)	2.448
	3 Be(1)	2.452

	Ligancy	Distance
Be(3)	1 Be(3)	2.141
	1 Rh(1)	2.274
	6 Be(4)	2.290
	3 x 1/3 Rh(2)	2.646
	3 x 1/3 Be(6)	2.646
Be(4)	4 Be(4)	2.093
	1 Be(2)	2.103
	2 Be(3)	2.290
	1 x 1/3 Rh(2)	2.332
	1 x 1/3 Be(6)	2.332
	1 Be(1)	2.432
	2 Rh(1)	2.493
Be(5)	2 Be(5)	2.039
	2 Be(2)	2.134
	2 Be(5)	2.152
	2 Rh(1)	2.411
	4 Be(1)	2.490
Be(6)	6 Be(4)	2.333
	3 x 1/3 Rh(2)	2.420
	6 Be(3)	2.646

Discussion

There are isomorphous compounds in the Co-, Fe-, and Ir-Be systems; extended ranges of composition for all phases are possible (Fig. 7).

Material and Details of analysis

Samples were prepared by melting the elements in BeO crucibles. Lattice parameters were derived from powder patterns (CrK_α, $\lambda_{\alpha 1}$ = 2.2896 Å) by a least-squares analysis. Oscillation, Weissenberg and precession photographs gave 6/*mmm* symmetry. Intensities for 140 reflexions were recorded using a stationary-crystal stationary-counter technique (MoK_α, Zr filter) from a fragment (0.06 mm max. dimension); no absorption correction was applied. A trial structure based on the $CaZn_5$ type was refined by difference Fouriers and least squares; dispersion corrections were applied, final *R* was 5.3%.

1. J.V. FLORIO, R.E. RUNDLE and A.I. SNOW, 1952. *Acta Cryst.*, 5, 449.
2. F.W. von BATCHELDER and R.F. RAEUCHLE, 1957. *Ibid.*, 10, 648; 1958. 11, 122.
3. D.E. SANDS, Q.C. JOHNSON, O.H. KRIKORIAN and K.L. KROMHOLTZ, 1962. *Ibid.*, 15, 1191.

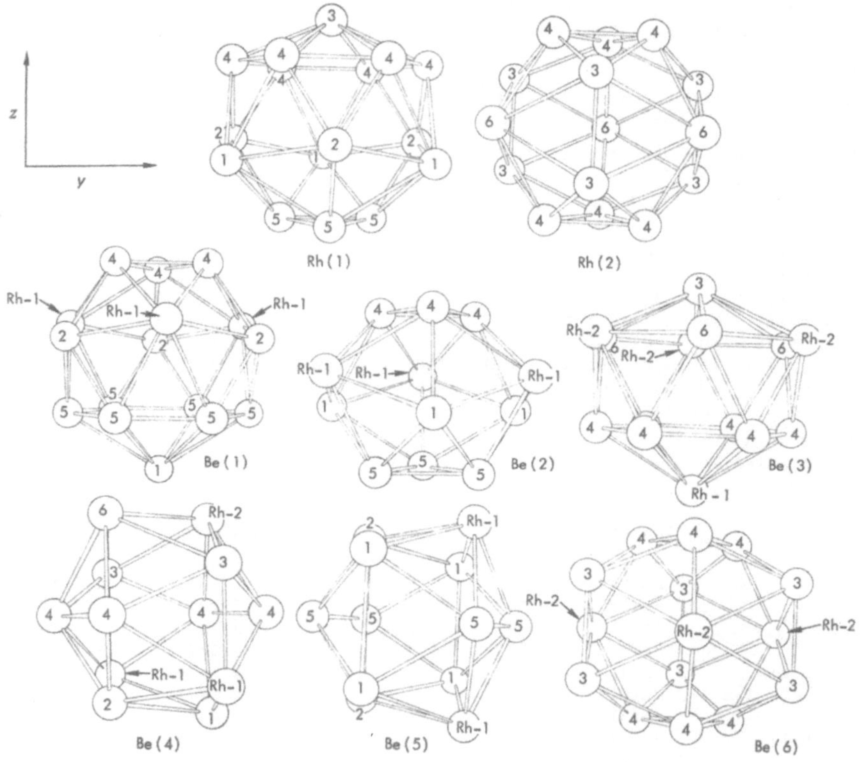

Fig. 7. The coordination polyhedra of the atoms in the $RhBe_{6.6}$ structure, the central atoms are omitted. Note that the Rh(2) and Be(6) sites are partially occupied.

BISMUTH COPPER LEAD SULPHUR

I. A redetermination of the crystal structure of aikinite ($BiS_2[S]Cu^{IV}Pb^{VII}$). M. OHMASA and W. NOWACKI, 1970. *Z. Kristallogr.*, 132, 71-86.

$BiCuPbS_3$ (aikinite) see also 1, 2. F.W. = 575.9

Orthorhombic, *a* = 11.638 ± 3, *b* = 4.039 ± 1, *c* = 11.319 ± 2 Å, [*U* = 532.06 Å^3], *Z* = 4, [D_x = 7.19].

Space group *Pnma* (D_{2h}^{16}) (piezoelectric test ruled out $Pn2_1a$)

Atomic positions

			x	*y*	*z*	*B* (Å^2)
4 Bi	in	4(*c*)	0.0189±1	1/4	0.6827±1	2.53±4
4 Cu	in	4(*c*)	0.2338±4	1/4	0.2085±4	3.75±8
4 Pb	in	4(*c*)	0.8329±1	1/4	0.0105±1	3.13±4
4 S(1)	in	4(*c*)	0.7160±5	1/4	0.6986±5	1.87±9
4 S(2)	in	4(*c*)	0.0467±5	1/4	0.1400±5	2.11±10
4 S(3)	in	4(*c*)	0.3775±5	1/4	0.0559±5	2.10±9

Interatomic distances (Å)

Bi	- 1 S(1)	2.658	S(1)	- 1 Bi	2.658
	- 2 S(2)	2.948		- 2 Cu	2.350
	- 1 S(3)	3.163		- 2 Pb	2.989
	- 2 S(3)	2.755	S(2)	- 2 Bi	2.948
Cu	- 2 S(1)	2.350		- 1 Cu	2.312
	- 1 S(2)	2.312		- 1 Pb	2.888
	- 1 S(3)	2.404		- 2 Pb	2.991
Pb	- 2 S(1)	2.989	S(3)	- 2 Bi	2.755
	- 2 S(2)	2.991		- 1 Cu	2.404
	- 1 S(2)	2.888			

Discussion

Each Pb atom has 6 S around it forming a prism. These prisms lie in a row parallel to the *b* axis; composite chains of $CuBiS_3$ along the *b* axis are formed by the infinite chains of Cu tetrahedra and Bi pyramids.

Aikinite shows a close relationship to stibnite (Sb_2S_3) and bismuthinite (Bi_2S_3) and is not isotypic with seligmannite and bournonite (p. 11, this volume).

Material and Details of analysis

A platy crystal of aikinite (0.141 x 0.063 x 0.021 mm) from Beresowsk, Ural, was used to record lattice parameters (back-reflexion Weissenberg photographs calibrated with Si) and intensities (diffractometer, Cu$K\alpha$, 573 *hkl*), which were corrected for Lorentz and polarization factors and absorption. The trial structure of 2 was refined by Fourier and difference syntheses and by block-diagonal least squares.

1. *Structure Reports*, 9, 192.
2. *Ibid.*, 17, 451.

BISMUTH GERMANIUM TELLURIUM

I. [Crystal structure determination for $Ge_3Bi_2Te_6$.] I.I. PETROV and R.M. IMAMOV, 1970. *Kristallografija, SSSR*, 15, 168-170 [*Soviet Physics - Crystallography*, 15, 134-136].

See also 1.

$Bi_2Ge_3Te_6$ F.W. = 1401.33

Hexagonal, *a* = 4.21 ± 2, *c* = 61.0 ± 5 Å, *c/a* = 14.49, [*U* = 1081 Å^3], D_m = 7.15, [*Z* = 3], D_x = 6.20 [6.46].

Space group *R*3*m* (C_{3v}^5) from ϕ^2 synthesis and absences

Atomic positions (hexagonal axes)

			x	*y*	*z*
3 Te(1)	in	3(a)	0	0	0.088
3 Te(2)	in	3(a)	0	0	0.188
3 Te(3)	in	3(a)	0	0	0.362
3 Te(4)	in	3(a)	0	0	0.637
3 Te(5)	in	3(a)	0	0	0.812
3 Te(6)	in	3(a)	0	0	0.912

Atomic positions (hexagonal axes)

			x	y	z
3 Bi(1)	in	3(*a*)	0	0	0
3 Bi(2)	in	3(*a*)	0	0	0.275
3 Ge(1)	in	3(*a*)	0	0	0.450
3 Ge(2)	in	3(*a*)	0	0	0.550
3 Ge(3)	in	3(*a*)	0	0	0.725

B = 0.3 Å^2 for all atoms.

[*Interatomic distances* (Å)]

Te(1)	- 3 Ge(1)	2.99	Bi(1)	- 3 Te(3)	2.99
	- 3 Ge(3)	3.03		- 3 Te(4)	3.03
Te(2)	- 3 Te(5)	3.56	Bi(2)	- 3 Te(4)	2.99
	- 3 Ge(2)	2.99		- 3 Te(6)	3.03
Te(3)	- 3 Bi(1)	2.99	Ge(1)	- 3 Te(1)	2.99
	- 3 Ge(3)	3.03		- 3 Te(5)	2.99
Te(4)	- 3 Bi(1)	3.03	Ge(2)	- 3 Te(2)	2.99
	- 3 Bi(2)	2.99		- 3 Te(6)	2.99
Te(5)	- 3 Te(2)	3.56	Ge(3)	- 3 Te(1)	3.03
	- 3 Ge(1)	2.99		- 3 Te(3)	3.03
Te(6)	- 3 Bi(2)	3.03			
	- 3 Ge(2)	2.99			

Material and Details of analysis

$Bi_2Ge_3Te_6$ was evaporated onto NaCl at 160°C and annealed for 3-4 hr. Oblique incidence electron diffraction photographs were used to obtain 150 *hkℓ* ($\sin \theta/\lambda \leq 0.92$ Å^{-1}) by microphotometer and visual estimates. A trial structure based on the 00z Patterson synthesis (ϕ^2) was refined by successive approximation; R = 19.6% or 17.8% after a dynamical correction was made.

Discussion

The structure can be described as an 18-layer rhombohedral stacking of Te atoms with Ge and Bi filling 5/6 of the octahedral holes; the sequence is *aBγ AβC Aβc aBc AβC αBC αBc AbC αBγ ABγ AbCa* ... where *ABC* are Te, *abc* are Bi and α,β,γ are Ge atoms.

1. N.K.H. ABRIKOSOV and G.T. DANILOVA-DOBRJAKOVA, 1965. *Izv. Akad. Nauk, SSSR, Neorg. Mater.*, 1, 57.

BISMUTH LEAD TELLURIUM

I. [X-ray structure determination of $PbBi_4Te_7$.] T.B. ŽUKOVA and A.I. ZASLAVSKIJ, 1970. *Ž. Strukt. Khim.*, 11, 462-468 [*J. Struct. Chem.*, 11, 423-428].

Bi_4PbTe_7 see also 1, 2, 3, 4. F.W. = 1936.31

Hexagonal, a = 4.451 ± 4, c = 41.532 ± 2 Å, (3), [c/a = 9.33], D_m = 7.73.
Rhombohedral cell, a = 14.070 Å, α = 18°12'. [A modified As_3Sn_4 structure. See ref. 6, p. 89.]

Space group $R\bar{3}m$ (D_{3d}^5)

Atomic positions (hexagonal coordinates)

				x	y	z
1.71 Pb + 6.86 Bi	} {	in	3(*a*)	0	0	0
		in	6(*c*)	0	0	0.429
6	Te(1)	in	6(*c*)	0	0	0.288
6	Te(2)	in	6(*c*)	0	0	0.136

B = 1.0 Å^2

Interatomic distances (Å)

Pb - 6 Te	3.19	Te(1) - 3 Pb	3.19
Bi - 3 Te(1)	3.31	- 3 Bi	3.31
- 3 Te(2)	3.07	Te(2) - 3 Bi	3.07
		- 3 Te(2)	3.62

Discussion

This structure differs from that of 2 and the possibility of polymorphism was tested by measuring lattice parameters over the range RT to 500°C; no changes were found.

Material and Details of analysis

Bi_4PbTe_7 was made by melting the elements in evacuated sealed SiO_2 ampoules. Debye-Scherrer photographs (FeK_α), oscillation photographs (CuK_α) and powder diffractometer patterns were used to establish the lattice parameters and the structure. Intensities for 00*l* reflexions (00.6 to 00.63) were measured on a linear diffractometer (MoK_α, Ross filters) and corrected for extinction (by comparison with a powder pattern). A trial model deduced from stacking considerations and the observed intensities was refined by systematic parameter variation (R = 16.8%). This model was confirmed by an electron density calculation which also served to fix the distribution of Bi and Pb atoms.

II. [Electron-diffraction analysis of $PbTe-Bi_2Te_3$ system phases.] I.I. PETROV and R.M. IMAMOV, 1970. *Kristallografija, SSSR*, 14, 699-703 [*Soviet Physics - Crystallography*, 14, 593-595].

Bi_4PbTe_7

Hexagonal, a = 4.42, c = 23.6 Å, $GeSb_4Te_7$ type (5), [c/a = 5.34, U = 399.28 Å^3, Z = 1, D_x = 8.05].

Space group $P\bar{3}m1$ (D^3_{3d})

Atomic positions

			x	*y*	*z*
1 Pb	in	1(*a*)	0	0	0
2 Bi(1)	in	2(*d*)	1/3	2/3	0.835
2 Bi(2)	in	2(*d*)	1/3	2/3	0.583
1 Te(1)	in	1(*b*)	0	0	1/2
2 Te(2)	in	2(*d*)	1/3	2/3	0.082
2 Te(3)	in	2(*c*)	0	0	0.248
2 Te(4)	in	2(*d*)	1/3	2/3	0.334

Interatomic distances (Å)

Pb - [6] Te(2)	3.20	[Te(1) - 6 Bi(2)	3.22
- [6] Pb	4.42		
Bi(1) - [3] Te(2)	3.22	Te(2) - 3 Pb	3.20
- [3] Te(3)	3.22	- 3 Bi(1)	3.22
Bi(2) - [3] Te(1)	3.22	Te(3) - 3 Bi(1)	3.22
- [3 Te(4)	3.22]	- 3 Te(4)	3.26
[given as Te(3)	3.28]	Te(4) - 3 Bi(2)	3.22
		- 3 Te(3)	3.26]

$Bi_2Pb_2Te_5$ F.W. = 1470.34

Hexagonal, $Pb_2Bi_2Se_5$ type (6), a = 4.46 ± 2 and c = 17.5 ± 2 Å, [c/a = 3.92, U = 301.46 Å^3, Z = 1, D_x = 8.10].

Space group $P\bar{3}m1$ (D^3_{3d})

Atomic positions

			x	y	z
1 Te(1)	in	1(*a*)	0	0	0
2 Te(2)	in	2(*d*)	1/3	2/3	0.782
2 Te(3)	in	2(*d*)	1/3	2/3	0.440
2 Pb	in	2(*c*)	0	0	0.329
2 Bi	in	2(*d*)	1/3	2/3	0.109

Interatomic distances (Å)

Te(1)	- 6 Bi(1)	3.20	Pb(1)	- 3 Te(2)	3.23
Te(2)	- 3 Pb	3.23		- 3 Te(3)	3.23
	- 3 Bi	3.20	Bi(1)	- 3 Te(1)	3.20
Te(3)	- 3 Te(3)	3.32		- 3 Te(2)	3.20
	- 3 Pb(1)	3.23			

Material and Details of analysis

$PbBi_4Te_7$ was prepared by sputtering the alloy onto an NaCl substrate and annealing at 130-140°C; $Pb_2Bi_2Te_5$ was prepared similarly but annealed at 180-200°C. $PbBi_4Te_7$: 190 reflexions ($\sin\theta/\lambda < 0.99$ Å^{-1}) were measured visually from an electron-diffraction photograph and used to calculate the Patterson (ϕ^2) and potential syntheses (ϕ) 00z and 1/3 2/3 z. The coordinates derived gave R = 18.1%.

Discussion

$PbBi_4Te_7$ is a 12-layer structure of form *aBcA BcAb CAbC* ...., where *a* = Pb, *b* and *c* = Bi and *A B* and *C* = Te, whereas the phase described in I. is a 21-layer structure; similarly, $Pb_2Bi_2Te_5$ is an *AbC aBC aBc* .... 9-layer structure.

1. E.I. ELAGINA and N. KH. ABRIKOSOV, 1959. *Ž. Neorg. Khim.*, 4, 1638.
2. A.G. TALYBOV and B.K. VAJNŠTEJN, 1961. *Kristallografija, SSSR*, 6, 541; *Idem*, 1961. *Ibid.*, 7, 43; A.G. TALYBOV, 1964. *Ibid.*, 9, 57.
3. A.I. ZASLAVSKIJ, 1968. *Ibid.*, 13, 232.
4. S. NAKAJIMA, 1963. *J. Phys. Chem. Solids*, 24, 479.
5. I.I. PETROV, R.M. IMAMOV and Z.G. PINSKER, 1968. *Kristallografija, SSSR*, 13, 417 [*Soviet Physics - Crystallography*, 13, 339].
6. K.A. AGEEV, A.G. TALYBOV and S.A. SEMILETOV, 1966. *Ibid.*, 11, 736 [*Ibid.*, 11, 630].

BORON CHROMIUM

I. The solubility of chromium in β-rhombohedral boron as determined in $CrB_{\sim 41}$ by single crystal diffractometry. S. ANDERSSON and T. LUNDSTROM, 1970. *J. Solid State Chem.*, 2, 603-611.

$B_{\sim 41}Cr$ see also 1. See $Al_{2.7}Cu_2B_{104}$ p. 3.

Rhombohedral, a = 10.162 Å, α = 65.295°, U = 827.5 Å^3.

Hexagonal axes, a = 10.9637 ± 2, c = 23.8477 ± 4 Å, U = 2482.5 Å^3 (single crystal). a = 10.9666 ± 3, c = 23.8514 ± 8 Å (powder).

Space group $R\bar{3}m$ (D_{3d}^5)

Atomic positions (hexagonal axes)

			x	y	z	B (Å^2)
6 Cr(1)*	in	6(*c*)	0	0	0.13451±6	0.146±31
18 Cr(2)†	in	18(*h*)	0.20547±19	0.41095±19	0.17364±14	0.944±61
3 B(1)	in	3(*b*)	0	0	1/2	0.79±17
6 B(2)	in	6(*c*)	0	0	0.38517±28	0.51±11
18 B(3)	in	18(*h*)	0.05543±20	0.11086±20	0.94409±15	0.469±64

* occupancy 71.9±6%

† occupancy 18.0±3%

Atomic positions (hexagonal axes)

			x	*y*	*z*	*B* (Å^2)
18 B(4)	in	18(*h*)	0.08635±20	0.17271±20	0.01332±15	0.316±61
18 B(5)	in	18(*h*)	0.11075±20	0.22150±20	0.88660±15	0.415±62
18 B(6)	in	18(*h*)	0.16993±20	0.33986±20	0.02782±15	0.454±63
18 B(7)	in	18(*h*)	0.12991±21	0.25981±21	0.76627±15	0.527±64
18 B(8)	in	18(*h*)	0.10239±20	0.20477±20	0.69813±15	0.432±63
18 B(9)	in	18(*h*)	0.05641±20	0.11283±20	0.32673±16	0.473±65
18 B(10)	in	18(*h*)	0.09018±21	0.18036±21	0.39859±15	0.501±63
18 B(11)§	in	18(*h*)	0.05642±31	0.11284±31	0.55495±23	0.88±10
36 B(12)	in	36(*h*)	0.17779±29	0.17576±29	0.17683±10	0.499±47
36 B(13)	in	36(*h*)	0.31909±28	0.29596±28	0.12884±10	0.504±47
36 B(14)	in	36(*h*)	0.26123±28	0.21692±28	0.41987±11	0.470±46
36 B(15)	in	36(*h*)	0.23635±28	0.25163±28	0.34685±10	0.426±46

§ occupancy 71.7±18%

Interatomic distances (Å)

B(1)	- 6 B(11)	1.693±5
	- 6 Cr(2)	2.434±2
B(2)	- 3 B(10)	1.742±2
	- 3 B(9)	1.758±6
	- 3 B(11)	1.785±7
B(3)	- B(5)	1.727±4
	- B(4)	1.752±5
	- 2 B(4)	1.761±4
	- 2 B(3)	1.823±3
	- Cr(1)	2.150±3
B(4)	- B(6)	1.624±3
	- B(3)	1.752±5
	- 2 B(4)	1.759±3
	- 2 B(3)	1.761±4
B(5)	- B(3)	1.727±4
	- B(7)	1.790±3
	- 2 B(13)	1.798±3
	- 2 B(12)	1.835±4
	- Cr(1)	2.163±2
B(6)	- B(4)	1.624±3
	- 2 B(15)	1.731±3
	- 2 B(14)	1.801±4
	- B(8)	1.828±4
B(7)	- B(8)	1.707±5
	- B(5)	1.790±3
	- 2 B(12)	1.832±4
	- 2 B(13)	1.849±3
B(8)	- B(7)	1.707±5
	- 2 B(9)	1.788±3
	- B(6)	1.828±4
	- 2 B(15)	1.839±4
B(9)	- B(2)	1.758±6
	- 2 B(8)	1.788±3
	- B(10)	1.830±5
	- 2 B(15)	1.854±3
	- 2 B(9)	1.855±3
B(10)	- B(2)	1.742±2
	- 2 B(14)	1.784±3
	- B(9)	1.830±5
	- 2 B(15)	1.857±4
	- 2 B(11)	1.864±5
	- Cr(2)	2.363±5
B(11)	- B(1)	1.693±5
	- B(2)	1.785±7
	- 2 B(14)	1.800±4
	- 2 B(11)	1.856±5
	- 2 B(10)	1.864±5
	- Cr(2)	2.005±5
	- 2 Cr(2)	2.398±4
B(12)	- B(13)	1.825±4
	- B(7)	1.832±4
	- B(5)	1.835±4
	- B(13)	1.845±4
	- B(12)	1.905±5
	- B(12)	1.972±5
	- Cr(1)	2.185±3
	- Cr(2)	2.442±3
B(13)	- B(14)	1.721±4
	- B(5)	1.798±3
	- B(13)	1.824±5
	- B(12)	1.825±4
	- B(12)	1.845±4
	- B(7)	1.849±3
	- Cr(2)	2.419±3
	- Cr(2)	2.431±3
B(14)	- B(13)	1.721±4
	- B(10)	1.784±3
	- B(11)	1.800±4
	- B(6)	1.801±4
	- B(15)	1.832±3
	- B(14)	1.892±5
	- Cr(2)	2.283±4
	- Cr(2)	2.446±4
B(15)	- B(15)	1.680±5
	- B(6)	1.731±3
	- B(14)	1.832±3
	- B(8)	1.839±4
	- B(9)	1.854±3
	- B(10)	1.857±4
	- B(15)	2.424±5
Cr(1)	- 3 B(3)	2.150±3
	- 3 B(5)	2.163±2
	- 6 B(12)	2.185±3

Interatomic distances (Å)

Cr(2) -	B(11)	2.005±5	Cr(2) - 2	B(13)	2.431±3
- 2	B(14)	2.283±4	-	B(1)	2.434±2
-	B(10)	2.363±5	- 2	B(12)	2.442±3
- 2	B(11)	2.398±4	- 2	B(14)	2.446±4
- 2	B(13)	2.419±3	- 2	Cr(2)	2.451±2

Material and Details of analysis

A triangular prismatic crystal was obtained from an arc-melted mixture of Cr (99.97%) and B (99.8%) and lattice parameters from Guinier photographs ($CrK\alpha_1$, λ = 2.28962 Å, Si standard a = 5.43054 Å) and also from the single crystal used. Intensities were taken from diffractometer data (θ-2θ scan, $CuK\alpha$, Ni filter, 1483 observations with 2θ < 130°, 553 independent *hkl*) and corrected for Lorentz and polarization factors. Scattering factors corrected for dispersion (f') were used in a full-matrix least squares. A preliminary investigation, using oscillation and Weissenberg photographs yielded the space group *C2/m* which was changed to $R\bar{3}m$ after analysing the diffractometer intensities; final R = 4.7% for all 553 *hkl*.

Discussion

The structure closely resembles that of $Al_{2.7}B_{104}Cu_2$ (p. 3) and both contain the three-dimensional B-atom network of β-rhombohedral boron (2) with the metal atoms occupying certain of the holes.

1. S. ANDERSSON and T. LUNDSTRÖM, 1968. *Acta Chem. Scand.*, 22, 3103.
2. J.L. HOARD and R.E. HUGHES, 1967. *The Chemistry of Boron and its Compounds*, Ed. E.L. Muetterties, p. 25. New York, Wiley.

BORON CHROMIUM YTTRIUM

I. [Crystal structure of the compound $YCrB_4$ and its analogues.] Ju.B. KUZ'MA, 1970. *Kristallografija, SSSR*, 15, 372-374 [*Soviet Physics - Crystallography*, 15, 312-314].

B_4CrY F.W. = 184.15

Orthorhombic, a = 5.972_5, b = 11.46_1, c = 3.461_4 Å, U = 236.9 Å^3, D_m = 5.14, Z = 4, D_x = 5.19. Lattice parameters of compounds with $YCrB_4$-type structure are given in Table I, p. 116.

Space group *Pbam* (D_{2h}^9) from absences and Patterson projection

Atomic positions

			x	*y*	*z*
4 Y	in	4(*g*)	0.125	0.150	0
4 Cr	in	4(*g*)	0.125	0.419	0
4 B(1)	in	4(*h*)	0.280	0.315	1/2
4 B(2)	in	4(*h*)	0.340	0.465	1/2
4 B(3)	in	4(*h*)	0.385	0.050	1/2
4 B(4)	in	4(*h*)	C.485	0.180	1/2

Interatomic distances (Å)

Y - 2	Y	3.76	Y - 2	B(4)	2.74
- 1	Y	3.75	- 2	B(3)	2.59
- 2	Y	3.46	Cr - 1	Y	3.08
- 1	Cr	3.08[3.09]	- 1	Y	3.04
- 1	Cr	3.04	- 2	Y	3.00[3.09]
- 2	Cr	3.00[3.10]	- 1	Cr	2.38
- 2	B(1)	2.97[2.72]	- 2	B(1)	2.30
- 2	B(4)	2.78	- 2	B(3)	2.29
- 2	B(1)	2.75[2.73]	- 2	B(3)	2.28
- 2	B(2)	2.75[2.76]	- 2	B(4)	2.23
- 2	B(2)	2.74	- 2	B(2)	2.22

Interatomic distances (Å)

B(1) -	2 Y	2.97[2.73]	B(3) -	2 Y	2.59
	2 Y	2.75[2.72]		2 Cr	2.29
	2 Cr	2.30		2 Cr	2.28
	1 B(4)	1.97		1 B(4)	2.09
	1 B(2)	1.76		1 B(3)	1.79
	1 B(4)	1.76		1 B(2)	1.66
B(2) -	2 Y	2.75[2.76]	B(4) -	2 Y	2.78
	2 Y	2.74		2 Y	2.74
	2 Cr	2.22		2 Cr	2.23
	1 B(2)	2.07		1 B(3)	2.09
	1 B(1)	1.76		1 B(1)	1.97
	1 B(3)	1.66		1 B(1)	1.76

Material and Details of analysis

Lattice parameters were determined from powder photographs (Cr*K*, Debye-Scherrer camera 142 mm diam.), whereas space group and intensities were derived from rotation photographs ([001] and [010], Mo and also Cu*K*). A trial structure based on the Patterson projection onto (001) was refined by Fourier syntheses and least squares; R = 8.8% (60 $hk0_{obs.}$) and 11.5% (79 $hkl_{obs.}$).

Discussion

The boron atoms lie at the centres of trigonal prisms composed of Y and Cr atoms and they also form a network of 7- and 5-membered rings. This structure is related to that of ScB_2C_2 and AlB_2 (Fig. 8).

1. G.S. SMITH, Q. JOHNSON and P.C. NORDINE, 1965. *Acta Cryst.*, 19, 668.
2. *Strukturbericht*, 3, 28.

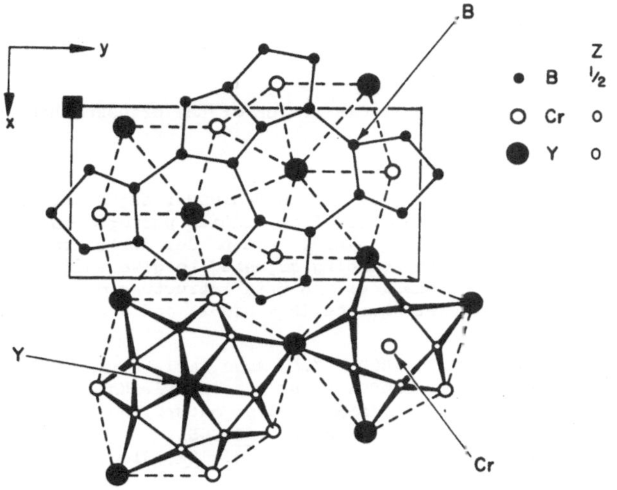

Fig. 8. The orthorhombic structure of B_4CrY projected onto (001); the unit cell is outlined and the coordination polyhedra of Y and Cr are also outlined [after I].

BORON MANGANESE

I. The crystal structure of MnB_4. S. ANDERSSON and J.-O. CARLSSON, 1970. *Acta Chem. Scand.*, 24, 1791-1799.

B_4Mn, see also 1, 2, 3, 4. F.W. = 98.18

Monoclinic, a = 5.5029 ± 3, b = 5.3669 ± 3, c = 2.9487 ± 2 Å, β = 122.71 ± 5°, U = 73.28 $Å^3$, Z = 2 [D_x = 4.45].

Space group $C2/m$ (C^3_{2h})

Atomic positions

			x	y	z	B ($Å^2$)
2 Mn	in	2(a)	0	0	0	1.52±41
						1.68±52*
8 B	in	8(j)	0.1948±57	0.3429±40	0.1967±95	1.67±64
			0.1994±71*	0.3460±49*	0.2033±119*	1.31±77*
Mean			0.1971	0.3445	0.2000	

* specimen with gum-arabic; the mean values of the positional parameters were used to calculate the interatomic distances.

Interatomic distances (Å)

Mn - 4 B	2.062	B - Mn	2.062
- 4 B	2.189	- Mn	2.189
- 4 B	2.210	- Mn	2.210
- 2 Mn	2.949	- B	2.474
B - B	1.669	- B	2.805
- B	1.825	- 2 B	2.920
- B	1.839	- 2 B	2.932
- B	1.871	- 2 B	2.949

Discussion

The MnB_4 structure is a small distortion of the orthorhombic CrB_4 structure (5).

Material and Details of analysis

Mn (99.99%) and B (99.8%) were arc-melted under argon and annealed between 1050° and 1400°C. Lattice parameters from Guinier photographs ($CuK\alpha_1$, Si internal standard, 5.43054 Å) and intensities from powder diffractometer data ($CrK\alpha$, LiF monochromator) with scanning adjusted to give uniform accuracy for 19 resolved reflexions (two samples). A trial structure derived from a Patterson was refined by least squares (R = 0.052) and confirmed by a difference Fourier.

1. R. FRUCHART and A. MICHEL, 1960. *C.R. Acad. Sci. Paris*, 251, 2953.
2. L.Ja. MARKOVSKIJ and E.T. BEZRUK, 1967. *Ž. Prikl. Khim.*, 40, 1199.
3. *Idem*, 1967. *Izv. Akad. Nauk, SSSR, Neorg. Mater.*, 3, 2165.
4. S. ANDERSSON, 1969. *Acta Chem. Scand.*, 23, 681.
5. S. ANDERSSON and T. LUNDSTRÖM, 1968. *Ibid.*, 22, 3103.

BORON NICKEL SILICON

I. A ternary W_5Si_3-type phase in the Ni-Si-B system. A.A. URAZ and S. RUNDQVIST, 1970. *Acta Chem. Scand.*, 24, 1843-1844.

$BNi_{4.6}Si_2$

Tetragonal, W_5Si_3 type structure, isostructural with $BCo_{4.7}Si_2$ (1), a = 8.632 ± 2, c = 4.290 ± 1 Å, c/a = 0.497.

Space group $I4/mcm$ (D^{18}_{4h}) from W_5Si_3 structure

Atomic positions

			x	*y*	*z*	*B* (Å^2)
	4 B	in 4(*a*)	0	0	1/4	1.5±8
0.58 x	4 Ni(1)	in 4(*b*)	0	1/2	1/4	-0.08±10*
	16 Ni(2)	in 16(*k*)	0.0787±2	0.2033±2	0	0.34±6
	8 Si	in 8(*h*)	0.1588±6	0.6588	0	0.58±11

* Became positive when a refinement was made with occupancy fixed at 0.61.

[*Interatomic distances* (Å)]

Ni(1)	- 2 Ni(1)	2.14	Ni(2)	- 2 B	2.17
	- 4 Si	2.22		- 2 Si	2.55
	- 8 Ni(2)	2.86		- 1 Si	2.30
Si	- 2 Ni(1)	2.22		- 1 Si	2.37
	- 2 Ni(2)	2.37		- 3 Ni(2)	2.66
	- 4 Ni(2)	2.55		- 2 Ni(2)	2.63
	- 2 Ni(2)	2.30		- 2 Ni(2)	2.54
B	- 2 B	2.15		- 2 Ni(1)	2.86
	- 8 Ni(2)	2.17			

Details of analysis

Lattice parameters were derived from Guinier photographs (Cu$K\alpha_1$, Si standard, *a* = 5.4305 Å) and intensities from Weissenberg photographs (Mo*K*, Zr, *hk*0-*hk*3, 117 observed *hkℓ*) were corrected for Lorentz and polarization factors. A trial structure based on 1 was refined by full-matrix least squares.

1. B. ARONSSON and G. LUNDGREN, 1959. *Acta Chem. Scand.*, 13, 433.

BORON SODIUM

I. The crystal structure of the ϕ phase in the boron-sodium system.
R. NASLAIN and J.S. KASPER, 1970. *J. Solid State Chem.*, 1, 150-151.

NaB_{15}

Orthorhombic, *a* = 5.847 ± 5, *b* = 8.415 ± 5, *c* = 10.298 ± 5 Å, [*U* = 506.69 Å^3], D_m = 2.44, *Z* = 4, D_x = 2.426.

Space group *Imam* (D_{2h}^{28}) [Non-standard setting; compare $AlB_{14}Mg$ p. 5]

Atomic positions

			x	*y*	*z*	*B* (Å^2)
16 B(1)	in	16(*j*): with	0.2475	0.0364	0.0820	0.89
16 B(2)	in	16(*j*): with	0.1609	0.8343	0.0484	0.81
8 B(3)	in	8(*h*): with	0.0000	0.1529	0.1011	0.98
8 B(4)	in	8(*h*): with	0.0000	-0.0472	0.1641	0.84
8 B(5)	in	8(*h*): with	0.0000	0.3941	0.1450	0.93
4 B(6)	in	4(*e*): with	0.0000	0.2541	0.2500	1.57
4 Na	in	4(*e*): with	0.0000	0.3386	0.7500	1.44

Material and Details of analysis

Crystals were prepared by thermal decomposition of NaB_6 at 1200°C. Laue and precession photographs (CuK_α and MoK_α) gave the symmetry *mmm*; 306 *hkl* reflexions were measured with a diffractometer (CuK_α) using peak height and background measurements. A trial structure based on $MgAlB_{14}$ (p. 5) was refined by Fourier and least-squares methods to *R* = 0.08.

Discussion

The structure is characterized by B_{12} icosahedra centred at 000, 1/2 1/2 0, 00 1/2, 1/2 1/2 1/2; the average B-B intraicosahedral distance is 1.807 Å; the average external B-B distance is 1.746 Å. The B(5)-B(6) separation of 1.599 Å is essentially the predicted distance for an electron-pair bond.

The Na atom occupies a cage formed by 16 B atoms; there are average B-Na distances of 2.798 (12 in all) and 2.551 Å (4 in all); the latter distance suggests Na bonding.

CADMIUM NICKEL

I. X-ray determination of the structure of primitive cubic gamma Ni,Cd phase. H. LJUNG and S. WESTMAN, 1970. *Acta Chem. Scand.*, 24, 611-617.

See 1, 2, 3, and also PtZn p. 90.

$NiCd_5$ F.W. = 620.7

Cubic, $NiCd_5$ type, a = 9.7878 ± 3 Å, [U = 937.7 $Å^3$], D_m = 8.90 ± 4 ($Ni_{0.162}Cd_{0.838}$), Z = 8, [D_x = 8.79].

Space group $P\bar{4}3m$ (T_d^1) from absences and structure

Atomic positions

			x	*y*	*z*	*B* ($Å^2$)
16/5Ni+4/5Cd)	in	4(*e*)	0.8169±22	0.8169±22	0.8169±22	2.0±7
24/5Ni+6/5Cd)	in	6(*f*)	0.2508±30	0	0	2.8±7
4 Cd(1)	in	4(*e*)	0.6066±15	0.6066±15	0.6066±15	2.9±5
4 Cd(2)	in	4(*e*)	0.3361±14	0.3361±14	0.3361±14	0.7±3
6 Cd(3)	in	6(*g*)	0.8631±17	1/2	1/2	0.9±2
12 Cd(4)	in	12(*i*)	0.2948±7	0.2948±7	0.0526±1	1.0±2
12 Cd(5)	in	12(*i*)	0.8096±9	0.8096±9	0.5403±14	2.6±3

Interatomic distances (Å)

σ's range from 0.008 to 0.041 Å.

A CLUSTER

Ni(1)*	- 3 Ni(2)	2.619
	- 3 Cd(4)	2.777
OT(*A*)	- 3 Cd(5)	2.709
Ni(2)*	- 2 Ni(1)	2.619
OH(*A*)	- 4 Ni(2)	3.471
	- 4 Cd(4)	2.962
	- 2 Cd(5)	3.336
Cd(4)	- 1 Ni(1)	2.777
CO(*A*)	- 2 Ni(2)	2.962
	- 2 Cd(4)	3.352
	- 1 Cd(2)	2.832
	- 1 Cd(3)	2.957
	- 2 Cd(5)	2.938
	- 1 Cd(3)	3.391
	- 2 Cd(5)	3.050

* Ni positions occupied by (4/5 Ni + 1/5 Cd)

B CLUSTER

Cd(1)	- 3 Cd(1)	2.950
IT(*B*)	- 3 Cd(2)	2.763
	- 3 Cd(3)	2.912
	- 3 Cd(5)	2.884
Cd(2)	- 3 Cd(4)	2.832
OT(*B*)	- 3 Cd(1)	2.763
	- 3 Cd(3)	2.991
	- 3 Cd(5)	2.839
Cd(3)	- 2 Cd(4)	2.957
OH(*B*)	- 2 Cd(1)	2.912
	- 2 Cd(2)	2.991
	- 4 Cd(5)	3.100
	- 2 Cd(4)	3.391
	- 1 Cd(3)	2.679
Cd(5)	- 1 Ni(1)	2.709
CO(*B*)	- 1 Ni(2)	3.336
	- 2 Cd(4)	2.938
	- 1 Cd(1)	2.884
	- 1 Cd(2)	2.839
	- 2 Cd(3)	3.100
	- 2 Cd(4)	3.050

Discussion

See Pt-Zn p. 90 for a description of the coordination nomenclature.

Material and Details of analysis

Ni (99.8%) and Cd were heated at 450 ± 10°C in sealed evacuated SiO_2 capsules, reground, reheated, slow-cooled to 400°C and quenched into water. Lattice parameters were derived from Guinier photographs (CuK_α, λ = 1.54050 Å, KCl *a* = 6.2919 Å as standard) and intensities from Weissenberg photographs and also from a diffractometer (MoK_α, 78 independent *hkl*). The trial structure, based on an ordered vacancy (at 0.11, 0.11, 0.11) in gamma-brass (4) was refined by block-diagonal and full-matrix least squares; *R* = 7.7% for 4 Ni + 2 Cd in 6(*g*) and *R* = 8.0% for 8 Ni in 4(*e*) + 6(*f*) not significantly worse, at the 25% probability level and chosen because temperature factors for Ni were physically more reasonable.

1. *Structure Reports*, 19, 79.
2. *Strukturbericht*, 2, 710.
3. O. VON HEIDENSTAM, A. JOHANSSON and S. WESTMAN, 1968. *Acta Chem. Scand.*, 22, 653.
4. *Strukturbericht*, 1, 497.

CADMIUM PHOSPHIDE

I. A note on the crystal structure of α-CdP_2. O. OLOFSSON and J. GULLMAN, 1970. *Acta Cryst.*, B 26, 1883-1884.

α-CdP_2, see also 1. F.W. = 174.35

Orthorhombic, a = 9.90 ± 1, b = 5.408 ± 5, c = 5.171 ± 5 Å, U = 276.9 Å^3, D_m = 4.19, Z = 4, D_x = 4.18. Data from 1.

Space group $Pna2_1$ (C_{2v}^9) from 1

Atomic positions

			x	y	z	B (Å^2)
4 Cd	in	4(a)	0.1529±2	0.1011±3	0.2606*	1.89±8
4 P(1)	in	4(a)	0.1185±7	0.4442±14	0.5957±23	1.43±12
4 P(2)	in	4(a)	0.9936±7	0.2693±12	0.8964±24	1.24±12

* Arbitrary value, held constant.

Interatomic distances (Å)

Cd - 1 P(1)	2.562±10	[P(1) - 1 Cd	2.562±10
- 1 P(1)	2.563±8	- 1 Cd	2.563±8]
- 1 P(2)	2.619±10	- 1 P(2)	2.200±14
- 1 P(2)	2.571±8	- 1 P(2)	2.167±12
		[P(2) - 1 Cd	2.571±8
		- 1 Cd	2.619±10
		- 1 P(1)	2.167±12
		- 1 P(1)	2.200±14]

Discussion

The P-P distances are now almost equal in agreement with the values found in other phosphides, contrary to 1, who found 2.386 Å and 2.050 Å.

Details of analysis

The structure given by 1 was refined using the data of 1. (Full-matrix least squares. R = 0.11 for 282 reflexions used in refinement; R = 0.12 for all 298 hkl).

1. J. GOODYEAR and G.A. STEIGMANN, 1969. *Acta Cryst.*, B 25, 2371.

CADMIUM PLATINUM ZINC

I. Über einige Strukturen im System Platin-Zink-Kadmium. Y. KHAN and K. SCHUBERT, 1970. *J. Less-Common Metals*, 20, 266-268.

Pt_2ZnCd

Tetragonal, a = 2.90, c = 7.24 Å, Z = 1.

Space group $P4/mmm$ (D_{4h}^1)

Atomic positions

			x	y	z
2 Pt	in	2(h)	0.5	0.5	0.23
1 Zn	in	1(a)	0	0	0
1 Cd	in	1(b)	0	0	1/2

[*Interatomic distances* (Å)]

Pt	- 4 Zn	2.64	Zn	- 8 Pt	2.64
	- 4 Cd	2.83		- 4 Zn	2.90
	- 4 Pt	2.90	Cd	- 8 Pt	2.83
				- 4 Cd	2.90

$Pt_3Zn_{4.4}Cd_{0.6}$

Trigonal, Th_3Pt_5 type (1), a = 7.050, c = 2.792 Å, Z = 1.

Space group $P31m$ (C^2_{3v}) [The more symmetrical space group $P\bar{6}2m$ (D^3_{3h}) can also be used]

Atomic positions

			$P31m$			[$P\bar{6}2m$]			
			x	y	z		x	y	z
3 Pt	in	3(c)	0.358	0	0.000	3(f)	0.358	0	0
2 (Zn,Cd)(1)	in	2(b)	1/3	2/3	0.500	2(d)	1/3	2/3	1/2
3 (Zn,Cd)(2)	in	3(c)	0.735	0	0.500	3(g)	0.735	0	1/2

[*Interatomic distances* (Å)]

Pt(1)	- 2 Pt	2.79	Zn	- 2 Pt	2.59	Cd	- 2 Pt	2.96
	- 2 Zn	2.74		- 2 Pt	2.74		- 2 Pt	3.00
	- 2 Cd	2.42*		- 2 Zn	2.79		- 1 Zn	2.40*
	- 2 Cd	3.00		- 1 Cd	2.40*		- 2 Cd	2.79
				- 1 Cd	2.87		- 1 Cd	2.64

* [Short distances]

$Pt_3Zn_2Cd_4$

Cubic, Ti_2Ni type (zeta carbide, $E9_3$, (2, 3)), a = 11.869 Å.

Space group $Fd3m$ (O^7_h)

[*Atomic positions*] [origin at $\bar{3}m$]

			x	y	z
(Zn,Cd)(1)	in	16(d)	1/2	1/2	1/2
(Zn,Cd)(2)	in	48(f)	0.940	1/8	1/8
Pt	in	32(e)	0.715	0.715	0.715

[*Interatomic distances* (Å)]

Zn(1)	- 6 Pt	2.62	Zn(2)	- 2 Pt	2.70
	- 6 Zn(2)	3.08		- 2 Pt	3.07
Pt	- 3 Zn(1)	2.62		- 2 Zn(1)	3.08
	- 3 Zn(2)	2.70		- 4 Zn(2)	3.11
	- 3 Pt	3.02		- 4 Zn(2)	3.16
	- 3 Zn(2)	3.07			

Material and Details of analysis

The alloys were prepared by melting the elements (99.9%) in evacuated silica tubes with subsequent annealing; Guinier powder patterns with qualitative agreement between observed (S, M, W, etc.) and calculated intensities.

1. J.R. THOMSON, 1963. *Acta Cryst.*, 16, 320.
2. *Structure Reports*, 23, 195.
3. *Ibid.*, 22, 187.

CALCIUM GERMANIUM LITHIUM

I. $Ca_{0.9}Li_{0.13}Ge_{1.97}$ - eine neue Stapelvariante des $CaSi_2$-Typs. W. MÜLLER, H. SCHÄFER and A. WEISS, 1970. *Z. Naturf.*, 25B, 431-432.

$Ca_{0.9}Li_{0.13}Ge_{1.97}$

Hexagonal, stacking variant of $CaSi_2$ type (1, 2), a = 3.98 ± 1, c = 10.19 ± 2 Å, c/a = 2.56, U = 140 Å^3, D_m = 4.20, Z = 2, D_x = 4.25.

Space group $P6_3mc$ (C_{6v}^4)

Atomic positions

			x	*y*	*z*
2 Ca*	in	2(*b*)	1/3	2/3	0.000
2 Ge(1)*	in	2(*b*)	1/3	2/3	0.688
2 Ge(2)*	in	2(*a*)	0	0	0.292

* Li content not located.

Interatomic distances (Å)

Ca - 3 Ge(1)	2.99	Ge(1) - 3 Ca	2.99	Ge(2) - 3 Ca	3.13
- 1 Ge(1)	3.18	- 1 Ca	3.18	- 3 Ge(1)	2.53
- 3 Ge(2)	3.13	- 3 Ge(2)	2.53		

Material and Details of analysis

Platy crystals were made by melting $CaGe_2$ and Li in a Ta crucible under argon. Lattice parameters, space group and intensities were derived from Weissenberg photographs. A trial structure based on a Patterson projection was refined by Fourier and trial and error methods (R = 10.3%).

1. K.H. JANZON, H. SCHAFER and A. WEISS, 1968. *Z. Naturf.*, 23B, 1544.
2. *Strukturbericht*, 1, 218.

CARBON TANTALUM VANADIUM

I. The crystal structure of Ta_2VC_2. E. RUDY, 1970. *J. Less-Common Metals*, 20, 49-55.

Ta_2VC_2 F.W. = 424.85

Rhombohedral, a = 7.481 Å, α = 23°27', *hexagonal cell*, a = 3.045, c = 21.81 Å, [U = 175.13 Å].

Space group $R\bar{3}m$ (D_{3d}^5) *from systematic absences*

Atomic positions

Hexagonal cell

			x	*y*	*z*
6 *T*	in	6(*c*)	0	0	2/9
3 *T*	in	3(*a*)	0	0	0
6 C	in	6(*c*)	0	0	7/18
		3(*b*)	0	0	0.5000

T = transition metal

[*Interatomic distances* (Å)]

T - 6 C	2.14	C - 6 *T*	2.14
- 6 *T*	2.99	- 1 C	2.42
- 6 *T*	3.05	- 3 C	2.99
		- 6 C	3.05
		C(*b*) - 6 *T*	2.14
		- 2 C	2.42
		6 C	3.05

Material and Details of analysis

Ta, VC and TaC were hot-pressed in graphite dies; powder photographs using CuK_{α} and CrK_{α} radiation; agreement for observed and calculated intensities; carbon positions not definite.

Discussion

The stacking of metal atoms is ABABCBCAC.

CERIUM IRON SILICON

I. [Crystal structures of CeFeSi and related compounds.] O.I. BODAK, E.I. GLADYŠEVSKIJ and P.I. KRIPJAKEVIČ, 1970. *Ž. Strukt. Khim.*, 11, 305-310.

CeFeSi F.W. = 224.05

Tetragonal, PbFCl(EO_1) (1) type, a = 4.062 ± 3, c = 6.752 ± 5 Å, c/a = 1.660, [U = 111.41 $Å^3$], D_m = 6.54, Z = 2, D_x = 6.64.

Space group $P4/nmm$ (D_{4h}^7)

Atomic positions

	x	y	z
2 Ce in 2(c)	1/4	1/4	0.672±1
2 Fe in 2(a)	3/4	1/4	0
2 Si in 2(c)	1/4	1/4	0.175±2

Interatomic distances (Å)

Ce - 4 Fe	3.005	Fe - 4 Si	2.349
- 4 Si	3.053	- 4 Fe	2.872
- 2 Si	3.356[3.356, 3.396]	- 4 Ce	3.005
- 4 Ce	3.693	Si - 4 Fe	2.349
		- 4 Ce	3.053
		- 1 Ce	3.356
		- 1 Ce	3.396

Material and Details of analysis

Ce (99.6%), carbonyl Fe (99.96%) and Si (99.99%) were arc-melted, and subsequently one alloy homogenized for 3 months at 800°C, yielded single crystals. Powder photographs (CrK) and rotation, oscillation and Weissenberg photographs (CuK); Patterson and Fourier projections; R = 13.5% for 27 visually estimated $0kl$, corrected for Lorentz and polarization factors.

1. *Strukturbericht*, 2, 45.

CHROMIUM LITHIUM SULPHUR

I. The crystal structure of lithium thiochromite, $LiCrS_2$. J.G. WHITE and H.L. PINCH, 1970. *Inorg. Chem.*, 9, 2581-2583.

$CrLiS_2$ see also 1. F.W. = 123.06

Trigonal, a = 3.456 ± 1, c = 6.020 ± 2 Å, U = 62.60 $Å^3$, D_m = 3.01, Z = 1, D_x = 3.26.

Space group $P\bar{3}m1$ (D_{3d}^3)

Atomic positions

	x	y	z	B ($Å^2$)	β_{11}	β_{22}	β_{33}	β_{12}
1 Cr in 1(a)	0	0	0	-	0.082±11	0.082	0.0038±12	0.041
1 Li in 1(b)	0	0	1/2	2.3±1.2	-	-	-	-
2 S in 2(d)	1/3	2/3	0.2247±7	-	0.069±10	0.069	0.0051±12	0.034

Interatomic distances (Å)

Li	- [6] S	2.597[2.594]		Cr	- [6] S	2.415[2.411]
	- [2] Cr	3.010		[Cr	- 2 Li	3.010]
[S	- 3 Cr	2.411				
	- 3 Li	2.594]				

Discussion

The structure is an ordered form of the NiAs structure.

Material and Details of analysis

Powders and crystals were grown by heating Li_2CO_3 and Cr_2O_3 with CS_2 saturated argon. Lattice parameters were taken from Debye-Scherrer photographs (CuK_α, Ni filter) and intensities were obtained visually from Weissenberg photographs (CuK_α, Ni filter, $h0\ell$-$h2\ell$, multiple-film) using a crystal 0.12 mm long x 0.053 mm diameter which was cut from a hexagonal plate. Lorentz, polarization and cylindrical absorption corrections were applied. A trial structure derived from packing arguments was refined by full-matrix least squares. Intensities of the pairs $hk\ell$ and $hk\bar{\ell}$ were affected by twinning and a twin factor (0.272) was used to correct these. The lithium atom was located from a difference Fourier and the final R = 0.068. Other space groups and structures were tested.

1. M. SERGENT and J. PRIGENT, 1965. *C.R. Acad. Sci., Paris,* 261, 5135.

CHROMIUM SELENIUM

I. Preparation, structure, and properties of some chromium selenides. Crystal growth with selenium vapor as a novel transport agent. F.H. WEHMEIER, E.T. KEVE and S.C. ABRAHAMS, 1970. *Inorg. Chem.,* 9, 2125-2131.

$Cr_{0.68}Se$-*T'* F.W. = 114.32

Trigonal, a = 3.612 ± 1, c = 5.775 ± 2 Å, c/a = 1.599, U = 65.24 Å^3, D_m = 6.0 ± 1, Z = 2, D_x = 5.94.

Space group $P3m1$ (C^1_{3v})

Atomic positions (ideal)

			x	y	z
1 Se(1)	in	1(*b*)	1/3	2/3	0.25
1 Se(2)	in	1(*c*)	2/3	1/3	0.75
1 Cr(1)	in	1(*a*)	0	0	0.50
0.36 Cr(2)	in	1(*a*)	0	0	0.00

[*Interatomic distances* (Å)]

Cr(1)	- 3 Se(1)	2.54		Se(1)	- 3 Cr(1)	2.54
	- 3 Se(2)	2.54			- 3 Cr(2)	2.54
	- 2 Cr(2)	2.89		Se(2)	- 3 Cr(1)	2.54
Cr(2)	- 3 Se(1)	2.54			- 3 Cr(3)	2.54
	- 3 Se(2)	2.54				
	- 2 Cr(1)	2.89				

Cr_2Se_3-*M* F.W. = 340.1

Monoclinic, a = 6.227 ± 1, b = 3.582 ± 1, c = 11.528 ± 2 Å, β = 90.77°, U = 257.1Å^3 D_m = 5.8 ± 1, Z = 8/3, D_x = 5.84. Alternative C-centred cell, a' = 13.101, b' = 3.582, c' = 11.528 Å and β = 151.62° by transformation $101/0\bar{1}0/00\bar{1}$.

Space group $I2/m$ (C^3_{2h}) non-standard setting for comparison with NiAs.

Atomic positions (ideal)

				x	y	z
4	Se(1)	in	4(*i*)	2/3	0	3/8
4	Se(2)	in	4(*i*)	1/3*	0	1/8*

[* Misprinted as 2/3 and 5/8. Private communication - author]

Atomic positions (ideal)

				x	y	z
4	Cr(1)	in	4(i)	0	0	1/4
1 1/3	Cr(2)	in	2(c)	0	0	1/2
	Vacancy		2(a)	0	0	0

[*Interatomic distances* (Å)]

Cr(1)	- 1 Se(1)	2.54	Se(1)	- 2 Cr(1)	2.51
	2 Se(1)	2.51		1 Cr(1)	2.54
	2 Se(2)	2.51		1 Cr(2)	2.51
	1 Se(2)	2.54	Se(2)	- 2 Cr(1)	2.51
	1 Cr	2.88		1 Cr(1)	2.54
Cr(2)	- 2 Se(1)	2.51		2 Cr(2)	2.53
	4 Se(2)	2.53			
	2 Cr(1)	2.88			

Cr_2Se_3-*T* F.W. = 340.1

Trigonal, *P*3 (C_3^1), a = 12.509 ± 2, c = 34.765 ± 5 Å, U = 4712 Å^3, Z = 48. There is a rhombohedral subcell with space group $R\bar{3}$, a' = 6.254, c' = 17.382 Å (hexagonal axes).

Space group $R\bar{3}$ (C_{3i}^2) for subcell

Atomic positions referred to $R\bar{3}$ subcell (idealized) with hexagonal axes

				x	y	z
3	Cr(1)	in	3(a)	0	0	0
3	Cr(2)	in	3(b)	0	0	1/2
6	Cr(3)	in	6(c)	0	0	1/3
18	Se	in	18(f)	1/3	0	1/4

[*Interatomic distances* (Å)]

Cr(1)	- 6 Se	2.54	Cr(3)	- 6 Se	2.54
Cr(2)	- 6 Se	2.54		- 1 Cr(2)	2.90
	- 2 Cr(3)	2.90	Se	- 1 Cr(1)	2.54
				- 1 Cr(2)	2.54
				- 2 Cr(3)	2.54

Material and Details of analysis

Phases were prepared from 99.9% elements in evacuated transparent SiO_2 ampoules with internal pressures between 3 and 10 atm. at the formation temperatures of 870° to 1031°C. *T*- and *M*-Cr_2Se_3 were prepared as single crystals by Se transport; *T'*-Cr_2Se_3 single crystals were prepared by iodine transport. Guinier powder photographs ($CuK\alpha_1$ = 1.540562 Å) were used to obtain the lattice parameters by a least-squares procedure; precession photographs were taken with MoK_α.

Discussion

The structures (Fig. 9) show that the Cr-Se phases (1-7) are all characterized by an alternation of completely filled and partially filled layers.

1. M. CHEVRETON and F. BERTAUT, 1961. *C.R. Acad. Sci., Paris*, 253, 145.
2. M. CHEVRETON, M. MURAT, C. EYRAUD and E.F. BERTAUT, 1963. *J. de Phys.*, 24, 443.
3. M. CHEVRETON and B. DUMONT, 1968. *C.R. Acad. Sci., Paris*, C 267, 884.
4. M. CHEVRETON, E.F. BERTAUT and S. BRUNIE, 1964. *Bull. Soc. Sci., Bretagne*, 39, 77.
5. M. CHEVRETON, 1967. *Bull. Soc. Franç. Minér. Crist.*, 90, 592.
6. F.K. LÖTGERING and E.W. GORTER, 1957. *J. Phys. Chem. Solids*, 3, 238.
7. A.W. SLEIGHT and T.A. BITHER, 1969. *Inorg. Chem.*, 8, 566.

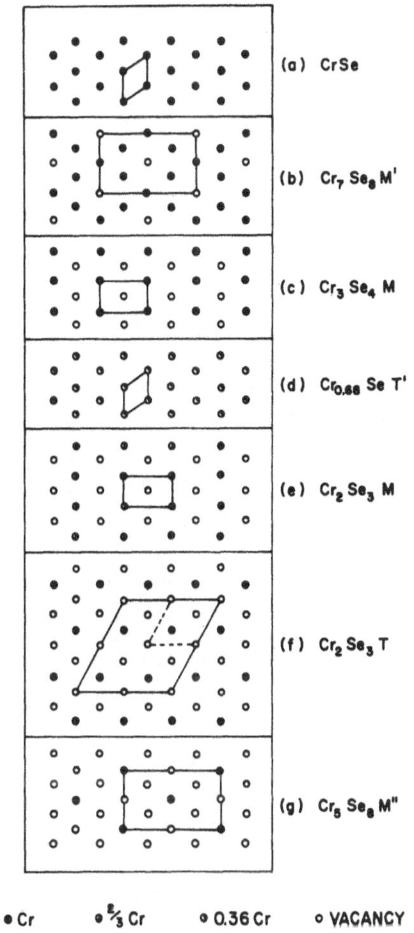

Fig. 9. Representation of the partly vacant, alternate, layers in the Cr-Se phases. The unit-cell is outlined as well as the subcell for *f*. [After I]

CHROMIUM TELLURIUM

I. The magnetic structure of Cr_2Te_3, Cr_3Te_4 and Cr_5Te_6. A.F. ANDRESEN, 1970. *Acta Chem. Scand.*, 24, 3495-3509.

Cr_2Te_3	see also 1, 2, 3, 4.	F.W. = 486.8
Cr_3Te_4	see also 2, 5.	F.W. = 666.4
Cr_5Te_6	see also 3, 6.	F.W. = 1026.0

Cr_2Te_3

Hexagonal, NiAs type cell ($a \approx \sqrt{3}a_o$, $c \approx 2c_o$), a = 6.829 ± 6, c = 11.922 ± 12 Å at 4.2°K, a = 6.814 ± 6, c = 12.073 ± 12 Å at 293°K, in agreement with 4.

Space group $P\bar{3}1c$ (D^2_{3d}) from absences and 4

Atomic positions

			x	*y*	*z*	*B* (Å^2)
2 □	in	2(*a*)	0	0	1/4	
2 □	in	2(*d*)	2/3	1/3	1/4	{1.42±23
2 Cr(1)	in	2(*c*)	1/3	2/3	1/4	{2.71±52*
2 Cr(2)	in	2(*b*)	0	0	0	
4 Cr(3)	in	4(*f*)	1/3	2/3	0.000±2	"
					-0.007±2*	
12 Te	in	12(*i*)	0.340±2	0.003±2	0.372±1	0.82±13
			0.337±3*	0.000±2*	0.379±2*	0.76±18*

*At 4.2°K. The moments at 4.2°K Cr(1) 0.20±18, Cr(2) 2.49±29, Cr(3) 2.61±23 μ_B, all parallel to *c*.

Cr_5Te_6

Monoclinic, *a* = 6.857, *b* = 3.954, *c* = 12.272 Å, β = 91.17° (at 4.2°K), [*U* = 332.7 Å^3].

Space group *I*2/*m* (C_{2h}^3) from Cr_3Te_4 (5)

Atomic positions (from Cr_3Te_4 (5))

				x	*y*	*z*
2/3	Cr(1)	in	2(*a*)	0	0	0
2	Cr(1)	in	2(*c*)	0	0	1/2
4	Cr(2)	in	4(*i*)	0.022	0	0.240
4	Te(1)	in	4(*i*)	0.336	0	0.866
4	Te(2)	in	4(*i*)	0.329	0	0.379

Average ferromagnetic moment per atom is 2.40 μ_B, and the paramagnetic value 3.92 μ_B. Cr_3Te_4 has a similar structure, modified so that the anti-ferromagnetic moments pointing along *b* in Cr_5Te_6 point along the (*b* + *c*) direction.

[*Interatomic distances* (Å)]

Cr_2Te_3		293°K	4.2°K
Cr(1)	- 2 Cr(3)	3.02	3.06
	- 6 Te	2.71	2.74
Cr(2)	- 6 Te	2.78	2.72
Cr(3)	- 3 Te	2.75	2.73
	- 3 Te	2.72	2.64
Te	- 1 Cr(1)	2.71	2.74
	- 1 Cr(2)	2.78	2.72
	- 1 Cr(3)	2.75	2.73
	- 1 Cr(3)	2.72	2.64

Cr_5Te_6		4.2°K
Cr(1)	- 2 Cr(3)	2.95
	- 2 Te(1)	2.86
	- 4 Te(2)	2.72
Cr(2)	- 4 Te(1)	2.79
	- 2 Te(2)	2.73
Cr(3)	- 2 Te(1)	2.83
	- 1 Te(1)	2.75
	- 2 Te(2)	2.67
	- 1 Te(2)	2.68
Te(1)	- Cr(1)	2.86
	- 2 Cr(2)	2.79
	- 1 Cr(3)	2.75
	- 2 Cr(3)	2.83
Te(2)	- 2 Cr(1)	2.72
	- 1 Cr(2)	2.73
	- 2 Cr(3)	2.67
	- 1 Cr(3)	2.68

Material and Details of analysis

Samples prepared as in 3 were slowly cooled from 900°C in 20° steps. High resolution powder neutron diffraction patterns (λ = 1.864 ± 2 Å, Ge (111) monochromator; λ/2 and λ/3 corrections were made). Refinement was carried out by a profile refinement program which took into account both overlapped and resolved peaks. A special correction ($I_{obs} \cdot \exp(-G\alpha^2)$. *G* = a parameter included in least squares, α = angle between scattering vector and plate normal) was applied to account for the preferred orientation present in the Cr_2Te_3 samples. Final *R* values: Cr_2Te_3 7.9% at room temperature; 3.2% at 4.2°K.

1. A. BERG, 1950. *Thesis*, University of Oslo.
2. M. CHEVRETON, E.F. BERTAUT and F. JELLINEK, 1963. *Acta Cryst.*, 16, 431.
3. A.F. ANDRESEN, 1963. *Acta Chem. Scand.*, 17, 1335.
4. M. CHEVRETON, 1964. *Thesis*, University of Lyons.
5. E.F. BERTAUT, G. ROULT, R. ALEONARD, R. PAUTHENET, M. CHEVRETON and R. JANSEN, 1964. *J. Phys. Radium*, 25, 582.
6. F. GRØNVOLD and E.F. WESTRUM, 1964. *Z. anorg. Chem.*, 328, 272.

COBALT

I. Nachweis einer neuen Phase bei der funkenerosiven Behandlung von Kobalt. E. KRAINER and J. ROBITSCH, 1970. *Z. Metallk.*, 61, 350-354.

δ-Co

Hexagonal, similar to Fe_3Th_7(2), a = 8.288, c = 10.542 Å, c/a = 1.272, [U = 627.11 Å^3], A = 46.

Space group $P6_3mc$ (C_{6v}^4)

Atomic positions

			x	*y*	*z*
2 Co(1)	in	2(*a*)	0.0000	0.0000	0.1185
2 Co(2)	in	2(*a*)	0.0000	0.0000	0.3814
2 Co(3)	in	2(*b*)	0.3333	0.6666	0.9430
2 Co(4)	in	2(*b*)	0.3333	0.6666	0.6682
2 Co(5)	in	2(*b*)	0.3333	0.6666	0.3948
6 Co(6)	in	6(*c*)	0.1560	0.8440	0.9860
6 Co(7)	in	6(*c*)	0.8154	0.1846	0.0190
6 Co(8)	in	6(*c*)	0.1940	0.8060	0.2510
6 Co(9)	in	6(*c*)	0.8552	0.1448	0.2500
6 Co(10)	in	6(*c*)	0.4924	0.5076	0.0427
6 Co(11)	in	6(*c*)	0.5260	0.4740	0.3056

[*Interatomic distances* (Å)

Coordination ranges from 5 to 9 and interatomic distances from 2.35 Å (Co(7) to 2 Co(10)) to 2.65 Å.]

Material and Details of analysis

Technically pure Co (99.9%) treated by spark erosion or a plasma jet yields surfaces containing δ-Co, C-content 0.042 ± 2%; Vickers Hardness No. = 462 (Co = 232). Diffractometer powder patterns of mixtures (δ was never obtained pure) were interpreted on the basis of the structure given; R' = 0.746[*sic*] for intensities.

Discussion

This modification was earlier reported by Metcalfe (1, II) but Newkirk and Geisler (1, I) and Basinski and Christian (1, ref. 4) were unable to observe it as a stable phase. The proposed structure is similar to that of Fe_3Th_7(2).

1. *Structure Reports*, 17, 128-130.
2. *Ibid*, 20, 132.

COBALT GADOLINIUM

I. The structure of Gd_3Co. O.A.W. STRYDOM and L. ALBERTS, 1970. *J. Less-Common Metals*, 22, 511-515.

$CoGd_3$, see also 1, 2. F.W. = 530.68

Orthorhombic, Fe_3C type (3), a = 7.05 ± 1, b = 9.54 ± 1, c = 6.32 ± 1 Å, [U = 425.06 Å^3, Z = 4, D_x = 8.29]. The lattice parameters agree with 2, but not with 1.

Space group *Pnma* (D_{2h}^{16}) from absences

Atomic positions

			x	*y*	*z*	*B* (Å^2)
4 Co	in	4(*c*)	0.3880±12	1/4	0.4512±13	3.0462±1460
4 Gd(1)	in	4(*c*)	0.0416±4	1/4	0.6372±4	2.1076±530
8 Gd(2)	in	8(*d*)	0.1758±3	0.0651±2	0.1738±3	2.0780±407

Interatomic distances (Å)

Co	- 1 Gd(1)	2.710	Gd(1)	- 1 Co	2.710	Gd(2)	- 1 Co	2.802
	- [1 Gd(1)	2.818]		- [1 Co	2.818]		- 1 Co	2.902
	- 1 Gd(1)	3.873		- 1 Co	3.873		- 1 Co	3.349
	- 2 Gd(2)	3.349		- [2 Gd(1)	3.802]		- 1 Gd(1)	3.547
	- 2 Gd(2)	2.902		- 2 Gd(2)	3.547		- [1 Gd(1)	3.579]
	- 2 Gd(2)	2.802		- 2 Gd(2)	3.613		- 1 Gd(1)	3.613
				- 2 Gd(2)	3.691		- 1 Gd(1)	3.691
				- [2 Gd(2)	3.579]		- 1 Gd(1)	3.938
				- [2 Gd(2)	3.938]		- 1 Gd(2)	3.528
							- [1 Gd(2)	3.537]
							- 2 Gd(2)	3.553
							- 2 Gd(2)	3.654

Material and Details of analysis

The elements were arc-melted under argon and then annealed at 10°C below the peritectic temperature. A single crystal was cut out by spark-erosion and a cleavage fragment cut to spherical shape. Intensities (556) were measured on a diffractometer (ω-scan, Mo, Zr filter) and corrected for Lorentz, polarization and spherical absorption factors. Patterson and Fourier syntheses gave a trial structure which was refined by full-matrix least squares to R = 0.099; anisotropic thermal parameters did not improve this.

1. *Structure Reports*, 26, 119.
2. K.H.J. BUSCHOW and A.S. VAN DER GOOT, 1969. *J. Less-Common Metals*, 17, 249.
3. *Strukturbericht*, 3, 33.

X PHASE IN COBALT MANGANESE SILICON

I. [The crystal structure of the *X* phase in the Mn-Co-Si system.] Ja.P. JARMOLJUK, P.I. KRIPJAKEVIČ and E.I. GLADYŠEVSKIJ, 1970. *Kristallografija, SSSR*, 15, 268-274 [*Soviet Physics - Crystallography*, 15, 226-230].

$Mn_{14}(Co,Si)_{23}$ is ideal formula.

See also 1, 2.

Orthorhombic, a = 12.47$_1$, b = 15.50$_1$, c = 4.76$_1$ Å (for $R_{14}X_{23}$), U = 920.04 Å^3, D_m = 6.95, A = 74, D_x = 7.05.

Space group *Pnnm* (D_{2h}^{12}) from absences and structure

Atomic positions

			x	*y*	*z*
4 *R*(1)	in	4(*g*)	0.104	0.336	0
4 *R*(2)	in	4(*g*)	0.286	0.236	0
4 *R*(3)	in	4(*g*)	0.225	0.059	0
4 *R*(4)	in	4(*g*)	0.596	0.047	0
4 *R*(5)	in	4(*g*)	0.079	0.768	0
4 *R*(6)	in	4(*g*)	0.410	0.522	0
4 *R*(7)	in	4(*g*)	0.028	0.932	0
2 *X*(1)	in	2(*c*)	0	1/2	0
4 *X*(2)	in	4(*g*)	0.244	0.645	0
4 *X*(3)	in	4(*g*)	0.308	0.785	0
4 *X*(4)	in	4(*g*)	0.511	0.224	0

Atomic positions

			x	*y*	*z*
4 *X*(5)	in	4(*g*)	0.451	0.356	0
4 *X*(6)	in	4(*g*)	0.190	0.507	0
8 *X*(7)	in	8(*h*)	0.108	0.174	1/4
8 *X*(8)	in	8(*h*)	0.415	0.112	1/4
8 *X*(9)	in	8(*h*)	0.287	0.396	1/4

Overall temperature factor *B* is 1.95 Å^2.

Interatomic distances (Å)

R(1) - 1 *R*(2)	2.75
R(1) - 1 *R*(5)	2.79
R(1) - 2 *R*(4)	2.99
R(1) - 1 *X*(1)	2.85
R(1) - 2 *X*(3)	2.74
R(1) - 2 *X*(4)	2.81
R(1) - 1 *X*(6)	2.86
R(1) - 2 *X*(7)	2.78
R(1) - 2 *X*(8)	2.76
R(1) - 2 *X*(9)	2.74
R(2) - 1 *R*(1)	2.75
R(2) - 1 *R*(3)	2.85
R(2) - 2 *R*(5)	2.96
R(2) - 2 *X*(2)	2.79
R(2) - 2 *X*(3)	2.76
R(2) - 1 *X*(4)	2.81
R(2) - 1 *X*(5)	2.77
R(2) - 2 *X*(7)	2.70
R(2) - 2 *X*(8)	2.77
R(2) - 2 *X*(9)	2.75
R(3) - 1 *R*(1)	2.85
R(3) - 1 *R*(4)	2.77
R(3) - 2 *R*(6)	2.97
R(3) - 2 *R*(7)	3.15
R(3) - 2 *X*(2)	2.75
R(3) - 2 *X*(6)	2.73
R(3) - 2 *X*(7)	2.59
R(3) - 2 *X*(8)	2.77
R(3) - 2 *X*(9)	2.80
R(4) - 2 *R*(1)	2.99
R(4) - 1 *R*(3)	2.77
R(4) - 1 *R*(4)	2.80
R(4) - 2 *X*(1)	2.76
R(4) - 1 *X*(3)	2.87
R(4) - 1 *X*(4)	2.94
R(4) - 2 *X*(6)	2.78
R(4) - 4 *X*(8)	2.74
R(4) - 2 *X*(9)	2.81
R(5) - 1 *R*(1)	2.79
R(5) - 2 *R*(2)	2.96
R(5) - 1 *R*(7)	2.62
R(5) - 1 *X*(2)	2.80
R(5) - 1 *X*(3)	2.87
R(5) - 2 *X*(4)	2.72
R(5) - 2 *X*(5)	2.77
R(5) - 2 *X*(7)	2.77
R(5) - 2 *X*(8)	2.70
R(5) - 2 *X*(9)	2.85
R(6) - 2 *R*(3)	2.97
R(6) - 1 *R*(6)	2.35
R(6) - 2 *R*(7)	2.88
R(6) - 2 *R*(7)	2.87
R(6) - 1 *X*(2)	2.81
R(6) - 1 *X*(5)	2.62
R(6) - 1 *X*(5)	2.56
R(6) - 1 *X*(6)	2.75
R(6) - 2 *X*(2)	2.65
R(6) - 2 *X*(9)	2.75
R(7) - 2 *R*(3)	3.15
R(7) - 1 *R*(5)	2.62
R(7) - 2 *R*(6)	2.88
R(7) - 2 *R*(6)	2.87
R(7) - 1 *R*(2)	2.22
R(7) - 2 *X*(5)	2.67
R(7) - 2 *X*(7)	2.65
R(7) - 2 *X*(9)	2.66
X(1) - 2 *R*(1)	2.85
X(1) - 4 *R*(4)	2.76
X(1) - 2 *X*(6)	2.37
X(1) - 4 *X*(8)	2.36
X(2) - 2 *R*(2)	2.79
X(2) - 2 *R*(3)	2.75
X(2) - 1 *R*(5)	2.80
X(2) - 1 *R*(6)	2.81
X(2) - 1 *X*(3)	2.31
X(2) - 1 *X*(6)	2.24
X(2) - 2 *X*(7)	2.24
X(2) - 2 *X*(8)	2.37
X(3) - 2 *R*(1)	2.74
X(3) - 2 *R*(2)	2.76
X(3) - 1 *R*(4)	2.87
X(3) - 1 *R*(5)	2.87
X(3) - 1 *X*(2)	2.31
X(3) - 1 *X*(4)	2.26
X(3) - 2 *X*(7)	2.34
X(3) - 2 *X*(9)	2.40
X(4) - 2 *R*(1)	2.81
X(4) - 1 *R*(2)	2.81
X(4) - 1 *R*(4)	2.94
X(4) - 2 *R*(5)	2.72
X(4) - 1 *X*(3)	2.26
X(4) - 1 *X*(5)	2.18
X(4) - 2 *X*(2)	2.32
X(4) - 2 *X*(9)	2.42
X(5) - 1 *R*(2)	2.77
X(5) - 2 *R*(5)	2.77
X(5) - 1 *R*(6)	2.62
X(5) - 1 *R*(6)	2.56
X(5) - 2 *R*(7)	2.67
X(5) - 1 *X*(4)	2.18
X(5) - 2 *X*(7)	2.34
X(5) - 2 *X*(9)	2.44
X(6) - 1 *R*(1)	2.86
X(6) - 2 *R*(3)	2.73
X(6) - 2 *R*(4)	2.78
X(6) - 1 *R*(6)	2.75
X(6) - 1 *X*(1)	2.37
X(6) - 1 *X*(2)	2.24
X(6) - 2 *X*(8)	2.40
X(6) - 2 *X*(9)	2.42
X(7) - 1 *R*(1)	2.78
X(7) - 1 *R*(2)	2.69
X(7) - 1 *R*(3)	2.59
X(7) - 1 *R*(5)	2.77
X(7) - 1 *R*(6)	2.65
X(7) - 1 *R*(7)	2.65
X(7) - 1 *X*(2)	2.24
X(7) - 1 *X*(3)	2.34
X(7) - 1 *X*(4)	2.32
X(7) - 1 *X*(5)	2.34
X(7) - 2 *X*(7)	2.38
X(8) - 1 *R*(1)	2.76
X(8) - 1 *R*(2)	2.77
X(8) - 1 *R*(3)	2.77
X(8) - 2 *R*(4)	2.74
X(8) - 1 *R*(5)	2.70
X(8) - 1 *X*(1)	2.36
X(8) - 1 *X*(2)	2.37
X(8) - 1 *X*(4)	2.42
X(8) - 1 *X*(6)	2.40
X(8) - 2 *X*(8)	2.38
X(9) - 1 *R*(1)	2.74
X(9) - 1 *R*(2)	2.75
X(9) - 1 *R*(3)	2.80
X(9) - 1 *R*(4)	2.81
X(9) - 1 *R*(5)	2.85
X(9) - 1 *R*(6)	2.75
X(9) - 1 *R*(7)	2.66
X(9) - 1 *X*(3)	2.40
X(9) - 1 *X*(5)	2.44
X(9) - 1 *X*(6)	2.42
X(9) - 2 *X*(9)	2.38

Material and Details of analysis

A prismatic crystal was obtained from a mixture containing 42.8 at.% Mn and 14.3 at.% Si which had been annealed at 800°C for 480 hr. Laue, rotation (FeK_α) and Weissenberg photographs (FeK_α, [001] axis) were used to determine the space group and lattice parameters. Intensities were estimated visually against a scale and corrected for Lorentz and polarization factors, scaled by Wilson's method and corrected for secondary extinction. A trial structure, derived from the Patterson section, *Pxy* 1/4, and packing considerations, was refined by Fourier and least squares; R = 15.5% for 135 $Fhk0_{obs}$.

Discussion

The structure (Fig. 10) consists of planar networks (z = 0 and 1/2 and 1/4 and 3/4) rotated with respect to each other and consisting solely of pentagons and triangles. The *R* atoms (Mn only) have coordination numbers 16, 15 and 14 while the *X* atoms (Co, Si + the excess Mn atoms) have a coordination of 12; the ideal formula is $Mn_{14}(Co,Si)_{23}$ and is a member of a series of structures Zr_4Al_3 - W_6Fe_7 (μ) - Nb_6(Ni,Al) (*M*) - $Mn_{14}(Co,Si)_{23}$ (*X*) - $MgZn_2$.

1. *Structure Reports*, 27, 165.
2. Ju.B. KUZ'MA and E.I. GLADYŠEVSKIJ, 1964. *Ž. Neorg. Khim.*, 9, 674.

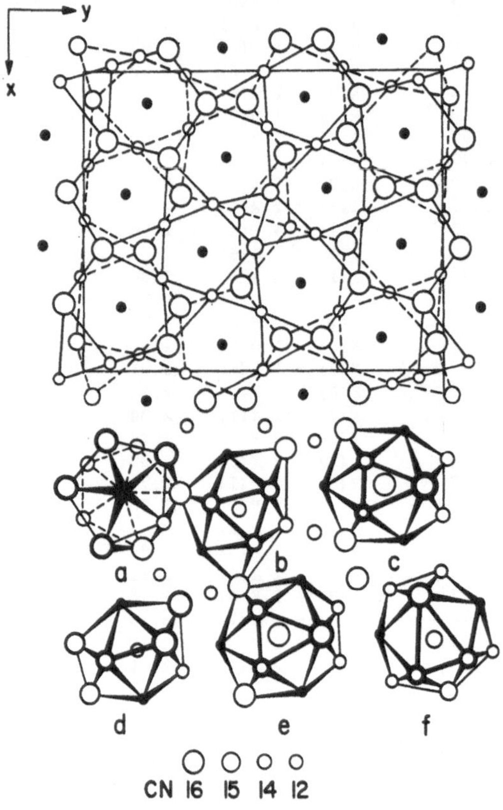

Fig. 10. The orthorhombic structure of the Co-Mn-Si *X*-phase projected onto (001). Two unit cells are shown; in the upper the networks at z = 0 (dashed lines) and z = 1/2 (solid lines) are shown with the intermediate layers at z = 1/4 and 3/4 (*X*(7), *X*(8) and *X*(9)) shown as solid dots; in the lower cell the different coordination polyhedra are outlined. *a* and *d* (CN = 12), *b* CN = 14, *f* CN = 15, *c* and *e* CN = 16 [after I].

COBALT THORIUM

I. The crystal structure of Th_2Co_7. K.H.J. BUSCHOW, 1970. *Acta Cryst.*, B 26, 1389-1392.

Co_7Th_2 see also 1, 2. F.W. = 878.3

α-Phase

Hexagonal, Ce_2Ni_7 (3) type, a = 5.030 ± 5, c = 24.62 ± 2 Å, c/a = 4.89, Z = 4.

β-Phase

Rhombohedral, β-Gd_2Co_7 type (4), a = 5.030 ± 8, c = 36.91 ± 2 Å (hexagonal axes).

Space group $P6_3/mmc$ (D_{6h}^4) from structure

Atomic positions α-phase

			x	*y*	*z*
2 Co(1)	in	2(*a*)	0	0	0
4 Co(2)	in	4(*e*)	0	0	0.167
4 Co(3)	in	4(*f*)	1/3	2/3	0.833
6 Co(4)	in	6(*h*)	0.833	0.666	1/4
12 Co(5)	in	12(*k*)	0.833	0.666	0.083
4 Th(1)	in	4(*f*)	1/3	2/3	0.030
4 Th(2)	in	4(*f*)	1/3	2/3	0.167

Temperature factor B = 0.60 $Å^2$ for all atoms.

[*Interatomic distances* (Å)]

Th(1)	- 3 Th(1)	3.26
	- 1 Th(2)	3.36
	- 3 Co(1)	3.00
	- 6 Co(5)	2.84
	- 3 Co(5)	3.15
Th(2)	- 1 Th(1)	3.36
	- 3 Co(2)	2.90
	- 3 Co(3)	2.90
	- 6 Co(4)	3.25
	- 6 Co(5)	3.25
Co(1)	- 6 Th(1)	3.00
	- 6 Co(5)	2.51
Co(2)	- 3 Co(3)	2.90
	- 3 Co(4)	2.51
	- 3 Co(5)	2.51
	- 3 Th(2)	2.90
Co(3)	- 3 Co(2)	2.90
	- 3 Co(4)	2.51
	- 3 Co(5)	2.51
	- 3 Th(2)	2.90
Co(4)	- 2 Co(2)	2.51
	- 2 Co(3)	2.51
	- 4 Co(4)	2.52
	- 4 Th(2)	3.25
Co(5)	- 1 Co(1)	2.51
	- 1 Co(2)	2.51
	- 1 Co(3)	2.51
	- 4 Co(5)	2.52
	- 2 Th(1)	2.84
	- 1 Th(1)	3.15
	- 2 Th(2)	3.25

Material and Details of analysis

Th (99.9%) and Co (99.9%) were arc-melted under argon and then vacuum annealed at 950°C. Powder patterns were obtained using a diffractometer (monochromatic CuK_α); intensities were corrected for Lorentz and polarization factors; R_I = 0.13.

1. *Structure Reports*, 20, 88.
2. J.R. THOMSON, 1966. *J. Less-Common Metals*, 10, 432.
3. *Structure Reports*, 23, 108.
4. E.F. BERTAUT, R. LEMAIRE and J. SCHWEIZER, 1965. *Bull. Soc. Franç. Minér. Crist.*, 88, 580.

COBALT ZIRCONIUM

I. [Crystal structure of the compound Zr_3Co.] P.I. KRIPJAKEVIČ, V.Ja. MARKIV, and V.V. BURNASOVA, 1970. *Dop. Akad. Nauk Ukr. RSR*, 32, 551-553.

$CoZr_3$ F.W. = 332.6

Orthorhombic, $PuBr_3$, (BRe_3) structure type (1, 2), a = 3.27, b = 10.84, c = 8.95 Å, [U = 317 $Å^3$], D_m = 7.07, Z = 4, D_x = 6.91.

Space group *Cmcm* (D_{2h}^{17}) from absences and type structure

Atomic positions

			x	y	z
4 Co	in	4(c)	0	0.740	1/4
4 Zr(1)	in	4(c)	0	0.424	1/4
8 Zr(2)	in	8(f)	0	0.135	0.057

B = 1.92 $Å^2$

Interatomic distances (Å)

Co	- 2 Co	3.27	Zr(2)	- 2 Co	2.74[2.64]
	- 2 Zr(1)	2.58		- 1 Co	3.06
	- 1 Zr(1)	3.43		- 2 Zr(1)	3.30
	- 4 Zr(2)	2.74[2.64]		- 2 Zr(1)	3.26
	- 2 Zr(2)	3.06		- 1 Zr(2)	3.10
Zr(1)	- 2 Co	2.58		- 2 Zr(2)	3.15
	- 1 Co	3.43		- 2 Zr(2)	3.27
	- 2 Zr(1)	3.27		- 1 Zr(2)	3.45
	- 4 Zr(2)	3.26		- 1 Zr(1)	3.58
	- 4 Zr(2)	3.30			
	- 2 Zr(2)	3.58			

Material and Details of analysis

Final R = 13.2% for 93 observed $0kl$ (17.5% for all reflexions); reciprocal lattice photographs.

1. *Structure Reports*, 11, 282.
2. *Ibid.*, 24, 73.

COPPER GADOLINIUM GERMANIUM

I. Die Kristallstruktur von $Gd_6Cu_8Ge_8$ und isotypen Phasen. W. RIEGER, 1970. *Mh. Chem.*, 101, 449-462.

$Gd_6Cu_8Ge_8$ F.W. = 1582.54

Orthorhombic, related to AlB_2 (1) and $CaZn_5$ (2) structure types, a = 14.000 ± 8, b = 6.655 ± 3, c = 4.223 ± 4 Å, [U = 393.46 $Å^3$], D_m = 8.20, Z = 1, D_x = 8.44.

Space group *Immm* (D_{2h}^{25})

Atomic positions

			x	y	z	B ($Å^2$)
2 Gd(1)	in	2(d)	0.00000	0.50000	0.0000	1.967±163
4 Gd(2)	in	4(e)	0.13116±30	0.00000	0.0000	1.928±144
8 Cu	in	8(n)	0.33048±55	0.18933±137	0.0000	2.041±183
4 Ge(1)	in	4(f)	0.21567±86	0.50000	0.0000	1.583±255
4 Ge(2)	in	4(h)	0.00000	0.18902±210	0.5000	1.501±250

Interatomic distances (Å)

Gd(1)	- 4 Gd(2)	3.801	Gd(2)	- 2 Ge(1)	3.533
	- 8 Cu	3.417		- 4 Ge(2)	3.061
	- 2 Ge(1)	3.019	Ge(1)	- 1 Gd(1)	3.019
	- 4 Ge(2)	2.957		- 2 Gd(2)	3.009
Gd(2)	- 2 Gd(1)	3.801		- 2 Gd(2)	3.533
	- 1 Gd(2)	3.673		- 2 Cu	2.619
	- [4]2 Cu	3.005		- 4 Cu	2.542
	- [2]4 Cu	3.061		- 4 Ge(1)	4.057
	- 2 Ge(1)	3.009		- 2 Ge(2)	3.271[4.175]

Interatomic distances (Å)

Ge(2)	- 2 Gd(1)	2.937
	- 4 Gd(2)	3.068
	- 2 Cu	2.508
	- 2 Ge(1)	3.271[4.175]
	- 1 Ge(2)	2.517

Cu	- 2 Gd(1)	3.417
	- [2]1 Gd(2)	3.005
	- [1]2 Gd(2)	3.061
	- 1 Cu	2.521
	- 2 Cu	3.192
	- 1 Ge(1)	2.619
	- 2 Ge(1)	2.542
	- 1 Ge(2)	2.508

Material and Details of analysis

Gd (99.99%)[sic], Cu (99.9%) and Ge (99.9%) were arc-melted under argon and annealed at 800°C in evacuated SiO_2 tubes. Powder photographs (CrK_α), rotation and Weissenberg photographs (CuK_α); visual estimation of layers 0-4 against a scale yielded 239 reflexions which were corrected for Lorentz-polarization factors but not for absorption. Patterson syntheses and full-matrix least squares gave R = 0.103 for F_{obs} (R = 0.125 for all F).

Discussion

The structure (Fig. 11) is closely related to AlB_2 (1) and $CuZn_5$ (2) types. The compounds $Dy_6Cu_8Ge_8$, $Er_6Cu_8Ge_8$ and $Lu_6Cu_8Ge_8$ are isotypic. (see Table I).

1. W. RIEGER and E. PARTHÉ, 1969. *Mh. Chem.*, 100, 439; *Strukturbericht*, 3, 28.
2. J.H. WERNICK and S. GELLER, 1959. *Acta Cryst.*, 12, 662; *Structure Reports*, 11, 59; 23, 222.

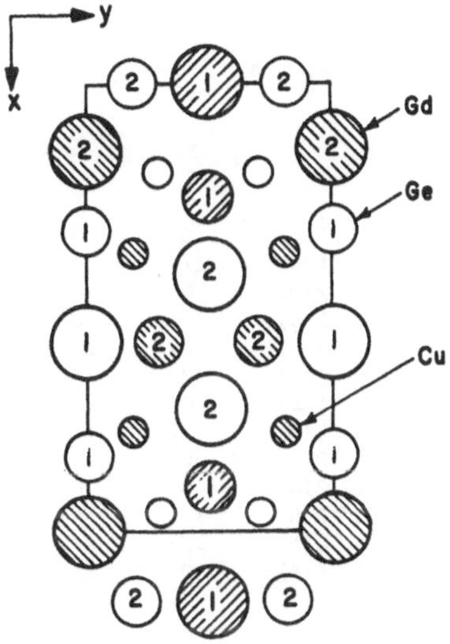

Fig. 11. Projection of the $Gd_6Cu_8Ge_8$ structure onto (001); atoms at z = 0 are cross hatched, the atoms are numbered as in the coordinate table [After I].

COPPER LITHIUM TIN

I. [Crystal structure of $LiCu_2Sn$.] P.I. KRIPJAKEVIČ and G.I. OLEKSIV, 1970. *Dop. Akad. Nauk Ukr. RSR, Ser. A*, 32, 63-65.

$LiCu_2Sn$ F.W. = 252.71

Hexagonal, ReB_3 type superstructure (1), a = 4.303, c = 7.637 Å, c/a = 1.774, [U = 122.46 Å^3], Z = 2.

Space group $P6_3/mmc$ (D^4_{6h})

Atomic positions

			x	y	z
2 Li	in	2(*a*)	0	0	0
4 Cu	in	4(*f*)	1/3	2/3	0.583
2 Sn	in	2(*c*)	1/3	2/3	1/4

[*Interatomic distances* (Å)]

Li	- 6 Cu	2.56	Cu - 3 Li	2.56
	- 6 Sn	3.13	- 3 Cu	2.79
Sn	- 6 Li	3.13	- 1 Cu	2.55
	- 6 Cu	2.79	- 1 Sn	2.54
	- 2 Cu	2.54	- 3 Sn	2.79

Material and Details of analysis

Li (98.9%), Cu (99.99%) and Sn (analysis grade) were melted in Al_2O_3 crucibles under a flame; powder photographs with Cr*K* radiation.

1. *Structure Reports*, 24, 71.

COPPER PLATINUM ZINC

I. Über die Struktur der Mischung Platin-Kupfer-Zink. Y. KHAN, B.V.R. MURTY and K. SCHUBERT, 1970. *J. Less-Common Metals*, 21, 293-303.

$Cu_3Pt_2Zn_5$

Tetragonal, $a = 3.7_7$, $c = 38.5_0$ Å, $c/a = 10.41$.

Space group *P4/mmm* (D^1_{4h})

Atomic positions Pt represents $Pt_{80}Zn_{20}$, Zn represents $Cu_{40}Zn_{60}$

			x	y	z
1 Pt(1)	in	1(*a*)	0	0	0
1 Pt(2)	in	1(*d*)	1/2	1/2	1/2
2 Pt(3)	in	2(*g*)	0	0	0.300
2 Pt(4)	in	2(*g*)	0	0	0.400
2 Pt(5)	in	2(*h*)	1/2	1/2	0.100
2 Pt(6)	in	2(*h*)	1/2	1/2	0.200
1 Zn(1)	in	1(*b*)	0	0	1/2
1 Zn(2)	in	1(*c*)	1/2	1/2	0
2 Zn(3)	in	2(*g*)	0	0	0.100
2 Zn(4)	in	2(*g*)	0	0	0.200
2 Zn(5)	in	2(*h*)	1/2	1/2	0.300
2 Zn(6)	in	2(*h*)	1/2	1/2	0.400
4 Zn(7)	in	4(*i*)	0	1/2	0.050
4 Zn(8)	in	4(*i*)	0	1/2	0.150
4 Zn(9)	in	4(*i*)	0	1/2	0.250
4 Zn(10)	in	4(*i*)	0	1/2	0.350
4 Zn(11)	in	4(*i*)	0	1/2	0.450

Discussion

This structure is based on a Cu_3Au sub-structure. [All atoms are 12 coordinated with interatomic distances either 2.67 Å or 2.69 Å]. The phase exists over a considerable range of composition, e.g. Pt_4Zn and Cu_2Zn_3 [The atomic distributions are not specified.]

COPPER SULPHUR

I. [On the structure of certain phases in the copper-sulphur system.] M.M. KAZINEČ, 1970. *Kristallografija, SSSR*, 14, 704-706 [*Soviet Physics - Crystallography*, 14, 599-600].

$Cu_{2-x}S$ see also 1.

Cubic, [β Ni_3S_2 type? (2)], *a* = 5.564 ± 16 Å (1).

Space group $F\bar{4}3m$ (T_d^2)

Atomic positions

			x	*y*	*z*
4 S	in	4(*a*)	0	0	0
4 Cu(1)	in	4(*c*)	1/4	1/4	1/4
3.8 Cu(2)	in	4(*b*)	1/2	1/2	1/2

B = 1.3 Å^2

[*Interatomic distances* (Å)]

Cu(1) - 4 S	2.41	S - 4 Cu(1)	2.41
- 4 Cu(2)	2.41	- 6 Cu(2)	2.78
Cu(2) - 6 S	2.78		

Material and Details of analysis

Cu and S evaporated and annealed [no details]. Multiple exposure electron diffraction photographs; microphotometer measurement of intensities; ϕ^2 synthesis; *R* = 19.2%; the occupancy of the 4(*b*) position was refined by trial and error.

CuS

Cubic, sphalerite (zinc blende) type, *a* = 5.387 ± 4 Å.

Space group $F\bar{4}3m$ (T_d^2)

Atomic positions

			x	*y*	*z*
4 S	in	4(*a*)	0	0	0
4 Cu	in	4(*c*)	1/4	1/4	1/4

B = 0.1 Å^2 [*sic*]

Interatomic distances (Å)

Cu - 4 S	2.33	S - 4 Cu	2.33

Details of analysis

Positions were obtained from a ϕ^2 synthesis; *R* = 15.4%.

II. The crystal structure of anilite. K. KOTO and N. MORIMOTO, 1970. *Acta Cryst.*, B 26, 915-924.

Cu_7S_4 (anilite) see also 3.

Orthorhombic, *a* = 7.89, *b* = 7.84, *c* = 11.01 Å, all ± 0.2%, [*U* = 681 Å^3], *Z* = 4, D_x = 5.68.

Space group *Pnma* (D_{2h}^{16}) from structure analysis and absences

Atomic positions

			x	*y*	*z*	*B* (Å^2)
4 S(1)	in	4(*c*)	0.2460±20	1/4	0.9873±11	0.25±23
4 S(2)	in	4(*c*)	0.7850±21	1/4	0.9841±12	0.41±25
8 S(3)	in	8(*d*)	0.9988±12	0.9980±15	0.2611±7	0.31±17
4 Cu(1)	in	4(*c*)	0.4832±13	3/4	0.0942±8	0.94±18
4 Cu(2)	in	4(*c*)	0.4258±13	1/4	0.3335±8	0.77±18
4 Cu(3)	in	4(*c*)	0.4790±13	3/4	0.3545±8	0.74±18
8 Cu(4)	in	8(*d*)	0.2371±10	0.4303±12	0.1553±6	1.35±15
8 Cu(5)	in	8(*d*)	0.1770±10	0.4767±12	0.4186±6	1.52±16

Interatomic distances (Å)

Cu(1)	- S(1)	2.317±18	Cu(4)	- S(1)	2.329±12
	- S(2)	2.285±19		- S(2)	2.944±11
	- 2 S(3)	2.517±12		- S(3)	2.282±12
Cu(2)	- S(1)	3.206±17		- S(3)	2.329±12
	- S(2)	2.295±17	Cu(5)	- S(1)	2.352±11
	- 2 S(3)	2.306±12		- S(2)	2.243±12
Cu(3)	- S(1)	2.300±17		- S(3)	2.241±11
	- S(2)	2.346±18		- S(3)	2.225±12
	- 2 S(3)	2.329±12			

Discussion

The anilite structure (Cu_7S_4) can be considered as a cubic face-centred arrangement of S atoms with the Cu atoms occupying the interstices in an ordered manner which fits the orthorhombic symmetry. Full details of the structure are given in II, and it is compared with other copper sulphide minerals.

Material and Details of analysis

A synthetic crystal [no details] (0.1 x 0.03 x 0.15 mm) with a small amount of twinning (ratio 40:1) and containing a very small amount of djurleite was used to record cell dimensions (precession photographs, CuK_α) and intensity data (multiple-film integrating Weissenberg photographs, CuK_α, levels 0-6, *a* and *b* rotation axes; 677 *b*-axis reflexions observed, 144 less than threshold, 450 corrected for twinning)). A trial structure derived from Patterson projections, was refined by Fourier and difference syntheses and by full-matrix least squares; R = 0.144 for 450 *hkl*. Anisotropic refinement and a refinement in $Pn2_1a$ were not successful; the final coordinates were used to produce a Fourier and difference maps. Synthetic and natural crystals gave the same X-ray patterns.

1. *Strukturbericht*, 4, 110.
2. G. LINÉ and M HUBER, 1963. *C.R. Acad. Sci., Paris*, 256, 3118.
3. N. MORIMOTO, K. KOTO and Y. SHIMAZAKI, 1969. *Amer. Min.*, 54, 1256.

EUROPIUM LEAD

I. On the crystal structure of Eu_5Pb_3. E. FRANCESCHI, 1970. *J. Less-Common Metals*, 22, 249-252.

Eu_5Pb_3 F.W. = 1381.37

See also 1.

Tetragonal, W_5Si_3 ($D8_m$) (2) type structure, $a = 13.184 \pm 5$, $c = 6.214 \pm 2$ Å, $c/a \simeq 0.47$, [U = 1080.1 Å^3], D_m = 8.35, Z = 4.

Space group $I4/mcm$ (D_{4h}^{18})

Atomic positions

			x	*y*	*z*
4 Eu(1)	in	4(*b*)	0	1/2	1/4
16 Eu(2)	in	16(*k*)	0.085	0.215	0
4 Pb(1)	in	4(*a*)	0	0	1/4
8 Pb(2)	in	8(*h*)	0.160	0.660	0

[*Interatomic distances* (Å)]

Eu(1)	- 2 Eu(1)	3.11	Eu(2)	- 1 Pb(2)	3.63
	- 4 Pb(2)	3.36		- 2 Eu(2)	3.83
	- 8 Eu(2)	4.22		- 2 Eu(2)	4.31
Eu(2)	- 2 Pb(1)	3.42		- 2 Eu(2)	3.94
	- 1 Pb(2)	3.44		- 1 Eu(2)	3.73
	- 2 Pb(2)	3.65		- 2 Eu(1)	4.22

Pb(1)	- 2 Pb(1)	3.11	Pb(2)	- 2 Eu(1)	3.36
	- 8 Eu(2)	3.42		- 4 Eu(2)	3.65
				- 2 Eu(2)	3.63
				- 2 Eu(2)	3.44

Material and Details of analysis

Eu (99.8%) and Pb (5N) were melted in Ta crucibles under argon and annealed at 900-1000°C; the parameters of 3 gave good qualitative agreement between observed and calculated intensities for powder patterns (Fe$K\alpha$, Cr$K\alpha$).

1. O.D. McMASTERS and K.A. GSCHNEIDNER, JR., 1967. *Trans. Met. Soc. AIME*, 239, 781.
2. *Structure Reports*, 19, 277.
3. E. FRANCESCHI and A. PALENZONA, 1969. *J. Less-Common Metals*, 18, 93.

EUROPIUM MAGNESIUM

I. Die Kristallstruktur von Mg_5Eu. W. MÜHLPFORDT, 1970. *Z. anorg. Chem.*, 374, 174-185.

$EuMg_5$ F.W. = 273.5

Hexagonal, a = 10.412 ± 6, c = 10.762 ± 6 Å, c/a = 1.03, U = 1010 Å^3, Z = 6, [D_x = 2.70].

Space group $P6_3/mmc$ (D_{6h}^4)

Atomic positions

			x	y	z
6 Eu	in	6(*h*)	0.195_1	0.390_2	1/4
2 Mg(1)	in	2(*a*)	0	0	0
4 Mg(2)	in	4(*f*)	1/3	2/3	-0.003_1
6 Mg(3)	in	6(*g*)	1/2	0	0
6 Mg(4)	in	6(*h*)	-0.431_5	-0.863_0	1/4
12 Mg(5)	in	12(*k*)	-0.161_5	-0.323_0	0.090_0

Interatomic distances (Å)

Eu	- [2]Eu	4.32	Mg(3)	- [2]Mg(4)	2.96
	- [2]Mg(1)	4.43		- [4]Mg(5)	3.20
	- [2]Mg(2)	3.69	Mg(4)	- [2 Eu	3.69
	- [4]Mg(3)	3.87		- 2 Mg(2)	3.19
	- [2]Mg(4)	3.69		- 2 Mg(3)	2.96]
	- [4]Mg(5)	3.69		- [2]Mg(4)	3.07
	- [2]Mg(5)	3.71		- [4]Mg(5)	3.20
Mg(1)	- [6 Eu	4.43]	Mg(5)	- [1 Eu	3.71
	- [6]Mg(5)	3.07		- 2 Eu	3.69
Mg(2)	- [3 Eu	3.69]		- 1 Mg(1)	3.07
	- [3]Mg(3)	3.01		- 1 Mg(2)	3.24
	- [3]Mg(4)	3.19		- 2 Mg(3)	3.20
	- [3]Mg(5)	3.24		- 2 Mg(4)	3.20]
Mg(3)	- [4 Eu	3.87]		- [2]Mg(5)	3.50
	- [2 Mg(2)	3.01]		- [1]Mg(5)	3.44

Discussion

The structure is very similar to that of Mg_4Sr (1).

Material and Details of analysis

Crystals were obtained from melts of the composition Mg_5Eu. Symmetry was deduced from Laue photographs and lattice parameters from Debye-Scherrer photographs. A trial structure derived from Patterson projections was refined by Fourier and difference syntheses; R = 0.08-0.10 for 552 observed reflexions.

1. F.E. WANG, F.A. KANDA, C.F. MISKELL and A.J. KING, 1965. *Acta Cryst.*, 18, 24.

GADOLINIUM IRON

I. Structures cristallines du composé intermetallique $(Gd_{2/3}2Fe_{1/3})Fe_5$.
F. GIVORD and R. LEMAIRE, 1970. *J. Less-Common Metals*, 21, 463-468.

$Fe_{17}Gd_2$ (quenched)

Hexagonal, $Ni_{17}Th_2$ (1) structure type, a = 8.496, c = 8.345 Å, [c/a = 0.982, U = 521.6 Å^3, Z = 2].

Space group $P6_3/mmc$ (D_{6h}^4)

Atomic positions

			x	*y*	*z*
2 Gd(1)	in	2(*b*)	0	0	1/4
2 Gd(2)	in	2(*d*)	1/3	2/3	3/4
4 Fe(1)	in	4(*f*)	1/3	2/3	0.105
6 Fe(2)	in	6(*g*)	1/2	0	0
12 Fe(3)	in	12(*j*)	1/3	0.969	1/4
12 Fe(4)	in	12(*k*)	0.167	1/3	0.985

For a sample annealed at 1050°C.

Rhombohedral, Th_2Zn_{17} (2,5) type structure, *hexagonal* cell, a = 8.540, c = 12.428 Å, [c/a = 1.455, U = 784.9 Å^3, Z = 3].

Space group $R\bar{3}m$ (D_{3d}^5)

Atomic positions (hexagonal axes)

			x	*y*	*z*
6 Gd	in	6(*c*)	0	0	1/3
6 Fe(1)	in	6(*c*)	0	0	0.096
9 Fe(2)	in	9(*d*)	1/2	0	1/2
18 Fe(3)	in	18(*f*)	0.286	0	0
18 Fe(4)	in	18(*h*)	1/2	-1/2	0.148

Annealed at 900°C.

Hexagonal, $ErCo_{5.75}$ type (3,4), a = 4.907, c = 4.168 Å, [c/a = 0.849, U = 86.91 Å^3].

Space group $P6/mmm$ (D_{6h}^1)

Atomic positions

			x	*y*	*z*
Gd	in	(*a*)*	0	0	0
2 Fe(1)	in	2(*c*)	1/3	2/3	0
2 Fe(2)	in	2(*e*)*	0	0	0.290
3 Fe(3)	in	3(*g*)	1/2	0	1/2

* Gd is replaced at random by the Fe in 2(*e*).

[*Interatomic distances* (Å)]

$Fe_{17}Gd_2$ (quenched)			$Fe_{17}Gd_2$ (1050°C)			$Fe_{17}Gd_2$ (900°C)		
Gd(1)	- 6 Fe(3)	2.97	Gd(1)	- 3 Fe(2)	3.22	Gd*	- 6 Fe(1)	2.83
	- 6 Fe(4)	3.14		- 6 Fe(3)	3.07		- 12 Fe(3)	3.22
	- 6 Fe(4)	3.30		- 6 Fe(4)	3.08	Fe(1)	- 6 Gd	2.83
Gd(2)	- 6 Fe(2)	3.22		- 3 Fe(4)	3.37		- 6 Fe(3)	2.52
	- 6 Fe(3)	2.97		- 1 Fe(1)	2.95		- 6 Fe(2)	3.08
	- 6 Fe(4)	3.14	Fe(1)	- 1 Gd	2.95	Fe(2)*	- 6 Fe(1)	3.08
	- 2 Fe(1)	2.96		- 3 Fe(2)	2.62		- 6 Fe(3)	2.60

* One or other of these is occupied. Distances between Gd and Fe(2) are not given.

[*Interatomic distances* (Å)]

$Fe_{17}Gd_2$ (quenched)			$Fe_{17}Gd_2$ (1050°C)			$Fe_{17}Gd_2$ (900°C)		
Fe(1)	- 1 Gd	2.96	Fe(1)	- 6 Fe(3)	2.72	Fe(3)	- 4 Gd	3.22
	- 3 Fe(2)	2.60		- 3 Fe(4)	2.70		- 4 Fe(1)	2.52
	- 6 Fe(3)	2.84		- 1 Fe(1)	2.39		- 4 Fe(3)	2.45
	- 3 Fe(4)	2.65	Fe(2)	- 2 Gd	3.22		- 4 Fe(2)	2.60
	- 1 Fe(1)	2.42		- 4 Fe(3)	2.43			
Fe(2)	- 2 Gd	3.22		- 4 Fe(3)	2.48			
	- 4 Fe(3)	2.46		- 2 Fe(1)	2.62			
	- 4 Fe(4)	2.46	Fe(3)	- 2 Gd	3.07			
	- 2 Fe(1)	2.60		- 2 Fe(2)	2.43			
Fe(3)	- 2 Gd	2.97		- 2 Fe(3)	2.44			
	- 2 Fe(2)	2.46		- 2 Fe(4)	2.63			
	- 2 Fe(3)	2.57		- 2 Fe(4)	2.59			
	- 1 Fe(3)	3.36		- 2 Fe(1)	2.72			
	- 2 Fe(4)	2.57	Fe(4)	- 2 Gd	3.08			
	- 2 Fe(4)	2.58		- 1 Gd	3.37			
	- 2 Fe(1)	2.84		- 2 Fe(2)	2.48			
Fe(4)	- 1 Gd(1)	3.14		- 2 Fe(3)	2.59			
	- 1 Gd(1)	3.30		- 2 Fe(3)	2.63			
	- 1 Gd(2)	3.14		- 2 Fe(4)	2.51			
	- 2 Fe(2)	2.46		- 1 Fe(1)	2.70			
	- 2 Fe(3)	2.58						
	- 2 Fe(3)	2.57						
	- 2 Fe(4)	2.47						
	- 1 Fe(1)	2.65						

Material and Details of analysis

Rare-earth elements (99.9%) and Fe(99.99%) were melted while levitated by induction heating and the alloys annealed *in vacuo* (900-1050°C). Debye-Scherrer photographs (CrK_α, 360 mm diam.), were used to obtain intensities which were compared with calculated values; the measurements were qualitative.

1. J.V. FLORIO, N.C. BAENZIGER and R.E. RUNDLE, 1965. *Acta Cryst.*, 9, 371.
2. E.S. MAKAROV and S.P. VINOGRADOV, 1956. *Kristallografija, SSSR*, 1, 634.
3. J. SCHWEIZER and F. TASSET, 1969. *J. Less-Common Metals*, 18, 245.
4. *Structure Reports*, 11, 59.
5. *Ibid.*, 20, 196.

GALLIUM GERMANIUM LITHIUM

I. Die Kristallstruktur der Verbindung LiGaGe. W. BOCKELMANN, H. JACOBS and H.-U. SCHUSTER, 1970. *Z. Naturf.*, 25 B, 1305-1306.

GaGeLi see also 1. F.W. = 149.25

Hexagonal, $a = 4.17_5$, $c = 6.78_3$ Å, $c/a = 1.62_5$, [U = 102 Å^3], $D_m = 4.54_6$, Z = 1.88, D_x = 4.84 (for Z = 2).

Space group $P6_3mc$ (C_{6v}^4) from intensity measurements

Atomic positions

				x	y	z
2 Li	in	2(*a*)		0	0	0.25
2 Ga	in	2(*b*)		1/3	2/3	0.06
2 Ge	in	2(*b*)		1/3	2/3	0.44

[*Interatomic distances* (Å)]

Li	- 3 Ga	2.73	Ga	- 3 Li	2.73	Ge	- 3 Li	2.73
	- 3 Ge	2.73		- 3 Ge	2.54		- 3 Ga	2.54
				- 1 Ge	[2.58]		- 1 Ga	[2.58]

Material and Details of analysis

LiGaGe was formed from the elements by heating under argon in sealed Ta crucibles at 850°C. Lattice parameters were derived from Debye-Scherrer photographs and diffractometer measurements, the space group from precession and Weissenberg photographs and the intensities from diffractometer measurements. A trial structure from packing considerations was refined by Fourier methods; $R = 0.105$. The Li atoms were placed from geometrical considerations.

Discussion

The structure is closely related to that of CdS in its wurtzite modification.

1. H.-U. SCHUSTER and W. BOCKELMANN, 1969. *Z. Naturf.*, 24 B, 1189.

GALLIUM MAGNESIUM

MAGNESIUM THALLIUM

I. Kristallstruktur von Mg_2Ga und Mg_2Tl. K. FRANK and K. SCHUBERT, 1970. *J. Less-Common Metals*, 20, 215-221.

Mg_2Ga F.W. = 118.34

Hexagonal, $a = 7.794 \pm 2$, $c = 6.893 \pm 1$ Å, $c/a = 0.884$, [$U = 362.6$ Å^3], $Z = 6$, [$D_x = 3.25$].

Space group $P\bar{6}2c$ (D_{3h}^4)

Atomic positions See Fig. 12.

			x	y	z
6 Mg(1)	in	6(*g*)	0.289±4	0	0
6 Mg(2)	in	6(*h*)	0.620±4	0.019±5	1/4
2 Ga(1)	in	2(*b*)	0	0	1/4
4 Ga(2)	in	4(*f*)	1/3	2/3	0.038±1

$B = 1.426$ Å^2

[*Interatomic distances* (Å)]

Mg(1)	- 2 Mg(2)	3.04	Ga(1)	- 6 Mg(1)	2.84
	- 2 Mg(2)	3.15		- 3 Mg(2)	3.04
	- 2 Ga(1)	2.84			
	- 2 Ga(2)	2.80			
Mg(2)	- 2 Mg(1)	3.04	Ga(2)	- 3 Mg(1)	2.80
	- 2 Mg(1)	3.15		- 1 Mg(2)	2.92
	- 1 Ga(1)	3.04		- 3 Mg(2)	3.03
	- 2 Ga(2)	2.92		- 2 Mg(2)	2.92
	- 2 Ga(2)	3.03		- 1 Ga(2)	2.92

Mg_2Tl F.W. = 253.0

Hexagonal, Fe_2P type (1), $a = 8.0829 \pm 4$, $c = 3.6796 \pm 4$ Å, $c/a = 0.4552$, [$U = 208.2$ Å^3, $Z = 3$, $\bar{D}_x = 6.05$].

Space group $P\bar{6}2m$ (D_{3h}^3)

Atomic positions (based on Fe_2P)

			x	y	z
3 Mg(1)	in	3(*f*)	0.256	0	0
3 Mg(2)	in	3(*g*)	0.594	0	1/2
1 Tl(1)	in	1(*b*)	0	0	1/2
2 Tl(2)	in	2(*c*)	1/3	2/3	0

[*Interatomic distances* (Å)]

Mg	- 2 Tl(1)	2.77	Tl(1)	- 6 Mg	2.77
	- 2 Tl(2)	3.06	Tl(2)	- 3 Mg	3.06

[*Interatomic distances* (Å)]

Mg	- 2 Tl(2)	3.07	Tl(2) - 3 Mg	3.07
	- 2 Tl(2)	3.07	- 3 Mg	3.07

Mg_5Tl_2 F.W. = 530.3

Orthorhombic, Mg_5Ga_2 type (2), a = 14.285 ± 3, b = 7.328 ± 2, c = 6.197 ± 2 Å, [U = 648.7 Å^3, Z = 4, D_x = 5.43].

Space group *Ibam* (D_{2h}^{26})

Atomic positions (from Mg_5Ga_2)

			x	*y*	*z*
4 Mg(1)	in	4(*b*)	1/2	0	1/4
8 Mg(2)	in	8(*f*)	0.26	0	1/4
8 Mg(3)	in	8(*j*)	0.077	0.155	0
8 Tl	in	8(*j*)	0.122	0.755	0

[*Interatomic distances* (Å)]

Mg(1)	- 4 Tl	2.99	Mg(3)	- 2 Mg(1)	3.16
	- 2 Mg(1)	3.10		- 2 Mg(2)	3.24
	- 2 Mg(2)	3.43		- 1 Mg(3)	3.16
	- 4 Mg(3)	3.16		- 1 Tl(1)	2.92
Mg(2)	- 1 Mg(1)	3.43		- 1 Tl(1)	3.00
	- 2 Mg(2)	3.10		- 2 Tl(1)	3.23
	- 2 Mg(3)	3.24	Tl	- 2 Mg(1)	2.99
	- 2 Tl(1)	3.08		- 2 Mg(2)	2.96
	- 2 Tl(1)	2.96		- 2 Mg(2)	3.08
				- 1 Mg(3)	2.92
				- 2 Mg(3)	3.23
				- 1 Mg(3)	3.00

Material and Details of analysis

Mg_2Ga: Mg and Ga were melted in Al_2O_3 crucibles under argon. Weissenberg photographs, Cu$K\alpha$. Mg_5Tl_2 and Mg_2Tl: Mg and Tl were melted in Al_2O_3 crucibles inside evacuated SiO_2 ampoules. The reactive alloy was handled under nitrogen; Guinier photographs.

1. *Structure Reports*, 23, 68.
2. K. SCHUBERT, F. GAUZZI and K. FRANK, 1963. *Z. Metallk.*, 54, 422.

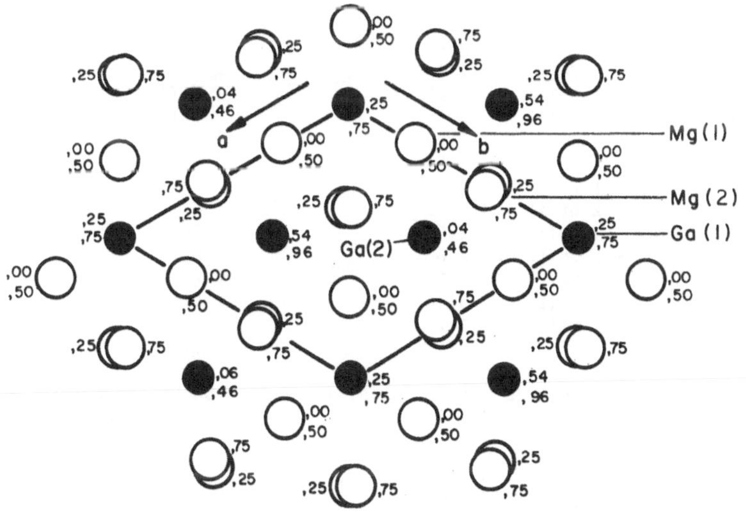

Fig. 12. The hexagonal $GaMg_2$ structure projected onto (00.1) [After I].

GERMANIUM LITHIUM

I. Die Kristallstruktur der Phase Li_9Ge_4. V. HOPF, H. SCHÄFER, and A. WEISS, 1970. *Z. Naturf.*, 25 B, 653.

Li_9Ge_4 F.W. = 352.81

Orthorhombic, a = 4.49 ± 1, b = 7.87 ± 2, c = 24.44 ± 6 Å, U = 861.3 Å^3 [863.62 Å^3], D_m = 2.62, Z = 4, D_x = 2.63[2.71].

Space group *Cmcm* (D_{2h}^{17}) from absences

Atomic positions

			x	y	z
8 Ge(1)	in	8(*f*)	0	0.166	0.200
8 Ge(2)	in	8(*f*)	0	0.500*	0.050
8 Li(1)	in	8(*f*)	0	0.182	0.410
8 Li(2)	in	8(*f*)	0	0.500	0.184
8 Li(3)	in	8(*f*)	0	0.900	0.140
8 Li(4)	in	8(*f*)	0	0.850	0.025
4 Li(5)	in	4(*c*)	0	0.820	0.250

*[Misprinted as 0.0 in paper. Private communication - Author]

[*Interatomic distances* (Å)]

Ge(1)	- 1 Ge(1)	2.44
	- 1 Li(1)	2.69
	- 2 Li(2)	2.63
	- 1 Li(2)	2.66
	- 1 Li(3)	2.56
	- 2 Li(5)	2.83
	- 2 Li(3)	3.25
	- 1 Li(5)	2.98
Ge(2)	- 1 Ge(2)	2.44
	- 2 Li(1)	2.84
	- 1 Li(1)	2.69
	- 1 Li(2)	3.27
	- 2 Li(3)	3.24
	- 2 Li(4)	2.61
	- 2 Li(4)	3.13
	- 1 Li(4)	3.31
	- 1 Li(4)	2.82
Li(1)	- 1 Ge(1)	2.69
	- 2 Ge(2)	2.84
	- 1 Ge(2)	2.69
	- 2 Li(3)	3.08
	- 1 Li(3)	2.53
	- 1 Li(4)	3.06
	- 1 Li(4)	2.82
	- 2 Li(4)	3.05
Li(2)	- 2 Ge(1)	2.63
	- 1 Ge(1)	2.66
	- 1 Ge(2)	3.27
	- 1 Li(1)	3.40
	- 1 Li(2)	3.23
Li(2)	- 2 Li(3)	2.61
	- 1 Li(3)	3.33
	- 2 Li(5)	3.11
	- 1 Li(5)	2.99
Li(3)	- 1 Ge(1)	2.56
	- 2 Ge(1)	3.25
	- 2 Ge(2)	3.24
	- 2 Li(1)	3.08
	- 1 Li(1)	2.53
	- 2 Li(2)	2.61
	- 1 Li(2)	3.33
	- 1 Li(4)	2.84
	- 1 Li(5)	2.76
Li(4)	- 1 Ge(2)	3.13
	- 2 Ge(2)	2.61
	- 1 Ge(2)	3.31
	- 1 Ge(2)	2.82
	- 1 Ge(2)	3.13
	- 2 Li(1)	3.05
	- 1 Li(1)	2.82
	- 1 Li(1)	3.06
	- 1 Li(3)	2.84
	- 2 Li(4)	3.00
	- 1 Li(4)	2.66
Li(5)	- 4 Ge(1)	2.83
	- 2 Ge(1)	2.98
	- 4 Li(2)	3.11
	- 2 Li(2)	2.99
	- 2 Li(3)	2.76

Discussion

The structure resembles that of Li_2Si (1) and is characterized by Ge-Ge pairs.

Material and Details of analysis

The elements when melted at 1000°C under argon in an iron crucible and slow-cooled, yielded deep-violet platelets. The Ge atoms were located from a

Patterson synthesis and the coordinates refined by Fourier and difference-Fourier syntheses and least squares; final *R* = 0.13; no absorption or temperature factors included.

1. H. AXEL, H. SCHÄFER and A. WEISS, 1965. *Angew. Chem.*, 77, 379.

GERMANIUM STRONTIUM SULPHUR

I. Structure cristalline de l'orthothiogermanate de strontium, Sr_2GeS_4. M. RIBES, É. PHILIPPOT and M. MAURIN, 1970. *C.R. Acad. Sci., Paris*, C 270, 1873-1874.

See also 1.

Sr_2GeS_4 F.W. = 376.09

Monoclinic, *a* = 8.231, *b* = 6.729, *c* = 6.672 Å, β = 108°21', [*U* = 348.66 Å^3], *Z* = 2, [D_x = 3.58].

Space group $P2_1m$ (C_{2h}^2)

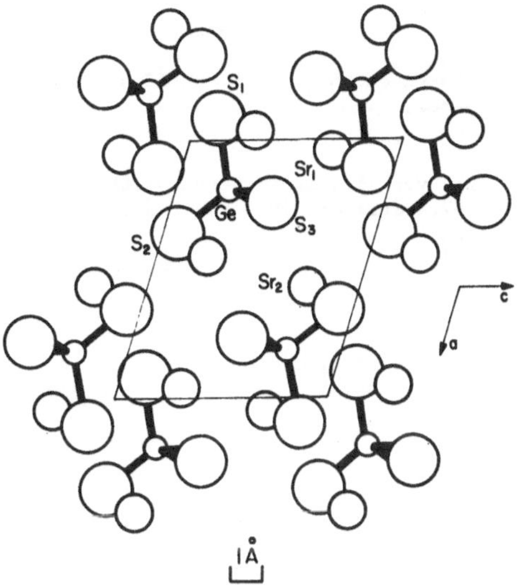

Fig. 13. Projection of the structure of GeS_4Sr_2 onto the (010) plane. [After I].

Atomic positions

			x	*y*	*z*	*B* (Å^2)
2 Sr(1)	in	2(*e*)	0.0515±2	0.25	0.7197±3	0.429±78
2 Sr(2)	in	2(*e*)	0.5677±2	0.25	0.7681±3	0.495±76
2 Ge	in	2(*e*)	0.2005±2	0.25	0.2749±3	0.666±86
2 S(1)	in	2(*e*)	-0.0753±6	0.25	0.0984±8	1.051±108
2 S(2)	in	2(*e*)	0.3731±5	0.25	0.0791±6	0.377±93
4 S(3)	in	4(*f*)	0.2606±3	-0.0020±4	0.4943±5	0.353±90

Interatomic distances (Å)

Ge	-	S(1)	2.206 [2.204]	[S(1)	- Ge	2.204
	-	S(2)	2.210 [2.212]		- Sr(1)	3.022
	- 2	S(3)	2.193		- Sr(2)	3.068]
[Sr(1)	-	S(1)	3.022	[S(2)	- Ge	2.212
	- 2	S(1)	3.561		- Sr(2)	2.994
	-	S(2)	2.960		- Sr(1)	2.960]

Interatomic distances (Å)

[Sr(1)	- 2 S(3)	3.019	[S(3)	- Ge	2.192
	- 2 S(3)	3.119]		- Sr(1)	3.019
[Sr(2)	- S(2)	2.994		- Sr(1)	3.119
	- S(1)	3.068		- Sr(2)	3.067
	- 2 S(2)	3.505		- Sr(2)	3.114]
	- 2 S(3)	3.067			
	- 2 S(3)	3.114]			

Material and Details of analysis

Intensities for zones $h0\ell$ - $h6\ell$ were visually estimated (by comparison with a scale) from Weissenberg photographs of a crystal 0.162 x 0.125 x 0.33 mm and corrected for Lorentz-polarization factors. Patterson and Fourier syntheses and least squares gave $R = 0.11$. See Fig. 13.

1. M. RIBES, É. PHILIPPOT and M. MAURIN, 1970. *C.R. Acad. Sci., Paris*, C 270, 716.

GOLD MANGANESE

I. The structure of Mn_2Au and Mn_3Au. P. WELLS and J.H. SMITH, 1970. *Acta Cryst.*, A 26, 379-381.

$AuMn_2$

Tetragonal, $MoSi_2$ ($C11_b$) type (1), in agreement with 2, $a = 3.328$, $c = 8.539$ Å, $Z = 2$, for sample with 67.4 at.% Mn; [$U = 94.57$ Å^3].

Space group $I4/mmm$ (D_{4h}^{17})

Atomic positions

			x	y	z
2 Au	in	2(a)	0	0	0
4 Mn	in	4(e)	0	0	0.33333

[*Interatomic distances* (Å)]

Au	- 8 Mn	2.75	Mn	- 4 Au	2.75
	- 2 Mn	2.85		- 4 Mn	2.75
				- 1 Mn	2.85
				- 1 Au	2.85

Material and Details of analysis

Mn (4N5) and Au (5N) were arc-melted under argon, homogenized at 900°C and annealed at 500°C; fair agreement between observed and calculated I's for both neutron and X-ray powder patterns.

1. *Strukturbericht*, 1, 740.
2. *Structure Reports*, 27, 223; E. STOLZ and K. SCHUBERT, 1962. *Z. Metallk.*, 53, 433.

GOLD MERCURY

I. The crystal structure of Au_6Hg_5. T. LINDAHL, 1970. *Acta Chem. Scand.*, 24, 946-952.

II. The crystal structure of the Au_2Hg phase. A.F. BERNDT and J.D. CUMMINS, 1970. *Acta Cryst.*, B 26, 864-867.

There has been considerable discussion on this phase, usually called Au_2Hg. See 1 - 3.

Au_6Hg_5 F.W. = 2184.8

Hexagonal, (I), a = 6.9937 ± 2, c = 10.1480 ± 4 Å, in equilibrium with Hg; used for structure analysis, [U = 429.85 Å^3], D_m = 16.3, Z = 2, D_x = 16.875; a = 6.9838 ± 2, c = 10.1510 ± 5 Å, in equilibrium with Au_3Hg.
(II), a = 7.019 ± 17, c = 10.184 Å, A = 22, D_x = 16.67.

Space group $P6_3/mcm$ (D^3_{6h}) from structure

Atomic positions I

			x	y	z	B (Å^2)
12 Au	in	12(k)	0.2416±5	0	0.3902±4	0.52±9
6 Hg	in	6(g)	0.5864±7	0	1/4	0.63±11
4 Hg	in	4(d)	1/3	2/3	0	0.62±12

II

			x	y	z	B (Å^2)
atom type 1	in	12(k)	0.2435±1	0	0.1150±3	3.2±1
atom type 2	in	6(g)	0.5878±2	0	1/4	3.2±1
atom type 3	in	4(d)	1/3	2/3	0	3.2±1

Interatomic distances (Å)

	I	II		I	II
Au - 2 Au	2.797	2.900	Hg(1) - 2 Au	2.799	2.780
- 1 Hg*	2.799	2.780	- 4 Au	2.890	2.870
- 1 Au	2.844	2.750	- 4 Hg(2)	3.298	3.312
- 2 Hg(1)	2.890	2.870	Hg(2) - 6 Au	2.930	2.953
- 2 Au	2.926	2.960	- 6 Hg(1)	3.298	3.312
- 2 Hg(2)	2.930	2.953			

σ's for I range from 0.002 to 0.007 Å; for II from 0.007 to 0.009 Å.

* [Misprinted as Au in I]

Discussion

The structure may be built up with the 10-coordinated polyhedra of Fig. 14 (a); alternatively the 12-fold coordination of Hg(2) Fig. 14(c), similar to that in Hg metal, may be used to build up the structure.

Materials and Details of analysis

I. A phase analysis was carried out to determine the equilibrium diagram in this region. A bipyramidal hexagonal prism of Au_6Hg_5 (0.04 diam.) was obtained electrolytically (Hg cathode, Au anode, HCl) and recrystallized at room temperature for 3 months. Lattice parameters were obtained from Guinier photographs ($CuK\alpha_1$, λ = 1.54050 Å, KCl standard a = 6.2919 Å) and intensities from diffractometer measurements ($CuK\alpha$, Ni filter, θ-2θ scan). Intensities were corrected for absorption and scattering factors were corrected for anomalous dispersion. A trial structure in $P6_3cm$ was obtained by consideration of layer packings and refined by full-matrix least squares using Hg scattering factors (R = 7.6%). This refinement yielded a centro-symmetric structure with Au and Hg distinguishable from the interatomic distances.

Further refinement in $P6_3/mcm$ confirmed this centro-symmetric space group but gave no improvement with separate Au and Hg scattering factors.

II. Amalgams were prepared by coreduction and then heated to remove excess Hg (110°C, *in vacuo*). A needle (0.04 mm diam. x 0.5 mm long) was used to record Weissenberg data (multiple-film, equi-inclination, $hk0$-$hk5$, $CuK\alpha$, Ni filter). Intensities were estimated visually and corrected for Lorentz and polarization effects; an approximate cylindrical absorption correction was applied. A trial structure based on packing arguments and a Patterson projection was refined by least squares (R = 0.07) with mean scattering factors. No attempt was made to distinguish between Hg and Au.

1. H. WINTERHAGER and W. SCHLOSSER, 1960. *Metall.*, 14, 1.
2. C. ROLFE and W. HUME-ROTHERY, 1967. *J. Less-Common Metals*, 13, 1.
3. *Strukturbericht*, 1, 561; 3, 611.

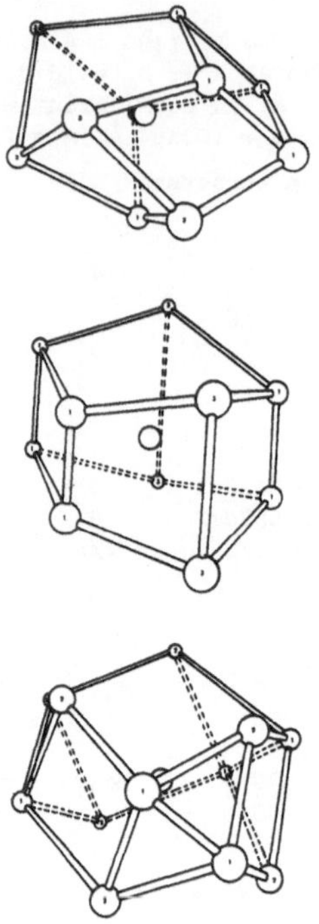

(a)

(b)

(c)

Fig. 14. The coordination around Au (a) and Hg(1) (b) and Hg(2) (c).

HAFNIUM NITROGEN

I. The crystal structures of Hf_3N_2 and Hf_4N_3. E. RUDY, 1970. *Metallurg. Trans.*, 1, 1249-1252.

See also 1

ε-Hf_3N_2 F.W. = 563.48

Rhombohedral, trigonal structure, a = 7.972 Å, α = 23°12', [U = 69.03 Å^3].
Hexagonal cell, a = 3.206, c = 23.26 Å, [U = 207.04 Å^3].

Space group $R\bar{3}m$ (D_{3d}^5) from systematic absences

Atomic positions (Hexagonal axes)

			x	y	z
3 Hf(1)	in	3(*a*)	0	0	0
6 Hf(2)	in	6(*c*)	0	0	2/9
6 N	in	6(*c*)	0	0	7/18
		3(*b*)	0	0	0.5000

[*Interatomic distances* (Å)]

Hf(1) - 6 N	2.26	N(*c*) - 6 Hf	2.26
		- N	2.58
Hf(2) - 6 N	2.26	N(*b*) - 6 Hf	2.26
		- 2 N	2.58

ζ-Hf_4N_3 F.W. = 755.98

Rhombohedral, trigonal structure isomorphous with ζ-V_4C_3 (1), a = 10.54 Å, α = 17°32', [U = 92.77 Å^3]. *Hexagonal* cell, a = 3.214, c = 31.12 Å, [U = 278.39 Å^3]. Other alloys with varying N contents gave powder photographs with parameters ranging from a = 3.206, c = 31.09 Å to a = 3.214, c = 31.12 Å.

Space group $R\bar{3}m$ (D^5_{3d}) from systematic absences

Atomic positions (Hexagonal axes)

			x	*y*	*z*
6 Hf(1)	in	6(*c*)	0	0	∿ 1/8
6 Hf(2)	in	6(*c*)	0	0	∿ 7/24
9 N	in	6(*c*)	0	0	∿ 5/12
		3(*a*)	0	0	0
		3(*b*)	0	0	0.5000

[*Interatomic distances* (Å)]

Hf(1) - 6 N	2.26	N(*c*) - 6 Hf	2.26
		- 1 N	2.59
Hf(2) - 6 N	2.26	N(*b*) - 6 Hf	2.26
		- 2 N	2.59
		N(*a*) - 6 Hf	2.26

Material and Details of analysis

Pellets were prepared by hot-pressing Hf (96+ %) and HfN at 1700-1900°C in graphite moulds; they were homogenized in argon (in the presence of HfN) at 1500-2100°C. Powder photographs (CuK_α and CrK_α) with agreement between observed and calculated intensities.

Discussion

The stacking sequence in Hf_3N_2 is ABABCBCAC and in Hf_4N_3 is ABABCACABCBC.

1. *Structure Reports*, 26, 73; E. RUDY and F. BENESOVSKY, 1961. *Mh. Chem.*, 92, 415.

HAFNIUM SULPHUR

ZIRCONIUM SULPHUR

I. Darstellung und Kristallstruktur der Phasen Zr_3S_4 und Hf_3S_4. K. STOCKS, G. EULENBERGER and H. HAHN, 1970. *Z. anorg. Chem.*, 374, 318-325.

S_4Zr_3 see also 1 F.W. = 401.92

Hf_3S_4 F.W. = 662.73

Cubic, NaCl type superstructure: S_4Zr_3, a = 10.25 ± 1 Å, [U = 1077 Å^3], D_m = 4.6 for $ZrS_{1.41}$. Hf_3S_4, *cubic*, a = 10.18 ± 1 Å, [U = 1055 Å^3], D_m = 7.79 for $HfS_{1.33}$.

Space group $Fd3m$ (O^7_h)

Atomic positions

S_4Zr_3

			x	*y*	*z*	*B* (Å^3)	Occupancy factor
Zr(1)	in	16(*c*)	0	0	0	0.399±15	0.944±10
Zr(2)	in	16(*d*)	1/2	1/2	1/2	2.59±32	0.293±18
S	in	32(*e*)	0.2513±10	0.2513±10	0.2513±10	0.99±10	0.977±29

Atomic positions

			Hf_3S_4			
		x	*y*	*z*	*B* (Å²)	Occupancy factor
Hf(1)	in 16(*c*)	0	0	0	1.0	1.0
Hf(2)	in 16(*d*)	1/2	1/2	1/2	1.0	0.40
S	in 32(*e*)	0.252	0.252	0.252	1.0	1.0

[*Interatomic distances* (Å)]

Zr(1)	- 6 Zr(1)	3.624	S	- 3 Zr(1)	2.576
	- 6 Zr(2)	3.624		- 3 Zr(2)	2.549
	- 6 S	2.576		- 3 S	3.586
Zr(2)	- 6 Zr(1)	3.624		- [6 S	3.624]
	- 6 Zr(2)	3.624		- 3 S	3.622
	- 6 S	2.549		- 3 Hf(1)	2.566
Hf(1)	- 6 S	2.566		- 3 Hf(2)	2.525
Hf(2)	- 6 S	2.525			

Material and Details of analysis

An octahedral crystal (185 μ) was prepared by heating $ZrS_{1.33}$ at 875°C. Rotation, Weissenberg and precession photographs were used for lattice parameters and space group; intensities were derived from integrated Weissenberg photographs (MoK_α radiation, Zr filter) and corrected for Lorentz and polarization effects. Scattering factors were corrected for the real part of the dispersion factor and the model of 1 refined by least squares. R = 7.6% for 57 non-zero *hkl*.

1. *Structure Reports*, 21, 176.

IRIDIUM SILICON

I. X-ray studies of silicon-rich iridium silicides. I. ENGSTRÖM and F. ZACKRISSON, 1970. *Acta Chem. Scand.*, 24, 2109-2116.

Ir_3Si_4 F.W. = 688.94

Orthorhombic, Rh_3Si_4 (1) type structure, a = 18.8701 ± 17, b = 3.6967 ± 3, c = 5.7742 ± 4 Å, U = 402.79 Å³, [Z = 4, D_x = 11.36]. For other phases in the Ir-Si system see Table I. The following lattice parameters are also given, a = 18.8741 ± 29, b = 3.6979 ± 8, c = 5.7717 ± 10 Å, U = 402.83 Å³.

Space group *Pnma* (D_{2h}^{16})

Atomic positions

	x	*y*	*z*	*B* (Å²)
4 Ir(1) in 4(*c*)	0.03154±10	1/4	0.19246±38	0.21±2
4 Ir(2) in 4(*c*)	0.17630±10	1/4	0.99915±34	0.21±2
4 Ir(3) in 4(*c*)	0.32570±10	1/4	0.10929±33	0.18±2
4 Si(1) in 4(*c*)	0.14371±93	1/4	0.38561±330	0.35±22
4 Si(2) in 4(*c*)	0.77035±98	1/4	0.78485±349	0.45±23
4 Si(3) in 4(*c*)	0.40882±100	1/4	0.41676±380	0.51±23
4 Si(4) in 4(*c*)	0.95882±90	1/4	0.52331±306	0.21±18

Interatomic distances (Å)†

Ir(1)	- Ir(2)	2.953	Ir(2)	- Ir(3)	2.893
	- 2 Ir(1)	3.131		- 2 Ir(3)	2.917
	- 2 Ir(1)	3.712		- Ir(1)	2.953
	- Si(1)	2.934*		- 2 Ir(2)	3.712
	- Si(3)	2.402		- 2 Ir(3)	3.981
	- Si(4)	2.420		- Si(1)	2.314
	- 2 Si(4)	2.495		- Si(2)	2.417
	- 2 Si(3)	2.692		- 2 Si(2)	2.453

*[Misprint for a value about 2.394 Å.] σ = 0.004 to 0.028 Å.

Ir(2)	- 2	Si(3)	2.501
	-	Si(1)	3.594
	- 2	Si(1)	3.929
Ir(3)	-	Ir(2)	2.893
	- 2	Ir(2)	2.917
	- 2	Ir(3)	3.712
	- 2	Ir(2)	3.981
	- 2	Si(1)	2.334
	-	Si(3)	2.369
	-	Si(2)	2.503
	-	Si(4)	2.520
	- 2	Si(2)	2.666
	-	Si(2)	3.694
	-	Si(1)	3.790
Si(1)	-	Ir(2)	2.314
	- 2	Ir(3)	2.334
	-	Ir(1)	2.394
	-	Ir(2)	3.594
	-	Ir(3)	3.790
	- 2	Ir(2)	3.929
	- 2	Si(4)	2.655
	- 2	Si(2)	2.655
	-	Si(2)	3.056
	- 2	Si(3)	3.428
	-	Si(4)	3.692
	- 2	Si(1)	3.712
	- 2	Si(3)	3.718
Si(2)	-	Ir(2)	2.417
	- 2	Ir(2)	2.435
	-	Ir(3)	2.505
Si(2)	- 2	Ir(3)	2.666
	-	Ir(3)	3.694
	- 2	Si(1)	2.655
	-	Si(1)	3.056
	-	Si(3)	3.132
	- 4	Si(2)	3.519
	- 2	Si(2)	3.712
	-	Si(4)	3.763
Si(3)	-	Ir(3)	2.369
	-	Ir(1)	2.402
	- 2	Ir(2)	2.501
	- 2	Ir(1)	2.692
	-	Si(4)	2.672
	-	Si(2)	3.132
	-	Si(4)	3.337
	- 2	Si(1)	3.428
	- 2	Si(3)	3.712
	- 2	Si(1)	3.718
	- 2	Si(4)	3.928
Si(4)	-	Ir(1)	2.420
	- 2	Ir(1)	2.495
	-	Ir(3)	2.520
	- 2	Si(4)	2.587
	- 2	Si(1)	2.655
	-	Si(3)	2.672
	-	Si(3)	3.337
	-	Si(1)	3.692
	- 2	Si(4)	3.712
	-	Si(2)	3.763
	- 2	Si(3)	3.928

† [These distances cannot be obtained from the coordinates given above using either of the two sets of axes given; the coordinations obtained are the same but the values differ by amounts of the order of 0.1 Å].

Materials and Details of analysis

Ir (spec. pure) and Si (99.9%) were arc-melted and then annealed *in vacuo*. Lattice parameters were derived from Guinier powder photographs (Cu$K\alpha_1$, λ = 1.54051 Å, Si standard a = 5.4305 Å) and intensities were visually estimated (multiple film Weissenberg; $h0\ell$, $h1\ell$, $h2\ell$; MoK (Zr) radiation, calibrated intensity scale; no absorption correction). A trial structure, based on 1, was refined by successive Fouriers and then by full-matrix least squares using scattering factors corrected for dispersion; final R = 0.127 for 582 $h0\ell$ and $h1\ell$; 11 strong $hk\ell$ omitted to reduce extinction effects. Space group $Pn2_1a$ was tested and rejected.

1. I. ENGSTRÖM and E. PERSSON, 1968. *Acta Chem. Scand.*, 22, 3120.

IRON NICKEL PHOSPHORUS
(RHABDITE)

I. Die Kristallstruktur des meteoritischen Rhabdits $(Fe,Ni)_3P$. F.-D. DOENITZ, 1970. *Z. Kristallogr.*, 131, 222-236.

Fe_2NiP (rhabdite) see also 1, 2.

Tetragonal, Ni_3P type structure (3), a = 9.040, c = 4.462 Å, [c/a = 0.494, U = 364.64 Å^3], Z = 8.

Space group $I\bar{4}$ (S_4^2)

Atomic positions

			x	*y*	*z*
M(1)	in	8(*g*)	0.0785±2	0.1081±2	0.2350±5
M(2)	in	8(*g*)*	0.3617±2	0.0324±2	0.9820±5
M(3)	in	8(*g*)*	0.1698±2	0.2199±2	0.7525±5
P	in	8(*g*)	0.2916±4	0.0481±2	0.4860±10

B = 0.40 Å^2 for all atoms.

* Ni atoms prefer these sites; the best model had 2/3 Fe and 1/3 Ni in M(2) and 1/3 Fe and 2/3 Ni in M(3).

[*Interatomic distances* (Å)]*

M(1)	- 1 *M*(1)	2.42	*M*(3)	- 1 *M*(1)	2.82
	- 2 *M*(1)	2.70		- 1 *M*(1)	2.76
	- 2 *M*(1)	2.92		- 1 *M*(1)	2.65
	- 1 *M*(2)	2.68		- 1 *M*(1)	2.76
	- 1 *M*(2)	2.88		- 1 *M*(1)	2.52
	- 1 *M*(3)	2.76		- 1 *M*(2)	2.53
	- 1 *M*(3)	2.82		- 1 *M*(2)	2.63
	- 1 *M*(3)	2.52		- 1 *M*(2)	2.56
	- 1 *M*(3)	2.65		- 2 *M*(3)	2.72
	- 1 *M*(3)	2.76		- 1 P	2.33
	- 1 P	2.37		- 1 P	2.24
	- 1 P	2.29		- 1 P	2.37
M(2)	- 1 *M*(1)	2.68	P	- 1 *M*(1)	2.37
	- 1 *M*(1)	2.88		- 1 *M*(1)	2.29
	- 2 *M*(2)	2.75		- 1 *M*(2)	2.32
	- 1 *M*(2)	2.57		- 1 *M*(2)	2.34
	- 2 *M*(2)	3.00		- 1 *M*(2)	2.31
	- 1 *M*(3)	2.53		- 1 *M*(2)	2.33
	- 1 *M*(3)	2.63		- 1 *M*(3)	2.33
	- 1 *M*(3)	2.56		- 1 *M*(3)	2.24
				- 1 *M*(3)	2.37

*[Values given in the paper are arranged in a special format and differ from these by 0.01 to 0.03 Å]

Discussion

The structure is of the Ni_3P type (3) characterized by a tetrakaidecahedral arrangement of metal atoms around each P atom.

Material and Details of analysis

A meteorite crystal from Buei Muerto, Chile, was used to record 151 *hkl* intensities (Weissenberg photographs, CuK_α) which were corrected for Lorentz, polarization and $\alpha_1\alpha_2$ splitting effects, absorption and extinction. The extinction correction, applied to F_o, is discussed in detail. Lattice parameters from Guinier photographs (NaCl standard). A trial structure was derived from the Patterson by using a minimum function and refined. The distribution of the Ni and Fe atoms was obtained by measuring 62 reflexions with CoK_α to obtain a strong anomalous scattering effect. Final R = 4.2%.

1. *Strukturbericht*, 2, 302.
2. G. HÄGG, 1929. *Nova Acta Reg. Soc. Sci. Upsaliensis* [IV, 7] No. 1.
3. *Structure Reports*, 19, 327.

IRON SELENIUM

I. The magnetic structure of Fe_3Se_4. A.F. ANDRESEN and B. van LAAR, 1970. *Acta Chem. Scand.*, 24, 2435-2439.

Fe_3Se_4 see also 1. F.W. = 483.4

Monoclinic, pseudo NiAs type structure, at 4.2°K, a = 6.159 ± 3, b = 3.493 ± 1, c = 11.385 ± 3 Å, β = 92.202 ± 3°, [U = 244.7 Å^3, Z = 2, D_x = 6.56]; at 293°K, a = 6.187 ± 2, b = 3.525 ± 1, c = 11.290 ± 5 Å, β = 91.98 ± 2°, [U = 246.1 Å^3, Z = 2, D_x = 6.52]; at 372°K, a = 6.208 ± 2, b = 3.541 ± 1, c = 11.281 ± 3 Å, β = 91.807 ± 4°, [U = 247.9 Å^3, Z = 2, D_x = 6.48].

Space group $I2/m$ (C^3_{2h})

Atomic positions

at 4.2°K

			x	*y*	*z*	N,N_B
2 Fe(1)	in	2(c)	0	0	1/2	2.89±12
4 Fe(2)	in	4(i)	0.0455±6	0	0.2402±4	2.27±8
4 Se(1)	in	4(i)	0.3358±8	0	0.8616±5	-
4 Se(2)	in	4(i)	0.3371±8	0	0.3890±5	-

at 372°K

			x	*y*	*z*
2 Fe(1)	in	2(c)	0	0	1/2
4 Fe(2)	in	4(i)	0.0431±6	0	0.2429±4
4 Se(1)	in	4(i)	0.3370±9	0	0.8623±4
4 Se(2)	in	4(i)	0.3366±8	0	0.3889±4

B (Å^2) for both sets of data were between 0.5 and 1.0.

[*Interatomic distances* (Å)]

		372°K	4.2°K
Fe(1)	- 2 Fe(2)	2.921*	2.982*
	- 4 Se(1)	2.544	2.536
	- 2 Se(2)	2.470	2.472
Fe(2)	- 1 Fe(1)	2.921	2.982
	- 2 Se(1)	2.587	2.601
	- 1 Se(1)	2.608	2.579
	- 2 Se(2)	2.444	2.412
	- 1 Se(2)	2.417	2.421
Se(1)	- 2 Fe(1)	2.544	2.536
	- 1 Fe(2)	2.608	2.579
	- 2 Fe(2)	2.587	2.601
	- 2 Se(1)	3.248	3.228
Se(2)	- 1 Fe(1)	2.470	2.472
	- 1 Fe(2)	2.417	2.421
	- 2 Fe(2)	2.444	2.412
	- 1 Se(2)	• 3.175	3.166

*[given as 2.79 Å in text].

Material and Details of analysis

The material of 1, contained in a 20 mm V cylinder, was examined using graphite monochromatized neutrons (λ = 2.582 ± 1 Å) with double Soller slits (15' divergence). The trial structure (1) was refined by a least-squares fitting of the entire diffraction pattern; refinement of four models for the spin moment direction ($\vec{a}$, $\vec{b}$, $\vec{a} + \vec{b}$ and ⊥ (101) clearly favoured the orientation $\vec{b}$ contrary to 1 and 2.

1. A.F. ANDRESEN, 1968. *Acta Chem. Scand.*, 22, 827.
2. B. LAMBERT-ANDRON and G. BERODIAS, 1969. *Solid State Comm.*, 7, 623.

IRON SILICON TITANIUM

I. The crystal structure of TiFeSi and related compounds. W. JEITSCHKO, 1970. *Acta Cryst.*, B 26, 815-822.

FeSiTi see also 1, 2, 3. F.W. = 131.83

Orthorhombic, (ordered Fe_2P type superstructure, (4)), a = 6.997 ± 2, b = 10.830 ± 5, c = 6.287 ± 2 Å, U = 476.4 Å^3, D_m = 5.57 (3), Z = 12, D_x = 5.51. Commonly occurs as a trilling with pseudo-hexagonal cell a = 6.29, c = 3.50 Å, (cf. 3).

Space group *Ima2* (C_{2v}^{22}) from least-squares refinement and symmetry of Fe_2P

Atomic positions

			x	y	z	B (Å^2)^b
4 Ti(1)	in	4(*b*)	1/4	0.2004±7	0.2964±14	0.5
4 Ti(2)	in	4(*b*)	1/4	0.7793±6	0.2707±14	0.5
4 Ti(3)	in	4(*b*)	1/4	0.9979±6	0.9178±15	0.5
8 Fe(1)	in	8(*c*)	0.0295±7	0.3764±4	0.1200*	0.5
4 Fe(2)	in	4(*a*)	0	0	0.2501±12	0.5
8 Si(1)	in	8(*c*)	0.0060±13	0.1675±9	0.9953±18	0.8
4 Si(2)	in	4(*b*)	1/4	0.9747±11	0.5055±23	0.8

* Held constant to fix origin of unit cell.

b Held fixed.

FeGeTi F.W. = 176.34

Orthorhombic, FeSiTi type, a = 7.155 ± 2, b = 11.025 ± 7, c = 6.405 ± 3 Å, U = 505.3 Å^3.

CoGeTi F.W. = 179.42

Hexagonal, Fe_2P type cell, a = 6.222 ± 2, c = 3.7267 ± 1 Å, c/a = 0.599 ± 2, U = 124.94 Å^3.

Discussion

The atomic distributions of some Fe_2P type structures with space group $P\bar{6}2m$ (D_{3h}^3) are given in the following table.

Compound	3(*g*)	3(*f*)	2(*c*) and 1(*b*)	Ref.
Fe_2P	Fe, x=0.594	Fe, x=0.256	P	4
ZrNiAl	Zr, x=0.600	Al, x=0.256	Ni	5
CeNiAl	Ce, x=0.580±1	Al, x=0.219±1	Ni	6
NbFeB	Nb	Fe	B	7
NbMnSi	Nb, x=0.589	Mn, x=0.242	Si	8
TiCoGe	Ti, x=0.57±1	Co, x=0.24$_5$±1	Ge	1

The structure is compared in detail with NiSiTi (9), CoGeTi (1) and $Cu_4Nb_5Si_4$ (10); all TM-Si(Ge) distances are shorter than the CN 12 radius whereas all TM-TM distances are longer (except two Ti-Fe distances).

Interatomic distances

Ti(1)	- 1 Ti(3)	3.237	Ti(2)	- 2 Fe(2)	2.965	Ti(3)	- 2 Si(1)	2.580
	- 2 Ti(1)	3.322		- 2 Fe(1)	3.123		- 1 Si(2)	2.604
	- 1 Ti(3)	3.355		- 2 Si(1)	2.557		- 1 Si(2)	3.703
	-[2 Ti(2)]	3.509		- 1 Si(2)	2.580	Fe(1)	- 1 Ti(1)	2.685
	- 2 Fe(1)	2.685		- 2 Si(1)	2.582		- 1 Ti(1)	2.691
	- 2 Fe(1)	2.691		-[1 Si(2)]	3.217		- 1 Ti(2)	2.751
	- 2 Fe(2)	2.803	Ti(3)	- 1 Ti(2)	3.141		- 1 Ti(3)	2.782
	- 2 Si(1)	2.554		- 1 Ti(1)	3.237		- 1 Ti(3)	3.010
	- 2 Si(1)	2.574		- 1 Ti(2)	3.245		- 1 Ti(2)	3.123
	- 1 Si(2)	2.775		- 1 Ti(1)	3.355		- 1 Fe(2)	2.691
Ti(2)	- 1 Ti(3)	3.141		-[2 Ti(3)]	3.499		- 1 Fe(1)	2.709
	- 2 Ti(2)	3.207		- 2 Fe(2)	2.725		-[1 Fe(1)]	3.086
	- 1 Ti(3)	3.245		- 2 Fe(1)	2.782		- 1 Si(2)	2.340
	-[2 Ti(1)]	3.509		- 2 Fe(1)	3.010		- 1 Si(2)	2.345
	- 2 Fe(1)	2.751		- 2 Si(1)	2.555		- 1 Si(1)	2.400
							- 1 Si(1)	2.413

Interatomic distances (Å)

Fe(2)	- 2 Ti(3)	2.725	Si(1)	- 1 Ti(3)	2.580	Si(2)	- 1 Ti(2)	2.580
	- 2 Ti(1)	2.803		- 1 Ti(2)	2.582		- 1 Ti(3)	2.604
	- 2 Ti(2)	2.965		- 1 Fe(1)	2.400		- 1 Ti(1)	2.775
	- 2 Fe(1)	2.691		- 1 Fe(1)	2.413		-[1 Ti(2)]	3.217
	- 2 Si(2)	2.390		- 1 Fe(2)	2.420		- 2 Fe(1)	2.340
	- 2 Si(1)	2.420		-[1 Si(1)]	3.415		- 2 Fe(1)	2.345
	- 2 Fe(2)	3.499		- 1 Si(1)	3.582		- 2 Fe(2)	2.390
Si(1)	- 1 Ti(1)	2.554		- 2 Si(1)	3.616		- 1 Ti(3)	3.703
	- 1 Ti(3)	2.555		- 2 Si(1)	3.629		- 2 Si(2)	3.499
	- 1 Ti(2)	2.557		- 1 Si(2)	3.779		- 2 Si(1)	3.779
	- 1 Ti(1)	2.574		- 1 Si(2)	3.881		- 2 Si(1)	3.881

The standard deviations range from 0.002 to 0.021 Å.

Materials and Details of analysis

Fe, Ti and Si (99.9 + %) were arc-melted under argon. Guinier powder patterns (CuK_α) and Weissenberg (CuK_α, MoK_α) and precession photographs (MoK_α); intensities were estimated visually (by comparison with an intensity scale) from precession photographs ($0k\ell$ - $3k\ell$, $h,2\ell,\ell$) on a trilled crystal containing 88% of one orientation plus 7 and 5 % of the other orientations; Lorentz and polarization corrections; trial parameters based on Fe_2P (4) were refined by full-matrix least squares with unit weights; scattering factors were corrected for dispersion; occupancy factors were tested and found to be essentially unity; final R = 0.094 for 344 observed F's.

1. F.X. SPIEGEL, D. BARDOS and P.A. BECK, 1963. *Trans. AIME*, 227, 575.
2. W. FREUNDLICH and N.F. MOCHAI, 1966. *C.R. Acad. Sci., Paris*, C 262, 1000.
3. V.Ja. MARKIV, E.I. GLADYŠEVSKIJ, R.V. SKOLOZDRA and P.I. KRIPJAKEVIČ, 1967. *Dopovidi: Akad. Nauk, Ukr. RSR*, A29, 266.
4. *Structure Reports*, 23, 68; S. RUNDQVIST and F. JELLINEK, 1959. *Acta Chem. Scand.*, 13, 425.
5. P.I. KRIPJAKEVIČ, V.Ja. MARKIV and Ja.V. MELNYK, 1967. *Dopovidi: Akad. Nauk, Ukr. RSR*, A29, 750.
6. A.E. DWIGHT, M.H. MUELLER, R.A. CONNER, JR., J.W. DOWNEY and H. KNOTT, 1968. *Trans. AIME*, 242, 2075.
7. Ju.B. KUZ'MA, 1967. *Dopovidi: Akad. Nauk, Ukr. RSR*, A29, 939.
8. B. DEYRIS, J. ROY-MONTREUIL, R. FRUCHART and A. MICHEL, 1968. *Bull. Soc. Chim. Fr.*, 1303.
9. C.B. SHOEMAKER and D.P. SHOEMAKER, 1965. *Acta Cryst.*, 18, 900.
10. E. GANGLBERGER, 1968. *Mh. Chem.*, 99, 549.

IRON SULPHUR
IRON TELLURIUM
COBALT TELLURIUM

I. Compounds with the marcasite type crystal structure V. The crystal structures of FeS_2, $FeTe_2$ and $CoTe_2$. G. BROSTIGEN and A. KJEKSHUS, 1970. *Acta Chem. Scand.*, 24, 1925-1940.

FeS_2 see also 1, 2, 3. F.W. = 120

$FeTe_2$ see also 4. F.W. = 311

$CoTe_2$ see also 5 F.W. = 314.1

Orthorhombic, marcasite-type (6), FeS_2, a = 4.4431_9, b = 5.4245_9, c = 3.3871_6 Å, [U = 81.63 Å^3, Z = 2, D_x = 4.88]. $FeTe_2$, a = 5.2655_8, b = 6.2679_8, c = 3.8738_7 Å, [U = 127.85 Å^3, Z = 2, D_x = 8.08]. $CoTe_2$, a = 5.3294_6, b = 6.3223_8, c = 3.9080_6 Å, [U = 131.68 Å^3, Z = 2, D_x = 7.92]. All for stoichiometric compositions.

Space group $Pnn2$ (C_{2v}^{10}) from structure refinement; contrary to 2, 3.

Atomic positions

	x	y	z	β_{11}	β_{22} ×10⁴	β_{33}	$2\beta_{12}$	$2\beta_{13}$	$2\beta_{23}$
FeS_2									
2 Fe in 2(*a*)	0	0	0	24±2	11±2	23±7	10±4	-	-
4 S in 4(*c*)	0.2008±4	0.3778±3	0.0089±22	42±4	26±3	67±13	-4±6	26±49	-234±35
$FeTe_2$									
2 Fe in 2(*a*)	0	0	0	25±3	12±2	6±36	-8±5	-	-
4 Te in 4(*c*)	0.2241±1	0.3620±1	0.0218±28	32±1	23±1	87±15	-5±2	61±11	-47±9
$CoTe_2$									
Co in 2(*a*)	0	0	0	45±4	24±3	58±23	3±6	-	-
4 Te in 4(*c*)	0.2204±1	0.3636±1	0.0007±39	36±1	24±1	103±10	-9±2	63±9	52±9

	FeS_2	$FeTe_2$	$CoTe_2$
R	0.0618	0.0737	0.0846

Interatomic distances (Å)

Compound	FeS_2	$FeTe_2$	$CoTe_2$	$FeSb_2$ (7)
T - 2 X^a	2.230±6	2.508±8	2.602±11	2.5782±11
- 2 X^a	2.235±2	2.5589±7	2.5815±6	2.5762±6
- 2 X^a	2.275±6	2.635±8	2.606±11	2.6164±11
X - 1 T	2.230±6	2.508±8	2.602±11	2.5782±11
- 1 T	2.235±2	2.5589±7	2.5815±6	2.5762±6
- 1 T	2.275±6	2.635±8	2.606±11	2.6164±11
- 1 T	2.223±3	2.9261±11	2.9143±11	2.8871±12

a nearest neighbours

Material and Details of analysis

FeS_2 crystals from Joplin, U.S.A. (as in 2, 3) and vapour grown crystals of $FeTe_2$ and $CoTe_2$. Guinier photographs (CuK_α, λ = 1.54050 Å, Si, a = 5.4305 Å) were used to obtain lattice parameters and intensities were obtained from integrated Weissenberg photographs (Mo*K*, Zr filter, microphotometer measurements except weakest which were compared visually with a scale), and corrected for Lorentz and polarization factors as well as absorption and secondary extinction (no dispersion correction). Trial structures were refined by full matrix least squares. All single crystals were found to be stoichiometric from their lattice parameters and occupancy refinement, although synthetic $CoTe_2$ and $FeTe_2$ show a small range of composition. Seven models, comprising the likely arrangements in *Pnnm* or *Pnn2* were refined with the result that the unrestrained model in *Pnn2* is clearly favoured at the 0.005 significance level.

Discussion

The structures are analysed in detail and it is concluded the Jahn-Teller effect does not account for the occurrence of the marcasite structure in preference to the pyrite structure.

1. M.L. HUGGINS, 1922. *Phys. Rev.*, 19, 369.
2. *Strukturbericht*, 2, 272.
3. *Ibid.*, 5, 52.
4. *Ibid.*, 6, 166.
5. *Structure Reports*, 18, 197.
6. *Strukturbericht*, 1, 495.
7. H. HOLSETH and A. KJEKSHUS, 1969. *Acta Chem. Scand.*, 23, 3043.

LANTHANUM MANGANESE SILICON SULPHUR

I. Structure cristalline de $La_6MnSi_2S_{14}$. G. COLLIN and P. LARUELLE, 1970. *C.R. Acad. Sci., Paris*, C 270, 410-412.

$La_6MnSi_2S_{14}$ F.W. = 1393.5

Hexagonal, $Ce_6Al_{10/3}S_{14}$ (1) type structure, a = 10.36, c = 5.73 Å, [c/a = 0.553, U = 615.0 Å^3, Z = 1, D_x = 3.76]. The compound can be considered as $La_6(Mn_1\square)Si_2S_{14}$, analogous to $L_6B_2C_2X_{14}$.

Space group $P6_3$ (C_6^6) from $Ce_6Al_{10/3}S_{14}$

Atomic positions

			x	y	z
1 Mn	in	2(a)	0	0	0.014
2 Si	in	2(b)	1/3	2/3	0.664
2 S(1)	in	2(b)	1/3	2/3	0.028
6 S(2)	in	6(c)	0.409	0.524	0.523
6 S(3)	in	6(c)	0.083	0.245	0.758
6 La	in	6(c)	0.123	0.358	0.250

Anisotropic thermal parameters are given;
σ values were ≈ 0.001 except for Mn where σ ≈ 0.01.

Interatomic distances (Å)

La	-	S(3)	2.87[2.88]	S(2)	-	La	2.91
	-	S(3)	2.90		-	La	3.01
	-	S(3)	3.00		-	La	3.06
	-	S(3)	3.09		-	Si	2.15
	-	S(2)	2.91[2.89]		-	S(3)	3.43
	-	S(2)	3.01		- 2	S(2)	3.45
	-	S(2)	3.04[3.06]		- 1	S(1)	3.46
	-	S(1)	3.10	S(3)	-	La	2.88
[Si	- 3	S(2)	2.15		-	La	2.90
	- 1	S(1)	2.09		-	La	3.00
Mn	- 2	Mn	2.87		-	La	3.09
	- 3	S(3)	2.64		-	Mn	2.64
	- 3	S(3)	2.67]		-	Mn	2.67
S(1)	- 3	La	3.10		-	S(2)	3.44
	-	Si	2.09				
	- 3	S(2)	3.46				

Material and Details of analysis

Red needles were obtained by heating La_2S_3 and MnS at 1350°C in H_2 containing SiS_2. Weissenberg photographs (MoK_α, $hk0$ - $hk9$) were used to record 489 independent hkl; a trial structure (1) was refined by 8 cycles of least squares [no details] to R = 0.056.

1. D. DE SAINT-GINIEZ, P. LARUELLE and J. FLAHAUT, 1963. *C.R. Acad. Sci., Paris*, C 270, 412.

LANTHANUM RHODIUM

I. The anti-Th_3P_4-type structure in the lanthanum-rhodium system: $La_4Rh_{\sim 3}$. A.V. VIRKAR, P.P. SINGH and A. RAMAN, 1970. *Inorg. Chem.*, 9, 353-356.

See also 1.

$La_4Rh_{\sim 3}$

Cubic, anti-Th_3P_4 type (2,3,4), a = 8.922 ± 3 Å*, U = 710.3 Å^3, A = 28, D_x = 8.006 for phase with 42.17 at.% Rh. D_m for $La_{60}Rh_{40}$ = 7.53 and for $La_{58}Rh_{42}$ = 7.85. La_4Rh_3 (16 La, 12 Rh) D_x = 8.086. For 10 2/3 Rh atoms, 40% Rh, D_x = 7.765.

*[Note the value of the Si parameter given below.]

Space group $I\bar{4}3d$ (T_d^6) from structure

Atomic positions

				x	y	z
∿11 2/3	Rh	in	12(*a*)	3/8	0	1/4
16	La	in	16(*a*)	0.0570(±5)	0.0570(±5)	0.0570(±5)

Interatomic distances (Å)

Rh - 4 La	2.851	La - 3 Rh	2.851
- 4 La	3.357	- 3 Rh	3.357
- 8 Rh	4.173	- 3 La	3.591
		- 5 La	4.122
		- 1 La	3.863

Material and Details of analysis

Alloys were prepared by 1 and annealed in Mo foil in sealed SiO_2 tubes at 900°; Guinier powder photographs (Si, a = 5.4199 Å[kX?] as internal standard) were used for visual estimation of 16 intensities; systematic variation of x gave R_{MIN} = 0.057.

1. P.P. SINGH and A. RAMAN, 1969. *Trans. Met. Soc. AIME*, 245, 1561.
2. K. MEISEL, 1939. *Z. anorg. Chem.*, 240, 300.
3. *Strukturbericht*, 7, 15.
4. D. HOHNKE and E. PARTHÉ, 1966. *Acta Cryst.*, 21, 435.

α-MANGANESE

I. A refinement of the atomic and thermal parameters of α-manganese from a single crystal. J.A. OBERTEUFFER and J.A. IBERS, 1970. *Acta Cryst.*, B 26, 1499-1504.

α-Mn see also 1, 2, 3, 4, 5.

Cubic, body-centred, a = 8.911 ± 2 Å, [U = 707.6 $Å^3$, A = 58, D_x = 7.463 (λMoK_{α_1} = 0.70930 Å, 20°C)].

Space group $I\bar{4}3m$ (T_d^3)

Atomic positions

			x	y	z	B ($Å^2$)
2 Mn(1)	in	2(*a*)	0	0	0	0.447±23
8 Mn(2)	in	8(*c*)	0.31787±10	0.31787±10	0.31787±10	0.454±12
24 Mn(3)	in	24(*g*)	0.35706±6	0.35706±6	0.03457±9	0.423±7
24 Mn(4)	in	24(*g*)	0.08958±6	0.08958±6	0.28194±9	0.391±7

Interatomic distances (Å)

All σ's = ± 0.001 Å

Mn(1) - 4 Mn(2)	2.811	Mn(3) - 1 Mn(2)	2.572	Mn(4) - 1 Mn(1)	2.754
- 12 Mn(4)	2.754	- 1 Mn(3)	2.930	- 2 Mn(2)	2.708
Mn(2) - 1 Mn(1)	2.811	- 2 Mn(3)	2.622	- 1 Mn(2)	2.895
- 3 Mn(3)	2.572	- 4 Mn(3)	2.661	- 1 Mn(3)	2.349
- 3 Mn(3)	2.930	- 1 Mn(4)	2.349	- 2 Mn(3)	2.524
- 6 Mn(4)	2.708	- 2 Mn(4)	2.524	- 2 Mn(3)	2.682
- 3 Mn(4)	2.895	- 2 Mn(4)	2.682	- 1 Mn(4)	2.258
				- 2 Mn(4)	2.424

Material and Details of analysis

A vapour grown single crystal was ground to a sphere (r = 0.2 mm) and then etched to a final ellipsoidal shape 0.11 x 0.084 x 0.068 mm; this shape was approximated by a 30-sided polygon to apply an absorption correction. Intensities for 1142 (650 independent) reflexions were collected with a diffractometer (θ-2θ scan, Mo radiation, Nb filter, h.w. = 0.1° 2θ at take-off angle of 1.1°; $R_{internal}$, low θ 8.9% before, 8.5% after absorption correction; $R_{internal}$, all *hkl* 7.9% before, 6.1% after correction). The trial structure (1-5) was refined by full-matrix least squares using isotropic temperature factors and dispersion corrections for the scattering factors (Mn^{2+}) and also an extinction parameter; anisotropic thermal parameters and Mn^{o} scattering factors gave no significant improvement; a final difference Fourier showed no interpretable features; R_F2 = 8.4%.

The mean Debye temperature (θ_D) was 394°K at mean r.m.s. displacement 0.0725 Å. The Debye temperature is close to that found by 4 (390°K) and contrary to 6; all atoms have relatively similar values for θ_D. No significant difference could be found between the scattering factors of different atoms although the coordinations are widely different.

1. *Strukturbericht*, 1, 64.
2. *Ibid.*, 1, 756.
3. *Ibid.*, 1, 757.
4. C.P. GAZZARA, R.M. MIDDLETON, R.J. WEISS and E.O. HALL, 1967. *Acta Cryst.*, 22, 859.
5. N. KUNITOMI, T. YAMADA, Y. NAKAI and Y. FUJII, 1969. *J. Appl. Phys.*, 40, 1265.
6. C.W. KIMBALL, W.C. PHILLIPS, M.C. NEVITT and R.S. PRESTON, 1966. *Phys. Rev.*, 146, 375.

MOLYBDENUM SULPHUR

I. Structure and phase transitions of molybdenum (III) sulfide and some related phases. R. DE JONGE, T.J.A. POPMA, G.A. WEIGERS and F. JELLINEK, 1970. *J. Solid State Chem.*, 2, 188-192.

See also 1

$Mo_{2.06}S_3$ (in agreement with 2, 3)

Monoclinic, *a* = 6.092, *b* = 3.208, *c* = 8.6335 Å, β = 102.43° at room temperature (1) for sub-cell used for structure refinement; true cell has all axes doubled. A phase transition occurs at low temperatures, the compound becoming triclinic with unit cell dimensions (at - 150°C), *a* = 6.06, *b* = 2 x 3.19, *c* = 8.60 Å, α = 89.6°, β = 102.5°, γ = 90.3°. The transition is weakly endothermic and occurs at -80°C when heating, showing a pronounced hysteresis. At +37°C the sub-cell becomes the true cell.

Space group $P2_1/m$ (C_{2h}^2) from structure refinement

Atomic positions

			x	*y*	*z*
2 Mo(1)	in	2(*e*)	0.310	1/4	0.009
2 Mo(2)	in	2(*e*)	0.108	1/4	0.632
2 S(1)	in	2(*e*)	0.505	1/4	0.803
2 S(2)	in	2(*e*)	0.963	1/4	0.161
2 S(3)	in	2(*e*)	0.729	1/4	0.516

e.s.d.'s = 0.001 for Mo and 0.005 for S (estimated from different refinements) overall temperature factor *B* = 1.27 Å^2.

Interatomic distances (Å)

Mo(1)	- 1 S(1)	2.34	Mo(2)	- 1 S(3)	2.31
	- 2 S(1)	2.39		- 2 S(3)	2.40
	- 2 S(2)	2.54		- 2 S(2)	2.51
	- 1 S(2)	2.72		- 1 S(1)	2.55
	- 2 Mo(1)	2.85		- 2 Mo(2)	2.87
	- 2 Mo(1)	3.21		- 2 Mo(2)	3.21
	- 1 Mo(2)	3.22		- 1 Mo(1)	3.22

Interatomic distances (Å)

[S(1) - 2 Mo(1)	2.39	[S(3) - 2 Mo(2)	2.40
- 1 Mo(1)	2.34	- 1 Mo(2)	2.31]
- 1 Mo(2)	2.55]		
[S(2) - 2 Mo(1)	2.54		
- 2 Mo(2)	2.51		
- 1 Mo(1)	2.72]		

Discussion

The metal atoms lie in octahedral holes of a *chh* stacking of sulphur atoms.

Material and Details of analysis

Mo and S were heated at 1300°C in evacuated SiO_2 ampoules, which were enclosed in an outer tube filled with argon, and quenched. Diffractometer and Guinier powder patterns, Weissenberg photographs (CuK_α); powder intensities by planimetering the strong lines and using peak intensities for weaker lines. Parameters were derived (*a*) from Fourier synthesis and (*b*) by the method of 7 and *c* by least squares; the parameters listed are the average values of *a*, *b* and *c*; $R = 0.25$ for *I*'s of 170 observed reflexions.

$MoNbSe_3$

Monoclinic, $a = 6.439$, $b = 3.414$, $c = 9.111$ Å, $\beta = 103.55°$, $P2_1/m$.

$Mo_{1.03}Nb_{1.03}S_3$ (room temperature)

Monoclinic, $a = 6.125$, $b = 3.292$, $c = 8.692$, $\beta = 102.67°$, $P2_1/m$. Both isotopic with $Mo_{2.06}S_3$.

In agreement with 4, 5, 6, the compounds previously regarded as Mo_2Se_3 and Mo_2Te_3 were found to have broad ranges of homogeneity, including the composition Mo_3X_4, but not Mo_2X_3.

Mo_3Se_4

Monoclinic, $a = 4.632$, $b = 4.762$, $c = 6.649$ Å, $\beta = 92.56°$, $U = 146.6$ Å^3.

Mo_3Te_4

Monoclinic, $a = 4.875$, $b = 5.092$, $c = 7.051$ Å, $\beta = 93.75°$, $U = 170.7$ Å^3.

1. *Structure Reports*, 26, 200; F. JELLINEK, 1961. *Nature, Lond.*, 192, 1065.
2. C.L. McCABE, 1955. *J. Metals*, 7, 61.
3. N. MORIMOTO and G. KULLERUD, *Carnegie Inst. Washington Year Book*, 61, 143, 1962.
4. A.A. OPALOVSKIJ and V.E. FEDOROV, 1966. *Izv. Akad. Nauk, SSSR, Neorg. Mater.*, 2, 443.
5. M. SPEISSER, C. MARCHAL and J. ROUXEL, 1968. *C.R. Acad. Sci., Paris*, C 266, 1583.
6. M. SPEISSER and J. ROUXEL, 1967. *Ibid.*, C 265, 92.
7. A.K. BHUIYA and E. STANLEY, 1963. *Acta Cryst.*, 16, 981.

NICKEL TERBIUM

I. Polymorphisme du composé TbNi. R. LEMAIRE and D. PACCARD, 1970. *J. Less-Common Metals*, 21, 403-413.

See also 1.

H.T. NiTb F.W. = 217.63

Orthorhombic, (quenched), $a = 21.09 \pm 3$, $b = 4.22 \pm 1$, $c = 5.45 \pm 1$ Å, [$U = 485.05$ Å^3, $Z = 12$, $D_x = 8.94$].

Space group *Pnma* (D_{2h}^{16}) from structure

Atomic positions

			x	*y*	*z*	*B* (Å^2)
Ni(1)	in	4(*c*)	0.0121±16	1/4	0.6403±35	0.8±4
Ni(2)	in	4(*c*)	0.1790±16	1/4	0.3911±35	0.8±4
Ni(3)	in	4(*c*)	0.3463±16	1/4	0.6209±35	0.8±4
Tb(1)	in	4(*c*)	0.0606±7	1/4	0.1393±12	0.5±2
Tb(2)	in	4(*c*)	0.3926±7	1/4	0.1323±12	0.5±2
Tb(3)	in	4(*c*)	0.2292±7	1/4	0.8792±12	0.5±2

Interatomic distances (Å)

Tb(1)	- 2 Ni(1)	2.87
	- [1 Ni(1)	2.91]
	- [1 Ni(1)	2.92]
	- 1 Ni(2)	2.85
	- 2 Ni(3)	2.89
	- 2 Tb(1)	3.65
	- 2 Tb(2)	3.56
	- [2 Tb(2)	3.61]
	- 1 Tb(3)	3.79[3.83]
Tb(2)	- 2 Ni(1)	2.92[2.91]
	- [1 Ni(1)	2.93]
	- 2 Ni(2)	2.91
	- 1 Ni(3)	2.83[2.84,2.95]
	- [2 Tb(1)	3.56
	- 2 Tb(1)	3.61
	- 1 Tb(1)	3.76]
	- 2 Tb(3)	3.62[3.59]
	- [1 Tb(3)	3.71]
Tb(3)	- 2 Ni(2)	2.86
	- [1 Ni(2)	2.86
	- 1 Ni(2)	2.98]
	- 1 Ni(3)	2.87[2.84]
	- [2 Ni(3)	2.95]
Tb(3)	- [1 Tb(1)	3.83
	- 2 Tb(2)	3.59
	- 1 Tb(2)	3.71]
	- 4 Tb(3)	3.56
[Ni(1)	- 2 Tb(1)	2.87
	- 1 Tb(1)	2.92
	- 1 Tb(1)	2.91
	- 2 Tb(2)	2.91
	- 1 Tb(2)	2.93]
	- 2 Ni(1)	2.65
[Ni(2)	- 1 Tb(1)	2.85
	- 2 Tb(2)	2.91
	- 2 Tb(3)	2.86
	- 1 Tb(3)	2.86
	- 1 Tb(3)	2.98]
	- 2 Ni(3)	2.63
[Ni(3)	- 2 Tb(1)	2.88
	- 1 Tb(2)	2.84
	- 1 Tb(2)	2.95
	- 2 Tb(3)	2.95
	- 1 Tb(3)	2.84
	- 2 Ni(2)	2.63

Details of analysis

A crystal 1 mm x 0.02 mm (prepared by melting the elements at 1400°C in an Al_2O_3 crucible) was used to obtain 44 *h0l* reflexions (Weissenberg camera, FeK_α). Intensities were measured photometrically and corrected for absorption. Patterson projection plus least squares refinement; final R = 10.3%.

L.T. NiTb

Monoclinic, (annealed at 950°C), a = 21.26 ± 3, b = 4.21 ± 1, c = 5.45 ± 1 Å, β = 97°25' ± 20', [U = 483.75 Å^3, Z = 12, D_x = 8.96].

Space group $P2_1/m$ (C^2_{2h}) from structure

Atomic positions

			x	*y*	*z*
Ni(1)	in	2(*e*)	0.1632±13	1/4	0.4675±29
Ni(2)	in	2(*e*)	0.3298±13	1/4	0.8000±29
Ni(3)	in	2(*e*)	0.4939±13	1/4	0.6325±29
Ni(4)	in	2(*e*)	0.6605±13	1/4	0.9450±29
Ni(5)	in	2(*e*)	0.8288±13	1/4	0.8000±29
Ni(6)	in	2(*e*)	0.9946±13	1/4	0.1325±29
Tb(1)	in	2(*e*)	0.1074±6	1/4	0.9214±11
Tb(2)	in	2(*e*)	0.2708±6	1/4	0.2382±11
Tb(3)	in	2(*e*)	0.4406±6	1/4	0.1050±11
Tb(4)	in	2(*e*)	0.6064±6	1/4	0.4175±11
Tb(5)	in	2(*e*)	0.7708±6	1/4	0.2853±11
Tb(6)	in	2(*e*)	0.9394±6	1/4	0.6090±11

Overall temperature factor B = 0.8±4 Å^2 for Ni and 0.5±2 Å^2 for Tb.

Interatomic distances (Å)

Tb(1)	- 1 Ni(1)	2.88	Ni(1)	- [1 Tb(1)	3.06
	- [1 Ni(1)	3.06]		- 1 Tb(1)	2.88
	- [2 Ni(5)	2.84]		- 1 Tb(2)	2.74
	- 1 Ni(6)	2.79		- 2 Tb(5)	2.78
	- [2 Ni(6)	3.00]		- 2 Tb(6)	3.02]
Tb(2)	- 1 Ni(1)	2.74		- 2 Ni(5)	2.59[2.58]
	- 1 Ni(2)	2.83[2.84]	[Ni(2)	- 1 Tb(2)	3.16
	- [1 Ni(2)	3.16]		- 1 Tb(2)	2.84
	- 2 Ni(4)	2.82		- 1 Tb(3)	2.70
	- 2 Ni(5)	2.98[2.97]		- 2 Tb(4)	2.84
Tb(3)	- 1 Ni(2)	2.70		- 2 Tb(5)	2.99
	- 2 Ni(3)	2.81		- 2 Ni(4)	2.52]
	- [1 Ni(3)	2.94]	[Ni(3)	- 2 Tb(3)	2.81
	- [1 Ni(3)	2.95]		- 1 Tb(3)	2.94
	- 2 Ni(4)	2.99[3.00]		- 1 Tb(3)	2.95
Tb(4)	- 2 Ni(2)	2.84		- 1 Tb(4)	2.80
	- 1 Ni(3)	2.79[2.80]		- 2 Tb(4)	2.98]
	- [2 Ni(3)	2.98]		- 2 Ni(3)	2.59[2.58]
	- 2 Ni(4)	2.95[2.95,2.96]	[Ni(4)	- 2 Tb(2)	2.82
Tb(5)	- 2 Ni(1)	2.78		- 2 Tb(3)	3.00
	- 2 Ni(2)	2.99		- 2 Tb(4)	2.95
	- 1 Ni(4)	2.79[2.80]		- 1 Tb(5)	2.80
	- 1 Ni(5)	2.91[2.92]		- 2 Ni(2)	2.52]
	- [1 Ni(5)	3.06]	Ni(5)	- [2 Tb(1)	2.84
Tb(6)	- [2 Ni(1)	3.02]		- 2 Tb(2)	2.97
	- 1 Ni(5)	2.69		- 1 Tb(5)	2.92
	- [2 Ni(6)	2.81		- 1 Tb(5)	3.06
	- 1 Ni(6)	2.94		- 1 Tb(6)	2.69]
	- [1 Ni(6)	2.98]		- 2 Ni(1)	2.59[2.58]
			[Ni(6)	- 2 Tb(1)	3.00
				- 1 Tb(1)	2.79
				- 2 Tb(6)	2.81
				- 1 Tb(6)	2.94
				- 1 Tb(6)	2.98
				- 2 Ni(6)	2.58]

[Many Tb-Tb distances ranging from 3.55 to 3.77 are omitted for brevity].

Material and Details of analysis

Ni and Tb (99.9%) were induction-melted in Al_2O_3 crucibles under helium. A needle, (1.2 mm x 0.016 mm) was used to record 78 reflexions (Weissenberg camera, FeK_α). Intensities were measured photometrically and corrected for absorption. A Patterson projection was used to obtain a trial structure refined by least squares to R = 14%. Powder pattern (CrK_α) is listed.

Discussion

Both structures of NiTb (Fig. 15) can be derived from that of DyNi (2) (FeB type (3)) by choosing new axes.

[Both structures show 7-fold coordination of Ni around Tb and of Tb around Ni; in addition each Ni is linked to 2 Ni, forming endless chains parallel to *b*; 6 Tb atoms around each Ni are arranged on the corners of a prism with the extra Tb opposite one face; the arrangement of Ni atoms around each Tb is similar but the Tb atom lies almost in one face of the prism and the prism axes are perpendicular to *b*.]

1. R. LEMAIRE and D. PACCARD, 1969. *Bull. Soc. Franç. Minér. Crist.*, 92, 9.
2. N.C. BAENZIGER and J.L. MORIARTY, 1961. *Acta Cryst.*, 14, 946.
3. *Strukturbericht*, 3, 12.

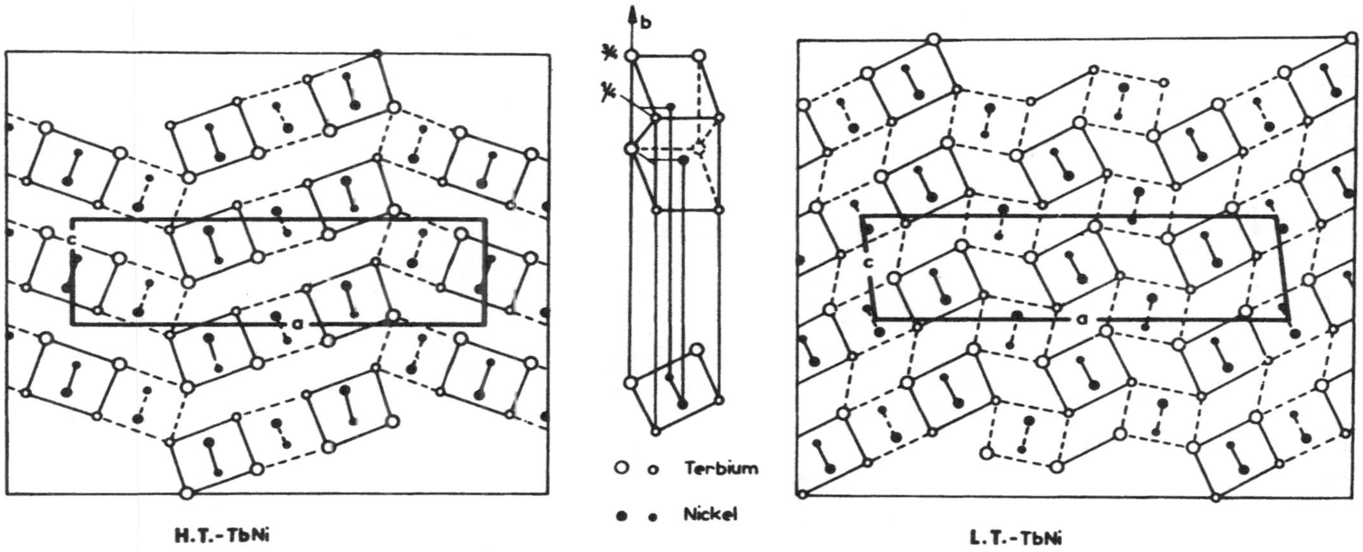

Fig. 15. The orthorhombic (1058°C) and monoclinic (950°C) structures of NiTb projected onto (010).

NIOBIUM SILICON TIN

I. The crystal structure of Ni_5Sn_2Si. R. HORYŃ and K. LUKASZEWICZ, 1970. *Bull. acad. polon. sci., ser. sci. chim.*, 18, 59-64.

Nb_5SiSn_2 F.W. = 730

Tetragonal, isostructural with W_5Si_3 (1), a = 10.541 ± 3, c = 5.138 ± 2 Å, c/a = 0.487, [U = 570.9 Å^3], D_m = 8.38 ± 2, Z = 4, D_x = 8.492.

Space group *I4/mcm* (D_{4h}^{18})

Atomic positions

			x	y	z	B (Å^2)
4 Nb(1)	in	4(*b*)	0	1/2	1/4	0.282±162
16 Nb(2)	in	16(*k*)	0.07038±54	0.21346±53	0	0.253±92
4 Si	in	4(*a*)	0	0	1/4	-0.343±495
8 Sn	in	8(*h*)	0.16830±37	0.66830±37	0	-0.041±92

Interatomic distances (Å)

Nb(1)	- 2 Nb(1)	2.569	Nb(2)	- 1 Sn	2.808
	- 8 Nb(2)	3.365		- 2 Sn	3.036
	- 4 Sn	2.819		- 2 Si	2.695
Nb(2)	- 2 Nb(1)	3.365	Sn	- 2 Nb(1)	2.819
	- 2 Nb(2)	2.967		- 2 Nb(2)	2.795
	- 1 Nb(2)	3.236[3.222]		- 2 Nb(2)	2.808
	- 2 Nb(2)	3.339		- 4 Nb(2)	3.036
	- 2 Nb(2)	3.351	Si	- 8 Nb(2)	2.695
	- 1 Sn	2.795		- 2 Si	2.569

Material and Details of analysis

This phase was prepared by vapour transport of Nb_3Sn (HCl). Intensities for *h0l* and *hkl* were obtained by photometering precession films (MoK_α) and corrected for Lorentz and polarization factors. Trial parameters (1) were refined by difference syntheses and least squares; scattering factors (TFD) were corrected for dispersion and R = 0.102 for observed reflexions.

1. *Structure Reports*, 19, 277.

OSMIUM SILICON

I. The crystal structure of $OsSi_2$. I. ENGSTRÖM, 1970. *Acta Chem. Scand.*, 24, 2117-2125.

$OsSi_2$ see also 1. F.W. = 246.37

Orthorhombic, a = 10.1496 ± 15, b = 8.1168 ± 11, c = 8.2230 ± 11 Å, U = 677.43 Å^3, Z = 16, D_x = 9.66.

Space group *Cmca* (D_{2h}^{18}) from absences and structure

Atomic positions

			x	y	z	B (Å^2)
8 Os(1)	in	8(*d*)	0.2142±3	0	0	0.21±2
8 Os(2)	in	8(*f*)	0	0.1881±3	0.1812±4	0.15±3
16 Si(1)	in	16(*g*)	0.3699±16	0.2208±13	0.0591±18	0.34±14
16 Si(2)	in	16(*g*)	0.1280±16	0.0534±13	0.7252±18	0.54±16

Interatomic distances (Å)

Os(1)	- 2 Os(2)	3.046	Os(1)	- 2 Si(1)	2.439	Os(1)	- 2 Si(2)	4.164
	- 2 Os(1)	4.123		- 2 Si(2)	2.462		- 2 Si(2)	4.167
	- 2 Os(2)	4.128		- 2 Si(1)	2.471		- 2 Si(1)	4.194
	- 2 Os(1)	4.175		- 2 Si(2)	2.486			
	- 2 Os(2)	4.197		- 2 Si(1)	4.130			

Interatomic distances (Å)

Os(2)	- 2 Os(1)	3.046	Si(1)	- Si(2)	2.548	Si(2)	- Os(2)	2.495
	- 2 Os(1)	4.128		- Si(2)	2.609		- Os(2)	4.116
	- 2 Os(1)	4.197		- Si(1)	2.640		- Os(1)	4.164
	- 2 Si(2)	2.474		- Si(1)	2.667		- Os(1)	4.167
	- 2 Si(1)	2.492		- Si(2)	2.708		- Si(2)	2.509
	- 2 Si(2)	2.495		- Si(2)	3.353		- Si(1)	2.548
	- 2 Si(1)	2.521		- Si(2)	3.654		- Si(2)	2.599
	- 2 Si(1)	3.894		- Si(2)	3.697		- Si(1)	2.609
	- 2 Si(2)	4.116		- Si(1)	3.717		- Si(1)	2.708
	- 2 Si(1)	4.138		- Si(2)	3.756		- Si(1)	3.353
Si(1)	- Os(1)	2.439		- Si(2)	3.768		- Si(1)	3.654
	- Os(1)	2.471		- Si(2)	3.854		- Si(1)	3.697
	- Os(2)	2.492		- Si(2)	3.929		- Si(1)	3.756
	- Os(2)	2.521		- Si(1)	3.966		- Si(1)	3.768
	- Os(2)	3.894		- 2 Si(1)	4.139		- Si(2)	3.803
	- Os(1)	4.130	Si(2)	- Os(1)	2.462		- Si(1)	3.854
	- Os(2)	4.138		- Os(2)	2.474		- Si(1)	3.929
	- Os(1)	4.194		- Os(1)	2.486		- 2 Si(2)	4.079

σ's for Os-Os range from 0.001-0.003 Å, for Si-Si from 0.017-0.033 Å.

Discussion

The structure (Fig. 16) can be described in terms of the approximately planar stacking of the Si atoms perpendicular to *a*, as an *AA BB AA BB* sequence. This structure can be derived from the CaF_2 structure.

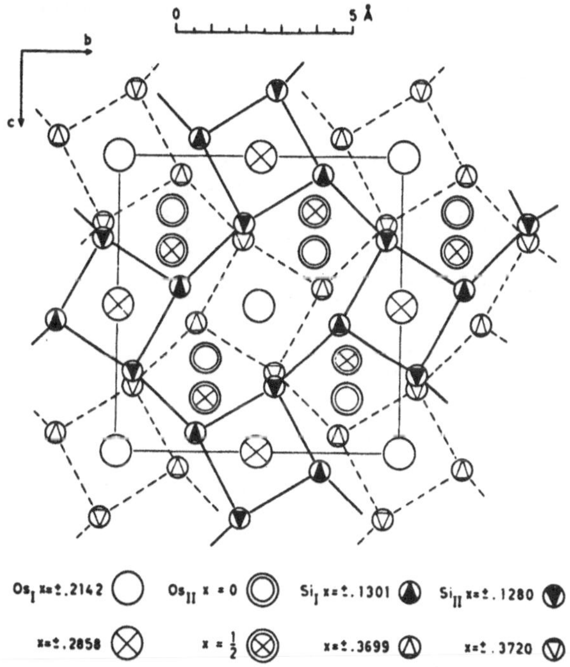

Fig. 16. Projection of the orthorhombic $OsSi_2$ structure onto (100).

Material and Details of analysis

Os (99.9%) and Si (99.9%) were arc-melted and then annealed *in vacuo*. Lattice parameters were obtained by least-squares analysis of Guinier photographs (CuK_α, silicon standard, a = 5.4305 Å), intensities were estimated visually by comparison with a scale (Weissenberg photographs, *hk*0, *hk*1, *hk*2, 0*kℓ*, Mo*K* radiation, Zr filter, multiple films, no absorption or extinction corrections). A trial structure based on Patterson projections was refined by difference Fouriers and full-matrix least squares. $R = 0.083$ for 270 observed *hkl*; the space group *C2ca* was tested and rejected.

1. *Structure Reports*, 15, 25; 21, 161; 27, 372.

PALLADIUM SILICON

I. Note on the crystal structures of Ru_5Si_3 and PdSi. I. ENGSTRÖM, 1970. *Acta Chem. Scand.*, 24, 1466-1468.

PdSi F.W. = 134.49

Orthorhombic, MnP(*B*31) type structure (1), $a = 5.6173 \pm 10$, $b = 3.3909 \pm 6$, $c = 6.1534 \pm 12$ Å, $U = 117.21$ Å^3, $Z = 4$, $D_x = 7.62$.

Space group *Pnma* (D_{2h}^{16})

Atomic positions

	x	*y*	*z*	*B* (Å^2)
4 Pd in 4(*c*)	0.0043±3	1/4	0.1906±3	0.27±3
4 Si in 4(*c*)	0.1770±14	1/4	0.5722±15	0.33±9

Interatomic distances (Å)

PdSi		NiSi	
Pd - 2 Pd	2.895	[Ni - 2 Ni	2.684
- 2 Pd	2.902	- 2 Ni	2.688
- 2 Pd	3.391	-	-
- 1 Si	2.449	- 1 Si	2.300
- 2 Si	2.458	- 2 Si	2.304
- 1 Si	2.541	- 1 Si	2.395
- 2 Si	2.571	- 2 Si	2.422]
- 1 Si	3.927	-	-
Si - 1 Pd	2.449	[Si - 1 Ni	2.300
- 2 Pd	2.458	- 2 Ni	2.304
- 1 Pd	2.541	- 1 Ni	2.395
- 2 Pd	2.571	- 2 Ni	2.422
- 1 Pd	3.927	-	-
- 2 Si	2.760	2 Si	2.635]
- 2 Si	3.391		
- 2 Si	3.560		
- 4 Si	3.607		

Material and Details of analysis

Pd (99.9%) and Si (99.9%) were arc-melted. Weissenberg photographs (Mo*K*, Zr filter, multiple film) were used to record 128 *h0l* and intensities were visually estimated by comparison with a scale; least squares analysis gave $R = 0.12$.

NiSi, *orthorhombic*, MnP type (1)

The data of 2 were refined by least squares to $R = 0.11$.

Space group *Pnma* (D_{2h}^{16})

Atomic positions

	x	*y*	*z*
4 Ni in 4(*c*)	0.0061	1/4	0.1873
4 Si in 4(*c*)	0.1741	1/4	0.5844

1. *Strukturbericht*, 3, 17.
2. *Structure Reports*, 15, 107.

PALLADIUM TELLURIUM

I. [The palladium telluride $Pd_{4-x}Te$ crystal structure.] V.S. KHAR'KIN, R.M. IMAMOV and S.A. SEMILETOV, 1970. *Kristallografija, SSSR*, 14, 907-910 [*Soviet Physics - Crystallography*, 14, 779-781.]

See also 1

$Pd_{4-x}Te$

Cubic, a = 12.69 Å.

Space group $F\bar{4}3m$ (T_d^2) from absences and structure

Atomic positions

				x	*y*	*z*
16	Pd(1)	in	16(*e*)	3/8	3/8	3/8
16	Pd(2)	in	16(*e*)	7/8	7/8	7/8
24	Pd(3)	in	24(*f*)	1/4	0	0
24	Pd(4)	in	24(*g*)	0	1/4	1/4
4	Pd(5)	in	16(*e*)	1/8	1/8	1/8
16	Te(1)	in	16(*e*)	5/8	5/8	5/8
4	Te(2)	in	4(*a*)	0	0	0
4	Te(3)	in	4(*c*)	1/4	1/4	1/4

Temperature factor B = 0.6 Å^2.

[*Interatomic distances* (Å)]

Pd(1)	- 3 Pd(3)	2.75
	- 3 Pd(4)	2.75
	- 1 Te(3)	2.75
Pd(2)	- 3 Pd(3)	2.75
	- 3 Pd(4)	2.75
	- 1 Te(2)	2.75
Pd(3)	- 2 Pd(5)	2.75
	- 2 Te(1)	2.75
	- 2 Pd(1)	2.75
	- 2 Pd(2)	2.75
Pd(4)	- 2 Pd(1)	2.75
	- 2 Pd(2)	2.75
	- 2 Pd(5)	2.75
	- 2 Te(1)	2.75
Pd(5)	- 3 Pd(3)	2.75
	- 3 Pd(4)	2.75
	- 1 Te(2)	2.75
	- 1 Te(3)	2.75
Te(1)	- 3 Pd(3)	2.75
	- 3 Pd(4)	2.75
Te(2)	- 4 Pd(2)	2.75
	- 4 Pd(5)	2.75
Te(3)	- 4 Pd(1)	2.75
	- 4 Pd(5)	2.75

Material and Details of analysis

Pd and Te were evaporated onto freshly cleaved NaCl at 5 x 10^{-5} mm Hg and then annealed 1.5 hr at 200°C. Intensities of 210 reflexions were determined by measurement of electron-diffraction photographs (multiple exposures; microphotometric measurement for zero layer, visual estimates for other reflexions). Trial structures based on $Fm3m$ were not satisfactory; ϕ^2 and ϕ projections gave a trial structure in $F\bar{4}3m$ which gave R = 0.22 when strong hkl were corrected for dynamical effects.

1. V.S. KHAR'KIN, R.M. IMAMOV and S.A. SEMILETOV, 1968. *Izv. Akad. Nauk SSSR, Ser. Neorg. Mater.*, 4, 10.

PHOSPHORUS TIN

I. The synthesis, structure and superconducting properties of new high-pressure forms of tin phosphide. P.C. DONOHUE, 1970. *Inorg. Chem.*, 9, 335-337.

II. X-ray investigations of the tin-phosphorus system. O. OLOFSSON, 1970. *Acta Chem. Scand.*, 24, 1153-1162.

See also 1, 2, 3.

SnP F.W. = 149.66

I. *Tetragonal*, GeP type (4), a = 3.831 ± 1, c = 5.963 ± 1 Å, c/a = 1.56, [U = 87.52 Å^3], Z = 2, D_x = 5.68.

Space group $I4mm$ (C_{4v}^9)

Atomic positions

	x	y	z	B (Å^2)
2 Sn in 2(a)	0	0	0	0.9±3
2 P in 2(a)	0	0	0.428±10	2.6±1.6

Interatomic distances (Å)

Sn - 1 P	2.55	P - 1 Sn	2.55
- 4 P	2.74	- 4 Sn	2.74
- 1 P	3.41	- 1 Sn	3.41

Cubic phase, NaCl type structure, a = 5.5359 ± 1 Å, [U = 169.57 Å^3], Z = 4, D_x = 5.86. Contrary to 5.

Space group $Fm3m$ (O_h^5)

Atomic positions

	x	y	z
4 Sn in 4(a)	0	0	0
4 P in 4(b)	1/2	1/2	1/2

[*Interatomic distances* (Å)]

Sn - 6 P	2.768	P - 6 Sn	2.768

Material and Details of analysis

Sn and P powder were compressed (15-65 kbar), heated (600-1300°C), then quenched to room temperature before pressure was released. Peaks on powder diffractometer (CuK_α, LiF monochromator) traces were cut out and weighed to yield intensities. A least squares refinement of weighted intensities gave R = 7.8% for the tetragonal form. For the NaCl form R = 18.4%. Scattering factors were corrected for dispersion.

Sn_4P_3 F.W. = 567.68

See also 2, 3.

II. *Rhombohedral*, As_3Sn_4 type, (6), a = 11.998 ± 1 Å, α = 19.036°, U = 160.6 Å^3, Z = 1, D_x = 5.870. Hexagonal cell, a = 3.9677 ± 3, c = 35.331 ± 4 Å, c/a = 8.905, U = 481.7 Å^3, Z = 3.

Space group $R\bar{3}m$ (D_{3d}^5) from Patterson sections and absences

Atomic positions (Hexagonal axes)

	x	y	z	B (Å^2)
6 Sn(1) in 6(c)	0	0	0.13405±6	0.30±3
6 Sn(2) in 6(c)	0	0	0.28951±7	0.62±3
3 P(1) in 3(a)	0	0	0	0.62±17
6 P(2) in 6(c)	0	0	0.4289±2	0.36±11

Interatomic distances (Å)

Sn(1) - 3 P(1)	2.664	P(1) - 3 Sn(1)	2.664
- 3 Sn(1)	3.250	- 3 Sn(2)	2.931
Sn(2) - 3 P(2)	2.765	P(2) - 6 Sn(2)	2.765
- 3 P(1)	2.931		

Material and Details of analysis

Sn and P were prereacted at 400°C in sealed SiO_2 tubes, then crushed and pressed into pellets which were annealed. Guinier powder films (CuK_α or α_1; Si, a = 5.43054 Å as internal standard) and Weissenberg photographs (Zr filtered MoK, multiple films with Fe foils). Intensities were obtained by visual comparison with a scale and corrected for Lorentz and polarization factors and absorption. Scattering factors were corrected for dispersion. A crystal 0.07 mm in size was used.

1. G. KATZ, J. KOHN and J. BRODER, 1957. *Acta Cryst.*, 10, 607.
2. O. OLOFSSON, 1967. *Acta Chem. Scand.*, 21, 1659.
3. P. ECKERLIN and W. KISCHIO, 1968. *Z. anorg. Chem.*, 363, 1.
4. P.C. DONOHUE and H.S. YOUNG, 1970. *J. Solid State Chem.*, 1, 143.
5. J. OSUGI, R. NAMIKAWA and Y. TANAKA, 1966. *Nippon Kag. Zass.*, 87, 1169; *Idem*, 1967. *Rev. Phys. Chem. Japan*, 37, 81.
6. G. HÄGG and A.G. HYBINETTE, 1935. *Phil. Mag.*, 20, 913; *Strukturbericht*, 3, 650. See also this volume p. 111.

PHOSPHORUS VANADIUM

I. The crystal structure of $V_{12}P_7$. O. OLOFSSON and E. GANGLBERGER, 1970. *Acta Chem. Scand.*, 24, 2389-2396.

P_7V_{12} see also 1, 2. F.W. = 828.12

Hexagonal, anti-Th_7S_{12} type structure (3,4), a = 9.299 ± 1, c = 3.2790 ± 7 Å, U = 245.5 Å^3, Z = 1, D_x = 5.60.

Space group $P6_3/m$ (C_{6h}^2)

Atomic positions

			x	y	z	B (Å^2)
3 V(1)	in	6(h)*	0.0143±5	0.2646±6	1/4	0.65±7
3 V(2)	in	6(h)*	0.0055±5	0.2046±6	1/4	0.71±7
6 V(3)	in	6(h)	0.3771±3	0.5095±3	1/4	0.74±4
6 P(1)	in	6(h)	0.7133±5	0.1640±4	1/4	0.82±7
1 P(2)	in	2(a)*	0	0	1/4	0.68±19

* Half filled position.

Interatomic distances (Å)

V(1)	- 2 V(1)	3.279	V(3)	- 1 V(1)	2.980 or	V(3)	- 1 V(2)	3.191
	- 2 V(2)	2.696		- 1 V(1)	2.940 or		- 1 V(2)	3.454
	- 2 V(2)	2.766		- 2 V(2)	2.994 or		- 2 V(1)	2.659
	- 1 V(3)	2.980		- 2 V(3)	3.279			
	- 1 V(3)	2.940		- 2 V(3)	2.948			
	- 2 V(3)	2.659		- 2 V(3)	2.889			
	- 1 P(1)	2.468		- 1 P(1)	2.485			
	- 2 P(1)	2.282		- 2 P(1)	2.415			
	- 1 P(2)	2.397		- 2 P(1)	2.410			
V(2)	- 2 V(1)	2.696	P(1)	- 1 V(1)	2.468 or	P(1)	- 1 V(2)	2.549
	- 2 V(1)	2.766		- 2 V(2)	2.592 or		- 2 V(1)	2.282
	- 2 V(2)	3.279		- 1 V(3)	2.485			
	- 2 V(2)	3.252		- 2 V(3)	2.410			
	- 2 V(3)	2.994		- 2 V(3)	2.415			
	- 1 V(3)	3.454		- 2 P(1)	3.279			
	- 1 V(3)	3.191		- 2 P(1)	3.171			
	- 2 P(1)	2.592	P(2)	- 3 V(1)	2.397			
	- 1 P(1)	2.549		- 6 V(2)	2.493			
	- 2 P(2)	2.493		- 2 P(2)	3.279			

Discussion

The structure is disordered (Fig. 17) and a detailed discussion is given. The interatomic distances given correspond to one or the other of the static possibilities; if all sites were fully occupied the interatomic distances would be too short.

Material and Details of analysis

V and P were sintered together and then arc-melted. Lattice parameters were derived from Guinier photographs (CuK_{α_1} λ = 1.54051 Å, Si standard *a* = 5.4305 Å) and *hk*0, *hkl* intensities from Weissenberg photographs (Mo*K*, Zr Filter, Fe foils, visual comparison with a scale) were corrected for Lorentz, polarization and absorption (approximately cylindrical, diam. = 0.02 mm). A trial structure, derived from Patterson sections, was refined by least squares and difference syntheses; final *R* = 0.070 for 156 F_{obs} (*hk*0 and *hkl*).

1. E. GANGLBERGER, 1968. *Mh. Chem.*, 99, 557; *Idem*, 1968. *Ibid.*, 99, 566.
2. T. LUNDSTRÖM, 1967. *2nd Intern. Conf. Solid Compounds of Transition Elements*, Twente; *Idem*, 1969. *Ark. Kemi*, 31, 227.
3. W.H. ZACHARIASEN, 1949. *Acta Cryst.*, 2, 288. *Structure Reports*, 12, 184.
4. I. ENGSTRÖM, 1965. *Acta Chem. Scand.*, 21, 1773.

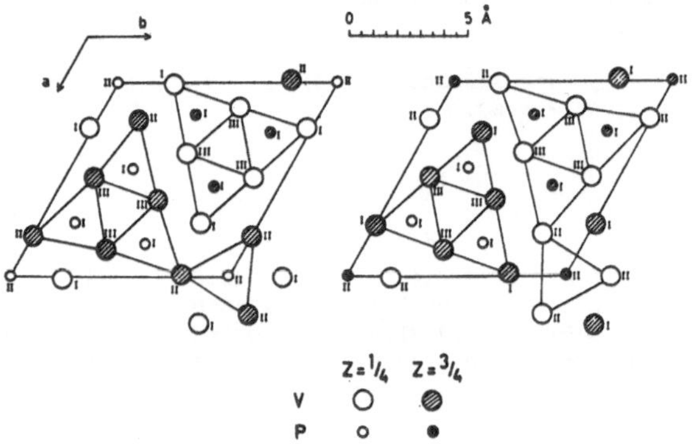

Fig. 17. Two possible arrangements in the disordered hexagonal $V_{12}P_7$ structure; (010) projection.

PLATINUM ZINC

I. Determination of the structure of cubic gamma-Pt, Zn; a phase of gamma brass type with an 18 Å superstructure. A. JOHANSSON and S. WESTMAN, 1970. *Acta Chem. Scand.*, 24, 3471-3479.

γ-Pt_3Zn_{10} (approx.) see also 1 and Cd-Ni, p. 40.

Cubic face-centred, *a* = 18.1128 ± 6 Å, D_m = 10.1, D_x = 10.1. Some crystals had *a* ≈ 9 Å.

Space group $F\bar{4}3m$ (T_d^2)

Atomic positions

The coordinates are chosen to centre clusters on 000, 1/2 1/2 1/2, 1/4 1/4 1/4 and 3/4 3/4 3/4.

Cluster A		*x*	*y*	*z*	*B* (Å^2)	Occupancy
OT(Pt) in	16(*e*)	-0.0896±4	-0.0896±4	-0.0896±4	1.2±2	
OH(Pt,Zn)	vac.					
IT(Zn) in	16(*e*)	0.0577±11	0.0577±11	0.0577±11	0.2±5	
CO(Zn) in	48(*h*)	0.0216±10	0.1550±6	0.1550±6	0.5±3	
Cluster B						
OT(Pt) in	16(*e*)	0.4350±11	0.4350±11	0.4350±11	1.1±6	1/2 Pt + 1/2 Zn*†
OH(Pt,Zn)	24(*f*)	0.6806±5	1/2	1/2	0.01±20	2/3 Pt + 1/3 Zn†
IT(Zn) in	16(*e*)	0.5506±18	0.5506±18	0.5506±18	0.2±9	1/12 Pt + 11/12 Zn*†
CO(Zn) in	48(*h*)	0.5204±20	0.6499±14	0.6499±14	0.8±6	
Cluster C						
OT(Pt) in	16(*e*)	0.1611±10	0.1611±10	0.1611±10	1.6±3	
OH(Pt,Zn)	24(*g*)	0.4296±6	1/4	1/4	0.3±3	2/3 Pt + 1/3 Zn†
IT(Zn) in	16(*e*)	0.3024±10	0.3024±10	0.3024±10	0.5±5	
CO(Zn) in	48(*h*)	0.2701±13	0.4065±8	0.4065±8	1.8±5	
Cluster D						
OT(Pt) in	16(*e*)	0.6655±5	0.6655±5	0.6655±5	1.0±2	
OH(Pt,Zn)	24(*g*)	0.9263±12	3/4	3/4	0.2±4	1/12 Pt + 11/12 Zn†
IT(Zn) in	16(*e*)	0.8251±17	0.8251±17	0.8251±17	3.1±9	
CO(Zn) in	48(*h*)	0.7591±10	0.9112±6	0.9112±6	-0.4±2	

* Either OT(B) or IT(B) is occupied by the atomic distribution shown.

† Occupancy only approximately determined.

The atomic sites are as follows: IT(Zn), inner tetrahedral; OT(Pt), outer tetrahedral; CO(Zn), cubo-octahedral; OH(Pt,Zn), octahedral.

Interatomic distances (Å)

IT(A)	- IT(A)	Zn	- 3 Zn	2.958±44
	- OT(A)		- 3 Pt	2.792±14
	- CO(A)		- 3 Zn	2.576±27
	- OT(C)		- Pt	3.242±37
OT(A)	- IT(A)	Pt	- 3 Zn	2.792±14
	- CO(A)		- 3 Zn	2.619±17
	- IT(D)		- Zn	2.675±42
	- CO(D)		- 3 Zn	2.740±19
CO(A)	- IT(A)	Zn	- Zn	2.576±27
	- OT(A)		- Pt	2.619±17
	- OT(C)		- Pt	2.532±20
	- OH(C)		- (Pt,Zn)	2.590±16
	- CO(C)		- 2 Zn	2.699±21
	- OH(D)		- Zn(Pt)	2.983±21
	- CO(D)		- 2 Zn	2.803±16
IT(B)	- IT(B)	Zn(Pt)	- 3 Zn(Pt)	2.592±92
	- OH(B)		- 3 (Pt,Zn)	2.688±37
	- CO(B)		- 3 Zn	2.601±58
OT(B)	- OT(B)	(Pt,Zn)	- 3 (Pt,Zn)	3.331±42
	- OH(B)		- 3 (Pt,Zn)	2.675±26
	- CO(B)		- 3 Zn	2.668±45
	- CO(C)		- 3 Zn	3.074±32

Interatomic distances (Å)

OH(B)	- IT(B)	(Pt,Zn)	- 2	Zn(Pt)	2.688±37
	- OT(B)		- 2	(Pt,Zn)	2.675±26
	- CO(B)		- 4	Zn	2.795±20
	- CO(C)		- 2	Zn	2.557±23
	- CO(D)		- 2	Zn	2.683±16
CO(B)	- IT(B)	Zn	-	Zn(Pt)	2.601±58
	- OT(B)		-	(Pt,Zn)	2.668±45
	- OH(B)		- 2	(Pt,Zn)	2.795±20
	- OH(C)		-	(Pt,Zn)	3.047±37
	- OT(D)		-	Pt	2.659±38
	- OH(D)		-	Zn(Pt)	2.741±37
	- CO(D)		- 2	Zn	2.583±24
	- CO(C)		- 2	Zn	2.720±32
IT(C)	- IT(C)	Zn	- 3	Zn	2.683±42
	- OT(C)		- 3	Pt	2.725±14
	- OH(C)		- 3	(Pt,Zn)	2.666±10
	- CO(C)		- 3	Zn	2.730±32
OT(C)	- IT(A)	Pt	-	Zn	3.242±37
	- CO(A)		- 3	Zn	2.532±20
	- IT(C)		- 3	Zn	2.725±14
	- OH(C)		- 3	(Pt,Zn)	2.808±8
	- CO(C)		- 3	Zn	2.626±23
OH(C)	- CO(A)	(Pt,Zn)	- 2	Zn	2.590±16
	- CO(B)		- 2	Zn	3.042±37
	- IT(C)		- 2	Zn	2.666±10
	- OT(C)		- 2	Pt	2.808±8
	- CO(C)		- 4	Zn	2.888±14
	- OH(D)		-	Zn(Pt)	2.610±25
CO(C)	- CO(A)	Zn	- 2	Zn	2.699±21
	- OT(B)		-	(Pt,Zn)	3.074±34
	- OH(B)		-	(Pt,Zn)	2.557±23
	- CO(B)		- 2	Zn	2.720±32
	- IT(C)		-	Zn	2.730±32
	- OT(C)		-	Pt	2.626±23
	- OH(C)		- 2	(Pt,Zn)	2.888±14
	- CO(D)		- 2	Zn	3.346±19
IT(D)	- OT(A)	Zn	-	Pt	2.675±42
	- OT(D)		- 3	Pt	2.900±31
	- OH(D)		- 3	Zn(Pt)	2.657±42
	- CO(D)		- 3	Zn	2.508±45
OT(D)	- CO(B)	Pt	- 3	Zn	2.659±38
	- IT(D)		- 3	Zn	2.900±31
	- OH(D)		- 3	Zn(Pt)	2.730±14
	- CO(D)		- 3	Zn	2.595±16
OH(D)	- CO(A)	Zn(Pt)	- 2	Zn	2.983±21
	- CO(B)		- 2	Zn	2.741±37
	- OH(C)		-	(Pt,Zn)	2.610±25
	- IT(D)		- 2	Zn	2.657±42
	- OT(D)		- 2	Pt	2.730±14
	- CO(D)		- 4	Zn	2.937±9
CO(D)	- OT(A)	Zn	-	Pt	2.740±19
	- CO(A)		- 2	Zn	2.803±16
	- OH(B)		-	(Pt,Zn)	2.683±16
	- CO(B)		- 2	Zn	2.583±24
	- CO(C)		- 2	Zn	3.346±19
	- IT(D)		-	Zn	2.508±45
	- OT(D)		-	Pt	2.595±16
	- OH(D)		- 2	Zn(Pt)	2.937±9

[The distances given above generally differ by 0.001 or 0.002 Å from those calculated with the given coordinates - presumably rounding-off errors.]

Material and Details of analysis

Pt and Zn were melted in sealed evacuated SiO_2 capsules, quenched into water, and then homogenized at 700°C and again quenched into water. Lattice parameters were derived from Guinier photographs ($CuK\alpha_1$, λ = 1.54050 Å, KCl standard *a* = 6.2919 Å). Intensities were obtained both photographically (Weissenberg, $CuK\alpha$, visual comparison with scale, ∿200 *hkl* yielding 54 independent values) and by diffractometer (0.05 x 0.05 x 0.06 crystal fragment, 800 *hkl* measured, 218 independent, absorption correction). Intensities were corrected for Lorentz-polarization factors and scattering factors were corrected for dispersion and the trial structure (2,3) refined by full-matrix least squares; *R* = 9.3% for 205 obs. *hkl* (218 measured); the selected model had the best combination of thermal parameters and can be considered an idealized ordering scheme with other arrangements not necessarily excluded.

Discussion

The structure may be described as a variant of the ordered gamma Pd,Zn structure and consists of four clusters, each containing an inner tetrahedral (IT), outer tetrahedral (OT), an octahedral (OH) and a cubo-octahedral (CO) position. The OH position in one cluster is vacant, in another cluster either the IT or the OT position is vacant. The two versions of this cluster are distributed statistically over the structure. The OT positions in all clusters are filled with Pt; some excess Pt together with Zn is distributed over the OH positions in the three clusters in which this position is occupied. [See also Cd-Ni, p. 40]

1. *Strukturbericht*, 2, 710.
2. O. von HEIDENSTAM, A. JOHANSSON and S. WESTMAN, 1968. *Acta Chem. Scand.*, 22, 653.
3. V.-A. EDSTRÖM and S. WESTMAN, 1969. *Ibid.*, 23, 279.

RARE-EARTH ARSENIDES

I. Rare-earth arsenides. S. ONO, J.G. DESPAULT, L.D. CALVERT and J.B. TAYLOR, 1970. *J. Less-Common Metals*, 22, 51-59.

As_3Yb_5 Data for other rare-earth arsenides are given in Table I. F.W. = 1090

Hexagonal, Mn_5Si_3 ($D8_8$) structure type (1), *a* = 8.480, *c* = 6.671 Å, *c/a* = 0.787 (Yb rich), *U* = 415.5 Å^3, *Z* = 2, *a* = 8.445, *c* = 6.618 Å, *c/a* = 0.784 (As rich), *U* = 408.7 Å^3, *Z* = 2.

Space group $P6_3/mcm$ (D^3_{6h})

Atomic positions

Yb-rich				*x*	*y*	*z*	*B* (Å^2)
3.4	Yb(1)	in	4(*d*)	1/3	2/3	0	1
6	Yb(2)	in	6(*g*)	0.2516±10	0	1/4	1
6	As	in	6(*g*)	0.6115±10	0	1/4	1

As_3Yb_4

Cubic, anti-Th_3P_4 type (2, 3), *a* = 8.791 ± 1 Å, *U* = 679.4 Å^3, *Z* = 4, D_x = 8.977.

Space group $I\bar{4}3d$ (T^6_d)

Atomic positions

	x	*y*	*z*	*B* (Å^2)
16 Yb in 16(*c*)	0.069±1	0.069±1	0.069±1	1
12 As in 12(*a*)	3/8	0	1/4	1

Interatomic distances (Å)

(Yb rich As_3Yb_5)

As(1)	- 4 Yb(1)	3.109	Yb(2)	- 2 As(1)	2.894
	- 2 Yb(2)	2.894		- 1 As(1)	3.052
	- 1 Yb(2)	3.052		- 2 As(1)	3.532
	- 2 Yb(2)	3.532		- 4 Yb(1)	3.635
	- 2 As(1)	3.834		- 2 Yb(2)	3.695
Yb(1)	- 6 As(1)	3.109		- 4 Yb(2)	3.960
	- 2 Yb(1)	3.336			
	- 6 Yb(2)	3.635			

(in cubic A_4B_3 phases)

		Ce_4As_3	Pr_4As_3	Yb_4As_3
As(1)	- 4 *B*(1)	2.998	2.978	2.911
	- 4 *B*(1)	3.279	3.257	3.184
B(1)	- 3 As(1)	2.998	2.978	2.911
	- 3 As(1)	3.279	3.257	3.184
	- 3 *B*(1)	3.507	3.484	3.406
	- 2 *B*(1)	3.920	3.895	3.807

Material and Details of analysis

As (6-9's) and Yb (99.9% nominal) were crushed, made into pellets and reacted under argon and then recrushed and pelletized and annealed *in vacuo* in sealed SiO_2 ampoules. Powder photographs in Guinier cameras ($CuK_{\alpha 1}$, λ = 1.54050 Å, Si *a* = 5.43052 Å) were used for lattice parameters; peak areas were obtained by micro-densitometer measurements and converted to intensity measurements by adding the equiatomic arsenide as an internal standard; trial structures refined by systematic variation of the parameters; R_I for As_3Yb_5 = 0.11.

1. *Strukturbericht*, 4, 24.
2. *Ibid.*, 7, 15.
3. O.D. McMASTERS and K.A. GSCHNEIDNER, *Trans. AIME, Inst. Metals Div. Spec. Rept. Ser. 10*(13), (1964), 93.

RARE-EARTH COPPER COMPOUNDS

I. The crystal structure of some copper compounds of the type RCu_6. K.H.J. BUSCHOW and A.S. VAN DER GOOT, 1970. *J. Less-Common Metals*, 20, 309-313.

See also 1.

$SmCu_6$ F.W. = 531.59

Orthorhombic, $CeCu_6$ (1) type structure, *a* = 8.060, *b* = 5.034, *c* = 10.049 Å, [U = 407.7 Å^3, Z = 4, D_x = 8.66].

Lattice parameters for other RCu_6 compounds are given in Table I.

Space group *Pnma* (D_{2h}^{16}) from close similarity to $CeCu_6$

Atomic positions (from 1)

			x	*y*	*z*
4 Cu(1)	in	4(*c*)	0.1470	1/4	0.1419
4 Cu(2)	in	4(*c*)	-0.1828	1/4	-0.2448
4 Cu(3)	in	4(*c*)	-0.4383	1/4	-0.4021
4 Cu(4)	in	4(*c*)	-0.0980	1/4	-0.4845
8 Cu(5)	in	8(*d*)	0.4356	0.0046	0.1904
4 Sm	in	4(*c*)	0.2604	1/4	0.4353

B = 0.800 Å^2 for all atoms.

[*Interatomic distances* (Å)]

Cu - Cu	2.42	to	3.06
Cu - Sm	2.88	to	3.35

Material and Details of analysis

Cu (99.99%) and the rare-earth element (99.9%) were arc-melted and then vacuum annealed in Al_2O_3 crucibles (700-800°C). Powder diffractometer patterns, (monochromatic CuK_α) gave good agreement with intensities based on the coordinates of 1; R = 0.12 for intensities for $SmCu_6$; other RCu_6 gave similar values for R.

1. *Structure Reports*, 24, 99.

RARE-EARTH NICKEL COMPOUNDS

I. The crystal structure of rare-earth nickel compounds of the type R_2Ni_7. K.H.J. BUSCHOW and A.S. VAN DER GOOT, 1970. *J. Less-Common Metals*, 22, 419-428.

La_2Ni_7 see also 1, 2. F.W. = 688.8

Hexagonal, Ce_2Ni_7 type (3) structure, a = 5.058, c = 24.71 Å, [c/a = 4.89, U = 547 Å^3, Z = 4, D_x = 8.36]. For lattice constants of other Ce_2Ni_7 type structure, and of Gd_2Co_7 types, see Table I.

Space group $P6_3/mmc$ (D_{6h}^4) from (3)

Atomic positions

			x	*y*	*z*
2 Ni(1)	in	2(*a*)	0	0	0
4 Ni(2)	in	4(*e*)	0	0	0.167
4 Ni(3)	in	4(*f*)	1/3	2/3	0.833
6 Ni(4)	in	6(*h*)	0.833	0.666	1/4
12 Ni(5)	in	12(*k*)	0.833	0.666	0.088
4 La(1)	in	4(*f*)	1/3	2/3	0.035
4 La(2)	in	4(*f*)	1/3	2/3	0.175

B (Å^2) for all atoms is 0.60.

[*Interatomic distances* (Å)]

Ni(1)	- 6 Ni(5)	2.62	Ni(5)	- 1 Ni(1)	2.62
	- 6 La(1)	3.05		- 1 Ni(2)	2.44
Ni(2)	- 3 Ni(3)	2.92		- 1 Ni(3)	2.44
	- 3 Ni(4)	2.52		- 2 Ni(5)	2.52
	- 3 Ni(5)	2.44		- 2 Ni(5)	2.53
	- 3 La(2)	2.93		- 2 La(1)	2.85
Ni(3)	- 3 Ni(2)	2.92		- 1 La(1)	3.37
	- 3 Ni(4)	2.52		- 2 La(2)	3.32
	- 3 Ni(5)	2.44	La(1)	- 3 Ni(1)	3.05
	- 3 La(2)	2.93		- 6 Ni(5)	2.85
Ni(4)	- 2 Ni(2)	2.52		- 3 Ni(5)	3.37
	- 2 Ni(3)	2.52		- 3 La(1)	3.39
	- 2 Ni(4)	2.52		- 1 La(2)	3.46
	- 2 Ni(4)	2.53	La(2)	- 3 Ni(2)	2.93
	- 4 La(2)	3.14		- 3 Ni(3)	2.93
				- 6 Ni(4)	3.14
				- 6 Ni(5)	3.32
				- 1 La(1)	3.46
				- 1 La(2)	3.71

Material and Details of analysis

Rare-earth metals (99.9%) and Ni (99.99%) were arc-melted under argon and annealed in Al_2O_3 crucibles in sealed SiO_2 tubes. Diffractometer powder intensities (CuK_α and CoK_α) were corrected for Lorentz and polarization factors (adapted to monochromatization when required) and the calculated values for the real dispersion term; R = 0.16 for intensities.

Discussion

The RE_2Ni_7 (RE = rare earth) compounds occur in two forms, the hexagonal Ce_2Ni_7 type being favoured by large RE atoms (Lu, Ce) whereas the Gd_2Co_7 (4) type is favoured by small RE atoms (Ho, Er, Y); both forms are observed for the medium size RE atoms.

1. A.V. VIRKAR and A. RAMAN, 1969. *J. Less-Common Metals*, 18, 59.
2. R. LEMAIRE, D. PACCARD and R. PAUTHENET, 1967. *C.R. Acad. Sci., Paris*, 265, 1280.
3. *Structure Reports*, 23, 108.
4. E.F. BERTAUT, R. LEMAIRE and J. SCHWEIZER, 1965. *Bull. Soc. Franç. Minér. Crist.*, 88, 580.

RARE-EARTH - NICKEL SILICIDES

I. [Crystal structure of $CeNiSi_2$ and kindred compounds.] O.I. BODAK and E.I. GLADYŠEVSKIJ, 1970. *Kristallografija, SSSR*, 14, 990-994 [*Soviet Physics - Crystallography*, 14, 859-862].

$CeNiSi_2$ F.W. = 255.0

Orthorhombic, a = 4.141 ± 5, b = 16.418 ± 5, c = 4.068 ± 5 Å, [U = 276.57 Å^3], D_m = 5.98, Z = 4, D_x = 6.09 [6.12]. [Note the resemblance to the Mo_2BC structure, (4)].

Space group *Cmcm* (D_{2h}^{17}) from absences and Patterson projections

Atomic positions

			x	y	z	B (Å^2)
4 Ce	in	4(c)	0	0.1070±2	1/4	0.85±5
4 Ni	in	4(c)	0	0.3158±8	1/4	2.95±17
4 Si(1)	in	4(c)	0	0.4566±20	1/4	2.53±37
4 Si(2)	in	4(c)	0	0.7492±4	1/4	0.79±1

Interatomic distances (Å)

Ce(1)	- 4 Ni	3.163[3.167]	Si(1)	- 4 Ce	3.086[3.085]
	- 1 Ni	3.433[3.428]		- 2 Ce	3.217[3.222]
	- 4 Si(1)	3.086[3.085]		- 2 Si(1)	2.476[2.484]
	- 2 Si(1)	[3.222]		- 1 Ni	2.299[2.312]
	- 2 Si(2)	3.095[3.116]	Si(2)	- 4 Si(2)	2.903
	- 2 Si(2)	3.141[3.120]		- 2 Ni	2.299[2.297]
	- 2 Ce	4.060		- 2 Ni	2.345[2.342]
	- 2 Ce	4.068		- 2 Ce	3.095[3.116]
	- 2 Ce	4.141		- 2 Ce	3.141[3.120]
			Ni	- 1 Si(1)	2.299[2.312]
				- 2 Si(2)	2.299[2,297]
				- 2 Si(2)	2.345[2.342]
				- 4 Ce	3.163[3.167]

Material and Details of analysis

Ce (99.56%), Ni (99.8%) and Si (99.99%) were arc-melted in argon. Powder photographs (CrK_α) were used for lattice parameters and Laue and Weissenberg photographs were used for obtaining the space group, 49 $hk0$ and 47 $0k\ell$ reflexions. Intensities were measured by comparison with a scale and corrected for absorption (crystal was .03 x .35 x .25 mm), Lorentz and polarization factors. A trial structure based on Patterson and Fourier projections was refined by least squares (R_{hk0} = 15.8%, $R_{0k\ell}$ = 13.3%).

Discussion

The structure may be described as being built up of elements of the $CeGa_2Al_2$ (1) and AlB_2 (2) structures. Detailed comparisons are given of the relationships in the various ternary Ce-Ni-Si structures (3) and related type structures.

The $RCeSi_2$ compounds with R = La, Pr, Nd, Sm, Eu, Gd, Tb, Dy, Ho, Er, Tm, Yb and Lu are isotypic with $CeNiSi_2$ (see Table I).

1. O.I. BODAK, E.I. GLADYŠEVSKIJ and P.I. KRIPJAKEVIČ, 1966. *Izv. Akad. Nauk SSSR, Neorg. Mater.*, 12, 2151.
2. *Strukturbericht*, 3, 28.
3. O.I. BODAK and E.I. GLADYŠEVSKIJ, 1969. *Dop. Akad. Nauk, Ukr.SSR*, A 5, 452.
4. W. JEITSCHKO, 1963. *Mh. Chem.*, 94, 565.

RARE-EARTH - ZINC PHASES

I. Rare-earth intermediate phases with zinc. G. BRUZZONE, M.L. FORNASINI and F. MERLO, 1970. *J. Less-Common Metals*, 22, 253-264.

See also Table I. for additional data.

$CeZn_3$ see also 1, 2.

Orthorhombic, YZn_3 type structure (1), a = 6.644, b = 4.627, c = 10.437 Å, [U = 320.85 Å^3, Z = 4.]

Space group *Pnma* (D_{2h}^{16})

Atomic positions

			x	*y*	*z*
4 Ce	in	4(*c*)	0.250	1/4	0.350
4 Zn(1)	in	4(*c*)	0.250	1/4	0.060
4 Zn(2)	in	4(*c*)	0.945	1/4	0.870
4 Zn(3)	in	4(*c*)	0.555	1/4	0.870

[*Interatomic distances* (Å)]

Ce	- 2 Zn(1)	3.45	Zn(2)	- 2 Ce	3.51
	- 2 Zn(1)	3.19		- 1 Ce	3.20
	- 1 Zn(1)	3.03		- 2 Ce	3.08
	- 2 Zn(2)	3.08		- 2 Zn(1)	2.75
	- 2 Zn(2)	3.51		- 1 Zn(1)	2.84
	- 1 Zn(2)	3.20		- 2 Zn(2)	3.64
	- 2 Zn(3)	3.08		- 1 Zn(3)	2.59
	- 2 Zn(3)	3.51		- 1 Zn(3)	2.61
	- 1 Zn(3)	3.20	Zn(3)	- 2 Ce	3.08
Zn(1)	- 2 Ce	3.45		- 2 Ce	3.51
	- 2 Ce	3.19		- 1 Ce	3.20
	- 1 Ce	3.03		- 2 Zn(1)	2.75
	- 2 Zn(2)	2.75		- 1 Zn(1)	2.84
	- 1 Zn(2)	2.84		- 1 Zn(2)	2.59
	- 2 Zn(3)	2.75		- 1 Zn(2)	2.61
	- 1 Zn(3)	2.84		- 2 Zn(3)	3.64

Gd_3Zn_{11}

Orthorhombic, La_3Al_{11} type structure (3), a = 4.423, b = 13.063, c = 8.842 Å, [U = 510.9 Å^3, Z = 2].

Space group *Immm* (D_{2h}^{25})

Atomic positions

			x	*y*	*z*
2 Gd(1)	in	2(*a*)	0	0	0
4 Gd(2)	in	4(*g*)	0	0.295	0
2 Zn(1)	in	2(*c*)	1/2	1/2	0
4 Zn(2)	in	4(*j*)	1/2	0	0.687
8 Zn(3)	in	8(*l*)	0	0.340	0.354
8 Zn(4)	in	8(*l*)	0	0.133	0.287

[*Interatomic distances* (Å)]

Gd(1)	- 4 Zn(2)	3.54	Zn(3)	- 2 Gd(1)	3.31
	- 8 Zn(3)	3.31		- 2 Gd(2)	3.11
	- 4 Zn(4)	3.08		- 1 Gd(2)	3.18
Gd(2)	- 2 Zn(1)	3.47		- 1 Zn(2)	2.56
	- 2 Zn(2)	3.15		- 1 Zn(3)	2.58
	- 4 Zn(3)	3.11		- 2 Zn(4)	2.56
	- 2 Zn(3)	3.18		- 1 Zn(4)	2.77
	- 4 Zn(4)	3.05	Zn(4)	- 1 Gd(1)	3.08
	- 2 Zn(4)	3.30		- 2 Gd(2)	3.05
Zn(1)	- 4 Gd(2)	3.47		- 1 Gd(2)	3.30
	- 4 Zn(2)	2.76		- 1 Zn(1)	2.56
	- 4 Zn(4)	2.56		- 2 Zn(2)	2.82
Zn(2)	- 2 Gd(1)	3.54		- 2 Zn(3)	2.56
	- 2 Gd(2)	3.15		- 1 Zn(3)	2.77
	- 2 Zn(1)	2.76		- 1 Zn(4)	3.47
	- 1 Zn(2)	3.31			
	- 2 Zn(3)	2.56			
	- 4 Zn(4)	2.82			

Material and Details of analysis

Rare-earth metal (99.5 - 99.9%) and Zn (99.99%) filings were heated in evacuated Pyrex ampoules (400°C) and then pressed and annealed in Ta crucibles sealed into SiO_2 ampoules with a small pressure of argon, (600-700°C). Powder (FeK_α) and single-crystal patterns (CuK_α, MoK_α); no other details.

1. B.G. LOTT and P. CHIOTTI, 1966. *Acta Cryst.*, 20, 733.
2. E. VELECKIS, R.V. SCHABLASKE, I. JOHNSON and H.M. FEDER, 1967. *Trans. Met. Soc. AIME*, 239, 58.
3. A.H. GOMES DE MESQUITA and K.H.J. BUSCHOW, 1967. *Acta Cryst.*, 22, 497.

SAMARIUM ZINC

I. The crystal structure of a samarium-zinc compound with approximate composition $SmZn_{11}$. J.T. MASON, K.S.S. HARSHA and P. CHIOTTI, 1970. *Acta Cryst.*, B 26, 356-361.

See also 1, 2, 3, 4.

$SmZn_{11}$ approximately

Hexagonal, a = 8.974 ± 2, c = 8.918 ± 3 Å, U = 622.1 Å^3.

Space group *P6/mmm* (D_{6h}^1)

Atomic positions

			x	*y*	*z*	occupancy, %
Sm(1)	in	1(*a*)	0	0	0	100
2 Sm(2)	in	2(*d*)	1/3	2/3	1/2	100
2 Sm(3)	in	2(*c*)	1/3	2/3	0	15
Sm(4)	in	1(*b*)	0	0	1/2	4
12 Zn(1)	in	12(*o*)	0.1671±2	0.3342	0.2415±3	100
6 Zn(2)	in	6(*j*)	0.3552±6	0	0	100

Atomic positions

			x	y	z	occupancy
6 Zn(3)	in	6(*k*)	0.2944±5	0	1/2	100
6 Zn(4)	in	6(*i*)	1/2	0	0.2742±5	100
2 Zn(5)	in	2(*e*)	0	0	0.3544±8	96
4 Zn(6)	in	4(*h*)	1/3	2/3	0.1457±7	85

Anisotropic β's are also given.

Interatomic distances (Å)

Sm(1)	- 2 Zn(5)	3.161	Zn(2)	- 4 Zn(6)	3.177
	- 6 Zn(2)	3.188		- 2 Zn(2)	3.188
	- 12 Zn(1)	3.374		- 1 Sm(1)	3.188
Sm(2)	- 2 Zn(6)	3.160	Zn(3)	- 1 Sm(4)	2.642*
	- 6 Zn(3)	3.180		- 2 Zn(3)	2.642
	- 6 Zn(4)	3.281		- 4 Zn(1)	2.675
	- 6 Zn(1)	3.463		- 2 Zn(4)	2.732
Sm(3)	- 6 Zn(2)	2.898*		- 2 Zn(5)	2.944
	- 6 Zn(1)	3.364		- 2 Sm(2)	3.180
	- 6 Zn(4)	3.562		- 1 Zn(3)	3.690
Sm(4)	- 6 Zn(3)	2.642*	Zn(4)	- 4 Zn(1)	2.604
	- 12 Zn(1)	3.473		- 2 Zn(3)	2.732
Zn(1)	- 2 Zn(1)	2.597		- 2 Zn(2)	2.770
	- 2 Zn(4)	2.604		- 6 Zn(6)	2.832
	- 2 Zn(3)	2.675		- 2 Sm(2)	3.282
	- 2 Zn(2)	2.685		- 1 Sm(3)	3.563
	- 1 Zn(6)	2.721	Zn(5)	- 1 Zn(5)	2.598
	- 1 Zn(5)	2.786		- 6 Zn(1)	2.786
	- 1 Sm(3)	3.364		- 6 Zn(3)	2.944
	- 1 Sm(1)	3.374		- 1 Sm(1)	3.161
	- 1 Sm(2)	3.463	Zn(6)	- 1 Zn(6)	2.601
	- 1 Sm(4)	3.473		- 3 Zn(1)	2.721
Zn(2)	- 1 Zn(2)	2.599		- 3 Zn(4)	2.832
	- 4 Zn(1)	2.685		- 1 Sm(2)	3.160
	- 2 Zn(4)	2.770		- 6 Zn(2)	3.177
	- 2 Sm(3)	2.898*			

* Short due to partial occupancy.

Sm(3) - Zn(6) and Sm(4) - Zn(5) pairs do not occur simultaneously (1.299, 1.298 Å)

Material and Details of analysis

Alloys were prepared as in 4 in Ta crucibles, annealed at 925°C and slowly cooled; crystals were obtained by dissolving the Zn matrix in NaOH. Debye-Scherrer photographs (CuK_α) plus oscillation, Weissenberg and precession photographs. A hexagonal needle 0.14 x 0.068 mm was used to record all accessible reflexions (MoK_α, diffractometer $2\theta < 70°$). Intensities (for 343 *hkl*) were corrected for absorption and a trial structure based on a Patterson map was refined by Fourier and least squares; scattering factors were corrected for dispersion; final refinement based on anisotropic thermal parameters, disorder and partial occupancy gave $R = 0.056$.

Discussion

The coordination of Sm(1) and (2) is given in Fig. 18; the disordered atoms Sm(3) and Sm(4) are both surrounded by similar polyhedra but 18-fold, instead of 20-fold. The structure is related to the $CaZn_5$ (5) family of structures (Fig. 19) as are also Pu_3Zn_{22}, Ce_5Mg_{24}, Th_2Zn_{17}, Th_2Ni_{17} and $SmZn_{12}$ ($ThMn_{12}$ type, 1, 2).

1. Yu.B. KUZ'MA, P.I. KRIPYAKEVIČ and N.S. UGRIN, 1966. *Inorg. Mater.*, 2, 544.
2. A. IANDELLI, A. PALENZONA, 1967. *J. Less-Common Metals*, 12, 333.
3. E. VELECKIS, R.V. SCHABLASKE and B.S. TANI, 1966. *Crystal Structure of the Zinc-Rich Phases in Lanthanon-Zinc Systems*, p. 162. Argonne National Lab. Report ANL-7225.

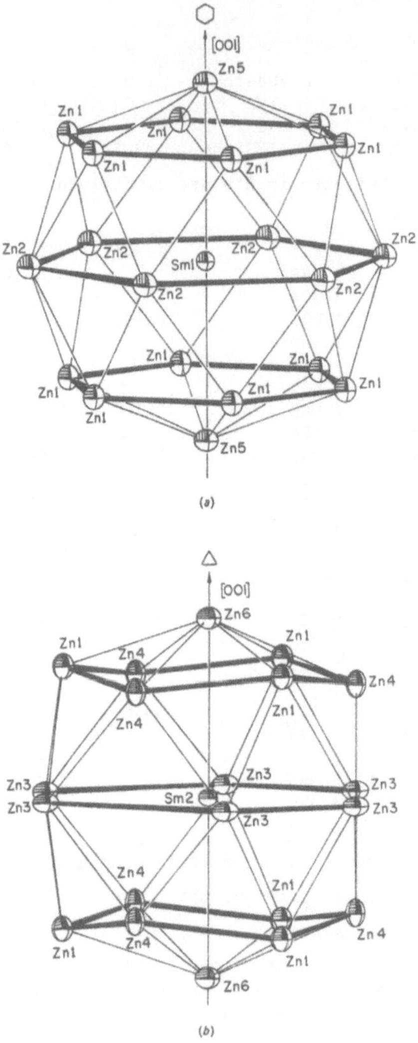

Fig. 18. $SmZn_{11}$ structure. (a) Coordination polyhedron about Sm(1). (b) Coordination polyhedron about Sm(2). The thermal ellipsoids in (a) and (b) are scaled to a 50% probability.

4. P. CHIOTTI and J.T. MASON, 1967. *Trans. Met. Soc. AIME,* 239, 547.
5. *Structure Reports,* 11, 59.

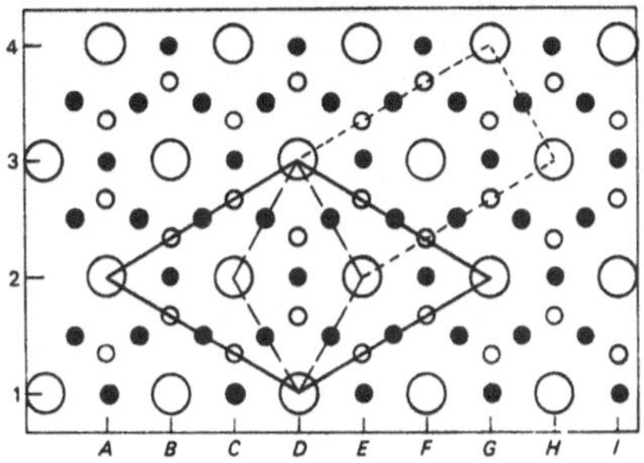

Fig. 19. Relationship of hexagonal $SmZn_{12}$ and tetragonal $SmZn_{12}$ type structures to the $D2_d$ structure. The grid represents the projection of an AB_5 structure of the $D2_d$ type onto the (001) plane. Large circles are *A* type atoms; small circles are *B* type atoms. Open circles are in the (001) plane; filled circles represent atoms in planes which are 1/2 above and below the (001) plane. The AB_5 unit cell is outlined by the long-dash lines, the $SmZn_{12}$ tetragonal unit cell by the short-dash lines, and the $SmZn_{12}$ hexagonal unit cell by the solid lines.

SCANDIUM SULPHUR YTTRIUM

I. Structure du sulfure mixte d'yttrium et de scandium $YScS_3$ et de certains composés isotypes. N. RODIER and P. LARUELLE, 1970. *C.R. Acad. Sci. Paris*, C 270, 2127-2130.

See also 1.

ScS_3Y F.W. = 230.05

Orthorhombic, a = 7.00, b = 6.36, c = 9.46 Å, (1), [U = 421.16 Å^3], Z = 4, [D_x = 3.63].

Space group $Pna2_1$ (C_{2v}^9) from Patterson and absences

Atomic positions

			x	*y*	*z*	*B* (Å^2)
4 Sc	in	4(*a*)	0.001	0.499	0.767	0.24
4 S(1)	in	4(*a*)	0.306	0.327	0.184	1.01
4 S(2)	in	4(*a*)	0.323	0.323	0.825	0.09
4 S(3)	in	4(*a*)	0.048	0.358	0.519	0.38
4 Y	in	4(*a*)	0.0993	0.0388	0	0.47

Interatomic distances (Å)

Y - S(1)	2.91	Sc - S(1)	2.54
- S(1)	2.82	- S(1)	2.61
- S(1)	3.34	- S(2)	2.46
- S(2)	2.69	- S(2)	2.58
- S(2)	2.91	- S(3)	2.57
- S(2)	3.41	- S(3)	2.53
- S(3)	2.73	S - S distances 3.40 to 3.97;	
- S(3)	2.73	average = 3.64	

Material and Details of analysis

A crystal was obtained by melting and then annealing at 600°C for several months. Integrated Weissenberg photographs (MoK_α) were used to record 486 independent reflexions. Intensities were measured photometrically and corrected for Lorentz and polarization factors and absorption. A trial structure based on Y positions derived from a Patterson synthesis was refined by Fourier and least squares. Allowance was made for anomalous scattering; R = 10% for all reflexions; R = 5.5% for 277 stronger reflexions which gave the coordinates listed.

1. N. RODIER, P. LARUELLE and J. FLAHAUT, 1969. *C.R. Acad. Sci., Paris*, C 269, 1391.

SILICON ZINC PHOSPHIDE

I. Crystal structure of luminescent $ZnSiP_2$. S.C. ABRAHAMS and J.L. BERNSTEIN, 1970. *J. Chem. Phys.*, 52, 5607-5613.

P_2SiZn F.W. = 154.936 [155.403]

Tetragonal, chalcopyrite type, (1), crystal 1 at 298°K, a = 5.398 ± 1, c = 10.434 ± 1 Å; crystal 2, a = 5.399 ± 1, c = 10.437 ± 1 Å (photoluminescent). Mean value, a = 5.399 ± 1, c = 10.435 ± 2 Å, U = 304.17 Å^3, D_m = 3.31 ± 8, Z = 4, D_x = 3.383[3.393].

Space group $I\bar{4}2d$ (D_{2d}^{12}) from absences and structure

Atomic positions

	x	y	z	B (Å^2)
4 Zn in 4(a)	0	0	0	0.600±15
4 Si in 4(b)	0	0	1/2	0.606±25
8 P in 8(d)				0.547±17
crystal 1	0.2686±6			
crystal 2	0.2699±6	1/4	1/8	
combined	0.2691±4			

Anisotropic β's are given.

Interatomic distances (Å) *and angles*

Zn - 4 P	2.375±1	Zn - Si	3.756±1
Si - 4 P	2.254±1		3.819±1
P - P	3.684±2	Zn - Zn	3.756±1
	3.832±2		
	3.968±1	Si - Si	3.756±1
P-Zn-P	113.3±1°	P-Si-P	109.6±1°
Zn-P-Zn	104.6±1°	Si-P-Si	112.8±1°
Zn-P-Si	108.4±1°	Zn-P-Si	111.2±1°

Discussion

Statistical analysis showed that crystal 1 did not differ detectably from crystal 2; a model for the combined data with 0.8% Zn on Si sites could not be ruled out at the 1% confidence level. A characteristic temperature of 343°K was calculated and the observed interatomic distances compared with various predicted values. See Fig. 20.

Material and Details of analysis

Dark red crystals were grown by vapour transport ($ZnCl_2$) from 6-9's Zn and P; for crystal 1 the Si was doped with As (ρ = 10^{-1} Ω cm), for crystal 2, 60 Ω cm p-type Si was used. Crystal 1, a sphere with radius 0.1435 ± 10 mm, was used to collect 2023 reflexions (MoK_α, sin θ/λ < 1.02 Å , Zr + Y balanced filters, diffractometer); crystal 2, a sphere of radius 0.1610 ± 15 mm, was similarly used to collect 1600 reflexions. Intensities (197 independent hkl for

1, 176 for 2, 158 *hkl* common to 1 & 2) were corrected for Lorentz, polarization and absorption factors. F_{obs} were corrected for extinction; relativistic Dirac-Slater scattering factors, corrected for dispersion, were used to refine the structure by full-matrix least squares; final *R* (anisotropic thermal parameters) was 0.049 for combined data. The two sets of data were compared and found to belong to the same population; replacement of Si by Zn was tested with the conclusion that departures from complete order are probably less than 1%. Half-normal probability plots were constructed for the atomic parameters leading to the conclusion that parameters from 1 and 2 belong to the same population and accordingly the data can be combined. Lattice parameters were derived from Debye-Scherrer photographs.

1. *Strukturbericht*, 2, 48.

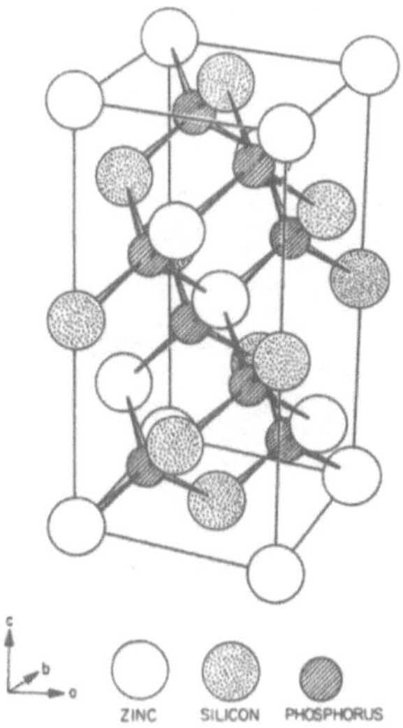

Fig. 20. The tetragonal $ZnSiP_2$ structure [After I].

SILVER YTTERBIUM

I. The ytterbium-silver system. A. PALENZONA, 1970. *J. Less-Common Metals*, 21, 443-446.

See also 1.

Ag_3Yb_5 F.W. = 1188.8

Tetragonal, Cr_5B_3 type structure, (3), *a* = 7.942, *c* = 14.881 Å, [*c/a* = 1.874, *U* = 938.62 Å, *Z* = 4, D_x = 8.41].

Space group *I*4/*mcm* (D_{4h}^{18})

Atomic positions

				x	*y*	*z*
4	Ag(1)	in	4(*a*)	0	0	1/4
8	Ag(2)	in	8(*h*)	0.378	0.878	0
4	Yb(1)	in	4(*c*)	0	0	0
16	Yb(2)	in	16(*l*)*	0.164	0.664	0.150

*[misprinted as 16(*e*)].

Ag_2Yb_3 F.W. = 734.86

Tetragonal, U_3Si_2 type structure, (2), a = 8.219, c = 4.218 Å, [c/a = 0.513, U = 284.93 Å^3, Z = 2, D_x = 8.56].

Space group *P4/mbm* (D_{4h}^5)

Atomic positions

			x	y	z
2 Yb(1)	in	2(*a*)	0	0	0
4 Yb(2)	in	4(*h*)	0.181	0.681	1/2
4 Ag	in	4(*g*)	0.389	0.889	0

H.T. AgYb F.W. = 280.91

Cubic, CsCl type structure, a = 3.680 Å, [U = 49.836 Å^3, Z = 1, D_x = 9.36].

Space group *Pm3m* (O_h^1)

Atomic positions

			x	y	z
1 Yb	in	1(*a*)	0	0	0
1 Ag	in	1(*b*)	1/2	1/2	1/2

L.T. AgYb

Orthorhombic, FeB type structure (6), a = 7.590, b = 4.670, c = 6.013 Å, [U = 213.13 Å^3, Z = 4, D_x = 8.75].

Space group *Pnma* (D_{2h}^{16})

Atomic positions

			x	y	z
4 Ag	in	4(*c*)	0.167	1/4	0.147
4 Yb	in	4(*c*)	0.041	1/4	0.603

Ag_2Yb F.W. = 388.78

See also 1.

Orthorhombic, $CeCu_2$ type structure, (4), a = 4.671, b = 7.204, c = 8.178 Å, [U = 275.19 Å^3, Z = 4, D_x = 9.38].

Space group *Imma* (D_{2h}^{28})

Atomic positions

			x	y	z
4 Yb	in	4(*e*)	0	1/4	0.640
8 Ag	in	8(*h*)	0	0.060	0.170

Ag_7Yb_2 F.W. = 1101.17

Hexagonal, Ag_7Ca_2 (5) type structure, a = 5.498, c = 14.059 Å, [c/a = 2.557, U = 424.98 Å^3].

[*Interatomic distances* (Å)]

Ag_3Yb_5

Yb(1)	- 4 Ag(2)	3.15
	- 2 Ag(1)	3.72
	- 8 Yb(2)	3.71
Yb(2)	- 2 Ag(1)	3.32
	- 2 Ag(2)	3.20
	- 1 Ag(2)	3.28
	- 1 Yb(2)	3.55
	- 2 Yb(2)	3.96

Ag_2Yb_3

Yb(1)	- 4 Ag(1)	3.32
	- 8 Yb(2)	3.68
	- 2 Yb(1)	4.22
Yb(2)	- 4 Ag(1)	3.25
	- 2 Ag(1)	3.21
	- 4 Yb(1)	3.68
	- 2 Yb(2)	4.22
	- 1 Yb(2)	4.21

H.T. AgYb

Yb	- 8 Ag	3.19
	- 6 Yb	3.68
Ag	- 8 Yb	3.19
	- 6 Ag	3.68

L.T. AgYb

Yb	- 2 Ag	3.19
	- 2 Ag	3.23

[*Interatomic distances* (Å)]

Ag_3Yb_5

Yb(2)	- 1 Yb(2)	3.68
	- 2 Yb(1)	3.71
	- 4 Yb(2)	4.20
Ag(1)	- 2 Yb(1)	3.72
	- 8 Yb(2)	3.32
Ag(2)	- 1 Ag(2)	2.74
	- 2 Yb(1)	3.15
	- 4 Yb(2)	3.20
	- 2 Yb(2)	3.28

Ag_2Yb_3

Ag(1)	- 1 Ag(1)	2.58
	- 2 Ag(1)	4.22
	- 2 Yb(1)	3.25
	- 2 Yb(2)	3.21
	- 4 Yb(2)	3.25

Ag_2Yb

Yb	- 2 Yb	2.95
	- 2 Ag	2.72
	- 2 Ag	3.24
	- 2 Ag	3.71
	- 2 Ag	3.24
	- 2 Ag	3.71
Ag	- 2 Ag	2.68
	- 1 Ag	2.74
	- 1 Ag	2.91
	- 1 Yb	2.72
	- 2 Yb	3.24
	- 2 Yb	3.71

L.T. AgYb

Yb	- 1 Ag	2.90
	- 1 Ag	3.21
	- 1 Ag	3.41
	- 2 Yb	3.87
	- 2 Yb	3.99
	- 4 Yb	4.01
Ag	- 2 Ag	2.72
	- 2 Yb	3.19
	- 1 Yb	3.21
	- 2 Yb	3.23
	- 1 Yb	2.90
	- 1 Yb	3.41

Material and Details of analysis

Yb (99.9%) and Ag (99.99%) were melted in sealed Mo crucibles under an argon atmosphere. Powder photographs (Cu, Cr) and single-crystal photographs; intensity calculations are based on literature values for the type structures.

1. A. IANDELLI and A. PALENZONA, 1968. *J. Less-Common Metals*, 15, 273.
2. *Structure Reports*, 11, 284.
3. *Ibid.*, 17, 67.
4. *Ibid.*, 26, 107.
5. W.A. ALEXANDER, L.D. CALVERT, A.DESAULNIERS, H.S. DUNSMORE and D.F. SARGENT, 1969. *Can. J. Chem.*, 47, 611.
6. *Strukturbericht*, 3, 12.

SULPHUR TANTALUM

I. The crystal structure of Ta_6S. H.F. FRANZEN and J.G. SMEGGIL, 1970. *Acta Cryst.*, B 26, 125-129.

Ta_6S F.W. = 1117.75

Monoclinic, a = 14.1576 ± 40, b = 5.2835 ± 13, c = 14.7885 ± 49 Å, β = 118.010°, [U = 976.56 Å^3], D_m = 15.21 ± 15, Z = 8, D_x = 15.18. Crystals are very commonly twinned.

Space group $C2/c$ (C_{2h}^6) from absences

Atomic positions

			x	y	z	B (Å^2)
8 Ta(1)	in	8(f)	0.2480±3	0.4067±7	0.2518±3	0.04±5
8 Ta(2)	in	8(f)	0.3401±2	0.6568±9	0.4547±3	0.26±5
8 Ta(3)	in	8(f)	0.4364±2	0.1628±8	0.4051±2	0.13±5
8 Ta(4)	in	8(f)	0.0358±2	0.1460±8	0.1740±3	0.12±5
8 Ta(5)	in	8(f)	0.2111±2	0.1475±8	0.3969±3	0.06±5
8 Ta(6)	in	8(f)	0.1063±2	0.6798±7	0.3173±3	0.06±5
8 S	in	8(f)	0.3920±15	0.9189±41	0.0437±16	0.51±30

Interatomic distances (Å)*

Ta(1)	- Ta(1)	2.644
	- Ta(1)	2.644
	- Ta(5)	2.795
	- Ta(5)	2.828
	- Ta(3)	2.873
	- Ta(3)	2.890
	- Ta(6)	2.953
	- Ta(2)	2.964
	- Ta(6)	2.979
	- Ta(4)	3.001
	- Ta(4)	3.004
	- Ta(2)	3.008
Ta(2)	- S(1)	2.525
	- Ta(1)	2.964
	- Ta(6)	2.966
	- Ta(1)	3.008
	- Ta(3)	3.010
	- Ta(5)	3.055
	- Ta(5)	3.071
	- Ta(4)	3.138
	- Ta(5)	3.140
	- Ta(3)	3.188
	- Ta(6)	3.208
	- Ta(3)	3.237
Ta(3)	- S(1)	2.448
	- S(1)	2.529
	- Ta(1)	2.873
	- Ta(1)	2.890
Ta(3)	- Ta(4)	2.911
	- Ta(2)	3.010
	- Ta(6)	3.055
	- Ta(3)	3.055
	- Ta(4)	3.069
	- Ta(5)	3.138
	- Ta(2)	3.188
	- Ta(6)	3.227
	- Ta(2)	3.237
Ta(4)	- S(1)	2.503
	- Ta(4)	2.869
	- Ta(3)	2.911
	- Ta(1)	3.001
	- Ta(1)	3.004
	- Ta(5)	3.058
	- Ta(3)	3.069
	- Ta(6)	3.094
	- Ta(2)	3.138
	- Ta(5)	3.143
	- Ta(6)	3.222
	- Ta(6)	3.387
Ta(5)	- S(1)	2.485
	- S(1)	2.489
	- Ta(1)	2.795
	- Ta(1)	2.828
	- Ta(6)	2.839
	- Ta(5)	2.921
	- Ta(2)	3.055
Ta(5)	- Ta(4)	3.058
	- Ta(2)	3.071
	- Ta(3)	3.138
	- Ta(2)	3.140
	- Ta(6)	3.140
	- Ta(4)	3.143
Ta(6)	- S(1)	2.465
	- Ta(6)	2.722
	- Ta(5)	2.839
	- Ta(1)	2.953
	- Ta(2)	2.966
	- Ta(1)	2.979
	- Ta(3)	3.055
	- Ta(4)	3.094
	- Ta(5)	3.140
	- Ta(2)	3.208
	- Ta(4)	3.222
	- Ta(3)	3.227
	- Ta(4)	3.387
S(1)	- Ta(3)	2.448
	- Ta(6)	2.465
	- Ta(5)	2.485
	- Ta(5)	2.489
	- Ta(4)	2.503
	- Ta(2)	2.525
	- Ta(3)	2.529

*[There appear to be trivial differences between these values and those derived from the listed coordinates.] The average e.s.d. in Ta - Ta distances is 0.005 Å; in Ta - S it is 0.020 Å.

Material and Details of analysis

Ta_6S was prepared from Ta (99.99%) and S (5N) at 1620°C in a closed crucible. Weissenberg and rotation photographs were used for space group; Guinier photographs for the lattice parameters (KCl standard a = 6.29300 Å, CuK_α, λ = 1.5405 Å). Intensities of 668 reflexions by diffractometer measurements (MoK_α, Zr filter, $2\theta < 60°$) were corrected for absorption, Lorentz and polarization factors. A Fourier based on signs determined by the triple product technique gave a trial structure which was refined by least squares; dispersion corrections were applied; R = 6.6%.

SULPHUR THALLIUM

I. [Determination of the structure of Tl_2S by the electron diffraction method.] L.I. MAN, 1970. *Kristallografija, SSSR*, 15, 471-476 [*Soviet Physics - Crystallography*, 15, 399-403].

STl_2 see also 1 F.W. = 440.8

Trigonal, hexagonal cell, based on CdI structure type, a = 12.20 ± 4, c = 18.17 ± 6 Å, [c/a = 1.49, U = 2342 Å^3], Z = 27, [D_x = 8.44]; $a \approx 3a_{CdI_2}$, $c \approx 3c_{CdI_2}$.

Space group $R3$ (C_3^4)

Atomic positions Obverse setting (contrary to 1, who used reverse setting), hexagonal axes.

				x	y	z
3	S(1)	in	3(a)	0	0	0.195
3	S(2)	in	3(a)	0	0	0.519
3	S(3)	in	3(a)	0	0	0.853
9	S(4)	in	9(b)	0.665	0.657	0.226
9	S(5)	in	9(b)	0.333	0.337	0.195
9	Tl(1)	in	9(b)	0.119	0.191	0.957
9	Tl(2)	in	9(b)	0.122	0.201	0.321
9	Tl(3)	in	9(b)	0.118	0.209	0.656
9	Tl(4)	in	9(b)	0.233	0.103	0.140
9	Tl(5)	in	9(b)	0.227	0.093	0.474
9	Tl(6)	in	9(b)	0.233	0.096	0.808

B = 0.84 Å^2. Accuracy for Tl 0.001, S 0.003.

Interatomic distances (Å)

Tl(1)	- S(3)	2.78	[S(1)	- 3 Tl(4)	2.66
	- S(4)	2.92		- 3 Tl(2)	3.13
	- S(5)	2.91	S(2)	- 3 Tl(5)	2.55
Tl(2)	- S(1)	3.13		- 3 Tl(3)	3.33
	- S(4)	3.28	S(3)	- 3 Tl(1)	2.78
	- S(5)	3.22		- 3 Tl(6)	2.61
Tl(3)	- S(2)	2.78[3.33]	S(4)	- Tl(1)	2.92
	- S(4)	3.09		- Tl(2)	3.28
	- S(5)	3.23		- Tl(3)	3.09
Tl(4)	- S(1)	2.66		- Tl(4)	2.67
	- S(4)	2.67		- Tl(5)	2.68
	- S(5)	2.67		- Tl(6)	2.50
Tl(5)	- S(2)	2.55	S(5)	- Tl(1)	2.91
	- S(4)	2.51[2.68]		- Tl(2)	3.22
	- S(5)	2.69		- Tl(3)	3.23
Tl(6)	- S(3)	2.61		- Tl(4)	2.67
	- S(4)	2.50		- Tl(5)	2.69
	- S(5)	2.70		- Tl(6)	2.70]

Discussion

The sulphur atoms have 6 Tl near neighbours, roughly arranged octahedrally; the Tl atoms have 3 near S neighbours lying roughly in the same z plane. The structure differs from that of 1 which contains errors.

Material and Details of analysis

Specimens were prepared by fast sublimation of Tl_2S onto collodion at room temperature followed by annealing at 30-40°C for 30 min. Intensities were obtained by microphotometering electron diffraction films using zero level intensities as a scale and were corrected for dynamical effects. The trial structure was derived from Patterson projections and potential syntheses (ϕ); overlap of reflexions ($hk0$ + $kh0$ etc.) was allowed for by systematically canvassing possible variations and thus deriving the proportions of the individual components in overlapped reflexions; six cycles of refinement (dynamical correction and absolute scale, temperature factor, B = 0.84 Å) gave $R_{hk\ell}$ = 0.30.

1. *Strukturbericht*, 7, 92; J.A. KETELAAR and E.W. GORTER, 1939. *Z. Kristallogr.*, 101, 24.

SULPHUR TITANIUM

I. Refinement of the crystal structure of non-stoichiometric $Ti_{2+x}S$. L.-J. NORRBY and H.F. FRANZEN, 1970. *J. Solid State Chem.*, 2, 36-41.

S_4Ti_{2+x} see also 1, 2.

Hexagonal, a = 3.4198 ± 4, c = 11.444 ± 2 Å, [U = 115.9 Å^3], D_m = 3.5 ± 1, Z = 1, D_x = 3.522.

Space group $P6_3mc$ (C_{6v}^4)

Atomic positions

			x	*y*	*z*	*B* (Å^2)
2	S(1)	in 2(*b*)	1/3	2/3	0.37711±7	0.504±9
2	S(2)	in 2(*a*)	0	0	0.12461±9	0.573±9
2	Ti(1)	in 2(*b*)	1/3	2/3	0.75000	0.631±7
0.455±9	Ti(2)	in 2(*b*)	1/3	2/3	0.01551±18	0.403±31

Interatomic distances (Å)

Ti(1)	- 3 S(2)	2.441±1	S(1)	- 3 Ti(1)	2.452±1
	- 3 S(1)	2.452±1		- 3 Ti(2)	2.531±1
	- Ti(2)	3.038±2		- 6 S(1)	3.420±0
	- 3 Ti(2)	3.332±2		- 3 S(2)	3.453±1
	- 6 Ti(1)	3.420±0	S(2)	- 3 Ti(2)	2.336±1
Ti(2)	- 3 S(2)	2.336±1		- 3 Ti(1)	2.441±1
	- 3 S(1)	2.531±1		- 6 S(2)	3.420±0
	- Ti(1)	3.038±2		- 3 S(1)	3.453±1
	- 3 Ti(1)	3.332±2			
	- 6 Ti(2)	3.420±0			

Material and Details of analysis

Lattice parameters were derived from Guinier photographs. A hexagonal prism, 0.131 mm long x 0.060 mm diam., was grown by vapour transport of TiS/NH_4Cl and used to collect 297 *hkℓ* reflexions (diffractometer, MoK_α, Nb filter, θ-2θ scale, sin θ/λ < 1.15 Å^{-1}) which were corrected for absorption, Lorentz and polarization factors. The trial structures of 1 and 2 were refined by full-matrix least squares; that of 2 refined satisfactorily. The five strongest reflexions were omitted (extinction); anisotropic refinement gave no better results; anomalous dispersion corrections yielded no improvement; final R = 2.6%.

Discussion

The structure can be considered as a stacking sequence *ABCBABCBA* .... or *chchch* ... with the *A* layers only about 25% occupied. Each Ti is at the centre of a slightly distorted octahedron of S Atoms with an additional Ti at 3.04 Å; S(1) is located at the centre of a slightly distorted trigonal prism of Ti atoms; each S(2) similarly in an octahedron; these two kinds of polyhedra alternate along *c* sharing triangular faces.

1. *Structure Reports*, 21, 173, Ref. II.
2. *Ibid.*, Ref. III.

TELLURIUM THALLIUM

I. Kristallstruktur von Tl_5Te_3 und Tl_2Te_3. S. BHAN and K. SCHUBERT, 1970. *J. Less-Common Metals*, 20, 229-235.

See also 1.

Te_3Tl_5 F.W. = 1404.65

Tetragonal, Cr_5B_3 ($D8_l$) type (2), a = 8.92$_9$, c = 12.62$_0$ Å, [U = 1004.13 Å^3, Z = 4, D_x = 9.29].

Space group $I4/mcm$ (D_{4h}^{18})

Atomic positions

			x	y	z
4 Tl(1)	in	4(*c*)	0	0	0
16 Tl(2)	in	16(*l*)	0.144	0.644	0.157
4 Te(1)	in	4(*a*)	0	0	1/4
8 Te(2)	in	8(*h*)	0.341	0.841	0

[*Interatomic distances* (Å)]

Tl(1)	- 2 Te(1)	3.16	Te(1)	- 2 Tl(1)	3.16
	- 4 Te(2)	3.36		- 8 Tl(2)	3.62
Tl(2)	- 2 Tl(2)	3.48	Te(2)	- 2 Tl(1)	3.36
	- 1 Tl(2)	3.64		- 2 Tl(2)	3.18
	- 1 Tl(2)	3.56		- 4 Tl(2)	3.36
	- 2 Te(1)	3.62			
	- 1 Te(2)	3.18			
	- 2 Te(2)	3.36			

Te_3Tl_2 F.W. = 791.54

Monoclinic, $a = 17.41_3$, $b = 6.55_2$, $c = 7.91_0$ Å, $\beta = 133.16°$, [U = 658.02 Å^3, Z = 4, D_x = 7.988]. [This structure appears to be the Ga_2S_3 type (3).]

Space group *Cc* (C_s^4)

Atomic positions

			x	y	z
4 Tl(1)	in	4(*a*)	0	0.839	0
4 Tl(2)	in	4(*a*)	0.080	0.358	0.025
4 Te(1)	in	4(*a*)	0.213	0.861	0.952
4 Te(2)	in	4(*a*)	0.281	0.643	0.481
4 Te(3)	in	4(*a*)	0.390	0.128	0.481

[*Interatomic distances* (Å)]

Tl(1)	- Tl(2)	3.40	Te(1)	- Te(2)	3.41
	- Tl(2)	3.59		- Te(3)	2.93
	- Tl(2)	3.63		- Te(3)	3.52
	- Te(1)	3.50	Te(2)	- Tl(1)	3.71
	- Te(2)	3.71		- Tl(1)	3.73
	- Te(2)	3.73		- Tl(2)	3.38
	- Te(3)	3.41		- Tl(2)	3.75
	- Te(3)	3.56		- Te(1)	3.76
Tl(2)	- Tl(1)	3.63		- Te(1)	3.46
	- Tl(1)	3.59		- Te(1)	3.41
	- Tl(1)	3.40		- Te(3)	3.35
	- Te(1)	2.85		- Te(3)	3.70
	- Te(2)	3.75	Te(3)	- Tl(1)	3.56
	- Te(2)	3.38		- Tl(1)	3.41
	- Te(3)	3.62		- Tl(2)	3.62
	- Te(3)	3.08		- Tl(2)	3.08
Te(1)	- Tl(1)	3.50		- Te(1)	3.52
	- Tl(2)	2.85		- Te(1)	2.93
	- Te(2)	3.46		- Te(2)	3.70
	- Te(2)	3.76		- Te(2)	3.35

Material and Details of analysis

Tl and Te (99.9%) were reacted in evacuated SiO_2 ampoules. Guinier and Weissenberg photographs; no absorption or temperature factor corrections; Te_3Tl_5, R = 0.20; Te_3Tl_2, R = 0.20; observed and calculated powder patterns are given; a three-dimensional Patterson synthesis was used to establish the Te_3Tl_2 structure; no details given.

1. A. RABENAU, A STEGHERR and P. ECKERLIN, 1960. *Z. Metallk.*, 51, 295.
2. *Structure Reports*, 17, 67.
3. J. GOODYEAR and G.A. STEIGMANN, 1963. *Acta Cryst.*, 16, 946.

TELLURIUM URANIUM

I. The crystal structure of stoichiometric uranium ditelluride. A.J.K. HANEVELD and F. JELLINEK, 1970. *J. Less-Common Metals*, 21, 45-49.

See also 1, 2.

Te_2U F.W. 493.23

Orthorhombic, body-centred cell, a = 4.1617, b = 6.1276, c = 13.965 Å, [U = 356.12 Å^3], D_m = 8.91 (4), Z = 4, D_x = 9.20.

Space group *Immm* (D_{2h}^{25}) from intensity agreement

Atomic positions

	x	*y*	*z*
4 Te(1) in 4(*h*)	0	0.251±2	1/2
4 Te(2) in 4(*j*)	1/2	0	0.2977±8
4 U in 4(*i*)	0	0	0.1348±6

Interatomic distances (Å)

U - 2 Te(1)	3.08	Te(1) - 4 Te(2)	3.83
- 4 Te(2)	3.19	- 4 Te(1)	3.94
- 2 Te(1)	3.205	Te(2) - 1 Te(2)	3.05
- 1 U	3.77	- 1 Te(2)	3.08
		- 4 Te(1)	3.83

Material and Details of analysis

U (99.7%) powder, prepared by dehydriding UH_3, was heated in sealed, evacuated SiO_2 ampoules at 1000° and 800°C and slow-cooled. Powder diffractometer intensities corrected for preferred orientation, were used in a least-squares refinement (74 unique lines) R = 8.9%.

Discussion

The structure is probably isomorphous with that of β-$ErSe_2$ (5) and related to that of NbPS (6).

1. R. FERRO, 1954. *Z. anorg. Chem.*, 275, 320.
2. A.J.K. HANEVELD and F. JELLINEK, 1964. *J. Inorg. Nucl. Chem.*, 26, 1127.
3. V.K. SLOVJANSKIKH, E.I. JAREMBAS, G.V. ELLERT and A.A. ELISEEV, 1968. *Izv. Akad. Nauk, SSSR, Neorg. Mater.*, 4, 624 [*Inorg. Mater.*, 4, 543].
4. L.K. MATSON, J.W. MOODY and R.C. HIMES, 1963. *J. Inorg. Nucl. Chem.*, 25, 795.
5. D.J. HAASE, H. STEINFINK and E.J. WEISS, 1965. *Inorg. Chem.*, 4, 538.
6. P.C. DONOHUE and P.E. BIERSTEDT, 1969. *Ibid.*, 8, 2690.

TRANSITION-METAL CARBIDES

I. On the crystal chemistry of the close packed transition-metal carbides. I. The crystal structure of the ζ-V, Nb and Ta carbides. K. YVON and E. PARTHÉ, 1970. *Acta Cryst.*, B 26, 149-153.

See also 1, 2, 3.

C_3V_4

Rhombohedral, a = 9.428 ± 7 Å, α = 17°48' ± 1', hexagonal cell, a = 2.917 ± 1, c = 27.83 ± 1 Å, c/a = 9.541. The ζ-phases in the systems Nb-C and Ta-C have the same crystal structure and the following lattice constants.
C_3Nb_4, a = 3.14 ± 1, c = 30.1 ± 1 Å, c/a = 9.58; C_3Ta_4, a = 3.116 ± 5, c = 30.00 ± 5 Å, c/a = 9.62.

Space group $R\bar{3}m$ (D_{3d}^5) from systematic absences

Atomic positions (hexagonal axes)

			x	*y*	*z*
6 V(1)	in	6(*c*)	0	0	0.1265±6
6 V(2)	in	6(*c*)	0	0	0.291±1
(1)	in	3(*a*)	0	0	0
8 C(2)	in	3(*b*)	0	0	1/2
(3)	in	6(*c*)	0	0	0.417±5

An overall temperature factor of 0.5 Å^2 was assumed.

[*Interatomic distances* (Å)]

V(1) - 3 C(3)	2.06	C(3) - 3 V(1)	2.06
- 3 C(2)	2.02	- 3 V(2)	2.04
V(2) - 3 C(3)	2.04	C(2) - 6 V(1)	2.02
- 3 C(1)	2.06	C(1) - 6 V(2)	2.06

Discussion

The stacking sequence is *ABABCACABCBC* (*hhcc*)$_3$; the carbon atoms occupy octahedral interstices. Sn_4P_3 (4) and Sn_4As_3 are probably isostructural with these zeta-carbides. See this volume p. 88.

1. E.K. STORMS and R.J. McNEAL, 1962. *J. Phys. Chem.*, 66, 1401.
2. H. RASSAERTS, F. BENESOVSKY and H. NOWOTNY, 1966. *Planseeber. Pulvermet.*, 14, 178.
3. E. RUDY, S. WINDISCH and C.E. BRUKL, 1968. *Ibid.*, 16, 3.
4. O. OLOFSSON, 1967. *Acta Chem. Scand.*, 21, 1659. See p. 88.

TRANSITION METALS - NIOBIUM SULPHIDE

I. The crystal structure and magnetic susceptibilities of $MnNb_3S_6$, $FeNb_3S_6$, $CoNb_3S_6$ and $NiNb_3S_6$. K. ANZENHOFER, J.M. VAN DEN BERG, P. COSSEE and J.N. HELLE, 1970. *J. Phys. Chem. Solids*, 31, 1057-1067.

See also 1.

Hexagonal, ordered variants of the 2*s* - NbS_2 structure (1) with formula MNb_3S_6, z = 2.

M	*a* (Å)	*c* (Å)	*c/a*	*U* (Å^3)	F.W.	D_x
Mn	5.782	12.629	2.18	365.6	526.04	4.78
Fe	5.761	12.201	2.12	350.7	526.95	4.99
Co	5.768	11.886	2.06	342.5	530.05	5.14
Ni	5.758	11.897	2.07	341.6	529.81	5.15

Space group $P6_322$ (D_6^6) from structure

Atomic positions

			x	*y*	*z*
λ *M*	in	2(*c*)	1/3	2/3	1/4
2 Nb(1)	in	2(*a*)	0	0	0
4 Nb(2)	in	4(*f*)	1/3	2/3	z_{Nb}
12 S	in	12(*i*)	x_S	y_S	z_S

M	λ	z_{Nb}	x_S	y_S	z_S
Mn	1.96±36	-0.0006±3	0.33299±19	0.00394±21	0.37601±7
Fe	2.03±108	-0.0011±2	0.33261±180	0.00222±160	0.37213±60
Co	1.91±72	-0.0013±8	0.33258±46	0.00075±48	0.36780±18
Ni	2.02±96	-0.0021	0.33260±89	0.00098±87	0.36875±34

Interatomic distances (Å)

		$2s.NbS_2$	MNb_3S_6 M=Mn	Fe	Co	Ni	$2s.MoS_2$
M	- 6 S	(2.42)	2.52	2.44	2.38	2.39	(2.38)
	- 2 Nb(2)	(2.97)	3.17	3.07	2.99	3.00	(3.58)
Nb(1)	- 6 S	2.47	2.47	2.47	2.47	2.47	-
Nb(2)	- 3 S	2.47	2.47	2.46	2.47	2.46	-
	- 3 S	2.47	2.51	2.49	2.50	2.50	-
S	- 1 S	3.14	3.13	3.12	3.14	3.12	3.19
	- 4 S	3.31	3.34	3.32	3.33	3.32	3.16
			[3.32]	[3.31,3.32]	[3.32,3.33]	[3.31,3.32]	
	- 2 S	3.40	3.73	3.55	3.40	3.42	3.47
	[- 2 S	-	3.38	3.35	3.34	3.34]	-
	[- 1 S	-	3.70	3.53	3.39	3.41]	-
	α, °	83	78	79	81	81	79

α is the octahedral angle, equal to 84.5° for an undistorted octahedron.

Material and Details of analysis

Powders of the ternary sulphides, MNb_3S_6, were prepared by heating the elements at 850°C; crystals were obtained by iodine vapour transport. Hexagonal prismatic crystals, rotated about the $10\bar{1}0$ axis were used to record intensities (θ-2θ scan, MoK_α, Zr filter, automatic diffractometer) for each compound. An approximate spherical absorption correction was applied and the observed intensities were corrected for primary extinction [secondary extinction, absorption ?] by correcting I_{obs} to that which would have been observed at the rotation angle giving I_{max}.

Trial structures (1) were refined by full-matrix least squares with anisotropic temperature factors; the Nb scattering factors were corrected for dispersion. Final values of R and the number of $hk\ell$ used in the refinements are given below.

	R (%)	$hk\ell$
$MnNbS_6$	2.7	236
$FeNbS_6$	9.2	204
$CoNbS_6$	4.5	278
$NiNbS_6$	10.3	248

1. J.M. VAN DEN BERG and P. COSSEE, 1968. *Inorg. chim. Acta*, 2, 143.

TRANSITION-METAL STANNIDES

I. Transition metal stannides with MgAgAs and $MnCu_2Al$ type structure. W. JEITSCHKO, 1970. *Metallurg. Trans.*, 1, 3159-3162.

NiSnZr F.W. = 268.6

Cubic, MgAgAs (1), ordered CaF_2 (2) type structure, a = 6.113 ± 1 Å, U = 228.4 Å^3, Z = 4, D_x = 7.81.

Space group $F\bar{4}3m$ (T_d^2)

Ni_2SnZr F.W. = 327.3

Cubic, $MnCu_2Al$, ordered BiF_3 (3) Heusler (4) type structure, a = 6.270 ± 1 Å, U = 246.5 Å^3, Z = 4, D_x = 8.82. For data on other ternary stannides with MgAgAs and $MnCu_2Al$ type structures, see Table I.

Space group $Fm3m$ (O_h^5)

Atomic positions

			x	*y*	*z*	
4 Sn	in	4(*a*)	0	0	0	NiSnZr, Ni_2SnZr
4 Zr	in	4(*b*)	1/2	1/2	1/2	NiSnZr, Ni_2SnZr
4 Ni(1)	in	8(*c*)	1/4	1/4	1/4	NiSnZr
8 Ni(2)	in	8(*c*)	1/4	1/4	1/4	Ni_2SnZr

Interatomic distances (Å)

NiSnZr		Ni_2SnZr	
Zr - 6 Sn	3.056	Zr - 6 Sn	3.135
- 4 Ni	2.647	- 8 Ni	2.715
Ni - 4 Zr	2.647	Ni - 4 Zr	2.715
- 4 Sn	2.647	- 4 Sn	2.715
Sn - 6 Zr	3.056	Sn - 6 Zr	3.135
- 4 Ni	2.647	- 8 Ni	2.715

Material and Details of analysis

Alloys were arc-melted from 99.9+% elements in purified argon. Powder patterns (Guinier camera, Si standard, a = 5.430 Å, CuK_α) were used for lattice parameters and powder diffractometer measurements for intensities (CuK_α, Ni Filter). Intensities were calculated with dispersion corrections; R = 0.03 for NiSnZr and R = 0.085 for Ni_2SnZr. Alternative structures could be rejected.

1. *Structure Reports*, 8, 21.
2. *Strukturbericht*, 1, 148.
3. *Ibid.*, 2, 22
4. *Ibid.*, 1, 488.

URANIUM AND THORIUM NITRIDES WITH ANTIMONY, BISMUTH AND TELLURIUM

I. Crystal structure of the compounds U_2N_2X and Th_2N_2X with x = Sb, Te and Bi. R. BENZ and W.H. ZACHARIASEN, 1970. *Acta Cryst.*, B 26, 823-827.

See also 1.

N_2SbU_2 F.W. = 625.02

Tetragonal, body-centred cell, $(Na_{1/4}Bi_{3/4})_2O_2Cl$ type (2), a = 3.8937 ± 2, c = 12.3371 ± 7 Å, [c/a = 3.168, U = 187.04 Å^3], Z = 2, D_x = 11.116 ± 2.

Space group *I*4/*mmm* (D_{4h}^{17}) from intensity observations

Atomic positions

			x	*y*	*z*
2 Sb	in	2(*a*)	0	0	0
4 N	in	4(*d*)	0.500	0	0.250
4 U	in	4(*e*)	0	0	0.344±3

Interatomic distances (Å)

	M - 4*Y*	*M* - 4*X*
U_2N_2Sb	2.27±2	3.36±3
U_2N_2Te	2.31	3.42
U_2N_2Bi	2.29*	3.40*
Th_2N_2Sb	2.39*	3.56*
$Th_2(N_{1/2}O_{1/2})_2Sb$	2.35*	3.49*
Th_2N_2Te	2.38	3.54
Th_2N_2Bi	2.41*	3.58*
$Th_2(N,O)_2Bi$	2.40*	3.57*

(oxygen saturated) M = U or Th; Y = N or (N,O); X = Sb, Te, Bi

* Uncertainty 2 to 3 times larger than for U_2N_2Sb.

U_2N_2Sb is representative of a series of isostructural compounds, the lattice parameters of which are as follows,

	a (Å)	c (Å)	[c/a]	D_x	z for 4(e)
U_2N_2Te	3.9631±2	12.561±2	3.169	10.64	0.344±3
N_2N_2Bi	3.9292±5	12.548±2	3.194	12.23	0.344±10
Th_2N_2Sb	4.049±1	13.57±1	3.35	9.17	0.344±3
$Th_2(N_{3/4}O_{1/4})_2Sb$	4.045±3	13.18±4	3.26	9.47	
$Th_2(N_{1/2}O_{1/2})_2Sb$ (oxygen saturated)	4.041±1	12.84±1	3.18	9.76	0.344±5
Th_2N_2Te	4.0939±4	13.014±1	3.179	9.44	0.344±3
Th_2N_2Bi	4.075±1	13.620±1	3.342	10.30	0.344±10
$Th_2(N,O)_2Bi$ (oxygen saturated)	4.074±2	13.53±3	3.32	10.37	

Discussion

This structure is observed with U_2N_2X compounds when X = Sb, Bi or Te, in which case there is room for 8U around each X. However, for the smaller atoms P, As, S or Se the Ce_2O_2S (1) structure is found.

It is noteworthy that increasing oxygen content for the $Th_2(N,O)_2Sb$ and $Th_2(N,O)_2Bi$ reduces the c axis thus also decreasing the M-Y distance, whereas an increase would be expected since the oxygen atom is larger. However, the Th-Sb bond strength in $Th_2(N_{1/2}O_{1/2})_2Sb$ is 0.375 and for Th_2N_2Sb it has decreased to 0.25 thus explaining the apparent discrepancy.

1. R. BENZ and W.H. ZACHARIASEN, 1969. *Acta Cryst.*, B 25, 294.
2. L.G. SILLÉN, 1929. *Z. anorg. Chem.*, 242, 41; *Strukturbericht*, 7, 159.

VANADIUM CARBON

I. A neutron diffraction investigation of V_8C_7. A.W. HENFREY and B.E.F. FENDER, 1970. *Acta Cryst.*, B26, 1882-1883.

V_8C_7 F.W. = 491.61

Cubic, a = 8.3307 ± 9 Å for $VC_{0.865}$ and a = 8.3303 ± 7 Å for $VC_{0.863}$. See also 1, 2.

Space group $P4_132$ (O^7) or enantiomorph. In agreement with the X-ray results of 2.

Atomic positions and occupation numbers

The coordinates given in $P4_132$ contain internal inconsistancies, which are avoided by using $P4_332$ as chosen by 1. A full report on the positions and occupancies is given in *Structure Reports*, 38.

Discussion

The occupation numbers agree well with those expected assuming a perfect spiral array, the C atoms being closely located on the special positions.

Details of analysis

Samples were prepared at 1650°C, and analysed for V and C content. Oxygen was determined by vacuum fusion as 0.02 to 0.06 wt%. Lattice parameters from X-ray powder photographs. Intensities (2θ ∿55°) from neutron powder photographs (λ = 1.053 Å). Least-squares refinement in terms of a scale factor, a temperature factor for C, and three C occupation numbers (the V scattering length for neutrons is very small).

1. D. KORDES, 1968. *Phys. Stat. Sol.*, 6, K 103.

2. C.H. DE NOVION, R. LORENZELLI and P. COSTA, 1966. *C.R. Acad. Sci. Paris*, 263B, 775.

ZIRCONIUM NITRIDE-HYDRIDE

I. [Neutron diffraction investigation of zirconium nitride-hydride.] V.N. BYKOV, V.S. GOLOVKIN, V.A. LEVDIK, V.P. KALININ and N.F. MIRON, 1970. *Kristallografija, SSSR*, 15, 376 [*Soviet Physics - Crystallography*, 15, 317].

$H_{0\cdot80}N_{0\cdot36}Zr$ see also 1.

Trigonal, *a* = 3.274, *c* = 5.529 Å; ZrCH type (anti-CdI_2) (2).

Space group $P\bar{3}m1$ (D^3_{3d})

Atomic positions

	x	*y*	*z*
0.72 N in 1(*b*)	0	0	1/2
0.32 H(1) in 2(*d*)	1/3	2/3	0.418
1.28 H(2) in 2(*d*)	1/3	2/3	0.082
2 Zr in 2(*d*)	1/3	2/3	0.235

[*Interatomic distances* (Å)]

N	- 6 H	1.94	H(2)	- 1 H(1)	1.86
	- 6 Zr	2.39		- 3 H(2)	2.10
H(1)	- 3 N	1.94		- 3 Zr	2.58
	- 3 H(1)	2.10		- 1 Zr	0.84*
	- 1 H(2)	1.86	Zr	- 3 N	2.39
	- 1 Zr	1.01*		- 3 H(1)	2.69
				- 1 H(1)	1.01*
				- 3 H(2)	2.58
				- 1 H(2)	0.84*

*[These distances are short].

Material and Details of analysis

Neutron diffraction (λ = 0.90 Å, peak half-width ∿ 30') intensities; trial and error gave *R* = 0.17.

1. R.A. ANDRIEVSKIJ, E.B. BOYKO and V.P. KALININ, 1967. *Kristallografija, SSSR*, 12, 1068 [*Soviet Physics - Crystallography*, 12, 930].
2. K. YVON, H. NOWOTNY and R. KIEFFER, 1967. *Mh. Chem.*, 98, 2164.

TABLE I CRYSTAL DATA

Phase	Space group	Structure type	a (Å)	b (Å)	c (Å)	c/a, or α,β,γ	D_m	Z	D_x	Notes	Ref.
Ag_2BaGe_2	$I4/mmm$	Cr_2Si_2Th	4.58		10.69	2.36		2			20
Ag_2CaGe_2	$I4/mmm$	Cr_2Si_2Th	4.33		10.93	2.52		2			20
$Ag_{51}Ce_{14}$	$P6/m$	$Ag_{51}Gd_{14}$	12.883		9.455					p.p.	1
$Ag_{51}Dy_{14}$	$P6/m$	$Ag_{51}Gd_{14}$	12.635		9.271					p.p.	1
$Ag_{51}Er_{14}$	$P6/m$	$Ag_{51}Gd_{14}$	12.596		9.236					p.p.	1
Ag_5Eu	$P6/mmm$	$CaCu_5$ ($D2_d$)	5.6201		4.6439					p.p.	1
$Ag_{51}Gd_{14}$	$P6/m$	$Ag_{51}Gd_{14}$	12.681		9.289					p.p.	1
Ag_2Ge_2Sr	$I4/mmm$	Cr_2Si_2Th	4.45		10.85	2.44		2			20
$Ag_{51}Ho_{14}$	$P6/m$	$Ag_{51}Gd_{14}$	12.609		9.257					p.p.	1
$Ag_{51}La_{14}$	$P6/m$	$Ag_{51}Gd_{14}$	12.955		9.525					p.p.	1
Ag_5La (L.T.)	$P6_3/mmc$	$MgZn_2$ ($C14$)	5.5690		9.0775					p.p.	1
Ag_4Lu	$I4/m$	$MoNi_4$ ($D1_a$)	6.6696		4.1581					p.p.	1
$Ag_{51}Nd_{14}$	$P6/m$	$Ag_{51}Gd_{14}$	12.814		9.432					p.p.	1
$Ag_{16}P_{12}Pd_{12}$	$P4/mmm$	$AsPd_5Tl$	3.917		6.895	1.760		1		p.p.	14
$Ag_{14}P_{14}Pt_{72}$	$P4/mmm$	$AsPd_5Tl$	3.926		7.014	1.787		1		p.p.	14
$Ag_{51}Pr_{14}$	$P6/m$	$Ag_{51}Gd_{14}$	12.846		9.446					p.p.	1
$Ag_{14}Pt_{72}Si_{14}$	$P4/mmm$	$AsPd_5Tl$	3.961		6.993	1.765		1		p.p.	14
Ag_4Sc	$I4/m$	$MoNi_4$ ($D1_a$)	6.5740		4.0686					p.p.	1
$Ag_{51}Sm_{14}$	$P6/m$	$Ag_{51}Gd_{14}$	12.750		9.381					p.p.	1
$Ag_{51}Tb_{14}$	$P6/m$	$Ag_{51}Gd_{14}$	12.650		9.280					p.p.	1
Ag_3Tm	$Pmmn$	Cu_3Ti (DO_a)	6.075	4.948	5.163					p.p.	1
Ag_3Tm	$Pm3m$	$AuCu_3$ ($L1_2$)	4.2117							p.p.	1
$Ag_{51}Y_{14}$	$P6/m$	$Ag_{51}Gd_{14}$	12.637		9.300					p.p.	1
Ag_7Yb_2	$P6\ 22$	Ag_7Ca_2	5.463		14.084					p.p.	1
$Al_{14}As_{14}Pt_{72}$	$P4/mmm$	$AsPd_5Tl$	3.895		6.934	1.780		1		p.p.	14
Al_3GeNb	$P23$ or $Fm3m$	ordered Cr_3Si	5.174_8x2					A=64		d.p.p.	9
$Al_{14}P_{14}Pt_{72}$	$P4/mmm$	$AsPd_5Tl$	3.896		6.818	1.750		1		p.p.	14
Am_2C_3	$I\bar{4}3d$	$D5_c$	8.2757							p.p.	6
$As_{13}Cd_{15}Pd_{72}$	$P4/mmm$	$AsPd_5Tl$	3.989		6.971	1.748		1		p.p.	14
$As_{14}Cd_{14}Pt_{72}$	$P4/mmm$	$AsPd_5Tl$	3.955		6.976	1.739		1		p.p.	14

As_3Ce_4	$I\bar{4}3d$	anti-Th_3P_4	9.053					4	7.028	p.p., s.c.d.	28
As_2Ce	$P2_1/c$	As_2Nd	4.165	6.871	10.561	β=106.72°	6.54	4	6.694	p.p., s.c.d.	28
$As_{15}Ga_{10}Pd_{75}$	$P4/mmm$	$AsPd_5Tl$	3.949		6.876	1.741		1		p.p.	14
$As_{10}Hg_{20}Pd_{70}$	$P4/mmm$	$AsPd_5Tl$	3.990		6.950	1.741		1		p.p.	14
$As_{12}In_{16}Pd_{72}$	$P4/mmm$	$AsPd_5Tl$	3.966		6.931	1.748		1		p.p.	14
$As_{14}In_{14}Pt_{72}$	$P4/mmm$	$AsPd_5Tl$	3.997		7.055	1.765		1		p.p.	14
As_2La (above 750°)	$B2/b$	As_2La	12.891	9.140	14.450	γ=135.16°	6.19	16	6.389	p.p., s.c.d.	28
As_2La (<750°)	$P2_1/c$	As_2Nd	4.212	6.935	10.647	β=106.60°	6.33	4	6.475	p.p., s.c.d.	28
$As_{14}Mg_{14}Pt_{72}$	$P4/mmm$	$AsPd_5Tl$	3.912		6.890	1.761		1		p.p.	14
As_2Nd	$P2_1/c$	As_2Nd	4.109	6.819	10.449	β=106.68°	6.94	4	7.006	p.p., s.c.d.	28
As_3Pr_4	$I\bar{4}3d$	anti-Th_3P_4	8.994					4	7.196	p.p., s.c.d.	28
As_2Pr	$P2_1/c$	As_2Nd	4.139	6.844	10.509	β=106.69°	6.76	4	6.812	p.p., s.c.d.	28
$As_{14}Pt_{72}Tl_{14}$	$P4/mmm$	$AsPd_5Tl$	4.009		7.161	1.786		1		p.p.	14
$As_{14}Pt_{72}Zn_{14}$	$P4/mmm$	$AsPd_5Tl$	3.880		7.000	1.804		1		p.p.	14
As_3Yb_4	$I\bar{4}3d$	anti-Th_3P_4	8.791					4	8.977	p.p., s.c.d.	28
As_3Yb_4	$R3$	deformed anti-Th_3P_4	8.784			α=90.80°		4	8.988	p.p., s.c.d.	28
B_4CrDy	$Pbam$	B_4CrY	5.792	11.48	3.451			4		p.p.	29
B_4CrEr	$Pbam$	B_4CrY	5.774	11.44	3.433			4		p.p.	29
B_4CrGd	$Pbam$	B_4CrY	5.872	11.55	3.485			4		p.p.	29
B_4CrHo	$Pbam$	B_4CrY	5.774	11.48	3.444			4		p.p.	29
B_4CrLu	$Pbam$	B_4CrY	5.726	11.38	3.404			4		p.p.	29
B_4CrTb	$Pbam$	B_4CrY	5.832	11.51	3.463			4		p.p.	29
$BaGe_2Mg_2$	$I4/mmm$	Cr_2Si_2Th	4.67		11.33	2.43		2			20
$Ba_3Li_4Sb_4$	$Immm$	$Li_4Sb_4Sr_3$	17.18	4.91	7.70		4.95	2	4.80	s.c.d.	15
Be_xCo	$P\bar{6}m2$	$CaZn_2$	4.114		10.66	2.591			2.929	s.c.d.	26
Be_xFe	$P\bar{6}m2$	$CaZn_2$	4.137		10.720	2.591			2.805	s.c.d.	26
Be_xIr	$P\bar{6}m2$	$CaZn_2$	4.197		10.842	2.583			5.887	s.c.d.	26
$CaCu_2Ge_2$	$I4/mmm$	Cr_2Si_2Th	4.139		10.232	2.48		2			20
$CaCu_2Si_2$	$I4/mmm$	Cr_2Si_2Th	4.04		10.00	2.48		2			20
$CaGe_2Zn_2$	$I4/mmm$	Cr_2Si_2Th	4.21		10.85	2.58		2			20
$Cd_{10}P_{15}Pd_{75}$	$P4/mmm$	$AsPd_5Tl$	3.984		6.982	1.753		1		p.p.	14
$Cd_{14}P_{14}Pt_{72}$	$P4/mmm$	$AsPd_5Tl$	3.922		6.783	1.730		1		p.p.	14
$Cd_{14}Pd_{72}Se_{14}$	$P4/mmm$	$AsPd_5Tl$	3.999		6.997	1.750		1		p.p.	14
Ce_2Co_7	$P6_3/mmc$	Ce_2Ni_7	4.94		24.46			4		d.p.p.	21
Ce_2Co_7	$R\bar{3}m$	Gd_2Co_7	4.940		36.52					d.p.p.	21
$CeCoSi$	$P4/nmm$	$PbFCl$	4.036		6.935	1.718		2		s.c.d.	18
$CeGe_2Ni$	$Cmcm$	$CeNiSi_2$	4.244	16.747	4.199			4		s.c.d.	23

Phase	Space group	Structure type	a (Å)	b (Å)	c (Å)	c/a, or α,β,γ	D_m	Z	D_x	Notes	Ref.
Ce_2Ni_7	$P6_3/mmc$	Ce_2Ni_7	4.927		24.45			4		d.p.p.	22
Co_7Dy_2	$R\bar{3}m$	Gd_2Co_7	4.992		36.13					d.p.p.	21
Co_7Er_2	$R\bar{3}m$	Gd_2Co_7	4.960		36.07					d.p.p.	21
Co_7Gd_2	$P6_3/mmc$	Ce_2Ni_7	5.022		24.19			4		d.p.p.	21
Co_7Gd_2	$R\bar{3}m$	Gd_2Co_7	5.022		36.24					d.p.p.	21
CoGeTi	$P\bar{6}2m$	Fe_2P	6.222		3.7276	0.599		3		p.p., s.c.d.	25
Co_2HfSn	$Fm3m$	$AlCu_2Mn$	6.201					4	11.56	d.p.p.	17
Co_7Ho_2	$R\bar{3}m$	Gd_2Co_7	4.977		36.10					d.p.p.	21
Co_7La_2	$P6_3/mmc$	Ce_2Ni_7	5.101		24.51			4		d.p.p.	21
Co_7La_2	$R\bar{3}m$	Gd_2Co_7	5.101		36.69					d.p.p.	21
CoLaSi	$P4/nmm$	PbFCl	4.069		7.180	1.765		2		s.c.d., p.p.	18
$Co_{1/3}NbSe_2$	$P6_3/mmc$	$Co_{1/4}NbSe_2$	2x3.456		12.384			8		d.p.p.	10
$Co_{1/4}NbSe_2$	$P6_3/mmc$	$Co_{1/4}NbSe_2$	2x3.452		13.379			8		d.p.p.	10
Co_2NbSn	$Fm3m$	$AlCu_2Mn$	6.146					4	9.43	d.p.p.	17
CoNbSn	$F\bar{4}3m$	MgAgAs	5.947					4	8.54	d.p.p.	17
Co_7Nd_2	$P6_3/mmc$	Ce_2Ni_7	5.059		24.39			4		d.p.p.	21
Co_7Nd_2	$R\bar{3}m$	Gd_2Co_7	5.059		36.43					d.p.p.	21
CoNdSi	$P4/nmm$	PbFCl	4.032		6.872	1.704		2		s.c.d., p.p.	18
Co_7Pr_2	$P6_3/mmc$	Ce_2Ni_7	5.060		24.43			4		d.p.p.	21
Co_7Pr_2	$R\bar{3}m$	Gd_2Co_7	5.060		36.52					d.p.p.	21
CoPrSi	$P4/nmm$	PbFCl	4.035		6.934	1.718		2		s.c.d., p.p.	18
CoSiSm	$P4/nmm$	PbFCl	4.010		6.776	1.690		2		s.c.d., p.p.	18
Co_7Sm_2	$P6_3/mmc$	Ce_2Ni_7	5.041		24.33			4		d.p.p.	21
Co_7Sm_2	$R\bar{3}m$	Gd_2Co_7	5.022		36.24					d.p.p.	21
Co_2SnTi	$Fm3m$	$AlCu_2Mn$	6.060					4	8.49	d.p.p.	17
Co_2SnZr	$Fm3m$	$AlCu_2Mn$	6.233					4	8.99	d.p.p.	17
Co_7Tb_2	$R\bar{3}m$	Gd_2Co_7	5.008		36.18					d.p.p.	21
$Cr_{1/3}NbSe_2$	$P6_3/mmc$	$Co_{1/4}NbSe_2$	2x3.452		12.567			8		d.p.p.	10
$Cr_{1/4}NbSe_2$	$P6_3/mmc$	$Co_{1/4}NbSe_2$	2x3.452		12.570			8		d.p.p.	10
$Cu_8Dy_6Ge_8$	$Immm$	$Cu_8Gd_6Ge_8$	13.845	6.618	4.173			1		d.p.p., s.c.d.	27
$Cu_8Er_6Ge_8$	$Immm$	$Cu_8Gd_6Ge_8$	13.79	6.617	4.153			1		d.p.p., s.c.d.	27
Cu_6Gd	$Pnma$	$CeCu_6$	8.040	5.009	9.995			4		d.p.p.	16
$Cu_8Ge_8Lu_6$	$Immm$	$Cu_8Gd_6Ge_8$	13.661	6.581	4.113			1		d.p.p., s.c.d.	27
Cu_3GeMg_2	*hex.*	Cu_3Mg_2Si	5.058		8.051	1.592		2	5.45	s.c.d.	5

CuGeMg	*P4/nmm*	AsCuMg	4.007		6.297	1.57	5.026	2	s.c.d.	5
Cu_6Nd	*Pnma*	$CeCu_6$	8.092	5.062	10.105			4	d.p.p.	16
Cu_6Pr	*Pnma*	$CeCu_6$	8.101	5.081	10.140			4	d.p.p.	16
$Cu_3Pt_{14}Zn_{83}$	*cubic*	Ni_4Zn_{22}	9.089						p.p.	19
$Cu_3Pt_2Zn_5$	*P4/mmm*	$Cu_3Pt_2Zn_5$	3.77		38.50			4	p.p.	19
$CuPt_{37}Zn_{62}$	*ortho.*	$_8Pt_7Zn_{12}$	41.095	6.969	2.775				p.p.	19
Cu_2Si_2Sr	*I4/mmm*	Cr_2Si_2Th	4.20		10.00	2.48[2.38]		2		20
Cu_6Th	*Pnma*	$CeCu_6$	8.115	5.078	10.122			4	d.p.p.	16
Dy_6Fe_{23}	*Fm3m*	Th_6Mn_{23}	12.055					4	d.p.p.	8
$DyFe_3$	*R$\bar{3}$m*	$PuNi_3$	5.123		24.570	4.795		3	d.p.p.	8
DyFeSi	*P4/nmm*	PbFCl	3.961		6.745	1.703		2	s.c.d., p.p.	18
Dy_2Ni_7	*P6$_3$/mmc*	Ce_2Ni_7	4.928		24.1			4	d.p.p.	22
Dy_2Ni_7	*R$\bar{3}$m*	Gd_2Co_7	4.928		36.18				d.p.p.	22
$DyNi_3$	*R$\bar{3}$m*	$PuNi_3$	4.966		24.35				d.p.p.	22
$DyNiSi_2$	*Cmcm*	$CeNiSi_2$	3.971	16.025	3.949			4	s.c.d., p.p.	23
$Dy_{13}Zn_{58}$	*P6$_3$mc*	$Gd_{13}Zn_{58}$	14.24		13.99	0.98		2	d.p.p.	12
Dy_3Zn_{11}	*Immm*	$Al_{11}La_3$	4.395	12.922	8.830			2	d.p.p.	12
Er_2Ni_7	*R$\bar{3}$m*	Gd_2Co_7	4.909		36.07				d.p.p.	22
$ErNi_3$	*R$\bar{3}$m*	$PuNi_3$	4.941		24.25				d.p.p.	22
$ErNiSi_2$	*Cmcm*	$CeNiSi_2$	3.917	15.935	3.917			4	s.c.d., p.p.	23
$Er_{13}Zn_{58}$	*P6$_3$mc*	$Gd_{13}Zn_{58}$	14.20		13.98	0.98		2	s.c.d., p.p.	12
$ErZn_3$	*Pnma*	YZn_3	6.678	4.350	10.024			4	d.p.p.	12
$EuNiSi_2$	*Cmcm*	$CeNiSi_2$	4.137	16.562	4.002			4	s.c.d., p.p.	23
FeGdSi	*P4/nmm*	PbFCl	4.001		6.798	1.699		2	s.c.d., p.p.	18
FeGeTi	*Ima2*	FeSiTi	7.155	11.025	6.405				p.p., s.c.d.	25
FeHoSi	*P4/nmm*	PbFCl	3.937		6.774	1.720		2	s.c.d., p.p.	18
FeLaSi	*P4/nmm*	PbFCl	4.062		7.179	1.767		2	s.c.d., p.p.	18
$Fe_{1/2}NbSe_2$	*P6$_3$22*	$Fe_{1/2}NbSe_2$	$\sqrt{3}$x3.468		12.773				d.p.p.	10
$Fe_{1/3}NbSe_2$	*P6$_3$/mmc*	$Co_{1/4}NbSe_2$	2x3.459		12.671			8	d.p.p.	10
$Fe_{1/4}NbSe_2$	*P6$_3$/mmc*	$Co_{1/4}NbSe_2$	2x3.454		12.623			8	d.p.p.	10
FeNdSi	*P4/nmm*	PbFCl	4.057		6.893	1.699		2	s.c.d., p.p.	18
$Fe_{0.50}Ni_{0.50}Sb_3$	*Im3*	As_3Co	9.0902				7.411	8	p.p.	4
FePrSi	*P4/nmm*	PbFCl	4.072		6.927	1.701		2	s.c.d., p.p.	18
FeSiSm	*P4/nmm*	PbFCl	4.031		6.820	1.692		2	s.c.d., p.p.	18
FeSiTb	*P4/nmm*	PbFCl	3.984		6.751	1.695		2	s.c.d., p.p.	18
FeSiYb	*P4/nmm*	PbFCl	3.972		6.784	1.708		2	s.c.d., p.p.	18
$Ga_{14}P_{14}Pd_{72}$	*P4/mmm*	$AsPd_5Tl$	3.871		6.735			1	p.p.	14
$Ga_{14}P_{14}Pt_{72}$	*P4/mmm*	$AsPd_5Tl$	3.897		6.651	1.707		1	p.p.	14
Gd_2Ni_7	*P6$_3$/mmc*	Ce_2Ni_7	4.953		24.21			4	d.p.p.	22

Phase	Space group	Structure type	a (Å)	b (Å)	c (Å)	c/a, or α,β,γ	D_m	Z	D_x	Notes	Ref.
Gd_2Ni_7	$R\bar{3}m$	Gd_2Co_7	4.953		36.41					d.p.p.	22
$GdNi_3$	$R\bar{3}m$	$PuNi_3$	4.993		24.49					d.p.p.	22
$GdNiSi_2$	$Cmcm$	$CeNiSi_2$	4.025	16.095	3.976			4		s.c.d., p.p.	23
Gd_3Zn_{22}	$I4_1/amd$	Ce_3Zn_{22}	8.831		21.118	2.39				d.p.p.	12
Gd_3Zn_{11}	$Immm$	$Al_{11}La_3$	4.423	13.063	8.842			2		d.p.p.	12
$GdZn_3$	$Pnma$	YZn_3	6.718	4.439	10.158			4		d.p.p.	12
$HfNi_2Sn$	$Fm3m$	$AlCu_2Mn$	6.237					4	11.35	d.p.p.	17
$HfNiSn$	$F\bar{4}3m$	$AgAsMg$	6.083					4	10.50	d.p.p.	17
$Hg_{14}P_{14}Pd_{72}$	$P4/mmm$	$AsPd_5T\ell$	3.983		6.957	1.747		1		p.p.	14
$Hg_{14}P_{14}Pt_{72}$	$P4/mmm$	$AsPd_5T\ell$	3.925		7.082	1.804		1		p.p.	14
$Hg_{14}Pd_{72}Se_{14}$	$P4/mmm$	$AsPd_5T\ell$	3.999		6.974	1.744		1		p.p.	14
Ho_2Ni_7	$R\bar{3}m$	Gd_2Co_7	4.921		36.09					d.p.p.	22
$HoNi_3$	$R\bar{3}m$	$PuNi_3$	4.951		24.31					d.p.p.	22
$HoNiSi_2$	$Cmcm$	$CeNiSi_2$	3.943	15.955	3.933			4		s.c.d., p.p.	23
$Ho_{13}Zn_{58}$	$P6_3mc$	Gd_3Zn_{58}	14.21		13.97	0.98		2		d.p.p.	12
$In_{14}P_{14}Pd_{72}$	$P4/mmm$	$AsPd_5T\ell$	3.928		6.917	1.761		1		p.p.	14
$In_{14}P_{14}Pt_{72}$	$P4/mmm$	$AsPd_5T\ell$	3.930		6.950	1.768		1		p.p.	14
$In_{14}Pt_{72}Si_{14}$	$P4/mmm$	$AsPd_5T\ell$	3.941		6.910	1.753		1		p.p.	14
Ir_4Si_5	$P2_1/m$	Rh_4Si_5	12.359	3.6181	5.8805					p.p.	11
Ir_3Si_4	$Pnma$	Rh_3Si_4	18.8741	3.6979	5.7717					s.c.d.	11
$IrSi_3$	$hex.?$	$IrSi_3$	4.3538		6.6277	1.52				p.p.	11
$IrSi_{\sim1.5}$	$mono.$	$IrSi_{\sim1.5}$	5.542	14.166	12.426					s.c.d.	11
$IrSi$	$Pnma$	MnP	5.5579	3.2213	6.2673			4		p.p.	11
$K_{2/3}NbS_2$	$P6_3/mmc$	$LiNbS_2$	3.345		16.22	4.849				d.p.p.	31
	or $P\bar{6}m2$	$KNbSe_2$									
$K_{2/3}NbSe_2$	$P\bar{6}m2$	$KNbSe_2$	3.480		17.04	4.879				d.p.p.	31
$K_{2/3}TaS_2$	$P\bar{6}m2$	$KTaS_2$	3.332		16.20	4.862				d.p.p.	31
	or $P6_3mc$										
$K_{2/3}TaSe_2$	$P\bar{6}m2$	$KNbSe_2$	3.463		17.05	4.923				d.p.p.	31
$LaNi_3$	$R\bar{3}m$	$PuNi_3$	5.083		25.09					d.p.p.	22
$LaNiSi_2$	$Cmcm$	$CeNiSi_2$	4.193	16.581	4.073			4		s.c.d., p.p.	23
$Li_{2/3}NbS_2$	$P6_3/mmc$	$LiNbS_2$	3.331		12.90	3.873				d.p.p.	31
$Li_{2/3}TaS_2$	$P6_3/mmc$	$LiNbS_2$	3.333		12.89	3.867				d.p.p.	31

$LuNiSi_2$	*Cmcm*	$CeNiSi_2$	3.851	15.810	3.851			4	s.c.d., p.p.	23
$Lu_{13}Zn_{58}$	$P6_3mc$	$Gd_{13}Zn_{58}$	14.09		13.89	0.99		2	d.p.p.	12
$LuZn_3$	*Pnma*	YZn_3	6.650	4.300	9.974			4	d.p.p.	12
$Mg_{14}P_{14}Pd_{72}$	$P4/mmm$	$AsPd_5T\ell$	3.844		6.715	1.739		1	p.p.	14
$Mg_{14}P_{14}Pt_{72}$	$P4/mmm$	$AsPd_5T\ell$	3.858		6.860	1.778		1	p.p.	14
MnB_x	$R\bar{3}m$	CrB_{41}	10.99		24.00		2.18		d.p.p.	7
$Mn_{1/2}NbSe_2$	$P6_322$	$Fe_{1/2}NbSe_2$	$\sqrt{3}$x3.476		13.165				d.p.p.	10
$Mn_{1/3}NbSe_2$	$P6_3/mmc$	$Co_{1/4}NbSe_2$	2x3.469		13.033			8	d.p.p.	10
$Mn_{1/4}NbSe_2$	$P6_3/mmc$	$Co_{1/4}NbSe_2$	2x3.462		12.899			8	d.p.p.	10
$Mo_{1.03}Nb_{1.03}S_3$	$P2_1/m$	$Mo_{2.06}S_3$	6.125	3.292	8.692	β=102.67°			p.p., s.c.d.	30
$MoNbSe_3$	$P2_1/m$	$Mo_{2.06}S_3$	6.439	3.414	9.111	β=103.55°			p.p., s.c.d.	30
$Mo_{2.06}S_3$	*triclinic*		6.06	2x3.19	8.60	α= 89.60°			p.p., s.c.d,	30
			(at -150°C)			β=102.5°			d.p.p.	
						γ= 90.3°				
Mo_3Se_4	*mono.*		4.632	4.762	6.649	β= 92.56°			p.p., s.c.d.,	30
									d.p.p.	
Mo_3Te_4	*mono.*		4.875	5.092	7.051	β= 93.75°			p.p., s.c.d.,	30
									d.p.p.	
NTa	$Fm3m$	NaCl	4.385				15.63			2
(30-100kb)										
∿1800°C										
NTa	$P\bar{6}m2$	WC	2.993		2.880	0.962				2
(30-100kb)										
1800-500°C										
$Na_{2/3}NbS_2$	$P6_3/mmc$	$LiNbS_2$	3.366		14.52	4.314			d.p.p.	31
$Na_{2/3}NbS_2$	$P\bar{6}m2$	$KTaS_2$	3.363		14.525	4.319			d.p.p.	31
$Na_{2/3}NbSe_2$	$P\bar{6}m2$	$KTaS_2$	3.476		15.366	4.421			d.p.p.	31
	or $P6_3mc$									
$Na_{2/3}TaS_2$	$P6_3/mmc$	$LiNbS_2$	3.345		14.54	4.347			d.p.p.	31
	or $P\bar{6}m2$	or $KNbSe_2$								
$Na_{2/3}TaS_2$	$P\bar{6}m2$	$KTaS_2$	3.337		14.59	4.372			d.p.p.	31
$Na_{2/3}TaSe_2$	$P\bar{6}m2$	$KTaS_2$	3.458		15.403	4.454			d.p.p.	31
	or $P6_3mc$									
$NbNi_{1/3}Se_2$	$P6_3/mmc$	$Co_{1/4}NbSe_2$	2x3.456		12.413			8	d.p.p.	10
$NbNi_{1/4}Se_2$	$P6_3/mmc$	$Co_{1/4}NbSe_2$	2x3.453		12.419			8	d.p.p.	10
$NbNi_2Sn$	$Fm3m$	$AlCu_2Mn$	6.162					4	9.34 d.p.p.	17
NbRhSn	$F\bar{4}3m$	AgAsMg	6.132					4	9.06 d.p.p.	17
$NbSe_2V_{1/3}$	$P6_3/mmc$	$Co_{1/4}NbSe_2$	2x3.454		12.673			8	d.p.p.	10
$NbSe_2V_{1/4}$	$P6_3/mmc$	$Co_{1/4}NbSe_2$	2x3.451		12.660			8	d.p.p.	10
Nd_2Ni_7	$P6_3/mmc$	Ce_2Ni_7	4.983		24.40			4	d.p.p.	22
Nd_2Ni_7	$R\bar{3}m$	Gd_2Co_7	4.983		36.60				d.p.p.	22

Phase	Space group	Structure type	a (Å)	b (Å)	c (Å)	c/a, or α,β,γ	D_m	Z	D_x	Notes	Ref.
$NdNi_3$	$R\bar{3}m$	$PuNi_3$	5.030		24.72					d.p.p.	22
$NdNiSi_2$	$Cmcm$	$CeNiSi_2$	4.105	16.450	4.012			4		p.p., s.c.d.	23
$Nd_{13}Zn_{58}$	$P6_3mc$	$Gd_{13}Zn_{58}$	14.53		14.09	0.97		2		d.p.p.	12
Nd_3Zn_{22}	$I4_1/amd$	Ce_3Zn_{22}	8.886		21.251	2.39				d.p.p.	12
Ni_7Pr_2	$P6_3/mmc$	Ce_2Ni_7	5.015		24.44			4		d.p.p.	22
Ni_7Pr_2	$R\bar{3}m$	Gd_2Co_7	5.015		36.64					d.p.p.	22
Ni_3Pr	$R\bar{3}m$	$PuNi_3$	5.035		24.82					d.p.p.	22
$NiPrSi_2$	$Cmcm$	$CeNiSi_2$	4.126	16.470	4.032			4		s.c.d., p.p.	23
$NiSi_2Sm$	$Cmcm$	$CeNiSi_2$	4.073	16.423	4.002			4		s.c.d., p.p.	23
$NiSi_2Tb$	$Cmcm$	$CeNiSi_2$	3.982	16.030	3.952			4		s.c.d., p.p.	23
$NiSi_2Tm$	$Cmcm$	$CeNiSi_2$	3.898	15.905	3.898			4		s.c.d., p.p.	23
$NiSi_2Yb$	$Cmcm$	$CeNiSi_2$	3.862	16.070	3.808			4		s.c.d., p.p.	23
Ni_7Sm_2	$P6_3/mmc$	$CeNi_7$	4.969		24.35			4		d.p.p.	22
Ni_7Sm_2	$R\bar{3}m$	Gd_2Co_7	4.969		36.53					d.p.p.	22
Ni_3Sm	$R\bar{3}m$	$PuNi_3$	5.005		24.60					d.p.p.	22
Ni_2SnTi	$Fm3m$	$AlCu_2Mn$	6.089					4	8.35	d.p.p.	17
$NiSnTi$	$F\bar{4}3m$	$AgAsMg$	5.941					4	7.14	d.p.p.	17
Ni_7Tb_2	$P6_3/mmc$	Ce_2Ni_7	4.948		24.12			4		d.p.p.	22
Ni_7Tb_2	$R\bar{3}m$	Gd_2Co_7	4.948		36.23					d.p.p.	22
Ni_3Tb	$R\bar{3}m$	$PuNi_3$	4.975		24.42					d.p.p.	22
Ni_7Y_2	$R\bar{3}m$	Gd_2Co_7	4.949		36.23					d.p.p.	22
Ni_3Y	$R\bar{3}m$	$PuNi_3$	4.973		24.42					d.p.p.	22
$P_{14}Pd_{72}Sn_{14}$	$P4/mmm$	$AsPd_5T\ell$	3.921		6.897	1.759		1		p.p.	14
$P_{14}Pd_{72}T\ell_{14}$	$P4/mmm$	$AsPd_5T\ell$	3.935		6.979	1.774		1		p.p.	14
$P_{14}Pd_{72}Zn_{14}$	$P4/mmm$	$AsPd_5T\ell$	3.861		6.727	1.742		1		p.p.	14
$P_7Pt_{72}Sn_{21}$	$P4/mmm$	$AsPd_5T\ell$	3.940		6.950	1.764		1		p.p.	14
$P_{14}Pt_{72}T\ell_{14}$	$P4/mmm$	$AsPd_5T\ell$	3.952		7.079	1.791		1		p.p.	14
$P_{14}Pt_{72}Zn_{14}$	$P4/mmm$	$AsPd_5T\ell$	3.877		6.871	1.772		1		p.p.	14
$PSn_{0.7}$	*hex.*		4.433		28.394	c/a=6.405				p.p., s.c.d.	13
	rhomb. cell		9.805			α= 26.132°					
$PSn_{0.3}$	*hex.*		7.3642		10.528	c/a=1.430				p.p., s.c.d.	13
	rhomb. cell		5.5130			α= 83.81°					
$Pb_{14}Pt_{72}Si_{14}$	$P4/mmm$	$AsPd_5T\ell$	3.984		7.121	1.787		1		p.p.	14
$Pd_{75}Se_{13}T\ell_{12}$	$P4/mmm$	$AsPd_5T\ell$	4.029		7.003	1.738		1		p.p.	14

$Pd_{72}Se_{14}Zn_{14}$	*P4/mmm*	$AsPd_5T\ell$	3.952		6.914	1.750	1		p.p.	<u>14</u>
PdSnZr	*F$\bar{4}$3m*	AgAsMg	6.321				4	8.32	d.p.p.	<u>17</u>
$Pr_{13}Zn_{58}$	*P6$_3$mc*	$Gd_{13}Zn_{58}$	14.57		14.15	0.97	2		d.p.p.	<u>12</u>
$Pt_{72}Sb_{14}Si_{14}$	*P4/mmm*	$AsPd_5T\ell$	3.951		7.027	1.779	1		p.p.	<u>14</u>
$Pt_{72}Si_{14}Sn_{14}$	*P4/mmm*	$AsPd_5T\ell$	3.950		6.959	1.762	1		p.p.	<u>14</u>
$Pt_{72}Si_{14}T\ell_{14}$	*P4/mmm*	$AsPd_5T\ell$	3.979		6.974	1.753	1		p.p.	<u>14</u>
Ru_5Si_3	*Pbam*	Ge_3Rh_5	5.246	9.815	4.023		2		d.p.p.	<u>24</u>
Se_2Zr	*P$\bar{3}$m1*	CdI_2	3.7715		6.1275	1.62			s.c.d.	<u>3</u>
$Sm_{13}Zn_{58}$	*P6$_3$mc*	$Gd_{13}Zn_{58}$	14.40		14.07	0.98	2		d.p.p.	<u>12</u>
Sm_3Zn_{22}	*I4$_1$/amd*	Ce_3Zn_{22}	8.856		21.175	2.39			d.p.p.	<u>12</u>
Sm_3Zn_{11}	*Immm*	$Al_{11}La_3$	4.452	13.179	8.838		2		d.p.p.	<u>12</u>
$SmZn_3$	*Pnma*	YZn_3	6.725	4.489	10.199		4		d.p.p.	<u>12</u>
$Tb_{13}Zn_{58}$	*P6$_3$mc*	$Gd_{13}Zn_{58}$	14.31		14.07	0.98	2		d.p.p.	<u>12</u>
Tb_3Zn_{11}	*Immm*	$Al_{11}La_3$	4.408	12.989	8.825		2		d.p.p.	<u>12</u>
$TbZn_3$	*Pnma*	YZn_3	6.690	4.411	10.104		4		d.p.p.	<u>12</u>
$Tm_{13}Zn_{58}$	*P6$_3$mc*	$Gd_{13}Zn_{58}$	14.14		13.94	0.99	2		d.p.p.	<u>12</u>
$TmZn_3$	*Pnma*	YZn_3	6.661	4.330	9.997		4		d.p.p.	<u>12</u>
$Yb_{13}Zn_{58}$	*P6$_3$mc*	$Gd_{13}Zn_{58}$	14.32		14.15	0.99	2		d.p.p.	<u>12</u>
Yb_3Zn_{17}	*Im3*	Ru_3Be_{17}	14.291				8		d.p.p.	<u>12</u>

<u>Abbreviations</u>

p.p. = powder photographs
d.p.p. = diffractometer powder patterns
s.c.d. = single crystal data

REFERENCES FOR TABLE I

1. O.D. McMASTERS, K.A. GSCHNEIDNER and F.R. VENTEICHER, 1970. *Acta Cryst.*, B26, 1224-1229.
2. L.G. BOIKO and S.V. POPOVA, 1970. *ZETF, Pis. Red.*, 12, 101-102 [*JETP Letters*, 12, 70-71].
3. A. GLEIZES and Y. JEANNIN, 1970. *J. Solid State Chem.*, 1, 180-184.
4. E. BJERKELUND and A. KJEKSHUS, 1970. *Acta Chem. Scand.*, 24, 3317-3325.
5. H.-U. SCHUSTER, W. BOCKELMANN and J. CAPTULLER, 1970. *Z. Naturf.*, 25B, 1304.
6. A.W. MITCHELL and D.J. LAM, 1970. *J. Nucl. Mater.*, 36, 110-112.
7. S. ANDERSSON and J.-O. CARLSSON, 1970. *Acta Chem. Scand.*, 24, 1791-1799.
8. M.P. DARIEL and G. EREZ, 1970. *J. Less-Common Metals*, 22, 360-362.
9. N.V. AGEEV, N.E. ALEKSEEVSKIJ and V.F. SAMRAJ, 1970. *Izv. Akad. Nauk, SSSR, Metally*, 3, 171-176.
10. J.M. VOORHOEVE and M. ROBBINS, 1970. *J. Solid State Chem.*, 1, 134-137.
11. I. ENGSTRÖM and F. ZACKRISSON, 1970. *Acta Chem. Scand.*, 24, 2109-2116.
12. G. BRUZZONE, M.L. FORNASINI and F. MERLO, 1970. *J. Less-Common Metals*, 22, 253-264.
13. O. OLOFSSON, 1970. *Acta Chem. Scand.*, 24, 1153-1162.
14. M. EL-BORAGY and K. SCHUBERT, 1970. *Z. Metallk.*, 61, 579-584.
15. O. LIEBRICH, H. SCHÄFER and A. WEISS, 1970. *Z. Naturf.*, 25B, 650-651.
16. K.H.J. BUSCHOW and A.S. VAN DER GOOT, 1970. *J. Less-Common Metals*, 20, 309-313.
17. W. JEITSCHKO, 1970. *Met. Trans.*, 1, 3159-3162.
18. O.I. BODAK, E.I. GLADYŠEVSKIJ and P.I. KRIPJAKEVIČ, 1970. *Ž. Strukt. Khim., SSSR*, 11, 305-310 [*J. Struct. Chem.*, 11, 283-288].
19. Y. KHAN, B.V.R. MURTY and K. SCHUBERT, 1970. *J. Less-Common Metals*, 21, 293-303.
20. B. EISENMANN, N. MAY, W. MÜLLER, H. SCHÄFER, A. WEISS, J. WINTER and G. ZIEGLEDER, 1970. *Z. Naturf.*, 25B, 1350-1352.
21. K.H.J. BUSCHOW, 1970. *Acta Cryst.*, B 26, 1389-1392.
22. K.H.J. BUSCHOW and A.S. VAN DER GOOT, 1970. *J. Less-Common Metals*, 22, 419-428.
23. O.I. BODAK and E.I. GLADYŠEVSKIJ, 1970. *Kristallografija, SSSR*, 14, 990-994 [*Soviet Physics - Crystallography*, 14, 859-862].
24. I. ENGSTRÖM, 1970. *Acta Chem. Scand.*, 24, 1466-1468.
25. W. JEITSCHKO, 1970. *Acta Cryst.*, B 26, 815-822.
26. Q. JOHNSON, G.S. SMITH, O.H. KRIKORIAN and D.E. SANDS, 1970. *Ibid.*, B 26, 109-113.
27. W. RIEGER, 1970. *Mh. Chem.*, 101, 449-462.
28. S. ONO, J.G. DESPAULT, L.D. CALVERT and J.B. TAYLOR, 1970. *J. Less-Common Metals*, 22, 51-59.
29. Ju.B. KUZ'MA, 1970. *Kristallografija, SSSR*, 15, 372 [*Soviet Physics - Crystallography*, 15, 312].
30. R. DE JONGE, T.J.A. POPMA, G.A. WEIGERS and F. JELLINEK, 1970. *J. Solid State Chem.*, 2, 188-192.
31. W.P.F.A.M. OMLOO and F. JELLINEK, 1970. *J. Less-Common Metals*, 20, 121-129.

STRUCTURE REPORTS

SECTION II

INORGANIC COMPOUNDS

EDITED BY
I. D. BROWN
C. CALVO

WITH THE ASSISTANCE OF THE FOLLOWING REPORTERS

H. D. GRUNDY
J. J. KIM

ARRANGEMENT

To find particular inorganic compounds the subject index or formula index should be used. The general arrangement is approximately: elements, hydrides, silicides, nitrides, sulphides, halides, oxides, borates, aluminates, carbonates, silicates, germanates, nitrates, phosphates, arsenates, sulphates, selenates, chromates, chlorates, iodates, and miscellaneous compounds and minerals. No strict classification is attempted, however; the indexes provide also a means of locating inorganic substances that have been dealt with in the Metals or the Organic sections.

BORON

I. Verfeinerung des β-rhomboedrischen Bors. D. GEIST, R. KLOSS and H. FOLLNER, 1970. *Acta. Cryst.*, B26, 1800-1802.

Rhombohedral, $a = 10.17 \pm 5$ Å, $\alpha = 65°12' \pm 20'$, $Z = 105$.

Space group $R\bar{3}m$ (D^5_{3d})

Atomic positions

	x	y	z	B(Å^2)
B(1)	0.0025± 8	0.0025± 8	0.1680±14	0.58±0.20
B(2)	0.1008± 9	0.1008± 9	0.8374±15	0.78±0.21
B(3)	0.9933± 8	0.9933± 8	0.6698±14	0.43±0.19
B(4)	0.1032± 9	0.1032± 9	0.4921±15	0.61±0.21
B(5)	0.1777±10	0.3473±10	0.0033±10	0.85±0.15
B(6)	0.1673±10	0.5521±10	0.8921±10	0.58±0.14
B(7)	0.1983± 9	0.1983± 9	0.6874±15	0.55±0.20
B(8)	0.3765±10	0.6826±10	0.2024±10	0.73±0.15
B(9)	0.3622±10	0.5811±10	0.0976±10	0.54±0.14
B(10)	0.1991± 9	0.1991± 9	0.5061±16	0.96±0.22
B(11)	0.3873±10	0.3873±10	0.5690±18	2.54±0.35
B(12)	0.4895± 9	0.4895± 9	0.2178±15	0.60±0.20
B(13)	0.3843± 8	0.3843± 8	0.2131±14	0.55±0.20
B(14)	0.3848±10	0.3848±10	0.3848±10	0.61±0.39
B(15)	0.5000	0.5000	0.5000	0.68±0.55

Interatomic distances (Å)

Nearest neighbours:

Intraicosahedra within the inner icosahedra system
1.78 - 1.91, average 1.83; 1.69 - 1.99, average 1.80

Intericosahedra 1.61 - 1.86, average 1.72

Average for whole structure 1.80

Discussion

A model of the structure is shown in Fig. 21. It is built up of nearly regular B icosahedra. Icosahedra are centered about 000 and ½00 on the cell axes. Inside the cell there are two icosahedra systems formed by the union of three icosahedra (Fig. 22) and having symmetry $3m$. These are joined through the central atom (B(15)) at ½½½.

Details of analysis

Single crystal, (approximately 99.999% B) 0.7 by 0.35 mm. Lattice parameters from rotation and Weissenberg photographs. Intensities of 862 independent reflexions obtained by the equi-inclination Weissenberg method using crystal-monochromated Ag$K\alpha$ radiation. Starting from the structure proposed by 1, refinement was by least squares using 501 reflexions with individual weights and individual isotropic temperature factors. $R_1 = 10.6\%$. ($R_1 = 51.9\%$ from non-observed reflexions). The structure was confirmed with a Fourier synthesis using all 862 reflexions.

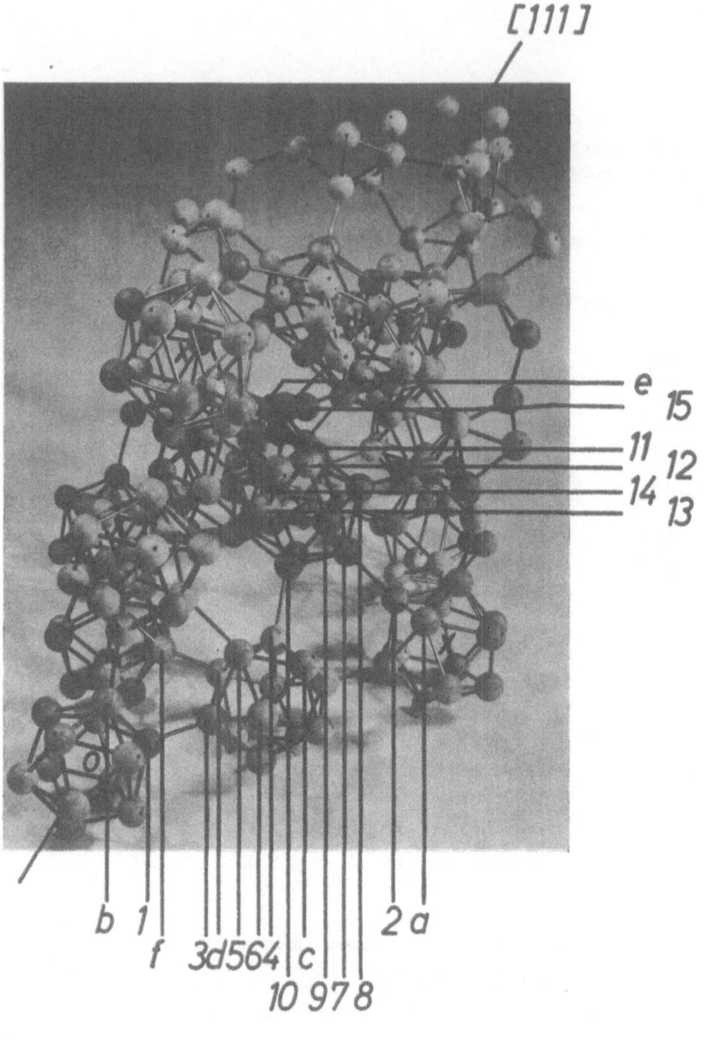

Fig. 21. Model of the structure of β-B. O indicates the origin and X the ends of the rhombohedral axes. Numbers correspond to those of the *Atomic positions*. Ignore letters.

1. J.L. HOARD, R.E. HUGHES, C.H.L. KENNARD, D.E. SANDS, D.B. SULLENGER and H.A. WEAKLIEM, 1963. *J. Amer. Chem. Soc.*, 85, 361.

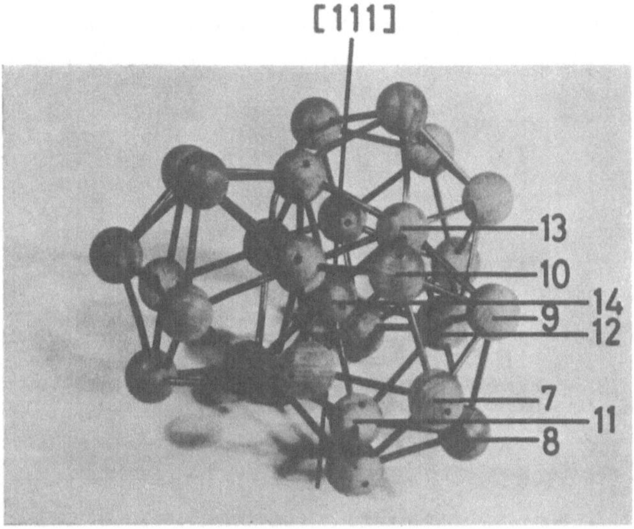

Fig. 22. Model of the inner icosahedra system. Numbers correspond to those of the *Atomic positions*.

α-FLUORINE

I. The crystal structure of α-fluorine. L. PAULING, I. KEAVENY and A.B. ROBINSON, 1970. *J. Solid State Chem.*, 2, 225-227.

The results of 1 are reinterpreted in terms of space group *C2/c* with refined F parameters of (0.287, 0.319, 0.100) leading to R = 0.20 (c.f. R = 0.33 in 1). The structure then consists of a cubic close packed array of F_2 molecules with each molecule (F-F = 1.49Å) tilted 18° from [001] towards the *b* axis.

1. L. MEYER, C.S. BARRETT and S.C. GREER, 1968. *J. Chem. Phys.*, 49, 1902.

NITROGEN

I. Crystal structures of the three modifications of nitrogen 14 and nitrogen 15 at high pressure. A.F. SCHUCH and R.L. MILLS, 1970. *J. Chem. Phys.* 52, 6000-6008.

α-N_2 (3785 atm, 19.6°K) F.W. 28.01

Cubic, a = 5.433 Å, U = 24.15 $Å^3$, Z = 2, D_x= 3.86. (See also 1)

β-N_2 (4125 atm, 49°K)

Hexagonal a = 3.861, c = 6.265 Å, U = 24.36 $Å^3$, Z = 2, D_x = 3.82.

Space group $P6_3/mmc$ (D^4_{6h})

The structure at this pressure is the same as that at low pressure (2).

γN_2 (4015±145 atm, 20.5±17°K)

Tetragonal a = 3.957, c = 5.109 Å, U = 24.09 Å^3, Z = 2, D_x = 3.86.

Space group $P4_2/mnm$ (D_{4h}^{14})

Atomic positions

N_2 in 4(f) with x = 0.098

Discussion

The structure consists of 001 layers of N_2 directed along [110], alternating with similar layers with the molecules oriented along [$1\bar{1}0$].

Details of analysis

X-ray powder photographs obtained from a special high pressure cryostat described in I. For α-N_2 14 lines were observed and listed. These agree with both models in space group $P2_13$ and $Pa3$. For β-N_2 10 lines were observed and listed and for γ-N_2 10 lines. A model was proposed using the molecular shape calculated by 3.

1. T.H. JORDAN, H.W. SMITH, W.E. STREIB and W.N. LIPSCOMB, 1964. *J. Chem. Phys.*, 41, 756.
2. W.E. STREIB, T.H. JORDAN and W.N. LIPSCOMB, 1962. *J. Chem. Phys.*, 37, 2962.
3. R.F.W. BADER, W.H. HENNEKER and P.E. CADE, 1967. *J. Chem. Phys.*, 46, 3341.

HEXADECABORANE (20)

I. The crystal and molecular structure of hexadecaborane (20). L.B. FRIEDMAN, R.E. COOK and M.D. GLICK, 1970. *Inorg. Chem.*, 9, 1452-1458.

$B_{16}H_{20}$ F.W. = 193.12

Monoclinic a = 5.849±9, b = 13.67±2, c = 16.75±3 Å, β = 100.83±5°, U = 1315 Å^3, D_m = 0.97±3, Z = 4, D_x = 0.975.

Space group $P2_1/c$ (C_{2h}^5) systematic absences

Atomic positions

	x	y	z	B (Å^2)
B(1)	0.4455±6	0.0830±3	0.1514±2	*
B(2)	0.5193±7	0.1999±3	0.1858±2	*
B(3)	0.4773±7	0.1014±3	0.2541±3	*
B(4)	0.5371±7	-0.0160±3	0.2074±2	*
B(5)	0.6696±6	0.1479±3	0.0936±2	*
B(6)	0.8110±6	0.2175±2	0.1512±2	*
B(7)	0.7175±7	0.1807±3	0.2529±2	*
B(8)	0.7206±8	0.0348±3	0.2699±3	*
B(9)	0.8294±7	-0.0359±3	0.1819±3	*
B(10)	0.6806±7	0.0090±3	0.1061±3	*
B(1')	0.9238±6	0.2042±2	0.0498±2	*
B(2')	0.8862±6	0.3127±3	-0.0065±2	*
B(4')	0.9375±7	0.3221±3	0.0972±2	*
B(5')	0.7457±7	0.4028±3	0.0564±3	*
B(6')	0.5953±7	0.3343±3	-0.0030±3	*
B(7')	0.6891±7	0.2096±3	-0.0036±2	*

	x	*y*	*z*	*B* (Å^2)
H(1)	0.272±5	0.074±2	0.129±2	2.2±7
H(2)	0.393±4	0.264±2	0.192±1	0.4±5
H(3)	0.346±5	0.109±2	0.199±2	2.6±7
H(4)	0.414±4	-0.077±2	0.226±1	1.6±6
H(7)	0.742±5	0.228±2	0.308±2	3.1±7
H(8)	0.734±5	0.008±2	0.334±2	3.7±8
H(9)	0.938±5	-0.103±2	0.181±2	3.0±7
H(10)	0.677±5	-0.037±2	0.049±2	2.6±7
H(1')	0.085±4	0.161±2	0.026±1	0.9±6
H(2')	0.017±4	0.333±2	0.939±2	1.6±6
H(4')	0.095±5	0.356±2	0.111±1	1.7±6
H(5')	0.771±5	0.482±2	0.053±2	2.5±7
H(6')	0.512±4	0.365±2	0.948±2	2.5±7
H(7')	0.677±5	0.167±2	0.911±2	2.0±6
H(B(6)B(7))	0.918±5	0.178±2	0.199±2	2.6±7
H(B(8)B(9))	0.918±5	0.026±2	0.223±2	2.2±7
H(B(9)B(10)	0.883±5	0.015±2	0.114±2	1.7±6
H(B(4')B(5'))	0.765±5	0.370±2	0.132±2	2.7±7
H(B(5')B(6'))	0.529±5	0.379±2	0.065±2	2.5±7
H(B(6')B(7'))	0.505±5	0.246±2	0.024±2	2.4±7

* Anisotropic temperature factors given in I. The numbering scheme used for the atoms is consistent with the proposed nomenclature rules for boron compounds (1).

Interatomic distances (Å)

B(1)	B(2)	1.780±5
B(1)	B(4)	1.786±6
B(2)	B(3)	1.754±6
B(3)	B(4)	1.790±6
B(2)	B(5)	1.777±5
B(2)	B(7)	1.780±6
B(4)	B(8)	1.775±6
B(4)	B(10)	1.779±6
B(1)	B(5)	1.722±5
B(1)	B(10)	1.760±6
B(3)	B(7)	1.772±6
B(3)	B(8)	1.751±6
B(5)	B(6)	1.680±5
B(6)	B(7)	1.763±6
B(8)	B(9)	1.779±7
B(9)	B(10)	1.778±6
B(1)	B(3)	1.782±6
B(2)	B(6)	1.713±6
B(4)	B(9)	1.704±6
B(5)	B(10)	1.914±6
B(7)	B(8)	2.015±7
B(5')	B(6')	1.723±6
B(2')	B(5')	1.725±6
B(2')	B(6')	1.717±6
B(2')	B(4')	1.821±6
B(2')	B(7')	1.816±6
B(1')	B(4')	1.806±5
B(1')	B(7')	1.775±6
B(4')	B(5')	1.797±6
B(6')	B(7')	1.791±6
B(4')	B(6)	1.779±5
B(5)	B(7')	1.818±6
B(1')	B(2')	1.794±5
B(1')	B(5)	1.712±5
B(1')	B(6)	1.712±6

B(2) - B(1) - B(3)	59.0±2	B(3) - B(8) - B(4)	61.0±3
B(2) - B(1) - B(5)	61.0±2	B(3) - B(8) - B(7)	55.6±2
B(3) - B(1) - B(4)	60.2±2	B(4) - B(8) - B(9)	57.3±2
B(4) - B(1) - B(10)	60.2±2	B(7) - B(8) - B(9)	115.6±3[a]
B(5) - B(1) - B(10)	66.7±2	B(4) - B(9) - B(8)	61.2±2
B(1) - B(2) - B(3)	60.6±2	B(4) - B(9) - B(10)	51.4±2
B(1) - B(2) - B(5)	57.9±2	B(8) - B(9) - B(10)	104.9±3[a]
B(3) - B(2) - B(7)	60.2±2	B(1) - B(10)- B(4)	60.6±2
B(5) - B(2) - B(6)	57.5±2	B(1) - B(10)- B(5)	55.7±2
B(6) - B(2) - B(7)	60.6±2	B(4) - B(10)- B(9)	57.3±2
B(1) - B(3) - B(2)	60.4±2	B(5) - B(10)- B(9)	116.9±3[a]
B(1) - B(3) - B(4)	60.0±2	B(5) - B(1')- B(6)	58.8±2
B(2) - B(3) - B(7)	60.6±2	B(5) - B(1')- B(7')	62.8±2
B(4) - B(3) - B(8)	60.2±2	B(6) - B(1')- B(4')	60.7±2
B(7) - B(3) - B(8)	69.8±3	B(2')- B(1')- B(4')	60.8±2
B(1) - B(4) - B(3)	59.8±2	B(2')- B(1')- B(7')	61.2±2
B(1) - B(4) - B(10)	59.2±2	B(1')- B(2')- B(4')	59.9±2
B(3) - B(4) - B(8)	58.8±2	B(1')- B(2')- B(7')	58.9±2
B(8) - B(4) - B(9)	61.5±3	B(4')- B(2')- B(5')	60.8±2
B(9) - B(4) - B(10)	61.3±3	B(5')- B(2')- B(6')	60.1±2
B(1) - B(5) - B(2)	61.1±2	B(6')- B(2')- B(7')	60.8±2
B(1) - B(5) - B(10)	57.6±2	B(6) - B(4')- B(1')	57.1±2
B(2) - B(5) - B(6)	59.3±2	B(6) - B(4')- B(5')	117.2±3[a]
B(6) - B(5) - B(1')	60.6±2	B(1')- B(4')- B(2')	59.3±2
B(10)- B(5) - B(7')	124.2±3[a]	B(2')- B(4')- B(5')	56.9±2
B(1')- B(5) - B(7')	60.3±2	B(2')- B(5')- B(4')	62.2±2
B(2) - B(6) - B(5)	63.2±2	B(2')- B(5')- B(6')	59.7±2
B(2) - B(6) - B(7)	61.6±2	B(4')- B(5')- B(6')	108.0±3
B(5) - B(6) - B(1')	60.6±2	B(2')- B(6')- B(5')	66.2±2
B(7) - B(6) - B(4')	138.2±3[a]	B(2')- B(6')- B(7')	62.3±2
B(1')- B(6) - B(4')	62.3±2	B(5')- B(6')- B(7')	109.4±3
B(2) - B(7) - B(3)	59.2±2	B(5) - B(7')- B(1')	56.9±2
B(2) - B(7) - B(6)	57.8±2	B(5) - B(7')- B(6')	118.0±3a
B(3) - B(7) - B(8)	54.6±2	B(1')- B(7')- B(2')	59.9±2
B(6) - B(7) - B(8)	114.2±3[a]	B(2')- B(7')- B(6')	56.8±2

a External angle.

B(1) - H(1)	1.15±3	B(6) - H(B(6)B(7))	1.23±3
B(2) - H(2)	1.13±2	B(7) - H(B(6)B(7))	1.34±3
B(3) - H(3)	1.08±3	B(8) - H(B(8)B(9))	1.28±3
B(4) - H(4)	1.11±3	B(9) - H(B(8)B(9))	1.26±3
B(7) - H(7)	1.16±3	B(9) - H(B(9)B(10))	1.31±3
B(8) - H(8)	1.15±3	B(10)- H(B(9)B(10))	1.22±3
B(9) - H(9)	1.11±3	B(4')- H(B(4')B(5'))	1.36±3
B(10)- H(10)	1.16±3	B(5')- H(B(4')B(5'))	1.25±3
B(1')- H(1')	1.12±2	B(5')- H(B(5')B(6'))	1.29±3
B(2')- H(2')	1.11±3	B(6')- H(B(6')B(7'))	1.36±3
B(4')- H(4')	1.09±3	B(6')- H(B(5')B(6'))	1.29±3
B(5')- H(5')	1.09±3	B(7')- H(B(6')B(7'))	1.20±3
B(6')- H(6')	1.12±3	Av	1.28±5
B(7')- H(7')	1.10±3		
Av	1.12±3		

Discussion

To a good approximation this molecule can be thought of as being composed of a $B_{10}H_{16}$ (2) and a B_8H_{12} (3) molecule fused along the B(5)-B(6) edge of $B_{10}H_{14}$ and the B(3)-B(8) edge of B_8H_{12}. See Fig. 23.

Details of analysis

Space group from Weissenberg photographs. Lattice parameters from least-squares refinement of 16 diffractometer measurements. Intensities measured on four-circle diffractometer (θ-2θ scan, Mo Kα radiation). Lorentz and polarisation corrections but not absorption (μ = 0.0384 mm^{-1}, crystal size = 0.5×0.5×1.0 mm^3). Structure solved by Sayre's method and 225 parameters refined by least squares to give R = 0.054 for 1095 reflexions listed.

1. 1968. *Inorg. Chem.*, 7, 1945.
2. *Structure Reports*, 13, 237; *Ibid.*, 15, 339; *Ibid.*, 21, 197.
3. R.D. DOBROTT, L.B. FRIEDMAN and W.N. LIPSCOMB, 1964. *J. Chem. Phys.*, 40, 866.

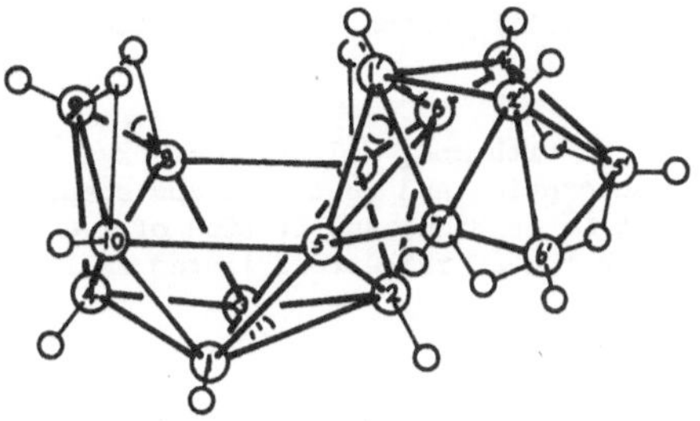

Fig. 23. The numbering scheme and molecular structure for $B_{16}H_{20}$ projected down the *a* axis.

TANTALUM DEUTERIDE

I. [The structure of Ta_2D]. V.F. PETRUNIN, V.A. SOMENKOV, S.Š. ŠIL'ŠTEIN and A.A. ČERTKOV, 1970. *Kristallografija*, 15, 171-173 [*Soviet Physics - Crystallography*, 15, 137-139].

Ta_2D F.W. 263.9

Orthorhombic [a = 4.77, b = 4.77, c = 3.40 inferred from I and 1]

Space group A222 (D_2^6)

Atomic positions

4Ta in 4(k) x = 0.012 (at room temperature)

= 0.017 (at liquid N_2 temperature)

2D in 2(a)

B calculated from data given in 2 using Einstein's model.

Discussion

The ordered phase of Ta_2D consists of layers of D atoms normal to *a* (orthorhombic) sandwiched between two layers of Ta atoms (D in tetrahedral interstices). The Ta layers are displaced from their f.c.c. positions away from the D atoms.

Details of analysis

Neutron [powder] diffraction, 16 lines listed. *R* (intensities) = 0.043.

1. *Structure Reports*, 26, 139.
2. M. SAKAMOTO, 1964. *J. Phys. Soc. Japan*, 19, 1862.
3. V.A. SOMENKOV, A.V. GURSKAJA, M.G. ZEMLYANOV, M.E. KOST, N.A. ČERNOPLEKOV and A.A. ČERTKOV, 1968. *Fiz. Tverd. Tela*, 10, 2697, [*Soviet Physics - Solid State*, 10, 2123.]

LITHIUM HYDRIDE: CARBON

LiH:C

I. Crystal structure of $LiH_{0.984}C_{0.016}$. M. ATOJI and M. KIKUCHI, 1970. *J. Chem. Phys.*, 52, 6434-6436.

Cubic (*B*1 type, 1), *a* = 4.096±2 Å.

Space group *Fm3m* (O_h^5)

The C atoms are found to substitute for H atoms in the lattice.

Details of analysis

Precession photographs with MoK_α radiation of crystals 0.4×0.4×0.3 mm^3. Lorentz, polarisation, absorption and spot size and shape corrections applied. Temperature factor of 1.5 Å^2 assumed, subsitution of 1.6 atomic % gave best agreement (*R* = 0.039) with the 12 reflexions listed in I which were each measured 27 times.

1. *Strukturbericht*, 1, 72.

MANGANESE SILICON CARBIDE

I. Structure de Mn_5SiC. P. SPINAT, R. FRUCHART, M. KABBANI and P. HERPIN, 1970. *Bull. Soc. fr. Minér Crist.*, 93, 171-184.

Mn_5SiC F.W. 314.8

Orthorhombic, *a* = 10.198, *b* = 8.035, *c* = 7.630 Å, *U* = 625.2 Å^3, D_m = 6.62, *Z* = 8, D_x = 6.67.

Space group *Cmc*2$_1$ (C_{2v}^{12}) Systematic absences, statistical tests and anomalous scattering.

Atomic positions

8 Mn(1)	in	8(*b*)	0.1478	0.3352	0.3810
4 Mn(2)	in	4(*a*)	0	0.4546	0.6769
4 Mn(*2)	in	4(*a*)	0	0.1386	0.1861
8 Mn(3)	in	4(*b*)	0.1252	0.0056	0.4389
8 Mn(4)	in	8(*b*)	0.1211	0.1796	0.6894
8 Mn(5)	in	8(*b*)	0.1469	0.3408	0
8 Si	in	8(*b*)	0.2520	0.0782	0.1869
4 C	probably in	4(*a*)	0	0.188	0.470
4 C*	probably in	4(*a*)	0	0.188	0.906

Anisotropic temperature factors given.

$MoMn_4SiC$ F.W. 355.8

Orthorhomic, a = 10.20, b = 8.30, c = 7.70 Å [numbers taken from graph], Z = 8.

Space group $Cmc2_1$ (C^{12}_{2v})

Atomic positions

			x	y	z
8 (Mn(1), Mo)	in	8(b)	0.1404	0.3326	0.3468
4 Mn(2)	in	4(a)	0	0.4456	0.6284
4 Mn(*2)	in	4(a)	0	0.1156	0.1195
8 Mn(3)	in	8(b)	0.1269	0.0046	0.4143
8 Mn(4)	in	8(b)	0.1227	0.1792	0.6194
8 (Mn(5), Mo)	in	8(b)	0.1466	0.3373	0
8 Si	in	8(b)	0.2575	0.0834	0.2160

z coordinates are particularly unreliable.

Interatomic distances (Å) (in Mn_5SiC)

Mn(1)	-	11 Mn	2.65 - 3.05
		3 Si	2.64 - 2.74
Mn(2)	-	10 Mn	2.53 - 3.02
		2 Si	2.54
Mn(*2)	-	10 Mn	2.55 - 2.84
		2 Si	2.62
Mn(3)	-	9 Mn	2.37 - 2.83
		2 Si	2.38 - 2.40
Mn(4)	-	9 Mn	2.36 - 2.84
		2 Si	2.34 - 2.47
Mn(5)	-	11 Mn	2.60 - 3.02
		3 Si	2.60 - 2.77
Si	-	12 Mn	2.34 - 2.77
C	-	6 Mn	about 2.04

Distances in $MoMn_4SiC$ are similar.

Discussion

Both structures are isotypic and consist of 8 layers of the following composition perpendicular to a:

Mn_4C_4: Mn_8: Si_4: Mn_8: Mn_4C_4: Mn_8: Si_4: Mn_8: Mn_4C_4

In $MoMn_4SiC$, the Mo is found in the Mn_8 layers.

Details of analysis

Lattice parameters from powder photographs with CuK_α radiation [λ not given]. Intensities for $MoMn_4SiC$ measured by microdensitometer from Weissenberg and precession photographs (MoK_α radiation). Structure solved by Patterson and electron-density syntheses and refined ($hk0$ and $hk1$ layers only) to give R = 0.18. [reflexions not listed]. Intensities for Mn_5SiC measured with an automatic diffractometer (MoK_α radiation). Refined by least squares to give R = 0.063 for 565 reflexions [not listed].

CALCIUM STRONTIUM SILICIDE

$CaSrSi_4$ F.W. 240.04

I. Zur Kenntnis ternärer Disilicide der Erdakalimetalle. B. EISENMANN, CH. RIEKEL, H. SCHÄFER and A. WEISS, 1970. *Z. anorg. Chem.*, 372, 325-331.

Trigonal, $a = 3.91 \pm 1$, $c = 5.15 \pm 1$ Å, $U = 68.18$ Å^3, $D_m = 2.93$, $Z = 1/2$, $D_x = 2.97$.

Space group $P\bar{3}m1$ (D^3_{3d})

Atomic positions

1/2 Ca + 1/2 Sr	in 1(*a*)	
2 Si	in 2(*d*)	$z = 0.568$

Interatomic distances (Å)

The crystallographic site symmetry of Ca(Sr) is D_{3d}

Ca,Sr - 6 Si	3.17	(×6)
Si - Si	2.36	

[σ not quoted but at least 0.01Å]

Si-Si-Si	111.5°

Discussion

The crystal has a graphite-like structure of condensed puckered Si_6 rings with the Ca and Sr atoms randomly distributed over octahedral holes between the layers.

Details of analysis

Weissenberg and precession photographs refined by trial and error to give $R = 0.08$ for 42 reflexions listed.

TERNARY NITRIDES

I. Structure de $MgSiN_2$ et $MgGeN_2$. J. DAVID, Y. LAURENT and J. LANG, 1970. *Bull. Soc. fr. Minér. Crist.*, 93, 153-159.

$MgSiN_2$ F.W. = 80.4

Orthorhombic, $a = 5.27_9$, $b = 6.47_6$, $c = 4.99_2$ Å, $U = 170.7$ Å^3, $D_m = 2.96 \pm 1$, $Z = 4$, $D_x = 3.12$.

Space group $Pna2_1$ (C^9_{2v}) systematic absences

Atomic positions All atoms in 4(*a*)

	x	*y*	*z*
4 Mg	0.083	0.600	0
4 Si	0.070	0.130	0.000
4 N(1)	0.065	0.125	0.385
4 N(2)	0.083	0.650	0.400

$MgGeN_2$ F.W. = 124.9

Orthorhombic, $a = 5.49_4$, $b = 6.61_1$, $c = 5.16_6$ Å, $U = 187.6$ Å^3, $D_m = 3.84 \pm 10$, $Z = 4$, $D_x = 4.41$.

Space group $Pna2_1$ (C^9_{2v}) systematic absences

Atomic positions All atoms in 4(*a*)

	x	*y*	*z*
4 Mg	0.083	0.625	0
4 Ge	0.083	0.125	0.000
4 N(1)	0.083	0.125	0.380
4 N(2)	0.083	0.625	0.400

Discussion

Isostructural with $BeSiN_2$ (1). The structures are derived from that of wurtzite by the ordering of the Mg and Si(Ge) atoms. Mean distances: Mg-N = 1.98, Si-N = 1.87, Ge-N = 1.95 Å. [The individual distances are unreliable because of the very small number of observations].

Details of analysis

Powder photographs and powder diffractometer indexed by analogy with $BeSiN_2$ (1). Lattice parameters determined by least squares. Intensities measured photometrically from powder photographs (CuK_α radiation) corrected for Lorentz, polarisation and multiplicity. Structure of $BeSiN_2$ assumed and refined to give R_1 ($MgSiN_2$) = 0.085 for 25 reflexions listed and R_1 ($MgGeN_2$) = 0.064 for 17 reflexions listed.

1. P. ECKERLIN, 1967. *Z. anorg. Chem.*, 353, 225.

TRIMERIC PHOSPHONITRILIC ISOTHIOCYANATE

$N_3P_3(NCS)_6$ F.W. = 483.4

I. The crystal and molecular structure of trimeric phosphonitrilic isothiocyanate. J.B. FAUGHT, T. MOELLER and I.C. PAUL, 1970. *Inorg. Chem*, 9, 1656-1660.

Triclinic, a = 11.79±3, b = 8.02±2, c = 10.31±1 Å, α = 96°0'±30', β = 99°4'±30', γ = 97°28'±30', U = 947 Å^3, D_m = 1.62, Z = 2, D_x = 1.70.

Space group P1 (C_i^1)

Atomic positions

	x	y	z
P(1)	0.8163	0.0847	1.0127
N(2)	0.8750	0.0200	0.8923
P(3)	0.8069	0.0347	0.7440
N(4)	0.7004	0.1311	0.7313
P(5)	0.6368	0.1801	0.8450
N(6)	0.7017	0.1543	0.9923
N(7)	0.9110	0.2313	1.1060
C(8)	0.9973	0.3327	1.1335
S(9)	1.1037	0.4664	1.1667
N(10)	0.8088	-0.0602	1.1154
C(11)	0.7914	-0.1929	1.1378
S(12)	0.7796	-0.3779	1.1828
N(13)	0.9053	0.1191	0.6679
C(14)	0.9280	0.1844	0.5856
S(15)	0.9783	0.2692	0.4675
N(16)	0.7773	-0.1530	0.6608
C(17)	0.7231	-0.2581	0.5996
S(18)	0.6528	-0.4187	0.5007
N(19)	0.5014	0.0895	0.8212
C(20)	0.4395	0.0110	0.7343
S(21)	0.3464	-0.1047	0.6182
N(22)	0.6201	0.3779	0.8465
C(23)	0.5854	0.5020	0.8519
S(24)	0.5389	0.6773	0.8704

Interatomic distances (Å) *and angles*

Mean	P-N(endo)	= 1.58
Mean	P-N(exo)	= 1.63
Mean	N-C	= 1.12
Mean	C-S	= 1.59

Mean	N(endo)-P-N(endo)	= 119°
Mean	P-N(endo)-P	= 121
Mean	N(endo)-P-N(exo)	= 109
Mean	N(exo)-P-N(exo)	= 100
Mean	P-N-C	= 152
Mean	N-C-S	= 175

Structure

See Figs. 24 and 25. The nitrogen atoms of two isothiocyanate groups are bonded to each phosphorus atom in the six-membered phosphorus-nitrogen ring. Five of the atoms of the phosphonitrilic ring are planar, with the sixth atom, a nitrogen, 0.15 Å out of the plane.

Details of analysis

Cu radiation, visual intensities from multiple-film equi-inclination Weissenberg photographs for 1246 observed reflexions. Structure by symbolic addition, electron-density and least-squares methods, $R = 0.12$. σ(bond lengths) = 0.02 - 0.04 Å.

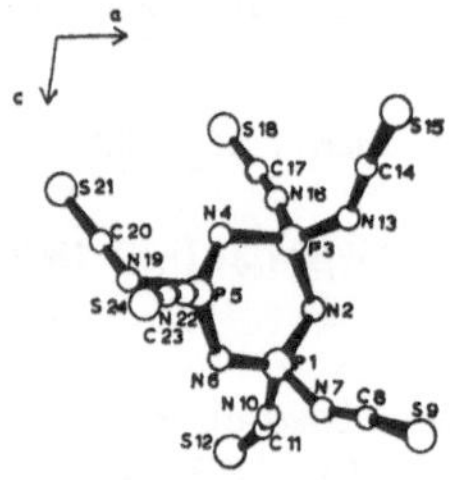

Fig. 24. View of a single $N_3P_3(NCS)_6$ molecule perpendicular to the *a-c* plane.

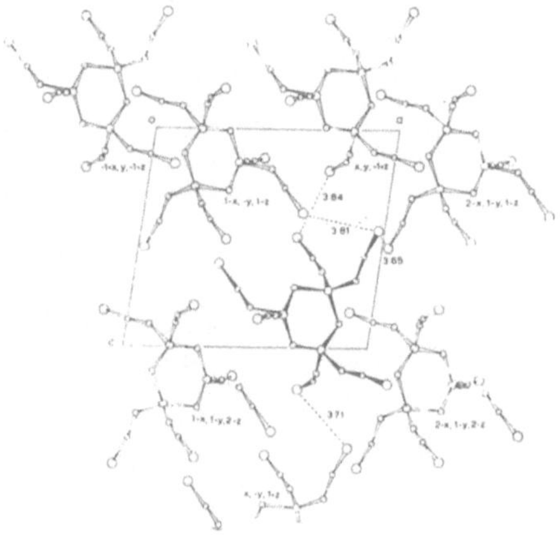

Fig. 25. Packing diagram of trimeric phosphonitrilic isothiocyanate. Some sulphur-sulphur inter-molecular contacts are shown.

IRON HEXATHIODIPHOSPHATE

$Fe_2P_2S_6$ F.W. = 366.00

I. Über Hexachalkogeno-hypodiphosphate vom Typ $M_2P_2X_6$. W. KLINGEN, G. EULENBERGER and H. HAHN, 1970. *Naturwissenschaften*, 57, 88.

Monoclinic, $a = 5.93$, $b = 10.28$, $c = 6.72$Å, $\beta = 107.16°$, $U = 390.0$Å^3, $Z = 2$, $D_x = 3.10$.

Space group, $C2/m$ (C^3_{2h})

Atomic positions

			x	*y*	*z*
4 Fe	in	4(*g*)	0	0.333	0
4 P	in	4(*i*)	0.057	0	0.171
4 S(1)	in	4(*i*)	0.751	0	0.247
8 S(2)	in	8(*j*)	0.248	0.166	0.248

Interatomic distances (Å)

P_2S_6 ion (crystallographic symmetry C_s, approximate symmetry D_{3d})

P - S(1)	2.02	
P - S(2)	2.03	(×2)
P - P	2.20	

Angles in degrees subtended at P by

	S(2)'	P
S(1)	114.2	104.0
S(2)	114.5	104.1

Environment of Fe (crystallographic symmetry C_2)

Fe - S(1)	2.54	(×2)
Fe - S(2)	2.54	(×2)
Fe - S(2)'	2.55	(×2)

Angles in degrees subtended at Fe by

	S(2)	S(1)	S(2)'
S(1)	95	85	85
S(2)		180	95
S(1)		95	95
S(1)			85
S(2)'			180

σ(distances) > 0.01Å σ(angles) > 0.5°.

Discussion

The P_2S_6 ions have approximate D_{3d} symmetry and are arranged perpendicular to the (001) plane with the sulphur atoms forming two close packed layers sandwiching the phosphorus and iron atoms. The layers are then stacked along *c* in such a way that the sulphur atoms lie in a cubic close packed array. Compare the similar structure of $Fe_2P_2Se_3$ where the selenium atoms lie in a hexagonal close packed array (2).

Details of analysis

Preparation described in 1 and correct unit cell and space group first reported in 2. Structure solved from three-dimensional Patterson and Fourier syntheses. Refined by least squares to give *R* = 0.127. No absorption correction applied. No details of intensity measurement nor list of structure factors given in I but are presumably given in 3.

1. H. HAHN and W. KLINGEN, 1965. *Naturwissenschaften*, 52, 494.
2. W. KLINGER, G. EULENBERGER and H. HAHN, 1968. *Ibid.*, 55, 229.
3. W. KLINGER, 1969. *Dissertation*, Stuttgart-Hohenheim.

TROILITE

I. Lunar troilite: crystallography. H.T. EVANS, 1970. *Science*, 20, 621-623.

FeS F.W. = 87.91

Hexagonal, a = 5.962±2, c = 11.750±3 Å, U = 361.7 Å^3, Z = 12, D_x = 4.83.

Space group $P\bar{6}2c$ (D^4_{3h})

Atomic positions

			x	y	z
12 Fe	in	12(i)	0.3797±3	0.0551±3	0.1230±1
2 S(1)	in	2(a)	0	0	0
4 S(2)	in	4(f)	1/3	2/3	0.0199 4
6 S(3)	in	6(h)	0.6652±9	-0.0031±8	1/4

Anisotropic temperature factors given.

Interatomic distances (Å)

Fe	-	S(3)	2.359±4
Fe	-	S(2)	2.379±3
Fe	-	S(3)	2.416±5
Fe	-	S(2)	2.503±3
Fe	-	S(1)	2.565±2
Fe	-	S(3)	2.722±2
Fe	-	Fe	2.919±2
Fe	-	Fe	2.920±2
Fe	-	Fe	2.946±2
Fe	-	Fe	2.984±2

Discussion

This work confirms and improves upon earlier work (1).

Details of analysis

Troilite from lunar sample 10050 from the Sea of Tranquility. Lattice parameters from powder photographs and X-ray goniostat [no calibration given]. Intensities measured on single crystal automatic diffractometer (MoK_α radiation) corrected for Lorentz and polarisation effects but not absorption (μ = 13.52 mm^{-1}, crystal size ∿ 0.15 mm). Structure refined by least squares to give R_1 = 0.084 for 600 reflexions [not listed but available from author].

1. *Structure Reports*, 20, 129, *Ibid.*, 24, 166.

α-DIMORPHITE

As_4S_3 F.W. = 395.85

I. The crystal structure of tetra-arsenic trisulphide. H.J. WHITFIELD, 1970. *J. Chem. Soc. Lond. A*, 1800-1807.

Orthorhomic, a = 9.12±2, b = 7.99±2, c = 10.10±2 Å, U = 736.0 Å^3, D_m = 3.55±5, Z = 4, D_x = 3.57.

Space group *Pnma* (D^{16}_{2h}) from systematic absences and statistical tests.

Atomic positions

			x	y	z
4 As(1)	in	4(c)	0.3425±5	1/4	0.4413±5
4 As(2)	in	4(c)	-0.0188±4	1/4	0.5697±5
8 As(3)	in	8(d)	0.1466±3	0.0966±6	0.7162±4
8 S(1)	in	8(d)	0.3326±7	0.0402±9	0.5851±8
4 S(2)	in	4(c)	0.1079±10	1/4	0.3830±10

Anisotropic temperature factors given.

Interatomic distances (Å)

The As_4S_3 molecule (Fig. 26) has crystallographic point symmetry C_s, but does not differ significantly from C_{3v} point symmetry.

As(1) - S(1)			2.220±9	(×2)
As(1) - S(2)			2.216±11	
As(2) - S(2)			2.213±12	
As(3) - S(1)			2.200±8	(×2)
As(2) - As(3)			2.445±6	(×2)
As(3) - As(3)'			2.450±10	
S(1) - As(1)	- S(2)		97.8±3°	(×2)
S(1) - As(1)	- S(1)		98.0±5°	
As(3) - As(2)	- As(3)'		60.2±2°	
As(3) - As(2)	- S(2)		101.1±3°	(×2)
As(2) - As(3)	- As(3)'		59.9±1°	(×2)
As(2) - As(3)	- S(1)		102.4±3°	(×2)
As(3)'- As(3)	- S(1)		101.8±3	(×2)
As(1) - S(1)	- As(3)		105.7±3°	(×2)
As(1) - S(2)	- As(2)		106.1±4°	

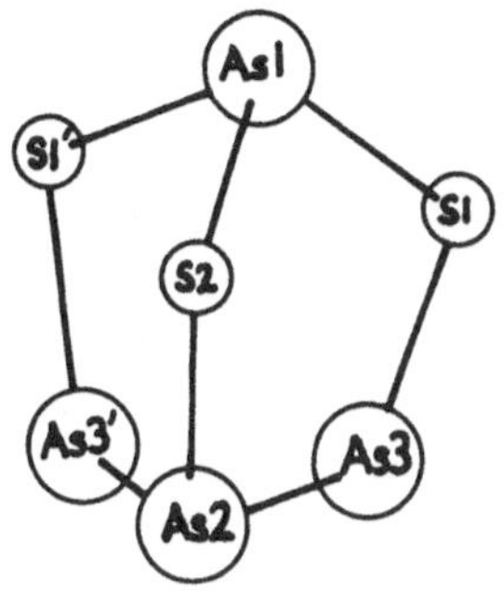

Fig. 26. The As_4S_3 molecule. The non-crystallographic three-fold axis passes through As(1) and the centers of the two triangles formed respectively by the S atoms and the remaining As atoms.

Details of analysis

Synthetic α-As_4S_3. Intensities measured visually from multiple-film equi-inclination Weissenberg photographs (CuK_α) of layers with k = 0 to 4 and l = 0 to 3. Lorentz and polarisation corrections. Structure solved by Sayre's method and refined by least squares to give R = 0.11 for 575 observed reflexions listed.

TITANIUM SULPHIDES

I. The system titanium-sulphur. II. The structures of Ti_3S_4 and Ti_4S_5. G.A. WEIGERS and F. JELLINEK, 1970. *J. Solid State Chem.*, 1, 519-525.

TRITITANIUM TETRASULPHIDE

Ti_3S_4 F.W. = 271.9

Rhombohedral, a = 3.441, c = 60.48 Å, U = 620.2 Å^3, Z = 5.25, D_x = 3.82.

Space group $R\bar{3}m$ (D_{3d}^{5})

Atomic positions Hexagonal setting

			x	y	z
3 Ti(1)	in	3(a)	0	0	0
3.1 Ti(2)	in	6(*c*)	0	0	0.0518±7
4.3 Ti(3)	in	6(*c*)	0	0	0.1879±7
5.7 Ti(4)	in	6(*c*)	0	0	0.2382±7
6 S(1)	in	6(*c*)	0	0	0.1190
6 S(2)	in	6(*c*)	0	0	0.3094
6 S(3)	in	6(*c*)	0	0	0.4028
3 S(4)	in	3(*b*)	0	0	1/2

B(Ti) = 2.08 Å^2 , B(S) = 1.64 Å^2.

TETRATITANIUM PENTASULPHIDE

Ti_4S_5 F.W. = 351.9

Hexagonal, a = 3.439, c = 28.29 Å, U = 296.3 Å^3, Z = 2, D_x = 3.93.

Space group $P6_3/mmc$ (D_{6h}^{4})

Atomic positions

			x	y	z
2 Ti(1)	in	2(*a*)	0	0	0
2.5 Ti(2)	in	4(*e*)	0	0	0.105
3.5 Ti(3)	in	4(*f*)	1/3	2/3	0.195
4 S(1)	in	4(*f*)	1/3	2/3	0.053
4 S(2)	in	4(*f*)	1/3	2/3	0.650
2 S(3)	in	2(*b*)	0	0	1/4

B = 1.02 Å^2

Interatomic distances (Å)

	Ti_3S_4		Ti_4S_5	
Ti(1)-S_h	2.46	(×6)	2.51	(×6)
Ti(2)-S_c	2.26	(×3)	2.37	(×3)
Ti(2)-S_h	2.61	(×3)	2.49	(×3)
Ti(3)-S_c	2.36	(×3)	2.38	(×3)
Ti(3)-S_h	2.55	(×3)	2.54	(×3)
Ti(4)-S_c	2.52	(×3)		
Ti(4)-S_h	2.45	(×3)		

S_c is a sulphur atom in a cubic layer

S_h is a sulphur atom in a hexagonal layer

Discussion

The relationship between Ti_3S_4, Ti_4S_5 and other sulphides of titanium (Ti_2S_3, Ti_8S_9 and TiS) are discussed in I. All the structures involve close packed layers of S atoms with Ti atoms partially filling the octahedral interstices. The Ti atoms tend to cluster so that their occupation numbers form waves along the *c* direction with repeats of 2 (Ti_2S_3), 2.33 (Ti_3S_4), 2.5 (Ti_4S_5), 3.0 (Ti_8S_9) and ∞ (TiS) layers. Hexagonal close packing of the S atoms occurs in regions where the occupation numbers are close to unity, cubic close packing when they are close to zero. This arrangement tends to bring pairs of Ti atoms (Ti-Ti = 3.09 Å) together along the *c* axis.

Details of analysis

Polycrystalline samples prepared by heating titanium sponge and sulphur at 1000°C. Integrated intensities measured from powder diffractometer. 30 reflexions (listed) used to refine the structure of Ti_3S_4 to give *R*(intensities) = 0.054, and 25 reflexions (listed) used to refine the structure of Ti_4S_5 to give *R* (intensities) = 0.13.

RATHITE-II

I. Die Kristallstruktur von Rathit-II $[As_{25}S_{56}[Pb^{vii}_{6.5}Pb^{ix}_{12}]$. P. ENGEL and W. NOWACKI, 1970. *Z. Kristallogr.*, 131, 356-375.

$Pb_{37}As_{50}S_{112}$ (See also 1, 2) F.W. = 15.003

Monoclinic, a = 8.371±5, b = 70.49±5, c = 7.914±4 Å, β = 90.13±3°, U = 4669±Å^3, D_m = 5.45, Z = 1, D_x = 5.32.

Space group $P2_1$ (C_2^2)

Atomic positions

Positional and anisotropic temperature coordinates for 19 Pb and 26 As atoms (two statistically distributed over 2 sites) and positional and isotropic temperature coordinates for 56 S atoms are given in I.

Discussion

The structure consists of a sequence of layers: ACABA$\overline{ACABA}$ along the *b* direction, the layers $\overline{C}$ etc. are related to C etc. by the screw axis. Layer A has the composition PbS_4 with Pb in ninefold coordination (Pb-S ranging from 2.7-3.5 Å). Layer B has the composition $As_3S_9PbAs_3S_7$ and layer C $As_3S_7Pb_3As_6S_{13}$ with seven-coordinate Pb (Pb-S again in the range 2.7-3.5) and three-coordinate (non-planar) As (As-S in the range 2.2-2.5 Å). The interatomic distances given in I are accurate to 0.01 to 0.02 Å.

Details of analysis

Rathite-II from Lengenbach, Canton Wallis, Switzerland. Lattice parameters by least squares from measurements on a back-reflexion camera calibrated with silicon [assumed parameters not given]. Intensities measured on automatic diffractometer with CuK_α radiation were corrected for Lorentz, polarisation and absorption effects (μ = 88.5 mm^{-1}, crystal ground to a regular sphere of radius 0.12 mm). Structure from Patterson and Buerger minimum functions and consideration of related structures (rathite I, baumhauerite and dufrenoysite) refined by difference electron synthesis and least squares to give R = 0.071 for 630 parameters and 8690 reflexions [not listed].

1. *Structure Reports*, 26, 428.
2. *Ibid.*, 27, 403, 405.

CUPROUS AMMONIUM THIOMOLYBDATE

$CuNH_4MoS_4$ F.W. = 305.73

I. On the preparation, properties and structure of cuprous ammonium thiomolybdate. W.P. BINNIE, M.J. REDMAN and W.J. MALLIO, 1970. *Inorg. Chem.*, 9, 1449-1452.

Tetragonal, a = 8.000±4, c = 5.409±3 Å, U = 346.2 Å^3 , D_m = 2.94±2, Z = 2, D_x = 2.93.

Space group $I\bar{4}$ (S_4^2) from systematic absences (electron diffraction) and final structure.

Atomic positions

			x	y	z	B (Å^2)
2 Cu	in	2(*b*)	0	0	1/2	3.5±8
2 N	in	2(*c*)	0	1/2	1/4	3.3±10
2 Mo	in	2(*a*)	0	0	0	2.0±4
8 S	in	8(*g*)	0.195±2	0.112±2	0.232±7	3.8±8

Interatomic distances (Å) *and angles*

Mo-S	2.19±3	(×4)	S-Mo-S	=	110±1°	(×2),	109±1°	(×4)
Cu-S	2.31±3	(×4)	S-Cu-S	-	102±1°	(×2),	113±1°	(×4)
N-S	3.48±3	(×4)	(equatorial)					
N-S'	3.68±3	(×4)						
N-S"	3.82±3	(×4)						

Discussion

All atoms except S and H lie in positions with S_4 crystallographic point symmetry. Edge sharing CuS_4 and MoS_4 tetrahedra alternate in a chain along c. The NH_4 ions occupy holes between the chains, the S coordination around N being that of a cubooctahedron. The four S atoms nearest to N lie only 0.1 Å above and below the (001) plane containing the N atom suggesting that any N-H...S hydrogen bonding is very weak.

Details of analysis

Sample prepared only as powder of crystallite size <5μ. X-ray powder pattern measured photographically (lattice parameters) and with diffractometer (intensities) using CoK_α radiation. Corrections for Lorentz, polarisation, multiplicity and absorption (μ = 44.18 mm^{-1}). No preferred orientation. Systematic absences observed by single-crystal electron diffraction. Structure by trial and error refined by least squares to give R = 0.15 (for $|F|^2$) for the 13 unique and 27 overlapped reflexions listed.

HEXABROMOCYCLOTRIPHOSPHAZENE

I. The crystal structure of compounds with (N-P)*n* rings. VII refinement of the crystal structure of hexabromocyclotriphosphazene. H. ZOER and A.J. WAGNER, 1970. *Acta Cryst.*, B26, 1812-1819.

$N_3P_3Br_6$ F.W. = 374.66

Orthorhombic, a = 14.463±2, b = 13.410±3, c = 6.601±1Å, [U = 1280.3Å^3], Z = 4, [D_x = 1.944].

Space group *Pnma* (D_{2h}^{16}) $Pn2_1a$ was not investigated

Atomic positions

(average over two sets)

				x	*y*	*z*
4 N(1)	in	4(*c*)	with	0.1432±10	1/4	0.5953±30
8 N(2)	in	8(*d*)	with	-0.0079±2	0.1503±8	0.4547±19
4 P(1)	in	4(*c*)	with	-0.0606±3	1/4	0.4119±9
8 P(2)	in	8(*d*)	with	0.0908±2	0.1478±3	0.5555±6
4 Br(1)	in	4(*c*)	with	-0.1895±2	1/4	0.5786±4
4 Br(2)	in	4(*c*)	with	-0.1078±2	1/4	0.1017±4
8 Br(3)	in	8(*d*)	with	0.0823±1	0.0714±1	0.8444±3
8 Br(4)	in	8(*d*)	with	0.1806±1	0.0506±1	0.3845±3

Anisotropic thermal parameters are given.

Interatomic distances (Å) *and average angles*

N(1)	-	P(2)	1.588±6	P - N - P	120.8°
N(2)	-	P(1)	1.565±8	N - P - N	118.6
N(2)	-	P(2)	1.575±8	Br - P - Br	102.0
P(1)	-	Br(1)	2.165±4	N - P - Br	108.8
P(1)	-	Br(2)	2.158±4		
P(2)	-	Br(3)	2.168±3		
P(2)	-	Br(4)	2.159±3		

Discussion of the structure

The structure contains non-planar rings of $(NP)_3$ very slightly distorted towards the chair configuration with a mirror plane through the molecule. A more accurate refinement of the structure reported by Giglio and Puliti (1).

Details of analysis

Data taken twice and refined separately. 1176 reflexions, Mo radiation, with diffractometer, least-squares refinement with extinction corrections, R = 0.096 and 0.094 for the two sets.

1. E. GIGLIO and R. PULITI, 1967. *Acta Cryst.*, 22, 304.

AMMONIUM HALIDES

I. Neutron diffraction study of NH_4Br and NH_4I. R.S. SEYMOUR, 1970. *Acta Cryst.*, B26, 1487-1491.

NH_4Br

50 reflexions with λ = 1.05Å and 30 with λ = 1.55Å were measured at room temperature. At 120°C 50 reflexions with λ = 1.05Å and at 136°C 50 reflexions with λ = 1.17 were measured. A single crystal of NH_4Br was used.

Parameters

	23°C	120°C	136°C
B factor for NH_4 group (Å^2)	2.31±3	2.62±9	3.48±6
B factor for Br (Å^2)	2.01±4	2.56±10	3.10±6
c,N-H bond length (Å^2)	1.046±5	1.049±10	1.040±5
R.M.S. angular vibration of NH_4 group (°)	11.0±5	12.4±9	12.1±6
R.M.S. vibration of N-H bond length (Å)	0.10±3	0.11±5	0.10±2

Refined with half hydrogen atoms at $c/\sqrt{3}(\pm1,\pm1,\pm1)$ and with the thermal motion of the hydrogens as a convolution of gaussians describing displacements of NH_4, libration of NH_4 and oscillation of H along the N-H bond.

NH_4I

42 reflexions with λ = 1.17Å and 80 reflexions with λ = 1.55Å were measured from a single crystal of NH_4I.

B factor for I (Å^2)	3.64±10
B factor for NH_4 group (Å^2)	4.29±10
N-H bond length (Å)	1.053±6

Hydrogen positions refined in terms of a distribution function in cubic harmonics leading to the conclusion that the hydrogen atom distribution has broad peaks (FWHM = 70°) in the six <100> directions.

II. The wurtzite *z* parameter and linear compressibilities for NH_4F. B. MOROSIN, 1970. *Acta Cryst.*, B26, 1635-1636.

NH_4F F.W. = 37.04

Hexagonal, *a* = 4.4389±4, *c* = 7.1635±4Å, [*U* = 122.24 Å^3], *Z* = 3, [D_x = 2.870].

Space group $P6_3mc$ (C^4_{6v})

Atomic parameters

				x	*y*	*z*	*B* (Å^2)
2 F	in	2(*b*)	with	1/3	2/3	0.0	2.53±10
2 N	in	2(*b*)	with	1/3	2/3	0.3781±7	2.61±8
2 H(1)	in	2(*b*)	with	1/3	2/3	0.217 ±12	0.36±14
6 H(2)	in	6(*c*)	with	0.440±17	-0.440	0.421 ±7	4.1 ±8

Interatomic distances (Å) *and bond angles*

N-F	2.709±5,	2.707±5 (3×)
mean N-H	1.0	
F-N-F	110.1±2° (3×),	108.2±2° (3×)

Discussion of the structure

The structure is the same as wurtzite as found by Zachariasen (1) with N atoms slightly displaced towards base of a tetrahedron of flourine atoms.

Details of analysis

108 reflexions measured with Mo radiation. Least-squares refinement to a value of 0.046. Linear compressibilities are 13.6±5 and 14.5±5 x 10^{-13}cm^2 dyne^{-1} along *a* and *c* respectively.

1. *Structure Reports*, 19, 316; 22, 532.

III. Refinement of the structure of ammonium tri-iodide, NH_4I_3. G.H. CHEESMAN and A.J.T. FINNEY, 1970. *Acta Cryst.*, B26, 904-906.

NH_4I_3 F.W. = 398.75

Orthorhombic, a = 10.819, b = 6.640, c = 9.662 Å, [U = 694.1 Å^3], Z = 4, [D_x = 3.817].

Space group *Pnma* (D_{2h}^{16})

Atomic positions

			x	*y*	*z*
4N	in	4(*c*)	0.835 ±6	1/4	0.470 ±6
4I(1)	in	4(*c*)	0.1569±3	1/4	0.3471±3
4I(2)	in	4(*c*)	0.3812±3	1/4	0.5489±3
4I(3)	in	4(*c*)	0.5784±3	1/4	0.7351±3

Anisotropic thermal parameters are given.

Interatomic distances (Å)

I(1)	-	I(2)	3.113(4)	
I(2)	-	I(3)	2.791(4)	
I(1)	-	N	3.62(7),	3.68(7)
I(3)	-	N	3.78(7),	3.87(7)
I(1)-I(2)-I(3)			180.00(2)°	

Discussion of the structure

Essentially the same as reported by Mooney (1) with greater accuracy. The triiodide ion is asymmetric but linear.

Details of analysis

Mo radiation, visually estimated from multiple-film Weissenberg layer-line photographs, without absorption corrections. 516 independent reflexions from three separate crystals (since they decomposed), were refined by least squares to an R of 0.183.

1. *Strukturbericht*, 3, 321.

ANTIMONY TRIFLUORIDE

I. Fluoride crystal structures. Part XIV. Antimony trifluoride: a redetermination. A.J. EDWARDS, 1970. *J. Chem. Soc. Lond.*, A, 2751-2753.

SbF_3 F.W. = 178.7

Orthorhombic a = 7.26±1, b = 7.46±1, c = 4.95±1Å, U = 268±Å^3, D_m = 4.385, Z = 4, D_x = 4.42.

Space group *Ama*2 [converted from *C*2*cm*] (C_{2v}^{16}) systematic absences and structure

Atomic positions

			x	*y*	*z*	B(Å^2)
4 Sb	-	4(*b*)	1/4	0.2138±1	0	0.77±1
8 F(1)	-	8(*c*)	0.0685±24	0.2540±17	-0.2602±45	2.25±23
4 F(2)	-	4(*b*)	1/4	0.4474±31	0.1530±55	2.18±32

Interatomic distances (Å)

Sb	-	F(1)	1.94	(×2)
Sb	-	F(2)	1.90	
Sb	-	$F(1)_I$	2.60	(×2)
Sb	-	$F(2)_I$	2.63	

Angles in degrees subtended at Sb by:

	F(1)'	$F(1)'_I$	$F(2)_I$
F(2)	89	81	163
F(1)	86	83	78
F(1)'		156	78
$F(1)_I$		125	106
Sb - F(1) - Sb		154	
Sb - F(2) - Sb		163	

σ(distances) = 0.02 Å, σ (angles) = 1°

Discussion

The structure of 2 is confirmed.

Details of analysis

Crystals of diameter 0.2 mm elongated along *c* mounted in Pyrex capillary. Lattice parameters from precession and Weissenberg photographs with MoK_α (λ = 0.7107 Å) and CuK_α (λ = 1.5418 Å). Intensities measured photometrically from integrating Weissenberg photographs of layers with *l* = 0 to 4 taken with MoK_α radiation, corrected for Lorentz and polarisation effects but not absorption (μ = 10.8 mm^{-1}). Structure from Patterson function refined by least-squares to give *R* = 0.072 for 370 reflexions listed in 1.

1. *Supplementary publication SUP* 20040.
2. *Structure Reports*, 9, 152.

ZINC CHLORIDE

I. Zur Kristallstruktur des $ZnCl_2.1/2HCl.H_2O$. H. FOLLNER, 1970. *Acta Cryst.*, B26, 1544-1547.

$ZnCl_2.\frac{1}{2}HCl.H_2O$ F.W. = 172.51

Orthorhombic, *a* = 9.26±5, *b* = 22.90±12, *c* = 8.91±5 Å, [*U* = 1889.4 $Å^3$], *Z* = 16, [D_x = 1.461].

Space group *Fdd*2 (C_{2v}^{19})

Atomic positions

			x	*y*	*z*	*B*($Å^2$)
16 Zn	in	16(*b*)	0.0143±8	0.0807±3	0	1.39±23
16 Cl(1)	in	16(*b*)	0.2434±19	0.0904±9	0.094±3	2.16±38
16 Cl(2)	in	16(*b*)	-0.1506±24	0.0750±8	0.179±3	2.26±35
8 Cl(3)	in	8(*a*)	0	0	-0.153±4	2.23±54
16 H_2O	in	16(*b*)	0.088 ± 7	0.037 ±2	0.468±7	2.9 ±1.1

Interatomic distances (Å) *and angles*

Zn - Cl(1)	2.29±2		Cl(1) - Zn - Cl(2)	112.4±1.0°
Zn - Cl(2)	2.21±3		Cl(1) - Zn - Cl(3)	110.3±6
Zn - Cl(3)	2.30±2		Cl(1) - Zn - Cl(1')	102.9±8
Zn - Cl(1')	2.29±2		Cl(2) - Zn - Cl(3)	109.9±8
H_2O - H_2O	2.35±9		Cl(2) - Zn - Cl(1')	115.2±8
H_2O - Cl	3.83±7, 3.60±7		Cl(3) - Zn - Cl(Cl')	105.7±10

Discussion of the structure

The chlorine atoms are in a hexagonally close-packed array with zinc atoms in tetrahedral site such as to form $Zn_2Cl_5^-$ groups by sharing a chlorine atom. The water molecules form pairs which are related by a center of symmetry and hydrogen bond to the chloride ions.

Details of analysis

504 independent reflexions were measured from integrated Weissenberg photographs taken with Ag radiation. Least-squares refinement gave an *R* value of 0.097.

II. Die Kristallstruktur des $ZnCl_2$.1 1/3 H_2O. H. FOLLNER and B. BREHLER, 1970. *Acta Cryst.*, B26, 1679-1682.

$ZnCl_2 . 4/3\ H_2O$ F.W. = 160.28

Orthorhombic, a = 6.29 , b = 12.82 , c = 15.94 Å, U = 1285.4 Å^3, Z = 12, [D_x = 3.314].

Space group *Pbca* (D_{2h}^{15})

Atomic positions

			x	*y*	*z*	*B*(Å^2)
4 Zn(1)	in	4(*a*)	0	0	0	1.48±18
8 Zn(2)	in	8(*c*)	0.0457±6	0.2251±4	0.1810±2	0.92±18
8 Cl(1)	in	8(*c*)	0.0319±12	0.4000±8	0.1796±5	0.12±25
8 Cl(2)	in	8(*c*)	0.1868±11	0.1636±8	0.0623±5	2.00±23
8 Cl(3)	in	8(*c*)	0.1972±13	0.1628±8	0.3011±4	1.68±26
8 H_2O(1)	in	8(*c*)	-0.0429±32	-0.0586±16	0.1164±14	2.32±63
8 H_2O(2)	in	8(*c*)	-0.2895±32	0.0675±15	0.0122±10	0.73±52

Interatomic distances (Å) *and angles*

Zn(2)	-	Cl	2.23±1, 2.24±2, 2.28±1, 2.35±1
Zn(1)	-	Cl(2)	2.60±1 (×2)
Zn(1)	-	H_2O(1)	2.02±1 (×2)
Zn(1)	-	H_2O(2)	2.03±1 (×2)

Cl(1) - Zn(2) - Cl(2)	111.1 ±4					Cl(2) - Zn(2) - Cl(3)	114.9±5°		
Cl(1) - Zn(2) - Cl(3)	112.0 ±4					Cl(2) - Zn(2) - Cl(4)	110.7±4		
Cl(1) - Zn(2) - Cl(4)	107.7 ±4					Cl(3) - Zn(2) - Cl(4)	99.7±4		

H_2O(1) - Zn(1) - H_2O(2)	87.2 ±8
H_2O(1) - Zn(1) - Cl(2)	90.5 ±7
H_2O(2) - Zn(1) - Cl(2)	91.4 ±6

Discussion of the structure

The structure has nearly close packed double layers of chlorine atoms normal to (010) with Zn(2) in tetrahedral sites forming chains by sharing one corner of each $ZnCl_4$ group. Zn(1) lies in octahedrally coordinated sites between the double layers sharing axial chlorine atoms with the layers. The four water molecules are bonded to Zn(1) only, completing the octahedron.

Details of analysis

Data were taken from single crystals in capillaries with Ag radiation and the structure was solved by direct methods. 333 reflexions, least-squares refinement, *R* = 0.065.

ZIRCONIUM TETRACHLORIDE

I. Die Kristallstruktur von Zirkoniumtetrachlorid. B. KREBS, 1970. *Z. anorg. Chem.*, 378, 263-272.

$ZrCl_4$ F.W. = 233.03

Monoclinic, a = 6.361±4, b = 7.407±4, c = 6.256±4 Å, β = 109.30±4°, U = 278.2 Å^3, D_m = 2.80, Z = 2, D_x = 2.78.

Space group *P*2/*c* (C_{2h}^6)

Atomic positions

			x	*y*	*z*	$B(Å^2)$
2 Zr	in	2(*f*)	0	0.1641±1	1/4	2.37
4 Cl(1)	in	4(*g*)	0.2263±3	0.1076±2	-0.0022±3	2.83
4 Cl(2)	in	4(*g*)	-0.2552±3	0.3629±2	0.0205±3	3.79

Anisotropic temperature factors also given.

Interatomic distances (Å) *and bond angles*

Zr - Cl(2)	2.308±1	(×2)
Zr - Cl(1)	2.498±2	(×2)
Zr - Cl(1)*	2.655±2	(×2)

Angles in degrees subtended at Zr by

	Cl(2)	Cl(1)	Cl(1)'	Cl(1)*	Cl(1)*'
Cl(2)	100.7±1	93.8±1	98.5±1	168.9±1	89.2±1
Cl(1)			160.7±1	79.5±1	85.8±1
Cl(1)*					81.5 1

Discussion

Distorted $ZrCl_6$ octahedra are linked along the *c* axis through asymmetrical double Cl bridges. Alternatively, the structure can be thought of as composed of very distorted $ZrCl_4$ tetrahedra with two long Zr-Cl bonds to adjacent tetrahedra along *c*.

Details of analysis

Crystals grown by sublimation and mounted in dry Lindemann capillaries. Lattice parameters from photographs and refined from single crystal diffractometer measurements. Intensities measured on automatic diffractometer using $\theta - 2\theta$ scan [and MoK_α radiation?]. Corrected for Lorentz and polarisation effects but not absorption ($\mu = 3.75$ mm^{-1}, crystal size = 0.02 × 0.04 × 0.2 mm^3). Structure solved from Patterson function and refined by least squares to give $R = 0.050$ for 614 reflexions listed.

GADOLINIUM SESQUICHLORIDE

I. Gadolinium sesquichloride, an unusual example of metal-metal bonding. D.A. LOKKEN and J.D. CORBETT, 1970. *J. Amer. Chem. Soc.*, 92, 1799-1800.

Gd_2Cl_3 F.W. = 440.86

Monoclinic, $a = 15.237±4$, $b = 3.896±1$, $c = 10.179±3$ Å, $\beta = 117.66±3°$, $U = 535.2$ Å^3, $D_m = 5.14±29$, $Z = 4$, $D_x = 5.21$.

Space group *Cm* (C_s^3)

Atomic positions All atoms in 2(*a*)

	x	*y*	*z*	$B(Å^2)$
Gd(1)	0	0	0	0.78±5
Gd(2)	0.2737±3	0	0.7568±4	0.72±5
Gd(3)	0.5463±3	0	0.3217±4	0.64±5
Gd(4)	0.8207±1	0	0.1029±2	0.60±4
Cl(5)	0.1786±16	0	0.9336±22	1.41±28
Cl(6)	0.0712±12	0	0.5642±17	0.70±20
Cl(7)	0.4582±14	0	0.7794±20	1.15±26
Cl(8)	0.3590±13	0	0.3169±18	0.80±22
Cl(9)	0.7532±15	0	0.5390±23	1.42±28
Cl(10)	0.6405±12	0	0.1682±16	0.53±18

Discussion

The gadolinium atoms are arranged at the corners of octahedra which are linked by a shared edge into a chain along *b*. (Gd–Gd = 3.35 Å on shared edge, 3.74 – 3.90 Å on unshared edges). The chlorine atoms lie above the faces of the octahedra (Gd – Cl = 2.71 – 2.88 Å) and serve to link the chains into sheets.

Details of analysis

Intensities measured with automatic four-circle diffractometer using MoK_α radiation and corrected for absorption. Crystal was a pentagonal prism 0.07 × 0.48 mm. Structure solved from Patterson and electron-density maps and refined by least squares to give *R* = 0.052 for about 700 reflexions [not listed] using anisotropic temperature factors [not given].

AMERICIUM TRICHLORIDE

I. Refinement of the crystal structure of $AmCl_3$. J.H. BURNS and J.R. PETERSON, 1970. *Acta Cryst.*, B26, 1885-1887.

$AmCl_3$

Hexagonal, *a* = 7.382±1, *c* = 4.214±1 Å, [*U* = 198.87 $Å^3$], *Z* = 2.

Space group $P6_3/m$ (C^2_{6h})

Atomic parameters

2 Am in 2(*c*) and 6 Cl in 6(*h*) with *x* = 0.3877±4, *y* = 0.3019±4

Anisotropic thermal parameters are given.

Interatomic distances (Å)

Am–Cl (6×) 2.874±2, (3×) 2.915±2

Discussion of the structure

Isotypic with UCl (1).

Details of analysis

1490 reflexions measured with diffractometer (MoK_α). Absorption corrections. Refined by least squares to an *R* value of 0.035.

1. *Structure Reports*, 11, 278.

PHOSPHORUS PENTABROMIDE

I. Refinement of the crystal structure of phosphorus pentabromide, PBr_5. W. GABES and K. OLIE, 1970. *Acta Cryst.*, B26, 443-444.

PBr_5 F.W. = 430.49

Orthorhombic, *a* = 5.663±3, *b* = 17.031±5, *c* = 8.247±5 Å, [*U* = 795.40 $Å^3$], Z = 4, [D_x = 3.596].

Space group *Pbcm* (D^{11}_{2h}) from previous studies (1).

Atomic positions

			x	*y*	*z*
4 P	in	4(*d*)	0.036 ±2	0.1340±6	1/4
8 Br(1)	in	8(*e*)	0.2591±6	0.1274±2	0.0384±5
4 Br(2)	in	4(*d*)	–0.2093±9	0.0390±3	1/4
4 Br(3)	in	4(*d*)	–0.1564±9	0.2440±3	1/4
4 Br^-	in	4(*d*)	0.603 ±1	0.4050±3	1/4

Anisotropic temperature factors given.

Interatomic distances (Å) *and angles*

P	-	Br(1)	2.16±2	Br(1) -	P -	Br(2)	110±1
P	-	Br(2)	2.13±3	Br(1) -	P -	Br(1)	108±2
P	-	Br(3)	2.17±3	Br(1) -	P -	Br(3)	110±1
				Br(2) -	P -	Br(3)	109±2

Discussion of the structure

Structure consists of PBr_4^+ tetrahedra and Br^- as previously reported (1).

Details of analysis

607 reflexions collected with diffractometer, taken with Cu radiation. Refined by least squares to *R* of 0.124. Absorption correction applied.

1. *Structure Reports*, 9, 156.

POTASSIUM TETRAHYDROGEN PENTAFLUORIDE

I. The structure of potassium tetrahydrogen pentafluoride. B.A. COYLE, L.W. SCHROEDER and J.A. IBERS, 1970. *J. Solid State Chem.*, 1, 386-393.

KH_4F_5 F.W. = 138.1

Tetragonal (Scheelite type, 1), a = 6.384±3, c = 13.227±7 Å, U = 539.1 Å^3, Z = 4, D_x = 1.70.

Space group $I4_1/a$ (C_{4h}^6) systematic absences

Atomic positions (origin at $\bar{1}$)

			x	*y*	*z*
4 K	in	4(*a*)	0	1/4	1/8
4 F(1)	in	4(*b*)	0	1/4	5/8
16 F(2)	in	16(*f*)	0.1837±4	0.0249±4	0.2805±2
16 H(2)	in	16(*f*)	0.13 ±1	0.45 ±1	-0.309 ±3

Anisotropic temperature factor given in I for all atoms except H.

Interatomic distances (Å)

$H_4F_5^-$ ion (crystallographic point symmetry S_4)

F(2) -	H		0.70 ±8	
F(2) -	F(1)		2.453±2	
F(1) -	H		1.76 ±8	(×4)
F(1) -	H -	F(2)	176 ±8°	
F(2) -	F(1) -	F(2)	118.7 ±1°	(×2)
			105.0 ±1°	(×4)
H -	F(1) -	H	120 ±2°	(×2)
			104 ±3°	(×4)

Eightfold coordination around K (crystallographic site symmetry S_4)

K -	F(2)	2.770±3	(×4)
K -	F(2)	2.776±2	(×4)

Discussion

The crystal contains K^+ and $H_4F_5^-$ ions, the latter being formed by the tetrahedral attachment through hydrogen bonds of four HF molecules to a central F atom.

Details of analysis

Crystal prepared by sublimation in Kel-F tube. Lattice parameters refined by least squares from single-crystal diffractometer measurements with MoK_α radiation (λ = 0.7093 Å). 243 intensities measured from a rather large [size not given] crystal on a single-crystal diffractometer using Zr filtered Mo radiation. Lorentz, polarisation, extinction and absorption corrections applied. Structure solved by inspection and difference Fourier syntheses and refined by least squares to give R_2 = 0.069 for 221 observed reflexions listed.

1. *Strukturbericht*, 1, 347.

MAGNESIUM NITRIDE FLUORIDES

I. Magnesium nitride fluorides. S. ANDERSSON, 1970. *J. Solid State Chem.*, 1, 306-309.

TRIMAGNESIUM NITRIDE TRIFLUORIDE

Mg_3NF_3 F.W. = 143.9

Cubic, a = 4.216 Å, U = 74.94 Å^3, D_m = 3.16, Z = 1, D_x = 3.18.

Space group $Pm3m$ (O_h^1)

Atomic positions

			site symmetry
Mg	in	3(c)	T_d
F	in	3(d)	O_h
N	in	1(b)	O_h

Interatomic distances (Å)

Mg	-	F	2.108	(×6)
Mg	-	N	2.108	(×6)

DIMAGNESIUM NITRIDE FLUORIDE

Mg_2NF F.W. = 81.6

Tetragonal, a = 4.186, c = 10.042 Å, U = 176.0Å^3, D_m = 3.05, Z = 4, D_x = 3.07.

Space group $I4_1/amd$ (D_{4h}^{19})

Atomic positions

				site symmetry
Mg	in	8(e)	z = 0.1595	C_{2v}
N	in	4(b)		T_d
F	in	4(a)		T_d

Interatomic distances (Å)

Mg	-	F			2.121	(×2)
Mg	-	N			2.121	(×2)
Mg	-	N'			2.164	
F	-	Mg	-	F	161.2°	
N	-	Mg	-	N	161.2	
F	-	Mg	-	N	88.5	(×4)
N	-	Mg	-	N'	99.4	(×2)
F	-	Mg	-	N'	99.4	(×2)

Discussion

Both structures are based on cubic close packing of anions with Mg in the octahedral holes. In Mg_3NF_3 some cation sites are vacant whereas in Mg_2NF the Mg ions are displaced from the centre of the octahedron to give square pyramidal coordination. At high pressure Mg_2NF appears to be isostructural with MgO.

Details of analysis

Intensities measured by weighing cut out powder diffractometer traces (CuK_α radiation). Structures refined by least squares to give R = 0.022 for Mg_3NF_3 for 19 reflexions listed. Random distribution of N and F atoms gave R = 0.06. For Mg_2NF, R = 0.044 for 41 reflexions listed. Interchanging the positions of N and F gave R = 0.095.

MOLYBDENUM NITRIDE TRICHLORIDE

I. Die Kristallstruktur von Molybdännitridchlorid, $MoNCl_3$. J. STRÄHLE, 1970. *Z. anorg. Chem.*, 375, 238–253.

$MoNCl_3$ F.W. = 214.3

Triclinic a = 9.14±1, b = 7.67±1, c = 8.15±1Å, α = 108.8±3, β = 99.3±3, γ - 108.6±3°, U = 490 Å^3, Z = 4, D_x = 2.93.

Space group $P\bar{1}$ (C_i^1) final structure

Atomic positions

All atoms in 2(i)

	x	y	z	B(Å^2)
Mo(1)	0.22568±8	0.51635±11	0.55554±13	0.90±3
Mo(2)	0.36395±8	0.36237±11	0.13369±13	0.91±3
Cl(1)	0.1179 ±3	0.1755 ±4	0.4546 ±5	2.23±8
Cl(2)	0.0619 ±3	0.5840 ±4	0.7349 ±4	2.02±8
Cl(3)	0.2996 ±3	0.8110 ±4	0.5188 ±5	2.25±8
Cl(4)	0.3393 ±3	0.0484 ±4	0.1021 ±5	2.15±8
Cl(5)	0.1345 ±3	0.2741 ±5	-0.0788 ±5	2.54±9
Cl(6)	0.4909 ±3	0.6928 ±3	0.1490 ±4	1.74±7
N (1)	0.3900 ±9	0.5649 ±12	0.7082 ±13	1.62±24
N (2)	0.3070 ±9	0.4320 ±11	0.3210 ±12	1.18±21

Anistropic temperature factors given for Mo and Cl atoms.

Interatomic distances (Å)

Mo(1)	-	N(1)	1.639
Mo(1)	-	N(2)	2.144
Mo(1)	-	Cl(3)	2.276
Mo(1)	-	Cl(1)	2.278
Mo(1)	-	Cl(2)	2.317
Mo(1)	-	Cl(2)	2.939

Angles in degrees subtended at Mo(1) by

	N(2)	Cl(3)	Cl(1)	Cl(2)	Cl(2)
N(1)	97.8	99.4	98.7	101.4	176.6
N(2)		80.7	78.7	160.9	78.9
Cl(3)			154.1	96.1	81.1
Cl(1)				98.3	79.7
Cl(2)					81.9

Mo(2)	-	N(2)	1.672
Mo(2)	-	N(1)	2.200
Mo(2)	-	Cl(5)	2.242
Mo(2)	-	Cl(4)	2.270
Mo(2)	-	Cl(6)	2.379
Mo(2)	-	Cl(6)'	2.823

Angles in degrees subtended at Mo(2) by

	N(1)	Cl(5)	Cl(4)	Cl(6)	Cl(6)'
N(2)	92.2	100.2	100.0	95.5	170.0
N(1)		167.6	81.1	81.2	80.7
Cl(5)			97.5	96.6	86.9
Cl(4)				156.8	86.0
Cl(6)					76.5

Mo(1)	-	N(1)	-	Mo(2)	167.3°
Mo(1)	-	N(2)	-	Mo(2)'	178.1°

σ(Mo-Cl) = 0.003; σ(Mo-N) = 0.010Å; σ(angles) = 0.5°

Discussion

Mo and N atoms alternate around an almost square eight membered ring which encircles a center of symmetry. The Cl atoms complete the sixfold coordination around Mo and also form asymmetric bridges between one tetramer and the next to form sheets in the *ac* plane.

Details of analysis

Lattice parameters from powder diagram calibrated with α-quartz [assumed values of *a* and *c* not given]. Intensities measured photometrically from integrated equi-inclination Weissenberg photographs with l = 0 to 5 and precession photographs with h = 0 and 1 taken with MoK_α radiation, corrected for Lorentz and polarisation but not absorption (μ = 4.06 mm^{-1}, crystal size 0.4 × 0.2 × 0.07 mm^3). Structure solved by Patterson and electron-density functions and refined by least squares to give R_1 = 0.062 for 1596 reflexions (225 *hk*0 reflexions listed).

RARE EARTH CHALCOGENIDE FLUORIDES

I. Über Chalkogenidfluoride einiger Seltener Erden. R. SCHMID and H. HAHN, 1970. *Z. anorg. Chem.*, 373, 168-175.

LaSF F.W. = 190.0

CeSF F.W. = 191.2

EuSF F.W. = 203.0

Tetragonal (*E*01 type 1), *Z* = 2.

	a (Å)	*c* (Å)	*U* (Å^3)	D_m	D_x	*u*	*v*
LaSF	4.02_4	6.97_9	113.0	5.5_4	5.58_1	0.180	0.626
CeSF	4.01_0	6.95_1	111.8	5.6_2	5.67_8	0.180	0.615
EuSF	3.87_4	6.73_5			6.66_9		

See 2 for results for other members of the series.

Interatomic distances (Å)

	M-F (×4)	M-S	M-S (×4)
LaSF	2.37	3.11	3.15
CeSF	2.36	3.02	3.18

Details of analysis

Prepared from binary salts heated *in vacuo* at 500-600°C. Powder diffractometer. Refined by trial and error. About forty reflexions listed for each of LaSF and CeSF.

1. *Strukturbericht*, 2, 45.
2. CH. DRAGON and F. THERET, 1969. *C.R. Acad. Sci. Paris*, C268, 1867.

WOLFRAM(VI) SULPHIDE TETRAHALIDES

I. Crystal and molecular structure of tungsten(VI) sulphide tetrachloride and tungsten(VI) sulphide tetrabromide. M.G.B. DREW and R. MANDYCZEWSKY, 1970. *J. Chem. Soc. Lond.*, A, 2815-2818.

WOLFRAM SULPHIDE TETRACHLORIDE

$WSCl_4$ F.W. = 357.7

Triclinic *a* = 8.278±7, *b* = 6.092±6, *c* = 6.923±9 Å, α = 94.04±12*, β = 104.30±7, γ = 68.76±16°*, *U* = 315.2 Å^3, *Z* = 2, D_x = 3.76 [*α and γ given incorrectly as 85.96 and 111.24° respectively in I].

Space group $P\bar{1}$ (C_i^1)

Atomic positions All atoms in 2(*i*)

	x	*y*	*z*
W	−0.2173±2	0.1353±2	0.2467±2
Cl(1)	0.0823±13	0.1162±16	0.3475±14
Cl(2)	−0.2865±13	0.4502±17	0.4478±15
Cl(3)	−0.4701±14	0.0807±21	0.2646±16
Cl(4)	−0.1037±13	−0.2493±16	0.1564±13
S	−0.2933±14	0.3079±17	−0.0315±14

Anisotropic temperature factors given.

WOLFRAM SULPHIDE TETRABROMIDE

$WSBr_4$ F.W. = 528.5

Monoclinic a = 9.642±10, b = 6.024±7, c = 12.776±14 Å, β = 109.17±9, U = 700.9 Å^3, Z = 4, D_x = 5.06.

Space group $P2_1/c$ (C^5_{2h}) systematic absences

Atomic positions

	x	*y*	*z*
W	−0.2346±4	0.0504±7	−0.0460±3
Br(1)	−0.0041±10	0.2410±15	0.0786±8
Br(2)	−0.1943±10	0.2349±16	−0.2041±8
Br(3)	−0.3923±10	−0.2165±15	−0.1715±8
Br(4)	−0.2073±10	−0.2133±15	0.1068±8
S	−0.3901±23	0.2603±38	−0.0183±18

Anisotropic temperature factors given.

Interatomic distances (Å)

	$WSCl_4$	$WSBr_4$
W - S	2.10	2.08
W - X(1)	2.37	2.54
W - X(2)	2.27	2.45
W - X(3)	2.27	2.42
W - X(4)	2.29	2.46
W - X(1)'	3.05	3.03
S - W - X(1)	99.7	98.7
S - W - X(2)	100.3	100.6
S - W - X(3)	102.4	100.6
S - W - X(4)	100.5	98.8
S - W - X(1)'	176.3	177.1
X(1) - W - X(2)	87.5	88.2
X(1) - W - X(3)	158.0	160.7
X(1) - W - X(4)	86.2	87.2
X(1) - W - X(1)'	76.6	78.4
X(2) - W - X(3)	88.9	89.5
X(2) - W - X(4)	158.9	160.6
X(2) - W - X(1)'	79.9	79.4
X(3) - W - X(4)	89.5	88.7
X(3) - W - X(1)'	81.3	82.3
X(4) - W - X(1)'	79.0	81.2

σ(distances) = 0.01 Å, σ(angle) = 0.4°.

Discussion

Both the structures consist of centrosymmetric dimers in which two *X*(1) atoms form asymmetric bridges. Each W atom is surrounded by a nearly regular octahedron of ligands, but is displaced about 0.4 Å away from one of the bridging halogens, *X*(1)', towards S.

Details of analysis

Intensities measured by stationary crystal–stationary counter method on manual diffractometer with MoK_α radiation and corrected for [Lorentz, polarisation and] absorption effects (WCl_4 crystal size 0.032 × 0.05 × 0.024 mm³, μ = 21.2 mm⁻¹; WSBr crystal size 0.07 × 0.21 × 0.22 mm³, μ = 44.3 mm⁻¹). Structure determined by Patterson and Fourier techniques and refined by least squares to give R = 0.089 for 844 significant [observed] reflexions for WCl_4 and R = 0.121 for 530 significant [observed] reflexions for $WSBr_4$. Structure factors listed in 1.

1. *Supplementary publications SUP* 20037.

TETRAPHOSPHORUS TRISELENODIIODIDE

I. The crystal and molecular structure of tetraphosphorus triselenodiiodide, G.J. PENNEY and G.M. SHELDRICK, 1970. *Acta Cryst.*, B26, 2092-2096.

$P_4Se_3I_2$ F.W. = 693.14

Orthorhombic, a = 9.783±9, b = 16.320±22, c = 6.800±6 Å, U = 1086Å³, D_m = 3.76, Z = 4, D_x = 3.76.

Space group *Pnma* (D_{2h}^{16}) confirmed by structure analysis

Atomic positions

			x	y	z	B(Å²)
8 I	in	8(*d*)	0.3426±8	0.0392±3	0.2745±11	
8 Se(1)	in	8(*d*)	0.3920±10	0.1416±5	0.7549±12	
4 Se(2)	in	4(*c*)	0.1694±15	1/4	0.4598±17	
8 P(1)	in	8(*d*)	0.484 ±2	0.1476±12	0.445 ±3	0.0033±6
4 P(2)	in	4(*c*)	0.250 ±4	1/4	0.765 ±5	0.0058±9
4 P(3)	in	4(*c*)	0.369 ±4	1/4	0.308 ±4	0.0035±7

Anisotropic thermal parameters are given for I, Se(1) and Se(2).

Interatomic distances (Å) *and angles*

I	-	P(1)	2.47±2
P(1)	-	Se(1)	2.29±2
Se(1)	-	P(2)	2.25±3
P(2)	-	Se(2)	2.22±4
Se(2)	-	P(3)	2.21±4
P(3)	-	P(1)	2.22±3

P(3)	-	P(1)	-	Se(1)	102.8±11
I	-	P(1)	-	Se(1)	101.4±8
P(3)		P(1)	-	I	94.6±10
Se(1)	-	P(2)	-	Se(2)	101.0±12
Se(1)		P(2)	-	Se(1)	103.6±13
P(1)	-	P(2)	-	P(1)	97.8±15
P(1)	-	Se(1)	-	P(2)	103.6±11
P(1)	-	Se(2)	-	P(3)	97.1±15

Materials

CS_2 and P_4Se_3 and iodine in solution for 3 days at 0°C.

Discussion of the structure

Molecular units of $P_4Se_3I_2$ contain a mirror plane as shown in Fig. 27. Molecules are approximately hexagonally close-packed with layers perpendicular to [010].

Details of analysis

Visual estimation of 337 reflexions recorded on Weissenberg photographs. Absorption correction. Least-squares refinement lead to an R value of 0.112.

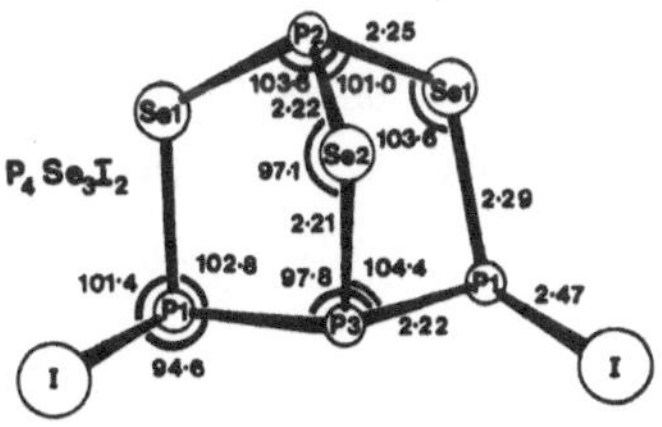

Fig. 27. The molecular dimensions of $P_4Se_3I_2$.

RUBIDIUM NICKEL FLUORIDE

I. The crystal structure of hexagonal $RbNiF_3$ (6H). R.J. ARNOTT and J.M. LONGO, 1970. *J. Solid State Chem.*, 2, 416–420.

II. Note on the crystal structure of $RbNiF_3$. D. BABEL, 1970. *Ibid.*, 2, 582–584.

III. Structures of ferrimagnetic fluorides of ABF_3 type. I $RbNiF_3$. J.E. WEIDENBORNER and A.L. BEDNOWITZ, 1970. *Acta Cryst.*, B26, 1464–1468.

$RbNiF_3$ F.W. = 201.17

Hexagonal, I, a = 5.843, c = 14.309 Å, U = 423.1 Å^3, Z = 6, D_x = 4.74.
II, a = 5.840±2, c = 14.308±4 Å, U = 422.6 Å^3, Z = 6, D_x = 4.742.

Space group $P6_3/mmc$ (D^4_{6h})

Atomic positions

				I	II	III
2 Rb(1)	in	2(b)				
4 Rb(2)	in	4(f)	z	= 0.0954±2	0.09535±5	0.0948±1
2 Ni(1)	in	2(a)				
4 Ni(2)	in	4(f)	z	= 0.8462±5	0.8443 ±7	0.8453±1
6 F(1)	in	6(h)	x	= 0.517 ±1	0.5155 ±37	0.5172±7
12 F(2)	in	12(k)	x	= 0.830 ±3	0.8351 ±27	0.8328±7
			z	= 0.081 ±1	0.0742 ±13	0.0798±3

Interatomic distances (Å)

			I	II	III
Ni(1)	- F(2)	(6×)	2.074±15	1.978±15	2.041±8
Ni(2)	- F(1)	(3×)	2.047±6	2.065±16	2.036±7
Ni(2)	- F(2)	(3×)	1.954±15	2.040±15	1.993±8
Rb(1)	- F(1)	(6×)	2.927±3	2.926±10	2.925±7
Rb(1)	- F(2)	(6×)	2.968±15	3.019±17	2.965±8
Rb(2)	- F(1)	(3×)	2.889±4	2.880±12	2.896±7
Rb(2)	- F(2)	(6×)	2.929±9	2.937±8	2.928±1
Rb(2)	- F(2)	(3×)	3.017±15	2.965±7	3.011±8

Materials

RbF and NiF_2 grown from the melt.

Discussion of the structure

The octahedra about Ni(1) form pairs of face sharing Ni_2F_9 groups which link Ni(2)F_6 octahedra by edge sharing. Rb ions are each coordinated to twelve fluorine atoms. The structure is not appreciably different at -150°C (I).

Details of analysis

I. Integrated intensities measured on scanning powder diffractometer (monochromated CuK_α radiation) and refined by least squares to give R = 0.058 (intensities) for 35 reflexions listed.

II. Intensities measured from $h0l$ precession photographs and rotation photographs of layers with l = 0 to 5 taken with CuK_α radiation. Structure refined by least squares to give R = 0.077 for 89 reflexions listed.

III. 327 measured reflexions using a diffractometer and MoK_α radiation. Refined by least-squares method to R = 0.042.

DICAESIUM SODIUM INDIUM HEXAFLUORIDE

I. Die Kristallstruktur von Cs_2NaInF_6. S. SCHNEIDER and R. HOPPE, 1970. *Z. anorg. Chem.*, 376, 277–281.

Cs_2NaInF_6 F.W. = 457.53

Cubic (K_2NaCrF_6 type 1), a = 8.905 Å, U = 706.2 Å^3, D_m = 4.74, Z = 4, D_x = 4.87.

Space group $Fm3m$ (O_h^5)

Atomic positions

8 Cs	in	8(c)	
4 Na	in	4(b)	
4 In	in	4(a)	
24 F	in	24(e)	z = 0.229

Interatomic distances (Å)

Cs - F	3.154	(×12)
Na - F	2.413	(× 6)
In - F	2.039	(× 6)

Details of analysis

Lattice parameters from Guinier powder photographs [no wavelengths given]. Intensities measured photometrically from Weissenberg and precession photographs of layers with $k = 2n-h$ for n = 0 to 7 using MoK_α radiation. Structure refined by least squares to give R = 0.111.

1. *Structure Reports*, 26, 309.

CAESIUM TRICHLOROSTANNATE (II)

$CsSnCl_3$ F.W. = 358.0

I. Crystal structure and phase transition of caesium trichlorostannate (II). F.R. POULSEN and S.E. RASMUSSEN, 1970. *Acta Chem. Scand.*, 24, 150-156.

Monoclinic, a = 16.10, b = 7.425, c = 5.748 Å, β = 93.2°, U = 686.1 Å^3, D_m = 3.45, Z = 4, D_x = 3.46.

Space group $P2_1/n$ (C_{2h}^5) systematic absences

Atomic positions

All atoms in 4(e) ($x, y, z; \bar{x}, \bar{y}, \bar{z}$; 1/2−$x$, 1/2+$y$, 1/2−$z$; 1/2+$x$, 1/2−$y$, 1/2+$z$.

	x	y	z
Cs	0.1534±1	0.5002±3	0.7504±3
Sn	0.3917±1	0.4713±3	0.2610±3
Cl(1)	0.2532±4	0.3134±9	0.2705±11
Cl(2)	0.4565±5	0.1660±11	0.2673±13
Cl(3)	0.3784±6	0.4763±12	0.8165±12

Anisotropic temperature factors given.

Interatomic distances (Å) *and bond angles*

Standard errors in distances = 0.01Å, in angles = 0.3°.

Environment of Sn

Sn - Cl(1)	2.52
Sn - Cl(2)	2.50
Sn - Cl(3)'	2.55
Sn - Cl(3)	3.21
Sn - Cl(1)"	3.45
Sn - Cl(2)"	3.77

Angles in degrees subtended at Sn by

	Cl(2)	Cl(3)'
Cl(1)	86.9	90.2
Cl(2)		92.3

Environment of Cs

Cs - Cl(2)	3.41	Cs - Cl(2)	3.59
Cs - Cl(1)	3.55	Cs - Cl(3)	3.63
Cs - Cl(2)	3.59	Cs - Cl(1)	3.85
Cs - Cl(3)	3.59	Cs - Cl(3)	3.94
Cs - Cl(1)	3.59	Cs - Cl(1)	4.14

Discussion

The crystal is composed of Cs^+ and pyramidal Sn Cl_3^- ions and is normally twinned on (100) plane. At 120°C the crystals pass through an irreversible phase transition to a form whose powder pattern can be partially indexed using two cubic cells (a_1 = 3.67, a_2 = 3.92Å).

Details of analysis

Prepared from CsCl and $SnCl_2.2H_2O$ in glycerol. Lattice parameters from Weissenberg, precession and retigraph photographs with Co and CuK_α radiation. Intensities measured from crystal (0.5 × 1.0 × 5.0 mm^3) using linear diffractometer and balanced filters (Mo radiation). Lorentz and polarisation corrections. Corrections also applied for twinning. Structure solved from three-dimensional Patterson and Fourier functions and refined by Fourier and least-squares methods to give R = 0.109 for 1139 reflexions listed.

CAESIUM MAGNESIUM CHLORIDE

$CsMgCl_3$ F.W. = 263.57

I. Single-crystal paramagnetic resonance studies of V(II), Mn(II) and Ni(II) in $CsMgCl_3$ and the crystal structure of $CsMgCl_3$. G.L. McPHERSON, T.J. KISTENMACHER and G.D. STUCKY, 1970. *J. Chem. Phys.*, 52, 815-824.

Hexagonal, a = 7.269±6, c = 6.187±5Å, U = 283.18Å^3, D_m = 3.07±3, Z = 2, D_x = 3.09.

Space group $P6_3/mmc$ (D_{6h}^4) from systematic absences and structure.

Atomic positions

			x	y	z
2 Cs	in	2(*d*)	1/3	2/3	3/4
2 Mg	in	2(*a*)	0	0	0
6 Cl	in	6(*h*)	0.1556	0.3112	1/4

Anisotropic temperature factors given.

Interatomic distances (Å) *and bond angles*

Six-fold coordination around Mg (crystallographic symmetry D_{3d})

Mg - Cl	2.496[±3]	
Cl - Mg - Cl	85.64°	(Cl atoms related by 3-fold axis)
Cl - Mg - Cl	94.36	(Cl atoms related by center of symmetry)
Mg - Cl - Mg	76.59°	

Environment of Cu (crystallographic symmetry D_{3h})

Cu - Cl	3.637 [±3]	(×6) in equatorial plane
Cu - Cl	3.818 [±3]	(×6) above and below plane

Discussion

$MgCl_6$ octahedra share faces to form columns running along the c axis of the crystal. The copper atoms lie between three such chains and are surrounded by twelve chlorine atoms. It is isotypic with $CsNiCl_3$ (1).

Details of analysis

Prepared by heating equimolar mixtures of CsCl and $MgCl_2$. Crystals grown by the vertical Bridgeman method. Lattice constants measured on Picker diffractometer. 994 measurements of the intensities of 481 reflexions made on diffractometer with Zr filtered Mo radiation from a cylindrically ground crystal, 0.34 mm. high by 0.15 mm. diameter. Lorentz, polarisation and absorption corrections. Assumed to be isostructural with $CsNiCl_3$ (1) and refined by least squares to give weighted R = 0.041. Structure factors given.

1. *Structure Reports*, 19, 332.

POTASSIUM ZINC TRIHALIDES

I. Die Kristallstruktur von $KZnBr_3.2H_2O$ und $KZnI_3.2H_2O$. R. HOLINSKI and B. BREHLER, 1970. *Acta Cryst.*, B26, 1915-1919.

$KZnBr_3.2H_2O$ F.W. = 380.20

$KZnI_3.2H_2O$ F.W. = 521.19

Orthorhombic: β-$KZnBr_3.2H_2O$, $a = 9.327_3$, $b = 13.067_4$, $c = 6.786_2$ Å, [U = 827.0 Å^3], $D_m = 3.00$, $Z = 4$, [$D_x = 3.054$].
$KZnI_3.2H_2O$, $a = 9.950_3$, $b = 13.726_4$, $c = 7.072_2$ Å, [U = 965.9 Å^3], $D_m = 3.55$, $Z = 4$, [$D_x = 3.585$].

Space group $P2_12_12_1$ (D_2^4)

Atomic positions

β-$KZnBr_3.2H_2O$

	x	y	z
Br(1)	0.4121±3	0.0129±1	0.8939±4
Br(2)	0.4193±3	0.3194±1	0.8794±4
Br(3)	0.0791±3	0.1750±2	0.0846±4
Zn	0.2785±3	0.1674±2	0.8661±4
$H_2O(1)$	0.2024±1	0.1626±1	0.5861±2
$H_2O(2)$	0.2873±2	0.9982±1	0.3853±3
K	0.2911±6	0.3407±5	0.3595±1

Anisotropic thermal parameters are given.

$KZnI_3.2H_2O$

	x	y	z	B(Å^2)
I(1)	0.4154±3	0.0063±3	0.9003±5	2.66±7
I(2)	0.4189±3	0.3223±3	0.8854±5	2.54±7
I(3)	0.0748±3	0.1688±3	0.0958±4	2.56±7
Zn	0.2812±5	0.1670±5	0.8727±8	1.83±9
$H_2O(1)$	0.2060±3	0.1580±2	0.6005±4	1.51±9
$H_2O(2)$	0.0280±4	0.0044±3	0.4092±7	3.7 ±8
K	0.2899±1	0.3377±1	0.3802±2	2.8 ±2

Interatomic distances (Å) *and bond angles*

	X=Br	X=I
Zn - X(1)	2.37	2.58
Zn - X(2)	2.38	2.53
Zn - X(3)	2.38	2.59
Zn - H_2O	2.03	2.07
7 K - X	3.41-3.77	3.63-3.93
K - $H_2O(1)$	2.91	3.03
K - $H_2O(2)$	2.78	2.81

	X=Br	X=I
X(1)- Zn - X(3)	113.2°	111.7°
X(1)- Zn - X(2)	114.5	115.7
X(2)- Zn - X(3)	111.8	113.5
H_2O - Zn - X(1)	103.4	101.8
H_2O - Zn - X(2)	104.7	106.2
H_2O - Zn - X(3)	105.4	106.3

Discussion of the structure

Nearly hexagonal closed-packed halogens with Zn in tetrahedrally coordinated sites as shown in Fig. 28.

Details of analysis

1110 reflexions of $KZnBr_3.2H_2O$ and 1169 of $KZnI_3.2H_2O$ recorded with scintillation counter and Ag radiation. Refined by least-squares methods to R = 0.064 and R = 0.084 respectively.

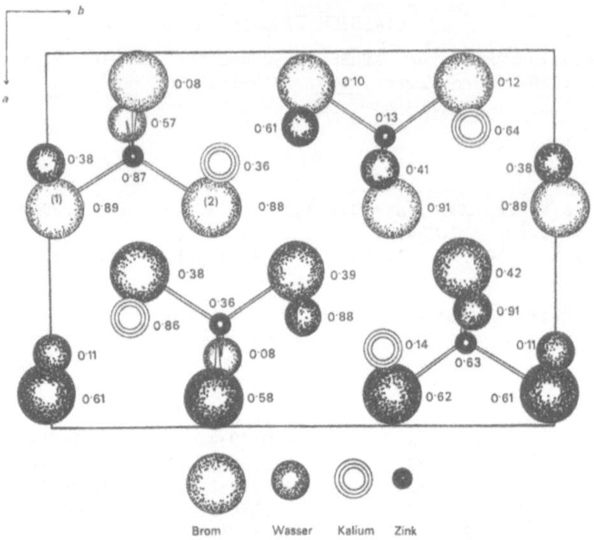

Fig. 28. The structure of $\beta KZnBr_3.2H_2O$ projected onto the *ab* plane. The approximate *y* coordinates are shown.

SILVER AND GOLD FLUORIDES

I. Neue Untersuchungen an Fluorkomplexen mit dreiwetigem Silber und Gold. R. HOPPE and R. HOMANN, 1970. *Z. anorg. Chem.*, 379, 193-198.

SODIUM TETRAFLUOROARGENTATE (III)
POTASSIUM TETRAFLUOROARGENTATE (III)
SODIUM TETRAFLUOROAURATE (III)
POTASSIUM TETRAFLUOROAURATE (III)
RUBIDIUM TETRAFLUOROAURATE (III)

Tetragonal, $KBrF_4$ Type (1), Z = 4.

	F.W.	*a* (Å)	*c* (Å)	*U* (Å^3)	D_x
$NaAgF_4$	206.85	5.54	10.56	324.1	4.23
$KAgF_4$	222.96	5.90	11.15	388.1	3.81
$NaAuF_4$	295.95	5.64	10.49	333.7	5.88
$KAuF_4$	312.06	5.99	11.38	408.3	5.07
$RbAuF_4$	358.43	6.18	11.85	452.6	5.25

Space Group I4/mcm (D_{4h}^{18})

Atomic positions

			x	*y*	*z*
Na, K, Rb	in	4(*a*)	0	0	1/4
Ag, Au	in	4(*d*)	0	1/2	0
F	in	16(*l*)	*x*	1/2 + *x*	*z*

[given at *x*, 1/2 + *y*, *z*, in I]

	x	*z*
$NaAgF_4$	0.17_1	0.12_7
$KAgF_4$	0.16_1	0.12_0
$NaAuF_4$	0.17_7	0.13_5
$KAuF_4$	0.16_7	0.12_4
$RbAuF_4$	0.16_2	0.11_9

Interatomic distances (Å)

	Ag,Au-F (×4)	Na,K,Rb-F (×8)
$NaAgF_4$	1.90*	2.43
$KAgF_4$	1.90*	2.65
$NaAuF_4$	2.00*	2.40
$KAuF_4$	2.00*	2.65
$RbAuF_4$	2.00*	2.79

* Assumed.

Details of analysis

Powder photographs, 10 lines listed for each compound, CuK_α radiation. Only reflexions from the subcell defined by the metals could be seen ($\sqrt{2}a_{sub} = a$ [not $2a_{sub} = a$ as given in I], $2c_{sub} = c$). The atomic parameters were derived by assuming suitable values for Ag-F and Au-F distances. The results gave qualitative agreement for the strong reflexions which are not sensitive to the fluorine positions.

1. *Structure Reports*, 21, 275.

TERNARY STRONTIUM FLUORIDES

I. Deux nouvelles structures difluorine et trifluorine. R. VON DER MÜHLL, D. DUMARA, J. RAVEZ and P. HAGENMULLER, 1970. *J. Solid State Chem.*, 2, 262-268.

STRONTIUM CHROMIUM (III) TETRAFLUORIDE

$SrCrF_4$ F.W. = 215.61

Tetragonal, a = 5.673+3, c = 10.920+6 Å, U = 351.4 Å^3, D_m = 4.11 5, Z = 4, D_x = 4.07.

Space group $I\bar{4}c2$ (D_{2d}^{10})

Atomis positions

			x	y	z	B(Å^2)
4 Sr	in	4(c)	0	0.5	0.25	0.67
4 Cr	in	4(b)	0	0	0	1.70
16 F	in	16(i)	0.3186	0.3186	0.1238	1.66

Interatomic distances (Å)

Sr	-	F	2.50	(×4)	Cr	-	F		1.99	(×4)
Sr	-	F	2.50	(×4)	F	-	Cr	- F	118°	(×4)
									94°	(×2)

DISTRONTIUM COPPER (II) HEXAFLUORIDE

Sr_2CuF_6 F.W. = 352.78

Tetragonal a = 5.71+3, c = 16.458+6 Å, U = 536.6 Å^3, D_m = 4.29+4, Z = 4, D_x = 4.36.

Space group $P\bar{4}b2$ (D_{2d}^{7})

Atomic positions

			x	y	z	B(Å^2)
4 Sr(1)	in	4(f)	0	0.5	0.162	0.86
4 Sr(2)	in	4(e)	0	0	0.337	0.86
2 Cu(1)	in	2(a)	0	0	0	0.86
2 Cu(2)	in	2(d)	0	0.5	0.5	0.86
8 F(1)	in	8(i)	0.357	0.323	0.075	1.40
8 F(2)	in	8(i)	0.235	0.259	0.251	1.20
8 F(3)	in	8(i)	0.323	0.142	0.422	1.40

Interatomic distances (Å)

Sr(1)	-	F(2)			2.42	(×2)
Sr(1)	-	F(1)			2.47	(×2)
Sr(1)	-	F(2)			2.57	(×2)
Sr(1)	-	F(1)			2.69	(×2)
Cu(1)	-	F(1)			1.79	(×4)
F(1)	-	Cu(1)	-	F(1)'	93°	(×2)
					118°	(×4)
Sr(2)	-	F(2)			2.45	(×2)
Sr(2)	-	F(3)			2.45	(×2)
Sr(2)	-	F(2)			2.49	(×2)
Sr(2)	-	F(3)			2.68	(×2)
Cu(2)	-	F(3)			1.82	(×4)
F(3)	-	Cu(2)	-	F(3)'	90°	(×4)
					171°	(×2)

Discussion

The structures of both compounds are derived from that of fluorite. In $SrCrF_4$ the Sr and Cr atoms alternate in planes along the *c* axis, in Sr_2CuF_6 there are two Sr planes followed by a Cu plane. The F atoms are displaced to give (distorted) tetrahedral coordination around Cr and Cu.

Details of analysis

Powder patterns given. $SrCrF_4$: intensities estimated visually from multiple-film Weissenberg photographs of layers with $h = 0$ to 3 taken with CuK_α radiation. Corrections for Lorentz and polarisation effects. Structure solved from Patterson function and refined by least squares to give $R = 0.084$ for 90 reflexions listed. Sr_2CuF_6: intensities measured with a planimeter from powder diffractometer trace. Structure solved by similarity with $SrCrF_4$ and refined by least squares to give $R = 0.057$ for 44 reflexions listed.

BARIUM MANGANESE FLUORIDE

I. Crystal structure of pyroelectric paramagnetic barium manganese fluoride, $BaMnF_4$. E.T. KEVE, S.C. ABRAHAMS and J.L. BERNSTEIN, 1970. *J. Chem. Phys.*, 51, 4928–4936.

$BaMnF_4$ F.W. = 268.3

Orthorhombic $a = 5.9845\pm3$, $b = 15.098\pm2$, $c = 4.2216\pm3$ Å, $U = 381.44$ Å^3, $D_m = 4.59\pm5$, $Z = 4$, $D_x = 4.63[4.66]$.

Space group $A2_1am$ (C_{2v}^{12}) Systematic absences piezoelectric effect and final structure.

Atomic positions

	x	*y*	*z*
Ba	0.4537	0.34383±5	0.5
Mn	0.0000±4	0.4160 ±1	0
F(1)	0.196±2	0.2982 ±6	0
F(2)	-0.275±3	0.3363 ±7	0
F(3)	0.337±3	0.4651 ±7	0
F(4)	0.016±3	0.4217 ±10	0.5

Ansiotropic temperature factors given.

Interatomic distances (Å)

Ba	-	F(1)	2.589±10	
Ba	-	F(2)	2.665±11	(×2)

Ba	-	F(1)	2.703+7	(×2)
Ba	-	F(4)	2.871+17	
Ba	-	F(3)	2.880+8	(×2)
Ba	-	F(2)	3.045+12	
Mn	-	F(2)	2.039+16	
Mn	-	F(3)	2.043+13	
Mn	-	F(4)	2.115+1	(×2)
Mn	-	F(1)	2.131+10	
Mn	-	F(3)	2.149+17	

Angles in degress subtended at Mn by

	F(3)	F(4)'	F(1)	F(3)
F(2)	97.7+6	93.5+5	87.2+5	164.0+5
F(3)		89.2+4	175.1+6	98.3+6
F(4)		173.0+6	90.5+4	86.8+5
F(1)				76.8+4

Discussion

The structure consists of MnF_6 octahedra sharing corners to form puckered sheets perpendicular to *b*. The Ba atoms lie between the sheets. The bridging F atoms show particularly large and anisotropic temperature factors which may be related to the ferroelectric switching observed in isomorphic compounds (but not in $BaMnF_4$). The assignment of the space group by 1 is wrong.

Details of analysis

Crystal grown from melt was ground to a sphere 0.119±3 mm radius and mounted inside a capillary. Lattice parameters measured on precission diffractometer [no calibration given]. Intensities measured on automatic diffractometer with MoK_α radiation and balanced filters, corrected for Lorentz, polarisation, absorption and extinction effects (μ = 14.05 mm^{-1}). Structure solved by Patterson and electron-density functions and refined by least squares to give R_2 = 0.062 for 455 reflexions listed. The absolute configuration was determined.

1. J.C. COUSSEINS and M. SAMOUËL, 1967. *C. R. Acad. Sci. Paris*, C265, 1121.

BARIUM COBALT FLUORIDE

I. Ferroelectric paraelastic paramagnetic barium cobalt fluoride, $BaCoF_4$, crystal structure. E.T. KEVE, S.C. ABRAHAMS and J.L. BERSTEIN,1970. *J. Chem. Phys.*, 53, 3279-3287.

$BaCoF_4$ F.W. = 272.3

Orthorhombic, *a* = 4.2102±3, *b* = 14.628±2, *c* = 5.8519±3 Å, *U* = 360.4 Å^3, D_m = 4.98, Z = 4, D_x = 5.034.

Space group *Cmc*2_1 (C^{12}_{2v}) [converted from *A*2_1*am*] from systematic absences, ferroelastic effect and similarity to $BaMnF_4$ (1).

Atomic positions

All atoms in 4(*a*)

	x	*y*	*z*
Ba	1/2	-0.35216±4	0.4605±2
Co	0	-0.41274±8	0
F(1)	0	-0.3018 ±4	0.1981±12
F(2)	0	-0.3336 ±6	0.2729±15
F(3)	0	-0.4718 ±5	0.3207±14
F(4)	1/2	-0.4206 ±8	0.0112±20

Anisotropic temperature factors given.

Interatomic distances (Å)

Ba	-	F(2)	2.635±5	(×2)
Ba	-	F(1)	2.647±7	
Ba	-	F(1)'	2.707±4	(×2)
Ba	-	F(4)	2.813±11	
Ba	-	F(3)	2.857±6	(×2)
Ba	-	F(2)'	3.041±9	
Ba	-	F(4)'	3.337±12	
Ba	-	F(4)"	3.375±12	
Co	-	F(2)	1.972±8	
Co	-	F(3)	1.989±7	(bridging)
Co	-	F(1)	1.994±7	
Co	-	F(3)'	2.066±7	(bridging)
Co	-	F(4)	2.109±1	(×2) (bridging)

Angles in degrees subtended at Co by

	F(3)	F(1)	F(3)'	F(4)
F(2)	94.1	89.6	168.8	93.3
F(3)		176.3	97.1	88.3
F(1)			79.2	91.5
F(3)'				87.1
F(4)'				172.8

σ(angles) = 0.3°

Co	-	F(3)	-	Co	146.6 4°
Co	-	F(4)	-	Co	172.8 6°

Discussion

The structure is derived from that of $BaMnF_4$ (1). The adjacent CoF_6 octahedra in the puckered sheets perpendicular to *b* are rotated in the opposite sense about the *a* axis. Ferroelectric switching results from the sense of rotation changing.

Details of analysis

Lattice parameters measured on a precision diffractometer [wavelength not given]. Intensities measured on single crystal diffractometer from switched crystals 0.37 × 0.36 × 0.42 mm^3 and 0.21 × 0.23 × 0.29 mm^3 [with MoK_α radiation] corrected for Lorentz, polarisation and absorption (μ = 15.93 mm^{-1}). Assumed isostructural with $BaMnF_4$, refined by least squares to give R_2 = 0.091 for 1261 reflexions listed. Absolute configuration of both crystals determined.

1. E.T. KEVE, S.C. ABRAHAMS and J.L. BERNSTEIN, 1969. *J. Chem. Phys.*, 51, 4928.

LITHIUM TETRAFLUOROYTTERBATE

$LiYbF_4$ F.W. = 255.97

I. Equilibrium reactions and crystal structure of lithium fluorolanthanate phases. R.E. THOMA, G.D. BRUNTON, R.A. PENNEMAN and T.K. KEENAN, 1970. *Inorg. Chem.*, 9, 1096-1101.

Tetragonal (Scheelite (*H*4) type, 1), *a* = 5.1335±2, *c* = 10.588±2 Å, *U* = 279.02 Å^3, *Z* = 4, D_x = 6.08.

Space group $I4_1/a$ (C^6_{4h}) from systematic absences

Atomic positions

			x	*y*	*z*
4 Yb	in	4(*b*)	0	1/4	5/8
4 Li	in	4(*a*)	0	1/4	1/8
16 F	in	16(*f*)	0.2166±6	0.4161±6	0.4564±3

Anisotropic temperature factors given.

Interatomic distances (Å)

Eightfold coordination around Yb (crystallographic site symmetry S_4)

Yb	-	F	2.217±3	(×4)
Yb	-	F'	2.270±3	(×4)

Fourfold coordination around Li (crystallographic site symmetry S_4)

L	-	F	1.894±3	(×4)
Li	-	F	2.871±4	(×4)

Details of analysis

Single crystals isolated from quenched LiF - YbF_3 mixtures ground into ellipsoids 0.184 × 0.184 × 0.292 mm^3. Intensities measured with 2θ scan on four-circle diffractometer with MoK_α ($\lambda K_{\alpha 1}$ = 0.70926 Å) and corrected for absorption (μ = 35.1 mm^{-1}). Lattice parameters measured on diffractometer from 30 high angle reflexions. Crystals assumed to be isostructural with scheelite (1) and structure refined by least squares to give R = 0.031 for 503 independent reflexions listed. Powder pattern also given.

1. *Strukturbericht*, 1, 347.

POTASSIUM PENTAFLUOROTELLURATE (IV)

I. Crystal structure of potassium pentafluorotellurate ($KTeF_5$). S.H. MASTIN, R.R. RYAN and L.B. ASPREY, 1970. *Inorg. Chem.*, 9, 2100-2103.

$KTeF_5$ F.W. = 261.8

Orthorhombic, a = 4.735±1, b = 9.209±2, c = 11.227±2 Å, U = 489.55 Å^3, D_m = 3.55, Z = 4, D_x = 3.54.

Space group *Pbcm* (C_{2h}^{11}) systematic absences and final structure

Atomic positions

	x	y	z
K	0.4973±4	0.25	0
Te	0.1124±2	0.05563±7	0.25
F(1)	0.3700±8	-0.0146 ±4	0.1270±3
F(2)	-0.0337±9	0.1844 ±4	0.1265±3
F(3)	0.3799±12	0.2037 ±5	0.25

Anisotropic temperature factors given.

Interatomic distances (Å)

K	-	F(1)	2.670±4	(×2)
K	-	F(2)	2.705±4	(×2)
K	-	F(1)	2.807±4	(×2)
K	-	F(3)	2.893±1	(×2)
K	-	F(2)	2.951±4	(×2)

TeF_5^- (Crystallographic symmetry C_s)

Te	-	F(3)	1.861±5	
Te	-	F(2)	1.951±4	(×2)
Te	-	F(1)	1.953±4	(×2)

Angles in degrees subtended at Te_2 by

	F(2)'	F(1)'
F(3)	78.2±2	79.5±2
F(2)	90.6±1	157.7±2
F(2)		85.4±1
F(1)		90.0±1

Discussion

This is a refinement of the structure first reported by 1.

Details of analysis

Lattice parameters and intensities measured on four-circle diffractometer from crystal 0.165 × 0.253 × 0.155 mm^3 using filtered MoK_α radiation (λ = 0.70926 Å). Lorentz, polarisation, absorption and extinction corrections (μ = 7.12 mm^{-1}). Structure of 1 refined by least squares to give R_2 = 0.025 for 1266 observed reflexions listed.

1. A.J. EDWARDS and M.A. MOUTY, 1969. *J. Chem. Soc. London*, A, 703.

POTASSIUM SODIUM THORIUM HEXAFLUORIDE

I. The crystal structure of $KNaThF_6$. G. BRUNTON, 1970. *Acta Cryst.*, B26, 1185-1186.

$KNaThF_6$ F.W. = 408.12

Hexagonal, a = 6.3073±2, c=7.8907±2 Å, [U = 271.83 Å^3], D_m = 4.985, Z = 2, [D_x = 4.987].

Space group $P\bar{3}$ (C^1_{3i}) P3 was not investigated.

Atomic parameters

2 Th	in	2(d)	with z = 0.1221±2,	β_{11} = 0.0034±4,	β_{33} = 0.0039±3
2 K	in	2(d)	with z = 0.608 ±2,	β_{11} = 0.014,	β_{33} = 0.004 ±2
2 Na	in	2(c)	with z = 0.236 ±3,	β = 0.014 ±3,	
6 F(1)	in	6(g)	with x = 0.104 ±3,	y = 0.381 ±3,	z = 0.322±2, β = 0.011±2
6 F(2)	in	6(g)	with x = 0.395 ±3,	y = 0.319 ±3,	z = 0.097±2, β = 0.010±2

Interatomic distances (Å)

Th	-	F(1)	2.28±2	(×3)	K	-	F(1)	2.68±2	(×3)
Th	-	F(2)	2.40±2	(×3)		-	F(1)	2.79±2	(×3)
Th	-	F(2)	2.42±2	(×3)		-	F(2)	2.87±2	(×3)
Na	-	F(1)	2.26±2	(×3)	Na	-	F(2)	2.54±2	(×3)

Discussion of the structure

ThF_9 and KF_9 polyhedra are trigonal prisms with a pyramid on each face. These polyhedra alternate along c. NaF_6 polyhedra share edges with Th and K polyhedra.

Details of analysis

365 reflexions were collected using silver radiation. Absorption and extinction corrections were applied.

POTASSIUM HEXACHLORORHENATE
POTASSIUM HEXABROMORHENATE
POTASSIUM HEXABROMOPLATINATE

I. A refinement of the crystal structures of K_2ReCl_6, K_2ReBr_6 and K_2PtBr_6. H.D. GRUNDY and I.D. BROWN, 1970. *Can. J. Chem.*, 48, 1151-1154.

Cubic, ($J1_1$-type, 1), Z = 4.

Space group $Fm3m$ (O^5_h)

	K_2ReCl_6	K_2ReBr_6	K_2PtBr_6
F.W.	477.12	743.83	752.72
a (Å)	9.840±1 (2)	10.385±2 (4)	10.293±1 (3)
U (Å^3)	952.8	1120.0	1090.5
D_x	3.32	4.40	4.57

Atomic positions

				K_2ReCl_6	K_2ReBr_6	K_2PtBr_6
8 K	in	8(*c*)				
4 Re, Pt	in	4(*a*)				
24 Cl, Br	in	24(*e*)	with *x*	0.2391±3	0.2391±4	0.2393±3

Interatomic distances (Å)

Re	-	Cl	(×6)	2.353±4		
Re	-	Br	(×6)		2.483±5	
Pt	-	Br	(×6)			2.463±3
K	-	Cl	(×12)	3.479±1		
K	-	Br	(×12)		3.672±1	3.640±1

Anisotropic temperature factors given.

Discussion

Refinement of earlier structures of K_2ReCl_6 (6) and K_2ReBr_6 (5). K_2PtBr_6 assumed to be isotypic. Anisotropic temperature factors discussed in terms of phase transformations.

Details of analysis

Spherical crystals ($\mu R < 1.0$) used to measure intensities on single-crystal diffractometer, $\theta - 2\theta$ scan, with Mo radiation. [Lorentz, polarisation and] absorption corrections. Refined by least squares to give R = 0.026 to 0.054 for 75 to 90 reflexions listed.

1. *Strukturbericht*, 1, 429.
2. H.E. SWANSON, M.C. MORRIS, R.P. STINCHFIELD and E.H. EVANS, 1963. *NBS monograph* 25, sect. 2, 28.
3. *Structure Reports*, 23, 500.
4. H. MUELLER, 1963. *Z. anorg. Chem.*, 321, 124; K. SCHWOCHAU, 1964. *Z. Naturforsch*, 19a, 1237; R. IKEDA, A. SASANE, D. NAKAMURA and M. KUBO, 1966. *J. Phys. Chem.*, 70, 2926.
5. *Structure Reports*, 15, 155.
6. *Strukturbericht*, 4, 189.

AMMONIUM HEPTAFLUOROZIRCONATE

I. The crystal structure of ammonium heptafluorozirconate and the disorder of the heptafluorozinconate ion. H.J. HURST and J.C. TAYLOR, 1970. *Acta Cryst.*, B26, 417-421.

II. A neutron diffraction analysis of the disorder in ammonium heptafluorozirconate. *Idem*, 1970. *Ibid.*, B26, 2136-2137.

$(NH_4)_4ZrF_7$ F.W. = 242.25

Cubic, a = 9.419±1 Å, U = 835.6 Å^3, D_m = 2.20, Z = 4, D_x = 2.21.

Space group $Fm3m$ (O_h^5)

Atomic positions

4	Zr	in	4(*a*)	y_{xr} = 0.0383±18, z_{xr} = 0.2086±5; y_{ne} = 0.0462±12, z_{ne} = 0.2062±8
20.8	F(1)	in	96(*j*)	y_{xr} = 0.1297±27, z_{xr} = 0.1771±22; y_{ne} = 0.1317±23, z_{ne} = 0.1880±21
8.3	F(2)	in	96(*j*)	
4	N(1)	in	4(*b*)	
8	N(2)	in	8(*c*)	
16.7	H(1)	in	96(*j*)	y_{ne} = 0.0329±22, z_{ne} = 0.3914±15
11.1	H(2)	in	192(*l*)	x_{ne} = 0.2342±57, y_{ne} = 0.2104±73, z_{ne} = 0.1560±54
11.1	H(3)	in	96(*k*)	x_{ne} = 0.1680±58, z_{ne} = 0.2259±74

Subscripts xr or ne refer to refinements with X-ray or neutron data respectively.

Thermal parameters

Anisotropic thermal parameters used in both studies but not all were varied in either case. In general the agreement is satisfactory.

Interatomic distances (Å) *and angles*

Zr	-	F(1)	(5)	2.00±1	nearest fluorine pairs in pentagonal plane
Zr	-	F(2)	(2)	2.07±2	F - Zr - F 64.2±12 to 79.8±4°
					others
					F - Zr - F 81.6±4 to 98.4±4
N(1)	-	H(1)		1.07	
N(2)	-	H(2)		0.97	
N(2)	-	H(3)		1.12	

Discussion of the structure

ZrF_7^{3-} forms a pentagonal bipyramid which is dynamically disordered. There are two distinct NH_4^+ groups which also are disordered.

Details of analysis

78 and 76 reflexions were measured in the X-ray (MoK_α) and neutron (λ = 1.17Å) diffraction experiments. Least-squares refinement with site populations determined from difference syntheses. Final *R* values of 0.023 and 0.033 respectively.

AMMONIUM OCTAFLUOROURANATE

I. The crystal structure of $(NH_4)_4UF_8$. A. ROSENZWEIZ and D.J. CROMER, 1970. *Acta Cryst.*, B26, 39-44.

$(NH_4)_4UF_8$ F.W. = 430.07

Monoclinic, *a* = 13.126±5, *b* = 6.692±3, *c* = 13.717±5 Å, β = 121°19', [*U* = 1029.3 $Å^3$], D_m = 2.96, *Z* = 4, D_x = 2.982.

Space group *C2/c* (C_{2h}^6) from structure solution

Atomic positions

			x	*y*	*z*
4 U	in	4(*e*) with	0	0.1682±2	1/4
8 F(1)	in	8(*j*)	0.0243±9	0.0071±18	0.1166±9
8 F(2)	in	8(*j*)	0.1597±9	-0.0317±18	0.3580±9
8 F(3)	in	8(*j*)	0.1599±8	0.3237±17	0.2528±8
8 F(4)	in	8(*j*)	0.0784±9	0.3952±15	0.3921±9
8 N(1)	in	8(*j*)	0.3824±12	0.1531±24	0.3921±12
8 N(2)	in	8(*j*)	0.2115±13	0.1111±21	0.0836±12

Interatomic distances (Å) *and angles*

U - F	= 2.28,	2.27,	2.33,	2.25	all ±0.01; each twice
N(1) - F	= 2.70,	2.82,	2.77,	2.75	all ±0.02
N(2) - F	= 2.81,	2.85,	2.79,	2.77	all ±0.02
F - N(1) - F	angles 87 - 122°				
F - N(2) - F	angles 85 - 128°				

Discussion of the structure

Tetragonal antiprisms of UF_8 interconnected by NH_4^+ ions as shown in Fig. 29.

Details of analysis

908 reflexions measured with diffractometer using Mo radiation. Absorption correction. Refined by least-squares method to *R* = 0.048.

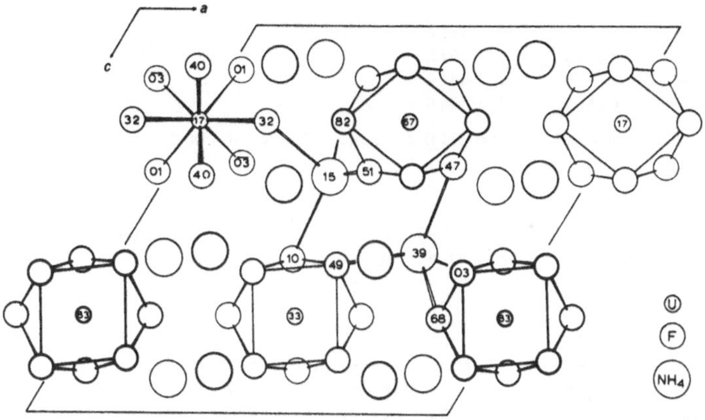

Fig. 29. Projection of the structure of $(NH_4)_4UF_8$ on the *ac* plane, *y* coordinates in hundredths of *b* are indicated. UF bonds are shown for one of the U atoms and the UF_8 polyhedra in the other cases.

CAESIUM OCTABROMODIRHENATE (III)

I. Some reactions of the octahalodirhenate (III) ions. VIII. Definitive structural characterization of the octabromodirhenate (III) ion. F.A. COTTON, B.G. DEBOER and M. JEREMIC, 1970. *Inorg. Chem.*, 9, 2143-2146.

$Cs_2Re_2Br_8$ F.W. = 1277.4

Orthorhombic, *a* = 12.625±4, *b* = 12.953±4, *c* = 10.141±3 Å, *U* = 1658.37 Å^3, D_m = 5.10, Z = 4, D_x = 5.11.

Space group *Pbca* (D_{2h}^{15}) systematic absences

Atomic positions

	x	*y*	*z*
Re	0.0154±1	0.0704±1	0.0600±1
Cs	-0.3072±2	0.1359±2	-0.1608±2
Br(1)	0.0640±3	0.0029±3	0.2801±3
Br(2)	-0.1585±3	0.1146±3	0.1558±4
Br(3)	-0.0155±3	0.2189±3	-0.0899±3
Br(4)	0.2068±3	0.1052±3	0.0247±4

Anisotropic temperature factors given.

Interatomic distances (Å)

Re	-	Re	2.227±2
Re	-	Br(2)	2.469±4
Re	-	Br(1)	2.474±3
Re	-	Br(3)	2.483±4
Re	-	Br(4)	2.485±4

Angles in degrees subtended at Re by

	Br(2)	Br(1)	Br(3)	Br(4)
Re	104.5±1	104.3±1	105.8±1	103.9±1
Br(2)		87.0±1	85.5±1	151.6±2
Br(1)			149.9±1	87.3±1
Br(3)				85.6±1

Cs	-	Br(4)B	3.631±5
Cs	-	Br(4)A	3.642±5
Cs	-	Br(4)B	3.716±4
Cs	-	Br(2)A	3.730±4
Cs	-	Br(1)B	3.756±4

Cs	-	Br(1)A	3.759±4
Cs	-	Br(2)B	3.765±5
Cs	-	Br(3)B	3.804±4
Cs	-	Br(3)A	3.904±4

The A atoms belong to one Re_2Br_8 ion. The B atoms to a second Re_2Br_8 ion.

Discussion

The $Re_2Br_8^{2-}$ ion has effectively D_{4h} symmetry, the Br atoms lying almost at the corners of a cube. Four caesium atoms lie opposite the four equivalent faces of the cube.

Details of analysis

Lattice parameters found from least-squares refinement of $h00$, $0k0$ and $00l$ $(0 \leqslant h, k, l \leqslant 14)$ reflexions measured on a single crystal diffractometer with MoK_α radiation ($\lambda\alpha_1$ = 0.70926, $\lambda\alpha_2$= 0.71354 Å). Intensities measured with diffractometer from crystal 0.018 × 0.06 × 0.06 mm^3 using $\theta - 2\theta$ scan and Zr filtered Mo radiation and corrected for absorption (μ = 40.2 mm^{-1}). Structure solved by Patterson and refined by least-squares methods to give R_2 = 0.054 for 770 reflexions listed.

BARIUM THULIUM FLUORIDE

I. [The crystal structure of $BaTm_2F_8$]. O.E. IZOTOVA and V.B. ALEKSANDROV, 1970. *Dokl. Akad. Nauk SSSR*, 192, 1037-1039 [*Soviet Physics - Doklady*. 15, 525-526].

$BaTm_2F_8$ F.W. = 627.19

Monoclinic, a = 6.935±1, b = 10.457±2, c = 4.243±1 Å, β = 99.67±3°, U = 303.3 Å^3, Z = 2, D_x = 6.85.

Space group $C2/m$ (C^3_{2h})

Atomic positions

			x	y	z	B(Å^2)
2 Ba	in	2(*a*)	0	0	0	-0.58
4 Tm	in	4(*h*)	0.5	0.1761	0.5	-0.35
8 F(1)	in	8(*j*)	0.187	0.140	0.560	-0.75
4 F(2)	in	4(*i*)	0.393	0	0.227	-0.70
4 F(3)	in	4(*g*)	0.5	0.239	0	-0.83

Interatomic distances (Å)

Ba	-	F(3)	2.73	(×2)
Ba	-	F(2)	2.73	(×2)
Ba	-	F(1)	2.85	(×4)
Ba	-	F(1)'	2.90	(×4)
Tm	-	F(3)	2.22	(×2)
Tm	-	F(2)	2.24	(×2)
Tm	-	F(1)	2.26	(×2)
Tm	-	F(1)'	2.31	(×2)

Errors uncertain but probably greater than 0.02 Å.

Discussion

The Ba atoms all lie at z = 0, the Tm atoms at z = 1/2, the F atoms lie between and within these layers to give 8 coordination around Tm and 12 coordination around Ba.

Details of analysis

Intensities measured on diffractometer using spherical crystal 0.35 mm in diameter. Structure solved from Patterson function and refined by least squares to give R = 0.06 for 416 zero layer reflexions [not listed]. [No correction appears to have been made for absorption which probably accounts for the negative temperature factors].

TETRASODIUM DECAFLUOROTRISTANNATE

I. Die Kristallstruktur des Tetranatrium-dekafluorotristannat(II), $Na_4Sn_3F_{10}$.
G. BERGERHOFF and L. GOOST, 1970. *Acta Cryst.*, B26, 19-23.

$Na_4Sn_3F_{10}$ F.W. = 661.00

Monoclinic, a = 12.18, b = 18.47, c = 5.50 Å, β = 91°, [U = 1237.1 Å^3], D_m = 3.34, Z = 4, D_x = 3.42.

Space group *C*2/*c* (C_{2h}^6) confirmed by structure solution

Atomic positions

				x	y	z	B(Å^2)
8 Sn(1)	in	8(*f*)	with	0.2522±1	0.0814±1	0.6690±5	1.17±5
4 Sn(2)	in	4(*e*)		0	0.1569±2	1/4	1.92±7
8 Na(1)	in	8(*f*)		0.1714±8	0.2712±6	0.7497±3	1.2 ±2
4 Na(2)	in	4(*e*)		0	0.342 ±1	1/4	2.9 ±4
4 Na(3)	in	4(*e*)		0	0.546 ±1	1/4	1.7 ±3
8 F(1)	in	8(*f*)		0.120 ±1	0.377 ±1	0.541 ±4	1.3 ±3
8 F(2)	in	8(*f*)		0.165 ±2	0.180 ±1	0.428 ±4	2.0 ±3
8 F(3)	in	8(*f*)		0.107 ±1	0.452 ±1	0.118 ±4	2.2 ±4
8 F(4)	in	8(*f*)		0.232 ±1	0.332 ±1	0.120 ±4	2.7 ±4
8 F(5)	in	8(*f*)		0.041 ±1	0.238 ±1	0.016 ±4	2.1 ±4

Interatomic distances (Å)

Sn	-	F	1.99 - 2.49
Na	-	F	2.24

Structure

Sn_3F_{10} consists of connected distorted tetragonal pyramids of SnF_4. Sodium ions interconnect Sn_3F_{10} leaving large channels parallel to c.

Details of analysis

Data from Weissenberg photographs taken with Mo radiation. Least-squares refinement, from coordinates determined from Patterson function, to an R of 0.092.

DIPOTASSIUM LEAD DIFLUOROBERYLLATE

I. Fluoroberyllates de structure palmiérite. Y. LE FUR and S. ALÉONARD, 1970. *Bull. Soc. Fr. Miner. Crist.*, 93, 260-262.

$K_2Pb(BeF_4)_2$ F.W. = 455.4

Rhombohedral (Palmierite type, 1), a = 5.455±2, c = 20.500±9 Å, U = 528.3 Å^3, Z = 3, D_x = 4.29.

Space group *R*$\bar{3}$*m* (D_{3d}^5)

Atomic positions

			x	y	z
3 Pb	in	3(*a*)	0	0	0
6 K	in	6(*c*)	0	0	0.206±2
6 Be	in	6(*c*)	0	0	0.407
6 F(1)	in	6(*c*)	0	0	0.330
6 F(2)	in	18(*h*)	0.846	0.154	0.431

Interatomic distances (Å)

Be	-	F(1)	1.58
Be	-	F(2)	1.54 (×3)
K	-	F(1)	2.54
K	-	F(2)	2.79 (×3)
K	-	F(2)'	2.80 (×6)

F(1)	-	Be	-	F(2)	109°
F(2)	-	Be	-	F(2)	110°
Pb	-	F(2)			2.62 (×6)
Pb	-	F(1)			3.15 (×6)

Errors uncertain but about 0.03 - 0.08 Å.

Discussion

$Rb_2Pb(BeF_4)_2$, $(NH_4)_2Pb(BeF_4)_2$, and $Tl_2Pb(BeF_4)_2$ are isostructural.

Details of analysis

Crystals prepared from aqueous solutions of K_2CO_3, $Pb(CH_3CO)_2$, BeF_2 and HF. Parameters checked against powder photograph to give qualitative agreement ($R \sim$ 0.10 to 0.20) for reflexions listed.

1. *Structure Reports*, 11, 390.

TRIFLUOROSELENIUM (IV) HEXAFLUORONIOBATE (V)

I. Fluoride crystal structures. Part XI. Trifluoroselenium (IV) hexafluoroniobate (V) and hexafluorotantalate (V). A.J. EDWARDS and G.R. JONES, 1970. *J. Chem. Soc. Lond. A*, 1891-1894.

SeF_3NbF_6 F.W. = 342.85

Rhombohedral, a = 8.99±1 Å, α = 92.9±2°, U = 724 Å^3, Z = 4, D_x = 3.15.

Space group $R3$ (C_3^4) from systematic absences and Patterson function.

Atomic positions

	x	y	z	U(Å^2)
3 Nb(1) in 3(*b*)	1.000	1.2196±5	0.6664±5	0.0307±8
1 Nb(2) in 1(*a*)	0.6030±5	0.6030±5	0.6030±5	0.0279±14
3 Se(1) in 3(*b*)	0.5478±6	1.0637±6	0.7556±5	0.0275±9
1 Se(2) in 1(*a*)	1.1218±5	1.1218±5	1.1218±5	0.0310±17
3 F(1) in 3(*b*)	0.5542±63	0.6304±64	0.4110±64	0.106 ±19
3 F(2) in 3(*b*)	0.5656±37	0.8052±37	0.6753±36	0.051 ±8
3 F(3) in 3(*b*)	1.1964±31	1.2614±30	0.6327±31	0.035 ±6
3 F(4) in 3(*b*)	0.9657±30	1.4077±31	0.7316±31	0.036 ±6
3 F(5) in 3(*b*)	0.5269±29	1.2437±33	0.8211±31	0.041 ±6
3 F(6) in 3(*b*)	0.3628±43	1.0053±38	0.7910±38	0.062 ±10
3 F(7) in 3(*b*)	0.9597±47	1.2350±47	0.4812±48	0.075 ±1
3 F(8) in 3(*b*)	1.0291±83	1.0038±81	0.6202±85	0.135 ±24
3 F(9) in 3(*b*)	1.0774±41	1.3007±44	1.1746±41	0.060 ±9
3 F(10) in 3(*b*)	0.7734±57	1.1615±61	0.7074±62	0.093 ±16
3 F(11) in 3(*b*)	1.0618±35	1.1800±37	0.8649±33	0.046 ±7
3 F(12) in 3(*b*)	0.5081±43	1.0865±44	0.5727±44	0.068 ±10

Interatomic distances (Å)

Sixfold coordination around Nb(1)

Nb(1) - F(3)	1.83±3
Nb(1) - F(4)	1.81±3
Nb(1) - F(7)	1.70±4
Nb(1) - F(8)*	2.00±8
Nb(1) - F(10)*	2.13±5
Nb(1) - F(11)*	1.90±3

Angles in degrees subtended at Nb(1) by

	F(4)	F(7)	F(8)*	F(10)*	F(11)*
F(3)	95	88	89		88
F(4)		99		87	88
F(7)			88	92	
F(8)*				88	85
F(10)*					91

σ(F-Nb-F) = 4°

Sixfold coordination around Nb(2) (crystallographic site symmetry C_3)

Nb(2) - F(1)	1.79±6	(×3)
Nb(2) - F(2)*	1.95±3	(×3)

Angles in degrees subtended at Nb(2) by

	F(1)'	F(2)"*
F(1)	89	96
F(2)'*	88	88

σ(F-Nb-F) = 4°

Sixfold coordination around Se(1)

Se(1) - F(5)	1.72±3	
Se(1) - F(6)	1.77±3	
Se(1) - F(12)	1.69±4	
Se(1) - F(2)*	2.41±3	
Se(1) - F(10)*	2.24±6	
Se(1) - F(8)*	2.33±7	

Angles in degrees subtended at Se(1) by

	F(6)	F(12)	F(2)*	F(10)*	F(8)*
F(5)	93	98		81	83
F(6)		94	84		90
F(12)			84	84	
F(2)*				102	95
F(10)*					93

σ(F–Se–F) = 4°

Sixfold coordination around Se(2) (crystallographic site symmetry C_3)

Se(2) - F(9)	1.74±4	(×3)
Se(2) - F(11)*	2.43±3	(×3)

Angles in degrees subtended at Se(2) by

	F(9)'	F(11)"*
F(9)	94	88
F(11)'*	83	95

* Bridging F atoms.

σ(F–Se–F) = 4°

Discussion

The crystal contains discrete tetrameric molecules consisting of four NbF_6 and four SeF_3 ions arranged at the alternate corners of a cube in such a way that each SeF_3 ion makes three long Se–F bonds with adjacent NbF_6 ions. The tetramer has nearly T_d symmetry. SeF_3TaF_6 is isostructural with virtually the same parameters throughout.

Details of analysis

Lattice parameters from precession and Weissenberg photographs with CuK_α (λ = 1.5418 Å) and MoK_α (λ = 0.7107 Å) radiations. Intensities measured by densitometer from integrated Nonius [Weissenberg?] photographs of layers with k = 0 to 4 taken with MoK_α radiation. Lorentz and polarisation corrections, but not absorption (μ = 7.27 mm^{-1}, crystal size ∿ 0.1 mm). Structure solved by determining Ta positions in isostructural SeF_3TaF_6 from Patterson function and calculating corresponding electron-density map for Nb compound. [68? parameters] refined by least squares to give R = 0.09 for 724 reflexions listed in 1.

1. *Supplementary Publications, UK National Lending Library SUP* 20014.

TRIFLUOROSELENIUM (IV) μ-FLUOROBIS (PENTAFLUORONIOBATE (V))

I. Fluoride crystal structures. Part X. Trifluoroselenium (IV) μ-fluorobis [pentafluoroniobate (V)]. A.J. EDWARDS and G.R. JONES, 1970. *J. Chem. Soc. Lond., A*, 1491-1497.

$SeF_3.Nb_2F_{11}$ F.W. = 516.74

Monoclinic, a = 7.60±1, b = 17.38±2, c = 8.80±1 Å, β = 103.4±2°, U = 1113 Å^3, Z = 4, D_x = 3.17.

Space group $P2_1/c$ (C^5_{2h}) from systematic absences

Atomic positions All atoms in 4(e)

	x	y	z	B(Å^2)
Nb(1)	-0.2205±2	0.3201±1	0.2719±3	5.78±10
Nb(2)	0.2158±2	0.4611±1	0.3287±2	0.51±12
Se	0.1694±3	0.1380±1	0.3694±3	5.82±12
F(1)	-0.0299±20	0.2471±9	0.3351±19	0.97±8
F(2)	-0.3744±27	0.3999±11	0.2068±24	1.38±10
F(3)	-0.3877±26	0.2462±12	0.2378±24	1.46±10

	x	y	z	$B(\text{Å}^2)$
F(4)	−0.1492±29	0.3244±12	0.0842±25	1.54±12
F(5)	−0.2196±23	0.3393±10	0.4754±21	1.14±8
F(6)	−0.0120±19	0.3982±8	0.3203±17	0.77±6
F(7)	0.0493±20	0.5412±9	0.2470±20	0.97±8
F(8)	0.3440±22	0.3727±9	0.3930±20	1.05±8
F(9)	0.4082±25	0.5218±10	0.3241±23	1.24±10
F(10)	0.1796±21	0.4245±9	0.1229±20	1.05±8
F(11)	0.2015±21	0.4860±9	0.5239±19	0.99±8
F(12)	0.3204±22	0.0681±9	0.3724±20	10.66±79
F(13)	0.3190±24	0.1991±10	0.4647±22	1.20±10
F(14)	0.1818±22	0.1695±9	0.1936±20	1.01±8

Interatomic distances (Å)

Trifluoroselenium (IV) ion

Se	- F(12)	1.67
Se	- F(13)	1.64
Se	- F(14)	1.66

longer Se-F distances

Se	- F(7)	2.42
Se	- F(10)	2.47
Se	- F(1)	2.40

σ(Se-F) = 0.02 Å

Angles in degrees subtended at Se by

	F(13)	F(14)	F(7)	F(10)	F(1)
F(12)	94	94	84	78	172
F(13)		95	174	88	84
F(14)			91	172	78
F(7)				87	98
F(10)					110

σ(F-Se-F) = 2°

Nb_2F_{11} ion

Nb(1)	- F(1)	1.91
Nb(1)	- F(2)	1.82
Nb(1)	- F(3)	1.78
Nb(1)	- F(4)	1.85
Nb(1)	- F(5)	1.82
Nb(1)	- F(6)	2.06

Angle in degrees subtended at Nb(2) by

	F(2)	F(3)	F(4)	F(5)	F(6)
F(1)	171	92	86	90	83
F(2)		96	89	92	88
F(3)			103	98	175
F(4)				159	78
F(5)					81

Nb(2)	- F(6)	2.03
Nb(2)	- F(7)	1.91
Nb(2)	- F(8)	1.84
Nb(2)	- F(9)	1.81
Nb(2)	- F(10)	1.88
Nb(2)	- F(11)	1.80

Angle in degrees subtended at Nb(2) by

	F(7)	F(8)	F(9)	F(10)	F(11)
F(6)	84	87	176	81	86
F(7)		170	92	87	90
F(8)			97	88	93
F(9)				97	96
F(10)					167

σ(Nb-F) = 0.02 Å

σ(F-Nb-F) = 2°

Nb(1)	-	F(6)	-	Nb(2)	166±2°
Se	-	F(7)	-	Nb(2)	177±1°
Se	-	F(1)	-	Nb(1)	167±2
Se	-	F(10)	-	Nb(2)	170±2

Discussion

The structure is composed of SeF_3 and $F_5NbFNbF_5$ ions connected by three fluorine bridges so as to give the selenium atom a trigonally distorted octahedral environment. The Nb-F distances occur in three ranges: 1.78 - 1.85 Å for terminal F atoms, 1.88 - 1.91 Å for F atoms bonded to Se and 2.03 - 2.06 Å for F atoms bonded to another Nb.

Details of analysis

Crystals grown in capillary by sublimation. Lattice parameters from Weissenberg and precession photographs ($\lambda(CuK_{\alpha 1})$ = 1.5418 Å). Intensities measured with photometer from integrated Weissenberg photographs of layers with k = 0 to 2 and l = 0 to 6. Lorentz and polarisation corrections but not absorption ($\mu(CuK_\alpha)$ = 23.5 mm^{-1}, crystal elongated along a with cross section 0.02 mm). Structure solved from Patterson and electron-density syntheses and refined by least squares to give R = 0.093 for the 1330 reflexions listed.

XENON DIFLUORIDE: IODINE PENTAFLUORIDE

I. The crystal structure of 1:1 molecular addition compound xenon difluoride - iodine pentafluoride, $XeF_2.IF_5$. G.R. JONES, R.D. BURBANK and N. BARTLETT, 1970. *Inorg. Chem.*, 9, 2264-2268.

$XeF_2.IF_5$ F.W. = 391.2

Tetragonal, a = 7.65±1, c = 10.94±1 Å, U = 640 Å^3, D_m = 3.8, Z = 4, D_x = 4.05.

Space group $I4/m$ (C^5_{4h}) systematic absences and structure

Atomic positions

			x	y	z
4 Xe	in	4(c)	0	1/2	0
4 I	in	4(e)	0	0	0.2859±1
4 F(1)	in	4(e)	0	0	0.1198±9
16 F(2)	in	16(i)	0.0596±8	0.2343±7	0.2589±5
8 F(3)	in	8(h)	0.2439±12	0.4035±13	0

Anisotropic temperature factors given.

Interatomic distances (Å)

		Xe	-	F(3)	2.009±9	(×2)
		I	-	F(1)	1.817±10	
		I	-	F(2)	1.874±5	(×4)
		I	-	F(3)	3.142±7	(×4)
F(2)	-	I	-	F(1)	80.9±2	
F(2)	-	I	-	F(2)'	88.6±3	
F(2)	-	I	-	F(2)"	161.9±5	

Discussion

The IF molecules consist of a square pyramid of F atoms with the I atom a little below the basal plane. They occur in layers perpendicular to c with alternate molecules having their unique axis inverted. XeF_2 molecules lie in planes between the layers in such a way that their F atoms form additional long bonds to the exposed side of the I atom.

Details of analysis

Lattice parameters measured on diffractometer with MoK_α radiation (λ = 0.7107 Å). Intensities, measured on manual diffractometer (stationary counter - stationary crystal) with MoK_α radiation from crystal of maximum dimension 0.1 mm mounted in a capillary corrected for Lorentz and polarisation effects but not absorption. Structure solved by Patterson function and systematic elimination of models for light atoms. 35 parameters refined by least squares to give R_1 = 0.034 for 403 unobserved reflexions listed.

DIFLUOROCHLORINE (III) HEXAFLUOROANTIMONATE (V)

I. Fluoride crystal structures. Part XIII. Difluorochlorine (III) hexafluoroantimonate (V). A.J. EDWARDS and R.J.C. SILLS, 1970. *J. Chem. Soc. (London), A*, 2697-2699.

$ClF_2.SbF_6$ F.W. = 309.2

Triclinic, a = 5.60±1, b = 10.55±1, c = 5.30±1 Å, α = 92.1±2, β = 91.8±2, γ = 91.5±2°, U = 312.64 Å^3, Z = 2, D_x = 3.28.

Space group $P\bar{1}$ (C_i^1)

Atomic positions

	x	y	z	B(Å^2)
Sb(1)	0	0	0	1.49±10
Sb(2)	0	1/2	0	1.41±9
Cl	-0.4022±14	0.2491±9	0.4211±17	2.08±15
F(1)	-0.5413±50	0.1527±29	0.5857±43	4.2 ±6
F(2)	-0.4722±46	0.3719±27	0.5772±49	3.8 ±5
F(3)	-0.2680±47	0.0665±27	0.1783±51	4.0 ±5
F(4)	-0.2030±41	0.3833±25	0.1523±45	3.3 ±5
F(5)	0.1861±48	0.1242±28	0.1419±53	4.4 ±6
F(6)	0.2711±40	0.4214±24	0.1150±44	3.1 ±3
F(7)	0.0189±45	0.3976±25	-0.2950±49	3.7 ±5
F(8)	-0.0617±54	0.1009±30	-0.2686±58	5.2 ±7

Interatomic distances (Å)

$Sb(1)F_6$ ion (crystallographic site symmetry C_i)

Sb(1) - F(5)	1.78	
Sb(1) - F(8)	1.84	
Sb(1) - F(3)	1.93	

Angles in degrees subtended at Sb(1) by

	F(8)	F(3)
F(5)	89	90
F(8)		91

Sb(2)F ion (crystallographic site symmetry C_i)

Sb(2) - F(6)	1.85
Sb(2) - F(7)	1.88
Sb(2) - F(4)	1.88

Angles in degrees subtended at Sb(2) by

	F(7)	F(4)
F(6)	87	93
F(7)		93

Cl - F(1)	1.57	planar with angles 87 - 96° between ligands
Cl - F(2)	1.58	
Cl - F(4)	2.33	
Cl - F(3)	2.43	

Cl - F(5)	2.96	both distances make angles of 70° with Cl - F(1) and Cl - F(3)
Cl - F(8)	2.99	

F(1) - Cl - F(2) 96°

σ(distance) = 0.03 Å, σ(angles) = 1° [Probably rather too low.]

Discussion

The structure consists of two crystallographically equivalent ClF_2 ions and two nonequivalent centrosymmetric SbF_6 ions. Two fluorine bridges (F(3) and F(4)) are coplanar with the ClF_2 ion and link the ions into a chain along b.

Details of analysis

Prepared from ClF_3 and SbF_5. Lattice parameters from precession and Weissenber photographs with MoK_α radiation, (λ = 0.7107 Å). Intensities measured photometrically from integrated (Weissenberg) photographs of layers with k = 0 - 9 using MoK_α radiation. Lorentz and polarisation corrections but not absorption (μ = 5.03 mm^{-1}, crystal size ∿ .0.03 mm). Structure solved from Patterson and electron-density functions and refined by least squares to give R = 0.096 for 625 reflexions listed in 1.

1. *Supplementary Publication Scheme SUP* 20035, National Lending Library, Boston Spa, U.K.

MOLYBDENUM FLUORO-BRONZES

I. Preparation and properties of molybdenum fluoro-bronzes. J.W. PIERCE, H.L. MCKINZIE, M. VLASSE and A. WOLD, 1970. *J. Solid State Chem.*, 1, 332-338.

$Mo_4O_{11.2}F_{0.8}$

Orthorhombic, a = 3.878±5, b = 13.96±1, c = 3.732±5 Å.

Space group *Cmcm* (D_{2h}^{17})

Structure similar to MoO_3 (1); full report to follow.

$MoO_{2.4}F_{0.6}$ F.W. = 145.74

Cubic, (ReO_3 type 2), a = 3.842±5 Å, U = 56.71 Å^3, D_m = 4.1±1, Z = 1, D_x = 4.27.

Space group *Pm3m* (O_h^1)

Atomic positions

1 Mo	in	1(a)
2.4 O + 0.6 F	in	3(d)

Interatomic distances (Å)

Mo	-	O	1.921 (×6)

Details of analysis

The proposed structure gives R(intensities) = 0.12 for reflexions measured from single crystal photographs.

1. L. KIHLBORG, 1963. *Ark. Kemi*, 21, 357.
2. K. MEISEL, 1932. *Z. anorg. Chem.*, 207, 121.

MAGNESIUM NIOBIUM OXYFLUORIDE

I. The crystal structure of $MgNb_{14}O_{35}F_2$. M. LUNDBERG, 1970. *J. Solid State Chemistry*, 1, 463-368.

$MgNb_{14}O_{35}F_2$ F.W. = 1922.94

Monoclinic, a = 20.628±10, b = 3.825±2, c = 19.098±10 Å, β = 107.75±2, U = 1435 Å^3, D_m = 4.41±0, Z = 2, D_x = 4.45.

Space group *C2/m* (C_{2h}^3)

Atomic positions

All atoms in 4(i) unless otherwise noted.

	x	y	z
Nb(1)	0.0630	0	0.0787
Nb(2)	0.1935	0.5	0.0791
Nb(3)	0.3736	0.5	0.0718
Nb(4)	0.1208	0	0.2893
Nb(5)	0.2515	0.5	0.2838
Nb(6)	0.4413	0.5	0.2971

	x	y	z
Nb(7)	0.1841	0	0.4991
Nb(8) in 2(c)	0	0	0.5000
O(1)	0.0857	0.5	0.0625
O(2)	0.1565	0	0.0429
O(3)	0.2700	0.5	0.0681
O(4)	0.0878	0	0.1926
O(5)	0.2252	0.5	0.1873
O(6)	0.4105	0.5	0.1665
O(7)	0.1428	0.5	0.2790
O(8)	0.3413	0.5	0.2788
O(9)	0.4271	0	0.2664
O(10)	0.1588	0	0.3895
O(11)	0.2904	0.5	0.3924
O(12)	0.2261	0	0.2916
O(13)	0.4692	0.5	0.0482
O(14)	0.3598	0	0.0470
O(15)	0.0896	0	0.5068
O(16)	0.4647	0.5	0.3897
O(17)	0.0321	0	0.2829
O(18)	0.2046	0.5	0.5228
O(19) in 2(d)	0	0.5	0.5

Interatomic distances (Å)

Nb(1)	-	O(13)	1.84		Nb(5)	-	O(5)	1.76	
Nb(1)	-	O(1)	2.02	(×2)	Nb(5)	-	O(8)	1.88	
Nb(1)	-	O(4)	2.08		Nb(5)	-	O(11)	1.98	
Nb(1)	-	O(2)	2.24		Nb(5)	-	O(12)	2.00	(×2)
Nb(1)	-	O(13)	2.31		Nb(5)	-	O(7)	2.22	
Nb(2)	-	O(3)	1.66		Nb(6)	-	O(16)	1.68	
Nb(2)	-	O(5)	1.97		Nb(6)	-	O(17)	1.97	
Nb(2)	-	O(2)	2.10	(×2)	Nb(6)	-	O(8)	1.98	
Nb(2)	-	O(1)	2.15		Nb(6)	-	O(9)	2.00	(×2)
Nb(2)	-	O(14)	2.32		Nb(6)	-	O(6)	2.38	
Nb(3)	-	O(6)	1.74		Nb(7)	-	O(11)	1.98	
Nb(3)	-	O(14)	1.97	(×2)	Nb(7)	-	O(18)	1.98	(×2)
Nb(3)	-	O(2)	2.09		Nb(7)	-	O(10)	2.00	
Nb(3)	-	O(3)	2.12		Nb(7)	-	O(15)	2.00	
Nb(3)	-	O(13)	2.15		Nb(7)	-	O(18)	2.46	
Nb(4)	-	O(4)	1.76		Nb(8)	-	O(15)	1.81	(×2)
Nb(4)	-	O(17)	1.80		Nb(8)	-	O(19)	1.91	(×2)
Nb(4)	-	O(10)	1.83		Nb(8)	-	O(16)	2.01	(×2)
Nb(4)	-	O(7)	1.99	(×2)					
Nb(4)	-	O(12)	2.16						

Standard error ∿ 0.15 Å

Discussion

The Mg and Nb atoms are disordered over the metal sites and the O and F atoms over the anion sites. The crystals are composed of blocks of ReO_3 structure of extent 5 × ∞ × 3 linked together in the *a* and *c* direction by edge shared octahedra.

Details of analysis

Material prepared by heating MgF_2 and H - Nb_2O_5 in 1:7 ratio at 100°C. Lattice parameters from powder photograph with Cu$K_{\alpha 1}$ radiation (powder pattern given). Intensities of layers with k = 0, 1 and 2 measured visually from multiple-film Weissenberg photographs taken with CuK_α radiation from a small twinned crystal. Structure determined by packing and 61 parameters (including individual isotropic temperature factors for the metal atoms) refined by least squares to give R = 0.064 for 216 reflexions listed.

ZIRCONIUM OXYGEN FLUORIDE

I. The crystal structure of $Zr_7O_9F_{10}$. B. HOLMBERG, 1970. *Acta Cryst.*, B26, 830-835.

$Zr_7O_9F_{10}$ F.W. = 972.54

Orthorhombic, a = 6.443±1, b = 26.851±1, c = 4.071±1 Å, [U = 704.28 Å^3], D_m = 4.55, Z = 2, D_x = 4.59.

Space group *Pbam* (D_{2h}^9) other possibility not tried

Atomic positions

	x	y	B(Å^2)
2 Zr(1) in 2(a)			1.12±10
4 Zr(2) in 4(*g*)	0.4484±4	0.0634±1	1.01±8
4 Zr(3) in 4(*g*)	0.0028±5	0.1442±1	1.16±8
4 Zr(4) in 4(*g*)	0.4401±5	0.2229±2	1.90±9
4 O, F(1) in 4(*g*)	0.6879±29	0.0095±8	2.4 ±4
4 O, F(2) in 4(*g*)	0.1230±34	0.0722±9	2.9 ±5
4 O, F(3) in 4(*g*)	0.7308±29	0.1040±7	1.9 ±4
4 O, F(4) in 4(*g*)	0.3573±29	0.1429±8	1.9 ±4
4 O, F(5) in 4(*g*)	0.7560±26	0.1969±7	1.7 ±4
4 O, F(6) in 4(*g*)	0.1332±30	0.2167±8	0.9 ±4
2 O, F(7) in 2(*b*)			3.2 ±16
4 O, F(8) in 4(*h*)	0.4569±44	0.0681±13	2.5 ±10
4 O, F(9) in 4(*h*)	0.9933±46	0.1437±10	2.9 ±13
4 O, F(10) in 4(*h*)	0.4439±41	0.2188±1	2.4 ±10

Interatomic distances (Å)

Zr(1) - O, F	2.03±2	2.09±3	2.04±1	(2 of each)			
Zr(2) - O, F	2.12±2	2.15±2	2.11±3	2.12±2	2.21±2	(2×)	2.04±1
Zr(3) - O, F	2.08±3	2.06±2	2.28±2	2.13±2	2.12±2	(2×)	2.04±1
Zr(4) - O, F	2.21±2	2.15±2	1.98±2	2.46±2	2.04±3	(2×)	2.04±1

Materials

Crystals obtained from ZrF_4 + ZrO_2 heated to 800°C for several days in a sealed platinum tube.

Discussion of the structure

Structure is built upon corner-shared and edge-shared polyhedra with some similarities to that of α-U_3O_8. Part of the structure is shown in Fig. 30.

Details of analysis

259 reflexions were measured from Weissenberg photographs taken with Cu radiation. Absorption corrected. Refined by least-squares method without distinction between fluorine and oxygen atoms to a final R value of 0.082. The wide distribution in temperature factors is not discussed.

IRON OXYCHLORIDE

I. Refinement of the crystal structure of iron oxychloride. M.D. LIND, 1970. *Acta Cryst.*, B26, 1058-1062.

FeOCl F.W. = 107.3

Orthorhombic, a = 3.780±5, b = 7.917±5, c = 3.302±5 Å, U = 98.82 Å^3, D_m = 3.55, Z = 2, D_x = 3.606.

Space group *Pmnm* (D_{2h}^{13})

Fig. 30. Part of the structure of $Zr_7O_9F_{10}$ viewed along the *c* axis. The zirconium atoms are represented by filled circles.

Atomic parameters

2 Fe	in	2(*b*)	with	*y* = 0.11568±13
2 O	in	2(*a*)	with	*y* =-0.0483±6
2 Cl	in	2(*a*)	with	*y* = 0.3300±2

Anisotropic thermal parameters are given.

Interatomic distances (Å)

Fe	-	O	1.964±8, 2.100±10 each twice
Fe	-	Cl	2.368±7

Interlayer distances

Cl	-	Cl	3.680±6 (4×)
Fe	-	Fe	3.107±6

Material preparation

Crystals grown by vapour transport using a mixture of $FeCl_3$ and Fe_2O_3. A temperature of 350°C was maintained at one end of an evacuated sealed tube and 325°C at the other end. Crystals grew in two weeks.

Description of the structure

Except for corrected *b* value and greater accuracy the structure is the same as that Goldsztaub (1) proposed. $Fe(OCl_2)_2$ octahedra form layers by sharing O - O edges, with the layers held by interactions between the Cl ions protruding from the layer as shown in Fig. 31.

Details of analysis

500 intensities were measured with an automatic diffractometer with MoK_α radiation and corrected for absorption. Structure refined to *R* = 0.055 by least squares.

1. *Strukturbericht*, 3, 376.

YTTRIUM OXYFLUORIDE

I. The crystal structure of stoichiometric yttrium oxyfluoride, YOF. A.W. MANN and D.J.M. BEVAN, 1970. *Acta Cryst.*, B26, 2129-2131.

YOF F.W. = 123.90

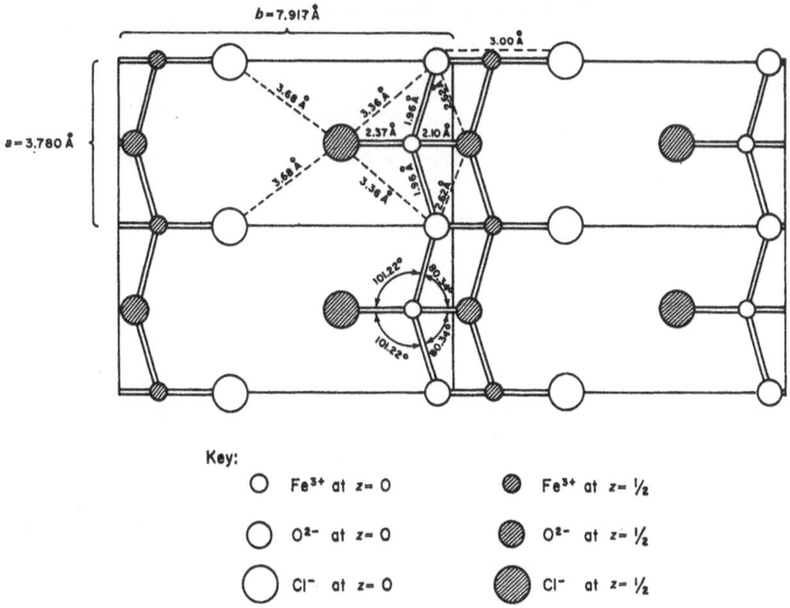

Fig. 31. Projection of the structure of FeOCl onto (001).

Rhombohedral, a = 6.666±2 Å, α = 33.09±1°, (hexagonal cell a = 3.797±1, c = 18.89±1 Å), U = 235.9 Å^3, D_m = 5.1, Z = 2, D_x = 5.23.

Space group $R\bar{3}m$ (D_{3d}^5) confirmed by refinement

Atomic positions (hexagonal)

6 Y	in	6(c)	with	z = 0.2412±1,	B = 0.3 (Å^2)
6 O	in	6(c)	with	z = 0.117±1,	B = 0.4
6 F	in	6(c)	with	z = 0.372±1,	B = 0.4

Interatomic distances (Å) and angles

Y – O	(2×)	2.24±2		F' – Y – F'		103.7±3°
Y – O'	(2×)	2.34±2		O' – Y – O'		116.2±4
Y – F	(2×)	2.41±2		F – Y – F'		65.3±4
Y – F'	(2×)	2.47±2		O – Y – O'		77.8±5

Material

YF_3 and Y_2O_3 reacted under nitrogen in sealed platinum crucibles at 1100°C.

Discussion of the structure

The structure is the same as proposed by Zachariasen (1) with oxygen and fluorine interchanged.

Details of analysis

Data (35 reflexions) were obtained from a series of integrated Hägg-Guinier patterns and measured with a microdensitometer. Least-squares refinement, excluding variation of thermal parameters, lead to an R of 0.076.

1. *Structure Reports*, 15, 167.

MOLYBDENUM (V) OXIDE TRICHLORIDE

$MoOCl_3$ F.W. = 218.30

I. Crystal and molecular structure of molybdenum (V) oxide trichloride. M.G.B. DREW and I.B. TOMKINS, 1970. *J. Chem. Soc. Lond. A*, 22–25.

Monoclinic, a = 5.732±6, b = 13.340±12, c = 6.029±6 Å, β = 93.79±8°, U = 459.9 Å^3, Z = 2, D_x = 3.151.

Space group $P2_1/c$ (C^5_{2h}) from systematic absences

Atomic positions

All atoms in 4(*e*)

	x	*y*	*z*
Mo	-0.0260±9	0.1608±4	0.2019±10
Cl(1)	-0.2977±33	0.0631±12	0.3575±31
Cl(2)	0.1784±26	0.1921±10	0.5585±25
Cl(3)	-0.2767±26	0.3156±10	0.3881±25
O	0.1544±66	0.0808±26	0.1022±61

Anisotropic temperature factors given.

Interatomic distances (Å)

Sixfold coordination around Mo

Mo – Cl(1)	2.26	[2.28]
Mo – Cl(2)	2.44	[2.42]
Mo – Cl(3)	2.78	[2.79]
Mo – Cl(2)'	2.46	[2.47]
Mo – Cl(3)'	2.34	[2.32]
Mo – O	1.63	

σ(Mo–Cl) = 0.02, σ(Mo–O) = 0.04

Mo – Cl(2) – Mo'	103.2±6°
Mo – Cl(3) – Mo'	96.7±5°

Angles in degrees subtended at Mo by

	Cl(2)	Cl(3)	Cl(2)'	Cl(3)'	O
Cl(1)	93	162	90	83	104
Cl(2)		76	88	160	99
Cl(3)			79	84	172
Cl(2)'				84	94
Cl(3)'					99

σ(O–Mo–Cl) = 1°, σ(Cl–Mo–Cl) = 1°

Discussion

The crystal is composed of $MoOCl_5$ octahedra linked through Cl(2) and Cl(3) into chains running along *c*. The terminal Cl(1) and O atoms are mutually *cis*. An independent determination of the structure of comparable accuracy is given in 1.

Details of analysis

Single crystal (0.01 × 0.02 × 0.3 mm^3) sealed in a capillary under nitrogen. Intensities measured on manual diffractometer with stationary crystal – stationary counter method (MoK_α radiation). Structure solved from Patterson function and refined by least squares to give R = 0.106 for 276 non-zero reflexions listed.

1. G. FERGUSON, M. MERCER and D.W.A. SHARP, 1969. *J. Chem. Soc. Lond. A*, 2415.

MOLYBDENUM OXYTRIBROMIDE

I. The crystal and molecular structure of molybdenum (V) oxytribromide. M.G.B. DREW and I.B. TOMKINS, 1970. *Acta Cryst.*, B26, 1161.

$MoOBr_3$ F.W. = 351.64

Tetragonal, a = 11.360±8, c = 3.948±4 Å, U = 509.48 Å^3, Z = 4, D_x = 3.508.

Space group $P4_2/mnm$ (D^{14}_{4h})

Atomic positions

4 Mo	in	8(*j*)	with	x = 0.1201±5,	z = 0.0846±10,	B = 0.0011±1 Å^2
4 Br(1)	in	4(*g*)	with	x = 0.1075±6,	B = 0.0018±1 Å^2,	
8 Br(2)	in	8(*i*)	with	x = 0.3300±3,	y = 0.1088±3 Å^2,	B = 0.0022±1 Å^2
4 O	in	4(*g*)	with	x = 0.390 ±4,	B = 0.0018±5 Å^2,	

Interatomic distances (Å)

Mo – O(1) 1.65±1; Mo – O(1') 2.31±1, Mo – Br(1) 2.61±1, Mo – Br(2) 2.41±1

Preparation of materials

Crystals prepared by the method of Colton and Tomkins (1).

Description of the structure

The Mo_2Br_6 entities with two bromine atoms as bridges are extended through the crystal by oxygen atom bridges. The short Mo–O bonds are regarded as ordered within domains yielding individual cells conforming to the space group symmetry.

Details of analysis

149 reflexions measured with a diffractometer using MoK_α radiation gave better agreement for a model involving disordered Mo positions than two ordered models investigated.

1. R. COLTON and I.B. TOMKINS, 1965. *Aust. J. Chem.*, 18, 447.

TECHNETIUM OXIDE TETRAFLUORIDE

I. Fluoride crystal structures. Part XII. Trimeric technetium oxide tetrafluoride. A.J. EDWARDS, G.R. JONES and R.J.C. SILLS, 1970. *J. Chem. Soc. Lond., A*, 2521-2523.

$TcOF_4$ F.W. = 190.9

Hexagonal, a = 9.00±1, c = 7.92±1 Å, U = 555.6 Å^3, Z = 6, D_x = 3.42.

Space group $P6_2/m$ (C^2_{6h})

Atomic positions

			x	y	z	B* (Å^2)
6 Tc	in	6(*h*)	0.0768±3	0.3992±3	1/4	0.48±1
6 O	in	6(*h*)	-0.1323±32	0.3272±29	1/4	1.21±13
6 F(1)	in	6(*h*)	0.0988±47	0.2079±45	1/4	1.76±17
6 F(2)	in	6(*h*)	0.3646±21	0.5139±22	1/4	0.59±7
12 F(3)	in	12(*i*)	0.1224±26	0.4217±25	0.0267±31	1.41±11

*Converted from isotropic U.

Interatomic distances

Tc	–	O	1.66±3
Tc	–	F(3)	1.80±3 (×2)
Tc	–	F(1)	1.82±5
Tc	–	F(2)	1.89±2
Tc	–	F(2)'	2.26±2
Tc	–	F(2) – Tc'	161±2°

Angles in degrees subtended at Tc by

	F(3)'	F(1)	F(2)	F(2)'
O	101	106	97	176
F(3)	157	89	86	79
F(1)			157	78
F(2)				79

σ(angles) = 1°

Discussion

The molecule consists of a cyclic trimer linked through the asymmetric F(2) bridge.

Details of analysis

Lattice parameters from oscillation and Weissenberg photographs with MoK_α radiation (λ = 0.7107 Å) or CuK_α radiation (λ = 1.5418 Å). Intensities measured photometrically from integrated [Weissenberg?] photographs of layers with h = 0 and l = 0 - 6 taken with MoK_α radiation (μ = 3.8 mm^{-1}) of a crystal 0.3 mm thick. Lorentz and polarisation, but no absorption correction. Structure determination from Patterson and electron-density projection refined by least squares to give R = 0.108 for 368 reflexions listed in 1.

1. *Supplementary Publications Scheme SUP* 20020, National Lending Library, Boston Spa, U.K.

BARIUM OXYCHLORIDE

I. Structure cristalline de l'oxychlorure de baryum, Ba_4OCl_6. B. FRIT, B. HOLMBERG and J. GALY, 1970. *Acta Cryst.*, B26, 16-19.

Ba_4OCl_6 F.W. = 707.17

Hexagonal, a = 9.97±1, c = 7.49±1 Å, [U = 644.7 Å^3], D_m = 3.98, Z = 2, D_x = 4.01.

Space group $P6_3mc$ (C^4_{6v}) other choices not considered

Atomic positions

6 Ba(1)	in	6(c)	x = 0.1955, z = 0	B = 1.2 Å^2
6 Cl(1)	in	6(c)	x = 0.1456, z = 0.3926	B = 1.8
6 Cl(2)	in	6(c)	x = 0.4704, z = 0.7044	B = 1.7
2 Ba(2)	in	2(b)	z = 0.4366	B = 1.5
2 O	in	2(b)	z = 0.0975	B = 0.8

Interatomic distances (Å)

7 Ba(1)	- Cl between	3.06 and 3.38
Ba(1)	- O	2.49
Ba(2)	- Cl(1) (3×)	3.26
Ba(2)	- Cl(2) (3×)	3.10
Ba(2)	- Cl(2) (3×)	3.81
Ba(2)	- O	2.54

Material

Direct reaction of BaO and $BaCl_2$ in a gold cell at 900°C in a nitrogen atmosphere.

Discussion of the structure

Six corner and the three face positions of a triangular prism about Ba(2), but only six corners and two face positions about Ba(1) are occupied.

Details of analysis

Least-squares refinement, 136 structure factors, R = 0.052. Data measured with a microdensitometer from integrated Weissenberg photographs taken with Cu radiation and corrected for absorption. No errors given for atomic parameters, but errors of ±0.02 Å specified for all bond lengths. Despite absorption corrections vibrational amplitude of oxygen smaller than that of the other atoms.

BIS(DICHLOROPHOSPHATE) BIS(PHOSPHORYL CHLORIDE) MAGNESIUM

$Mg(PO_2Cl_2)_2(POCl_3)_2$ F.W. = 598.7

I. Direct determination of the crystal structure of bis(dichlorophosphoryl chloride) magnesium: $Mg(PO_2Cl_2)_2(POCl_3)_2$. J. NYBORG and J. DANIELSEN, 1970. *Acta Chem. Scand.*, 24, 59-71.

Triclinic, $a = 11.21$, $b = 10.97$, $c = 9.75$ Å, $\alpha = 116°44'$, $\beta = 83°51'$, $\gamma = 114°5'$, $Z = 2$, $D_x = 2.04$.

Space group $P\bar{1}$ (C_i^1) from final structure

Atomic positions

All atoms in 2(*i*)

	x	*y*	*z*
Mg	0.0586±4	0.9926±5	0.2322±4
O(1)	0.2155±11	0.1602±13	0.1864±11
O(2)	0.1058±11	0.1431±12	0.4542±10
O(3)	0.0804±10	0.1658±12	0.7229±10
O(4)	0.2063±12	0.9278±14	0.2566±12
O(5)	0.0438±10	0.8512±12	0.0064±9
O(6)	0.0693±10	0.9335±12	0.7975±10
P(1)	0.3281±4	0.2839±6	0.1918±5
P(2)	0.2490±5	0.8261±6	0.2679±5
P(3)	0.1521±5	0.2176±4	0.6151±4
P(4)	0.0840±5	0.8418±5	0.8579±4
Cl(1)	0.3540±6	0.4713±6	0.3751±6
Cl(2)	0.3288±8	0.3215±8	0.0200±7
Cl(3)	0.4906±6	0.2605±9	0.1964±10
Cl(4)	0.4385±6	0.8913±10	0.2693±9
Cl(5)	0.1770±8	0.6265±7	0.0975±6
Cl(6)	0.2054±7	0.7998±7	0.4511±6
Cl(7)	0.3366±5	0.2374±8	0.6309±6
Cl(8)	0.1816±6	0.4312±5	0.6971±6
Cl(9)	0.2735±5	0.8719±7	0.8511±6
Cl(10)	0.0023±6	0.3710±6	0.2993±5

Interatomic distances (Å)

Standard error ∿ 0.010 Å for distances and ∿ 0.5° for angles.

Sixfold coordination around Mg

Mg	-	O(1)	2.146
Mg	-	O(2)	2.003
Mg	-	O(3)	1.993
Mg	-	O(4)	2.122
Mg	-	O(5)	2.019
Mg	-	O(6)	2.010

Angles in degrees subtended at Mg by

	O(2)	O(3)	O(4)	O(5)	O(6)
O(1)	85.5	177.1	85.1	87.7	89.9
O(2)		94.1	87.7	170.4	92.6
O(3)			92.1	92.4	93.0
O(4)				85.0	174.9
O(5)					94.2

Fourfold coordination around P(1)

P(1)	-	O(1)	1.410
P(1)	-	Cl(1)	1.955
P(1)	-	Cl(2)	1.895
P(1)	-	Cl(3)	1.948

Angles in degrees subtended at P(1) by

	Cl(1)	Cl(2)	Cl(3)
O(1)	114.3	113.1	113.0
Cl(1)		106.1	103.7
Cl(2)			105.7

Fourfold coordination around P(2)

P(2)	-	O(4)	1.429
P(2)	-	Cl(4)	1.944
P(2)	-	Cl(5)	1.944
P(2)	-	Cl(6)	1.917

Angles in degrees subtended at P(2) by

	Cl(4)	Cl(5)	Cl(6)
O(4)	112.1	113.7	114.0
Cl(4)		105.4	105.5
Cl(5)			105.5

Fourfold coordination around P(3)

P(3)	-	O(2)	1.440
P(3)	-	O(3)'	1.434
P(3)	-	Cl(7)	2.005
P(3)	-	Cl(8)	1.995

Angles in degrees subtended at P(3) by

	O(3)	Cl(7)	Cl(8)
O(2)	122.0	108.3	108.0
O(3)		108.6	107.3
Cl(7)			100.8

Fourfold coordination around P(4)

P(4)	-	O(5)	1.441
P(4)	-	O(6)	1.440
P(4)	-	Cl(9)	2.010
P(4)	-	Cl(10)	1.997

Angles in degrees subtended at P(4) by

	O(6)	Cl(9)	Cl(10)
O(5)	122.3	108.7	108.1
O(6)		107.0	107.9
Cl(9)			100.8

Angles at O atoms

Mg	-	O(1)	-	P(1)	167.0
Mg	-	O(4)	-	P(2)	152.6
Mg	-	O(2)	-	P(3)	162.0
Mg	-	O(5)	-	P(4)	144.2

Discussion

Each Mg atom is octahedrally surrounded by six oxygen atoms, two *cis* O atoms belonging to $POCl_3$ groups and the remainder belonging to PO_2Cl_2 groups arranged in two centrosymmetrically related pairs. Each pair bridges between a different pair of Mg atoms. In this way, the Mg atoms are strung together by double bridges into chains along the *c* axis. The chains linked by Cl–Cl van der Waals bonds of 3.4 Å or more.

Details of analysis

Crystals 0.25 × 0.03 × 0.03 cm^3 sealed in a Lindemann glass tube. Cell constants from oscillation and Weissenberg photographs (CuK_α radiation) and precession and Rimsky photographs (MoK_α radiation). Intensities (l = 0 to 12) measured on linear diffractometer with balanced filters, but accurate values difficult to obtain because of reflexions from small crystallites attached to the main crystal. 2271 non-zero reflexions (listed) corrected for Lorentz and polarisation effects. Structure solved by symbolic addition and Fourier synthesis, and refined by least-squares methods to give R = 0.10.

POTASSIUM BORATOHEXAFLUOROBERYLLATE

I. [Crystal structure of potassium boratohexafluoroberyllate, $KBe_2(BO_3)F_2$]. L.P. SOLOV'EVA and V.V. BAKAKIN, 1970. *Kristallografija*, 15, 922-925 [*Soviet Physics - Crystallography*, 15, 802-805].

$KBe_2(BO_3)F_2$ F.W. = 153.93

Monoclinic, a = 7.718±6, b = 4.444±2, c = 6.790±3 Å, β = 112.8±1°, U = 214.7 Å^3, D_m = 2.40, Z = 2, D_x = 2.38.

Space group $C2$ (C_2^3)

Atomic positions

			x	y	z	B(Å^2)
2 K	in	2(*a*)	0	0	0	2.0±1
2 B	in	2(*b*)	0	0.519±9	0.5	1.7±3
4 Be	in	4(*c*)	0.198±2	0.025±7	0.590±3	1.6±3
2 O(1)	in	2(*b*)	0	0.190±4	0.5	1.8±2
4 O(2)	in	4(*c*)	0.344±1	0.157±3	0.498±2	2.0±2
4 F	in	4(*c*)	0.222±1	0.496±4	0.165±1	2.6±2

Interatomic distances (Å)

K	-	F	2.75	(×2),	2.78 (×2), 2.78 (×2), σ(K-F) = 0.01				
B	-	O(1)	1.46±4		O(1)	-	B	- O(2)	117±2°
B	-	O(2)	1.35±2	(×2)	O(2)	-	B	- O(2)'	126±3°
Be	-	F	1.54±2						
Be	-	O(1)	1.59±2						
Be	-	O(2)	1.60±3						
Be	-	O(2)'	1.73±3						

Angles in degrees subtended at Be by

	O(1)	O(2)	O(2)'
F	112	114	104
O(1)		114	106
O(2)			106

Discussion

BO_3 triangles are linked into sheets in the *ab* plane by sharing each oxygen with two BeO_3F tetrahedra, the beryllium atoms lying alternately above and below the oxygen plane. The sheets are separated by K atoms through K-F interactions.

Details of analysis

Lattice parameters from single crystal diffractometer measurements. Intensities estimated visually from Weissenberg photographs of layers with k = 0 to 4 taken with MoK_α radiation. No correction for absorption (crystal size 0.5 × 0.5 × 0.2 mm^3). Structure determined from Patterson function refined by least squares to give R = 0.131 for 340 reflexions [not listed].

SODIUM HYDROGEN TRIFLUOROBORATE

$NaBF_3OH$ F.W. = 107.81

I. The crystal and molecular structure of $NaBF_3OH$. M.J.R. CLARK and H. LYNTON, 1970. *Can. J. Chem.*, 48, 405–409.

Hexagonal, a = 8.08_4, c = 7.95_8 Å, U = 450.4 $Å^3$, D_m = 2.38, Z = 6, D_x = 2.38_5.

Space group $P6_3$ (C_6^6) from systematic absences, pyroelectric effect and final structure

Atomic positions

				x	*y*	*z*
2	Na(1)	in	2(*b*)	1/3	2/3	0.4495±19
2	Na(2)	in	2(*b*)	1/3	2/3	0.0581±20
2	Na(3)	in	2(*a*)	0	0	0
6	B	in	6(*c*)	0.3485±22	0.3086±21	0.2607±34
6	F(1)	in	6(*c*)	0.1539±14	0.2268±13	0.2380±29
6	F(2)	in	6(*c*)	0.4108±17	0.2357±17	0.3781±18
6	F(3)	in	6(*c*)	0.3789±17	0.2384±17	0.0946±18
6	F(4)	in	6(*c*)	0.4379±14	0.5117±15	0.2563±30

Anisotropic (B and F) and isotropic (Na) temperature factors given.

Interatomic distances (Å)

BF OH ion (treated as BF_4)

B	-	F(1)	1.38
B	-	F(2)	1.33
B	-	F(3)	1.51
B	-	F(4)	1.43

Angles in degrees subtended at B by

	F(1)	F(2)	F(3)
F(2)	118		
F(3)	94	106	
F(4)	110	116	109

Sixfold coordination around Na(1) (crystallographic site symmetry C_3)

Na(1)	-	F(3)	2.36	(×3)
Na(1)	-	F(4)	2.39	(×3)

Sixfold coordination around Na(2) (crystallographic site symmetry C_3)

Na(2)	- F(2)	2.31	(×3)
Na(2)	- F(4)	2.42	(×3)

Coordination around Na(3) (crystallographic site symmetry C_3)

Na(3)	- F(1)	2.49	(×3)
Na(3)	- F(1)	2.64	(×3)
Na(3)	- F(3)	2.79	(×3)
Na(3)	- F(2)	3.05	(×3)

Standard error in distances is 0.02 Å, in angles is 2°.

Discussion

It was found impossible to identify which "fluorine" atom was the hydroxyl group. The BF_3OH ions lie on hexagonal close packed sites with Na in the octahedral interstices.

Details of analysis

Prepared by method of 1. Lattice constants measured from Weissenberg photographs calibrated with aluminum powder. Equi-inclination Weissenberg photographs of layers l = 0 to 6 (crystal 0.2 × 0.2 × 0.8 mm^3) and h = 0 (crystal 0.2 × 0.2 × 0.3 mm^3) used for visual estimation of intensities which were corrected for Lorentz, polarisation and absorption effects. Phases determined by direct methods and structure refined by Fourier and least-squares techniques. Final R = 0.090 for 195 observed reflexions listed in 2.

1. I.G. ROSS and M.M. SLUCKAJA, 1952. *J. Gen. Chem. USSR.*, 22, 45.
2. Depository of Unpublished Data, National Science Library, National Research Council of Canada, Ottawa, Canada.

TRISILVER (I) IODIDE DINITRATE

I. Strukturbestimmung von AgJ: $2AgNO_3$. R. BIRNSTOCK and D. BRITTON, 1970. *Z. Kristallogr.*, 132, 87-98.

$Ag_3I(NO_3)_2$ F.W. = 574.52

Orthorhombic, a = 7.53±1, b = 12.47±2, c = 7.86±1 Å, U = 738 Å³, D_m = 5.04, Z = 4, D_x = 5.16.

Space group $P2_12_12_1$ (D_2^4)

Atomic positions

All atoms in 4(a)

	x	y	z	B(Å²)
I	0.7235±4	0.6112±3	0.1510±5	1.34±11*
Ag(1)	0.0557±6	0.0634±4	0.0283±8	1.71±15*
Ag(2)	0.4453±6	0.0461±5	0.0242±9	2.39±25*
Ag(3)	0.5878±8	0.2230±7	0.4500±7	3.71±102*
O(11)	0.936 ±5	0.253 ±3	0.043 ±5	1.82±70
O(12)	0.815 ±8	0.150 ±5	0.234 ±9	5.25±144
O(13)	0.712 ±5	0.320 ±3	0.240 ±5	1.59±73
O(21)	0.253 ±7	0.457 ±4	0.084 ±6	2.49±77
O(22)	0.083 ±5	0.398 ±4	0.290 ±6	2.04±79
O(23)	0.391 ±8	0.401 ±5	0.319 ±9	4.84±132
N(1)	0.838 ±8	0.246 ±6	0.140 ±10	4.10±140
N(2)	0.234 ±5	0.417 ±3	0.230 ±5	0.44±62

* Anisotropic temperature factors also given.

Interatomic distances (Å)

Ag(1) - O(21)	2.46		Ag(2) - O(11)	2.56
Ag(1) - O(11)	2.53		Ag(2) - O(22)	2.77
Ag(1) - O(12)	2.66		Ag(2) - I	2.922
Ag(1) - O(22)	2.72		Ag(2) - Ag(1)	2.946
Ag(1) - O(13)	2.82		Ag(2) - I	2.966
Ag(1) - I	2.886		Ag(2) - N(2)	2.99
Ag(1) - N(1)	2.94		Ag(2) - N(1)	3.01
Ag(1) - Ag(2)	2.946		Ag(3) - O(13)	2.25
Ag(1) - O(23)	3.03		Ag(3) - O(22)	2.54
Ag(1) - I	3.079		Ag(3) - O(12)	2.58
Ag(1) - N(2)	3.17		Ag(3) - I	2.843
Ag(2) - O(21)	2.47		Ag(3) - O(23)	2.86
Ag(2) - O(23)	2.51		Ag(3) - I	2.966
			Ag(3) - N(1)	3.10
N(1) - O(11)	1.06		N(2) - O(22)	1.26
N(1) - O(12)	1.42		N(2) - O(21)	1.26
N(1) - O(13)	1.54		N(2) - O(23)	1.39

Angles in degrees subtended at N(1) by

	O(12)	O(13)
O(11)	127	138
O(12)		100

Angles in degrees subtended at N(2) by

	O(21)	O(23)
O(22)	121	124
O(21)		115

Standard errors in Ag-Ag, Ag-I = 0.006;
Ag-O , Ag-N = 0.05;
N-O = 0.07 Å
O-N-O = 6°

Discussion

The structure can best be considered as composed of edge and corner sharing IAg_6 triangular prisms (prism axis *a*) with the NO_3 groups lying between.

Details of analysis

Lattice parameters from Weissenberg and precession photographs taken with MoK_α radiation (λ = 0.7107 Å). Intensities estimated visually from precession photographs of layers with l = 0 to 3 and k = 0 to 4 taken with MoK_α radiation, corrected for Lorentz and polarisation effects, but not absorption (μ = 11.8 mm^{-1}, crystal size = 0.05 × 0.06 × 0.3 mm^3). Structure determined by Patterson and refined by least squares to give R = 0.106 for 524 observed reflexions listed.

AMMONIUM FLUOROSULPHATE

NH_4SO_3F F.W. = 117.1

I. Crystal structure of ammonium fluorosulphate. R. O'SULLIVAN, R.C. THOMPSON and J. TROTTER, 1970. *J. Chem. Soc. Lond. A*, 1814-1817.

Orthorhombic, a = 8.972±10, b = 5.996±10, c = 7.542±10 Å, U = 405.7 Å^3, D_m ∿ 2.0, Z = 4, D_x = 1.92.

Space group *Pnma* (D_{2h}^{16}) systematic absences and final structure

Atomic positions

			x	*y*	*z*
4 N	in	4(*c*)	0.1751±10	1/4	0.1638±13
4 S	in	4(*c*)	0.0755±3	1/4	-0.3076±2
4 O(1)	in	4(*c*)	0.1933±8	1/4	-0.4443±8
8 O(2)	in	8(*d*)	0.0828±5	0.0492±5	-0.2026±5

4 F	in	4(*c*)	−0.0700±7	1/4	−0.4061±8
4 H(1)	in	4(*c*)	0.19 ±15	1/4	0.29 ±15
4 H(2)	in	4(*c*)	0.13 ±15	1/4	0.15 ±15
8 H(3)	in	8(*d*)	0.20 ±15	0.13 ± 15	0.09 ±15

Anisotropic temperature factors given for N, S, O and F. B(H) = 2.02 Å^2 (assumed).

Interatomic distances (Å)

Fourfold coordination around S (crystallographic point symmetry C_s)

S	−	O(1)	1.476
S	−	O(2)	1.443 (×2)
S	−	F	1.502

Angles in degrees at S subtended by

	O(2)'	F
O(1)	110.5	106.1
O(2)	113.2	108.1

σ(distance) = 0.008 but thermal motion corrections of order 0.01 Å should be applied

σ(angle) at least 0.5° and probably higher because of disorder.

Fourfold coordination around N (crystallographic site symmetry C_s)

N − H (mean) 0.8±1 Å (standard deviation of individual bonds 0.08 Å)

N − N − H (mean) 108±15° (standard deviation of individual angles 16°)

Discussion

The deviation of the SO_3F ion from strict C_{3v} symmetry is assumed to mean that 25% of the ions are disordered with F occupying the position of O(1). The partial ordering (as against the complete disorder in KSO_3F (1)) is probably the result of hydrogen bond formation, N–H(1)...O(1) = 2.96 Å being the one best defined.

Details of analysis

Single crystals prepared from NH_4Cl and HSO_3F and recrystallized from ethanol. Lattice parameters from X-ray photographs and diffractometer measurements. Intensities measured on single-crystal diffractometer (GE-XRD6, CuK_α radiation, θ − 2θ scan). [Lorentz, polarisation] and absorption (μ = 6.3 mm^{-1}, crystal size 0.2 × 0.2 × 0.05 mm^3) corrections. Structure assumed from isotypic KSO_3F (1) and 41 parameters refined by least squares to give R = 0.078 for 293 observed reflexions listed.

1. K. O'SULLIVAN, R.C. THOMPSON and J. TROTTER, 1967. *J. Chem. Soc. Lond. A*, 2024.

SYNTHETIC SPODIOSITES

I. The crystal structures of synthetic spodiosites. Ca_2VO_4Cl and Ca_2AsO_4Cl.
F. BANKS, W. GREENBLATT and B. POST, 1970. *Inorg. Chem.*, 9, 2259-2263.

VANADIUM SPODIOSITE

Ca_2VO_4Cl F.W. = 230.55

Orthorhombic, a = 6.311±5, b = 7.140±5, c = 11.052±5 Å, U = 498.0 Å^3, D_m = 3.06, Z = 4, D_x = 3.07.

Space group *Pbcm* (C_{2h}^{11}) isostructural with Ca_2CrO_4Cl (1).

Atomic positions Anisotropic temperature factors given.

			x	*y*	*z*
4 Ca(1)	in	4(*c*)	0.6218±1	0.25	0
4 Ca(2)	in	4(*d*)	0.1395±1	0.4717±1	0.25
4 Cl	in	4(*d*)	0.4886±1	0.1944±1	0.25
4 V	in	4(*c*)	0.1266±1	0.25	0
8 O(1)	in	8(*e*)	0.0345±2	0.7308±2	0.3762±1
8 O(2)	in	8(*e*)	0.7103±2	0.5652±2	0.5285±1

ARSENIC SPODIOSITE

Ca_2AsO_4Cl F.W. = 254.53

Orthorhombic, a = 6.318±2, b = 7.108±2, c = 11.058±3 Å, U = 496.6 Å^3, D_m = 3.42, Z = 4, D_x = 3.40.

Space group *Pbcm* (C_{2h}^{11})

Atomic positions

			x	y	z
4 Ca(1)	in	4(*c*)	0.6252±1	0.25	0
4 Ca(2)	in	4(*d*)	0.1384±1	0.4764±1	0.25
4 Cl	in	4(*d*)	0.4958±1	0.2007±1	0.25
4 As	in	4(*c*)	0.12909±3	0.25	0
8 O(1)	in	8(*e*)	0.0297±2	0.7352±2	0.3774±1
8 O(2)	in	8(*e*)	0.7100±2	0.5669±2	0.5272±1

Anisotropic temperature factors given.

Interatomic distances (Å)

		Ca_2VO_4Cl	Ca_2AsO_4Cl
V, As - O(1)	(×2)	1.711	1.690
V, As - O(2)	(×2)	1.703	1.679
Ca(1) - O(2)	(×2)	2.340	2.335
Ca(1) - O(2)'	(×2)	2.498	2.504
Ca(1) - O(1)	(×2)	2.570	2.570
Ca(1) - Cl	(×2)	2.915	2.904
Ca(2) - O(1)	(×2)	2.410	2.417
Ca(2) - O(1)'	(×2)	2.472	2.460
Ca(2) - O(2)	(×2)	2.639	2.661
Ca(2) - Cl		2.837	2.808
Ca(2) - Cl'		2.963	2.990
O(1) - V - O(2)	(×2)	116.5°	117.1°
O(1) - V - O(1)'		107.1	107.1
O(1) - V - O(2)'	(×2)	105.9	105.3
O(2) - V - O(2)'		105.6	105.4

σ(distance) = 0.001 Å, σ(angles) = 0.1°

Discussion

Ca_2VO_4Cl and Ca_2AsO_4Cl are isostructural and consist of alternating planes perpendicular to c of Ca(1)VO_4 and Ca(2)Cl, the former groups lying on a two-fold axis along a, the latter lying on the mirror plane in the layer.

Details of analysis

Crystals grown from melt. Lattice parameters for Ca_2AsO_4Cl from least squares refinement of 12 diffractometer measured reflexions. Lattice parameters for Ca_2VO_4Cl from powder diffractometer measurements. Intensities for Ca_2VO_4Cl measured from a spherical crystal (r = 0.18 mm) on an equi-inclination automatic diffractometer with crystal monochromated MoK_α radiation (layers k = 0 - 9). Intensities for Ca_2AsO_4Cl measured from a spherical crystal (r = 0.16 mm) on an equiatorial automatic diffractometer with filtered MoK_α radiation using θ - 2θ scan. Intensities corrected for Lorentz, polarisation and absorption effects (μ = 4.48 and 9.6 mm^{-1} respectively). The structure of Ca_2CrO_4Cl was assumed and refined by least squares (including isotropic extinction correction) to give R_2 = 0.032 for 799 observed reflexions listed for Ca_2VO_4Cl and R_2 = 0.024 for 1147 observed reflexions listed for Ca_2AsO_4Cl.

1. M. GREENBLATT, E. BANKS and B. POST, 1967. *Acta Cryst.*, 23, 166.

HYDRATES OF HYDROGEN BROMIDE

I. Hydrogen bond studies, XL. The crystal structure of three hydrates of hydrogen bromide, $HBr.nH_2O$, n + 1, 2 and 3. J.- O. LUNDGREN, 1970. *Acta Cryst.*, B26, 1893-1899.

$HBr.H_2O$ F.W. = 98.93

Rhombohedral (hexagonal parameters) at − 182°C, a = 5.058±1, b = 8.906 4 Å, U = 197.3 Å^3, Z = 3, D_x = 2.497.

Space group $R3m$ (C^5_{3v}) only space group tested

Atomic parameters

3 Br	in	3(*a*)	with	z = 0,	β_{11} = 0.075±1,	β_{33} = 0.004±3
3 O	in	3(*a*)	with	z = 0.451±3,	β_{11} = 0.03 ±1,	β_{33} = 0.004±3

$HBr.2H_2O$ F.W. = 116.97

Monoclinic, at − 190°C, a = 4.164±6, b = 12.422±3, c = 6.946±5 Å, β = 101.21±12°, U = 352.4 Å^3, Z = 2, D_x = 2.204.

Space group $P2_1/c$ (C^5_{2h})

Atomic parameters

	x	y	z
Br	0.0337±3	0.3302±1	0.1561±1
O(1)	0.570 ±2	0.1241±6	0.042 ±1
O(2)	0.310 ±3	0.0575±7	0.300 ±1

Anisotropic temperature factors are given.

$HBr.3H_2O$ F.W. = 134.97

Orthorhombic, at − 62°C, a = 6.228±3, b = 5.961±1, c = 11.722±1 Å, U = 435.2 Å^3, Z = 4, D_x = 2.060.

Space group $Aba2$ (C^{17}_{2v}) [others rejected as unreasonable but no details were given.]

Atomic parameters

4 Br	in	4(*a*)	with	z = 0		
4 O(1)	in	4(*a*)	with	z = 0.363±2		
4 O(2)	in	8(*b*)	with	x = 0.320±2,	y = 0.087±2,	z = 0.2163±6

$HBr.2H_2O$

The water molecules are associated forming $H_5O_2^+$ ions. Each water molecule is bonded to two bromine ions through hydrogen bonds and each bromine is bonded to four water molecules. One end of the H_5O_2 group bonds with Br ions in one layer, and the other end with Br ions in the next adjacent layer. The compound is isostructural with $H_5O_2^+Cl^-$ (1).

$HBr.3H_2O$

Layers of bromide ions parallel to the *ab* plane are interleaved with layers of water molecules with the latter forming chains by hydrogen bonding parallel to *a*.

Details of analysis

Data recorded with Weissenberg camera, CuK_α radiation, intensities visually estimated and corrected for absorption. Parameters refined by least-squares method using 64, 543 and 162 reflexions to R values of 0.067, 0.092 and 0.069 for $HBr.nH_2O$ with n = 1, 2 and 3 respectively. Diffuse streaks along reciprocal lattice rows were observed in the trihydrate and a change to space $Bba2$ (once) indicates a new phase between −180 and −190°C.

1. J.O. LUNDGREN and I. OLOVSSON, 1967. *Acta Cryst.*, 23, 966.

BASIC SAMARIUM CHLORIDE

I. [The structure of basic samarium chloride]. T.N. TARKHOVA, N.N. MIRONOV and I.A. GRIŠIN, 1970. *Ž. Strukt. Khim.*, 11, 556-557 [*J. Struct. Chem.*, 11, 515-516].

$SmCl(OH)_2$ F.W. = 219.86

Monoclinic, a = 6.18, b = 6.75, c = 3.82 Å, γ = 112°, U = 148 Å^3, D_m = 4.74, Z = 4, D_x = 4.93.

Space group $P2_1/m$ (c axis setting) (C^2_{2h})

Atomic positions

All atoms in 2(e) ±(x, y, 1/4 or 3/4)

	x	y	z	B(Å^2)
Sm	0.129	0.332	0.25	0.35
Cl	0.559	0.306	0.25	1.53
O(1)	0.907	0.307	0.75	1.57
O(2)	0.157	0.088	0.75	2.63

Interatomic distances (Å)

Errors uncertain but at least 0.02 Å.

Sm	-	O(1)	2.332±0	(×2)
Sm	-	O(2)	2.525±0	(×2)
Sm	-	O(1)*	2.596±0	
Sm	-	O(2)*	2.668±0	
Sm	-	Cl	2.700±1	

Angles in degrees subtended at Sm by

	O(1)'	O(2)	O(2)'	O(1)*	O(2)*	Cl
O(1)	110	62	140	77	79	125
O(2)			98	131	62	71
O(1)*					138	119
O(2)*						104

Discussion

The structure is similar to that of $YCl(OH)_2$ (1).

Details of analysis

Crystals grown hydrothermally. Lattice parameters from Laue, oscillation and goniometric photographs. Intensities estimated visually from Weissenberg photographs of layers with l = 0 to 3 and k = 0 taken with Mo radiation. Corrected for Lorentz and polarisation, but not for absorption (crystal size 1.0 × 0.15 × 0.05 mm^3). Structure determined from Patterson and electron-density functions and refined to give R = 0.15.

1. R.F. KLEVTSOVA and P.V. KLEVSTOV, 1966. *Ž. Strukt. Khim.*, 7, 556.

SODIUM IODIDE DIHYDRATE

I. Structure cristalline et protonique de l'iodure de sodium dihydraté $NaI.2H_2O$. J. VERBIST, P. PIRET and M. VAN MEERSCHE, 1970. *Bull. Soç. fr. Minér, Cristallogr.*, 93, 509-514.

$NaI.2H_2O$ F.W. = 185.92

Triclinic, $a = 7.161 \pm 1$, $b = 6.035 \pm 5$, $c = 7.282 \pm 5$ Å, $\alpha = 81.2° \pm 2$, $\beta = 118.7° \pm 2$, $\gamma = 115.3 \pm 2°$, $U = 249.0$ Å^3, $D_m = 2.48$, $Z = 2$, $D_x = 2.47$.

Space group $P\bar{1}$ (C_i^1)

Atomic positions

All atoms in 2(*i*)

	x	*y*	*z*
I	0.3764±2	0.2393±2	0.2228±1
Na	0.1430±15	0.3243±11	0.4805±10
O(1)	0.0129±28	0.6267±20	0.2865±18
O(2)	0.7795±26	0.9938±22	0.2879±17
H(11)	-0.1264	0.5894	0.1603
H(12)	0.1370	0.7890	0.2945
H(21)	0.7578	0.9580	0.1514
H(22)	0.6633	0.0352	0.2860

Anisotropic temperature factors are given.

Interatomic distances (Å) *and angles*

Na	-	O(1)	2.37 ±2
Na	-	O(2)	2.37 ±1
Na	-	O(2)'	2.42 ±1
Na	-	O(1)'	2.54 ±2
Na	-	I	3.210±6
Na	-	I	3.260±11

Angles in degrees subtended at Na by

	O(2)	O(2)'	O(1)'	I	I'
O(1)	93.6±5	167.2±6	84.4±7	88.2±4	101.0±6
O(2)		82.7±5	89.0±6	174.6±5	92.9±6
O(2)'			83.3±6	94.6±4	91.4±6
O(1)'				86.1±3	174.1±6
I					91.8±3

Na - I - Na 88.2±3°

O(1)	-	H(12)	-	I	3.62±1	167°
O(1)	-	H(11)	-	I'	3.64±1	147°
O(2)	-	H(22)	-	I	3.59±2	166°
O(2)	-	H(21)	-	I'	3.65±1	167°

H(11) - H(12)	1.58	H(11) - O(1) - H(12)	112°
H(21) - H(22)	1.64	H(21) - O(2) - H(22)	115°

Discussion

The structure consists of sheets in the *ab* plane composed of an hexagonal array of octahedrally coordinated Na atoms. Each Na is coordinated to its three neighbouring Na atoms by a centro-symmetric pair of I atoms or water molecules. Sheets are bonded by O-H...I hydrogen bonds.

Details of analysis

Lattice parameters from Weissenberg photographs with CuK_α and CoK_α radiation. Intensities estimated visually from Weissenberg photographs of layers with $h = 0$ to 5 taken with MoK_α radiation. Corrected for Lorentz, polarisation and spot shape, but not for absorption effects ($\mu = 6.36$ mm^{-1}, crystal size not given). Structure solved by heavy-atom method and electron-density functions, and refined by least squares to give $R = 0.085$ for 954 reflexions [not listed]. Proton positions estimated from single crystal NMR spectra (56.4 MHz, 14,200 G) assuming the hydrogen atoms are symmetrically displaced 0.96 Å from the oxygen atoms.

COPPER(II) CHLORIDE DIHYDRATE

I. An X-ray refinement of the crystal structure of copper (II) chloride dihydrate. A. ENGBERG, 1970. *Acta Chem. Scand.*, 24, 3510-3526.

$CuCl_2.2H_2O$ F.W. = 170.48

Orthorhombic, a = 8.126±8, b = 3.764±5, c = 7.440±7 Å, U = 227.6 Å^3, D_m = 2.49, Z = 2, D_x = 2.48.

Space group *Pmna* (D_{2h}^7)

Atomic positions

			x	*y*	*z*
2 Cu	in	2(*a*)	0	0	0
4 Cl	in	4(*h*)	0	0.3800±5	0.24037±17
4 O	in	4(*e*)	0.2408±4	0	0
8 H	in	8(*i*)	0.3065	0.1295	0.0822

Anisotropic temperature factors given

Interatomic distances (Å)

Cu	-	O	1.959±3	(×2)
Cu	-	Cl	2.290±1	(×2)
Cu	-	Cl	2.940±2	(×2)

Symmetry around Cu is 2/*m*

O - H		0.95	(×2)
O - H ... Cl		3.197±2	(×2)
H - O - Cl		7°	

Discussion

This work, which confirms and improves that of previous workers (1), was undertaken primarily to study thermal motions which are discussed in some detail.

Details of analysis

Experimental procedures are described in considerable detail. Lattice parameters measured by precession photographs and diffractometer. Intensities were measured on a diffractometer and corrected for a variety of effects [explained in detail]. Structure refined by least squares to give R = 0.053 for 800 reflexions listed.

1. *Strukturbericht*, 4, 104; *Structure Reports*, 21, 397.

POTASSIUM COPPER CHLORIDE DIHYDRATE

I. Neutron diffraction refinement of the crystal structure of potassium copper chloride dihydrate, $K_2CuCl_4.2H_2O$. R. CHIDAMBARAM, Q.O. NAVARRO, A. GARCIA, K. LINGGOATMODJO, LIN SHI-CHEIN, IL-HWAN SUH, A. SEQUEIRA and S. SRIKANTA, 1970. *Acta Cryst.*, B26, 827-830.

$K_2CuCl_4.2H_2O$ F.W. = 280.45

Tetragonal, a = 7.477±13, c = 7.935±19 Å, [U = 455.5 Å^3], Z = 2, [D_x = 2.045].

Space group $P4_2/mnm$ (D_{4h}^{14}) other possibilities not considered

Atomic parameters Anisotropic thermal parameters are given

2 Cu	in	2(*a*)	
4 K	in	4(*d*)	
4 Cl(1)	in	4(*g*)	with x = 0.2161±3
4 Cl(2)	in	4(*g*)	with x = 0.7262±3
4 O	in	4(*e*)	with z = 0.2484±5
4 H	in	4(*g*)	with x = 0.0738±10, z = 0.3178±6

Interatomic distances (Å) *and angles*

Cu - O (2×)	1.971±6			
Cu - Cl(1) (2×)	2.285±3			
Cu - Cl(2) (2×)	2.895±4			
K - Cl(2) (4×)	3.315±4			
K - Cl(1) (4×)	3.325±4			
O - H	0.955±6		H - O - H	109.7±7°
	0.966±6			

Discussion of the structure

The structure is essentially that proposed by Hendricks and Dickinson (1) with the proton-proton vector of the water molecule along the *ab* face diagonal. The potassium ions are surrounded by eight chlorine atoms at the corners of a cube.

Details of analysis

201 reflexions. λ = 1.17Å. Refined by least-squares methods to an *R* value of 0.039.

1. *Structure Reports*, 17, 343; 21, 401.

RUBIDIUM MANGANESE (II) CHLORIDE DIHYDRATE

I. Neutron diffraction study of β-$RbMnCl_3.2H_2O$. S.J. JENSEN and M.S. LEHMANN, 1970. *Acta Chem. Scand.*, 24, 3422-3424.

β-$RbMnCl_3.2H_2O$ F.W. = 282.80

Triclinic, *a* = 6.65, *b* = 7.01, *c* = 9.03 Å, α = 92.3°, β = 109.4, γ 112.9°, *U* = 359 Å^3, *Z* = 2, D_x = 2.61.

Space group $P\bar{1}$ (C_i^1)

Atomic positions

All atoms in 2(*i*)

	x	*y*	*z*	*B*(Å^2)
Re	0.256±2	0.934±1	0.177±1	*
Mn	0.998±3	0.331±2	0.329±2	*
Cl(1)	0.242±1	0.713±1	0.495±1	*
Cl(2)	0.771±1	0.959±1	0.187±1	*
Cl(3)	0.244±1	0.426±1	0.162±1	*
O(1)	0.743±2	0.749±1	0.491±1	*
O(2)	0.746±2	0.438±1	0.168±1	*
H(1)	0.581±4	0.685±3	0.461±3	4.8±4
H(2)	0.778±4	0.817±3	0.414±3	4.0±4
H(3)	0.640±4	0.459±3	0.190±2	4.4±4
H(4)	0.735±3	0.464±3	0.069±2	3.4±4

*Anisotropic temperature factors given in I.

Interatomic distances (Å) *and angles*

σ(distances) = 0.02 Å; σ(angles) = 1°

These errors, if derived from the least-squares refinement, are liable to be rather low because of the small numbers of reflexions used.

Rb - Cl(2)	3.33
Rb - Cl(1)	3.33
Rb - Cl(2)'	3.33
Rb - Cl(2)"	3.36
Rb - Cl(3)	3.49
Rb - H(4)	3.52

Rb	-	Cl(3)'	3.53
Rb	-	H(4)'	3.53
Rb	-	Cl(1)'	3.53
Rb	-	H(3)	3.54
Rb	-	Cl(3)"	3.59
Rb	-	O(1)	3.66
Mn	-	O(1)	2.21
Mn	-	O(2)	2.22
Mn	-	Cl(2)	2.47
Mn	-	Cl(3)	2.50
Mn	-	Cl(1)	2.55
Mn	-	Cl(1)'	2.59

Angles subtended at Mn by:

	O(2)	Cl(2)	Cl(3)	Cl(1)	Cl(1)'
O(1)	174°	91°	92°	91°	86°
O(2)		94	91	86	89
Cl(2)			94	93	176
Cl(3)				172	88
Cl(1)					85

Water (1)

O(1)	-	H(1)	0.92	O(1)	-	Cl(1)	3.25
O(1)	-	H(2)	0.90	O(1)	-	Cl(3)	3.40
O(1)	-	Mn	2.21	O(1)	-	Cl(2)	3.19
H(1)	-	Cl(1)	2.45				
H(1)	-	Cl(3)	2.83				
H(2)	-	Cl(2)	2.31				

O(1)	-	H(1)	-	Cl(1)	145°
O(1)	-	H(1)	-	Cl(3)	127°
O(1)	-	H(2)	-	Cl(2)	166°
Mn	-	O(1)	-	H(1)	127°
Mn	-	O(1)	-	H(2)	122°
H(1)	-	O(1)	-	H(2)	109°

Water (2)

O(2)	-	H(3)	0.85		H(3)	-	Cl(3)	2.48	
O(2)	-	H(4)	0.90		H(4)	-	Cl(3)	2.28	
O(2)	-	Mn	2.22		O(2)	-	Cl(3)	3.29	
Mn	-	O(2) - H(3)	125°		O(2)	-	Cl(3)	3.19	
Mn	-	O(2) - H(4)	127°		O(2)	-	H(4)	- Cl(3)	159°
H(3)	-	O(2) - H(4)	108°		O(2)	-	H(4)	- Cl(3)	172°

Discussion

The structure agrees with that proposed in 1.

Details of analysis

Intensities measured with 1.025 Å neutrons using ω scan. No absorption correction applied (μ = 1.31 cm^{-1}, crystal size = 0.4 × 0.05 × 0.05 cm). Structure of 1 assumed and H atoms found by nuclear density maps. 80 parameters refined to give R = 0.056 for 326 reflexions listed.

1. S.J. JENSEN, 1967. *Acta Chem. Scand.*, 21, 889.

HEPTASODIUMFLUORIDEBISARSENATE-19-HYDRATE

$Na[Na_6F(OH_2)_{18}](AsO_4)_2.H_2O$ F.W. = 800.10

I. A new type of polycation in heptasodiumfluoride-bisarsenate-19-hydrate.
E. TILLMANNS and W.H. BAUR, 1970. *Naturwissenschaften*, 57, 242.

Cubic, a = 28.12 Å, U = 22.235 Å^3, D_m = 1.89, Z = 32, D_x = 1.91.

Space group $Fd3c$ (O_h^8)

Preliminary atomic coordinates are given.

The $Na_6F(OH_2)_{18}$ ion has $\bar{3}$ symmetry with F at the centre octahedrally surrounded by 6 Na atoms. Each Na atom is further surrounded by 5 water molecules, four of which are shared with adjacent Na atoms to form a compact hydrated ion. Na - F = 2.42 Å, Na - OH_2 = 2.39 - 2.48 Å. One of the two AsO_4 ions is disordered. As - O = 1.68 - 1.70 Å.

Details of analysis

Solved by Patterson and Fourier methods and refined to R = 0.12 for 572 observed reflexions [not listed].

RUBIDIUM PALLADO-CYANIDE

I. La structure cristalline du palladocyanure de rubidium monohydrate. L. DUPONT, 1970. *Acta Cryst.*, B26, 964-971.

$Rb_2Pd(CN)_4.H_2O$ F.W. = 381.42

Orthorhombic, a = 10.01±2, b = 13.74±2, c = 7.44±2 Å, [U = 1023.3 Å^3], D_m = 2.56, Z = 4, [D_x = 2.46].

Space group $Pncn$ (D_{2h}^6)

Atomic positions

				x	y	z
4 Pd	in	4(*a*)	at	origin		
4 Rb(1)	in	4(*d*)	at	1/4	1/4	0.8169±4
4 Rb(2)	in	4(*c*)	at	1/2	0.0962±2	1/4
8 C(1)	in	8(*e*)	at	0.030±3	0.141 ±1	0.505±3
8 C(2)	in	8(*e*)	at	0.291±2	-0.021 ±2	0.472±3
8 N(1)	in	8(*e*)	at	0.046±2	0.225 ±1	0.507±3
8 N(2)	in	8(*e*)	at	0.308±2	-0.031 ±2	0.449±3
4 H_2O	in	4(*d*)	at	1/4	1/4	0.211±4

Anistropic temperature factors are listed in the paper.

Interatomic distances (Å)

Pd	-	C(1)	1.97±2	(2×)
Pd	-	C(2)	2.04±2	(2×)
C(1)	-	N(1)	1.16±3	(4×)
C(2)	-	N(2)	1.09±3	(4×)
Rb(1)	-	N(1)	3.10±2, 3.26±2	(2×)
Rb(1)	-	N(2)	3.22±2	(2×)
Rb(1)	-	H_2O	2.94±3	(2×)
Rb(2)	-	N(1)	3.14±2	(2×)
Rb(2)	-	N(2)	2.99±2, 3.08±2	(2×)
Rb(2)	-	H_2O	3.228±5	(2×)

Description of the structure

The structure contains chains of planar $Pd(CN)_4^{2-}$ groups parallel to c with the plane of the anion perpendicular to the chain. The chains are held together through interactions with Rb ions as seen in Fig. 32.

Details of analysis

814 reflexions were measured with a microdensitometer from integrated Weissenberg photographs taken with CuK_α radiation. Least-squares refinement lead to an R value of 0.092.

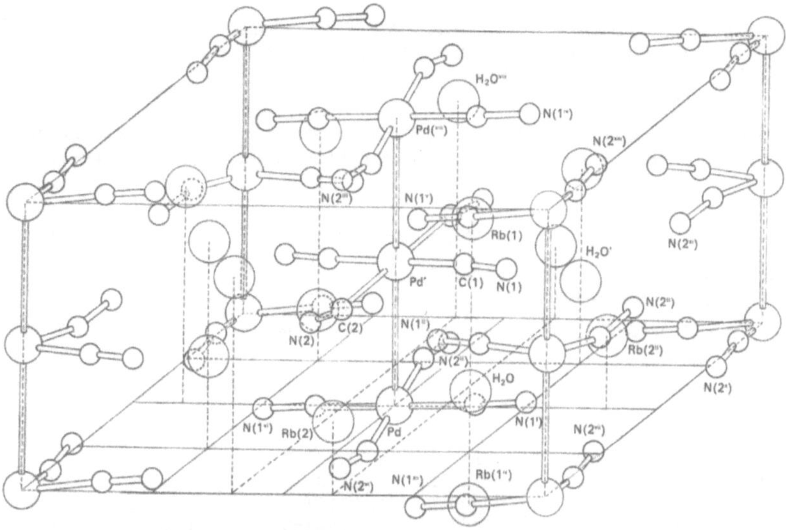

Fig. 32. A unit cell of $Rb_2Pd(CN)_4H_2O$.

POTASSIUM AURICYANIDE MONOHYDRATE

I. Structure cristalline de l'auricyanure de potassium monohydrate par le diffraction des neutrons. C. BERTINOTTI and E. BERTINOTTI, 1970. *Acta Cryst.*, B26,422-428.

$K[Au(CN)_4].H_2O$ F.W. = 358.16

Orthorhombic, a = 6.65, b = 7.12, c = 17.35 Å, [U = 821.5 Å^3], Z = 4, [D_x = 2.89].

Space group $P2_12_12_1$ (D_2^4)

Atomic positions

	x	y	z	B(Å^2)
Au	0.7589±9	-0.0714±8	0.7605±3	1.81
K	0.2594±34	-0.0364±25	0.5178±8	3.3
O	0.6317±18	0.1603±16	0.5190±5	3.7
N(1)	0.5846±11	0.0855±8	0.9151±3	3.1
N(2)	0.1436±25	-0.1922±9	0.8480±4	4.0
N(3)	0.9269±10	-0.2142±8	0.6031±3	3.1
N(4)	0.3643±10	0.0449±9	0.6753±3	3.5
C(1)	0.6453±11	0.0274±9	0.8580±3	2.0
C(2)	0.0082±12	-0.1458±10	0.8146±3	2.4
C(3)	0.8559±24	-0.1644±9	0.6612±3	2.0
H(1)	0.6754±34	0.1844±25	0.5681±9	4.8
H(2)	0.7168±49	0.0917±33	0.4926±12	5.5

Interatomic distances (Å) *and bond angles*

Au - C(1), -C(2), -C(3), -C(4)	1.982±8,	1.98±1,	1.955±9,	1.99±1
C - N	1.12 -	1.17		
O - H(1), -H(2)	0.92±2,	0.88±3		
O - H(1)... N(2)	2.94±1			
O - H(2)... N(1)	3.14±1			
Au - C - N	175 - 178°			
H(1) - O - H(2)	113±2			
O - H(1)... N(2)	156±2			
O - H(2)... N(1)	175±2			

Description of the structure

The $Au(CN_4)^-$ ion has square planar geometry. Each of these groups is bonded to two others through two water molecules forming links in a chain winding around the two-fold screw axes parallel to *a*. The K^+ ions are in channels forming chains by bonding to five nitrogen atoms and two oxygen atoms.

Details of analysis

The structure was determined using 360 symmetry independent reflexions obtained with neutrons of wavelength 1.1375 Å. Absorption and extinction corrections were applied.

BORON SESQUIOXIDE

I. The crystal structure of trigonal diboron trioxide. G.E. GURR, P.W. MONTGOMERY, C.D. KNUTSON and B.T. GORRES, 1970. *Acta Cryst.*, B26, 906-914.

Trigonal, a = 4.3368±5, c = 8.3397±18 Å, [U = 135.8 Å^3], Z = 3, D_x = 2.56.

Space group $P3_1$ (C_3^2)

Atomic positions

			x	y	z	B(Å^2)
3 O(1)	in	3(*a*)	0.547±3	0.397±3	0	0.82±16
3 O(2)	in	3(*a*)	0.148±2	0.600±3	0.0775±9	0.75±16
3 O(3)	in	3(*a*)	0.005±2	0.161±4	-0.129 ±2	0.58±12
3 B(1)	in	3(*a*)	0.223±5	0.393±5	-0.020 ±3	1.28±39
3 B(2)	in	3(*a*)	0.828±4	0.603±4	0.092 ±2	0.31±36

Interatomic distances (Å) *and bond angles*

B(1) - O(1)	1.404±19
B(1) - O(2)	1.366±23
B(1) - O(3)	1.337±18
B(2) - O(1)	1.336±22
B(2) - O(2)	1.401±19
B(2) - O(3)	1.384±17

Angles in degrees subtended at B(1)

	O(2)	O(3)
O(1)	119.0	114.7
O(2)		126.1

Angles in degrees subtended at B(2)

	O(2)	O(3)
O(1)	121.5	113.8
O(2)		121.5

Description of structure

The structure consists of infinite chains of interconnected BO_3 triangles. The two distinct BO_3 triangles are significantly distorted, but very similar forming ribbons, as indicated in Fig. 33. These ribbons are similar to the $(BO_2)_\infty^{-\infty}$ ribbons found in metaborate structures. The previously reported structure is in error (1).

Materials

Crystals were grown at high pressure in a graphite cell upon slow cooling from 700°C.

1. *Structure Reports*, 16, 216.

VANADIUM OXIDES

I. On the crystal structure of a new vanadium oxide, V_4O_9. K.-A. WHILHELMI and K. WALTERSSON, 1970. *Acta Chem. Scand.*, 24, 3409-3411.

V_4O_9 F.W. = 347.76

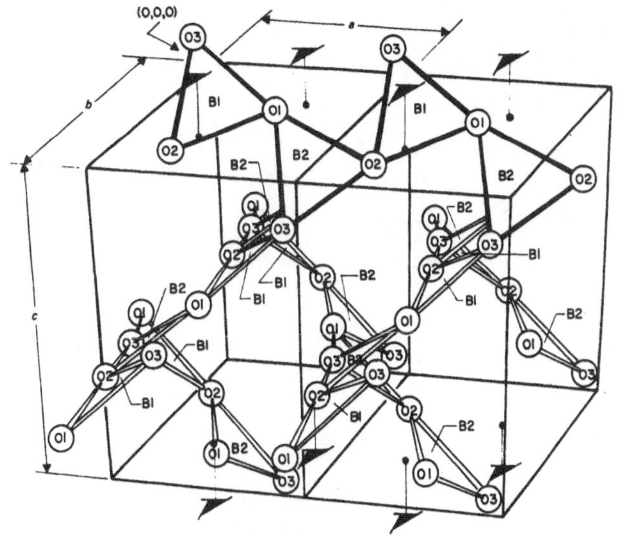

Fig. 33. Two unit-cells of the B_2O_3 I structure.

Orthorhombic, a = 17.926±4, b = 3.631±1, c = 9.396±2 Å, U = 611.6 Å^3, Z = 4, D_x = 3.77.

Space group *Pnma* (D_{2h}^{16})

Atomic positions All atoms in 4(c)

	x	y	z	B(Å^2)
V(1)	-0.0059±2	0.25	0.2287±3	1.16±13
V(2)	0.0780±2	0.25	0.5385±3	1.26±14
V(3)	0.1675±2	0.25	0.2053±3	1.21±13
V(4)	0.3169±2	0.25	0.4566±3	1.13±13
O(1)	-0.0233±6	0.75	0.2740±10	3.06±78
O(2)	0.0480±7	0.75	0.5477±11	2.67±64
O(3)	0.1543±6	0.75	0.1579±11	2.06±64
O(4)	0.2969±6	0.75	0.5108±11	1.82±43
O(5)	0.0879±6	0.25	0.3401±11	0.65±40
O(6)	0.0435±8	0.25	0.0833±12	2.70±65
O(7)	0.3924±6	0.25	0.3551±12	1.92±56
O(8)	0.1639±7	0.25	0.5843±11	1.80±53
O(9)	0.2393±6	0.25	0.3124±11	2.02±54

Interatomic distances (Å) *and bond angles*

V(1) - O(6)	1.628±13	
V(1) - O(1)	1.891±3	(×2)
V(1) - O(5)	1.981±11	
V(1) - O(7)	1.986±11	
V(1) - O(2)	2.232±11	

Angles in degrees subtended at V(1) by

	O(1)'	O(5)	O(7)	O(2)
O(6)	106.2±3	88.9±6	99.6±6	166.8±6
O(1)	147 6±4	91.2±3	86.4±3	74.5±3
O(5)			171.5±5	77.9±4
O(7)				93.6±5

V(2) - O(8)	1.559±13		
V(2) - O(5)	1.873±11		
V(2) - O(2)	1.895±4	(×2)	
V(2) - O(1)	2.016±10		
V(2) - O(2)	2.400±13		

Angles in degrees subtended at V(2) by

	O(5)	O(2)'	O(1)	O(2)
O(8)	100.2±5	105.14±4	103.5±5	175.9±8
O(5)		94.1±3	156.3±5	75.7±4
O(2)		146.6±5	79.8±5	75.4±4
O(1)				80.6±4

V(3) - O(9)	1.634±11	
V(3) - O(3)	1.884±3	(×2)
V(3) - O(5)	1.908±11	
V(3) - O(4)	1.936±11	

Angles in degrees subtended at V(3) by

	O(3)'	O(5)	O(4)
O(9)	104.2±3	100.4±5	108.8±5
O(3)	148.9±5	93.6±3	79.5±3
O(5)			150.8±5

V(4) - O(7)	1.656±11	
V(4) - O(4)	1.919±3	(×2)
V(4) - O(9)	1.942±11	
V(4) - O(3)	1.961±11	

Angles in degress subtended at V(4) by

	O(4)'	O(9)	O(3)
O(7)	107.8±3	100.6±5	109.9±5
O(4)	142.1±4	92.9±3	78.1±3
O(9)			149.5±5

Discussion

V_4O_9 contains double chains of edge sharing VO_6 octahedra and single chains of edge sharing VO_5 square pyramids running along *b* (all atoms lying in the mirror planes perpendicular to *b*). The chains are linked into a network by means of corner shared O atoms.

Details of analysis

Prepared by decomposition of V_3O_7 in supercritical water at 600°C and 2Kb. Lattice parameters from Guinier-Hägg powder photograph. Composition determined from a consideration of oxygen packing densities. Intensities measured from Weissenberg photographs of layers with k = 0 to 2 taken with CuK_α radiation. Structure determined from Patterson function and refined by least squares to give R = 0.07 for 316 reflexions (k = 0,1) not listed.

VANADIUM DIOXIDE

VO_2 F.W. = 82.94

II. A refinement of the structure of VO_2. J.M. LONGO and P. KIERKEGAARD, 1970. *Acta Chem. Scand.*, <u>24</u>, 420-426.

Monoclinic, a = 5.7517±30, b = 4.5378±25, c = 5.3825±25 Å, β = 122.65±10°, U = 118.28 Å^3, D_m = 4.65(<u>1</u>), Z = 4, D_x = 4.65.

Space group $P2_1/c$ (C^5_{2h})

Atomic positions All atoms in 4(*e*).

	x	*y*	*z*	$B(Å^2)$
V	0.23947±5	0.97894±5	0.02646±6	0.299±2
O(1)	0.10616±24	0.21185±25	0.20859±27	0.396±11
O(2)	0.40051±24	0.70258±26	0.29884±27	0.441±11

Interatomic distances (Å) *and bond angles*

V - O(2)	1.765±1
V - O(1)	1.865±2
V - O(1)	1.895±1
V - O(1)	2.017±1
V - O(2)	2.026±1
V - O(2)	2.065±2
V - V	2.620±0

Angles in degrees subtended at V by

	O(1)	O(1)	O(1)	O(2)	O(2)
O(2)	98.4±1	97.0±1	171.1±5	91.3±1	90.4±1
O(1)		91.6±1	88.6±1	94.6±1	169.0±2
O(1)			88.3±1	168.8±8	93.8±1
O(1)				82.6±1	82.0±1
O(2)					78.5±1
O(2)					

Details of analysis

Lattice parameters from focusing powder camera with monochromated CuK_α radiation, KCl (a = 6.29288 Å) used as internal standard. Intensities measured on four-circle goniometer, MoK_α radiation, Lorentz, polarisation, absorption (μ = 8.12 mm^{-1}, crystal size not given) and extinction corrections. Parameters of 1 refined by least squares to give unweighted R = 0.026 for 864 reflexions listed.

VANADIUM SESQUIOXIDE

III. The crystal structure of V_2O_3 and $(V_{0.962}Cr_{0.035})_2O_3$ near the metal-insulator transition. P.D. DERNIER, 1970. *J. Phys. Chem. Solids*, 31, 2569-2575.

V_2O_3 F.W. = 144.88

Rhombohedral, a = 4.9515±3 (4.9985±5), c = 14.003±1 (13.912±1) Å, U = 297.32 $Å^3$, Z = 6, D_x = 5.01. (Figures in parentheses refer to $(V_{0.962}Cr_{0.038})_2O_3$.)

Space group $R\bar{3}c$ (D^6_{3d})

Atomic positions Hexagonal setting

			x	*y*	*z*
12 V	in	12(*c*)	0	0	0.34630±2 (0.34870±2)
18 O	in	18(*e*)	0.31164±18 (0.30745±22)	0	0.25

Anisotropic temperature factors given.

Interatomic distances (Å) *and angles*

		V_2O_3	$V(Cr)_2O_3$	
V - O	(×3)	1.971	1.975	
V - O'	(×3)	2.049	2.062	
O-V-O	(×3)	97.2°	99.0°	
O-V-O	(×3)	88.4	87.5	
	(×3)	168.6	166.5	
	(×3)	92.0	91.6	
O'-V-O'	(×3)	81.4	80.5	
V - V		2.697	2.746	(across face)
		2.882	2.918	(across edge)

σ(bonds) = 0.001 Å σ(angles) = 0.1°

Discussion

Refinement of 2. The onset of insulating properties is associated with the increase in the distance between V atoms across a shared octahedral face.

Details of analysis

Lattice parameters measured from back reflexion powder photographs with CrK_α radiation ($K_{\alpha 1}$, = 2.28962 Å). Intensities measured with diffractometer using MoK_α radiation and stationary crystal - stationary counter method. Crystals ground to spheres 0.16 mm (V_2O_3) and 0.171 mm ($(V_{0.962}Cr_{0.038})_2O_3$) intensities corrected for Lorentz, polarisation, absorption and extinction and structure refined by least squares to give R = 0.019 for 188 reflexions and R = 0.020 for 208 reflexions, both listed, of the pure and doped crystals respectively.

IV. Ordered phases in the monoxide region of the vanadium-oxygen system. B. ANDERSSON and J. GJØNNES, 1970. *Acta Chem. Scand.*, 24, 2250-2252.

$V_{13}O_{16}$ F.W. = 938.23

Tetragonal a = 11.72, c = 8.245 Å.

Space group $I4_1/amd$ (D_{4h}^{19}) Systematic absences in electron diffraction pattern.

Atomic positions

			x	*y*	*z*
4 V(1)	in	4(*a*)			
16 V(2)	in	16(*h*)	1/8		1/4
16 V(3)	in	16(*f*)	1/8		
16 V(4)	in	16(*f*)	5/8		
16 O(1)	in	16(*h*)	1/8		0
16 O(2)	in	16(*i*)	1/8		1/2
32 O(3)	in	32(*i*)	1/8	0	1/4

Discussion

$V_{13}O_{16}$ has an NaCl arrangement with intersticial V atoms in 4(*a*) surrounded by four metal vacancies in 16(*h*) with x = 1/8 and z = 3/4. A second ordered phase of composition $VO_{1.17}$ was also found to be tetragonal a = 26.08, c = 8.30 Å, space group $I4_1/a$ (C_{4h}^6).

Details of analysis

Space group determined by single-crystal electron diffraction, lattice parameters and structure determined from powder diffraction photograph (CuK_α radiation) to give R = 0.08 for 14 reflexions [not listed]. Neither disordering of the intersticial atom and vacancies, nor refining of the variable parameters resulted in better argreement.

1. *Structure Reports*, 20, 264.
2. R.E. NEWNHAM and Y.M. DE HAAN, 1965. *Z. Kristallogr.*, 117, 235.

CHROMIUM TRIOXIDE

I. The crystal structure of $(CrO_3)_x$. J.S. STEPHENS and D.W.J. CRUICKSHANK, 1970. *Acta Cryst.*, B26, 222-226.

CrO_3 F.W. = 151.99

Orthorhombic, a = 4.789±5, b = 8.557±5, c = 5.743±4 Å, U = 235.4 Å^3, Z = 4, D_x = 2.82.

Space group $C2cm$ (C_{2v}^{15}) as chosen previously (1, 2, 3).

Atomic positions

4 Cr	in	4(*c*)	with	*x* = 0.0	*y* = 0.4023±5
4 O(1)	in	4(*b*)	with	*x* = 0.1159±2	
4 O(2)	in	4(*c*)	with	*x* = 0.1245±14	*y* = 0.2323±5
4 O(3)	in	4(*c*)	with	*x* = 0.3284±9	*y* = 0.3922±5

Thermal parameters are listed in the paper.

Interatomic distances (Å) *and bond angles*

Cr - O(1) (2×)	1.748±2	O(1) - Cr - O(1')		110.5±1°
- O(2)	1.579±5	O(1) - Cr - O(2)	(2×)	108.6±1
- O(3)	1.576±5	O(1) - Cr - O(3)	(2×)	110.2±2
		O(2) - Cr - O(3)		108.7±3
		Cr - O(1) - Cr		143.0±2

Discussion

Linear chains of corner-sharing CrO_4 groups as reported previously (1, 2, 3) but with greater accuracy. Thermal corrections increase Cr-O(2) and Cr-O(3) bond lengths by 0.024 and 0.018 Å respectively.

Details of analysis

Crystals were grown from saturated solution of $Na_2Cr_2O_7$ in H_2SO_4. 334 reflexions measured partially with automatic diffractometer and partially by microdensitometer from precession photographs. Least-squares refinement to *R* = 0.0234.

1. A. BYSTROM and K.A. WILHELMI, 1950. *Acta Chem. Scand.*, 4, 1131.
2. F. HANIC and D. STEMPELOVA, 1960. *Chem. Zvest.*, 14, 156.
3. *Structure Reports*, 19, 368.

CUPRIC OXIDE

I. A refinement of the crystal structure of copper(II) oxide with a discussion of some exceptional esd's. S. ÅSBRINK and L.J. NORRBY, 1970. *Acta Cryst.*, B26, 8-15.

CuO F.W. = 79.54

Monoclinic, *a* = 4.6837±5, *b* = 3.4226±5, *c* = 5.1288±6 Å, β = 99.54±1°, *U* = 81.08 Å^3, *Z* = 4, D_x = 6.515.

Space group *C*2/*c* (C^6_{2h})

Atomic positions

4 Cu	in	4(*c*)	
4 O	in	4(*e*)	with *y* = 0.4184±13

Thermal parameters are listed in the paper.

Interatomic distances (Å)

Cu - O	1.9608±13	(2×)
	1.9509±26	(2×)
	2.7840±37	(2×)

Description of the structure

More accurate determination of that proposed by 1.

Details of analysis

Crystal grown from Na_2CO_3 flux. 267 reflexions measured with a diffractometer and Mo radiation. Least-squares refinement to *R* = 0.038. Absorption and extinction corrections.

1. G. TUNELL, E. POSNIAK and C.J. KSANDA, 1935. *Z. Kristallogr.*, 90, 120.

IODINE PENTOXIDE

I. Iodine Oxides. Part III. The crystal structure of I_2O_5. K. SELTE and A. KJEKSHUS, 1970. *Acta Chem. Scand.*, 24, 1912-1924.

I_2O_5 F.W. = 333.80

Monoclinic, a = 11.036±3, b = 5.063±1, c = 8.135±2 Å, β = 107.18±2°, U = 434.26 Å^3, D_m = 5.08, Z = 4, D_x = 5.09.

Space group $P2_1/c$ (C^5_{2h})

Atomic positions All atoms in 4(e)

	x	y	z	B(Å^2)
I(1)	0.1260±2	0.1143±6	0.2136±3	*
I(2)	0.3730±2	0.6825±5	0.1597±3	*
O(1)	0.015 ±3	0.850 ±7	0.154 ±3	1.70±40
O(2)	0.193 ±2	0.041 ±7	0.434 ±3	1.20±40
O(3)	0.486 ±2	0.862 ±6	0.333 ±3	1.24±4
O(4)	0.309 ±2	0.492 ±6	0.300 ±3	0.90±30
O(5)	0.250 ±2	0.968 ±6	0.116 ±3	1.10±30

* Anisotropic temperature factors given.

Interatomic distances (Å) *and bond angles*

I(1) - O(2)	1.76
I(1) - O(1)	1.78
I(1) - O(5)	1.92
I(1) - O(1)	2.45
I(1) - O(4)	2.72
I(1) - O(1)	2.94
I(1) - O(2)	3.12

Angles in degrees subtended at I(1) by

	O(1)	O(5)
O(2)	99.6	101.7
O(1)		96.6

I(2) - O(4)	1.79
I(2) - O(3)	1.82
I(2) - O(5)	1.94
I(2) - O(3)	2.24
I(2) - O(2)	2.54

Angles in degrees subtended at I(2) by

	O(3)	O(5)
O(4)	94.8	97.5
O(3)		93.0

I(1) - O(5) - I(2) = 139.2°

Standard errors in distances = 0.03 Å, in angles = 1.2°.

Discussion

Discrete O_2IOIO_2 molecules are linked by comparatively strong intermolecular I - O bonds.

Details of analysis

Crystals mounted in thin walled B-Li glass capillaries. Intensities measured on microdensitometer from multiple-film integrated Weissenberg photographs of layers with k = 0-3 taken with MoK_α radiation, corrected for Lorentz, polarisation, absorption (μR = 1.1, cylindrical shape assumed) and secondary extinction. Structure from Patterson and electron-density functions refined by least squares to give R_1 = 0.089 for 814 observed reflexions listed.

NIOBIUM PENTOXIDE

$(Nb_2O_5)_m$ F.W. = $(275.81)_m$

I. Über die Kristallstruktur von M - Nb_2O_5. W. MARTIN, S. ANDERSSON and R. GRUEHN, 1970. *J. Solid State Chem.*, 1, 419-424.

Tetragonal, a = 20.44, c = 3.832 Å, U = 1601 Å^3, D_m = 4.3-4.4, Z = 16, D_x = 4.4.

Space group $I4/mmm$ (D_{2h}^{17}) from systematic absence and comparing with the structure of N - Nb_2O_5.

Atomic positions

			x	y	z
Nb(1)	in	8(i)	0.1294±6	0	0
Nb(2)	in	8(i)	0.4047±9	0	0
Nb(3)	in	16(l)	0.2738±5	0.1344±5	0
O(1)	in	8(j)	0.3479±74	1/2	0
O(2)	in	8(j)	0.0731±64	1/2	0
O(3)	in	16(l)	0.2095±55	0.3402±55	0
O(4)	in	8(h)	0.0678±42	0.0678±42	0
O(5)	in	8(h)	0.2116±56	0.2116±56	0
O(6)	in	16(l)	0.2095±34	0.0727±33	0
O(7)	in	16(l)	0.3555±45	0.0630±44	0

No temperature factor is given.

Interatomic distances (Å) *and bond angles*

Environment of Nb(1) (crystallographic symmetry C_{2v})

Nb(1) - O(4)	1.87±9	(×2)
Nb(1) - O(1)	1.97±4	(×2)
Nb(1) - O(6)	2.21±7	(×2)

Angles in degrees subtended at Nb(1) by

	O(4)'	O(1)	O(1)'	O(6)	O(6)'
O(4)	96±4	99±3		175±3	90±3
O(1)			153±6	80±3	
O(6)					85±3

Environment of Nb(2) (crystallographic symmetry C_{2v})

Nb(2) - O(7)	1.63±9	(×2)
Nb(2) - O(2)	1.97±3	(×2)
Nb(2) - O'(2)	2.46±8	(×2)

Angles in degrees subtended at Nb(2) by

	O(7)'	O(2)	O(2)'	O'(2)	O'(2)'
O(7)	104±5	98±2		166±4	91±4
O(2)			153±5	80±3	

Environment of Nb(3) (crystallographic symmetry C_s)

Nb(3) - O(6)	1.82±7	
Nb(3) - O(3)	2.01±4	(×2)
Nb(3) - O(5)	2.03±12	
Nb(3) - O'(3)	2.01±11	
Nb(3) - O(7)	2.22±9	

Angles in degrees subtended at Nb(3) by

	O(3)	O(3)'	O(5)	O'(3)	O(7)
O(6)	108±3		95±4	175±4	95±3
O(3)		144±4	85±3	72±3	92±3
O(5)				80±5	170±4
O'(3)					90±4

Discussion

Nb-O octahedra are arranged in blocks by sharing the vertices of the octahedra. Perpendicular to the *c* axis, 4×4 octahedra blocks are linked in checker-board fashion. Along the *c* direction the blocks extend to infinite columns by joining the vertices of octahedra.

Details of analysis

Single crystals were obtained by chemical transport. Lattice parameter from powder photograph. Intensity data were from Weissenberg photographs (*hk*0 - *hk*3) using a 0.16×0.02×0.01 mm^3 crystal with MoK_α radiation. An overall temperature factor for each layer was used. Refinement of structure derived from that of N-Nb_2O_5(1), gave R = 0.157.

1. S. ANDERSSON, 1967, *Z. anorg. Chem.*, 351, 106.

BISMUTH TRIOXIDE

α-Bi_2O_3 F.W. = 465.96

I. The crystal structure of α-Bi_2O_3. G. MALMROS, 1970. *Acta Chem. Scand.*, 24, 384-396.

Monoclinic, a = 5.8486±5, b = 8.1661±10, c = 7.5097±8 Å, β = 113.00±1°, U = 330.2 Å^3, D_m = 9.20, Z = 4, D_x = 9.40.

Space group $P2_1/c$ (C^5_{2h}) from systematic absences

Atomic positions

	x	*y*	*z*	*B*(Å^2)
Bi(1)	0.5240±1	0.1831±2	0.3613±1	*
Bi(2)	0.0409±2	0.0425±1	0.7762±1	*
O(1)	0.780 ±6	0.300 ±3	0.710 ±3	0.9±2
O(2)	0.242 ±5	0.044 ±4	0.134 ±4	1.2±3
O(3)	0.271 ±4	0.024 ±3	0.513 ±3	0.8±2

* Anisotropic temperature factors given.

Interatomic distances (Å)

Fivefold coordination around Bi(1)

Bi(1) -	O(3)	2.08
Bi(1) -	O(2)	2.17
Bi(1) -	O(1)	2.21
Bi(1) -	O(3)'	2.54
Bi(1) -	O(1)'	2.63

Sixford coordination around Bi(2)

Bi(2) -	O(2)	2.14
Bi(2) -	O(1)	2.22
Bi(2) -	O(3)	2.29
Bi(2) -	O(2)	2.48
Bi(2) -	O(1)	2.54
Bi(2) -	O(3)	2.80

σ(Bi-O) = 0.03 Å

Discussion

The structure, more or less correctly described in 1, consists of layers of Bi and O alternating along the *a* axis. Each Bi atom is surrounded by a polyhedron of O atoms which approximates to a distorted octahedron with three short Bi-O distances and two or three longer distances. There are tunnels of diameter about 2.5 Å running along *c*.

Details of analysis

Crystals prepared by heating commercial Bi_2O_3 in KOH solution in an autoclave at 250°C for a week, were ground to a sphere of radius 0.0835 mm. Lattice parameters from focused powder photographs with monochromated CuK_α radiation (KCl, a = 6.2929 Å used as a standard). Intensities estimated visually from integrated Weissenberg photographs of layers with k = 0 to 9 taken with MoK_α radiation. Corrections for Lorentz, polarisation and absorption (μ = 100.8 mm^{-1}) effects. Structure from Patterson and Fourier syntheses. Refined by least squares to give R = 0.087 for about 1600 reflexions listed.

1. *Structure Reports*, 8, 142.

OSMIUM DIOXIDE

OsO_2 F.W. = 222.2

I. Precision determination of the crystal structure of osmium dioxide. C.-E. BOMAN, 1970. *Acta Chem. Scand.*, 24, 123-128.

Tetragonal (C4 type), a = 4.5003±5, c = 3.1839±4 Å, U = 64.48 $Å^3$, D_m = 11.38 (1), Z = 2, D_x = 11.44 (see 2 for previous work).

Space group $P4_2/mnm$ (D_{4h}^{14}) from systematic absences and similarity to rutile.

Atomic positions

			x	y	z
2 Os	in	2(a)	0	0	0
4 O	in	4(f)	0.308±2	0.308±2	0

Anisotropic temperature factors given.

Interatomic distances (Å)

The crystallographic site symmetry of Os is D_{2h}

Os - O	1.962±13	(×2)
Os - O'	2.006±8	(×4)
O - Os - O'	90°	(×8)
O' - Os - O'	75.0 ±5	(×2)
O' - Os - O'	105.0±5	(×2)

Details of analysis

Crystals of OsO_2 prepared by sublimation. Lattice parameters measured from Guinier photographs (CuK_α radiation) calibrated with KCl (a = 6.2919(4) Å). Intensities measured on PAILRED diffractometer (MoK_α radiation) from crystal (0.02×0.02×0.063 mm^3) and corrected for Lorentz, polarisation and absorption effects. Structure refined by least squares omitting 11 extinction affected reflexions to give R = 0.031. 145 independent reflexions listed.

1. F. KRAUSE and G. SCHRADER, 1928. *Z. anorg. Chem.*, 176, 385.
2. *Strukturbericht*, 1, 213, 158.

HIGH PRESSURE RHODIUM SESQUIOXIDE

Rh_2O_3 F.W. = 253.81

I. Synthesis and structure of a new high-pressure form of Rh_2O_3. R.D. SHANNON and C.T. PREWITT, 1970. *J. Solid State Chem.*, 2, 134-136.

Orthorhombic, a = 5.1686±3, b = 5.3814±4, c = 7.2426±4 Å, U = 201.45 $Å^3$, D_m = 8.28, Z = 4, D_x = 8.35.

Space group $Pbna$ (D_{2h}^{14})

Atomic positions

Atoms at $(x, y, z;\ 1/2-x, 1/2+y, 1/2-z;\ 1/2-x, -y, 1/2+z;\ x, 1/2-y, -z)$

			x	y	z	B(Å^2)
8 **Rh**	in	8(*d*)	0.7498±1	0.0312±2	0.1058±1	0.15±4
8 O(1)	in	8(*d*)	0.6037±11	0.1161±10	0.8494±8	0.08±8
4 O(2)	in	4(*c*)	0.0505±16	0.25	0	0.14±12

Interatomic distances (Å)

Rh	-	O(2)	1.986+4
Rh	-	O(1)	2.018+6
Rh	-	O(1)'	2.056+6
Rh	-	O(1)''	2.068+5
Rh	-	O(1)'''	2.077+6
Rh	-	O(2)'	2.095+6

Angles in degrees subtended at Rh by

	O(1)	O(1)'	O(1)''	O(1)'''	O(2)'
O(2)	103.5+2	90.7+2	161.8+2	81.2+2	84.3+2
O(1)		84.3+2	90.3+2	92.5+2	162.0+2
O(1)'			78.7+2	170.3+2	79.5+2
O(1)''				110.5+2	79.2+2
O(1)'''					104.8+2

Discussion

Two RhO_6 octahedra share a face as in the corundum structure, but the arrangements of these groups is different.

Details of analysis

Intensities measured on single crystal diffractometer using MoK_α radiation with spherical crystal (radius 0.08 mm). Structure refined by least squares to give R = 0.042. [Neither the number of reflexions nor intensities are given and no further details of the experimental method are reported.] Powder pattern listed.

RUTHENIUM DIOXIDE

I. Refinement of the crystal structure of ruthenium dioxide. C.-E. BOMAN, 1970. *Acta Chem. Scand.*, 24, 116-122.

Tetragonal (*C*4 type) a = 4.4919±8, c = 3.1066±7 Å, U = 62.28 Å^3, D_m = 7.2 (1), Z = 2, D_x = 7.05 (see 2, 3 and 4 for earlier work.)

Space group $P4_2/mnm$ (D_{4h}^{14}) systematic absences and assumed isomorphism with rutile.

Atomic positions			x	y	z	B(Å^2)
Ru	in	2(*a*)	0	0	0	0.384±12
O	in	4(*f*)	0.3058±16	0.3058±16	0	0.517±8

Interatomic distances (Å)

The crystallographic site symmetry of the Ru is D_{2h}

Ru - O = 1.942±10 (×2)
Ru - O' = 1.984±6 (×4)

O - Ru - O' = 90° (×8)
O' - Ru - O' = 76.9±4 (×2)
O' - Ru - O' = 103.1±4 (×2)

Discussion

This work is rather more accurate than that of 4 which suggested that the two types of Ru-O bond differed significantly by 4σ.

Details of analysis

Crystals grown by sublimation of Ru or RuO_2 in O_2. Lattice parameters from Guinier photographs (CuK_α radiation) calibrated with KCl (a = 6.2919(4) Å). Intensities from visually measured multiple-film equi-inclination Weissenberg photographs (MoK_α) of layers with l = 0 to 4 and k = 0 to 5. Lorentz and polarisation correction, no absorption correction (crystal 0.07×0.07×0.10 mm^3). Rutile structure assumed and refined by least squares to give R = 0.071 for 235 reflexions.

1. H. DEBRAY and A. JOLY, 1888. *C.R. Acad. Sci. Paris*, 106, 100, 328, 1494.
2. G. LUNDE, 1927 *Z. anorg. Chem.*, 163, 345; *Strukturbericht*, 1, 213, 158.
3. H. SCHÄFER, G. SCHNEIDEREIT and W. GERHARDT, 1963. *Z. anorg. Chem.*, 319, 327.
4. F.A. COTTON and J.T. MAGUE, 1966. *Inorg. Chem.*, 5, 317.

URANIUM OXIDE

I. The structure of β-U_3O_8. B.O. LOOPSTRA, 1970. *Acta Cryst.*, B26, 656-657.

U_3O_8 F.W. = 842.08

Orthorhombic, a = 7.069±1, b = 11.445±1, c = 8.303±1 Å, [U = 671.75 Å^3], Z = 4, [D_x= 8.324].

Space group *Cmcm* (D_{2h}^{17}) only possibility tried

Atomic positions

			x	*y*	*z*
4 U(1)	in	4(*c*)	0	0.989±1	1/4
4 U(2)	in	4(*c*)	0	0.350±1	1/4
4 U(3)	in	4(*c*)	0	0.668±1	1/4
4 O(1)	in	4(*a*)	0	0	0
8 O(2)	in	8(*f*)	0	0.352±1	0.023±1
4 O(3)	in	4(*c*)	0	0.165±1	1/4
8 O(4)	in	8(*g*)	0.318±1	0.024±1	1/4
8 O(5)	in	8(*g*)	0.319±1	0.312±1	1/4

Interatomic distances (Å)

U(1) - O	7 distances	2.021 - 2.398
U(2) - O	7 distances	1.888 - 2.295
U(3) - O	6 distances	2.087 - 2.275

Description of the structure

Interconnected chains formed of uranium octahedra and pentagonal bipyramids. Distortion from that of α-U_3O_8.

Details of analysis

From neutron diffraction using powders. Least-squares refinement from models based upon α-U_3O_8 structure. No temperature factors were given.

SODIUM INDIUM OXIDE

I. Zur Kenntnis des Systems $NaInO_2/Na_2SnO_3$. E. HUBBERT-PALETTA, R. HOPPE and G. KREUZBURG, 1970. *Z. anorg. Chem.*, 379, 255-261.

$NaInO_2$ F.W. = 169.81

Rhombohedral, α-$NaFeO_2$ type (Hexagonal setting), a = 3.23$_2$, c = 16.39 Å, U = 85.5 Å^3, Z = 3, D_x = 9.90.

Space group $R\bar{3}m$ (D_{3d}^5)

Atomic positions

3 Na	in	3(*b*)	
3 In	in	3(*a*)	
6 O	in	6(*c*)	*z* = 0.267±1

Interatomic distances

In – O	2.15	(×6)
Na – O	2.48	(×6)

σ ∿ 0.06 Å

Details of analysis

Lattice parameters from rotation photographs around [01.0] taken with CuK_α radiation. Intensities estimated visually and the photometer from *h0l* Weissenberg photograph taken with MoK_α radiation. *R* = 0.10 for 96 reflexions listed.

SAMARIUM LITHIUM OXIDE

I. Structure cristalline de δ-$SmLiO_2$. M. GONDRAND, 1970. *Bull. Soc. fr. Minér. Crist.*, 93, 421-425.

$SmLiO_2$ F.W. = 189.34

Monoclinic, *a* = 5.707, *b* = 6.024, *c* = 5.647 Å, β = 103.05, *U* = 189.1 Å^3, *Z* = 4, D_x = 6.64.

Space group $P2_1/c$ (C^5_{2h})

Atomic positions All atoms in 4(*e*)

	x	*y*	*z*
Sm	0.199	0.067	0.303
O(1)	-0.11	0.27	0.48
O(2)	0.46	0.15	0.72
Li	0.67	0.10	0.15

Interatomic distances (Å)

Sm – O(2)	2.37	Li – O(2)	1.90
Sm – O(1)	2.38	Li – O(1)	1.91
Sm – O(2)	2.38	Li – O(2)	2.02
Sm – O(1)	2.44	Li – O(1)	2.24
Sm – O(1)	2.48	Li – O(2)	2.47
Sm – O(1)	2.53		
Sm – O(2)	2.54	σ ∿ 0.05 to 0.10 Å	

Details of analysis

Intensities measured from powder diagram. Structure determined by trial and error and refined by least squares to give *R*(intensities) = 0.12 for 26 reflexions listed.

ALKALI METAL AURATES

I. Zur Kristallstruktur von Li_3AuO_3, Li_5AuO_4, $KAuO_2$ und $RbAuO_2$. H.-D. WASEL-NIELSEN and R. HOPPE, 1970. *Z. anorg. Chem.*, 375, 43-54.

HEXALITHIUM DIAURATE (III)

$Li_6Au_2O_6$ F.W. = 265.8

Tetragonal, *a* = 9.111, *c* = 3.576 Å, *U* = 296.8 Å^3, D_m = 5.75, *Z* = 4, D_x = 5.93.

Space group $P4_1/mnm$ (D^{14}_{4h}) final structure

Atomic positions

			x	y	z
8 Li(1)	in	8(*i*)	0.59	0.18	0
4 Li(2)	in	4(*f*)	0.17	x	0
4 Au	in	4(*f*)	0.38_0	x	0
8 O(1)	in	8(*i*)	0.15_2	0.38_4	0
4 O(2)	in	4(*g*)	0.39_2	$-x$	0

Interatomic distances (Å)

Au - O(1)	2.08	(×2)
Au - O(2)	2.08	(×2)
Li(1) - O(1)	1.89	
Li(1) - O(2)	1.94	
Li(1) - O(1)'	1.96	(×2)
Li(2) - O(1)	1.96	(×2)
Li(2) - O(2)	1.96	(×2)

Errors uncertain but about 0.10 Å.

Discussion

The structure is composed of tetrahedrally coordinated Li ions and planar $O_2AuO_2AuO_2^{6-}$ ions which have C_{2v} crystallographic symmetry.

Details of analysis

Powder photographs. Structure determined by trial and error and refined by least squares to give R = 0.054 for 28 observed reflexions (listed). Li positions assumed.

PENTALITHIUM AURATE (III)

Li_5AuO_4 F.W. = 295.7

Orthorhombic (Li_2PdO_2 type, 1), a = 3.673, b = 9.505, c = 2.940 Å, U = 102.6 $Å^3$, D_m = 4.36, Z = 1, D_x = 4.46.

Space group *Immm* (D_{2h}^{25}) systematic absences

Atomic positions

			x	y	z	$B(Å^2)$
1 Li(1) + 1 Au	in	2(*a*)	0	0	0	1.52
4 Li(2)	in	4(*g*)	0	0.287	0	Not given
4 O	in	4(*h*)	0	0.85_3	1/2	1.65

Interatomic distances (Å)

Li(1), Au - O	2.03	(×4)	(square planar)
Li(2) - O	1.94	(×2)	(tetrahedral)
	1.98	(×2)	

Errors uncertain, but about 0.10 Å.

[The tetrahedron of oxygen atoms around Li(2) is very distorted but could be made more regular by the adjustment of y_{Li}. More work is needed to confirm whether this structure is correct.]

Details of analysis

Powder photographs taken with CuK_α radiation calibrated with α quartz [assumed values of a_0 and c_0 not given]. Intensities measured photometrically. Structure determined by trial and error and five parameters refined by least squares to give R [intensities?] = 0.048 for 19 reflexions listed.

MONOPOTASSIUM AURATE (III)

$KAuO_2$ F.W. = 268.1

Orthorhombic, a = 3.585, b = 5.849, c = 3.005 Å, U = 63.01 Å^3, D_m = 6.94, Z = 1, D_x = 7.06.

Space group *Pmmm* (D_{2h}^1)

Atomic positions

			x	y	z
1 K	in	1(*f*)	1/2	1/2	0
1 Au	in	1(*a*)	0	0	0
2 O	in	2(*n*)	0	0.24_7	1.2

Interatomic distances (Å)

K - O	2.77	(×8)	(cubic)
Au - O	2.08	(×4)	(square planar)

Errors uncertain, but about 0.10 Å.

Details of analysis

Powder photographs. Structure refined by least squares to give R[intensities?] = 0.089 for 17 reflexions listed.

MONORUBIDIUM AURATE (III)

$RbAuO_2$ F.W. = 314.4

Orthorhombic ($KCuO_2$ type, 2), a = 4.56, b = 12.30, c = 5.96 Å, U = 334.3 Å^3, Z = 4, D_x = 6.2.

Space group *Cmcm* (D_{2h}^{17})

Atomic positions

			x	y	z
4 Rb	in	4(*c*)	0	0.31	1/4
4 Au	in	4(*c*)	0	0.00	1/4
8 O	in	8(*f*)	0	0.11_8	0.00

Interatomic distances (Å)

Rb - O	2.80	(×2)	
Rb - O	2.86	(×4)	
Rb - O	3.26	(×2)	
Au - O	2.08	(×4)	(square planar)

Errors uncertain, but about 0.10 Å.

$CsAuO_2$ is probably isotypic, a = 5.04, b = 12.76, c = 5.96 Å.

Details of analysis

Intensities taken from 3 and refined by trial and error to give qualitative agreement. Some reflexions listed.

1. J.J. SCHEER, 1956. *Disertation,* Leiden.
2. W. KLEMM and K. WAHL, 1952. *Z. anorg. Chem.*, 270, 69.
3. R. HOPPE and K.H. AREND, 1962. *Z. anorg. Chem.*, 314, 4.

LITHIUM CUPRATE

I. Die Kristallstruktur von Li_2CuO_2. R. HOPPE and H. RIECK, 1970. *Z. anorg. Chem.*, 379, 157-164.

Li_2CuO_2 F.W. = 109.41

Orthorhombic, $a = 3.66_3$, $b = 2.86_3$, $c = 9.39_6$ Å, $U = 98.54$ Å^3, $D_m = 3.62$, $Z = 2$, $D_x = 3.68$.

Space group *Immm* (D_{2h}^{25}) systematic absences

Atomic positions

			x	*y*	*z*
2 Cu	in	2(*b*)	0	0.5	0.5
4 Li	in	4(*j*)	0.5	0	0.293 *
4 O	in	4(*i*)	0	0	0.3565±11

* No standard error given.

Interatomic distances (Å) *and bond angles*

Cu - O	1.97±1	(×4)	O - Cu - O	93.4±3°, 86.6±3°
Li - O	1.93	(×2)	O - Li - O'	144°
Li - O*	2.01	(×2)	O - Li - O*	102°
			O* - Li - O*'	91°

Discussion

CuO_4 planar groups oriented perpendicular to *a* share edges to form chains along *b*. The Li atoms lie in between with distorted tetrahedral coordination. The proposed positions also give rise to 4 short (2.46 Å) Li-Li contacts.

Details of analysis

Rotation, Weissenberg and precession photographs around [100], [010] and [001] taken with MoK_α radiation. Structure solved from Patterson function and refined by least squares to give $R = 0.11$ for 137 reflexions [not listed]. 19 powder lines listed.

STRONTIUM COPPER OXIDE

I. Zur Kenntnis von $SrCu_2O_2$. CHR. L. TESKE and H. MÜLLER-BUSCHBAUM, 1970. *Z. anorg. Chem.*, <u>379</u>, 113-224.

$SrCu_2O_2$ F.W. = 246.71

Tetragonal, $a = 5.48$, $c = 9.82_5$ Å, $U = 295$ Å^3, $D_m = 5.51$, $Z = 4$, $D_x = 5.54$.

Space group $I4_1/amd$ (D_{4h}^{19})

Atomic positions Origin at $\bar{4}m2$

			x	*y*	*z*
4 Sr	in	4(*a*)	0	0	0
8 Cu	in	8(*d*)	0	1/4	5/8
8 O	in	8(*e*)	0	0	0.25 [given in I as 0.5]

Interatomic distances (Å)

Sr - O	2.46	(×3)
Sr - O	2.47	(×4)
Cu - O	1.84	(×2)

Discussion

$SrCu_2O_2$ consists of a cubic close packed array of O atoms with half the octahedral intensities containing Sr atom. The SrO_6 octahedra share four edges with other octahedra. The Cu atoms lie along the edges of the unfilled octahedral intensities.

Details of analysis

Lattice parameters from rotation, Weissenberg and precession photographs around [100], [010] and [001] taken with MoK_α radiation. Intensities of layers with k = 0 to 2 recorded. Structure determined from Patterson and elctron-density syntheses to give R = 0.07 for 107 reflexions listed.

RARE EARTH ORTHOFERRITES

I. The crystal chemistry of the rare earth orthoferrites. M. MAREZIO, J.P. REMEIKA and P.D. DERNIER, 1970. *Acta Cryst.*, B26, 2008-2022.

$PrFeO_3$	F.W. = 244.76
$NdFeO_3$	F.W. = 248.09
$SmFeO_3$	F.W. = 254.20
$EuFeO_3$	F.W. = 255.81
$GdFeO_3$	F.W. = 261.10
$TbFeO_3$	F.W. = 262.77
$DyFeO_3$	F.W. = 266.35
$HoFeO_3$	F.W. = 268.78
$ErFeO_3$	F.W. = 271.11
$TmFeO_3$	F.W. = 272.78
$YbFeO_3$	F.W. = 276.89
$LuFeO_3$	F.W. = 278.12

Orthorhombic

	a	b	c(Å)	U(Å^3)	Z	D_x
$PrFeO_3$	5.483	5.578	7.786	[238.13]	4	[6.820]
$NdFeO_3$	5.453	5.584	7.768	[236.53]	4	[6.960]
$SmFeO_3$	5.400	5.597	7.711	[233.06]	4	[7.239]
$EuFeO_3$	5.372	5.606	7.685	[231.44]	4	[7.338]
$GdFeO_3$	5.349	5.611	7.669	[230.17]	4	[7.531]
$TbFeO_3$	5.326	5.602	7.635	[227.80]	4	[7.657]
$DyFeO_3$	5.302	5.598	7.623	[226.26]	4	[7.816]
$HoFeO_3$	5.278	5.591	7.602	[224.33]	4	[7.956]
$ErFeO_3$	5.263	5.582	7.591	[223.01]	4	[8.069]
$TmFeO_3$	5.251	5.576	7.584	[222.06]	4	[8.155]
$YbFeO_3$	5.233	5.557	7.570	[220.13]	4	[8.353]
$LuFeO_3$	5.213	5.547	7.565	[218.75]	4	[8.460]

Space group *Pbnm* (D_{2h}^{16})

Atomic positions

4 RE	in	4(c)	$x, y, 1/4; \bar{x}, \bar{y}, 3/4; 1/2+x, 1/2-y, 3/4; 1/2-x, 1/2-y, 1/4.$
4 Fe	in	4(b)	1/2, 0, 0; 1/2, 0, 1/2; 0, 1/2, 0; 0, 1/2, 1/2.
4 O(1)	in	4(c)	
8 O(2)	in	8(a)	$\pm(x, y, z; 1/2-x, 1/2+y, 1/2-z; \bar{x}, \bar{y}, 1/2+z; 1/2+x, 1/2-y, \bar{z}).$

		Pr	Nd	Sm	Eu	Gd	Tb
RE	x	0.99097±4	0.98931±4	0.98688±3	0.98555±5	0.98444±5	0.98403±3
	y	0.04367±5	0.04881±5	0.05666±3	0.06012±4	0.06284±6	0.06408±3
O(1)	x	0.0817 ±7	0.0876 ±7	0.0947 ±6	0.0978 ±7	0.1005 ±10	0.1035 ±6
	y	0.4788 ±9	0.4759 ±8	0.4710 ±6	0.4680 ±7	0.4672 ±11	0.4640 ±6
O(2)	x	0.7075 ±5	0.7052 ±5	0.7004 ±4	0.6977 ±5	0.6957 ±6	0.6950 ±4
	y	0.2919 ±5	0.2936 ±5	0.2992 ±4	0.3006 ±5	0.3016 ±6	0.3026 ±4
	z	0.0437 ±5	0.0462 ±4	0.0497 ±3	0.0506 ±4	0.0506 ±4	0.0538 ±3

		Dy	Ho	Er	Tm	Yb	Lu
RE	*x*	0.98293±4	0.98219±4	0.98155±4	0.98104±5	0.98064±5	0.98003±4
	y	0.06648±4	0.04881±3	0.06913±4	0.06913±4	0.07076±5	0.07149±5
O(1)	*x*	0.1060 ±8	0.0876 ±7	0.1137 ±8	0.1148 ±9	0.1169 ±11	0.1199 ±10
	y	0.4624 ±7	0.4759 ±8	0.4594 ±7	0.4559 ±8	0.4537 ±12	0.4539 ±10
O(2)	*x*	0.6930 ±5	0.7052 ±5	0.6910 ±5	0.6907 ±6	0.6886 ±6	0.6893 ±6
	y	0.3052 ±6	0.2936 ±5	0.3057 ±6	0.3057 ±6	0.3071 ±7	0.3071 ±7
	z	0.0549 ±4	0.0462 ±4	0.0573 ±4	0.0587 ±5	0.0599 ±6	0.0621 ±5

Thermal parameters are given.

Interatomic distances (Å) *and bond angles*

	Pr	Nd	Sm	Eu	Gd	Tb
Fe-O(1)	2.001±1	2.004±1	2.001±1	2.000±1	2.000±2	1.997±1
Fe-O(2)	2.015±3	2.017±3	2.030±3	2.029±3	2.030±4	2.030±2
Fe-O(2)	2.010±3	2.010±3	2.007±2	2.010±3	2.013±4	2.008±2
O(1)-Fe-O(2)	89.4±2	89.2±1	88.8±1	88.4±1	88.6±1	88.2±1
O(1)-Fe-O(2)	88.8±2	88.8±1	88.7±1	88.7±1	88.5±2	88.4±1
O(2)-Fe-O(2)	89.3±1	89.5±1	89.8±1	90.1±1	90.2±1	90.3±1
RE-O(1)	2.371±4	2.343±4	2.310±3	2.297±4	2.284±6	2.267±4
RE-O(1)	2.478±5	2.444±5	2.391±4	2.365±4	2.352±6	2.329±4
RE-O(1)	3.160±4	3.172±4	3.177±4	3.176±4	3.175±5	3.179±4
RE-O(1)	3.190±5	3.244±5	3.329±4	3.374±4	3.399±6	3.422±4
RE-O(2)	2.395±3	2.380±3	2.342±2	2.331±3	2.317±4	2.302±3
RE-O(2)	2.629±3	2.603±3	2.573±2	2.561±3	2.543±4	2.530±3
RE-O(2)	2.730±3	2.730±3	2.706±2	2.692±3	2.692±4	2.684±3
RE-O(2)	3.386±3	3.424±3	3.487±2	3.510±3	3.535±4	3.539±3

	Dy	Ho	Er	Tm	Yb	Lu
Fe-O(1)	1.998±1	1.998±2	2.003±3	2.005±2	2.006±2	2.008±2
Fe-O(2)	2.033±3	2.031±3	2.029±3	2.026±3	2.025±4	2.024±4
Fe-O(2)	2.004±3	2.001±3	2.002±3	2.003±3	2.001±4	1.997±4
O(1)-Fe-O(2)	88.1±1	87.8±2	87.7±1	87.5±2	87.4±2	87.7±2
O(1)-Fe-O(2)	88.4±1	88.4±2	88.6±1	88.1±2	88.1±2	87.9±2
O(2)-Fe-O(2)	90.3±1	90.4±1	90.4±1	90.3±1	90.2±1	90.1±1
RE-O(1)	2.256±4	2.240±5	2.217±4	2.215±5	2.204±6	2.185±5
RE-O(1)	2.310±4	2.295±5	2.287±4	2.265±5	2.244±7	2.243±5
RE-O(1)	2.176±4	3.178±5	3.192±5	3.193±5	3.194±6	3.195±5
RE-O(1)	3.444±4	3.462±5	3.474±4	3.494±5	3.502±7	3.502±6
RE-O(2)	2.286±3	2.277±3	2.264±3	2.255±3	2.234±4	2.225±4
RE-O(2)	2.521±3	2.505±3	2.495±3	2.482±4	2.478±4	2.455±4
RE-O(2)	2.676±3	2.673±4	2.673±3	2.679±4	2.673±4	2.687±4
RE-O(2)	3.560±3	3.566±3	3.577±3	3.582±3	3.595±4	3.599±4

Materials

Crystals grown from a flux composed of K_2CO_3 and B_2O_3.

Structure and details of analysis

Structure is isotypic with $YFeO_3$ (1). Intensity data measured with diffractometer with MoK_α radiation and corrected for absorption and extinction. Refined by least-squares methods.

1. D.T. COPPENS and M. EIBSCHÜTZ, 1965. *Acta Cryst.*, 19, 524.

SODIUM URANATES

I. Preparation and crystal structure of $NaUO_3$ and $Na_{11}U_5O_{16}$. S.F. BARTRAM and R.E. FRYXELL, 1970. *J. Inorg. Nucl. Chem.*, 32, 3701-3706.

SODIUM URANATE (V)

$NaUO_3$ F.W. = 309.01

Orthorhombic ($CaTiO_3$ type, 1), a = 5.910±1, b = 8.293±1, c = 5.776±1 Å, U = 282.7 Å^3, Z = 4, D_x = 7.2.

Space group *Pnma* (D_{2h}^{16}) [converted from *Pbnm*]

Atomic positions

			x	y	z
4 Na	in	4(*c*)	0.03	1/4	0.00
4 U	in	4(*b*)	0	0	1/2
4 O(1)	in	4(*c*)	0.48	1/4	0.04
8 O(2)	in	8(*d*)	0.27	0.03	0.73

Interatomic distances (Å)

Na	- O(1)	2.7	(×2)
Na	- O(2)	2.7	(×2)
Na	- O(2)'	2.8	(×2)
Na	- O(2)''	2.9	(×2)
U	- O(1)	2.1	(×2)
U	- O(2)	2.1	(×2)
U	- O(2)'	2.1	(×2)

Errors about 0.2 Å.

Details of analysis

Powder mounted in oxygen free atmosphere. Lattice parameters from 25 back-reflexion lines. Powder diffractometer (CuK_α radiation). Structure assumed the same as $CaTiO_3$. Proposed coordinates give R(intensities) = 0.10 for 48 lines listed.

SODIUM URANATE (IV-V)

$Na_{11}U_5O_{16}$ F.W. = 1699.1

Cubic, a = 9.543±2 Å, U = 869.0 Å^3, Z = 2, D_x = 6.49.

Space group $P4_232$ (O^2) systematic absences

Atomic positions

4	U(1)	in	4(b)			
6	U(2)	in	6(*f*)			
4	Na(1)	in	4(*c*)			
6	Na(2)	in	6(*e*)			
12	Na(3)	in	12(*h*)	x = 0.25		
2	O(1)	in	2(*a*)			
6	O(2)	in	6(*d*)			
24	O(3)	in	24(*m*)	x = 0.75,	y = 0.25,	z = 0.

Discussion

The metal atoms are in octahedral holes of a f.c.c. lattice of oxygen atoms.

Details of analysis

29 observed powder diffractometer lines listed agree qualitatively in intensity with the proposed model.

1. *Structure Reports*, 21, 317.

BETA LITHIUM STANNATE

I. Die Kristallstruktur von Li_2SnO_3. G. KREUZBURG, F. STEWNER and R. HOPPE, 1970. *Z. anorg. Chem.*, 379, 242-254.

β-Li_2SnO_3 F.W. = 180.57.

Monoclinic, a = 5.295, b = 9.184, c = 10.032 Å, β = 100.13, U = 480.2 $Å^3$, D_m = 4.84, Z = 8, D_x = 4.98.

Space group $C2/c$ (C^6_{2h})

Atomic positions

			x	y	z	$B(Å^2)$
8 Li(1)	in	8(*f*)	0.239+33	0.078 +2	-0.001 +2	1.10
4 Li(2)	in	4(*d*)	0.25	0.25	0.5	1.50
4 Li(3)	in	4(*e*)	0	0.083*	0.25	3.80
4 Sn(1)	in	4(*e*)	0	0.4165+1	0.25	0.18
4 Sn(2)	in	4(*e*)	0	0.7508+1	0.25	0.15
8 O(1)	in	8(*f*)	0.1337+8	0.2597+16	0.1333+4	0.49+10
8 O(2)	in	8(*f*)	0.1102+9	0.5844+16	0.1342+5	0.68+14
8 O(3)	in	8(*f*)	0.1346+8	0.9092+20	0.1329+4	0.48+10

* chosen so that Li(3) lies at the centre of its coordination octahedron.

Interatomic distances (Å)

σ(Li-O) = 0.02 Å; σ(Sn-O) = 0.01 Å; σ(angles) = 1°.

Li(1)	-	O(1)	2.18
Li(1)	-	O(3)	2.18
Li(1)	-	O(2)	2.18
Li(1)	-	O(3)	2.18
Li(1)	-	O(2)	2.24
Li(1)	-	O(1)	2.28

Angles in degrees subtended at Li(1) by

	O(3)	O(2)	O(3)	O(2)	O(1)
O(1)	86	91	176	85	90
O(3)		175	96	93	92
O(2)			86	91	84
O(3)				93	92
O(2)					172

Li(2)	-	O(3)	2.14	(×2)
Li(2)	-	O(1)	2.23	(×2)
Li(2)	-	O(2)	2.24	(×2)

Angles in degrees subtended at Li(2) by

	O(1)	O(2)
O(3)	86	94
O(3)'	94	86
O(1)		88
O(1)'		92

Li(3)	-	O(3)	2.18	(×2)
Li(3)	-	O(2)	2.18	(×2)
Li(3)	-	O(1)	2.19	(×2)

Errors in Li(3)-O distances and angles uncertain.

Angles in degrees subtended at Li(3) by

	O(3)'	O(2)'	O(1)'
O(3)	86	95	95
O(3)'		86	179
O(2)		179	85
O(2)'			94
O(1)			84

Sn(1)	-	O(1)	2.06	(×2)
Sn(1)	-	O(2)	2.08	(×2)
Sn(1)	-	O(3)	2.08	(×2)

Angles in degrees subtended at Sn(1) by

	O(1)'	O(2)'	O(3)'
O(1)	91	92	85
O(1)'		176	92
O(2)		84	92
O(2)'			91
O(3)			176

Sn(2)	-	O(2)	2.06
Sn(2)	-	O(3)	2.07
Sn(2)	-	O(1)	2.08

Angles in degrees subtended at Sn(2) by

	O(2)	O(3)'	O(1)'
O(2)	84	92	92
O(2)'		176	91
O(3)		91	85
O(3)'			92
O(1)			176

Discussion

The O atoms are cubic close packed with Li and Sn ordered on the octahedral interstices.

Details of analysis

Lattice parameters and space group from rotation, Weissenberg, precession and powder photographs. Structure determined from Patterson and difference electron-density functions calculated with intensities measured from k = 0-7 layers. Refined by least squares to give R = 0.10 for 1462 reflexions listed. The position of Li(3) was inferred from chemical considerations.

POTASSIUM METAZIRCONATE

POTASSIUM METASTANNATE

I. The crystal structure of potassium metazirconate, K_2ZrO_3, and its tin analogue, K_2SnO_3. B.M. GATEHOUSE and D.J. LLOYD, 1970. *J. Solid State Chemistry*, 2, 410-415.

POTASSIUM METAZIRCONATE

K_2ZrO_3 F.W. = 217.42

Orthorhombic, a = 5.93±2, b = 10.48±2, c = 7.03±2 Å, U = 437 Å^3, Z = 4, D_x = 3.30.

Space group *Pnma* (D_{2h}^{16})

Atomic positions

			x	y	z
4 Zr	in	4(c)	0.2730±6	0.25	0.1598±4
8 K	in	8(d)	0.264 ±1	0.5857±5	0.1143±6
4 O(1)	in	4(c)	0.175 ±5	0.75	0.390 ±3
8 O(2)	in	8(d)	0.015 ±3	0.121 ±1	0.234 ±4

Anisotropic temperature factors given in I.

Interatomic distances (Å)

Zr	-	O(1)	1.92	
Zr	-	O(2)	2.11	(×2)
Zr	-	O(2)'	2.11	(×2)

Errors in distances = 0.02 Å
Errors in angles = 1°

Angles in degrees subtended at Zr by

	O(2)'	O(2)*'
O(1)	111	104
O(2)	80	90
O(2)'		145
O(2)*		80

K	-	O(1)	2.65
K	-	O(2)	2.76
K	-	O(2)	2.84
K	-	O(2)	2.98
K	-	O(1)	2.99

Other K-O distances are greater than 3.0 Å.

POTASSIUM METASTANNATE

K_2SnO_3 F.W. = 244.89

Orthorhombic, a = 5.74, b = 10.34, c = 7.14 Å, U = 42 Å^3, Z = 4, D_x = 3.83.

Space group *Pnma* (D_{2h}^{16})

Atomic positions

			x	y	z	B(Å^2)
4 Sn	in	4(c)	0.267±5	0.25	0.163±1	*
8 K	in	8(d)	0.266±9	0.588±2	0.113±2	3.2±5
4 O(1)	in	4(c)	0.19 ±2	0.75	0.40 ±1	1.0±15
8 O(2)	in	8(d)	0.02 ±2	0.113±5	0.23 ±1	2.2±18

* Anisotropic temperature factors given.

Interatomic distances (Å)

Errors in distances about 0.08 Å, in angles about 3°.

Sn - O(1)	1.89	
Sn - O(2)	2.06	(×2)
Sn - O(2)*	2.17	(×2)

Angles in degrees subtended at Sn by

	O(2)'	O(2)*'
O(1)	109	105
O(2)	87	86
O(2)'		146
O(2)*		82

K - O(2)	2.65
K - O(1)	2.68
K - O(2)	2.78
K - O(1)	2.96
K - O(2)	2.96

Other K-O distances are greater than 3.0.

Discussion

The structure proposed by 1 for K_2SnO_3 is shown to be incorrect. K_2SnO_3 and K_2ZrO_3 are isostructural with ZrO_5 square pyramids linked through shared basal edges into chains along a.

Details of analysis

K_2ZrO_3 prepared by dissolution of ZrO_2 in excess K_2O formed from the decomposition of KNO_3 at 1100°C. Crystal (0.07×0.03×0.03 mm^3) mounted in capillary. Lattice parameters from oscillation and Weissenberg photographs. Intensities estimated visually from multiple-film Weissenberg photographs of layers with h = 0 to 5 taken with CuK_α radiation. Structure determined from Patterson function and refined by least squares to give R_1 = 0.087 for 293 reflexions listed. Intensities for K_2SnO_3 measured from Weissenberg photographs and refined to give R = 0.14 for 119 reflexions listed.

1. M. TOURNOUX, 1964. *Ann. Chim. Paris*, 9, 576.

RUBIDIUM PLUMBATE(IV)

I. Über Oxoplumbate(IV). I Die Kristallstruktur von Rb_2PbO_3. K. SEEGER and R. HOPPE, 1970. *Z. anorg. Chem.*, 375, 255-263.

Rb_2PbO_3 F.W. = 426.1

Orthorhombic, a = 10.8, b = 7.49, c = 5.98 Å, U = 484 Å^3, D_m = 5.69, Z = 4, D_x = 5.84.

Space group *Pmcn* (D_{2h}^{16}) systematic absences

Atomic positions

Equivalent positions $\pm(x, y, z; 1/2-x, 1/2-y, 1/2+z; 1/2+x, \bar{y}, \bar{z}; \bar{x}, 1/2+y, 1/2-z)$

			x	y	z
8 Rb	in	8(d)	0.887	0.606	0.225
4 Pb	in	4(c)	0.25	0.159	0.247
8 O(1)	in	8(d)	0.119	0.225	0.008
4 O(2)	in	4(c)	0.25	-0.39_1	-0.15_4

Interatomic distances (Å)

Rb	- O(2)	2.85
Rb	- O(2)	2.86
Rb	- O(1)	2.88
Rb	- O(1)	2.91
Rb	- O(1)	3.02
Rb	- O(1)	3.15

Environment of Pb

Pb	- O(1)	2.07	(×2)
Pb	- O(2)	2.11	
Pb	- O(1)	2.28	(×2)

Errors uncertain but probably ∿ 0.05 Å.

Discussion

The structure consists of square pyramidal PbO_5 groups linked into chains along c by shared basal edges. Rb atoms lie between the chains.

Isotypic compounds (1)

	a	b	c
Rb_2ZrO_3	10.7	7.40	5.87 Å
Rb_2SnO_3	10.7	7.49	5.75 Å

[Powder patterns]

Details of analysis

Lattice parameters for rotation, Weissenberg and precession photographs. Intensities measured photometrically from precession photographs of layers with $k = 0$ to 4 (MoK_α radiation). Structure determined from Patterson function and refined by least squares and considerations of Madelung energy. Final $R = 0.116$ for 244 reflexions listed.

1. R. HOPPE and K. SEEGER, 1970. *Z. anorg. Chem.*, 375, 264.

STRONTIUM PALLADATE

I. Über die Systeme CaO/PdO und SrO/PdO. H.-D. WASEL-NIELSEN and R. HOPPE, 1970. *Z. anorg. Chem.*, 375, 209-213.

Sr_2PdO_3 F.W. = 329.6

Orthorhombic, $a = 3.977$, $b = 3.530$, $c = 12.82$ Å, $U = 180.0$ Å^3, $Z = 2$, $D_x = 6.06$.

Space group *Immm* (D_{2h}^{25})

Atomic positions

			x	y	z	Crystal site symmetry
4 Sr	in	4(i)	0	0	0.35_2	*mm*
2 Pd	in	2(a)	0	0	0	*mmm*
2 O(1)	in	2(b)	0	1/2	1/2	*mmm*
4 O(2)	in	4(i)	0	0	0.16_2	*mm*

Interatomic distances (Å)

Sr	-	O(2)	2.44±13	
Sr	-	O(1)	2.59±9	(×2)
Sr	-	O(2')	2.66±1	(×4)
Pd	-	O(1)	1.99±1	(×2)
Pd	-	O(2)	2.08±13	(×2)

Discussion

Square planar PdO_4 groups whose normals are directed along *b* are linked through shared O(1) atoms into chains running along *a*. The chains stack to form planes perpendicular to *c*. Sr atoms lie in positions of sevenfold coordination between the planes.

Details of analysis

Powder photographs taken with CuK_α, radiation calibrated with α-quartz. [No value for a_o given.] Structure refined by least squares to given R = 0.07 for 17 observed reflexions listed.

LITHIUM INDIUM OXIDE

I. Die Kristallstruktur von Li_3InO_3. F. STEWNER and R. HOPPE, 1970. *Z. anorg. Chem.*, 374, 239-258.

Li_3InO_3 F.W. = 183.64

Trigonal, a = 9.606 , c = 10.420 Å, U = 847 Å^3, Z = 12, D_x = 4.40.

Space group $P\bar{3}c1$ (D^4_{3d})

Atomic positions

			x	*y*	*z*	*B*(Å^2)
2 In(1)	in	2(*a*)	0	0	1/4	1.27±1
4 In(2)	in	4(*d*)	1/3	2/3	0.2683±1	0.24±1
6 In(3)	in	6(*f*)	0.3332±2	0	1/4	0.18±1
12 O(1)	in	12(*a*)	0.096 ±2	0.218±1	0.1267±7	0.4 ±1
12 O(2)	in	12(*g*)	0.439 ±3	0.879±2	0.1359±9	0.9 ±1
12 O(3)	in	12(*g*)	0.769 ±2	0.556±1	0.1259±8	0.5 ±1
12 Li(1)	in	12(*g*)	0.134 ±5	0.229±5	0.431 ±3	1.2 ±4
12 Li(2)	in	12(*g*)	0.431 ±5	0.911±5	0.443 ±3	2.8 ±5
12 Li(3)	in	12(*g*)	0.800 ±5	0.552±5	0.436 ±3	1.2 ±5

Interatomic distances (Å)

Standard error 0.02 Å

In(1)	-	O(1)	2.23	(×6)
In(2)	-	O(2)	2.24	(×3)
In(2)	-	O(3)	2.16	(×3)
In(3)	-	O(1)	2.18	(×2)
In(3)	-	O(2)	2.23	(×2)
In(3)	-	O(3)	2.19	(×2)

Li atoms at centres of O tetrahedra (coordinates different from those given above) give rise to Li-O bonds of 1.96 - 2.03 Å.

Discussion

The oxygen atoms are in hexagonal close packing with the In atoms in octahedral holes between one pair of layers, and the Li in tetrahedral holes between the next.

Details of analysis

Lattice parameters from rotation, Weissenberg and precession photographs ($CuK_{\alpha1}$, λ = 1.54051 Å). Intensities measured photometrically from integrated Weissenberg photographs (MoK_α radiation) of layers with k = 0 - 6, l = 0 - 2, and h = k. Structure solved from Patterson function and refined to give R = 0.098 for 1544 reflexions listed. Li atom positions were determined by least squares (quoted above), Fourier syntheses and finding the centers of the tetrahedra of O atoms. The latter mostly lie within 3 standard errors of the values given above.

BARIUM MOLBYDATE
BARIUM WOLFRAMATE

I. [Crystal structures of Ba molybate and Ba tungstate]. T.I. BYLIČKINA, L.I. SOLEVA, E.A. POBEDIMSKAJA, M.A. PORAJ-KOŠIC and N.V. BELOV, 1970. *Kristallografija*, 15, 165-167 [*Soviet Physics - Crystallography*, 15, 130-131].

Tetragonal, Scheelite Type 1.

	$BaMoO_4$	$BaWO_4$
F.W.	297.3	385.2
a (Å)	5.62±3	5.614±3
c	12.82±3	12.719±3
U (Å^3)	404.9	400.9
D_x	4.87	6.30

Space group $I4_1/a$ (C_{4h}^6) systematic absences

Atomic positions

				$BaMoO_4$	$BaWO_4$
Ba	in	4(*a*)			
Mo,W	in	4(*b*)			
O	in	16(*f*)	*x*	0.143±10	0.146±10
			y	0.016±10	0.008±20
			z	0.709±10	0.714±10

[The coordinates are only given in I for *x* and *z* but they are given for two O atoms, allowing *y*(O) to be computed].

Interatomic distances (Å)

	$BaMoO_4$	$BaWO_4$
Ba - O (×4)	2.72±11	2.64±12
Ba - O (×4)	2.73±7	2.71±7
Mo(W) - O (×4)	1.88±8	1.95±8

Angles at Mo(W) = 109±4°.

Details of analysis

Lattice parameters from oscillation and rotation photographs and layer line patterns, also by least squares refinement using the powder diffractometer pattern (CuK_α radiation). Intensities measured visually were not corrected for absorption [no crystal size given]. Oxygen coordinates refined by least squares to give *R* = 0.14 for $BaMoO_4$ and *R* = 0.16 for $BaWO_4$ for an unspecified number of *h0l* reflexions [not listed].

1. *Strukturbericht*, 1, 347.

COPPER WOLFRAMATE

I. $CuWO_4$, a distorted wolframite-type structure. L. KIHLBORG and E. GEBERT, 1970. *Acta Cryst.*, B26, 1020-1025.

$CuWO_4$ F.W. = 329.41

Triclinic, *a* = 4.7026±6, *b* = 5.8389±7, *c* = 4.8784±6 Å, α = 91.677±9, β = 92.469±7, γ = 82.805±10°, *U* = 132.73 Å^3, D_m = 7.61-7.65, *Z* = 2, D_x = 7.790.

Space group $P\bar{1}$ (C_i^1)

Atomic positions

	x	*y*	*z*
Cu	0.49533±16	0.65976±13	0.24481±5
W	0.02106±4	0.17348±3	0.25429±4
O(1)	0.2491 ±10	0.3535 ±9	0.4245 ±10
O(2)	0.2145 ±10	0.8812 ±7	0.4309 ±9
O(3)	0.7353 ±10	0.3803 ±8	0.0981 ±9
O(4)	0.7826 ±9	0.9079 ±8	0.0533 ±9

Thermal parameters are given.

Interatomic distances (Å)

Cu - O	1.961±4,	1.967±5,	1.978±5,	1.997±5,	2.340±5,	2.450±5
W - O	1.760±5,	1.816±5,	1.845±4,	1.988±4,	2.028±4,	2.208±5

Material

CuO and WO_3 oven heated for 4 days at 800°C in an evacuated Pt tube.

Discussion of the structure

Distorted wolframate structure as reported previously (1) with linked octahedra about each cation as shown in Fig. 34.

Details of analysis

1111 reflexions measured on a diffractometer using Mo radiation. Absorption corrected. The structure was refined using least-squares methods to an *R* value of 0.029. Oxygen atom occupancy parameters do not deviate significantly from full occupancy.

1. E. GEBERT and L. KIHLBORG, 1967. *Acta Chem. Scand.*, 21, 2575.

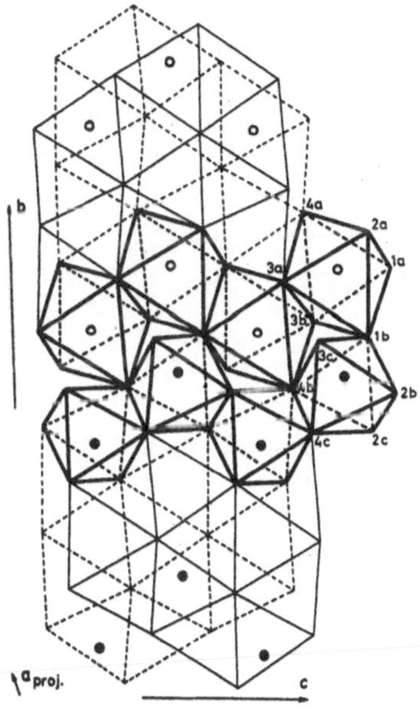

Fig. 34. The structure of $CuWO_4$ projected onto the *bc* plane. Three pseudo-hexagonal oxygen atom layers are indicated as triangular nets: in the upper part of the Figure two layers surrounding copper atoms (open circles) and in the lower part two layers enclosing wolfram atoms (filled circles).

LEAD WOLFRAMATE

I. [The crystal structure of $PbWO_4$]. G.F. PLAKHOV, E.A. POBEDIMSKAJA, M.A. SIMONOV and N.V. BELOV, 1970. *Kristallografija*, 15, 1067-1068 [*Soviet Physics - Crystallography*, 15, 928-929].

$PbWO_4$ F.W. = 455.04

Tetragonal, Scheelite Type (1), a = 5.50±1, c = 12.12±2 Å, U = 367 Å^3, Z = 4, D_x = 8.23.

Space group $I4_1/a$ (C^6_{4h})

Atomic positions Origin at $\bar{1}$

			x	y	z
4 Pb	in	4(b)	0.5	0.75	0.125
4 W	in	4(a)	0	0.25	0.125
16O	in	16(f)	0.221	0.401	0.389

Discussion

[The coordinates given are manifestly wrong as they give rise to square planar coordination around Pb and W with Pb-O = 2.28 Å and W-O = 1.75 Å, whereas I claim that the coordination is tetrahedral with W-O = 1.80 and Pb-O = 2.66.]

Details of analysis

Lattice parameters from high-angle rotation photograph reflexions. Intensities estimated visually from Weissenberg photographs of layers k = 0 to 3 taken with Mo radiation. Structure refined by least squares to give R = 0.05 (B = 0.1 Å^2) for 90 observed reflexions [not listed].

1. *Strukturbericht*, 1, 347.

STRONTIUM NEODYMIUM FERRATE

I. Sur quelques nouveaux oxydes mixtes de strontium et d'éléments de transition du type K_2NiF_4. J.C. JOUBERT, A. COLLOMB, D. ELMALEH, G. LEFLEM, A. DAOUDI and G. OLLIVIER, 1970. *J. Solid State Chem.*, 2, 343-346.

$SrNdFeO_4$ F.W. = 351.70

Tetragonal, a = 3.834+2, c = 12.360+3 Å, U = 181.7 Å^3, Z = 2, D_x = 5.38.

Space group $I4/mmm$ (D^{17}_{4h})

Atomic positions

			x	y	z(X-ray)	z(neutron)
2 Fe	in	2(a)	0	0	0	
2 Sr + 2 Nd	in	4(e)	0	0	0.360	0.359±2
4 O(1)	in	4(c)	0	0.5	0	
4 O(2)	in	4(e)	0	0	0.172	0.168±3

Interatomic distances (Å)

Fe	- O(1)	1.92±1	(×4)
Fe	- O(2)	2.13±4	(×2)
Sr,Nd	- O(2)	2.32±4	
Sr,Nd	- O(1)	2.58±1	(×4)
Sr,Nd	- O(2)	2.74±1	(×4)

Discussion

$SrNdFeO_4$, together with a range of other rare earth ferrates and chromates(III), is found to have the K_2NiF_4 structure (1).

Details of analysis

Prepared by heating $SrCO_3$, Nd_2O_3 and Fe_2O_3 at 1350°C. X-ray (CoK_α) and neutron powder diffraction. Structure determined by trial and error to give R = 0.07 for

28 X-ray reflexions and 0.04 for 10 neutron reflexions listed.

1. *Structure Reports*, 17, 232.

SODIUM YTTRIUM GERMANATE

I. [Crystal structure of synthetic Na, Y-orthogarmanate, $NaYGeO_4$]. E.A. KUZ'MIN, B.A. MAKSIMOV, V.V. ILJUKHIN and N.V. BELOV, 1970. *Ž. Strukt. Khim.*, 11, 159-161 [*J. Struct. Chem.*, 11, 152-154].

$NaYGeO_4$ F.W. = 248.48

Orthorhombic, γ-Ca_2SiO_4 type (1), $a = 5.32$ $b = 11.49$, $c = 6.49$ Å, $U = 396.71$ Å^3, $D = 4.4$, $Z = 4$, $D_x = 4.15$.

Space group $Pbn2_1$ (C_{2v}^9) absences and Patterson function

Atomic positions All atoms in 4(*a*)

	x	*y*	*z*	*B*(Å^2)
Y	0.003	0.222	0	0.0
Na	0.476	0.009	0.253	0.4
Ge	0.057	0.097	0.477	0.0
O(1)	0.221	0.162	0.692	0.1
O(2)	0.182	0.189	0.306	0.12
O(2)	0.333	0.462	0.473	0.0
O(3)	0.253	0.398	0.015	0.1

Interatomic distances (Å) *and bond angles*

Errors uncerain but at least 0.02 Å

Y - O(2)	2.24
Y - O(3)	2.31
Y - O(2)'	2.36
Y - O(1)	2.36
Y - O(1')	2.41
Y - O(4)	2.42

Angles in degrees subtended at Y by

	O(3)	O(2')	O(1)	O(1')	O(4)
O(2)	94	149	84	119	83
O(3)		94	108	82	170
O(2')			65	92	93
O(1)				155	82
O(1')					90

Na - O(3)	2.24
Na - O(4)	2.34
Na - O(4')	2.49
Na - O(1)	2.57
Na - O(2)	2.62
Na - O(3')	2.65

Angles in degrees subtended at Na by

	O(4)	O(4')	O(1)	O(2)	O(3')
O(3)	85	96	112	71	173
O(4)		172	79	102	89
O(4)			94	86	89
O(1)				177	63
O(2)					115

Ge - O(4)	1.64
Ge - O(3)	1.66
Ge - O(2)	1.67
Ge - O(1)	1.81

	O(3)	O(2)	O(1)
O(4)	113	118	110
O(3)		116	103
O(2)			93

Details of analysis

Crystals prepared hydrothermally. Intensities estimated visually from photographs of layers with l = 0 to 5 and h = 0 to 1 taken with MoK_α radiation and corrected for Lorentz and polarisation. Structure solved from Patterson and electron-density functions and refined by least squares to give R = 0.15 for 450 reflexions [not listed]. No correction for absorption ($\mu R \sim 2$).

1. W. EITEL and G. TRÖMEL, 1929. *Zbl. Mineral.*, A12, 415.

ALKALI EARTH RARE EARTH OXIDES

I. Untersuchungen an $SrYb_2O_4$, $CaYb_2O_4$ und $CaLu_2O_4$; ein Beitrag zur Kristallstruktur des Calciumferrat (III)-Typs. H. MÜLLER-BUSCHBAUM and R. VON SCHENK, 1970. *Z. anorg. Chem.*, 377, 70-78.

$SrYb_2O_4$ STRONTIUM YTTERBATE F.W. = 497.69

Orthorhombic, a = 9.98, b = 3.345, c = 11.73 Å, U = 391.5 Å^3, Z = 4, D_x = 8.42.

Space group *Pnma* (D_{2h}^{16}) [converted from *Pnam*]

Atomic positions All atoms in 4(c)

	x	y	z
Se	0.7540	1/4	0.6512
Yb(1)	0.4216	1/4	0.1092
Yb(2)	0.4253	1/4	0.6121
O(1)	0.208	1/4	0.177
O(2)	0.119	1/4	0.483
O(3)	0.519	1/4	0.781
O(4)	0.419	1/4	0.425

Interatomic distances (Å) *and bond angles*

Sr - O(4)	2.56	(×2)
Sr - O(2)	2.62	(×2)
Sr - O(1)	2.65	(×2)
Sr - O(3)	2.76	
Sr - O(3)'	2.80	
Yb(1) - O(3)	2.19	(×2)
Yb(1) - O(2)	2.25	
Yb(1) - O(2)*	2.27	(×2)
Yb(1) - O(1)	2.28	

Angles in degrees subtended at Yb(1) by

	O(3)'	O(2)	O(2)*	O(2)*'	O(1)
O(3)	99	93	173	82	93
O(2)			81	81	172
O(2)*				95	94

Yb(2) - O(3)	2.19	
Yb(2) - O(4)	2.20	
Yb(2) - O(1)	2.27	(×2)
Yb(2) - O(4)*	2.32	(×2)

Angles in degrees subtended at Yb(2) by

	O(4)	O(1)'	O(4)*'
O(3)	156	87	83
O(4)		109	80
O(1)		95	86
O(1)'			170
O(4)*			92

CALCIUM YTTERBATE

$CaYb_2O_4$ F.W. = 450.15

Orthorhombic, *a* = 9.74, *b* = 3.316, *c* = 11.594 Å, *U* = 374.5 Å^3, *Z* = 4, D_x = 7.97.

Space group *Pnma* (D_{2h}^{16}) [converted from *Pnam*]

Atomic positions All atoms in 4(*c*)

	x	*y*	*z*
Ca	0.7539	1/4	0.6502
Yb(1)	0.4246	1/4	0.1105
Yb(2)	0.4196	1/4	0.6124
O(1)	0.197	1/4	0.194
O(2)	0.131	1/4	0.484
O(3)	0.521	1/4	0.776
O(4)	0.412	1/4	0.428

Interatomic distances (Å) *and bond angles*

Ca - O(4)	2.49	(×2)	
Ca - O(1)	2.50	(×2)	
Ca - O(2)	2.54	(×2)	
Ca - O(3)	2.70		
Ca - O(3)	2.74		
Yb(1) - O(3)	2.18	(×2)	
Yb(1) - O(2)	2.28	(×2)	
Yb(1) - O(2)*	2.29		
Yb(1) - O(1)	2.42		

Angles in degrees subtended at Yb(1) by

	O(3)'	O(2)	O(2)'	O(2)*	O(1)
O(3)	99	177	84	94	89
O(2)			93	84	72
O(2)*					175

Yb(2) - O(4)	2.14		Yb(2) - O(3)	2.14
Yb(2) - O(1)	2.22	(×2)		
Yb(2) - O(4)*	2.38	(×2)		

Angles in degrees subtended at Yb(2) by

	O(4)	O(1)'	O(4)*'
O(3)	154	82	82
O(4)		114	80
O(1)		97	85
O(1)'			163
O(4)			88

CALCIUM LUTECIATE

$CaLu_2O_4$ F.W. = 454.01

Orthorhombic, *a* = 9.731, *b* = 3.304, *c* = 11.573 Å, *U* = 372.1 Å^3, *Z* = 4, D_x = 8.09

Space group *Pnma* (D_{2h}^{16}) [converted from *Pnam*]

Atomic positions All atoms in 4(*c*)

	x	*y*	*z*
Ca	0.7431	1/4	0.6500
Lu(1)	0.4238	1/4	0.1096
Lu(2)	0.4209	1/4	0.613
O(1)	0.204	1/4	0.191
O(2)	0.134	1/4	0.477
O(3)	0.514	1/4	0.778
O(4)	0.404	1/4	0.432

Interatomic distances (Å) *and bond angles*

Ca - O(4)	2.44	(×2)
Ca - O(2)	2.47	(×2)
Ca - O(1)	2.51	(×2)
Ca - O(3)	2.67	
Ca - O(3)	2.76	
Lu(1) - O(3)	2.19	(×2)
Lu(1) - O(2)	2.28	
Lu(1) - O(2)*	2.32	(×2)
Lu(1) - O(1)	2.34	

Angles in degrees subtended at Lu(1) by

	O(3)'	O(2)	O(2)*	O(2)*'	O(1)
O(3)	98	91	175	86	91
O(2)			86	86	178
O(2)*				91	93

Lu(2) - O(4)	2.10	
Lu(2) - O(3)	2.11	
Lu(2) - O(1)	2.24	(×2)
Lu(2) - O(4)*	2.43	(×2)

Angles in degrees subtended at Lu(2) by

	O(3)	O(1)'	O(4)*'
O(4)	159	111	81
O(3)		82	84
O(1)		95	88
O(1)'			166
O(4)*			86

Discussion

All three crystals have the $CaFe_2O_4$ structure (1). The errors in the bond lengths are uncertain, but are likely to be fairly large (∿ 0.05 Å).

Details of analysis

Rotation and Weissenberg photographs around [001] [*l* = 0, 1?] and precession photographs of 0*kl* and *h*0*l*. $CaFe_2O_4$ structure (1) assumed and refined by differential synthesis to give *R* ∿ 0.08 for $SrYb_2O_4$, 0.1$\bar{6}$ for $CaYb_2O_4$ and 0.13 for $CaLu_2O_4$, each for about 200 listed reflexions.

1. *Structure Reports*, 20, 271.

ALKALI METAL MOLYBDATES AND WOLFRAMATES

I. The structure of potassium, rubidium and cesium molybdate and tungstate. F.X.N.M. KOOLS, A.S. KOSTER and G.D. RIBEK, 1970. *Acta Cryst.*, B26, 1974-1977.

K_2MoO_4 F.W. = 238.14

K_2WO_4 F.W. = 344.08

Rb_2MoO_4	F.W. = 330.87
Rb_2WO_4	F.W. = 436.81
Cs_2MoO_4	F.W. = 421.73
Cs_2WO_4	F.W. = 527.67

Monoclinic

	a(Å)	*b*(Å)	*c*(Å)	β°	*U*(Å^3)	D_x	Z
K_2MoO_4	12.345(2)	6.078(1)	7.535(1)	115.73(1)	509.3	[3.105]	4
K_2WO_4	12.380	6.117	7.554	115.96	514.4	[4.443]	4
Rb_2MoO_4(I)	12.821	6.253	7.842	115.64	566.8	[3.877]	4
Rb_2WO_4	12.841	6.285	7.854	115.82	570.5	[5.086]	4

Orthorhombic

	a(Å)	*b*(Å)	*c*(Å)	*U*(Å^3)	D_x	Z
Rb_2MoO_4(II)	6.375(2)	11.112(3)	8.097(2)	573.6	[3.831]	4
Cs_2MoO_4	6.551(2)	11.586(2)	8.499(2)	645.2	[4.342]	4
Cs_2WO_4	6.598(2)	11.647(3)	8.513(2)	654.2	[5.358]	4

Space group

*C*2/*m* (C_{2h}^3) K_2MoO_4, K_2WO_4, Rb_2MoO_4(I), Rb_2WO_4

Pmcn (D_{2h}^{16}) Rb_2MoO_4(II), Cs_2MoO_4, Cs_2WO_4

Atomic positions

Rb_2MoO_4			*x*	*z*
4 Mo	in	4(*i*)	0.174(2)	0.227(4)
4 Rb(1)	in	4(*i*)	0.515(3)	0.237(6)
4 Rb(2)	in	4(*i*)	0.848(3)	0.269(5)

Rb_2WO_4			*x*	*z*
4 Wo	in	4(*i*)	0.175(1)	0.227(2)
4 Rb(1)	in	4(*i*)	0.517(3)	0.243(4)
4 Rb(2)	in	4(*i*)	0.846(3)	0.262(4)

Cs_2MoO_4			*y*	*z*
4 Mo	in	4(*c*)	0.416(2)	0.224(2)
4 Cs(1)	in	4(*c*)	0.418(1)	0.671(2)
4 Cs(2)	in	4(*c*)	0.707(1)	0.991(1)

Cs_2WO_4			*y*	*z*
4 W	in	4(*c*)	0.418(1)	0.226(1)
4 Cs(1)	in	4(*c*)	0.420(1)	0.671(2)
4 Cs(2)	in	4(*c*)	0.709(1)	0.990(2)

Discussion of the structure

Cs_2MoO_4 and Cs_2WO_4 have the β-K_2SO_4 structure (1). Monoclinic phases are isostructural with K_2WO_4 (2). The transition in Rb_2MoO_4 occurs at about 100°C.

Details of analysis

About 120 powder lines determined using CuK_α radiation with assumed oxygen positions and isotropic temperature factors.

1. *Structure Reports*, 22, 447.
2. A.S. KOSTER, F.X.N.M. KOOLS and G.D. RIECK, 1969. *Acta Cryst.*, B25, 1704.

LITHIUM EUROPIUM OXIDE

$LiEu_3O_4$ F.W. = 526.82

I. Die Kristallstruktur von $LiEu_3O_4$. H. BÄRNIGHAUSEN, 1970. *Z. anorg. Chem.*, 374, 201-224.

Orthorhombic, a = 11.535±2, b = 3.480±1, c = 11.565±2 Å, U = 465.04 Å^3, Z = 4, D_x = 7.51.

Space group *Pnma* (D_{2h}^{16}) [converted from *Pbnm*]

Atomic positions All atoms in 4(c) (y = 1/4)

	x	*z*
Eu(1)	0.35862±4	0.20426±4
Eu(2)	0.40230±4	0.61529±5
Eu(3)	0.13562±4	0.42543±4
Li	0.1555 ±19	0.9108 ±19
O(1)	0.2441 ±6	0.0450 ±6
O(2)	0.4369 ±6	0.3894 ±7
O(3)	0.0263 ±8	0.6297 ±8
O(4)	0.2575 ±7	0.7808 ±7

Anisotropic temperature factors given for Eu, isotropic for Li and O.

Interatomic distances (Å) *and bond angles*

Fourfold coordination around Li (crystallographic site symmetry C_s)

Li - O(1)	1.86		
Li - O(2)	2.06	(×2)	
Li - O(4)	1.91		σ(Li-O) = 0.02 Å

Angle in degrees subtended at Li by

	O(2)'	O(4)	
O(1)	112.7	108.7	
O(2)	115.6	103.0	σ(O-Li-O) = 1.0°

Sixfold coordination around Eu(1) = Eu^{3+} (crystallographic site symmetry C_s)

Eu(1) - O(1)	2.267		
Eu(1) - O(2)	2.324		
Eu(1) - O(3)	2.353	(×2)	
Eu(1) - O(4)	2.367	(×2)	σ(Eu-O) = 0.007 Å

Angle in degrees subtended at Eu(1) by

	O(3)'	O(4)'	
O(1)	91.8	88.5	
O(2)	96.8	82.9	
O(3)	95.4	85.0	
O(4)		94.6	σ(O-Eu-O) - 0.3°

Sevenfold coordination around Eu(2) = Eu^{2+} (crystallographic site symmetry C_s)

Eu(2) - O(1)	2.558	(×2)	
Eu(2) - O(2)	2.643		
Eu(2) - O(2)*	2.544	(×2)	
Eu(2) - O(3)	3.278		
Eu(2) - O(4)	2.540		σ(Eu-O) = 0.008 Å

Angles in degrees subtended at Eu(2) by

	O(1)'	O(2)*'	O(4)	
O(1)	85.7	90.5	78.8	
O(2)	77.6	82.5		
O(2)*		86.3		
O(3)		72.6	67.0	σ(O-Eu-O) = 0.2°

Sevenfold coordination around Eu(3) = Eu^{2+} (cyrstallographic site symmetry C_s)

Eu(3)	-	O(1)	2.620	(×2)	
Eu(3)	-	O(3)*	2.631	(×2)	
Eu(3)	-	O(3)	2.678		
Eu(3)	-	O(4)	2.711	(×2)	σ(Eu-O) = 0.008 Å

Angles in degrees subtended at Eu(3) by

	O(1)'	O(3)*'	O(4)'
O(1)	83.2	93.7	70.0
O(3)*		82.8	75.5
O(3)	77.5	83.1	
O(4)			79.9 σ(O-Eu-O) = 0.2°

Angles at O (given in I) range from 78.1 to 115.6°.

Discussion

The assignment of Eu^{2+} and Eu^{3+} is based on crystal radii of the two forms, and a powder analysis of the isostructural $LiSr_2EuO_4$ which showed Sr substituting for Eu(2) and Eu(3) (intensities of powder lines given in I). The structure is virtually identical to that of Yb_3S_4(2) with the addition of Li in a tetrahedral hole and is similar to that of Eu_3O_4 into which it converts on heating *in vacuo* at 900°C.

Details of analysis

Lattice parameters and preparation from 1. Intensities measured by photometer from integrated Weissenberg photographs of layers with l = 0, 1, and precession photographs of layers with h = 0, 1 and k = 0, 1, taken with a needle shaped crystal 0.035 mm in diameter. Corrected for Lorentz, polarisation, absorption (μ = 40.2 mm^{-1}) and extinction. Structure from Harker sections and difference Fourier syntheses refined by least squares to give R = 0.045 for 787 observed reflexions listed.

1. H. BÄRNIGHAUSEN, 1967. *Zeit. anorg. Chem.*, 349, 280.
2. R. CHEVALIER, P. LARUETTE and J. FLAHAUT, 1967. *Bull. Soc. franç. Minér. Crist.*, 90, 564.

LANTHANUM OXYSELENIDE

I. Structure cristalline de l'oxyséléniure de lanthane $La_4O_4Se_3$. J. DUGUE, C. ADOLPHE and P. KHODADAD, 1970. *Acta Cryst.*, B26, 1627-1628.

$La_4O_4Se_3$

Orthorhombic, a = 13.232±5, b = 5.590±8, c = 4.101±3 Å, [U = 303.33 $Å^3$], D_m = 6.19, Z = 2, D_x = 6.10.

Space group *B2mm* (C_{2v}^{11}) confirmed by the structure solution.

Atomic positions

	x	y	z	$B(Å^2)$
La(1)	0	0	0	*
La(2)	0.018	1/2	0	*
La(3)	0.0186	0.236	1/2	*
Se(1)	0.356	0.356	0	*
Se(2)	0.331	0	0	*
O(1)	0.092	0.244	0	0.33
O(2)	0.085	0	1/2	0.41
O(3)	0.123	1/2	1/2	0.24

*Anisotropic thermal parameters are given.

Interatomic distances (Å)

La	- O	2.34 - 2.47
La	- Se	3.21 - 3.33

Discussion of the structure

La(1) is coordinated to 4 oxygen and 2 selenium atoms, whereas La(2) and La(3) are both coordinated to 4 oxygen and 4 selenium atoms.

Details of analysis

Data measured by densitometer from Weissenberg photographs taken with Mo radiation. Refined by least-squares methods to an *R* value of 0.08. No absorption correction.

URANIUM VANADIUM PENTOXIDE

I. Structure cristalline de UVO_5. R. CHEVALIER and M. GASPERIN, 1970. *Bull. Soc. franç. Minér. Crist.*, 93, 18-22.

UVO_5 F.W. = 368.96

Orthorhombic, a = 4.12, b = 7.19, c = 12.31 Å, U = 364.7 Å^3, Z = 4, D_x = 6.71.

Space group *Pbcm* (D_{2h}^{11}) [converted from *Pbma*]

Atomic positions

			x	y	z	B(Å^2)
U	in	4(*d*)	0.0137	1/4	0.0273	*
V	in	4(*c*)+	-0.1134	0	1/4	*
O(1)	in	4(*d*)	0.511	1/4	0.029	3.48
O(2)	in	4(*c*)	0.513	0	1/4	4.08
O(3)	in	4(*d*)	0.026	1/4	0.717	0.80
O(4)	in	8(*e*)	0.008	0.538	0.599	0.82

* anisotropic temperature factor given.

+ These coordinates have converted from the values (-*x*, 0, 3/4) given in I.

Interatomic distances (Å)

Sevenfold coordination around U (crystallographic site symmetry C_s)

U - O(1)	2.05	
U - O(1)'	2.07	
U - O(3)	2.34	
U - O(4)	2.25	(×2)
U - O(4)'	2.18	(×2)

Fivefold coordination around V (crystallographic site symmetry C_2)

V - O(2)	1.54	
V - O(2)	2.58	
V - O(4)	1.92	(×2)
V - O(3)	1.87	(×2)

Errors not given but probably about 0.05 Å.

Discussion

All atoms except O(1) and O(2) lie close to the plane with x = 0, the U atoms being surrounded by 5, the V atoms by 4 oxygen atoms in the plane. O(1) and O(2) lie between the planes above and below U and V respectively, to complete a pentagonal bipyramid around U and a very distorted octahedron around V. A rather inaccurate structure for UVO_5 was first reported by 1. The present work confirms the general arrangement of atoms proposed by 1 but leads to a considerably better view of the structure.

Details of analysis

Lattice parameters from Weissenberg and precession cameras. Intensities from single crystal diffractometer (θ-2θ scan, CuK_α radiation). Lorentz, polarisation and spherical crystal absorption corrections (μR = 5.0, crystal size ∿ 0.05 mm, μ = 198 mm^{-1}.) Structure refined by difference syntheses to give R = 0.10 for 227 reflexions [not listed].

1. O.A. EFREMOVA, L.M. KOVBA and V.I. SPICYN, 1969. *Ž. Strukt. Khim.*, 10, 344.

CHROMIUM TITANIUM NEODYMIUM PENTOXIDE
IRON TITANIUM NEODYMIUM PENTOXIDE

I. Étude par rayons x et neutrons de la serie isomorphe ATiTO (*A* = Cr, Mn, Fe, *T* = terres rares). G. BUISSON, 1970. *J. Phys. Chem. Solids*, 31, 1171-1183.

$CrTiNdO_5$ F.W. = 324.13

Orthorhombic, a = 7.56, b = 8.67, c = 5.80 Å, U = 380 Å^3, Z = 4, D_x = 5.65.

Space group *Pbam* (D_{2h}^9) or *Pba*2 (C_{2v}^8)

Atomic positions (in *Pbam*)

			x	*y*	*z*
4 Nd	in	4(*g*)	0.142	0.172	0
4 (0.95 Ti+0.05 Cr)	in	4(*h*)	0.124	-0.134	1/2
4 (0.95 Cr+0.05 Ti)	in	4(*f*)	0	1/2	0.252
8 O(1)	in	8(*i*)	0.111	-0.285	0
4 O(2)	in	4(*g*)	0.186	0.433	0
4 O(3)	in	4(*h*)	0.156	0.444	1/2
4 O(4)	in	4(*e*)	0	0	0.300

$FeTiNdO_5$ F.W. = 327.98

Orthorhombic, a = 7.54, b = 8.71, c = 5.85 Å, U = 384 Å^3, Z = 4, D_x = 5.66.

Space group *Pbam* (D_{2h}^9) (isostructural with $CrTiNdO_5$)

Atomic positions

			x	*y*	*z*
4 Nd	in	4(*g*)	0.136	0.184	0
4 (0.57 Ti+0.43 Fe)	in	4(*h*)	0.088	-0.156	1/2
4 (0.57 Fe+0.43 Ti)	in	4(*f*)	0	1/2	0.238
8 O(1)	in	8(*i*)	0.107	-0.256	0.247
4 O(2)	in	4(*g*)	0.206	0.429	0
4 O(3)	in	4(*h*)	0.179	0.404	1/2
4 O(4)	in	4(*e*)	0	0	0.285

Discussion

The magnetic structure of both materials is given together with a crystal field interpretation of the distribution of the two ions between sites having an octahedral and a square pyramidal environment.

Details of analysis

X-ray (Cr radiation) and neutron powder patterns both given. Distribution of Ti and transition metal refined by trial and error. It is not clear how the other parameters were obtained. X-ray intensities are not given and only 13 neutron lines could be measured for each material. The structure was assumed to be isomorphous with $HoMn_2O_5$.

POTASSIUM DIZIRCONATE

I. The crystal structure of beta-potassium dizirconate: β-$K_2Zr_2O_5$. B.W. GATEHOUSE and D.J. LLOYD, 1970. *J. Solid State Chem.*, 1, 478-483.

β-$K_2Zr_2O_5$ F.W. = 340.6

Orthorhombic, a = 5.85, b = 10.79, c = 8.76 Å, U = 552.94 Å^3, Z = 4, D_x = 4.08.

Space group *Pnna* (D_{2h}^6) systematic absences

Atomic positions

			x	y	z	B(Å^2)
8 Zr	in	8(*e*)	0.9697±7	0.0763±3	0.1603±3	*
8 K	in	8(*e*)	0.438 ±2	0.3340±9	0.077 ±1	*
8 O(1)	in	8(*e*)	0.136 ±5	0.398 ±2	0.554 ±2	0.6±5
4 O(2)	in	4(*c*)	0.25	0	0.255 ±5	1.7±2
4 O(3)	in	4(*c*)	0.25	0	0.677 ±4	1.2±7
4 O(4)	in	4(*d*)	0.40 ±8	0.25	0.25	1.5±7

* Anisotropic temperature factors given.

Interatomic distances (Å) *and bond angles*

Zr - Zr	3.052±8	(across faces)
Zr - O(2)	2.01 ±2	
Zr - O(4)	2.07 ±1	
Zr - O(3)	2.09 ±2	
Zr - O(1)	2.13 ±2	
Zr - O(1)'	2.18 ±3	
Zr - O(1)"	2.22 ±2	

Angles in degrees subtended at Zr by

	O(4)	O(3)	O(1)	O(1)'	O(1)"
O(2)	93.0±12	93.4±10	92.5±13	163.1±6	92.6±8
O(4)		102.7±9	97.2±9	103.0±13	174.4±13
O(3)			158.9±9	77.9±8	76.9±8
O(1)				90.9±9	82.6±8
O(1)'					71.5±9

K - O(2)	2.57±3
K - O(3)	2.70±2
K - O(1)	2.77±2
K - O(4)	2.92±4
K - O(3)	3.06±3
K - O(4)	3.06±1

Discussion

Two ZrO_6 octahedra share faces to form pairs which are linked into chains along *a* by edge and corner sharing. The potassium atoms lie in holes between the chains at a distance 3.2 Å apart.

Details of analysis

Crystals (0.03×0.02×0.02 mm^3) grown in nitrogen furnace and mounted in oil in a capillary tube. Lattice parameters measured from oscillation and Weissenberg photographs. Intensities estimated visually from multiple-film Weissenberg photographs of layers with h = 0-5 (CuK_α radiation) and corrected for [Lorentz and polarisation effects and] absorption. Structure solved by three-dimensional Patterson and electron-density functions and refined by least squares to give R_1 = 0.097 for 267 reflexions listed.

NEODYMIUM WOLFRAMATE

I. [A new form of the scheelite structural type: crystal structure of Nd_2WO_6]. T.M. POLJANSKAJA, S.V. BORISOV and N.V. BELOV, 1970. *Dokl. Akad. Nauk SSSR*, 193, 83-86 [*Soviet Physics - Doklady*, 15, 636-639].

Nd_2WO_6 F.W. = 568.32

Monoclinic, a = 15.92, b = 11.39, c = 5.508 Å, β = 92°, U = 998.2 Å^3, Z = 8, D_x = 7.55.

Space group $I2/c$ (C_{2h}^6)

Atomic positions

Atoms in (0, 0, 0; 1/2, 1/2, 1/2)±(x, y, z; x, $\bar{y}$, 1/2+z)

			x	y	z	B(Å^2)
8 W	in	8(f)	0.1531±1	0.3526±2	0.2145±5	0.2
4 Nd(1)	in	4(e)	0	0.6314±4	0.25	0.4
8 Nd(2)	in	8(f)	0.1731±2	0.8845±3	0.2859±6	0.4
4 Nd(3)	in	4(e)	0	0.1110±4	0.25	0.3
8 O(1)	in	8(f)	0.407 ±3	0.519 ±5	0.485 ±10	0.3
8 O(2)	in	8(f)	0.420 ±3	0.261 ±5	0.493 ±11	0.3
8 O(3)	in	8(f)	0.422 ±3	0.035 ±5	0.396 ±11	0.7
8 O(4)	in	8(f)	0.272 ±3	0.053 ±4	0.117 ±11	0.4
8 O(5)	in	8(f)	0.270 ±3	0.205 ±5	0.487 ±11	0.1
8 O(6)	in	8(f)	0.390 ±3	0.222 ±5	0.036 ±10	0.7

Interatomic distances (Å) and bond angles

W	-	O(6)	1.77
W	-	O(5)	1.80
W	-	O(4)	1.83
W	-	O(3)	1.84
W	-	O(2)	2.06

Angles in degrees subtended at W by

	O(5)	O(4)	O(3)	O(2)
O(6)	126	99	109	84
O(5)		95	120	79
O(4)			99	174
O(3)				85

Nd(1)	-	O(2)	2.38	(×2)
Nd(1)	-	O(3)	2.42	(×2)
Nd(1)	-	O(3)	2.53	(×2)
Nd(1)	-	O(6)	2.61	(×2)
Nd(2)	-	O(1)	2.22	
Nd(2)	-	O(1)	2.38	
Nd(2)	-	O(2)	2.40	
Nd(2)	-	O(4)	2.47	
Nd(2)	-	O(5)	2.51	
Nd(2)	-	O(5)	2.55	
Nd(2)	-	O(4)	2.67	
Nd(2)	-	O(6)	2.73	
Nd(3)	-	O(1)	2.30	(×2)
Nd(3)	-	O(2)	2.38	(×2)
Nd(3)	-	O(1)	2.49	(×2)
Nd(3)	-	O(6)	2.82	(×2)

Discussion

The structure is similar to that of scheelite.

Details of analysis

Crystals prepared by hydrothermal synthesis. Intensities estimated visually from photographs of layers l = 0 to 6, k = 0 taken with Mo radiation. Intensities corrected for Lorentz and polarisation effects, but not absorption (crystal size 0.26×0.33×0.02 mm^3, μ = 47.7 mm^{-1}). Structure solved from Patterson function and refined by least squares to give R = 0.17 for 2400 reflexions [not listed].

CALCIUM NIOBATE

I. The crystal structure of calcium niobate ($CaNb_2O_6$). J.P. CUMMINGS and S.H. SIMONSEN, 1970. *Amer. Min.*, 55, 90-97.

$CaNb_2O_6$ F.W. = 321.88

Orthorhombic, a = 14.926±4, b = 5.752±4, c = 5.204±4 Å, U = 446.8 Å^3, D_m = 4.70, Z = 4, D_x = 4.78.

Space group *Pbcn* (D_{2h}^{14}) from systematic absences

Atomic positions

			x	*y*	*z*	*B*(Å^2)
4 Ca	in	4(*c*)†	0	0.7756	1/4	*
8 Nb	in	8(*d*)	0.1653	0.3166	0.2987	*
8 O(1)	in	8(*d*)	0.0893	0.0997	0.4040	1.50
8 O(2)	in	8(*d*)	0.1003	0.4280	0.0056	1.84
8 O(3)	in	8(*d*)	0.2576	0.1351	0.1266	0.46

* Anisotropic temperature factors given.

[† converted from atom at position 0, *y*, 3/4.]

σ(Nb) = 0.0013; σ(Ca) = 0.004; σ(O) = 0.009 Å.

Interatomic distances (Å) *and bond angles*

Sixfold coordination around Nb

Nb	- O(1)	1.774
Nb	- O(2)	1.919
Nb	- O(3)"	2.077
Nb	- O(3)	1.947
Nb	- O(2)'	2.065
Nb	- O(3)'	2.342

σ(Nb-O) = 0.009 Å

Angles in degrees subtended at Nb by

	O(2)	O(3)"	O(3)	O(2)'	O(3)'
O(1)	99.0	101.2	102.6	92.2	169.5
O(2)		152.8	99.9	86.5	71.5
O(3)"			93.3	74.7	86.4
O(3)				162.7	84.0
O(2)'					82.7

σ(O-Nb-O) = 0.4°

Eightfold coordination around Ca (crystallographic site symmetry C_2)

Ca	- O(1)	2.353	(×2)
Ca	- O(2)	2.319	(×2)
Ca	- O(1)'	2.427	(×2)
Ca	- O(2)"	2.322	(×2)

σ(Ca-O) = 0.009 Å

Discussion

The present work confirms and refines the structure described and illustrated in 2. The structure consists of chains of edge-sharing NbO_6 octahedra running along c. These are linked into corrugated sheets perpendicular to a by corner-shared oxygen atoms. The Ca atoms lie between the sheets.

Details of analysis

Lattice parameters from least-squares refinement of film and diffractometer measurements. Intensities measured with diffractometer (XRD-5) using stationary crystal-stationary counter method (MoK_α radiation, balanced filters). Lorentz, polarisation and absorption corrections (μ = 5.99 mm^{-1}), crystal dimensions 0.13 ×0.20×0.40 mm^3). Structure determined by Patterson and Fourier functions and refined by least squares to give R = 0.067 for 618 observed reflexions [eleven reflexions were omitted because of grossly bad agreement]. Intensities listed in 1.

1. *NAPS document* #00708, ASIS National Auxiliary Publications Service, c/o CCM Information Sciences Inc., 22 West 34th St., New York, N.Y. 10001.
2. *Structure Reports*, 24, 367.

LITHIUM NIOBIUM TITINATE

I. Synthèse de monocristeaux et étude structurale de $LiNbTiO_6$. D. FAUQUIER and M. GASPERIN, 1970. *Bull. Soc. franç Minér Crist.*, 93, 258-259.

$LaNbTiO_6$ F.W. = 375.7

Orthorhombic, a = 10.93, b = 7.572, c = 5.446 Å, U = 450.89 Å^3, Z = 4, D_x = 6.18.

Space group *Pnma* (D_{2h}^{16})

Atomic positions

			x	y	z	B(Å^2)
4 La	in	4(c)	0.0421	0.25	0.5403	*
8 Nb, Ti	in	8(d)	0.1445	-0.0057	0.0362	0.89
8 O(1)	in	8(d)	-0.026	0.033	0.228	0.19
8 O(2)	in	8(d)	0.210	0.051	0.374	0.33
4 O(3)	in	4(c)	0.144	0.25	-0.019	0.09
4 O(4)	in	4(c)	0.128	-0.25	0.126	0.92

* Anisotropic temperature factors given

Interatomic distances (Å) *and bond angles*

La - O(1)	2.48	(×2)		Nb,Ti - O(2)	1.85
La - O(1)'	2.49	(×2)		Nb,Ti - O(4)	1.92
La - O(2)	2.54	(×2)		Nb,Ti - O(1)	1.95
La - O(4)	2.60			Nb,Ti - O(3)	1.96
La - O(3)	2.65			Nb,Ti - O(2)'	2.02
				Nb,Ti - O(1)'	2.16

Angles in degrees subtended at Nb,Ti by

	O(4)	O(1)	O(3)	O(2)'	O(1)'
O(2)	91.4	101.5	96.4	99.7	177.1
O(4)		91.3	171.8	90.3	85.8
O(1)			89.4	158.6	78.3
O(3)				86.0	86.4
O(2)'					80.6

Errors not given, but at least 0.01 Å and 0.5°.

Discussion

Isostructural with $CaTa_2O_6$ (1). The Ti and Nb atoms are distributed randomly over the octahedrally coordinated sites.

Details of analysis

Intensities measured with automatic diffractometer (CuK_α radiation) corrected for [Lorentz, polarisation and] absorption effects (μ = 108 mm^{-1}, mean radius 0.07 mm). Assumed isostructural with $CaTa_2O_6$ (1), and refined by least squares to give R = 0.11 for 341 non-equivalent reflexions [not listed].

1. L. JAHNBERG, 1963. *Acta Chem. Scand.*, 17, 2548.

MAGNESIUM TELLURATE (VI)

I. A new structure type Mg_3TeO_6. H. SCHULZ and G. BAYER, 1970. *Naturwissenschaften*, 57, 393.

Mg_3TeO_6 F.W. = 296.5

Rhombohedral, a = 8.615, c = 10.315 Å, U = 663.0 Å^3, Z = 6, D_x = 4.45.

Space group $R\bar{3}$ (C_{3i}^2) from a "centre test" [*sic*]

Atomic positions Hexagonal setting

			x	*y*	*z*	*B*(Å^2)
3 Te(1)	in	3(*a*)	0	0	0	0.26±1
3 Te(2)	in	3(*b*)	0	0	0.5	*
18 Mg	in	18(*f*)	0.4013±3	0.1082±3	0.0448±2	0.55±2
18 O(1)	in	18(*f*)	0.2017±5	0.1720±5	0.0981±3	0.65±3
18 O(2)	in	18(*f*)	0.1851±5	0.1598±5	0.6162±3	0.58±3

* Not given [same as Te(1)?]

Interatomic distances (Å) *and bond angles*

Te(1) - O(1)	1.914±3 (×6)
O(1) - Te - O(1)	94.6±1°, 85.4±1°
Te(2) - O(2)	1.918±4 (×6)
O(2) - Te - O(2)	85.1±1°, 94.9±1°
Mg - O(1)	2.029±4
Mg - O(2)	2.051±5
Mg - O(2)	2.118±3
Mg - O(1)	2.123±6
Mg - O(2)	2.156±3
Mg - O(1)	2.267±4

Angles in degrees subtended at Mg by

	O(2)	O(2)	O(1)	O(2)	O(1)
O(1)	95.6±2	106.8±2	81.6±2	106.0±2	144.5±2
O(2)		149.9±2	122.6±1	80.0±1	79.3±1
O(2)			81.2±2	74.7±2	92.9±1
O(1)				155.9±2	72.4±1
O(2)					107.6±1

Details of analysis

Crystals grown from NaCl by cooling from 950°C. Intensities measured on automatic diffractometer. Final R_1 = 0.065 for 1182 reflexions [not listed].

STRONTIUM NICKEL TELLURATE

I. Das $Sr_2[NiTe]O_6$. P. KÖHL, E. SCHULTZE-RHONHOF and D. REINEN, 1970. *Z. anorg. Chem.*, 378, 129-143.

Sr_2NiTeO_6 F.W. = 457.54

Monoclinic, a = 9.64$_9$, b = 5.61$_3$, c = 5.58$_1$ Å, β = 54.67, U = 246.6 Å^3, Z = 2, D_x = 6.15.

Space group $C2/m$ (C_{2h}^3) See *Details of analysis*

Atomic positions

			x	*y*	*z*	*B*(Å^2)
2 Te	in	2(*a*)	0	0	0	0.12±4
2 Ni	in	2(*d*)	0	0	0.5	0.40±10
4 Sr	in	4(*i*)	0.25	0	0.25	0.77±4
8 O(1)	in	8(*j*)	0.0021±7	-0.2398±6	0.2415±7	1.10±20
4 O(2)	in	4(*i*)	0.2420±10	0	-0.2420±10	1.30±30

Transformed from pseudocubic setting. See *Details of analysis*.

Interatomic distances (Å) and bond angles

Te - O(2)	1.905±8	(×2)
Te - O(1)	1.913±5	(×4)

Angles in degrees subtended at Te by

	O(1)	O(1)"
O(2)	89.5±3	89.5±3
O(2)'	90.5±3	90.5±3
O(1)		89.4±2
O(1)'		90.6±2

Ni - O(2)	2.031±8	(×2)
Ni - O(1)	2.045±4	(×4)

Angles in degrees subtended at Ni by

	O(1)	O(1)
O(2)	90.5±2	89.5±2
O(2)	89.5±2	90.5±2
O(1)		88.8±2
O(1)		91.2±2
O(1)		

Sr - O(1)	2.769±6	(×2)
Sr - O(1)	2.791±3	(×2)
Sr - O(2)	2.791±8	(×2)
Sr - O(1)	2.793±3	(×2)
Sr - O(2)	2.807±6	(×2)
Sr - O(1)	2.814±6	(×2)

Because of ambiguities in the space group, these errors are probably too small.

Discussion

The structure is a distorted perovskite, the distortion being a reorientation of the TeO_6 and NiO_6 octahedra. The details of the distortion are not well characterised and there is some uncertainty about the correct space group.

Details of analysis

The assignment of a monoclinic space group is based only on splitting of pseudo cubic lines in the powder photograph. The face centering systematic absences of the cubic structure are apparently preserved leading to a *C* face centered monoclinic cell. Intensities for layers with *k* (pseudocubic setting), *h* (monoclinic setting) = 0 to 6 were measured from Weissenberg and precession photographs (AgK_α and MoK_α radiation). A variety of models in *C2/m*, *P*$\bar{1}$ and *Immm* were examined and not all could be rejected. The final coordinates are given in the pseudocubic setting [thereby implying a greater symmetry than is required by *C2/m*]. Refinement by least squares gave R = 0.068 for 183 strong reflexions and 0.125 for 147 weak reflexions listed.

CALCIUM FERRATE (IV)

I. The crystal structure of $Ca_2Fe_2O_6$ and its relation to the nuclear electric field gradient at the iron sites. A.F. COLVILLE, 1970. *Acta Cryst.*, B26, 1469-1473.

$Ca_2Fe_2O_6$ F.W. = 287.86

Orthorhombic, a = 5.599±1, b = 14.771±2, c = 5.429±1 Å, U = 448.99 Å^3, Z = 4, D_x = 4.02

Space group *Pcmn* (D_{2h}^{16})

Atomic positions

			x	*y*	*z*
8 Ca	in	8(*d*)	0.0233±2	0.1079±1	0.4806±2
4 Fe(1)	in	4(*a*)	0	0	0
4 Fe(2)	in	4(*c*)	0.9338±2	1/4	0.9459±2
8 O(1)	in	8(*d*)	0.2366±7	0.9839±2	0.2632±8
8 O(1)	in	8(*d*)	0.0716±7	0.1403±2	0.0234±8
4 O(3)	in	4(*c*)	0.8746±8	1/4	0.6005±9

Thermal parameters are given.

Interatomic distances (Å) *and bond angles*

Ca - O(1)	2.484±4,		2.483±4,	2.427±4,		2.741±2
Ca - O(2)	2.542±5,		2.330±4,	3.000±5		
Ca - O(3)	2.350					
Fe(1) - O(1)	1.963±4,	(×2)	1.970±4		(×2)	
Fe(1) - O(2)	2.115±3	(×2)				
Fe(2) - O(2)	1.844±3	(×2)				
Fe(2) - O(3)	1.904±6,		1.920±4			

Bond angles in degrees subtended at Fe(1)

	O(1)	O(2)
O(1)	92.6±1	86.9±1
O(1)'		89.1±1

Bond angles in degrees subtended at Fe(2)

	O(2)	O(3)	O(3)'
O(2)	123.0±2	107.3±2	106.1±2
O(3)			105.9±2

Description of the structure

The structure is essentially as reported by 1. Distortions found along the *b* axis, confirm those predicted from electric field gradient measurements.

Details of analysis

Crystals were grown by the Czochralski technique. 413 reflexions measured with MoK_α radiation using a diffractometer were refined to a final *R* value of 0.034. Absorption corrections were applied.

1. *Structure Reports*, 23, 387.

PYROCHLORES

I. Ein Beitrag zur Pyrochlorestruktur an $La_2Zr_2O_7$. H.-J. DIESEROTH and H. MÜLLER-BUSCHBAUM, 1970. *Z. anorg. Chem.*, 375, 152-156.

II. Röntgen-und Neutronenbeugungsuntersuchungen an $Y_2Ti_2O_7$. W.-J. BECKER and G. WILL, 1970. *Z. Kristallogr.*, 131, 278-288.

Cubic, (Pyrochlore type, 1), Z = 8.

	F.W.	a(Å)	U(Å^3)	D_m	D_x	x	Ref.
$La_2Zr_2O_7$	574.2	10.786	1254		6.07	-0.080	I
$Y_2Ti_2O_7$	385.61	10.0896±14	1206.4	5.03±3	4.24	-0.080±3	II (X-ray)
						-0.0788±5	II (neutron)

Space group *Fd*3*m* (O_h^7)

Atomic positions

Origin at centre ($\bar{3}m$)

			x	y	z
M^{III}	in	16(*c*):	0	0	0
M^{IV}	in	16(*d*):	1/2	1/2	1/2
O(1)	in	8(*b*):	3/8	3/8	3/8
O(2)	in	48(*f*):	x	1/8	1/8

Interatomic distances (Å) *and angles*

	M^{III}-O		M^{IV}-O	O-M^{IV}-O equitorial	O-M^{IV}-O axial
$La_2Zr_2O_7$	2.33	2.64	2.09	83.3	96.7°
$Y_2Ti_2O_7$	2.1845±2	2.483±3	1.954±2	83.6	96.0

Details of analysis

I. Intensities measured from integrated Weissenberg and precession photographs. Structure refined by differential synthesis to give *R*(for different layers) ranging from 0.03 to 0.11 for 172 reflexions listed.

II. Crystals grown from PbO-PbF_2 flux. Lattice parameters refined by least squares from powder photographs. (Cu$K_{\alpha 1}$ radiation, λ = 1.5405 Å). X-ray intensities from photometered powder photographs corrected for Lorentz, polarisation and multiplicity. x refined to give R = 0.067 for 24 observed reflexions. Neutron intensities measured with θ,2θ scan on four-circle gonimeter (λ = 0.95 Å) with single crystal of longest dimension 5 mm. Structure refined to give R = 0.066 for 52 good observed reflexions listed. Individual isotropic temperature factors given.

1. *Strukturbericht*, 7, 24.

STRONTIUM COPPER NICKEL OXIDE

I. Zur Kenntnis von $SrCu_{0.75}Ni_{0.25}O_2$. H. MÜLLER-BUSCHBAUM and H. MATTAUSCH, 1970. Z. *anorg. Chem.*, 377, 144-151.

$SrCu_{0.75}Ni_{0.25}O_2$ F.W. = 181.96

Orthorhombic, a = 3.565, b = 16.308 c = 3.916 Å, U = 227.7 Å^3, Z = 4, D_x = 5.29.

Space group *Cmcm* (D_{2h}^{17})

Atomic positions

			x	y	z
4 Sr	in	4(*c*)	0	0.3303	1/4
3 Cu + 1 Ni	in	4(*c*)	0	0.614	1/4
4 O(1)	in	4(*c*)	0	0.941	1/4
4 O(2)	in	4(*c*)	0	0.178	1/4

Interatomic distances (Å)

Sr - O(2)	2.48	
Sr - O(1)	2.54	(×2)
Sr - O(2)'	2.65	(×4)
Cu - O(2)	1.90	
Cu - O(1)	1.96	(×2)
Cu - O(1)'	1.96	

Errors uncerain but not less than 0.02 Å.

[The distances quoted by I differ slightly from those calculated above.]

Discussion

The Cu(Ni) atoms are surrounded by squares of O atoms which share edges to form chains along *c*. The chains are linked by Sr atoms.

Details of analysis

Intensities measured from layers *l* = 0 - 3 on two-circle automatic diffractometer. Structure from Patterson and electron-density projections refined [? by least squares] to give $R \sim 0.10$ for 136 reflexions listed.

ALKALINE EARTH CUPRATES

I. Zur Kenntnis von Ca_2CuO_3 und $SrCuO_2$. C.L. TESKE and H. MÜLLER-BUSCHBAUM, 1970. *Z. anorg. Chem.*, 379, 234-241.

DICALCIUM CUPRATE

Ca_2CuO_3 F.W. = 191.70

Orthorhombic, a = 12.239, b = 3.779, c = 3.259 Å, U = 150.73 Å^3, D_m = 4.28, Z = 2, D_x = 4.21.

Space group *Immm* (D_{2h}^{25})

Atomic positions

			x	*y*	*z*
4 Ca	in	4(*f*)	0.150	0	1/2
2 Cu	in	2(*d*)	0	1/2	0
4 O(1)	in	4(*f*)	0.340	0	1/2
2 O(2)	in	2(*a*)	0	0	0

Interatomic distances (Å)

Errors uncertain but about 0.02 Å.

Ca - O(1)	2.32	
Ca - O(2)	2.46	(×2)
Ca - O(1)	2.50	(×4)
Cu - O(2)	1.89	(×2)
Cu - O(1)	1.96	(×2)

STRONTIUM CUPRATE

$SrCuO_2$ F.W. = 183.16

Orthorhombic, a = 3.565, b = 16.326, c = 3.921 Å, U = 228.21 Å^3, Z = 4, D_x = 5.32.

Space group *CmCm* (D_{2h}^{17})

Atomic positions All atoms in 4(*c*)

	x	*y*	*z*
Sr	0	0.331_4	1/4
Cu	0	0.061	1/4
O(1)	0	0.945	1/4
O(2)	0	0.178	1/4

Interatomic distances (Å)

Sr - O(2)	2.50	
Sr - O(1)	2.57	(×2)
Sr - O(2)	2.65	(×2)
Cu - O(1)	1.89	
Cu - O(2)	1.91	
Cu - O(1)	1.96	(×2)

Errors uncertain but about 0.02 Å.

Discussion

In both structures the copper atoms are square coordinated. In the Sr compound Cu atoms approach within 2.80 Å of each other across a shared edge of the square.

Details of analysis

Crystals assumed to be isostructural with Sr_2CuO_3 (1) and $SrCu_{0.25}Ni_{0.25}O_2$ (2) and structure refined by differential synthesis and least squares to give R = 0.08 for 71 reflexions listed (Ca_2CuO_3) and R = 0.10 for 122 reflexions listed ($SrCuO_2$).

1. C.L. TESKE and H. MÜLLER-BUSCHBAUM, 1969. *Z. anorg. Chem.*, 371, 325.
2. See previous report.

LITHIUM ALUMINUM MOLYBDATE (VI)

I. [Crystal structure of $LiAl(MoO_4)_2$]. L.P. SOLOV'EVA and S.V. BORISOV, 1970. *Kristalografija*, 15, 577-580 [*Soviet Physics - Crystallography*, 15, 493-495].

$LiAl(MoO_4)_2$ F.W. = 353.80

Triclinic, a = 7.10+3, b = 7.25+5, c = 6.67+5 Å, α = 111.0±5°, β = 105.0±5°, γ = 90.0±5°, U = 308 Å^3, Z = 2, D_x = 3.81.

Space group $P\bar{1}$ (C_i^1)

Atomic positions All atoms in 2(i)

	x	y	z	B(Å^2)
Mo(1)	0.0335±5	0.2849±3	0.6776±5	0.67±4
Mo(2)	0.4232±3	0.2070±3	0.1678±5	0.71±4
Al	0.1002±13	0.8258±13	0.9132±22	0.60±14
Li	0.556 ±10	0.202 ±11	0.713 ±15	2
O(1)	0.4237±37	0.4619±39	0.2557±56	1.62±43
O(2)	0.0422±36	0.5435±37	0.7351±56	1.54±43
O(3)	0.2742±30	0.2317±31	0.7664±48	0.72±35
O(4)	0.6234±28	0.1507±29	0.0283±45	0.39±31
O(5)	0.9441±33	0.1555±33	0.3829±51	1.07±39
O(6)	0.1544±29	0.1104±29	0.0835±46	0.44±32
O(7)	0.1314±30	0.7795±31	0.1925±47	0.66±33
O(8)	0.5161±35	0.1559±35	0.4087±52	1.22±39

Interatomic distances (Å) and bond angles

Mo(1)	- O(3)	1.76
Mo(1)	- O(2)	1.77
Mo(1)	- O(7)	1.78
Mo(1)	- O(5)	1.78

Angles in degrees subtended at Mo(1) by

	O(2)	O(7)	O(5)
O(3)	107	115	109
O(2)		109	109
O(7)			108

MO(2)	- O(1)	1.73
Mo(2)	- O(8)	1.74
Mo(2)	- O(4)	1.87
Mo(2)	- O(6)	1.91
Mo(2)	- O(4)'	2.43
Mo(2)	- O(3)	2.68

Angles in degrees subtended at Mo(2) by

	O(8)	O(4)	O(6)	O(4)'	O(3)
O(1)	106	104	104	168	82
O(8)		103	102	86	172
O(4)			134	75	75
O(6)				69	75
O(4)'					86

σ(distances) = 0.02 (0.1 for Li-O) σ(angles) = 2° (4° for O-Li-O)

Al	- O(6)	1.88
Al	- O(4)	1.90
Al	- O(2)	1.94
Al	- O(6)	1.95
Al	- O(7)	1.97
Al	- O(5)	1.97

Angles in Degrees subtended at Al by

	O(4)	O(2)	O(6)	O(7)	O(5)
O(6)	161	99	80	89	90
O(4)		100	81	89	94
O(2)			178	92	82
O(6)				90	96
O(7)					174

Li	- O(8)	1.88
Li	- O(3)	2.13
Li	- O(7)	2.14
Li	- O(4)	2.20
Li	- O(1)	2.37
Li	- O(8)	2.44

Angles in Degrees subtended at Li by

	O(3)	O(7)	O(4)	O(1)	O(8)
O(8)	105	98	161	82	86
O(3)		156	82	89	88
O(7)			77	89	99
O(4)				116	77
O(1)					166

Discussion

The structure consists of one regular MoO_4 ion and one MoO_4 ion distorted by having long Mo-O bonds to each of the crystallographically distinct MoO_4 ions. Mo(2) is thus in a transitional state between tetrahedral and octahedral coordination.

Details of analysis

Crystals grown from melt. Intensities estimated visually from equi-inclination Weissenberg photographs of layers h = 0 and l = 0 to 4 taken with Mo radiation. Corrected for Lorentz and polarisation but not absorption. Structure solved from Patterson and electron-density functions and refined by least squares to give R = 0.10 for about 900 reflexions [not listed].

LITHIUM IRON WOLFRAMATE

I. [Crystallographic investigation of a double tungstate, $LiFe(WO_4)_2$.] P.V. KLEVCOV and R.F. KLEVCOVA, 1970. *Kristallografija*, 15, 294-298 [*Soviet Physics - Crystallography*, 15, 245-248].

$LiFe(WO_4)_2$ F.W. = 558.5

Monoclinic, a = 9.26, b = 11.38, c = 4.91 Å, β = 90.3, U = 517 Å^3, D_m = 7.14±2, Z = 4, D_x = 7.15.

Space group $C2/c$ (C^6_{2h}) systematic absences

Atomic positions

			x	y	z	B(Å^2)
8 W	in	8(f)	0.2460±5	0.0914±2	0.2470±15	0.42
4 Fe	in	4(e)	0	0.336 ±2	0.25	0.74
4 Li	in	4(e)	0.5	0.327 *	0.25	2.0
8 O(1)	in	8(f)	0.387 ±9	0.062 ±8	0.940 ±10	1.0
8 O(2)	in	8(f)	0.392 ±8	0.178 ±7	0.420 ±12	1.0
8 O(3)	in	8(f)	0.356 ±5	0.552 ±4	0.929 ±12	1.0
8 O(4)	in	8(f)	0.378 ±6	0.697 ±4	0.389 ±13	1.0

* not refined.

Interatomic distances (Å) *and bond angles*

W	- O(4)	1.79
W	- O(2)	1.87
W	- O(3)	1.91
W	- O(1)	2.03
W	- O(3)'	2.07
W	- O(1)'	2.37

Angles in degrees subtended at W by

	O(2)	O(3)	O(1)	O(3)'	O(1)'
O(4)	106	99	104	95	173
O(2)		96	88	158	79
O(3)			155	86	77
O(1)				82	80
O(3)'					81

				Angles in degrees subtended at Fe by	O(1)'	O(2)'	O(4)'
Fe - O(1)	1.82	(×2)		O(1)	101	101	90
Fe - O(2)	1.91	(×2)		O(1)'		85	167
Fe - O(4)	2.06	(×2)		O(2)		170	87
				O(2)'			86
				O(4)			80

				Angles in degrees subtended at Li by	O(3)'	O(4)'	O(2)'
Li - O(3)	2.11	(×2)		O(3)	99	96	93
Li - O(4)	2.11	(×2)		O(3)'		94	167
Li - O(2)	2.14	(×2)		O(4)		165	80
				O(4)'			89
				O(2)			75

σ(distances) = 0.06 Å   σ(angles) = 3°.

Discussion

An ordered variant of wolframite (1).

Details of analysis

Lattice parameters from powder photographs (CuK_α radiation). Intensities measured from equi-inclination Weissenberg photographs of layer with l = 0 to 5. Wolframite structure (1) assumed with an ordered arrangement of Fe and Li. R = 0.15 for about 800 reflexions [not listed]. Coordinates of Li not refined.

1. D. ÜKLÜ, 1967. *Z. Kristallogr.*, 124, 192.

LITHIUM YTTERBIUM WOLFRAMATE

I. [Variations on a single wolframite motif in the structures of $LiYb(WO_4)_2$, $LiFe(WO_4)_2$, and $NaFe(WO_4)_2$]. R.F. KLEVCOVA and N.V. BELOV, 1970. *Kristallografija*, 15, 43-46 [*Soviet Physics - Crystallography*, 15, 32-35].

$LiYb(WO_4)_2$ F.W. = 675.6

Monoclinic, a = 9.89, b = 5.77, c = 4.98 Å, β = 93.5°, U = 284 Å^3, Z = 2, D_x = 7.89.

Space group $P2/n$ (C^4_{2h}) systematic absences and absence of piezoelectric effect.

Atomic positions Equivalent positions (x, y, z; $\bar{x}$, $\bar{y}$, $\bar{z}$; 1/2-x, y, 1/2-z; 1/2+x, $\bar{y}$, 1/2+z)

			x	y	z	B(Å^2)
4 W	in	4(g)	0.0156±20	0.1808±4	0.2503±4	0.21
2 Yb	in	2(f)	0.25	0.6972±2	0.25	0.58
2 Li	in	2(f)	0.25	0.315 ±13	0.75	1.00
4 O(1)	in	4(g)	0.114 ±4	0.629 ±8	0.896 ±9	0.60
4 O(2)	in	4(g)	0.138 ±5	0.374 ±9	0.403 ±9	1.18
4 O(3)	in	4(g)	0.115 ±5	0.090 ±8	0.971 ±9	0.85
4 O(4)	in	4(g)	0.101 ±6	0.896 ±9	0.448 ±9	1.33

Interatomic distances (Å) *and bond angles*

			Angles in degrees subtended at W by	O(1)	O(3)	O(4)	O(4)	O(3)
W - O(2)	1.78		O(2)	104	97	103	93	172
W - O(1)	1.81		O(1)		106	91	159	82
W - O(3)	1.83		O(3)			150	85	76
W - O(4)	2.00		O(4)				73	83
W - O(4)	2.07		O(4)					84
W - O(3)	2.27							

Yb - O(4)	2.15	(×2)	
Yb - O(1)	2.19	(×2)	
Yb - O(2)	2.32	(×2)	

Angles in degrees subtended at Yb by

	O(4)'	O(1)'	O(2)'
O(4)	116	93	86
O(4)'		97	159
O(1)		159	83
O(1)'			81
O(2)			73

Li - O(2)	2.02	(×2)
Li - O(3)	2.20	(×2)
Li - O(1)	2.40	(×2)

Angles in degrees subtended at Li by

	O(2)'	O(3)'	O(1)'
O(2)	161	102	84
O(2)'		89	81
O(3)		108	86
O(3)'			165
O(1)			82

σ(distances) = 0.05 Å

σ(angles) = 2°

Discussion

The O atoms form an hexagonal close packed array with the metal atoms in octahedral interstices. The structure is related to that of wolframite (1).

Details of analysis

Intensities measured [visually] from equi-inclination Weissenberg photographs of layers with k = 0,1 and l = 1 - 6 taken with Mo radiation and corrected for Lorentz and polarisation, but not absorption which was negligible. Structure determined from Patterson and electron-density projections refined by least squares to give R = 0.112 for 900 reflexions [not listed].

1. D. ÜLKÜ, 1967. *Z. Kristallogr.*, 124, 192.

LITHIUM IRON MOLYBDATES

I. [Crystal structure of lithium-iron molybdates $Li_3Fe^{\cdot\cdot\cdot}(MoO_4)_3$ and $Li_2Fe_2^{\cdot\cdot}(MoO_4)_3$]. R.F. KLEVCOVA and S.A. MAGARILL, 1970. *Kristallografija*, 15, 710-715 [*Soviet Physics - Crystallography*, 15, 612-615].

$Li_3Fe(MoO_4)_3$ F.W. = 556.48

Orthorhombic, a = 5.07, b = 10.48, c = 17.64 Å, U = 937 Å^3, D_m = 3.97, Z = 4, D_x = 3.94.

Space group *Pnma* (D_{2h}^{16})

Atomic positions

				x	y	z	B(Å^2)
8	Mo(1)	in	8(*d*)	0.2784±4	0.52619±18	0.15620±11	0.70±3
4	Mo(2)	in	4(*c*)	0.7795±6	0.25	0.05712±15	0.55±4
1-1/3	Fe(1) + 2-2/3 Li	in	4(*c*)	0.1066±23	0.25	0.2513 ±8	0.69±16
5-1/3	Li(1) + 2-2/3 Fe	in	8(*d*)	0.7548±19	0.5731 ±9	0.0265 ±5	0.64±11
4	Li(2)	in	4(*c*)	0.2470	0.75	0.300	2.00
4	O(1)	in	4(*c*)	0.865 ±5	0.25	0.1549 ±16	0.08±3
4	O(2)	in	4(*c*)	0.055 ±5	0.25	-0.0053 ±16	0.90±40
8	O(3)	in	8(*d*)	0.580 ±4	0.1148 ±18	0.0368 ±10	0.60±20
8	O(4)	in	8(*d*)	0.082 ±3	0.4900 ±17	0.0739 ±10	0.50±20
8	O(5)	in	8(*d*)	0.078 ±4	0.6255 ±19	0.2118 ±11	1.00±30
8	O(6)	in	8(*d*)	0.358 ±3	0.3810 ±18	0.2070 ±10	0.70±20
8	O(7)	in	8(*d*)	0.561 ±4	0.6120 ±20	0.1271 ±11	1.00±30

Interatomic distances (Å) *and bond angles*

Mo(1) - O(5)	1.75			
Mo(1) - O(7)	1.77			
Mo(1) - O(4)	1.80			
Mo(1) - O(6)	1.81			

Angles in degrees subtended at Mo(1) by

	O(7)	O(4)	O(6)
O(5)	109	105	110
O(7)		109	113
O(4)			110

Mo(2) - O(3)	1.78	(×2)
Mo(2) - O(1)	1.78	
Mo(2) - O(2)	1.78	

Angles in degrees subtended at Mo(2) by

	O(3)'	O(1)	O(2)
O(3)	106	110	109
O(1)			114

(Fe(1), Li) - O(6)	2.00	(×2)
(Fe(1), Li) - O(6)*	2.03	(×2)
(Fe(1), Li) - O(1)	2.09	
(Fe(1), Li) - O(1)*	2.11	

Angles in degrees subtended at (Fe(1), Li) by

	O(6)'	O(6)*	O(6)*'	O(1)	O(1)*
O(6)	86	179	94	86	96
O(6)*			85	93	85
O(1)					177

(Fe, Li(1)) - O(4)	2.05
(Fe, Li(1)) - O(4)*	2.06
(Fe, Li(1)) - O(7)	2.07
(Fe, Li(1)) - O(3)	2.08
(Fe, Li(1)) - O(2)	2.12
(Fe, Li(1)) - O(3)*	2.17

Angles in degrees subtended at (Fe, Li(1) by

	O(4)*	O(7)	O(3)	O(2)	O(3)*
O(4)	84	97	166	94	85
O(4)*		172	86	87	87
O(7)			92	101	85
O(3)				95	84
O(2)					174

Li(2) - O(5)	2.14	(×2)
Li(2) - O(7)	2.15	(×2)
Li(2) - O(5)*	2.20	(×2)

Angles in degrees subtended at Li(2) by

	O(5)'	O(7)'	O(5)*'
O(5)	75	144	83
O(5)'		90	127
O(7)		84	82
O(7)'			130
O(5)*			73

DILITHIUM DIFERROUS MOLYBDATE

$Li_2Fe_2(MoO_4)_3$ F.W. = 605.38

Orthorhombic, a = 5.11, b = 10.55, c = 17.76 Å, U = 957 Å^3, Z = 4, D_x = 4.19.

Space group *Pnma* (D_{2h}^{16})

Atomic positions

			x	y	z	B(Å^2)
8 Mo(1)	in	8(d)	0.2793±6	0.52672±29	0.15690±17	0.75±6
4 Mo(2)	in	4(c)	0.7807±10	0.25	0.05806±24	0.65±7
2-2/3 Fe(1) + 1-1/3 Li	in	4(c)	0.1074±32	0.25	0.25040±100	1.86±22
2-2/3 Li(1) + 5-1/3 Fe	in	8(d)	0.7506±25	0.5741 ±10	0.02730±60	2.20±18
4 Li(2)	in	4(c)	0.2470	0.75	0.3000	2.00
4 O(1)	in	4(c)	0.851 ±8	0.25	0.1534 ±24	1.1 ±6
4 O(2)	in	4(c)	0.063 ±9	0.25	-0.0039 ±25	0.4 ±5
8 O(3)	in	8(d)	0.590 ±6	0.1184 ±28	0.0404 ±15	0.7 ±4
8 O(4)	in	8(d)	0.084 ±6	0.4890 ±27	0.0747 ±15	0.6 ±4
8 O(5)	in	8(d)	0.081 ±6	0.6261 ±29	0.2137 ±16	1.2
8 O(6)	in	8(d)	0.354 ±6	0.3817 ±28	0.2055 ±15	0.6 ±4
8 O(7)	in	8(d)	0.558 ±6	0.6160 ±31	0.1244 ±17	1.3 ±5

Interatomic distances (Å) *and bond angles*

Standard errors in distances = 0.04 Å, in angles = 2°

Mo(1)	- O(5)	1.77	
Mo(1)	- O(6)	1.80	
Mo(1)	- O(7)	1.80	
Mo(1)	- O(4)	1.81	

Angles in degrees subtended at Mo(1) by

	O(6)	O(7)	O(4)
O(5)	111	109	106
O(6)		116	108
O(7)			107

Mo(2)	- O(3)	1.72	(×2)
Mo(2)	- O(1)	1.73	
Mo(2)	- O(2)	1.81	

Angles in degrees subtended at Mo(2) by

	O(3)'	O(1)	O(2)
O(3)	107	107	110
O(1)			115

(Fe(1), Li)	- O(6)	2.03	(×2)
(Fe(1), Li)	- O(6)*	2.05	(×2)
(Fe(1), Li)	- O(1)	2.11	
(Fe(1), Li)	- O(1)*	2.16	

Angles in degrees subtended at (Fe(1), Li) by

	O(6)'	O(6)*	O(6)*'	O(1)	O(1)*
O(6)	86	179	94	87	94
O(6)*			85	94	86
O(1)					179

(Fe, Li(1))	- O(7)	2.03	
(Fe, Li(1))	- O(4)	2.10	
(Fe, Li(1))	- O(4)*	2.10	
(Fe, Li(1))	- O(2)	2.12	
(Fe, Li(1))	- O(3)	2.16	
(Fe, Li(1))	- O(3)*	2.20	

Angles in degrees subtended at (Fe, Li(1)) by

	O(4)	O(4)*	O(2)	O(3)	O(3)*
O(7)	98	173	101	92	86
O(4)		83	95	165	82
O(4)*			86	85	87
O(2)				94	173
O(3)					88

Li(2)	- O(5)	2.16	(×2)
Li(2)	- O(7)	2.18	(×2)
Li(2)	- O(5)*	2.19	(×2)

Angles in degrees subtended at Li(2) by

	O(5)'	O(7)'	O(5)*'
O(5)	74	144	82
O(5)'		92	126
O(7)		81	83
O(7)'			130
O(5)*			74

Discussion

The structure of $Li_3Fe(MoO_4)_3$, shows a pseudo-hexagonal motif centered around the $Fe(1)O_6$ octahedron which shares corners with 6 MoO_4 tetrahedra. These in turn are linked to similar groups through the $Li(1)O_6$ octahedra and $Li(2)O_6$ triangular prisms.

Details of analysis

Intensities for $Li_3Fe(MoO_4)_3$ estimated visually from Weissenberg photographs of layers with h = 0 to 6 and k = 0 taken with Mo radiation, corrected for Lorentz and polarisation effects, but not absorption (μ = 5.5 mm^{-1} crystal size = 0.4 × 0.05×0.05 mm^3). Structure determined from Patterson and electron-density functions and refined (except for Li(2)) by least squares to give R = 0.099 for 964 reflexions. Analysis of $Li_2Fe_2(MoO_4)_3$ leads to R = 0.121 for 692 reflexions [not listed].

SODIUM INDIUM WOLFRAMATE

I. Single-crystal synthesis and investigation of the double tungstates $NaR^{3+}(WO_4)_2$, where R^{3+} = Fe, Sc, Ga, and In. P.V. KLEVCOV and R.F. KLEVCOVA, 1970. *J. Solid State Chem.*, 2, 278-282.

$NaIn(WO_4)_2$ F.W. = 633.50

Monoclinic, a = 10.08+2, b = 5.81+1, c = 5.03+2 Å, U = 295 Å³, β = 90.0+5°, D_m = 7.13, Z = 2, D_x = 7.13.

Space group *P2/c* (C^4_{2h}) systematic absences and structure

Atomic positions

			x	*y*	*z*	*B*(Å³)
4 W	in	4(*g*)	0.2432±3	0.1785±3	0.2468±5	-0.13±4
2 In	in	2(*e*)	0	0.6856±11	0.25	0.42±7
2 Na	in	2(*f*)	0.5	0.655 ±11	0.25	0.3 ±4
4 O(1)	in	4(*g*)	0.120 ±4	0.626 ±7	0.598 ±10	0.3 ±4
4 O(2)	in	4(*g*)	0.357 ±6	0.374 ±10	0.365 ±13	1.1 ±6
4 O(3)	in	4(*g*)	0.337 ±4	0.095 ±8	0.939 ±10	0.3 ±5
4 O(4)	in	4(*g*)	0.131 ±5	0.101 ±9	0.536 ±12	0.7 ±6

Interatomic distances (Å) *and bond angles*

W - O(2)	1.72
W - O(1)	1.84
W - O(3)	1.88
W - O(4)	1.90
W - O(3)'	2.09
W - O(4)'	2.25

Angles in degrees subtended at W by

	O(1)	O(3)	O(4)	O(3)'	O(4)'
O(2)	101	97	107	92	168
O(1)		100	93	164	85
O(3)			151	88	71
O(4)				75	84
O(3)'					84

In - O(4)	2.11	(×2)
In - O(1)	2.16	(×2)
In - O(1)*	2.31	(×2)

Angles in degrees subtended at In by

	O(4)'	O(1)'	O(1)*'
O(4)	108	99	88
O(4)'		92	164
O(1)		162	84
O(1)'			81
O(1)*			77

Na - O(2)	2.26	(×2)
Na - O(3)	2.39	(×2)
Na - O(2)*	2.42	(×2)

Angles in degrees subtended at Na by

	O(2)'	O(3)'	O(2)*'
O(2)	87	84	97
O(2)'		169	77
O(3)		105	87
O(3)'			98
O(2)*			172

Errors about 0.05 Å for distances and 2° for angles

Discussion

The O atoms form an hexagonal close packed array ([100] is the hexagonal axis) with half of the octahedral intensities filled. Each layer is separated by a different metal, viz. Na, W, In and W respectively. Within a layer the metal-O_6 octahedra share edges to form chains along *c*.

Details of analysis

Prepared by heating oxides of constituent metals to 1010°C. Lattice parameters measured on powder diffractometer (CuK_α radiation) calibrated with Si. Intensities estimated visually from equi-inclination Weissenberg photographs of layers with k = 0 and l = 0 to 6 taken with MoK_α radiation; corrected for Lorentz and polarisation, but not absorption effects (crystal size = 0.25×0.25×0.80 mm³). Structure determined from Patterson and electron-density functions, and refined by least squares to give $R = 0.13$ for 650 reflexions [not listed].

DOUBLE MOLYBDATES

I. [Synthesis and crystal structure of double molybdates $KR(MoO_4)_3$ for R^{3+} = Al, Sc, and Fe and tungstate $KSc(WO_4)_2$]. R.F. KLEVCOVA and P.V. KLEVCOV, 1970. *Kristallografija*, 15, 953-959 [*Soviet Physics - Crystallography*, 15, 829-834].

POTASSIUM ALUMINUM MOLYBDATE

$KAl(MoO_4)_2$ F.W. = 385.96

Trigonal, a = 5.545+3, c = 7.070+5 Å, U = 188.3 Å^3, D_m = 3.42+2, Z = 1, D_x = 3.40.

Space group $P\bar{3}m1$ (D^3_{3d})

Atomic positions

			x	*y*	*z*	*B*(Å^2)
2 Mo	in	2(*d*)	1/3	2/3	0.23488±32	0.23
1 Al	in	1(*a*)	0	0	0	0.41
1 K	in	1(*b*)	0	0	1/2	1.45
2 O(1)	in	2(*d*)	1/3	2/3	0.4645 ±22	0.01
6 O(2)	in	6(*i*)	0.155	0.310	0.1561 ±24	1.53

POTASSIUM SCANDIUM MOLYBDATE

$KSc(MoO_4)_2$ F.W. = 403.93

Trigonal, a = 5.774+6, c = 7.219+7 Å, U = 208.5 Å^3, D_m = 3.22+2, Z = 1, D_x = 3.21

Space group $P\bar{3}m1$ (D^3_{3d})

Atomic positions

			x	*y*	*z*	*B*(Å^2)
2 Mo	in	2(*d*)	1/3	2/3	0.25225±45	0.16
1 Sc	in	1(*a*)	0	0	0	0.54
1 K	in	1(*b*)	0	0	1/2	1.55
2 O(1)	in	2(*d*)	1/3	2/3	0.4941 ±60	2.60
6 O(2)	in	6(*i*)	0.157	0.314	0.1738 ±35	1.77

POTASSIUM IRON MOLYBDATE

$KFe(MoO_4)_2$ F.W. = 414.82

Trigonal, a = 5.66+1, c = 14.24+2 Å, U = 395.1 Å^3, D_m = 3.36+4, Z = 2, D_x = 3.48.

Space group $P\bar{3}c1$ (D^4_{3d})

Atomic positions

			x	*y*	*z*	*B*(Å^2)
4 Mo	in	4(*d*)	1/3	2/3	0.12086±14	1.03
2 Fe	in	2(*b*)	0	0	0	0.90
2 K	in	2(*a*)	0	0	1/4	2.50
4 O(1)	in	4(*d*)	1/3	2/3	0.2383 ±12	2.00
12 O(2)	in	12(*g*)	0.1262±22	0.3199±21	0.0840 ±7	1.74

Interatomic distances (Å)

	$KAl(MoO_4)_2$	$KSc(MoO_4)_2$	$KFe(MoO_4)_2$
Mo - O(1)	1.62 ±2	1.75 ±4	1.67 ±2
Mo - O(2) (×3)	1.801±5	1.853±8	1.790±11
O(1) - Mo - O(2)	108.0±5°	107.8±7°	107.1±3°
O(2) - Mo - O(2)	110.9±3°	111.1±5°	111.8±5°
Al, Sc, Fe - O(2) (×6)	1.85 ±1	2.01 ±2	1.98 ±1
K - O(1) (×6)	3.211±1	3.336±1	3.275±1
K - O(2) (×6)	2.851±14	2.831±21	2.844±10

Discussion

In spite of the doubling of the *c* axis in $KFe(MoO_4)_2$, the three structures are essentially similar. $KSc(WO_4)_2$ is also isostructural.

Details of analysis

Lattice parameters from rotation and powder photographs. Intensities measured from Weissenberg photographs taken with MoK_α radiation of layers with $k = 0$, $l = 0$ to 8 ($KAl(MoO_4)_2$) and $k = 0$ to 7, $l = 0$ ($KFe(MoO_4)_2$). Structure determined from Patterson function refined by least squares to give $R = 0.088$ for 375 reflexions ($KAl(MoO_4)_2$), $R = 0.134$ for 307 reflexions ($KSc(MoO_4)_2$) and $R = 0.103$ for 870 reflexions ($KFe(MoO_4)_2$). No intensities listed.

PRASEODYMIUM OXYWOLFRAMITE

I. [The crystal structure of $Pr_2W_2O_9$.] S.V. BORISOV and R.F. KLEVCOVA, 1970. *Kristallografija*, 15, 38-42 [*Soviet Physics - Crystallography*, 15, 28-31].

$Pr_2W_2O_9$ F.W. = 793.5

Monoclinic, $a = 7.70$, $b = 9.84$, $c = 9.27$ Å, $\beta = 106.5°$, $U = 673.45$ Å^3, $Z = 4$, $D_x = 7.81$.

Space group $P2_1/c$ (C^5_{2h})

Atomic positions

	x	y	z	B(Å^2)
W(1)	0.4271±9	0.2729±6	0.0337±5	0.52±7
W(2)	0.0715±9	0.2505±6	0.2368±5	0.46±6
Pr(1)	0.2792±14	0.9555±9	0.0757±8	0.85±10
Pr(2)	0.2299±14	0.5522±9	0.1537±8	0.84±10
O(1)	0.257 ±16	0.183 ±10	0.142 ±8	1.4
O(2)	0.529 ±14	0.390 ±10	0.171 ±8	-0.3
O(3)	0.586 ±12	0.212 ±11	0.421 ±8	0.4
O(4)	0.556 ±12	0.397 ±9	0.605 ±7	0.7
O(5)	0.275 ±13	0.364 ±10	0.366 ±9	2.3
O(6)	0.240 ±14	0.123 ±10	0.424 ±8	1.3
O(7)	0.988 ±14	0.378 ±11	0.089 ±9	0.7
O(8)	-0.022 ±13	0.074 ±10	0.164 ±8	0.0
O(9)	-0.078 ±14	0.280 ±11	0.341 ±8	1.1

Interatomic distances (Å) *and bond angles*

W(1)	-	O(2)	1.73
W(1)	-	O(6)	1.82
W(1)	-	O(3)	1.83
W(1)	-	O(4)	1.96
W(1)	-	O(1)	2.06
W(1)	-	O(5)	2.14

Angles in degrees subtended at W(1) by

	O(6)	O(3)	O(4)	O(1)	O(5)
O(2)	99	97	104	98	174
O(6)		101	156	90	76
O(3)			84	159	87
O(4)				78	80
O(1)					78

W(2)	-	O(9)	1.72
W(2)	-	O(7)	1.83
W(2)	-	O(8)	1.93
W(2)	-	O(1)	1.99
W(2)	-	O(5)	2.02
W(2)	-	O(6)	2.23

Angles in degrees subtended at W(2) by

	O(7)	O(8)	O(1)	O(5)	O(6)
O(9)	99	96	168	96	90
O(7)		110	92	96	165
O(8)			78	149	81
O(1)				85	80
O(5)					71

Pr(1) - O(1)	2.34		Pr(2) - O(3)	2.35	
Pr(1) - O(4)	2.44		Pr(2) - O(4)	2.42	
Pr(1) - O(2)	2.48		Pr(2) - O(7)	2.48	
Pr(1) - O(4)	2.53		Pr(2) - O(7)	2.49	
Pr(1) - O(8)	2.54		Pr(2) - O(9)	2.53	
Pr(1) - O(9)	2.58		Pr(2) - O(8)	2.65	
Pr(1) - O(3)	2.61		Pr(2) - O(5)	2.65	
Pr(1) - O(5)	2.63		Pr(2) - O(6)	2.76	
Pr(1) - O(8)	2.91		Pr(2) - O(2)	2.77	

σ(distances) = 0.10 Å; σ(angles) = 4°

Discussion

WO_6 octahedra are joined in pairs through a shared edge (O(5) - O(6)). The pairs are linked through a shared corner (O(1)) into a chain running along *c*.

Details of analysis

Intensities [estimated visually] from Weissenberg photographs of layers with *h* = 0 to 3 and *k* = 0 to 4 taken with Mo radiation. Corrected for Lorentz and polarization effects. Structure from Patterson function and chemical considerations refined by least squares to give *R* = 0.17 for 800 reflexions [not listed].

NEODYMIUM RHENIUM(V) OXIDE

I. On the crystal structure of $Nd_4Re_2O_{11}$. K.-A. WILHELMI and E. LAGERVALL, 1970. *Acta Chem. Scand.*, 24, 3406-3408.

$Nd_4Re_2O_{11}$ F.W. = 1125.35

Tetragonal, *a* = 12.676±2, *c* = 5.601±1 Å, *U* = 899.97 Å^3, D_m = 8.14±0, *Z* = 4, D_x = 8.29.

Space group $P4_2/n$ (C^4_{4h})

Atomic positions Origin at $\bar{1}$

			x	*y*	*z*	*B*(Å^2)
8 Nd(1)	in	8(*g*)	0.1840±1	0.1182±1	0.9979±2	0.44±2
8 Nd(2)	in	8(*g*)	0.1107±1	0.8064±1	0.7030±2	0.53±2
8 Re(1)	in	8(*g*)	0.0247±1	0.9130±1	0.5707±1	0.35±2
8 O(1)	in	8(*g*)	0.0301±10	0.1938±9	0.2009±21	0.41±18
8 O(2)	in	8(*g*)	0.0013±9	0.4086±9	0.1937±22	0.32±18
8 O(3)	in	8(*g*)	0.0415±10	0.6291±9	0.2905±22	0.45±18
8 O(4)	in	8(*g*)	0.1484±10	0.8032±11	0.4943±26	1.03±22
8 O(5)	in	8(*g*)	0.4559±10	0.3400±10	0.6868±24	0.75±19
2 O(6)	in	2(*a*)	0.25	0.25	0.25	0.25±36
2 O(7)	in	2(*b*)	0.25	0.25	0.25	0.92±43

[There is manifestly an error in some of these coordinates which give bond distances typically 0.02 Å different from those quoted in I. In addition, Nd(2) is only 1.89 Å from Re and 1.26 Å from O(4), whereas the Nd(2) - O distances quoted in I range from 2.32 to 2.80 Å.]

Interatomic distances (Å) *and bond angles*

Nd(1) - O(7)	2.337	[2.328±1]
Nd(1) - O(6)	2.323	[2.342±1]
Nd(1) - O(5)	2.428	[2.422±13]
Nd(1) - O(3)	2.443	[2.436±12]
Nd(1) - O(1)	2.437	[2.453±12]
Nd(1) - O(5)	2.552	[2.542±13]
Nd(1) - O(1)	2.528	[2.548±12]
Nd(1) - O(2)	2.539	[2.561±12]

Angles in degrees subtended at Re(1) by

	O(5)	O(2)	O(1)	O(2)	O(4)
O(3)	161.7±5	97.3±5	91.9±5	90.9±5	79.4±5
O(5)		101.0±5	89.8±5	84.7±5	82.5±5
O(2)			83.6±5	105.3±5	169.6±5
O(1)				170.3±5	86.7±5
O(2)					84.7±5

[These angles are not given in I]

[Re(1)	-	No(2)		1.888±2]	See note above
Re(1)	-	O(3)	1.894	[1.897±12]	
Re(1)	-	O(5)	1.918	[1.923±13]	
Re(1)	-	O(2)	1.951	[1.955±11]	
Re(1)	-	O(1)	1.984	[1.988±12]	
Re(1)	-	O(2)	2.035	[2.040±12]	
Re(1)	-	O(4)	2.138	[2.140±13]	
Re(1)	-	Re(1)	2.421	[2.426±2]	

Discussion

$Nd_4Re_2O_{11}$ contains Re_2O_{10} clusters formed from two edge sharing ReO_6 octahedra. The remaining O surrounds Nd(1) atoms in 8 fold coordination. Nd(2) apparently has a two-face-capped trigonal prism of oxygen atoms around it.

Details of analysis

Lattice parameters from Guinier-Hägg powder photographs. Intensities measured from films (CuK_α radiation) and automatic diffractometer (MoK_α radiation) and corrected for Lorentz, polarisation, absorption and extinction effects. Structure determined from Patterson function and refined by least squares to give R = 0.026. No reflexions listed.

NIOBATES AND TANTALATES

I. Crystal structures of $Na_2Nb_4O_{22}$ and $CaTa_4O_{11}$. L. JAHNBERG, 1970. *J. Solid State Chem.*, 1, 454-462.

SODIUM NIOBATE

$Na_2Nb_4O_{11}$ F.W. = 592.62

Monoclinic, a = 10.840, b = 6.162, c = 12.745 Å, β = 106°22', D_m = 4.75, Z = 4, D_x = 4.82.

Space group $C2/c$ (C_{2h}^6)

Atomic parameters

			x	y	z
8 Nb(1)	in	8(*f*)	0.1825±2	0.5658±9	0.2499±2
4 Nb(2)	in	4(*e*)	0	0.1140±13	1/4
4 Nb(3)	in	4(*d*)	1/4	1/4	1/2
8 Na	in	8(*f*)	0.082 ±2	0.253 ±3	0.996 ±2
8 O(1)	in	8(*f*)	0.233 ±2	0.512 ±5	0.407 ±2
8 O(2)	in	8(*f*)	0.157 ±2	0.601 ±4	0.090 ±1
8 O(3)	in	8(*f*)	0.081 ±2	0.135 ±4	0.410 ±2
8 O(4)	in	8(*f*)	0.125 ±2	0.880 ±5	0.251 ±2
8 O(5)	in	8(*f*)	0.159 ±2	0.254 ±6	0.218 ±2
4 O(6)	in	4(*e*)	0	0.505 ±9	1/4

Isotropic temperature factors given in I.

Interatomic distances (Å)

Standard errors 0.02 Å.

Nb(1)	-	O(1)	1.96	
Nb(1)	-	O(2)	1.99	
Nb(1)	-	O(4)	2.04	
Nb(1)	-	O(4)'	2.38	
Nb(1)	-	O(5)	1.96	
Nb(1)	-	O(5)'	2.02	
Nb(1)	-	O(6)	2.01	
Nb(2)	-	O(3)	1.99	(×2)
Nb(2)	-	O(4)	1.98	(×2)
Nb(2)	-	O(5)	2.07	(×2)
Nb(2)	-	O(6)	2.41	
Nb(3)	-	O(1)	1.98	(×2)
Nb(3)	-	O(2)	1.95	(×2)
Nb(3)	-	O(3)	2.00	(×2)
Na	-	O(1)	2.52	
Na	-	O(1)'	2.66	
Na	-	O(2)	2.48	
Na	-	O(2)'	2.68	
Na	-	O(3)	2.51	
Na	-	O(3)'	2.63	
Na	-	O(5)	2.72	

CALCIUM TANTALATE

$CaTa_4O_{11}$ F.W. = 939.86

Hexagonal, a = 6.213+1, c = 12.265+1 Å, U = 410.0 Å^3, D_m = 7.58, Z = 2, D_x = 7.60.

Space group $P6_322$ (D_6^6)

Atomic positions

			x	*y*	*z*
6 Ta(1)	in	6(*g*)	0.3592±3	0	0
2 Ta(2)	in	2(*c*)	1/3	2/3	1/4
2 Ca	in	2(*d*)	1/3	2/3	3/4
12 O(1)	in	12(*i*)	0.375 ±4	0.945±3	0.156±2
6 O(2)	in	6(*g*)	0.754 ±4	0	0
4 O(3)	in	4(*f*)	1/3	2/3	0.966±2

Interatomic distances (Å)

Standard errors ∿ 0.02 Å

Ta(1)	-	O(1)	1.96	(×2)
Ta(1)	-	O(3)	1.98	(×2)
Ta(1)	-	O(2)'	2.46	
Ta(1)	-	O(3)	2.04	(×2)
Ta(2)	-	O(1)	1.98	(×6)
Ca	-	O(1)	2.46	(×6)
Ca	-	O(3)	2.65	(×2)

Discussion

Both structures consist of alternating layers along *c* of Nb(Ta) in edge sharing pentagonal bipyramids and Nb+Na(Ta+Ca) in edge sharing octahedra. The structures differ in that the Ca occupies only half of the Na octahedral sites. It is possible that there is some substitution of O by OH in the sodium compound with a corresponding drop in the sodium content.

Details of analysis

$Na_2Nb_4O_{11}$: Lattice parameters from Guinier photographs. Intensities measured from integrated Weissenberg photographs (CuK_α radiation) of layers with k = 0 to 3. Structure determined by similarity with $CaTa_4O_{11}$ and refined by least squares to give R = 0.072 for 292 reflexions listed.

$CaTa_4O_{11}$: Lattice parameters from Guinier powder photographs. Intensities estimated visually from integrated multiple-film Weissenberg photographs (CuK_α radiation) of layers with k = 0 to 4. Also a set of "integrated data around [100]" [*sic*] was measured and this was refined by least squares to give R = 0.039 for 134 reflexions (0.105 for 177 reflexions including some poor strong reflexions) listed.

CHROME PYROPE GARNET

I. Refinement of the crystal structure of a chrome pyrope garnet: an inclusion in natural diamond. G.A. NOVAK and H.O.A. MEYER, 1970. *Amer. Min.*, 55, 2124-2127.

$Mg_3(Al,Cr)_2Si_3O_{12}$ F.W. = 428.16

Cubic, a = 11.526(1) Å, U = 1531 Å^3, Z = 8, D_x = 3.71.

Space group $Ia3d$ (O_h^{10})

Atomic positions

24 Mg	in	24(*c*)	
16 (Al, Cr)	in	16(*a*)	
24 Si	in	24(*d*)	
96 O	in	96(*h*)	*x* = 0.03346±9, *y* = 0.0507±1, *z* = 0.65366±9

Anisotropic temperature factors given.

Interatomic distances (Å)

Si – O	1.639(1)	(×4)		O – Si – O	99.88(8), 114.47(5)°
Al,Cr – O	1.905(1)	(×6)		O – Al,Cr – O	88.00(5), 92.00(5)°
Mg – O	2.216(1)	(×4)			
Mg – O*	2.353(1)	(×4)			

Details of analysis

Claret coloured garnet formed as inclusion in a Venezualan diamond had composition $(Mg_{0.875}Fe_{0.089}Ca_{0.030}Mn_{0.005})_{3.006}(Al_{0.670}Cr_{0.286}Fe_{0.044}Ti_{0.0005})_{2.000}(Si_{0.996}Al_{0.004})_{3.000}O_{12}$. Lattice parameter from back reflexion Weissenberg photograph. Intensities measured with Nb filtered MoK_α radiation on an equi-inclination diffractometer. Crystal equidimensional diameter 0.12 mm. Structure refined by least squares to give R = 0.038 for 200 observed reflexions listed in 1.

1. NAPS 0113.

CALCIUM IRON VANADIUM GARNET

I. [Neutron diffraction study of the garnet ferrite $Ca_3Fe_{3.5}V_{1.5}O_{12}$]. E.L. DUKHOVSKAJA, J.V. LIPIN and J.Z. NOZIK, 1970. *Kristallografija*, 15, 1247-1248 [*Soviet Physics - Crystallography*, 15, 1088-1089].

$Ca_3Fe_{3.5}V_{1.5}O_{12}$

Cubic (Garnet structure), a = 12.465 Å.

Oxygen positions given as x = -0.0331, y = 0.0543, z = 0.1524, B = -1.19 $Å^2$ [? at above 220°C].

Interatomic distances (Å)

Ca – O	2.413	(×12)
Fe – O	2.058	(×6)
(1/2Fe+1/2V) – O	1.803	(×4)

Details of analysis

Neutron powder diffraction measurements made at -120°C and above 220°C. Oxygen parameters refined by least squares using 9 intensities [given in I] to give R = 0.032.

ZIRCONIUM YTTERBIUM OXIDE

I. Mixed oxides of the type MO_2(fluorite)-M_2O_3. IV. Crystal structures of the high- and low-temperature forms of $Zr_3Yb_4O_{12}$. M.R. THORNBER and D.J.M. BEVAN, 1970. *J. Solid State Chem.*, 1, 536-544.

$Zr_3Yb_4O_{12}$ F.W. = 1157.81

LOW - $Zr_3Yb_4O_{12}$

Rhombohedral (Hexagonal setting), a = 9.65±5, c = 9.02±1 Å, U = 727 $Å^3$, Z = 3, D_x = 9.03.

Space group $R\bar{3}$ (C^2_{3i})

Atomic positions

		x	y	z	$B(Å^2)$
2-1/4 Zr(1) + 3/4 Yb(1)	in 3(a)	0	0	0	0.3 ±2
6-3/4 Zr(2) + 11-1/4 Yb(2)	in 18(f)	0.2554±4	0.2126±4	0.3512±4	0.6[0]±6
O(1)	in 18(f)	0.197 ±4	0.163 ±4	0.112 ±4	0
O(2)	in 18(f)	0.031 ±4	0.217 ±4	0.407 ±4	0

[x and y interchanged from I to conform to standard setting].

Interatomic distances (Å)

Zr(1) - O(1)		2.03	(×6)
O(1) - Zr	- O(1)'	83°	
Yb(2) - O(2)		2.16	
Yb(2) - O(2)'		2.21	
Yb(2) - O(1)		2.22	
Yb(2) - O(2)"		2.25	
Yb(2) - O(2)"		2.26	
Yb(2) - O(1)'		2.27	
Yb(2) - O(1)"		2.59	

σ(distances) = 0.04 Å; σ(angles) = 1°.

HIGH - $Zr_3Yb_4O_{12}$

Rhombohedral (Hexagonal setting, a = 9.68±3, c = 8.96±7 Å, U = 727 Å^3, Z = 3, D_x = 9.03.

Space group $R\bar{3}$ (C_{3i}^2)

Atomic positions

			x	*y*	*z*
1-1/2 Zr(1) + 1-1/2 Yb(1)	in	3(*a*)	0	0	0
9 Zr(2) + 9 Yb(2)	in	18(*f*)	0.2567±3	0.2130±3	0.3516±6
O(1)	in	18(*f*)	0.021 ±4	0.205 ±4	0.381 ±7
O(2)	in	18(*f*)	0.205 ±5	0.180 ±5	0.098 ±9

$B = 0$

[*x* and *y* interchanged from I to conform to standard setting]

Interatomic distances (Å)

Zr(1) - O(2)		2.07	(×6)
O(2) - Zr(1)	- O(2)'	77°	
Zr(2) - O(1)		2.02	
Zr(2) - O(2)		2.08	
Zr(2) - O(1)'		2.27	
Zr(2) - O(1)"		2.27	
Zr(2) - O(2)'		2.31	
Zr(2) - O(1)"'		2.43	
Zr(2) - O(2)"		2.56	

σ(distances) = 0.05 Å; σ(angles) = 2°.

Discussion

Both structures consist of a fluorite arrangement with 1/7 of the anion sites vacant (those running along [111] of the fluorite cell). In the high-temperature phase the cations are randomly distributed, but in the low-temperature phase they are partly ordered.

Details of analysis

Crystals of the low-temperature form were mounted with *c* along the goniometer axis. Intensities were measured photometrically from 20° inclined beam multiple-film oscillation photographs taken with Cu$K\alpha$ radiation and corrected for Lorentz, polarisation, but not absorption effects. Absorption and extinction effects were accounted for by using powder intensities with overlapping reflexions partitioned according to the single-crystal measurements. Intensities of the high temperature form were measured as above from single crystals using only a visual estimate of intensities. Structures refined by Fourier and least-squares methods to give R = 0.13 for 133 reflexions listed for the low-temperature phase and R = 0.168 for 89 reflexions listed for the high-temperature phase.

BARIUM TANTALIUM OXIDE

I. A refinement of the structure of barium tantalum oxide, $Ba_5Ta_4O_{15}$, J. SHANNON and L. KATZ, 1970. *Acta Cryst.*, B26, 102.

$Ba_5Ta_4O_{15}$ F.W. = 595.23

Trigonal, a = 5.776±5, c = 11.82±1Å, [U = 341.50Å^3], Z = 1, [D_x = 2.89].

Space group $P\bar{3}m1$ (D_{3d}^3)

Atomic positions

1 Ba(1) in 1(*a*)
2 Ba(2) in 2(*d*) z = 0.8837±5
2 Ba(3) in 2(*d*) z = 0.4282±5
2 Ta(1) in 2(*c*) z = 0.6873±3
2 Ta(2) in 2(*d*) z = 0.1035±3
3 O(1) in 3(*e*)
6 O(2) in 6(*i*) x = 0.1695±23, z = 0.1916±26
6 O(3) in 6(*i*) x = 0.1632±18, z = 0.6136±24

Thermal parameters are given.

Interatomic distances (Å)

Ba (1) - O	2.89, 2.83
Ba (2) - O	2.64, 2.90 , 3.05
Ba (3) - O	2.76, 2.92, 3.24
Ta (1) - O	1.86±2, 2.22±3
Ta (2) - O	2.07±2, 1.94±3

Each distance occurs three times
Error or Ba-O distance are of the order of ±0.03 Å

Description of structure

A more accurate determination but basically the same as previously reported (1).

Details of analysis

$BaCO_3$ was reacted with Ta_2O_5 at 1100°C in air. Crystals were grown from PbO_2 flux. 253 reflexions were measured with a diffractometer using Mo radiation. Least-squares refinement to R = 0.057.

1. *Structure Reports*, 26, 421.

NEODYMIUM SESQUIWOLFRAMATE

I. [The structure of $Nd_4W_3O_{15}$]. T.M. POLJANSKAJA, S.V. BORISOV and N.V. BELOV, 1970. *Kristallografija*, 15, 1135-1139 [*Soviet Physics-Crystallography*, 15, 991-994].

$Nd_4W_3O_{15}$ F.W. = 1368.51

Tetragonal, a = 9.92, c = 12.5Å, U = 1230Å^3, D_m = 7.24, Z = 4, D_x = 7.37.

Space group $P4_2/nmc$ (D_{4h}^{15})

Atomic positions non-centrosymmetric setting

	x	y	z	B(Å^2)
4W(1) in 4(d)	0.5	0	0.0434±5	0.85
8W(2) in 8(g)	0.1974±4	0	0.2261±4	0.82
2Nd(1) in 2(a)	0	0	0	0.78
8Nd(2) in 8(f)	0.2775±10	0.2775±10	0	1.57
6Nd(3) in 8(g)	0	0.2691±11	0.2330±11	2.25
16 O(1) in 16(h)	0.162±5	0.128±5	0.114±5	-0.2
16 O(2) in 16(h)	0.143±5	0.199±5	0.671±5	-0.1
8 O(3) in 8(g)	0	0.313±9	0.431±9	0.7
8 O(4) in 8(g)	0	0.384±10	0.819±11	1.0
8 O(5) in 8(g)	0	0.357±9	0.027±8	0.5
4 O(6) in 4(c)	0	0	0.238±*	0.1

* no value given.

Interatomic distances (Å)

W(1) - O(5)	1.67±9 (×2)
W(1) - O(3)	1.88±9 (×2)
W(1) - O(4)	2.07±13 (×2)
W(2) - O(2)	1.92±6 (×2)
W(2) - O(1)	1.92±6 (×2)
W(2) - O(4)	1.93±10
W(2) - O(6)	1.96±6
Nd(1) - O(1)	2.50±5 (×8)
Nd(1) - O(6)	2.98±5 (×2)
Nd(2) - O(2)	2.29±6 (×2)
Nd(2) - O(1)	2.35±6 (×2)
Nd(2) - O(3)	2.53±5 (×2)
Nd(2) - O(5)	2.88±3 (×2)
Nd(3) - O(3)	2.51±11
Nd(3) - O(1)	2.60±6 (×2)
Nd(3) - O(2)	2.63±5 (×2)
Nd(3) - O(6)	2.67±1
Nd(3) - O(5)	2.72±10
Nd(3) - O(4)	2.78±7 (×2)

Details of analysis

Crystal 0.12 x 0.12 x 0.20 mm grown hydrothermally. Intensities measured visually from Weissenberg photographs of layers with l = 0 to 13 and k = 0. Structure determined from Patterson function and chemical considerations and refined by least squares to give R = 0.17 for 600 reflexions [not listed]. [The presence of disorder in this structure and the large R factor makes this determination somewhat unsatisfactory].

SODIUM COBALT MOLYBDENUM OXYCHLORIDE

I. The refinement of the crystal structure of $Na_2Co_5Mo_4Cl_4O_{16}$. G.W. SMITH and B.G.A. NELSON, 1970. *Acta Cryst.*, B26, 449-451.

$Na_2Co_5Mo_4Cl_4O_{16}$ F.W. = 1122.2

Monoclinic, a = 10.706±3, b = 8.852±2, c = 10.663±3Å, β = 109°45'±6', U = 951Å^3 , Z = 2, [D_x = 3.92].

Space group $C2/m$ (C^3_{2h})

Atomic positions

			x	*y*	*z*	*B*(Å^2)
4 Na	in	4(*i*)	0.3260±10	0	0.4918±10	
8 Co(1)	in	8(*j*)	0.2644±1	0.1790±1	0.8654±1	
2 Co(2)	in	2(*c*)	0	0	1/2	
4 Mo(1)	in	4(*i*)	0.0396±1	0	0.1913±1	
4 Mo(2)	in	4(*i*)	0.4466±1	0	0.1532±1	
8 Cl	in	8(*j*)	0.1650±3	0.1964±3	0.6242±2	
8 O(1)	in	8(*j*)	0.3523±7	0.1629±8	0.0845±7	0.934
8 O(2)	in	8(*j*)	0.1025±8	0.1621±9	0.1399±7	1.085
4 O(3)	in	4(*i*)	0.1404±10	0	0.8715±10	0.871
4 O(4)	in	4(*i*)	0.3936±10	0	0.8717±10	0.872
4 O(5)	in	4(*i*)	0.4767±12	0	0.3218±12	1.241
4 O(6)	in	4(*i*)	0.0955±12	0	0.3675±12	1.212

Anisotropic thermal parameters are listed.

Interatomic distances (Å)

Mo(1) - O	1.75±1,	1.81±1,	1.77±1,	1.75±1		
Mo(2) - O	1.77±1,	1.82±1,	1.72±1,	1.77±1		
Co(1) - O	2.06±1,	2.21±7,	2.02±1,	2.08±1,	2.09±1,	2.02±1
- Cl	2.43±3					

Co - O	(2×) 2.01,	(4×)	2.51	
Na - O	2.36,	2.80,	2.37	
- Cl	2.97,	3.10		

Details of analysis

An accurate refinement is based upon previously reported data (1). The data set consisted of 2132 reflexions. Block-diagonal least-squares were used. The final *R* value is 0.096.

1. G.W. Smith, 1965. *Acta Cryst.*, 18, 582.

LEAD GERMANIUM ALUMINATE

I. Die Kristallstruktur von $Pb_3GeAl_{10}O_{20}$ ($Pb_3SiAl_{10}O_{20}$). H. VINEK, H. VÖLLENKLE and H. NOWOTNY, 1970. *Mh. Chem.*, 101, 275-284.

$Pb_3GeAl_{10}O_{20}$ F.W. = 1284.0

Monoclinic, *a* = 14.39, *b* = 11.44, *c* = 5.004Å, β = 90.0°, *U* = 823.77Å^3, D_m = 5.30, *Z* = 2, D_x = 5.17.

Space group I2/m (C^3_{2h}) Intensities and systematic absences

Atomic positions *Atoms at* (0,0,0; ½,½,½)±(*x*,*y*,*z*; *x*,$\bar{y}$,*z*)

			x	*y*	*z*	*B*(Å^2)
2 Pb(1)	in	2(*a*)	0	0	0	2.26
4 Pb(2)	in	4(*i*)	0.2847±4	0	0.0315±14	0.52
2 Al(1)*	in	2(*b*)	0	0.5	0	0.96
4 Al(2)*	in	4(*h*)	0	0.6326±27	0.5	1.31
8 Al(3)*	in	8(*j*)	0.3548±22	0.3603±17	0.9979±70	1.37
8 Al(4)*	in	8(*j*)	0.1284±16	0.2905±16	0.0299±57	0.76
4 O(1)	in	4(*i*)	0.4387±72	0	0.1599±240	1.02
4 O(2)	in	4(*i*)	0.8983±71	0	0.5889±210	0.64
8 O(3)	in	8(*j*)	0.2411±59	0.3572±63	0.0772±193	2.09
8 O(4)	in	8(*j*)	0.4188±50	0.2485±59	0.1637±168	1.56
8 O(5)	in	8(*j*)	0.8643±59	0.1445±65	0.1190±180	2.05
8 O(6)	in	8(*j*)	0.9330±45	0.3860±47	0.1650±145	0.42

* The Ge atoms are randomly distributed over the Al sites.

Interatomic distances (Å)

Pb(1) - O(2)	2.52±10 (×2)	Al(Ge)(2) - O(6)	1.95±7 (×2)
Pb(1) - O(5)	2.63±8 (×4)	Al(Ge)(2) - O(4)	1.95±7 (×2)
Pb(2) - O(1)	2.31±10	Al(Ge)(3) - O(3)	1.68±9
Pb(2) - O(3)	2.58±9 (×2)	Al(Ge)(3) - O(2)	1.78±5
Pb(2) - O(5)	2.81±8 (×2)	Al(Ge)(3) - O(4)	1.78±8
Pb(2) - O(6)	3.10±7 (×2)	Al(Ge)(3) - O(5)	1.90±10
Pb(2) - O(2)	3.25±10	Al(Ge)(4) - O(6)	1.71±7
Al(Ge)(1) - O(6)	1.82±6 (×4)	Al(Ge)(4) - O(4)	1.74±8
Al(Ge)(1) - O(1)	1.92±12 (×2)	Al(Ge)(4) - O(3)	1.81±8
Al(Ge)(2) - O(1)	1.93±7 (×2)	Al(Ge)(4) - O(5)	1.83±8

Isostructural compound

$Pb_3SiAl_{10}O_{20}$: a = 14.34, b = 11.39, c = 4.96Å, β = 90.0°.

Discussion

The structure consists of chains of edge sharing $Al(Ge)O_6$ octahedra running along c linked by corner sharing to sheets of double $Al(Ge)O_4$ tetrahedra perpendicular to a. The resultant framework encloses the Pb atoms in channels which run along c both within and between the sheets. It was not possible to determine to what extent the Ge atoms were ordered on the Al sites. The crystals are pseudo-orthorhombic, space group *Immm*.

Details of analysis

Single crystal, powder and diffractometer records. No absorption correction. Structure determined from Patterson and electron-density syntheses and refined by least squares to give R = 0.16. Reflexions for l = 0 listed.

MONOCLINIC WOLFRAM VANADIUM OXIDE

I. The crystal structure of monoclinic wolfram vanadium oxide, $W_3V_5O_{20}$, an OD structure related to R-Nb_2O_5. M. ISREALSSON and L. KIHLBORG, 1970. *J. Solid State Chem.*, 1, 469-477.

$W_3V_5O_{20}$ F.W. = 1126.25

Monoclinic, a = 24.413±3, b = 7.446±2, c = 3.950±1Å, β = 91.31±2°, U = 717.9Å³, D_m = 5.18 2, Z = 2, D_x = 5.21.

Space group $C2/m$ (C^3_{2h})

Atomic positions

			x	y	z
2W(1) and 6V(1)	in	8(j)	0.0739±2	0.2494±9	0.0856±14
4W(2)	in	4(i)	0.1787±1	0.5	0.9291±7
4V(3)	in	4(i)	0.1715±4	0	0.9163±25
4O(1)	in	4(g)	0	0.262 ±6	0
4O(2)	in	4(i)	0.0902±14	0	0.026 ±9
8O(3)	in	8(j)	0.1589±13	0.255 ±5	0.033 ±9
4O(4)	in	4(i)	0.0886±14	0.5	0.036 ±9
4O(5)	in	4(i)	0.2493±12	0.5	0.995 ±8
8O(6)	in	8(j)	0.0795±13	0.250±5	0.536 ±9
4O(7)	in	4(i)	0.1787±14	0.5	0.511 ±9
4O(8)	in	4(i)	0.1822±14	0.5	0.556±10

Isotropic temperature factors given.

Interatomic distances (Å)

W,V(1) - O(6)	1.78±4
W,V(1) - O(1)	1.83±1
W,V(1) - O(4)	1.91±1
W,V(1) - O(2)	1.92±1
W,V(1) - O(3)	2.09±3
W,V(1) - O(6)	2.18±4
W(2) - O(7)	1.65±4
W(2) - O(5)	1.74±3
W(2) - O(3)	1.93±4 (×2)
W(2) - O(4)	2.25±3
W(2) - O(7)	2.30±3
V(3) - O(8)	1.45±4
V(3) - O(5)	1.96±3
V(3) - O(3)	1.98±4 (×2)
V(3) - O(2)	2.04±4
V(3) - O(8)	2.54±4

Discussion

The crystal has the R-Nb_2O_5 structure (3) and differs from that proposed by 1. A preliminary account appears in 2. The V and W atoms are partially ordered on the Nb sites. The stacking of layers is disordered along the *b* direction resulting diffuse reflexions for reciprocal lattice layers with $k=2n+1$.

Details of analysis

Lattice parameters from Guinier powder photograph ($CuK\alpha_1$, λ = 1.54051Å). Intensities estimated visually from multiple film Weissenberg photographs of layers with $k = 0$ to 6 taken with $CuK\alpha$ radiation. Structure determined from Patterson function and refined by least squares to give R = 0.073 for 316 observed reflexions listed.

1. S. MONDER, A. RIMSKY, J. BORÈNE and W. FREUNDLICH, 1968. *C.R. Acad. Sci. Paris*, C266, 1145.
2. L. KIHLBORG and M. ISREALSSON, 1968. *Acta Chem. Scand.*, 22, 1685.
3. R. GRUEHN, 1966. *J. Less-Common Metals*, 11, 119.

Y FERRITES

I. Refinement of the crystal structure of $Ba_2Zn_2Fe_{12}O_{22}$. W.D. TOWNES and J.H. FANG, 1970. *Z. Kritallogr.* 131, 196-205.

II. [A neutron diffraction study of the hexagonal Y ferrites $Ba_2Zn_2Fe_{12}O_{22}$ and $Ba_2Zn_{0.3}Co_{1.7}Fe_{12}O_{22}$]. I.I. JAMZIN and J. LECIEJEWICZ, 1970. *Kristallografija*, 15, 280-286 [*Soviet Physics - Crystallography*, 15, 235-238].

Zn_2 Y-FERRITE

$Ba_2Zn_2Fe_{12}O_{22}$ F.W. = 1423.6

Rhombohedral, a = 5.876, c = 43.558Å, U = 1302Å^3, Z = 3, D_x = 5.4, (from 1).

Space group $R\bar{3}m$ (D^5_{3d})

Atomic positions

The first value is from I, the second from II [converted in order to be directly comparable with that from I].

	x	*y*	*z*	$B(Å^2)$†
6 Ba in 6(*c*)	0	0	0.29997±2	0.75±1
6 Ba	0	0	0.2997*	
3 Fe(1) in 3(*a*)	0	0	0	0.47±3
3 Fe(1)	0	0	0	
3 Fe(2) in 3(*b*)	0	0	1/2	0.50±3
3 Fe(2)	0	0	1/2	
1.8 Fe(3)+4.2 Zn in 6(*c*)	0	0	0.06523±5	0.55±2†
6 Fe(3)	0	0	0.0656*	
18 Fe(4) in 18(*h*)	0.5035±1	-*x*	0.10967±3	0.55±1
18 Fe(4)	0.496	-*x*	0.110	
3 Fe+3Zn(1) in 6(*c*)	0	0	0.15233±5	0.53±2
6 (Fe+Zn)(1)	0	0	0.152*	
4.2Fe+1.8 Zn(2) in 6(*c*)	0	0	0.37619±5	0.65±2†
6 (Fe+Zn)(2)	0	0	0.3761*	
6 O(1) in 6(*c*)	0	0	0.4194±3	0.82±11
6 O(1)	0	0	0.417*	
6 O(2) in 6(*c*)	0	0	0.1968±3	0.87±11
6 O(2)	0	0	0.197*	
18 O(3) in 18(*h*)	0.1552±7	-*x*	0.0286±2	1.07±8
18 O(3)	0.150	-*x*	0.025	
18 O(4) in 18(*h*)	0.8300±6	-*x*	0.0849±1	0.74±7
18 O(4)	0.820	-*x*	0.086	
18 O(5) in 18(*h*)	0.1797±7	-*x*	0.1376±1	0.69±7
18 O(5)	0.178	-*x*	0.138	

† Values given before refinement of site occupations. Since these values are strongly correlated with the site occupation the values of *B* for Fe(3) and Fe, Zn(2) may be in error.

* Values given by 1 not refined.

Interatomic distances (Å) *and bond angles* from I

Fe(1) - O(3)	2.012±5	(×6)		
O(3)-Fe(1)-O(3)	85.7±2°	(×6),	94.3±2°	(×6)
Fe(2)-O(5)	2.012±3	(×6)		
O(5)-Fe(2)-O(5)	84.6±1°	(×6),	95.4±1°	(×6)
Fe(3)-O(4)	1.931±2	(×3)		
Fe(3)-O(3)	2.245±6	(×3)		
O(4)-Fe(3)-O(4)	101.8±2°	(×3)		
O(4)-Fe(3)-O(3)	90.0±2°	(×6),	161.1±2°	(×3)
O(3)-Fe(3)-O(3)	75.1±2°	(×3)		
Fe(4)-O(4)	1.981±3	(×2)		
Fe(4)-O(1)	2.014±7			
Fe(4)-O(2)	2.031±8			
Fe(4)-O(5)	2.053±4	(×2)		

Angles in degrees subtended at Fe(4) by

	O(4)'	O(1)	O(2)	O(5)	O(5)'
O(4)	93.2±1	94.7±2	88.4±2	174.6±1	92.1±1
O(1)			175.5±4	84.5±3	84.5±3
O(2)				92.1±2	92.1±2
O(5)					82.5±1

(Fe+Zn)(1)-O(2)	1.937±13	(Fe+Zn)(2)-O(1)	1.882±12	
(Fe+Zn)(1)-O(5)	1.938±5 (×3)	(Fe+Zn)(2)-O(3)	1.916±3	(×3)
O(2)-(Fe+Zn)(1)-O(5)	109.3±1° (×3)	O(1)-(Fe+Zn)(2)-O(3)	108.9±2°	(×3)
O(5)-(Fe+Zn)(1)-O(5)	109.6±2° (×3)	O(3)-(Fe+Zn)(2)-O(3)	110.0±2°	(×3)

Discussion

The structure is composed of mixed cubic and hexagonal close packed layers of O and Ba with Fe and Zn in octahedral and tetrahedral holes. The stacking sequence is C'BCABA'B'ABCAC'A'CABCB'. The primed layers have composition BaO_3; the others contain only oxygen atoms.

Details of analysis

I. Intensities measured with balanced filters on diffractometer using Mo*K*α radiation and stationary crystal-stationary counter. Corrected for Lorentz, polarisation, α_1, α_2 and absorption (μ = 17.6 mm^{-1}, crystal a sphere of radius 0.47 mm). Structure of 1 refined by least squares to give R = 0.071 for 1033 independent observed reflexions listed.

II. Powder neutron diffractometer (λ = 1.22 Å) at 78 and 293°K. Both chemical (coordinates of Fe(4), O(3), O(4) and O(5)) and magnetic structure refined by least squares to give R[intensities ?] = 0.052 (at 293°K) for 13 reflexions listed. The distribution of Fe and Zn atoms is inferred from the magnetic moments.

1. *Structure Reports*, 21, 281.

BARIUM SILICON TANTALUM OXIDE

I. The structures of the reduced and oxidized forms of barium silicon tantalum oxide, $Ba_3Si_4Ta_6O_{23}$ and $Ba_3Si_4Ta_6O_{26}$. J. SHANNON and L. KATZ, 1970. *J. Solid State Chem.*, 1, 399-408.

$Ba_3Si_4Ta_6O_{23}$ (reduced phase) F.W. = 1978.03

Hexagonal, a = 8.997±3, c = 7.745±5Å, U = 542.9Å^3, D_m = 5.7 to 6.3, Z = 1, D_x = 6.04.

Space group $P\bar{6}2m$ (D^3_{3h}) from Laue symmetry and structure

Atomic positions

			x	*y*	*z*
3 Ba	in	3(*g*)	0.6016±8	0	1/2
6 Ta	in	6(*i*)	0.2384±3	0	0.2428±4
4 Si	in	4(*h*)	1/3	2/3	0.2046±31
2 O(1)	in	2(*c*)	1/3	2/3	0
3 O(2)	in	3(*f*)	0.2919±111	0	0
6 O(3)	in	6(*i*)	0.8191±59	0	0.2363±68
12 O(4)	in	12(*l*)	0.4939±48	0.1773±46	0.2802±36

Isotropic temperature factors for all atoms are reported.

$Ba_3Si_4Ta_6O_{26}$ (oxidized phase) is nearly isostructural with $Ba_3Si_4Ta_6O_{23}$

Hexagonal, a = 8.99±1, c = 7.79±1Å. F.W. = 2026.03

Atomic positions

3 Ba	in	3(g)	with	x = 0.6022±10
6 Ta	in	6(i)		x = 0.2379±3, z = 0.2452±11
4 Si	in	4(h)		z = 0.2137±81
2 O(1)	in	2(c)		
3 O(2)	in	3(f)		x = 0.2767±86
6 O(3)	in	6(i)		x = 0.8332±97, z = 0.2112±171
12 O(4)	in	12(l)		x = 0.4881±39, y = 0.1809±38, z = 0.2903±62
3 O(5)	in	3(g)		x = 0.2217±111 y = 0, z = 1/2

Isotropic temperature factors reported.

Interatomic distances (Å)

		$Ba_3Si_4Ta_6O_{23}$ (Reduced phase)	$Ba_3Si_4Ta_6O_{26}$ (Oxidized phase)
Ba-O(3)	(×2)	2.83±4	3.06±11
Ba-O(4)	(×4)	2.82±4	2.84±4
Ba-O(4)'	(×4)	3.04±4	2.95±4
Ba-O(5)			3.11±1
Ba-O(5)'			3.42±10
Ta-O(2)		1.94±2	1.94±2
Ta-O(3)	(×2)	1.94±2	1.92±3
Ta-O(4)	(×2)	2.06±4	2. 4±3
Ta-O(5)			1.99±1
Si-O(1)		1.59±3	1.66±6
Si-O(4)	(×3)	1.60±4	1.61±4

Discussion

Previously reported as $Ba_{0.5-x}TaO_{3-x}$ (1). The structure of the reduced phase consists of layers composed of corner sharing Si_2O_7 double tetrahedra and Ta_2O_9 double square pyramids. The layers are separated by Ba ions. In the oxidized phase, the additional oxygen atoms lie between the layers in such a way as to complete the octahedral coordination around tantalum atoms in both layers.

Details of analysis

Crystals cut into plates. Lattice parameters and intensities measured with a single-crystal diffractometer using filtered Mo radiation and θ-2θ scan. Intensities corrected for absorption. Structure of 1 adapted and refined by Fourier methods and least squares to give a weighted R = 0.10 for 509 reflexions listed for the reduced phase and R = 0.096 for 258 reflexions listed for the oxidized phase.

1. *Structure Reports*, 24, 339.

BARIUM NIOBIUM SILICON OXIDE

I. The structure of barium silicon niobium oxide, $Ba_3Si_4Nb_6O_{26}$, A compound with linear silicon-oxygen-silicon groups. J. SHANNON and L. KATZ, 1970. *Acta Cryst.*, B26, 105.

$Ba_3Si_4Nb_6O_{26}$ F.W. = 1368.14

Hexagonal, a = 9.00±1, c = 7.89±1Å, [U = 553.5Å^3], Z = 1, D_x = 4.10.

Space group $P\bar{6}2m$ (D^3_{3h})

Atomic positions

			x	y	z	B(Å^2)
3 Ba	in	3(g)	0.59334±14(1)	0	1/2	*
6 Nb	in	6(i)	0.23809±12	0	0.23939±21	*
4 Si	in	4(h)	1/3	2/3	0.20309±60	0.25±7
2 O(1)	in	2(c)	1/3	2/3	0	0.64±28
3 O(2)	in	3(f)	0.2734±15	0	0	0.69±24
6 O(3)	in	6(i)	0.8188±11	0	0.2358±14	0.74±15
12 O(4)	in	12(l)	0.4888±9	0.1830±8	0.2841±8	0.48±10
3 O(5)	in	3(g)	0.2269±14	0	1/2	0.68±20

*Anisotropic thermal parameters are listed.

Interatomic distances (Å) *and angles*

Ba-O(3)		2.867±10		O(1)-Si-O(4)	113.0±3
Ba-O(4)		2.873±7,	2.989±7	O(4)-Si-O(4)	105.7±3
Ba-O(5)		3.132±4,	3.332±15	Si-O(1)-Si	180
Nb-O(2)		1.918±4		O(2)-Nb-O(3)	96.0±4
Nb-O(3)	(2×)	1.937±4		O(2)-Nb-O(4)	92.8±4
Nb-O(4)	(2×)	2.044±7		O(3)-Nb-O(3)	93.2±6
Nb-O(5)		2.056±3		O(3)-Nb-O(4)	88.7±3
Si-O(1)		1.599±6		O(4)-Nb-O(4)	88.2±4
Si-O(4)	(3×)	1.629±7			

Structure

Isotypic with $Ba_3Si_4Ta_6O_{26}$ (1). Si_2O_7 groups link together chains of NbO_6 octahedra which share corners. The barium ion has a distorted pentagonal environment of oxygen atoms.

Materials

Heating 5BaO + $4Nb_2O_5$ + 2Nb with 1% BaF_2 to 1100°C in an evacuated sealed silica capsule produced the crystals.

Details of analysis

1133 reflexions obtained using Mo radiation and diffractometer. Absorption corrections applied. Least-squares refinement to R = 0.046.

1. See previous report.

SILVERDECAMOLYBDATE

I. The crystal structure of silverdecamolybdate, $Ag_6Mo_{10}O_{33}$. B.M. GATEHOUSE and P. LEVERETT, 1970, *J. Solid State Chem.*, 1, 484-496.

$Ag_6Mo_{10}O_{33}$ F.W. = 2134.57

Triclinic, a = 7.59±2, b = 8.31±2, c = 11.42±2Å, α = 82.6±2, β = 102.9±2, γ = 106.4 2°, U = 671.9Å^3, D_m = 5.5±1, Z = 1, D_x = 5.26.

Space group $P\overline{1}$ (C_i^1) from structure refinement.

Atomic positions

			x	*y*	*z*
2 Ag(1)	in	2(*i*)	0.6519± 5	0.0065± 4	0.2515± 3
2 Ag(2)		"	0.8398± 6	0.4832± 4	0.2075± 3
2 Ag(3)		"	0.4438± 8	0.5227± 5	0.2388± 3
2 Mo(1)		"	0.5691± 5	0.2249± 3	0.4887± 3
2 Mo(2)		"	0.0731± 5	0.2334± 3	0.4909± 3
2 Mo(3)		"	0.1404± 5	0.8005± 3	0.0282± 3
2 Mo(4)		"	0.3011± 5	0.2048± 3	0.0144± 3
2 Mo(5)		"	0.1823± 5	0.9993± 3	0.3179± 3
1 O(1)		1(*a*)	0	0	0
2 O(2)		2(*i*)	0.8794±40	0.6701±28	0.0215±21
2 O(3)		"	0.3466±41	0.9952±29	0.0790±22
2 O(4)		"	0.4921±43	0.3519±31	0.0847±24
2 O(5)		"	0.2423±40	0.6484±28	0.1080±22
2 O(6)		"	0.8310±47	0.2029±35	0.1161±26
2 O(7)		"	0.6694±41	0.7948±28	0.1334±22
2 O(8)		"	0.2178±41	0.1738±29	0.2130±22
2 O(9)		"	0.0940±41	0.8312±28	0.2262±22
2 O(10)		"	0.5999±44	0.3322±31	0.3516±23
2 O(11)		"	0.0427±47	0.3725±35	0.3709±26
2 O(12)		"	0.9688±41	0.0067±29	0.3762±22
2 O(13)		"	0.4436±41	0.9884±29	0.3820±22
2 O(14)		"	0.8143±41	0.6537±28	0.3888±22
2 O(15)		"	0.3864±42	0.6202±29	0.4129±22
2 O(16)		"	0.2982±42	0.1742±30	0.4625±22
2 O(17)		"	0.1792±42	0.8214±30	0.4821±22

Anisotropic temperature factors for silver atoms and isotropic temperature factors for other atoms are given.

Interatomic distances (Å)

Ag(1) - O(12)	2.37±3	Mo(5) - O(9)	1.74±2
Ag(1) - O(7)	2.40±3	Mo(5) - O(8)	1.76±2
Ag(1) - O(6)	2.44±3	Mo(5) - O(12)	1.91±3
Ag(1) - O(12)	2.51±3	Mo(5) - O(12)	1.98±3
Ag(1) - O(3)	2.68±3	Mo(5) - O(16)	2.20±2
Ag(2) - O(11)	2.40±3	Mo(5) - O(17)	2.23±2
Ag(2) - O(2)	2.49±2	Mo(1) - O(10)	1.73±2
Ag(2) - O(6)	2.65±3	Mo(1) - O(15)	1.73±3
Ag(2) - O(10)	2.67±3	Mo(1) - O(16)	1.94±3
Ag(2) - O(4)	2.69±3	Mo(1) - O(17)	2.01±3
Ag(2) - O(14)	2.72±3	Mo(1) - O(12)	2.15±2
Ag(2) - O(9)	2.99±2	Mo(1) - O(12)	2.30±2
Ag(3) - O(5)	2.27±3	Mo(2) - O(14)	1.69±2
Ag(3) - O(10)	2.33±3	Mo(2) - O(11)	1.69±3
Ag(3) - O(15)	2.41±3	Mo(2) - O(17)	1.93±3
Ag(3) - O(4)	2.54±3	Mo(2) - O(16)	2.01±3
Ag(3) - O(7)	2.75±2	Mo(2) - O(12)	2.28±2
Ag(3) - O(8)	2.94±2	Mo(2) - O(12)	2.34±2
Ag(3) - O(14)	2.94±2		

Mo(3) - O(6)	1.71±3	Mo(4) - O(4)	1.73±2
Mo(3) - O(5)	1.72±3	Mo(4) - O(7)	1.75±3
Mo(3) - O(3)	1.95±2	Mo(4) - O(2)	1.89±3
Mo(3) - O(2)	1.97±3	Mo(4) - O(3)	1.90±2
Mo(3) - O(1)	2.17±1	Mo(4) - O(1)	2.42±1
Mo(3) - O(9)	2.42±3	Mo(4) - O(8)	2.45±3

Discussion

The structure is composed of sheets of edge and corner sharing MoO_6 octahedra. The silver atoms lie in holes within the sheets (Ag(1)) and in layers between the sheets (Ag(2) and Ag(3)).

Details of analysis

Intensities for layers with h = 0 to 5, k = 0 and l = 0 were measured visually from integrated Weissenberg photographs taken with Cu$K\alpha$ radiation from a crystal 0.10×0.02×0.02 mm^3. Intensities corrected for Lorentz and polarization effects but not for absorption. Structure solved from Patterson function and electron difference maps and refined by least squares to give R = 0.076 for 1267 reflexions listed. A refinement in $P1$ gave essentially the same structure.

BARIUM TITANIUM OXIDE

I. The crystal structure of hexabarium 17-Titanate. E. TILLMANNS and W.H. BAUR, 1970. *Acta Cryst.*, B26, 1645-1654.

$Ba_6Ti_{17}O_{40}$ F.W. = 2278.32

Monoclinic, a = 9.883±6, b = 17.08±1, c = 18.92±1Å, β = 98°42±2', U = 3156Å^3, Z = 4, D_x = 4.80.

Space group $C2/c$ (C_{2h}^6) Cc tested and rejected.

Atomic parameters

	x	y	z	B(Å^2)
8 Ba(1) in 8(f)	0.06961±12	0.08981±7	0.44274±6	0.77±3
8 Ba(2) in 8(f)	0.86257±12	0.42547±7	0.82157±6	0.82±3
8 Ba(3) in 8(f)	0.85369±12	0.28319±7	0.31496±6	0.85±3
4 Ti(1) in 4(e)	0	0.1040±3	¼	0.53±9
4 Ti(2) in 4(e)	0	0.9378±3	¼	0.35±8
8 Ti(3) in 8(f)	0.7227±3	0.9011±2	0.8618±2	0.26±6
4 Ti(4) in 4(e)	0	0.7476±3	¼	0.35±8
8 Ti(5) in 8(f)	0.7041±4	0.9257±2	0.3697±2	0.46±6
8 Ti(6) in 8(f)	0.7097±3	0.7527±2	0.3736±2	0.23±6
8 Ti(7) in 8(f)	0.9777±3	0.1599±2	0.8816±2	0.34±6
4 Ti(8) in 4(c)	¼	¼	0	0.27±8
8 Ti(9) in 8(f)	0.9669±3	0.3241±2	-0.0011±2	0.26±6
8 Ti(10) in 8(f)	0.2560±3	0.4199±2	0.0135±2	0.21±6
4 Ti(11) in 4(b)	0	½	0	0.49±9
O(1) in 8(f)	0.9310±13	0.0803±7	0.5594±6	0.5±3
O(2) in 8(f)	0.9390±13	0.2415±7	0.5530±7	0.8±3
O(3) in 8(f)	0.9395±13	0.4120±7	0.5542±7	0.7±3
O(4) in 8(f)	0.2049±12	0.0156±7	0.0674±6	0.2±3
O(5) in 8(f)	0.1883±12	0.1586±7	0.5748±6	0.2±3
O(6) in 8(f)	0.1852±12	0.3225±7	0.5674±6	0.3±3
O(7) in 8(f)	0.1775±12	0.4950±7	0.0605±6	0.7±3
O(8) in 8(f)	0.9263±14	0.4131±8	0.0582±7	1.1±3
O(9) in 8(f)	0.9300±13	0.2532±7	0.0698±7	0.7±3
O(10) in 8(f)	0.1772±12	0.3303±7	0.0558±7	0.6±3
O(11) in 8(f)	0.2074±12	0.1642±7	0.0616±6	0.1±3

O(12)	in 8(*f*)	0.9011±13	0.0183±8	0.1965±7	0.8±3
O(13)	in 8(*f*)	0.9085±12	0.1621±7	0.6900±6	0.1±3
O(14)	in 8(*f*)	0.8901±13	0.3272±8	0.6925±7	1.2±3
O(15)	in 8(*f*)	0.8713±13	0.4869±8	0.6821±7	0.7±3
O(16)	in 8(*f*)	0.1223±12	0.0802±7	0.6776±6	0.5±3
O(17)	in 8(*f*)	0.1293±13	0.2421±8	0.6801±7	0.8±3
O(18)	in 8(*f*)	0.8759±12	0.3410±7	0.1795±7	0.3±3
O(19)	in 8(*f*)	0.8842±12	0.1803±7	0.1899±6	0.4±3
O(20)	in 8(*f*)	0.1287±13	0.1137±8	0.1785±7	1.2±3

Interatomic distances (Å)

Average Ti(1) - O	1.977
Ti(2) - O	2.014
Ti(3) - O	1.972
Ti(4) - O	1.977
Ti(5) - O	1.974
Ti(6) - O	1.931
Ti(7) - O	1.941
Ti(8) - O	1.960
Ti(9) - O	1.995
Ti(10) - O	1.962
Ti(11) - O	1.986

Ba(1) - O twelve in range 2.77 - 3.10
Ba(2) - O eleven in range 2.75 - 3.35
Ba(3) - O eleven in range 2.64 - 3.22

Description of the structure

The structure features nearly hexagonally packed oxygen atom layers stacked along [103] with sequence ABCACABCABABCACABCBCAB. Ti in octahedral sites, Ba(1) in the middle of a BAB layer sequence and Ba(2) and Ba(3) in the middle of an ABC sequence.

Material

$BaO:TiO_2$ = 1:3 was heated to 1400°C in platinum and quenched.

Details of analysis

2981 unique reflexions measured with a diffractometer (Ag radiation). No absorption corrections. Least-squares refinement to an *R* value of 0.133 including unobserved reflexions.

MAGNESIUM GERMANATE

I. The crystal structure of magnesium germanate: A reformulation of Mg_4GeO_6 as $Mg_{28}Ge_{10}O_{48}$. R.B. VON DREELE, P.W. BLESS, E. KOSTINER and R.E. HUGHES, 1970. *J. Solid State Chem.*, 2, 612-618.

$Mg_{28}Ge_{10}O_{48}$ F.W. = 2174.39

Orthorhombic, a = 14.512±2, b = 10.219±2, c = 5.944±1 Å, U = 881.48 Å^3, D_m = 3.98±4, Z = 1, D_x = 4.09.

Space group Pbam (D^9_{2h}) systematic absences and final structure.

Atomic positions

	x	*y*	*z*	$B(Å^2)$
2 Ge(1) in 2(*a*)	0	0	0	0.30±1
4 Ge(2) in 4(*g*)	0.12563±3	0.50156±5	0	0.28±1
4 Ge(3) in 4(*h*)	0.18623±3	0.32509±5	0.5	0.39±1
2 Mg(1) in 2(*d*)	0	0.5	0.5	0.39±3
2 Mg(2) in 2(*b*)	0	0	0.5	0.42±3
4 Mg(3) in 4(*g*)	0.1759 ±1	0.1785 ±2	0	0.39±2
4 Mg(4) in 4(*h*)	0.3259 ±1	0.1466 ±2	0.5	0.36±2
8 Mg(5) in 8(*i*)	-0.0046 ±1	0.2517 ±1	0.2423±2	0.38±2
8 Mg(6) in 8(*i*)	0.3315 ±1	0.4192 ±1	0.2460±2	0.37±2
4 O(1) in 4(*g*)	0.0840 ±2	0.3374 ±3	0	0.38±4
4 O(2) in 4(*g*)	0.4217 ±2	0.3488 ±3	0	0.35±4
4 O(3) in 4(*g*)	0.2520 ±2	0.0034 ±4	0	0.43±4
4 O(4) in 4(*h*)	0.0677 ±2	0.3294 ±4	0.5	0.48±5
4 O(5) in 4(*h*)	0.4129 ±2	0.3315 ±3	0.5	0.37±4
4 O(6) in 4(*h*)	0.2562 ±2	-0.0231 ±3	0.5	0.45±4
8 O(7) in 8(*i*)	0.0757 ±2	0.0780 ±2	0.2234±4	0.42±3
8 O(8) in 8(*i*)	0.4142 ±2	0.0810 ±2	0.2482±4	0.38±3
8 O(9) in 8(*i*)	0.2426 ±2	0.2513 ±2	0.2731±4	0.45±3

Interatomic distances (Å) *and bond angles*

Ge(1) - O(7)	1.899±2	(×4)
Ge(1) - O(2)	1.918±3	(×2)

Angles in degrees subtended at Ge(1) by

	O(7)"	O(2)
O(7)	88.7±1	90.3±1
O(7)'	91.3±1	89.7±1

Ge(2) - O(3)	1.776±3	
Ge(2) - O(8)	1.780±2	(×2)
Ge(2) - O(1)	1.783±3	

Angles in degrees subtended at Ge(2) by

	O(8)'	O(1)
O(3)	108.7±1	110.4±2
O(8)	111.9±1	108.6±1

Ge(3) - O(4)	1.721±3	
Ge(3) - O(9)	1.748±2	(×2)
Ge(3) - O(6)	1.762±3	

Angles in degrees subtended at Ge(3) by

	O(9)'	O(6)
O(4)	118.6±1	116.8±2
O(9)	101.0±1	99.1±1

Mg(1) - O(4)	2.001±4	(×2)
Mg(1) - O(8)	2.116±2	(×4)

Angles in degrees subtended at Mg(1) by

	O(4)	O(4)'	O(8)	O(8)'
O(8)"	87.0±1	93.0±1	90.1±1	89.9±1

Mg(2) - O(7)	2.132±2	(×4)
Mg(2) - O(5)	2.136±3	(×2)

Angles in degrees subtended at Mg(2) by

	O(7)"	O(5)
O(7)	79.1±1	89.8±$\bar{1}$
O(7)'	100.9±1	90.2±$\bar{1}$

Mg(3) - O(9)	2.031±3	(×2)
Mg(3) - O(1)	2.101±3	
Mg(3) - O(3)	2.103±4	
Mg(3) - O(7)	2.221±3	(×2)

Angles in degrees subtended at Mg(3) by

	O(9)'	O(1)	O(3)	O(7)	O(7)'
O(9)	106.1±1	91.1±1	93.5±1	163.6±1	90.2±1
O(1)			172.3±2	86.7±1	86.7±1
O(3)				87.2±1	87.2±1
O(7)					73.4±1

Mg(4) - O(6)	2.008±3	
Mg(4) - O(8)	2.081±3	(×2)
Mg(4) - O(9)	2.104±3	(×2)
Mg(4) - O(5)	2.273±3	

Angles in degrees subtended at Mg(4) by

	O(8)'	O(9)	O(9)'	O(5)
O(6)	91.8±1	98.6±1	98.6±1	176.5±3
O(8)	92.0±1	168.2±1	93.2±1	85.7±1
O(9)			79.7±1	84.1±1

Mg(5) - O(4)	2.019±2
Mg(5) - O(2)	2.067±2
Mg(5) - O(8)	2.077±3
Mg(5) - O(1)	2.120±2
Mg(5) - O(5)	2.122±2
Mg(5) - O(7)	2.126±3

Angles in degrees subtended at Mg(5) by

	O(2)	O(8)	O(1)	O(5)	O(7)
O(4)	173.1±1	87.6±1	92.2±1	84.4±1	94.8±1
O(2)		97.3±1	92.6±1	90.7±1	80.3±1
O(8)			90.9±1	89.9±1	177.6±4
O(1)				176.5±2	88.7±1
O(5)					90.7±1

Mg(6) - O(6)	2.061±2
Mg(6) - O(3)	2.085±3
Mg(6) - O(2)	2.090±2
Mg(6) - O(7)	2.113±3
Mg(6) - O(5)	2.116±2
Mg(6) - O(9)	2.153±3

Angles in degrees subtended at Mg(6) by

	O(3)	O(2)	O(7)	O(5)	O(9)
O(6)	92.1±1	176.1±2	102.7±1	86.8±1	78.6±1
O(3)		90.9±1	90.5±1	178.4±4	91.9±1
O(2)			79.7±1	90.2±1	98.8±1
O(7)				90.8±1	177.2±3
O(5)					86.8±1

Discussion

The O atoms are close packed and stacked along **a** with sequence $(ABACBC)_2$. The sequence of metal atoms is (Ge(1), Mg(1), Mg(2), Mg(5) | O(1), O(2), O(4), O(5), O(7), O(8) | Ge(2), Ge(3), Mg(3), Mg(4), Mg(6) | O(3), O(6), O(9) | Ge(2), Ge(3), Mg(3), Mg(4), Mg(6) | O(1), O(2), O(4), O(5), O(7), $O(8))_{2n}$

Details of analysis

Prepared from 4:1 mixture of MgO and GeO_2 with 1% $MnCo_3$ and fired at 1200°C. Crystals grown from PbO flux. Analysis given. Lattice parameters measured on four-circle diffractometer at 21°C with MoK_α radiation [λ not given]. Intensities measured on four-circle diffractometer (MoK_α radiation) and corrected for Lorentz, polarization, extinction and absorption (μR = 0.89 for crystal ground as a sphere with diameter 0.19±1 mm). Structure determined from Patterson function assuming close packed oxygen atom array and refined by least squares to give R_2= 0.062 for 2096 reflexions listed in 1. There was no evidence of anisotropic thermal motions or partial occupancy.

1. Document 1108 A.D.I. Auxiliary Publications project, Library of Congress, U.S.A.

METAL(I) VANADATE - PHASES

I. Structure of the $M_xV_2O_5$-β and $M_xV_{2-y}T_yO_5$-β phases. J. GALY, J. DARRIET, A. CASALOR and J.B. GOODENOUGH, 1970. *J. Solid State Chem.*, 1, 339-348.

$Li_{0.48}V_2O_5$-β'

Sample prepared by heating LiV_2O_5 and V_2O_5 to 680°C. Structure very similar to $Li_{0.30}V_2O_5$-β (1), but with Li atoms in the tetrahedral M_3 site [coordinates not given]. R = 0.066 for Weissenberg measured intensities.

Interatomic distances (Å)

V(1) - O(2)	2.28	V(2) - O(1)	1.80	V(3) - O(5)	1.80
V(1) - O(3)	1.99	V(2) - O(2)	2.32	V(3) - O(7)	1.98
V(1) - O(4)	1.53	V(2) - O(5)	2.12	V(3) - O(8)	1.50
V(1) - O(5)	1.97	V(2) - O(6)	1.62	V(3) - O(7)'	1.91
V(1) - O(2)'	1.88	V(2) - O(3)	1.89	V(3) - O(6)	2.78

$Cu_{0.60}V_2O_5$

Monoclinic, a = 15.25, b = 3.624, c = 10.205 Å, β = 106°9'.

Space group Cm (C_s^3) (Wilson statistics)

Atomic positions

None given except:

$y(Cu_1)$ = ±0.108 (disordered)
$y(Cu_2)$ = ±0.08

With Cu atoms in M_3 positions (see above)

Interatomic distances (Å)

V(1) - O(2)	2.25	V(2) - O(1)	1.89	V(3) - O(5)	1.52
V(1) - O(3)	2.28	V(2) - O(2)	2.40	V(3) - O(7)	2.12
V(1) - O(4)	1.51	V(2) - O(5)	2.31	V(3) - O(8)	1.60
V(1) - O(5)	2.06	V(2) - O(6)	1.55	V(3) - O(9)	1.96
V(1) - O(15)	1.92	V(2) - O(14)	1.88	V(3) - O(6)	3.12
V(4) - O(12)	1.91	V(5) - O(15)	2.23	V(6) - O(1)	1.69
V(4) - O(9)	1.84	V(5) - O(14)	1.90	V(6) - O(15)	2.35
V(4) - O(10)	1.65	V(5) - O(11)	1.64	V(6) - O(12)	2.12
V(4) - O(7)	1.87	V(5) - O(12)	1.83	V(6) - O(13)	1.51
V(4) - O(13)	2.69	V(5) - O(2)	1.91	V(6) - O(3)	1.87

Discussion

The structure is the same as that of $Li_{0.48}V_2O_5$-β', but with a lower symmetry and hence twice the number of atoms in the asymmetric unit.

Details of analysis

Single crystal, 0.02 × 0.2 × 0.02 mm^3, prepared by heating to 800°C. Intensities measured from integrated Weissenberg photographs using filtered Cu radiation. Structure solved from Patterson function and refined by least squares to give R = 0.105 [no intensities given].

$Na_xV_{2-x}Mo_xO_5$

Fourier and least squares analysis of measurements from single crystals of compositions with x = 0.22, 0.30 and 0.40 show that up to x = 0.33, Na occupies site M_1 and Mo occupies V_2 (R = 0.057). For x > 0.33 Na occupies site M_1 (50%) and then M_2, whereas Mo occupies site V_2(50%) and then V_3.

Interatomic distances (Å) (for x = 0.40)

V(1) - O(2)	2.38	V(2) - O(1)	1.82	V(3) - O(5)	1.81
V(1) - O(3)	2.10	V(2) - O(2)	2.34	V(3) - O(7)	1.97
V(1) - O(4)	1.57	V(2) - O(5)	2.19	V(3) - O(8)	1.63
V(1) - O(5)	1.94	V(2) - O(6)	1.69	V(3) - O(7)'	1.91
V(1) - O(2)'	1.89	V(2) - O(3)	1.90	V(3) - O(6)	2.67

1. J. GALY, A. CASALOT and M. POUCHARD, 1965. *Bull. Soc. Chim. Fr.*, 6, 1056.

ALKALI TITANATE BRONZE

$K_3Ti_8O_{17}$ F.W. = 772.5

I. $K_3Ti_8O_{17}$, a new alkali titanate bronze. J.A. WATTS, 1970. *J. Solid State Chem.*, 1, 319-325.

Monoclinic, a = 15.68±1, b = 3.809±3, c = 12.06±1 Å, β = 95.0°, U = 717.5 Å^3, Z = 2, D_x = 3.57.

Space group C2/m (C_{2h}^3) from systematic absences and refinement

Atomic positions

	x	*y*	*z*
2 K(1) in 4(*i*)	0.4871±22	0	0.46102±6
4 K(2) in 4(*i*)	0.4429±9	0	0.1715 ±2
4 Ti(1)in 4(*i*)	0.1134±7	0	0.0736 ±8
4 Ti(2)in 4(*i*)	0.1484±7	0	0.3249 ±9
4 Ti(3)in 4(*i*)	0.1984±8	0	0.5769 ±9
4 Ti(4)in 4(*i*)	0.2457±7	0	0.8311 ±9
2 O(1) in 2(*a*)	0	0	0
4 O(2) in 4(*i*)	0.064 ±2	0	0.230 ±3
4 O(3) in 4(*i*)	0.103 ±3	0	0.461 ±4
4 O(4) in 4(*i*)	0.144 ±4	0	0.713 ±5
4 O(5) in 4(*i*)	0.172 ±3	0	0.931 ±4
4 O(6) in 4(*i*)	0.233 ±3	0	0.191 ±4
4 O(7) in 4(*i*)	0.279 ±2	0	0.433 ±3
4 O(8) in 4(*i*)	0.317 ±3	0	0.678 ±4
4 O(9) in 4(*i*)	0.361 ±2	0	0.913 ±3

Isotropic temperature factors are given.

Interatomic distances (Å) *and bond angles*

Environment of Ti(1)
(crystallographic symmetry C_s)

Ti(1) - O(1)	1.92±1
Ti(1) - O(9)	1.95±1
Ti(1) - O(9)'	1.95±1
Ti(1) - O(5)	2.02±5
Ti(1) - O(2)	2.10±4
Ti(1) - O(6)	2.25±5

Angles in degrees subtended at Ti(1) by

	O(9)	O(9)'	O(5)	O(2)	O(6)
O(1)	102±1	102±1	95±1	91±1	169±2°
O(9)		155±1	88±1	91±1	78±1
O(9)'			88±1	91±1	78±1
O(5)				175±2	97±2
O(2)					78±2

Environment of Ti(2)
(crystallogrpahic symmetry C_s)

Ti(2) - O(2)	1.67±3
Ti(2) - O(3)	1.85±5
Ti(2) - O(8)	1.98±1
Ti(2) - O(8)'	1.98±1
Ti(2) - O(6)	2.18±5
Ti(2) - O(7)	2.33±3

Angles in degrees subtended at Ti(2) by

	O(3)	O(8)	O(8)'	O(6)	O(7)
O(2)	105±2	101±1	101±1	89±2	171±2°
O(3)		98±2	98±2	165±2	84±2
O(8)			148±2	79±2	77±1
O(8)'				79±2	77±1
O(6)					82±2

Environment of Ti(3)
(crystallographic symmetry C_s)

Ti(3) - O(4)	1.92±6
Ti(3) - O(7)	1.94±1
Ti(3) - O(7)'	1.94±1
Ti(3) - O(3)	1.96±5
Ti(3) - O(7)	2.24±4

Angles in degrees subtended at Ti(3) by

	O(7)	O(7)'	O(3)	O(8)	O(7)
O(4)	99±1	99±1	104±2	87±2	172±2°
O(7)		157±2	95±1	83±1	80±1
O(7)'			95±1	83±1	80±1
O(3)				169±2	84±2
O(8)					85±2

Environment of Ti(4)
(crystallographic symmetry C_s)

Ti(4) - O(5)	1.74±5
Ti(4) - O(6)	1.96±1
Ti(4) - O(6)'	1.96±1
Ti(4) - O(9)	1.98±3
Ti(4) - O(4)	2.04±6
Ti(4) - O(8)	2.24±5

Angles in degrees subtended at Ti(4) by

	O(6)	O(6)'	O(9)	O(4)	O(8)
O(5)	103±1	103±1	107±2	88±2	168±2°
O(6)		154±2	85±1	92±1	78±2
O(6)'			85±1	92±1	78±2
O(9)				166±2	85±2
O(4)					81±2

Discussion

The compound possesses the limiting formula $K_3Ti_8O_{17}$ with a formal charge distribution $K_3Ti^{3+}Ti_7^{4+}O_{17}$. The basic framework of the structure is built up from four octahedra sharing edges at one level (b = 0,½), which combined with similar units above and below form zigzag ribbons of octahedra (1).

Details of analysis

Crystals prepared from electrolysis of melt of K_2O, TiO_2 and Nb_2O_5. Intensity data *h0l* to *h3l* were recorded on Weissenberg photographs using CuK_α radiation. No absorption corrections made.

1. A.D. Wadsley and W.G. Mumme, 1968. *Acta Cryst.*, B24, 392.

CAESIUM MOLYBDENUM BRONZE

I. The crystal structure of the molybdenum bronze Cs_xMoO_3 (x≈0.25). W.G. MUMME and J.A. WATTS, 1970. *J. Solid State Chem.*, 2, 16-23.

Cs_xMoO_3 F.W. = 199.31

Monoclinic, a = 6.425±5, b = 7.543±4, c = 8.169±5 Å, β = 96.50±5, U = 393.46 Å^3, Z = 6, D_x = 5.06 (for x = 0.25)

Space group $P2_1/m$ (C^2_{2h})

Atomic positions

		x	y	z	B(Å^2)
2 Cs	in 2(*e*)	0.1785±8	0.75	0.6113±6	1.25±8
2 Mo(1)	in 2(*e*)	0.3482±8	0.25	0.7753±7	1.74±9
4 Mo(2)	in 4(*f*)	0.2715±5	0.0047±5	0.1060±4	1.71±6
2 O(1)	in 2(*a*)	0	0	0	4.54±95
2 O(2)	in 2(*e*)	0.2991±65	0.25	0.0405±55	1.22±76
2 O(3)	in 2(*e*)	0.3043±82	0.75	0.0927±68	2.76±103
2 O(4)	in 2(*e*)	0.4635±77	0.25	0.5958±61	1.79±91
2 O(5)	in 2(*e*)	0.0942±93	0.25	0.7214±76	3.22±113
4 O(6)	in 4(*f*)	0.3970±53	0.0085±52	0.8542±47	2.02±65
4 O(7)	in 4(*f*)	0.2218±54	0.0397±68	0.3032±46	2.57±73

Interatomic distances (Å) and bond angles

σ(distances) = 0.04 Å; σ(angles) = 2°

Cs - O(4)	2.76[3.01]
Cs - O(6)	3.19[3.01] (×2)
Cs - O(5)	3.34[3.06]
Cs - O(7)	3.07[3.17] (×2)
Cs - O(7)	3.34[3.37] (×2)

Mo(1) - O(5)	1.73 [1.64]
Mo(1) - O(4)	1.59 [1.72]
Mo(1) - O(6)	1.99 [1.95] (×2)
Mo(1) - O(2)	2.16 [2.22]
Mo(1) - O(3)	2.55 [2.37]

Angles in degrees subtended at Mo(1) by

	O(4)	O(6)'	O(2)	O(3)
O(5)	106	101	91	169
O(4)		102	163	85
O(6)		139	73	75
O(2)				78

Mo(2) - O(7)	1.64 [1.70]
Mo(2) - O(1)	2.01 [1.86]
Mo(2) - O(3)	1.96 [1.94]
Mo(2) - O(2)	1.96 [1.94]
Mo(2) - O(6)	2.15 [2.12]
Mo(2) - O(6)'	2.19 [2.29]

Angles in degrees subtended at Mo(2) by

	O(1)	O(3)	O(2)	O(6)	O(6)'
O(7)	100	104	98	99	166
O(1)		93	90	161	89
O(3)			156	81	85
O(2)				88	72
O(6)					72

[The values of the angles are those calculated by the reporter. The angles are not given in I]

Discussion

The MoO_6 octahedra share edges with 3 or 4 other octahedra to form chains running along *b*. Corner sharing between chains results in sheets in the *ab* plane between which lie the Cs atoms.

Details of analysis

10 mg of copper coloured crystals isolated from an electrolysed melt of Cs_2MoO_4 and MoO_3. Lattice parameters from focussing powder camera calibrated with KCl [assumed parameters not given]. Intensities measured visually from integrated multiple-film Weissenberg photographs of layers with h=0 and k=0 to 6 taken with CuK_α radiation. Structure solved from Patterson function was refined by electron-density and least-squares methods to give R = 0.12 for 475 reflexions listed.

INDIUM OXIDE HYDROXIDE

I. Neutron and X-ray crystallographic studies on indium oxide hydroxide. M.S. LEHMANN, F.K. LARSEN, F.R. POULSEN, A.N. CHRISTENSEN and S.E. RASMUSSEN, 1970. *Acta Chem. Scand.*, 24, 1662-1670.

InOOH F.W. = 147.82

Orthorhombic, a = 3.27±1, b = 4.56±1, c = 5.26±1 Å, U = 78.4 Å^3, D_m = 7.[0]±7, Z = 2, D_x = 6.25.

Space group $Pmn2_1$ (C^7_{2v}) [converted from $P2_1nm$] systematic absences and structure.

Atomic positions All atoms in 2(*a*)

	X-ray			neutron		
	x	*y*	*z*	*x*	*y*	*z*
In	0	0.26438±7	0	0	0.2673±7	0
O(1)	0	0.478 ±1	0.362±1	0	0.4770±4	0.3608±5
O(2)	0	0.018 ±1	0.636±1	0	0.0186±5	0.6343±5
H				0	0.215 ±2	0.520 ±1

Anisotropic temperature factors are given.

Interatomic distances (Å) *and bond angles*

	X-ray	neutron
In - O(1)	2.140±6	2.128±7
In - O(1)'	2.140±4	2.138±3 (×2)
In - O(2)	2.201±4	2.208±2 (×2)
In - O(2)'	2.220±6	2.236±3
O(2) - H...O(1)	2.543±8	2.539±3
O(2) - H		1.079±8
O(1)...H		1.460±9
O(2) - H...O(1)		179.1±8°

Angles in degrees (neutron data) subtended at In by

	O(1)'*	O(2)*	O(2)'
O(1)	93.5	88.8	176.2°
O(1)'	99.8	82.3	88.9
O(1)'*		176.8	88.9
O(2)		95.6	88.7

σ(angles) = 0.3°

Discussion

The structure is similar to that proposed by 1, but in a different space group. InO_6 octahedra share edges to form chains along *a*. The chains are linked into a tight network through shared corners and hydrogen bonds.

Details of anlysis

Prepared hydrothermally. X-ray intensities measured on equi-inclination diffractometer with monochromated Mo$K\beta$ radiation using ω scan and corrected for Lorentz and polarization, but not absorption effects. Neutron intensities measured on automatic diffractometer (λ = 1.022 A) using θ-2θ scan and corrected for Lorentz and extinction, but not absorption effects (μ = 0.379 mm^{-1}, crystal size ∿ 0.4-0.7 mm). Structure determined from Patterson and electron-density functions and refined by least squares to give R_1 = 0.051 for 748 X-ray reflexions listed and R_2 = 0.053 for 144 neutron reflexions listed. The two refinements agree at the 5% significance level.

1. A.N. CHRISTENSEN, R. GRØNBACK and S.E. RASMUSSEN, 1965. *Acta Chem. Scand.*, 18, 1261.

DIRHENIUM HEPTOXIDE DIHYDRATE

I. Die Struktur von Dirheniumdihydratoheptoxid $Re_2O_7(OH_2)_2$ - Ein neuer Typ von Wasserbindung in einem Aquoxid -. H. BEYER, O. GLEMSER, B. KREBS and G. WAGNER, 1970. *Z. anorg. Chem.*, 376, 87-100.

$Re_2O_7(OH_2)_2$ F.W. = 520.4

Monoclinic, *a* = 8.82±2, *b* = 8.89±2, *c* = 5.03±1 Å, β = 112.0±2, *U* = 366 Å^3, D_m = 4.87, *Z* = 2, D_x = 4.74.

Space group $P2_1/m$ (C_{2h}^2) systematic absences and statistical tests

Atomic positions

	x	*y*	*z*	*B*(Å^2)
2 Re(1) in 2(*e*)	0.1030±1	0.25	0.9644±2	0.81±1
2 Re(2) in 2(*e*)	0.5546±1	0.25	0.0146±2	1.19±1
2 O(1) in 2(*e*)	0.346 ±2	0.25	0.989 ±4	2.3 ±3
2 O(2) in 2(*e*)	0.697 ±3	0.25	0.373 ±6	4.8 ±5
2 O(3) in 2(*e*)	0.942 ±2	0.25	0.091 ±4	2.8 ±3
4 O(4) in 4(*f*)	0.066 ±2	0.097±2	0.734 ±3	2.1 ±2
4 O(5) in 4(*f*)	0.582 ±2	0.092±2	0.836 ±3	2.2 ±2
4 O(6) in 4(*f*)	0.222 ±1	0.095±2	0.328 ±3	3.3 ±3

Interatomic distances (Å) *and bond angles*

Re(1) - O(4)	1.74 (×2)
Re(1) - O(3)	1.76
Re(1) - O(1)	2.10
Re(1) - O(6)	2.22 (×2)

Angles in degrees subtended at Re(1) by

	O(4)'	O(3)	O(1)	O(6)	O(6)'
O(4)	103.1	105.4	88.4	161.9	88.6
O(3)			157.3	84.2	84.2
O(1)				78.1	78.1
O(6)					76.9

Re(2) - O(5)	1.73 (×2)
Re(2) - O(2)	1.77
Re(2) - O(1)	1.80

Angles in degrees subtended at Re(2) by

	O(5)'	O(2)	O(1)
O(5)	108.3	110.1	107.5
O(2)			113.0

Re(1) - O(1) - Re(2) 179±1°

Possible hydrogen bonds

O(6)-H ... O(5)	2.74
O(6)-H ... O(4)	2.86
O(5)-O(6)-O(4)	138.3°
Re(1)-O(6)-O(5)	106.7°
Re(1)-O(6)-O(4)	112.3°

σ(distances) = 0.02 Å, σ(angles) = 0.8°

Discussion

The crystal contains discrete $O_3Re\text{-}O\text{-}ReO_3(OH_2)_2$ molecules linked by hydrogen bonds.

Details of analysis

Lattice parameters from rotation, Weissenberg and precession photographs with MoK_α radiation. Intensities measured photometrically from integrated Weissenberg photographs of layers with l=0-4 and integrated precession photographs of layers with k=0-2 taken with MoK_α radiation and corrected "in the usual way", as well as for absorption (μ = 35.0 mm^{-1}, crystal size = 0.09×0.10×0.10 mm^3). Structure found from Patterson and electron-density syntheses and refined by least squares to give R_2 = 0.057 for 1150 observed reflexions listed. Further refinement with anisotropic temperature factors gave little improvement.

MONOCLINIC TELLURIC ACID

I. The crystal structure of telluric acid, $Te(OH)_6$(*mon*). O. LINDQVIST, 1970. *Acta Chem. Scand.*, 24, 3178-3188.

$Te(OH)_6$ F.W. = 229.63

Monoclinic, a = 6.495±1, b = 9.320±1, c = 11.393±1 Å, β = 133.88 1°, U = 497.1 Å^3, D_m = 3.06 1, Z = 4, D_x = 3.05.

Space group $P2_1/n$ (C^5_{2h})

Atomic positions Atoms at ±(x,y,z;½-x,½+y,½-z)

	x	y	z	B(Å^2)
2 Te(1) in 2(*a*)	0	0	0	*
2 Te(2) in 2(*d*)+	0	0	0.5	*
4 O(1) in 4(*e*)	0.7291±1	0.1544±5	0.8928±6	*
4 O(2) in 6(*e*)	0.1255±11	0.0482±7	0.2037±6	*
4 O(3) in 4(*e*)	0.7739±10	0.3744±5	0.5373±7	*
4 O(4) in 4(*e*)	0.6895±9	0.0957±5	0.4544±6	*
4 O(5) in 4(*e*)	0.3241±10	0.4509±7	0.7845±6	*
4 O(6) in 4(*e*)	0.7168±10	0.3243±5	0.0774±6	*
4 H(1) in 4(*e*)	0.199 ±20	0.282 ±12	0.457 ±14	4.5
4 H(2) in 4(*e*)	0.268 ±23	0.118 ±12	0.275 ±13	4.5
4 H(3) in 4(*e*)	0.499 ±23	0.084 ±12	0.130 ±14	4.5
4 H(5) in 4(*e*)	0.994 ±20	0.080 ±14	0.264 ±11	4.5
4 H(6) in 4(*e*)	0.883 ±24	0.346 ±12	0.192 ±14	4.5

*Anisotropic temperature factors given.
+Incorrectly given as 2(*c*).

Interatomic distances (Å) *and bond angles*

Te(1) - O(2)	1.905±7	(×2)
Te(1) - O(3)	1.915±7	(×2)
Te(1) - O(1)	1.918±5	(×2)

Angles in degrees subtended at Te(1) by

	O(3)	O(1)
O(2)	87.8±3	91.6±3°
O(2)'	92.2±3	88.4±3
O(3)		90.9±2
O(3)'		89.1±2

Te(2) - O(5)	1.912±6	(×2)
Te(2) - O(4)	1.919±6	(×2)
Te(2) - O(6)	1.929±5	(×2)

Angles in degrees subtended at Te(2) by

	O(4)	O(6)
O(5)	92.3±3	89.9±3°
O(5)'	87.7±3	91.1±3
O(4)		88.5±2
O(4)'		91.5±2

Hydrogen bonds O(*A*) - H(*A*)...O(*B*)

A	*B*	O(*A*)-O(*B*)	O(*A*)-H(*A*)	H(*A*)...O(*B*)	O(*A*)-H(*A*)...O(*B*)	Te-O(*A*)-H(*A*)
1	6	2.678±10	1.06±17	1.63±16	169±10°	120±6°
2	4	2.696±6	0.95±9	1.99±10	130±9°	123±11°
3	5	2.694±6	1.12±10	1.58±10	168±8°	114±8°
4	3	2.687±7				
5	2	2.679±12	1.31±16	1.45±16	152±10°	117±3°
6	1	2.729±5	0.98±8	1.78±10	163±13°	102±7°

Discussion

Independent $Te(OH)_6$ molecules hydrogen bond together to form the crystal.

Details of analysis

Lattice parameters from Guinier powder photograph (CuK_α radiation) calibrated with $Pb(NO_3)_2$ (a = 7.8566 Å). Intensities estimated visually from multiple-film, equi-inclination Weissenberg photographs of layers with l = 0 to 9 taken with unfiltered CuK_α radiation corrected for Lorentz, polarization, extinction and absorption (μ = 61 mm^{-1}, crystal dimensions about 0.1 mm [described in detail in I]). Structure determined from electron-density function phased on Te positions and refined by least squares to give R = 0.066 for 589 observed reflexions listed. The position of H(4) could not be satisfactorily found.

LANTHANUM BORATES

High-$LaBO_3$ F.W. = 198.7

I. Hochtemperatur-Lanthanborate. R. BOHLOFF, H.U. BAMBAUER and W. HOFFMANN, 1970. *Naturwissenschaften*, 57, 129.

Monoclinic, a = 6.348, b = 5.084, c = 4.186 Å, β = 107.89°, U = 128.6 Å^3, D_m = 5.06, Z = 2, D_x = 5.12.

Space group $P2_1/m$ (C^2_{2h})

Atomic positions

	x	y	z
2 La in 2(e)	0.776	1/4	0.620
2 B in 2(e)	0.257	1/4	0.002
2 O(1) in 2(e)	0.380	1/4	0.319
4 O(2) in 4(f)	0.174	0.020	0.839

Interatomic distances (Å)

B-O(1)	1.32	O(1)-B-O(2)	121°
B-O(2)	1.38 (×2)	O(2)-B-O(2)	117°
La-O(1)	2.44		
La-O(2)	2.46 (×2)		
La-O(2)'	2.58 (×2)		
La-O(2)"	2.68 (×2)		
La-O(1)'	2.77 (×2)		

Errors uncertain but greater than 0.01 Å.

Discussion

The structure is composed of triangular BO_3 ions linked by nine-coordinated La atoms.

Details of analysis

Integrating Weissenberg photographs about b used to measure 663 non-zero reflexions. No correction made for a noticeably large absorption effect. Three-dimensional Patterson and Fourier synthesis gave with isotropic temperature factors [not listed], R = 0.118. No intensities given.

$La(B_3O_6)$ F.W. = 268.3

II. Die Kristallstruktur des $La(B_3O_6)$, ein neuer Ketten-Borat-Typ. J. ST. YSKER and W. HOFFMANN, 1970. *Naturwissenschaften*, 57, 129-130.

Monoclinic, a = 7.956, b = 8.161, c = 6.499 Å, β = 93.63°, U = 421.1 Å^3, Z = 4, D_x = 4.2.

Space group I2/a (C_{2h}^6)

Atomic positions $(0,0,0;\frac{1}{2},\frac{1}{2},\frac{1}{2}) \pm (x,y,z;\frac{1}{2}-x,y,\bar{z})$

		x	y	z
4 La	4(e)	1/4	0.051	0
4 B(1)	4(e)*	1/4	0.481	0
8 B(2)	8(f)	0.522	0.184	0.609
8 O(1)	8(f)	0.145	0.135	0.369
8 O(2)	8(f)	0.547	0.166	0.044
8 O(3)	8(f)	0.647	0.064	0.640

*coordinates transformed from those of atom at $(\frac{1}{4},\frac{1}{4}-y,\frac{1}{2})$

Overall isotropic temperature factor = 0.30 Å^2.

Interatomic distances (Å)

B(1)-O(3)	1.43 (×2)	O(3)-B(1)-O(3)'	124°
B(1)-O(1)	1.49 (×2)	O(3)-B(1)-O(1)	110° (×2)
		O(3)-B(1)-O(1)'	105° (×2)
		O(1)-B(1)-O(1)'	101°
B(2)-O(2)	1.31	O(2)-B(2)-O(3)	125°
B(2)-O(3)	1.40	O(2)-B(2)-O(1)	117°
B(2)-O(1)	1.40	O(3)-B(2)-O(1)	117°
La-O(2)	2.46 (×2)		
La-O(2)'	2.52 (×2)		
La-O(3)	2.62 (×2)		
La-O(1)	2.67 (×2)		
La-O(1)'	2.77 (×2)		

Errors uncertain, but at least 0.01 Å.

Discussion

The structure consists of an infinite chain B_3O_6 polyion running along *a*. Adjacent BO_4 tetrahedra in the polyion are bridged by two triangular BO_3 ions.

Details of analysis

Patterson projections of (001) and (100). *R* = 0.051. No further details given.

CALCIUM BORITE

I. [Crystal structure of calciborite $CaB_2O_4 = Ca_2(BO_3BO)_2$]. D.P. ŠAŠKIN, M.A. SIMONOV and N.V. BELOV, 1970. *Dokl. Akad. Nauk SSSR*, 195, 345-348. [*Soviet Physics-Doklady*, 15, 1003-1005].

CaB_2O_4 F.W. = 125.70

Orthorhombic, *a* = 8.38±1, *b* = 13.82±1, *c* = 5.006±2 Å, *U* = 580 Å^3, D_m = 2.88, *Z* = 8, D_x = 2.87.

Space group Pccn (D_{2h}^{10})

Atomic positions

All atoms in 8(*e*)

	x	*y*	*z*
Ca	0.386	0.143	0.123
B(1)	0.537	0.139	0.624
B(2)	0.742	0.052	0.365
O(1)	0.391	0.185	0.633
O(2)	0.742	-0.009	0.114
O(3)	0.596	0.112	0.365
O(4)	0.885	0.112	0.378

Interatomic distances (Å)

Ca - O(3)	2.18	B(1) - O(1)	1.38
Ca - O(4)	2.32	B(1) - O(3)	1.44 [given at 1.38 in I]
Ca - O(1)	2.38	B(1) - O(4)	1.48
Ca - O(1)'	2.40	B(2) - O(4)	1.46
Ca - O(2)	2.45	B(2) - O(3)	1.48
Ca - O(1)"	2.52	B(2) - O(2)	1.51
Ca - O(2)'	2.58	B(2) - O(2)'	1.51
Ca - O(1)"'	2.62		

Errors uncertain, but around 0.03 Å.

Discussion

B_2O_4 chains consisting of equal numbers of BO_4 tetrahedra and BO_3 triangles run along *c*. Each tetrahedron is linked to the next directly through a stated corner and indirectly through a BO_3 group.

Details of analysis

Intensities estimated visually from Weissenberg photographs of layers with h=0, k=0,1 and l=0 to 4 taken with MoK_α radiation. The structure determined from the Patterson function was refined by least squares to give R = 0.12 for 300 observed reflexions [not listed].

TRIZINC DIORTHOBORATE

I. The space group and crystal structure of trizinc diorthoborate. W.H. BAUR and E. TILLMANNS, 1970. *Z. Kristallogr.*, 131, 213-221.

$Zn_3(BO_3)_2$ F.W. = 307.7

Monoclinic, a = 23.406, b = 5.048, c = 8.381 Å, β = 97.53 (from 1), U = 939.8 Å^3, Z = 8, D_x = 4.36.

Space group $I2/c$ (C^6_{2h})

Atomic positions

All atoms in 8(*f*)±(0,0,0;½,½,½) ($x,y,z;\bar{x},y,½-z$)

	x	*y*	*z*	*B*(Å^2)
Zn(1)	0.0495±1	0.8291±4	0.3746±2	0.61±7
Zn(2)	0.1275±1	0.6841±4	0.7488±2	0.53±7
Zn(3)	0.2095±1	0.6921±4	0.4992±3	0.60±7
B(1)	0.0660±7	0.3213±31	0.5320±18	-0.3 ±2
B(2)	0.1870±7	0.1835±33	0.6683±19	-0.2 ±2
O(1)	0.0363±5	0.2018±24	0.6362±13	0.2 ±2
O(2)	0.0820±4	0.5885±23	0.5408±13	0.1 ±2
O(3)	0.0798±5	0.1874±24	0.4001±15	0.7 ±2
O(4)	0.1843±4	0.9117±25	0.6632±12	0.0 ±2

O(5)	0.2149±5	0.3132±32	0.5537±13	0.2±2
O(6)	0.1633±5	0.6783±21	0.2847±13	0.4±2

Interatomic distances (Å) and bond angles

Zn(1) - O(2)	1.928±11
Zn(1) - O(3)	1.944±12
Zn(1) - O(1)	1.987±11
Zn(1) - O(1)	2.004±12

Angles in degrees subtended at Zn(1) by

	O(3)	O(1)	O(1)
O(2)	113.8±5	132.2±5	106.4±4°
O(3)		100.9±5	115.2±5
O(1)			85.7±5

Zn(2) - O(3)	1.910±13
Zn(2) - O(4)	1.963±11
Zn(2) - O(2)	1.980±10
Zn(2) - O(6)	2.018±11

Angles in degrees subtended at Zn(2) by

	O(4)	O(2)	O(6)
O(3)	121.8±5	111.9±5	118.4±5°
O(4)		97.6±4	107.3±5
O(2)			94.3±4

Zn(3) - O(4)	1.918±11
Zn(3) - O(5)	1.967±11
Zn(3) - O(6)	1.974±11
Zn(3) - O(5)	1.977±12

Angles in degrees subtended at Zn(3) by

	O(5)	O(6)	O(5)
O(4)	114.3±5	119.3±4	110.5±4°
O(5)		100.8±4	108.7±5
O(6)			101.9±5

B(1) - O(1)	1.330±20
B(1) - O(3)	1.370±20
B(1) - O(2)	1.399±19

Angles in degrees subtended at B(1) by

	O(3)	O(2)
O(1)	120.3±13	124.2±14°
O(3)		115.4±13

B(2) - O(4)	1.374±21
B(2) - O(6)	1.374±21
B(2) - O(5)	1.394±20

Angles in degrees subtended at B(2) by

	O(6)	O(5)
O(4)	120.5±14	118.1±13°
O(6)		121.3±14

Discussion

Refinement in *I2/c* rather than *Ic* leads to chemically equivalent bonds being more nearly equal.

Details of analysis

Intensities of 1 re-refined in space group *I2/c* to give R = 0.133 for 800 reflexions listed. Refinement (with twice as many parameters) in the space group *Ic* proposed by 1 lead to R = 0.12 but the refinement did not converge satisfactorily and the results were appreciably different from those reported by 1.

1. S. GARCIA-BLANCO and T. FAYOS, 1968. *Z. Kristallogr.*, 127,145.

POTASSIUM METABORATE

I. Bond lengths and thermal parameters of potassium metaborate, $K_3B_3O_6$. W. SCHNEIDER and G.B. CARPENTER, 1970. *Acta Cryst.*, B26, 1189-1191.

$K_3B_3O_6$ F.W. = 245.74

Rhombohedral (hexagonal crystal data, a = 12.76±1, c = 7.34±1 Å, [U = 1034.9 Å^3], Z = 6).

Space group $R\bar{3}c$ (D_{3d}^6)

Atomic parameters

All atoms in 18(e)

	x
K	0.5613±1
B	0.8889±6
O(1)	0.7843±3
O(2)	0.1084±4

Anisotropic thermal parameters are listed

Interatomic distances (Å) *and angles*

B-O(1)	1.331±10	O(2)-B-O(2)	117.3±8°
-O(2)(2×)	1.398±5	O(1)-B-O(2)	121.3±4
		B-O(2)-B	122.6±8

K-O(1)	2.849±6
K-O(1)(2×)	2.801±2
-O(1)(2×)	2.835±3
-O(2)(2×)	2.775±5

Discussion

Structure is a more accurate refinement of $K_3B_3O_6$, previously reported (1) and isostructural with $Na_3B_3O_6$ (2).

Details of analysis

277 reflexions from precession photographs. No extinction or absorption corrections. Radiation not given. Least-squares refinement gave an *R* value of 0.061.

1. *Strukturbericht*, 5, 74.
2. M. MAREZIO, H.A. PLETTINGER and W.H. ZACHARIASEN, 1963. *Acta Cryst.*, 16, 594.

FABIANITE

I. Crystal structure of fabianite, $CaB_3O_5(OH)$, and comparison with the structure of its synthetic dimorph. J.A. KONNERT, J.R. CLARK and C.L. CHRIST, 1970. *Z. Kristallogr.*, 132, 241-254.

$CaB_3O_5(OH)$ F.W. = 169.51

Monoclinic, a = 6.593±1, b = 10.488±2, c = 6.365±1 Å, β = 113.3±2°, U = 401.1 Å^3, D_m = 2.72 1, Z = 4, D_x = 2.788.

Space group $P2_1/a$ (C_{2h}^5)

Atomic positions

All atoms in 4(*e*) $\pm(x,y,z;\frac{1}{2}-x,\frac{1}{2}+y,z)$

	x	*y*	*z*
Ca	0.3041±2	0.4371±2	0.6372±2
OH(1)	0.4243±6	0.1284±4	0.5278±6
O(2)	0.3930±6	0.2387±4	0.8489±6
O(3)	0.2161±6	0.0325±3	0.7225±6
O(4)	0.6207±6	0.0502±3	0.8902±6
O(5)	0.6739±6	0.1892±3	1.2175±6
O(6)	0.9845±6	0.1131±4	1.1527±6
B(1)	0.4139±9	0.1124±5	0.7556±9
B(2)	0.7562±10	0.0765±5	1.1259±9
B(3)	0.5174 9	0.2690±5	1.0739±9

Anisotropic temperature factors given in I.

Interatomic distances (Å) and bond angles

Ca - O(1)	2.417±4	B(2) - O(3)	1.460±6
Ca - O(4)'	2.478±4	B(2) - O(4)	1.436±6
Ca - O(2)	2.421±4	B(2) - O(5)	1.511±7
Ca - O(3)	2.570±4	B(2) - O(6)	1.496±7
Ca - O(3)'	2.453±4	B(3) - O(2)	1.376±6
Ca - O(4)	2.370±4	B(3) - O(5)	1.360±1
Ca - O(5)	2.796±4	B(3) - O(6)	1.383±7
Ca - O(6)	2.379±4	O(1)-H...O(5)	2.76 ±1
B(1) - O(1)	1.488±6		
B(1) - O(2)	1.481±7		
B(1) - O(3)	1.493±7		
B(1) - O(4)	1.446±6		

Angles in degrees subtended at B(1) by

	O(2)	O(3)	O(4)
O(1)	109.7±4	107.6±4	103.4±4°
O(2)		110.1±4	112.4±4
O(3)			113.3±4

Angles in degrees subtended at B(2) by

	O(4)	O(5)	O(6)
O(3)	113.7±4	109.1±4	105.7±4°
O(4)		112.6±4	110.4±4
O(5)			104.7±4

O(2) - B(3) - O(5)	122.5±5
O(2) - B(3) - O(6)	118.0±4
O(5) - B(3) - O(6)	119.5±4
B(1) - O(4) - B(2)	125.6±4
B(2) - O(5) - B(3)	121.0±4
B(1) - O(2) - B(3)	120.8±4
B(1) - O(3) - B(2)'	118.2±4
B(2) - O(6) - B(3)'	118.9±4

Discussion

The borate ions are arranged in rings of two tetrahedrally coordinated B and one trigonally coordinated B. Each ring is linked through two ^{IV}B-O-^{IV}B bridges to a second ring related by a centre of symmetry and the pairs are further linked into sheets through ^{IV}B-O-^{III}B linkages. The sheets are held togeher by Ca ions and hydrogen bonds.

Details of analysis

Fabianite from Diepholz, Germany (1). Crystal size, 0.18×0.40×0.40 mm^3. Cell parameters from 2. Intensities measured with filtered Mo radiation on automatic diffractometer (θ-2θ scan). Corrections for Lorentz, polarization, but not absorption effects (μ = 1.43 mm^{-1}). Structure solved from Patterson and electron-density functions and refined by least squares to give R = 0.075 for 1309 reflexions listed.

1. H. GARTNER, K.-L. ROSE and R. KUHN, 1962.
2. R.C. ERD, G.D. EBERLEIN and C.L. CHRIST, 1969. *Canad. Miner.*, 10, 108.

NICKEL ALUMINUM BORATE

I. Structure du boroaluminate $B_2O_3.Al_2O_3.4NiO$. A.M. SCHWAB and E.F. BERTAUT, 1970. *Bull. Soc. Franç, Minér. Crist.*, 93, 255-257.

$Al_2Ni_4B_2O_{10}$ F.W. = 470.4

Orthorhombic, a = 9.2, b = 12.2, c = 2.99 Å, U = 335.60 Å^3, Z = 2, D_x = 4.65.

Space group Pbam (D_{2h}^9)

Atomic positions

	x	y	z
4 O(1) in 4(g)	0.147	-0.047*	0
4 O(2) in 4(h)	0.123	0.140	0.5
4 O(3) in 4(g)	0.113	0.362	0
4 O(4) in 4(h)	0.389	0.083	0.5
4 O(5) in 4(g)	0.353	0.271	0
2 M(1) in 2(b)	0	0	0.5
2 M(2) in 2(c)	0.5	0	0
4 M(3) in 4(h)	0	0.283	0.5
4 M(4) in 4(g)	0.244	0.115	0
4 B in 4(g)	0.364**	0.363	0

*[given as +0.047 in I], **[Manifestly wrong. Probably meant to be 0.264]

Discussion

Al and Ni atoms are randomly distributed in the ratio 1:2 on the *M* site. The *M* atoms each have an octahedral environment (*M*-O ∿ 2.0 Å). B is trigonally co-ordinated.

Details of analysis

Lattice parameters measured with CoK_α radiation. R = 0.17 for 32 hk0 reflexions listed. Refinement promised.

BORACITES

I. Röntgen- und Neutronenbeugungsuntersuchungen an Ni-J-Boracit. W.-J. BECKER and G. WILL, 1970. *Z. Kristallogr.*, 131, 139-146.

II. Les systèmes B_2O_3-MO-MS boracites M-S (M=Mg, Mn, Fe, Cd) et sodalites M-S (M = Co, Zn). C. FOUASSIER, A. LEVASSEUR, J.-C. JOUBERT, J. MULLER and P. HAGENMULLER, 1970. *Z. anorg. Chem.*, 375, 202-208.

Cubic (Boracite type, 1)

	a	
$Ni_3B_7O_{13}I$	12.0368±8	(I)
$Mg_3B_7O_{12.65}S_{0.85}$	12.097±1	(II)
$Fe_3B_7O_{12.65}S_{0.85}$	12.146±3	(II)
$Mn_3B_7O_{12.65}S_{0.85}$	12.269±3	(II)
$Cd_3B_7O_{12.65}S_{0.85}$	12.484±4	(II)

Space group $F\bar{4}3c$ (T_d^5)

Atomic positions

			Ni (I)	Mg (II)
Ni, Mg	in 24(*c*)			
I,S,	in 8(*b*)			
O(1)	in 8(*a*)			
O(2)	in 96(*h*)	x =	0.0047±12	0.021
		y =	0.0934±13	0.098
		z =	0.1813±9	0.180
B(1)	in 24(*d*)			
B(2)	in 32(*e*)*	x =	0.0784±30	
			(0.085±3)+	
Temperature factor ($Å^2$)			0.18±4	1.0-2.0

*[given in I in error as 32 (*c*).]
+determined from neutron diffraction.

Interatomic distances in $Ni_3B_7O_{13}I$ (Å)

σ(O-O) = 0.02 σ(B-O) = 0.05

Sixfold coordination around Ni (crystallographic site symmetry S_4)

Ni-O(2)	2.06 (×4)
Ni-I	3.01 (×2)
I-O(2)	3.60
I-O(2)	3.69
O(2)-O(2)'	2.91(4) (×2)

Fourfold coordination around B(2) (crystallographic site symmetry C_3)

B(2)-O(1)	1.55
B(2)-O(2)	1.35 (×2)
O(1)-O(2)	2.45(5) (×12)

Fourfold coordination around B(1) (crystallographic site symmetry S_4)

B(1)-O(2)	1.39(6) (×4)
O(2)-O(2)	2.25
O(2)-O(2)	2.29(5) (×2)

Discussion

Neutron diffraction did not detect any magnetic ordering in $Ni_3B_7O_{13}I$ at 77 and 300°K (I).

Details of analysis

I. Lattice parameters and intensities from powder photographs taken with CuK_α (λ = 1.542 Å) and measured on microdensitometer. Structure of boracite (1) assumed and parameters refined by least squares to give $R(F^2)$ = 0.074 for 42 reflexions listed. Significant corrections made for anomalous dispersion. Neutron powder diffraction at 77 and 300°K gave similar results.

II. X-ray [powder] spectrum (listed) agrees with structure factors calculated from parameters of 1. [No refinement of these parameters was attempted].

1. *Structure Reports*, 15, 282.

CALCIUM DIALUMINATE

I. [Crystal structure of calcium dialuminate]. V.I. PONOMAREV, D.M. KHEIKER and N.V. BELOV, 1970. *Kristallografija*, 15, 1140-1143 [*Soviet Physics - Crystallography*, 15, 995-998].

$CaAl_4O_7$ F.W. = 260.00

Monoclinic, a = 12.866±6, b = 8.879±5, c = 5.440±4 Å, β = 106.75°, U = 595.18 Å^3, Z = 4, D_x = 2.90.

Space group C2/c (C^6_{2h})

Atomic positions

	x	y	z	B(Å^2)
4 Ca in 4(*e*)	0	0.1908±5	0.25	1.58±12
8 Al(1) in 8(*f*)	0.1647±3	0.9137±5	0.3029±6	1.48±12
8 Al(2) in 8(*f*)	0.1195±3	0.5589±6	0.2418±6	1.59±12
4 O(1) in 4(*e*)	0	0.4675±6	0.25	2.20±27
8 O(2) in 8(*f*)	0.1149±6	0.9493±12	0.5641±14	1.95±20
8 O(3) in 8(*f*)	0.1195±6	0.7476±10	0.1482±14	1.83±20
8 O(4) in 8(*f*)	0.1935±6	0.5581±11	0.5805±14	1.86±20

Interatomic distances (Å) *and bond angles*

Ca - O(3)	2.332±7 (×2)
Ca - O(2)	2.371±9 (×2)
Ca - O(1)	2.457±7
Ca - O(2)	2.872±10 (×2)

Al(1) - O(3)	1.715±9
Al(1) - O(2)	1.748±9
Al(1) - O(2)'	1.760±10
Al(1) - O(4)	1.769±8

Angles in degrees subtended at Al(1) by

	O(2)	O(2)'	O(4)
O(3)	113.7±5	103.4±4	117.8±5
O(2)		109.5±5	105.4±4
O(2)'			106.6±4

Al(2) - O(1)	1.751±5
Al(2) - O(3)	1.741±10
Al(2) - O(4)	1.800±10
Al(2) - O(4)'	1.814±7

Angles in degrees subtended at Al(2) by

	O(3)	O(4)	O(4)'
O(1)	121.6±4	109.7±4	100.4±3
O(3)		111.5±5	104.8±4
O(4)			107.2±4

Details of analysis

Lattice parameters and intensities measured with a diffractometer with Cu radiation. No correction for absorption (crystal size 0.07×0.17×0.07 mm^3) isostructural with $SrAl_4O_7$. Refined by least squares to give R = 0.090 for 400 observed reflexions [not listed].

TETRACALCIUM TRIALUMINATE

I. [Crystal structure of tetracalcium trialuminate - the aluminate analog of sodalite.] V.I. PONOMAREV, D.M. KHEIKER and N.V. BELOV, 1970. *Kristallografija*, 15, 918-921 [*Soviet Physics-Crystallography*, 15. 799-801].

$Ca_4(Al_2O_4)_3O$ F.W. = 530.20

Cubic, a = 8.86±1 Å, U = 695 Å^3, D_m = 2.50, Z = 2, D_x = 2.59.

Space group $I\bar{4}3m$ (T_d^3)

Atomic positions

8 Ca	in 8(*c*):	x = 0.160±1
12 Al	in 12(*d*)	
24 O(1)	in 24(*g*):	x = 0.365±2, z = 0.087±1
2 O(2)	in 2(*a*)	

Interatomic distances (Å)

Al-O	1.75 (×4)
Ca-O	2.46
Ca-O	2.62 (×3)

Discussion

The crystals are an aluminum analog of sodalite.

Details of analysis

Single crystal obtained by dehydration of $4CaO.3Al_2O_3.3H_2O$. Lattice parameter and intensities measured on DRON-1 diffractometer using CuK_α radiation and stationary counter method (θ<33) or θ-2θ method (θ>33). Crystal had large mosaic spread. No absorption correction (crystal size 0.015×0.030×0.080 mm^3). Structure determined by Patterson and electron-density functions, 8 parameters refined by least squares to give R = 0.069 for 21 reflexions [not listed].

TETRACALCIUM TRIHYDRO TRIALUMINATE

I. [Crystal structure of tetracalcium trihydrotrialuminate $C_4A_3H_3$]. V.I. PONOMAREV, D.M. KHEIKER and N.V. BELOV, 1970. *Dokl. Akad. Nauk SSSR*, 194, 1072-1075.

$Ca_2Al_3O_6(OH)(H_2O)$ F.W. = 584.24

Orthorhombic, a = 12.426±2, b = 12.809±2, c = 8.864±1 Å, U = 141 Å^3, Z = 4, D_x = 2.74.

Space group $Abma$ (D_{2h}^{18})

Atomic positions

Atoms in (0,0,0;0,½,½) ($x,y,z;\bar{x},y,\bar{z};$½+$x,y,$½$-z;$½$-x,y,$½$+z$)

	x	y	z	B(Å^2)
8 Ca(1) in 8(*d*)	0	0.1422	0	0.18
8 Ca(2) in 8(*f*)	0.3492	0	0.2107	0.12
16 Al(1) in 16(*g*)	0.1296	0.1229	0.3630	0.12
8 Al(2) in 8(*e*)	0.25	0.25	0.1106	0.14
16 O(1) in 16(*g*)	0.3663	0.2186	0.0026	0.54
16 O(2) in 16(*g*)	0.2222	0.1393	0.2195	0.64
16 O(3) in 16(*g*)	-0.0025	0.1310	0.2769	0.85
8 O(4) in 8(*f*)	0.1467	0	0.4463	0.56
8 O(5) in 8(*f*)	0.3913	0	0.4640	0.80

Interatomic distances (Å) *and bond angles*

Errors uncertain but about 0.005-0.010 Å and 1°

Ca(1) - O(5)	2.290 (×2)
Ca(1) - O(1)	2.437 (×2)
Ca(1) - O(3)	2.459 (×2)
Ca(2) - O(5)	2.305
Ca(2) - O(4)	2.344
Ca(2) - O(2)	2.383 (×2)
Ca(2) - O(3)	2.495 (×2)
Al(1) - O(2)	1.728
Al(1) - O(1)	1.743
Al(1) - O(4)	1.752
Al(1) - O(3)	1.813

Angles in degrees subtended at Al(1) by

	O(1)	O(4)	O(3)
O(2)	115	110	107
O(1)		109	107
O(4)			110

Possible hydrogen bonds

O(3)-H...O(1)	2.782
O(5)-H...O(5)	2.776

The second H bond from O(5) is not apparent from interatomic distances.

Al(2) - O(2)	1.750 (×2)
Al(2) - O(1)	1.780 (×2)

Angles in degrees subtended at Al(2) by

	O(2)	O(1)
O(2)	113	109
O(2)'		106
O(1)		115

Discussion

The crystal contains a cubic close packed array of oxygen atoms with (201) as the close packed plane. Ca occupies 1/4 of the octahedral and Al 3/16 of the tetrahedral holes. The AlO_4 tetrahedra share corners to form sheets perpendicular to *a*.

Details of analysis

Crystals grown by hydrothermal synthesis. Intensities 4 layers with $l = 0$ to 9 on equi-inclination automatic diffractometer with CuK_α radiation and corrected for Lorentz, polarization and absorption effects (crystal spherical $d = 0.302 \pm 5$, $\mu R = 2.64$). Structure solved by Patterson and electron-density maps and chemical considerations and refined by least squares to give $R = 0.084$ for 598 non-zero reflexions [not listed].

ALUMINUM BARIUM HYDRATES

I. Barium aluminate hydrates. II. The crystal structure of $Ba_2[Al_2(OH)_{10}]$. A.H.M. AHMED and L.S.D. GLASSER, 1970. *Acta Cryst.*, B26, 867-871.

$Ba_2O.Al_2O_3.5H_2O$ F.W. = 482.72

Triclinic, $a = 6.704$, $b = 5.758$, $c = 6.179$ Å, $\alpha = 90°4'$, $\beta = 98°25'$, $\gamma = 109°36'$, $U = 215.3$ Å^3, $D_m = 3.66$, $Z = 1$, $D_x = 3.723$.

Space group $P\bar{1}$ (C_i^1) from statistics

Atomic parameters All atoms in 2(*i*)

	x	*y*	*z*
Ba	0.0370±4	0.2761±3	0.2943±3
Al	0.318±1	0.346±2	0.826±2
O(1)	0.278±3	0.084±4	0.029±3
O(2)	0.306±3	0.141±4	0.591±3
O(3)	0.339±4	0.618±4	0.653±3
O(4)	0.018±4	0.294±4	0.768±4
O(5)	0.630±3	0.431±4	0.911±3

Thermal parameters are given.

Interatomic distances (Å) *and bond angles*

Al-O	1.84±2, 1.88±3, 1.91±3, 1.94±2, 1.97±2, 1.99±2
O-Al-O	angles between nearest oxygen atoms range from 85 to 95° except for O(5)-Al-O(5') = 79±1°
Ba-O	nine between 2.68 - 2.98

Description of analysis

Centrosymmetrically related pairs of shared edges forming $[Al_2(OH)_{10}]^{4-}$ groups are linked through Ba ions as shown in Fig. 35. Structure also reported by Louis and Moras (1).

Details of analysis

700 and 760 reflexions collected on a diffractometer using Mo radiation from two crystals. 14 discrepant reflexions in both sets dropped. No absorption correction. Least-squares refinement to an *R* value of 0.11. No discussion is offered of the large thermal parameter for O(4).

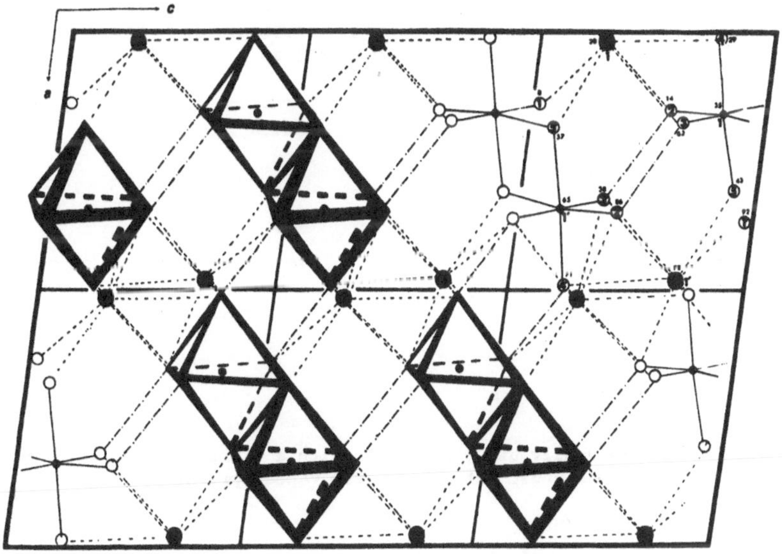

Fig. 35. The structure of six unit cells of $Ba_2[Al_2(OH)_{10}]$ viewed along *b*. The large solid circles represent Ba, small solid circles Al, open circles, O atoms. Dotted lines show Ba-O contacts; dash lines are possible hydrogen bonds.

II. Barium aluminate hydrates. III. The crystal structure of α-$Ba[AlO(OH)_2]_2$.
A.H.M. AHMED and L.S.D. GLASSER, 1970. *Acta Cryst.*, B26, 1686-1690.

$Ba[AlO(OH)_2]_2$ F.W. = 281.30

Orthorhombic, *a* = 12.99, *b* = 12.34, *c* = 7.19 Å, *U* = 1152.5 $Å^3$, D_m = 3.30, *Z* = 8, D_x = 3.356.

Space group Pbca (D_{2h}^{15})

Atomic positions All atoms in 8(*c*)

	x	*y*	*z*	*B*($Å^2$)
Ba	0.0733	0.3433	0.3947	1.54
Al(1)	0.149	0.015	0.387	1.1
Al(2)	0.326	0.150	0.226	1.1
O(1)	0.429	0.222	0.328	0.4
O(2)	0.094	-0.008	0.162	1.0
O(3)	0.223	0.242	0.180	1.7
O(4)	0.0795	0.120	0.493	1.3
O(5)	0.278	0.053	0.373	1.4
O(6)	0.370	0.103	0.017	1.1

Interatomic distances (Å) *and bond angles*

Al(1)-O(2)	1.78	Ba-O(1)	2.85
-O(4)	1.75	-O(3)	3.02
-O(5)	1.75	-O(1)	2.88
-O(6')	1.75	-O(3)	2.78
Al(2)-O(1)	1.77	-O(4)	2.85
-O(3)	1.78	-O(2)	2.81
-O(5)	1.72	-O(2)	2.88
-O(6)	1.71	-O(4)	2.92
		-O(5)	3.22

Angle in degrees subtended at Al(1) by

	O(4)	O(5)	O(6')
O(2)	108.3	111.8	107.2
O(4)		108.8	108.2
O(5')			112.5

Angle in degrees subtended at Al(2) by

	O(3)	O(5)	O(6)
O(1)	109.1	111.9	106.3
O(3)		106.7	107.5
O(5)			115.2

Materials

Prepared hydrothermally using materials with a starting molar ratio of 1 BaO to 1 Al_2O_3 at 80 - 215°C under saturated water vapour pressure.

Description of the structure

The structure contains infinite chains of AlO_4H_2 tetrahedra as shown in Fig. 36. Ba ions are bonded to eight non-bridging hydroxyl groups and one bridging oxygen atom.

Details of the analysis

943 symmetry independent reflexions, measured with MoK_α radiation, were refined by least squares to a final *R* value of 0.09. No absorption or extinction corrections were made. Errors in unit cell lengths were not given.

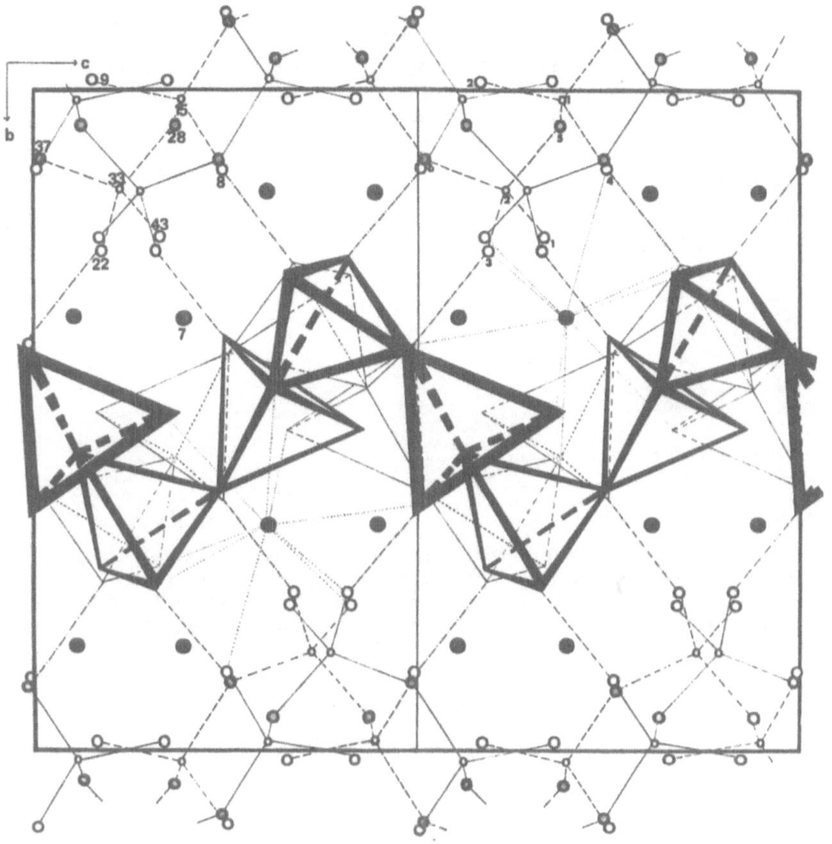

Fig. 36. Projection of two cells of α-Ba[AlO(OH)$_2$] on (100), showing the configuration of the [AlO(OH)$_2$]$_\infty$ chains and large solid circles represent barium atoms and small circles aluminum atoms. Medium circles represent oxygen atoms if shaded, and hydroxyl group if not. Dot - dash lines represent postulated hydrogen bonds.

1. R. Louis and D. Moras, 1969. *Bull. Soc. Chim. Fr.*, 3471.

STRONTIUM CARBONATE

I. Zur Kristallstruktur von Strontianit, $SrCO_3$. W. PANNHORST and J. LÖHN, 1970. *Z. Kristallogr.*, 131, 455-459.

$SrCO_3$ F.W. = 147.63

Orthorhombic, Aragonite type (1), a = 6.020±2, b = 5.093±2, c = 8.376±3 Å, U = 256.8 Å^3, D_m = 3.78, Z = 4, D_x = 3.81.

Space group Pnma (D_{2h}^{16})

Atomic positions

	x	y	z	$B(Å^2)$
4 Sr in 4(c)	0.7583±7	0.25	0.4159±3	0.77±2
4 O in 4(c)	0.920±6	0.25	0.758±3	0.81±36
4 O(1) in 4(c)	0.905±4	0.25	0.913±3	1.10±27
8 O(2) in 8(d)	0.912±3	0.469±2	0.681±1	0.76±19

Interatomic distances (Å)

Sr - O(1)	2.56
Sr-O(2)	2.58 (×2)
Sr - O(2)'	2.64 (×2)
Sr - O(2)"	2.65 (×2)
Sr - O(1)'	2.73 (×2)
σ(Sr-O) = 0.01	
C - O(2)	1.29±2 (×2)
C - O(1)	1.30±4
O(1)-C-O(2)	120±1°
O(2)-C-O(2)'	120±2°

Details of analysis

Lattice parameters from precession photographs. Intensities measured photometrically from Weissenberg photographs of layers with h = 0 to 5 taken with monochromated MoK_α radiation. Intensities corrected for distortion twinning and absorption (crystal size 0.12×0.03×0.34 mm³). Structure of aragonite (1) assumed and refined by least squares to give R = 0.073 for 203 observed reflexions listed.

1. *Strukturbericht*, 1, 295.

CALCIUM CARBONATE HEXAHYDRATE

$Ca(H_2O)_6CO_3$ F.W. = 208.15

I. The crystal structure of calcium carbonate hexahydrate at ∿ - 120°. B. DICKENS and W.E. BROWN, 1970. *Inorg. Chem.*, 9, 480-486.

Monoclinic, a = 8.87±1, b = 8.23±1, c = 11.02±1 Å, β = 110.2°, U = 755.0 Å³, D_m = 1.82 (at 0°C), Z = 4, D_x = 1.80 [1.83] (at -120°C).

Space group C2/c (C^6_{2h}) from systematic absences and statistical analysis of intensities.

Atomic positions

	x	y	z	$B(Å^2)$
4 Ca in 4(e)*	0	0.1472±1	1/4	0.10 (not refined)
4 C in 4(e)*	0	0.8067±1	1/4	0.37±5
4 O(1) in 4(e)*	0	0.6508±6	1/4	0.77±5
8 O(2) in 8(f)	0.5263±14	0.3849±3	0.1582±3	0.49±3
8 O(3) in 8(f)	0.6163±4	0.7229±4	0.0916±4	0.70±4
8 O(4) in 8(f)	0.7883±4	0.5576±4	0.3825±3	0.59±4
8 O(5) in 8(f)	0.6703±4	0.8842±3	0.3593±3	0.53±3

*converted from atomic position at 1/2, 1/2 + y, 1/4 given in I.

Proposed coordinates for the hydrogen atoms are given, but were not included in the refinement.

Interatomic distances (Å) and bond angles

CO_3 ion (crystallographic site symmetry C_2)

C-O(1)	1.283±7	O(1)-C-O(2)	120.0±2°
C-O(2)	1.287±5 (×2)	O(2)-C-O(2)'	119.9±4°

Eightfold coordination around Ca (crystallographic site symmetry C_2)

Ca-O(2)	2.429±3 (×2)
Ca-O(3)	2.397±4 (×2)
Ca-O(4)	2.576±3 (×2)
Ca-O(5)	2.506±3 (×2)

Environment of water O(3)

O(3)-Ca	2.397±4
O(3)-O(2)	2.749±5
O(3)-O(4)	2.867±4

Angles in degrees subtended by

	O(2)	O(4)
Ca	118.7±3	111.3±2
O(2)		115.3±2

Environment of water O(4)

O(4)-Ca	2.576±3
O(4)-O(1)	2.854±4
O(4)-O(5)	2.781±5
O(4)-O(5)'	2.864±4
O(4)-O(3)	2.867±4

Angles in degrees subtended at O(4) by

	O(1)	O(5)	O(5)'	O(3)
Ca	109.5±2	106.5±1	54.5±1	122.2±1
O(1)		126.8±1	89.2±2	94.9±2
O(5)			81.1±1	97.7±1
O(5)'				175.6±2

Environment of water O(5)

O(5)-Ca	2.506±3
O(5)-O(1)	2.700±5
O(5)-O(2)	2.764±6
O(5)-O(4)	2.781±5
O(5)-O(4)'	2.864±4

Angles in degrees subtended at O(5) by

	O(1)	O(2)	O(4)	O(4)'
Ca	105.4±2	114.1±2	118.0±2	56.9±2
O(1)		112.0±1	99.9±1	159.0±3
O(2)			106.4±1	70.9±2
O(4)				98.9±1

Discussion

Each calcium atom lies on a twofold axis and is chelated by one CO_3 ion lying on the same twofold axis. The six water molecules complete the coordination sphere around Ca, giving discrete molecules which are linked in the crystal only by hydrogen and van der Waals bonds.

Details of analysis

Crystals grown from $CaCO_3$ gels. A crystal of diameter less than 0.3 mm was sealed in borate glass tube and cooled to -120°C. Intensities measured visually from multiple-film oscillation photographs and corrected for Lorentz, polarization and Cox-Shaw (1) effects. Mo radiation. No correction for absorption. Structure determined by Patterson and Fourier syntheses and refined by least squares to give R = 0.10 for 1420 reflexions listed.

1. E.G. COX and W.F.B. SHAW, 1930. *Proc. Roy. Soc. Lond.* A 127, 71.

YTTRIUM SILICATE

I. [Crystal structure of yttrium oxyorthosilicate $Y_2O_3.SiO_2{=}Y_2SiO_5$. Dual function of yttrium.] B.A. MAKSIMOV, V.V. ILJUKHIN, JU.A. KHANTANOV and N.V. BELOV, 1970. *Kristallografija*, 15, 926-933 [*Soviet Physics - Crystallography*, 15, 806-812].

Y_2SiO_5 F.W. = 285.89

Monoclinic, a = 10.410±3, b = 6.721±2, c = 12.490±5 Å, β = 102.65±3°, U = 852.7 Å^3, D_m = 4.60, Z = 8, D_x = 4.44.

Space group I2/a (C_{2h}^6)

Atomic positions

All atoms in 8(*f*) (0,0,0;½,½,½)±(*x*,*y*,*z*;*x*+½,$\bar{y}$,*z*)

	x	*y*	*z*	*B*(Å^2)
Y(1)	-0.194	0.378	0.141	0.02
Y(2)	-0.071	-0.257	-0.037	0.03
Si	-0.127	-0.093	0.181	-0.30
O(1)	-0.200	-0.287	0.118	0.09
O(2)	-0.054	-0.002	0.089	0.29
O(3)	-0.032	-0.157	0.298	0.33
O(4)	-0.237	0.071	0.203	0.06
O(5)	-0.118	0.398	-0.015	0.40

[Note: the origin has been transformed by subtracting 0.25 from the *x* coordinates given in I. This may not be the correct transformation for the symmetry positions given above, but it gives a better agreement with the interatomic distances quoted in I, than does the coordinate set quoted by the authors.]

Interatomic distances (Å) *and bond angles*

Errors uncertain but probably 0.02 to 0.03 Å

Y(1)-O(5)	2.21 [2.23]
Y(1)-O(4)	2.29 [2.25]
Y(1)-O(5)'	2.33 [2.26]
Y(1)-O(1)	2.28 [2.27]
Y(1)-O(4)'	2.31 [2.28]
Y(1)-O(3)	2.30 [2.32]

Angles in degrees subtended at Y(1) by

	O(4)	O(5)'	O(1)	O(4)'	O(3)
O(5)	102	79	82	95	155
O(4)		168	87	74	103
O(5)'			81	118	77
O(1)				160	98
O(4)'					93

Y(2)-O(5)	2.15
Y(2)-O(3)	2.32 [2.26]
Y(2)-O(2)	2.33 [2.31]
Y(2)-O(2)'	2.37 [2.35]
Y(2)-O(1)	2.35 [2.39]
Y(2)-O(5)'	2.39 [2.40]
Y(2)-O(1)'	2.60

Angles in degrees subtended at Y(2) by

	O(3)	O(2)	O(2)'	O(1)	O(5)'	O(1)'
O(5)	80	101	83	149	75	109
O(3)		144	74	87	86	152
O(2)			71	106	129	62
O(2)'				120	153	132
O(1)					76	71
O(5)'						72

Si-O(1)	1.60 [1.62]
Si-O(2)	1.66 [1.63]
Si-O(3)	1.63
Si-O(4)	1.66

Angles in degrees subtended at Si by

	O(2)	O(3)	O(4)
O(1)	102	110	110
O(2)		116	108
O(3)			109

Discussion

The structure is composed of a network of edge sharing YO_6 octahedra linked by SiO_4 tetrahedra.

Details of analysis

Lattice parameters measured on diffractometer with CuK_α radiation. Intensities estimated visually from photographs of layers with $k = 0$ to 4 and $h = 0$ and 1 using MoK_α radiation. No correction for absorption ($\mu R = 0.9$). Structure solved from Patterson function and refined by electron density and least squares to give $R = 0.10$ for 800 observed reflexions [not listed].

SODIUM GADOLINIUM SILICATE

I. [Crystal structure of sodium gadolinium orthosilicate]. E.I. AVERTISJAN, A.V. ČIČAGOV and N.V. BELOV, 1970. *Kristallografija*, 15, 1066-1067 [*Soviet Physics - Crystallography*, 15, 926-927].

$NaGdSiO_4$

F.W. = 272.32

Tetragonal, a = 11.63, c = 5.41 Å, U = 732 Å^3, D_m = 4.9, Z = 8, D_x = 4.93.

Space group $I4/m$ (C^5_{4h}) from final structure and X-ray point symmetry.

Atomic positions

		x	y	z
8 Gd	in 8(h)	0.184	0.117	0
8 Si	in 8(h)	0.106	0.246	0.5
8 Na	in 8(h)	0.420	0.109	0.5
8 O(1)	in 8(h)	0.385	0.113	0
8 O(2)	in 8(h)	0.240	0.210	0.5
16 O(3)	in 16(i)	0.041	0.190	0.260

B = -0.68 Å^2

Interatomic distances (Å) *and bond angles*

Si-O(2)	1.61
Si-O(3)	1.64 (×2)
Si-O(1)	1.64

Angles in degrees subtended at Si by

	O(3)'	O(1)'
O(2)	110	101
O(3)	105	115

Gd-O(2)	2.20
Gd-O(3)	2.32 (×2)
Gd-O(3)'	2.34 (×2)
Gd-O(1)	2.34
Na-O(1)	2.25
Na-O(2)	2.40
Na-O(1)'	2.63
Na-O(1)"	2.74 (×2)
Na-O(3)	2.77 (×2)

Errors uncertain but probably at least 0.03 Å

Discussion

Isostructural with $NaSmSiO_4$ (1). There is good evidence for 1/4 molecule of H_2O per formula unit situated at 2(a) [0,0,0].

Details of analysis

Intensities estimated visually from KFOR photograph of layers with l = 0 to 3 taken with MoK_α radiation. Structure solved from Patterson and difference electron-density functions and refined to give R = 0.15 for 325 observed reflexions [not listed].

1. A.V. ČIČAGOV, V.V. ILJUKHIN and N.V. BELOV, 1967. *Dokl. Akad. Nauk SSSR.*, 173, 3.

RARE EARTH DISILICATES

I. The crystal structures of the rare earth pyrosilicates. JU.I. SMOLIN and JU.F. ŠEPELEV, 1970. *Acta Cryst.*, B26, 484-492.

$Yb_2Si_2O_7$ F.W. = 514.25

$Tm_2Si_2O_7$ F.W. = 506.04

Monoclinic: $Yb_2Si_2O_7$, a = 6.802±5, b = 8.875±10, c = 4.703 5 Å, β = 102°07±1', [U = 277.58 Å^3], D_m = 5.99, Z = 2, D_x = 6.15.

$Tm_2Si_2O_7$, a = 6.824±10, b = 8.91±1, c = 4.704±5 Å, β = 101°50'±10', [U = 279.87 Å^3], Z = 2.

Space group C2/m (C^3_{2h}) other choices tested but rejected on chemical grounds.

Atomic positions

		x	y	z	B(Å^2)
4 Yb	in 4(*g*)	1/2	0.80687±22	0	0.25
4 Si	in 4(*i*)	0.7189±3	1/2	0.4125±6	0.37
2 O(1)	in 2(*a*)	1/2	1/2	1/2	1.02
4 O(2)	in 4(*i*)	0.8831±5	1/2	0.7151±15	0.50
8 O(3)	in 8(*j*)	0.7361±5	0.6504±4	0.2197±11	0.54

Interatomic distances (Å) *and bond angles*

Si-O(1)	(2×)	1.626±3	O(1)-Si-O(2)	106°07'
-O(2)	(2×)	1.614±6	O(1)-Si-O(3)	108°28'
-O(3)	(4×)	1.632±5	O(2)-Si-O(3)	111°53'
			O(3)-Si-O(3')	109°49'

$Er_2Si_2O_7$ F.W. = 472.09

$Y_2Si_2O_7$ F.W. = 345.98

$Ho_2Si_2O_7$ F.W. = 499.03

Monoclinic: $Er_2Si_2O_7$, a = 4.683±5, b = 5.556±5, c = 10.79±1 Å, γ = 96°±10', [U = 279.11 Å^3], D_m = 5.85, Z = 2, D_x = 5.98.

$Y_2Si_2O_7$, a = 4.663±5, b = 5.536±5, c = 10.784±21 Å, γ = 96°06', U = 276.73 Å^3, Z = 2.

$Ho_2Si_2O_7$, a = 4.686±5, b = 5.58±5, c = 10.84±2 Å, γ = 95°58'±10', [U = 278.38 Å^3], Z = 2.

Space group $P2_1/b$ (C^5_{2h})

Atomic parameters

		x	y	z	B(Å^2)
4 Er	in 4(*e*)	0.88829±8	0.09318±6	0.34934±5	0.29
4 Si	in 4(*e*)	0.3601±4	0.6442±3	0.3871±3	0.33
2 O(1')	in 2(*b*)	1/2	1/2	1/2	0.91
4 O(2)	in 4(*e*)	0.2052±8	0.8653±7	0.4486±6	0.64
4 O(3)	in 4(*e*)	0.1235±9	0.4583±8	0.3191±6	0.63
4 O(4)	in 4(*e*)	0.6184±9	0.7522±7	0.2984±6	0.56

Interatomic distances (Å) and bond angles

Si-O(1)	1.632±3	O(1)-Si-O(2)	107°24'
-O(2)	1.631±5	O(1)-Si-O(3)	108°03'
-O(3)	1.611±5	O(1)-Si-O(4)	107°36'
-O(4)	1.609±5	O(2)-Si-O(4)	109°43'
		O(2)-Si-O(3)	109°44'
Er-O(2), O(2)'	2.312±5, 2.240±7	O(3)-Si-O(4)	114°03'
-O(3), O(3')	2.229±5, 2.290±6		
-O(4), O(4')	2.233±5, 2.270±6		

$Y_2Si_2O_7$ F.W. = 345.98

$Ho_2Si_2O_7$ F.W. = 499.03

$Dy_2Si_2O_7$ F.W. = 503.17

$Gd_2Si_2O_7$ F.W. = 482.67

Orthorhombic, $Y_2Si_2O_7$, a = 13.69±2, b = 5.020±5, c = 8.164±10 Å, U = 561.12 Å^3.
$Ho_2Si_2O_7$, a = 13.700±15, b = 4.980±5, c = 8.23±1 Å, U = 561.50 Å^3.
$Dy_2Si_2O_7$, a = 13.740±15, b = 5.012±5, c = 8.26±1 Å, U = 568.82 Å^3.
$Gd_2Si_2O_7$, a = 13.870±15, b = 5.073±5, c = 8.33±1 Å, U = 586.12 Å^3, D_m = 5.33, Z = 4, D_x = 5.46.

*Space group Pna*2$_1$ (C^9_{2v}) *Pnam* yields poor agreement and unsatisfactory variations in SiO bond lengths.

Atomic parameters All atoms in 4(*a*).

	x	*y*	*z*	*B*(Å^2)
4 Gd(1)	0.12551±4	0.33730±13	0.99831±7	0.38
4 Gd(2)	0.12564±4	0.33739±13	0.51409±7	0.36
4 Si(1)	0.3205±2	0.3744±7	0.2505±0	0.31
4 Si(2)	0.5390±2	0.6253±6	0.2498±8	0.17
4 O(1)	0.2715±7	0.4769±22	0.0876±13	0.58
4 O(2)	0.2658±6	0.4857±21	0.4130±12	0.34
4 O(3)	0.3457±6	0.0706±17	0.2465±15	0.44
4 O(4)	0.4211±5	0.5557±16	0.2448±11	0.23
4 O(5)	0.5472±8	0.7858±23	0.0866±14	0.61
4 O(6)	0.5456±8	0.7882±25	0.4206±15	0.73
4 O(7)	0.5988±5	0.3526±17	0.2560±19	0.60

Interatomic distances (Å) and bond angles

Si(1)-O(1)	1.604±13	O(1)-Si(1)-O(2)	112°50'
-O(2)	1.651±12	O(1)-Si(1)-O(3)	113°05'
-O(3)	1.581±9	O(1)-Si(1)-O(4)	98°40'
-O(4)	1.672±8	O(2)-Si(1)-O(3)	116°53'
		O(2)-Si(1)-O(4)	102°37'
Si(2)-O(4)	1.673±7	O(3)-Si(1)-O(4)	110°30'
-O(5)	1.588±13	O(4)-Si(2)-O(5)	99°
-O(6)	1.648±14	O(4)-Si(2)-O(6)	110°27'
-O(7)	1.614±9	O(4)-Si(2)-O(7)	108°48'
		O(5)-Si(2)-O(6)	118°32'
		O(5)-Si(2)-O(7)	115°29'
		O(6)-Si(2)-O(7)	111°55'
		Si(1)-O(4)-Si(2)	158°40'

Gd(-1)-O seven 2.27-2.47
Gd(2)-O seven 2.248-2.485

$Eu_2Si_2O_7$	F.W. = 472.09
$Sm_2Si_2O_7$	F.W. = 468.89
$Ce_2Si_2O_7$	F.W. = 448.41
$La_2Si_2O_7$	F.W. = 445.98
$Nd_2Si_2O_7$	F.W. = 456.65

Orthorhombic, $Eu_2Si_2O_7$, a = 5.374±5, b = 12.82±2, c = 8.65±1 Å, U = 593.93 Å^3.
$Sm_2Si_2O_7$, a = 5.384±5, b = 12.85±2, c = 8.69±1 Å, U = 601.21 Å^3.
$Ce_2Si_2O_7$, a = 5.400±5, b = 13.05±2, c = 8.73±1 Å, U = 615.20 Å^3.
$La_2Si_2O_7$, a = 5.410±5, b = 13.17±2, c = 8.76±1 Å, U = 624.14 Å^3.
$Nd_2Si_2O_7$, a = 5.39±5, b = 12.95±1, c = 8.72±1 Å, U = 608.66 Å^3, D_m = 5.04, Z = 4, D_x = 4.98.

Space group $P2_12_12_1$ (D^4)

Atomic positions All atoms in 4(*a*)

	x	y	z	B(Å^2)
4 Nd(1)	0.99736±22	0.23136±3	0.44325±7	0.45
4 Nd(2)	0.50719±16	0.40960±3	0.64617±7	0.45
4 Si(1)	0.4770±6	0.3218±2	0.2494±5	0.52
4 Si(2)	0.4767±6	0.4748±2	-0.0018±4	0.52
4 O(1)	0.4710±4	0.4340±5	0.1758±12	0.60
4 O(2)	0.2500±17	0.2593±6	0.1696±14	0.66
4 O(3)	0.7232±18	0.2487±7	0.2325±16	0.90
4 O(4)	0.3947±21	0.3301±8	0.4291±18	1.34
4 O(5)	0.2534±16	0.4205±6	-0.1070±13	0.54
4 O(6)	0.7302±19	0.4485±8	-0.0926±14	0.89
4 O(7)	0.3970±18	0.5941±7	-0.0040±15	0.96

Interatomic distances (Å) *and bond angles*

Si(1)-O(1)	1.590±8	O(1)-Si(1)-O(2)	105°32'
-O(2)	1.624±10	O(1) Si(1)-O(3)	120°34'
-O(3)	1.638±11	O(1)-Si(1)-O(4)	108°47'
-O(4)	1.632±15	O(2)-Si(1)-O(3)	106°31'
Si(2)-O(1)	1.636±11	O(2)-Si(1)-O(4)	103°50'
-O(5)	1.669±10	O(3)-Si(1)-O(4)	110°10'
-O(6)	1.616±9	O(1)-Si(2)-O(5)	111°45'
-O(7)	1.604±10	O(1)-Si(2)-O(6)	114°17'
		O(1)-Si(2)-O(7)	108°47'
		O(5)-Si(2)-O(6)	104°37'
		O(5)-Si(2)-O(7)	101°41'
		O(6)-Si(2)-O(7)	114°55'
		Si(1)-O(1)-Si(2)	132°36'

Nd(1)-O eight in the range 2.385-2.949
Nd(2)-O eight in the range 2.239-2.625

Materials

Single crystals of $M_2Si_2O_7$, M=Yb, Er, Gd, Nd, were grown by the Verneuil method. Powders were prepared of the others by sintering the oxides.

Discussion of the structures

The structures of $Yb_2Si_2O_7$, $Er_2Si_2O_7$ and $Gd_2Si_2O_7$ have anion sheets with the cations bonded to oxygen atoms bridging between sheets. $Yb_2Si_2O_7$ has the thortvietite structure (<u>1</u>) with linear Si-O-Si bonds perpendicular to the *b* axis as seen in Fig. 37. The structure of $Er_2Si_2O_7$ is related to that of $Yb_2Si_2O_7$ by rotating the anions so that in adjacent rows they subtend angles of +30° and -30° with the *c* axis, as can be seen in Fig. 38. In both of these structures the cations are bonded to terminal oxygen atoms only. In $Gd_2Si_2O_7$ adjacent sheets of anions no longer are translationally equivalent, and the cations now bond to bridging oxygen atoms in addition to terminal ones (see Fig. 39). The structure of $Nd_2Si_2O_7$, as shown in Fig. 40, consists of chains of Si_2O_7 anions with adjacent anions related by a two-fold screw axis. Each chain is surrounded by four others which are rotated 90° about the chain axis.

Details of analysis

Least-squares refinement of data collected using MoK_α radiation. For $Er_2Si_2O_7$, 1860 reflexions, $R = 0.062$, for $Ge_2Si_2O_7$, about 1300 reflexions, $R = 0.073$, and for $Nd_2Si_2O_7$, 1820 reflexions, $R = 0.077$.

1. *Structure Reports*, <u>27</u>, 702.

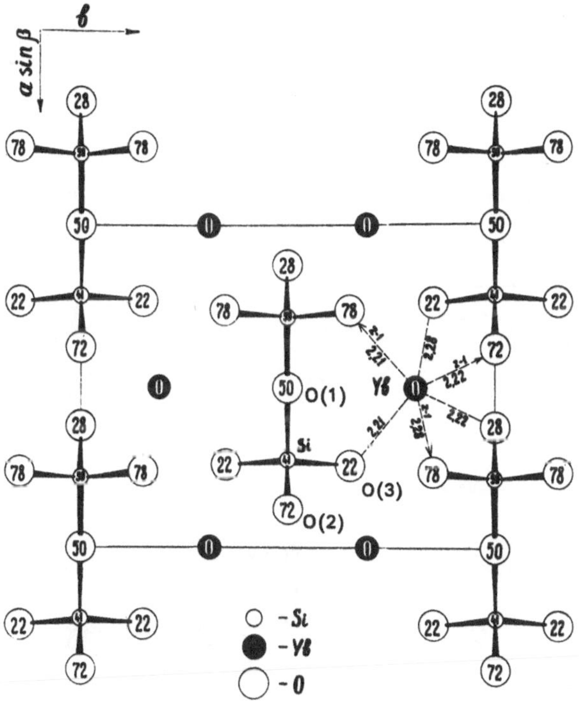

Fig. 37. The structure of $Yb_2Si_2O_7$ projected along the *c* axis.

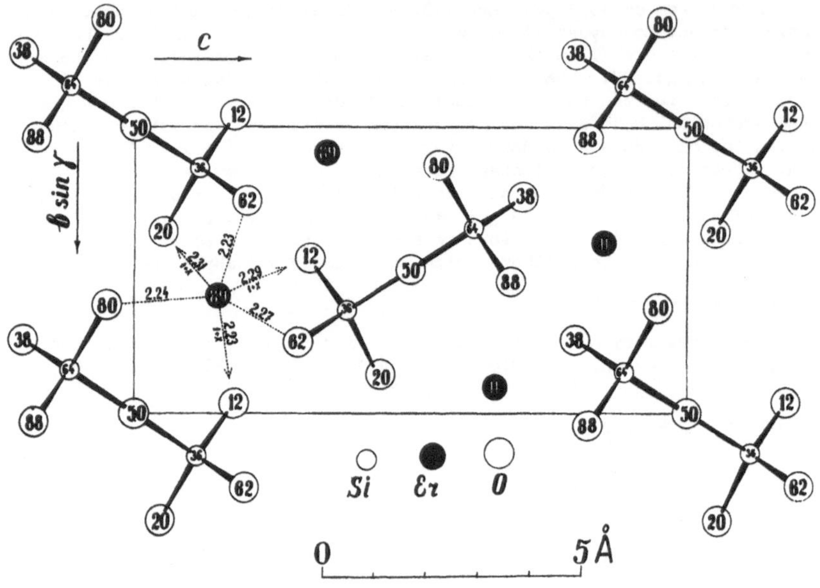

Fig. 38. The structure of $Er_2Si_2O_7$ projected along the *a* axis.

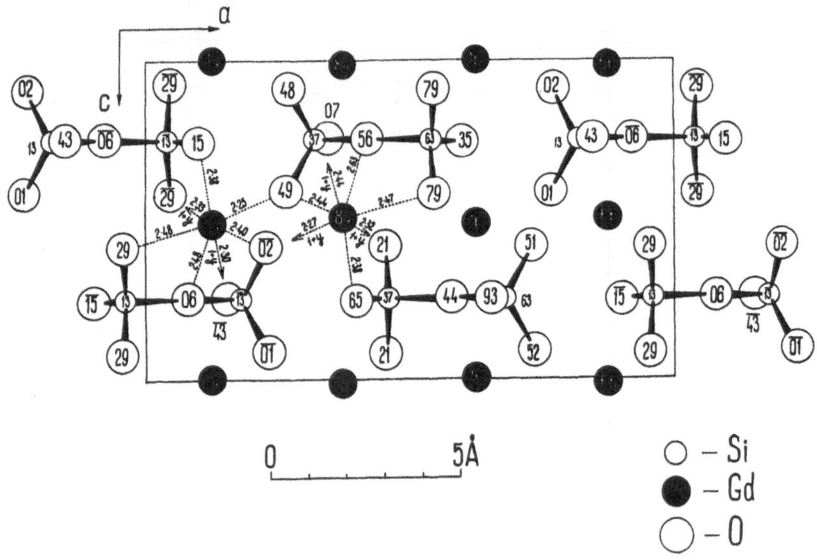

Fig. 39. The structure of $Gd_2Si_2O_7$ projected along the *b* axis.

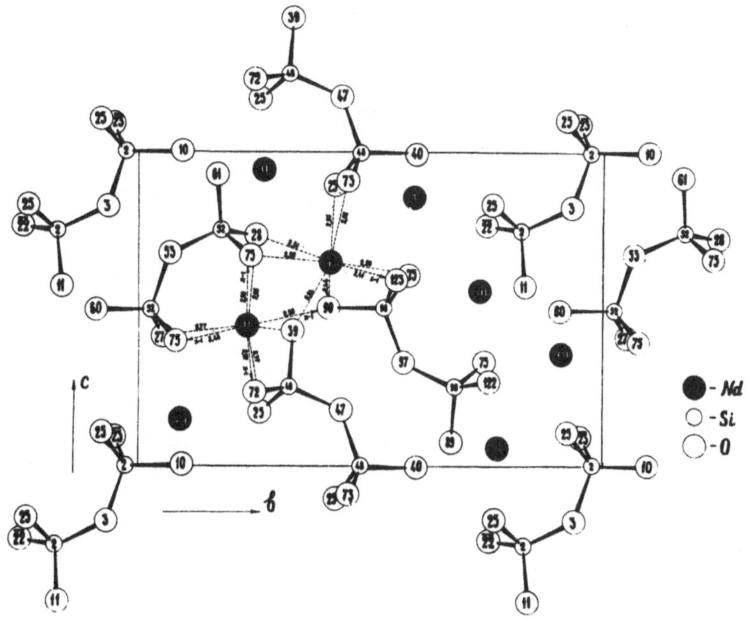

Fig. 40. The structure of $Nd_2Si_2O_7$ projected along the *a* axis.

SAMARIUM DISILICATE

I. [The crystal structure of the low temperature form of samarium pyrosilicate, $Sm_2Si_2O_7$.] JU.I. SMOLIN, JU.F. ŠEPELEV and I.K. BUTIKOVA, 1970. *Kristallografija*, 15, 256-261 [*Soviet Physics - Crystallography*, 15, 214-219].

$Sm_2Si_2O_7$ F.W. = 469.0

Tetragonal, a = 6.695±5, c = 24.40±2 Å, U = 1093.68 Å^3, Z = 8, D_m = 5.68.

Space group $P4_1$ (C_4^2)

Atomic positions

All atoms in 4(*a*)

	x	*y*	*z*	*B*(Å^2)
Sm(1)	0.7816±2	0.2821±2	0	0.38
Sm(2)	0.5201±2	0.1622±2	0.14618±6	0.39
Sm(3)	0.3463±2	0.9121±2	0.00052±6	0.35
Sm(4)	0.1294±2	0.7553±2	0.13778±6	0.47
Si(1)	0.8575±9	0.7529±9	0.0146±2	0.42
Si(2)	0.6043±9	0.6834±9	0.1135±2	0.36
Si(3)	0.2687±9	0.3692±10	0.0193±2	0.43
Si(4)	0.0151±10	0.2817±10	0.1202±2	0.40
O(1)	0.8956±23	0.5997±23	-0.0354±6	0.66
O(2)	0.7244±23	0.9386±23	-0.0115±6	0.72
O(3)	0.0480±20	0.8376±20	0.0462±5	0.38
O(4)	0.7252±19	0.6168±20	0.0570±5	0.51
O(5)	0.4753±19	0.5062±20	0.1375±5	0.81

O(6)	0.4441±19	0.8542±20	0.0945±5	0.41
O(7)	0.7659±19	0.7968±22	0.1549±5	0.78
O(8)	0.3327±20	0.5760±20	-0.0113±6	0.63
O(9)	0.4547±20	0.2319±20	0.0366±5	0.60
O(10)	0.1324±20	0.2363±20	-0.0228±5	0.54
O(11)	0.1230±21	0.4154±20	0.0809±5	0.56
O(12)	-0.0125±20	0.4463±20	0.1658±5	0.32
O(13)	0.1660±19	0.1105±19	0.1427±5	0.29
O(14)	-0.1844±20	0.1907±20	0.0940±5	0.86

Interatomic distances (Å) *and bond angles*

Si(1)-O(3)	1.595
Si(1)-O(1)	1.614
Si(1)-O(4)	1.628
Si(1)-O(2)	1.638

Angles in degrees subtended at Si(1) by

	O(1)	O(4)	O(2)
O(3)	117.7	109.9	110.5
O(1)		102.1	105.7
O(4)			111.7

Si(2)-O(5)	1.580
Si(2)-O(6)	1.635
Si(2)-O(7)	1.664
Si(2)-O(4)	1.660

Angles in degrees subtended at Si(2) by

	O(6)	O(7)	O(4)
O(5)	105.7	118.3	111.9
O(6)		106.3	105.8
O(7)			108.1

Si(1)-O(4)-Si(2) 129.8±9

Si(3)-O(9)	1.605
Si(3)-O(11)	1.623
Si(3)-O(8)	1.630
Si(3)-O(10)	1.637

Angles in degrees subtended at Si(3) by

	O(11)	O(8)	O(10)
O(9)	111.8	113.7	106.7
O(11)		110.6	104.8
O(8)			108.7

Si(4)-O(12)	1.577
Si(4)-O(14)	1.602
Si(4)-O(13)	1.624
Si(4)-O(11)	1.665

Angles in degrees subtended at Si(4) by

	O(14)	O(13)	O(11)
O(12)	116.7	109.1	100.7
O(14)		112.7	106.2
O(13)			110.7

Si(3)-O(11)-Si(4) 136.0±9°

Sm(1)-O(7)	2.345	Sm(13)-O(8)	2.270
Sm(1)-O(2)	2.348	Sm(3)-O(12)	2.275
Sm(1)-O(14)	2.385	Sm(3)-O(3)	2.342
Sm(1)-O(9)	2.389	Sm(3)-O(6)	2.416
Sm(1)-O(1)	2.419	Sm(3)-O(9)	2.426
Sm(1)-O(10)	2.435	Sm(3)-O(2)	2.556
Sm(1)-O(4)	2.662	Sm(3)-O(10)	2.662

Sm(2)-O(5)	2.332	Sm(4)-O(3)	2.365
Sm(2)-O(14)	2.362	Sm(4)-O(12)	2.337
Sm(2)-O(13)	2.399	Sm(4)-O(13)	2.394
Sm(2)-O(6)	2.470	Sm(4)-O(6)	2.449
Sm(2)-O(10)	2.571	Sm(4)-O(7)	2.487
Sm(2)-O(1)	2.572	Sm(4)-O(2)	2.508
Sm(2)-O(8)	2.610	Sm(4)-O(1)	2.773
Sm(2)-O(9)	2.749	Sm(4)-O(11)	2.801
Sm(2)-O(7)	2.957	Sm(4)-O(5)	2.855

σ(distances) = 0.15 Å, σ(angles) = 0.7°

Discussion

Isostructural with β-$Ca_2P_2O_7$ (1).

Details of analysis

Prepared from solution in Bi_2O_3. Lattice parameters from diffractometer [Mo$K\alpha$ radiation?, no calibration given]. Intensities measured with crystal-monochromated Mo$K\alpha$ radiation on single-crystal diffractomerer (ω scan) corrected for Lorentz effect, polarization and absorption (spherical crystal, diameter 0.27 mm). Structure assumed similar to β-$Ca_2P_2O_7$ (1) and refined by difference electron-density syntheses and least squares to give R = 0.046 for 1420 reflexions [not listed].

1. N.C. WEBB, 1967. *Acta Cryst.*, 21, 942.

PRASEODYMIUM DISILICATE

I. The crystal structure of α-$Pr_2Si_2O_7$. J. FELSCHE, 1970. *Naturwissenschaften*, 57, 452.

$Pr_2Si_2O_7$ F.W. = 450.0

Monoclinic, a = 5.40_5, b = 8.67_4, c = 14.07_5 Å, β = 112.59°, U = 609.25 Å^3, D_m = 4.86, Z = 4, D_x = 4.90.

[The lattice parameters are given erroneously in I as a = 8.67_4, b = 12.99_6, c = 5.40_5 Å, β[*sic*] = 90.00°, Space group $P2_1/n$].

Space group $P2_1/c$ (C^5_{2h})

Atomic positions All atoms in 4(*e*)

	x	y	z	B(Å^2)
Pr(1)	0.5212±1	0.8068±1	0.7685±1	0.40±1
Pr(2)	0.8481±1	0.6053±1	0.5901±1	0.40±1
Si(1)	0.7969±8	0.2476±4	0.0249±2	0.39±6
Si(2)	0.9086±7	0.4964±5	0.1788±2	0.44±6
O(1)	0.8329±21	0.4232±11	0.0627±7	0.64±19
O(2)	0.0727±20	0.1456±11	0.0789±7	0.32±18
O(3)	0.5726±20	0.1564±11	0.0536±7	0.36±18
O(4)	0.7350±20	0.2406±11	0.9057±7	0.37±17
O(5)	0.7469±20	0.4188±11	0.2413±7	0.40±18
O(6)	0.2170±19	0.4727±12	0.2469±7	0.33±17
O(7)	0.8183±21	0.6772±12	0.1722±7	0.50±18

Interatomic distances (Å) *and bond angles*

Si(1)-O(4)	1.581±10
Si(1)-O(1)	1.600±10
Si(1)-O(3)	1.623±12
Si(1)-O(2)	1.647±10

Angles in degrees subtended at Si(1) by

	O(1)	O(3)	O(2)
O(4)	109.8±5	111.8±5	104.7±6
O(1)		112.8±6	112.6±5
O(3)			104.7±5

Si(2)-O(6)	1.585±9
Si(2)-O(5)	1.607±12
Si(2)-O(7)	1.634±11
Si(2)-O(1)	1.651±10

Angles in degrees subtended at Si(2) by

	O(5)	O(7)	O(1)
O(6)	106.4±5	112.4±5	110.6±6
O(5)		102.7±6	113.6±5
O(7)			111.0±5

Pr(1)-O(4)	2.377±8	Pr(2)-O(7)	2.250±11
Pr(1)-O(5)	2.407±10	Pr(2)-O(3)	2.434±8
Pr(1)-O(2)	2.446±8	Pr(2)-O(4)	2.521±11
Pr(1)-O(6)	2.464±10	Pr(2)-O(2)	2.526±10
Pr(1)-O(7)	2.472±13	Pr(2)-O(6)	2.543±11
Pr(1)-O(3)	2.758±11	Pr(2)-O(5)	2.545±8
Pr(1)-O(5)	2.768±10	Pr(2)-O(2)	2.597±11
Pr(1)-O(6)	2.857±11	Pr(2)-O(3)	2.655±10

Si(1)-O(1)-Si(2) = 130.3±6°

Discussion

Isostructural with α-$Ca_2P_2O_7$(1).

Details of analysis

Crystals obtained by sintering $Pr_2Si_2O_7$ at 1550°C in air. Intensities measured with monochromatised MoK_α radiation on a single-crystal diffractometer with a spherical crystal (diameter = 0.06 mm) and corrected "in the usual way" [*sic*]. Structure solved from Patterson and electron-density functions are refined by least squares to give R = 0.046 for 1093 non-equivalent reflexions [not listed].

II. The crystal structure of β-$Pr_2Si_2O_7$. J. FELSCHE, 1970. *Naturwissenschaften*, 57, 669-670.

Tetragonal (β-$Ca_2P_2O_7$ type 2) a = 6.765$_7$, c = 24.60$_8$ Å, U = 1126 Å^3, Z = 8, D_x = 5.30.

Space group $P4_1$ (C_4^2)

Atomic positions

All atoms in 4(*a*)

	x	y	z	B(Å^2)
Pr(1)	-0.2340±8	0.2961±8	-0.0115±5	0.78±7
Pr(2)	0.5216±8	0.1671±8	0.1359±5	0.89±8
Pr(2)	0.3374±8	-0.0837±8	-0.0110±5	0.76±7
Pr(4)	0.1242±8	-0.2353±8	0.1277±5	0.91±8
Si(1)	-0.141±3	0.762±3	0.010±1	0.36±14
Si(2)	-0.391±3	0.689±3	0.109±1	0.48±15
Si(3)	0.267±3	0.374±3	0.016±1	0.43±15
Si(4)	0.012±3	0.289±3	0.115±1	0.35±14
O(1)	-0.106±8	0.620±9	-0.040±2	0.72±23
O(2)	-0.272±8	0.927±8	-0.015±2	0.63±23
O(3)	0.049±8	0.852±8	0.040±2	0.83±31
O(4)	-0.264±8	0.623±8	0.050±2	0.76±24
O(5)	0.491±9	0.524±9	0.130±2	1.10±36
O(6)	0.426±9	0.844±8	0.089±2	0.92±25
O(7)	-0.244±8	0.831±9	0.145±2	0.67±33
O(8)	0.308±8	0.567±9	-0.013±2	0.84±44
O(9)	0.471±8	0.235±8	0.034±2	0.77±24
O(10)	0.125±8	0.248±9	-0.035±2	0.94±25
O(11)	0.127±9	0.437±9	0.063±2	0.88±25
O(12)	-0.025±9	0.456±9	0.161±2	0.89±34
O(13)	0.178±9	0.123±9	0.141±2	1.11±25
O(14)	-0.183±9	0.212±9	0.088±2	0.64±24

Discussion

β-$Pr_2Si_2O_7$ has the β-$Ca_2P_2O_7$ structure (2). The interatomic distances are only accurate to 0.05 Å. The Si-O distances range from 1.47 to 1.80 with Si-O-Si angles at 129±4°. The Pr atoms are surrounded by between 7 and 9 oxygen atoms at distances between 2.2 and 2.9 Å.

Details of analysis

Intensities measured on single-crystal diffractometer with MoK_α radiation (θ-2θ scan) corrected in "the usual way" [*sic*]. Structure refined to give R_2 = 0.58 for 882 reflexions [not listed].

1. C. CALVO, 1968. *Inorg. Chem.*, 7, 1345.
2. N.C. WEBB, 1966. *Acta Cryst.*, 21, 942.

SODIUM ZIRCONIUM DISILICATE

I. [Crystal structure of a new natural modification of $Na_2Zr[Si_2O_7]$.] A.A. VORONKOV, N.G. SUMJACKAJA and JU.A. PJATENKO, 1970. *Ž. Strukt. Khim.*, 11, 932-933 [*Journal of Structural Chemistry*, 11, 866-867].

$Na_2Zr(Si_2O_7)$ F.W. = 305.36

Triclinic, a = 6.66±3, b = 8.83±4, c = 5.42±2 Å, α = 92.75, β = 94.25, γ = 72.33°, U = 303 Å^3, Z = 2, D_x = 3.34.

Space group $P\bar{1}$ (C_i^1)

Atomic positions All atoms in 2(i)

	x	*y*	*z*
Zr	0.289	0.271	0.220
Si(1)	0.642	0.146	0.772
Si(2)	0.935	0.336	0.689
Na(1)	0.874	0.099	0.264
Na(2)	0.299	0.492	0.738
O(1)	0.296	0.037	0.176
O(2)	0.884	0.169	0.745
O(3)	0.497	0.206	0.524
O(4)	0.557	0.252	0.023
O(5)	0.002	0.316	0.395
O(6)	0.123	0.340	0.886
O(7)	0.300	0.520	0.230

Interatomic distances (Å)

Zr	-O(1)	2.06
Zr	-O(3)	2.06
Zr	-O(6)	2.07
Zr	-O(4)	2.11
Zr	-O(5)	2.12
Zr	-O(7)	2.22
Si(1)	-O(1)	1.58
Si(1)	-O(3)	1.61
Si(1)	-O(4)	1.65
Si(1)	-O(2)	1.71
Si(2)	-O(6)	1.60
Si(2)	-O(2)	1.66
Si(2)	-O(5)	1.67
Si(2)	-O(7)	1.76
Na(1)	-O(5)	2.38
Na(1)	-O(2)	2.43
Na(1)	-O(4)	2.46

Na(1)-O(2)'	2.65
Na(1)-O(1)	2.78
Na(1)-O(3)	2.86
Na(1)-O(2)"	2.92
Na(1)-O(1)'	2.93
Na(2)-O(6)	2.24
Na(2)-O(5)	2.30
Na(2)-O(7)	2.64
Na(2)-O(7)'	2.67
Na(2)-O(3)	2.72
Na(2)-O(4)	2.75
Na(2)-O(7)"	2.78
Na(2)-O(4)'	2.93

Discussion

The Si_2O_7 groups (Si-O-Si = 127°) are linked together by sharing the terminal O atoms with Zr. The Na atoms lie in the voids in the resulting framework.

Details of analysis

Natural crystals found in conjunction with keldyshite. Intensities recorded on a KFOR camera. Structure solved by Patterson and refined by least squares to give R = 0.14 for $hk0$ and $0kl$ reflexions [not listed].

POTASSIUM ZIRCONIUM DISILICATE

[Crystalline structure of monoclinic modification of K,Zr-diorthosilicate ($K_2ZrSi_2O_7$).] A.N. ČERNOV, B.A. MAKSIMOV, V.V. ILJUKHIN and N.V. BELOV, 1970. *Dokl. Akad. Nauk SSSR*, 193, 1293-1296.

$K_2ZrSi_2O_7$ F.W. = 337.59

Monoclinic, a = 9.54, b = 14.26, c = 5.60 Å, γ = 116.51°, U = 682 Å^3, Z = 4, D_x = 3.28.

Space group $P2_1/b$ (C_{2h}^5)

Atomic positions c axis unique. All atoms in 4(e)

	x	y	z
Zr	0.237	0.232	0.253
Si(1)	0.172	0.369	0.744
Si(2)	0.336	0.114	0.741
K(1)	0.035	0.426	0.264
K(2)	0.482	0.592	0.260
O(1)	0.004	0.629	0.264
O(2)	0.309	0.492	0.795
O(3)	0.476	0.316	0.232
O(4)	0.217	0.342	0.464
O(5)	0.185	0.296	0.935
O(6)	0.245	0.111	0.007
O(7)	0.265	0.148	0.540

Interatomic distances (Å)

Si(1) - O(5)	1.54	K(1) - O(1)	2.74
Si(1) - O(2)	1.68	K(1) - O(4)	2.75
Si(1) - O(1)	1.69	K(1) - O(6)	2.81
Si(1) - O(4)	1.71	K(1) - O(6)	2.83
		K(1) - O(7)	2.86
Si(2) - O(7)	1.50	K(1) - O(1)	3.07
Si(2) - O(3)	1.62	K(1) - O(1)	3.03
Si(2) - O(2)	1.65	K(1) - O(5)	3.08
Si(2) - O(6)	1.72		
		K(2) - O(6)	2.72
Zr - O(4)	2.04	K(2) - O(2)	2.76
Zr - O(1)	2.05	K(2) - O(2)	3.08
Zr - O(3)	2.05	K(2) - O(3)	3.08
Zr - O(7)	2.09	K(2) - O(3)	2.99
Zr - O(5)	2.16	K(2) - O(4)	3.00
Zr - O(6)	2.24	K(2) - O(5)	3.08
		K(2) - O(7)	3.03

[The uncertainties in the bond lengths are probably very large. The wide variation in Si-O distances and the unusually large K-O distances suggest that this structure should be reexamined].

Details of analysis

Intensities estimated visually from Weissenberg photographs of layers with l = 0 to 5 and k = 0 and 1 taken with Mo radiation. Structure determined from Patterson and electron-density functions and refined by least squares to give R = 0.14 for 1200 observed reflexions [not listed].

TETRALITHIUM DIHYDROGEN DISILICATE

I. Darstellung und Kristallstrukur der Verbindung $Li_4H_2Si_2O_7$. H. VÖLLENKLE, A. WITTMANN and H. NOWOTNY, 1970. *Mh. Chem.*, 101, 684-692.

$Li_4H_2Si_2O_7$ F.W. = 197.9

Tetragonal, a = 7.595, c = 5.06 Å, U = 291 Å^3, Z = 2, D_x = 2.25.

Space group $P\bar{4}2_1/m$ (D^3_{2d})

Atomic positions

	x	y	z	B(Å^2)
4 Si in 4(e)	0.1425±11	½-x	0.1659±9	*
2 O(1) in 2(c)+	0	0.5	0.2914±38	*
4 O(2) in 4(e)	0.1426±31	½-x	0.8561±21	*
8 O(3) in 8(f)	0.0836±16	0.1754±13	0.2894±20	*
8 Li in 8(f)	0.0744±34	0.1382±39	0.6804±50	3.88±49

*Anisotropic temperature factors given.
+Given incorrectly in I as 4O(1) in 4(d).

Interatomic distances (Å)

Corrections for thermal motion increase all the distances by about 0.05 Å.

Si - O(2)	1.57
Si - O(3)	1.58 (×2)
Si - O(1)	1.66

σ(Si-O) = 0.03 Å

Angles in degrees subtended at Si by

	O(3)'	O(1)
O(2)	113	113
O(3)	110	104

(O-Si-O) = 2°

O(1) - Si - O(1) 135±3°

Li - O(3)	1.86
Li - O(3)	1.95
Li - O(2)	1.96
Li - O(3)	2.00

σ(Li-O) = 0.05 Å

Angles in degrees subtended at Li by

	O(3)	O(2)	O(3)
O(3)	102	129	101
O(3)		114	98
O(2)			109

σ(O-Li-O) = 2°

Discussion

The structure consists of Li^+ ions and $Si_2O_7^{6-}$ ions with C_{2v} symmetry. The hydrogen atoms could not be found.

Details of analysis

Lattice parameters from powder diagrams and Weissenberg photographs (CuK_α radiation). Intensities estimated visually from Weissenberg photographs of layers with h = 0 to 5 (CuK_α radiation) corrected for Lorentz and polarization effects. Structure based on that of $Li_6Si_2O_7$ (1) and 31 parameters refined by least squares to give R_2 = 0.134 for 119 reflexions listed.

PENTABARIUM DIHYDROXY TETRAMETASILICATE

I. [Crystal structure of the synthetic barium silicate $Ba_5(Si_4O_{12})(OH)_2$.] O.S. FILIPENKO, E.A. POBEDIMSKAJA, V.I. PONOMAREV, D.M. KHEIKER and N.V. BELOV, 1970. *Dokl. Akad. Nauk SSSR*, 194, 83-86 [*Soviet Physics - Doklady*, 15, 800-802].

$Ba_5(Si_4O_{12})(OH)_2$ F.W. = 1025.06

Tetragonal, a = 7.745±6, c = 11.68±6 Å, U = 700.6 Å³, D_m = 4.64, Z = 2, D_x = 4.86.

Space group P4/mnc (D_{4h}^6)

Atomic positions

	x	y	z
8 Ba(1) in 8(*g*)	0.3409	0.8409	0.25
2 Ba(2) in 2(*a*)	0	0	0
8 Si in 8(*h*)	0.114	0.275	0.5
16 O(1) in 16(*n*)	0.125	0.334	0.114
8 O(2) in 8(*h*)	0.409	0.279	0
4 O(3) in 4(*e*)	0	0	0.25

Interatomic distances (Å)

Si - O(1)	1.60 (×2)
Si - O(2)	1.64
Si - O(2)'	1.65
Ba(1) - O(1)	2.67 (×2)
Ba(1) - O(1)'	2.71 (×2)
Ba(1) - O(1)"	2.99 (×2)
Ba(1) - O(2)	3.11 (×2)
Ba(1) - O(3)	2.91 (×2)
Ba(2) - O(1)	3.06 (×8)
Ba(2) - O(3)	2.92 (×2)

Standard error 0.01 Å

Discussion

The structure consists of a body centered array of Si_4O_{12} rings with the rings in alternate layers rotated with respect to each other. The Ba(1) atoms lie in the plane of the rings, being bonded to four rings equatorially and two OH groups axially. The Ba(2) atoms lie between the planes of the rings.

Details of analysis

Crystals prepared hydrothermally. Lattice parameters from diffractometer. Intensities measured on automatic equi-inclination diffractometer with CuK_α radiation. Absorption correction for spherical crystal (d = 0.224 mm, μR = 12.2), Lorentz and polarization corrections. Structure determined from Patterson and electron-density functions and refined by least squares to give R = 0.073 for 310 non zero reflexions [not listed].

HIGH PRESSURE CALCIUM GERMANATE

I. The crystal chemistry of dense M_3O_4 polymorphs: high pressure Ca_2GeO_4 of K_2NiF_4 structure type. A.F. REID and A.E. RINGWOOD, 1970. *J. Solid State Chem.*, 1, 557-565.

Ca_2GeO_4 F.W. = 216.7

Tetragonal (K_2NiF_4 type, 1) a = 3.700±5, c = 11.88±1 Å, U = 162.6 Å^3, Z = 2, D_x = 4.42.

Space group I4/mmm (D_{4h}^{17})

Atomic positions

		x	y	z
2 Ge	in 2(a)	0	0	0
4 Ca	in 4(d)	0	0	0.353±5
4 O(1)	in 4(c)	1/2	0	0
4 O(2)	in 4(e)	0	0	0.162±15

B assumed to be 1 Å^2

Interatomic distances (Å)

Sixfold coordination around Ge (crystallographic site symmetry D_{4h})

Ge-O(1)	1.850±2 (×4)
Ge-O(2)	1.92 ±18 (×2)

Ninefold coordination around Ca (crystallographic site symmetry C_{4v})

Ca-O(2)	2.70±19
Ca-O(1)	2.54±4 (×4)
Ca-O(2)	2.62±1 (×4)

Details of analysis

$2CaCO_3$ and GeO_3 heated at 1200°C and reacted at 100-120 kbars and 900°C. Debye-Scherrer powder photographs taken at atmospheric pressure and room temperature with CuK_α radiation (λ = 1.5418 Å). Lattice and atom position parameters found by least-squares refinement to give R(intensities) = 0.26 for 31 powder lines listed.

1. *Structure Reports*, 17, 332.

WILLEMITE

I. [Crystal structures of willemite $Zn_2(SiO_4)$ and its germanium analog $Zn_2(GeO_4)$]. CHIN HANG, M.A. SIMONOV and N.V. BELOV, 1970. *Kristallografija*, 15, 457-460 [*Soviet Physics - Crystallography*, 15, 387-389].

ZINC ORTHOSILICATE

Zn_2SiO_4 F.W. = 218.82

Rhombohedral, a = 8.628 Å, α = 107.87°, U = 521.79 Å^3, Z = 6, D_x = 4.17.

Space group $R\bar{3}$ (C_{3i}^2)

Atomic positions rhombohedral setting All atoms in 6(*f*)

	x	*y*	*z*
Si	0.459	0.225	0.051
Zn(1)	0.793	0.562	0.385
Zn(2)	0.128	0.897	0.720
O(1)	0.329	0.019	-0.027
O(2)	0.614	0.307	0.258
O(3)	0.971	0.639	0.640
O(4)	0.553	0.229	-0.092

Interatomic distances (Å)

Errors uncertain but probably about 0.02-0.03 Å

Si - O(1)	1.56
Si - O(2)	1.63
Si - O(3)	1.66
Si - O(4)	1.66

Angles in degrees subtended at Si by

	O(2)	O(3)	O(4)
O(1)	112	116	103
O(2)		112	111
O(3)			101

Zn(1) - O(4)	1.90
Zn(1) - O(2)	1.96
Zn(1) - O(2)'	1.99
Zn(1) - O(3)	2.01

Angles in degrees subtended at Zn(1) by

	O(2)	O(2')	O(3)
O(4)	98	108	111
O(2)		116	114
O(2)'			109

Zn(2) - O(4)	1.91
Zn(2) - O(1)	1.97
Zn(2) - O(3)	1.98
Zn(2) - O(1)'	2.01

Angles in degrees subtended at Zn(2) by

	O(1)	O(3)	O(1)'
O(4)	99	108	100
O(1)		111	115
O(3)			114

ZINC ORTHOGERMANATE

Zn_2GeO_4 F.W. = 263.33

Rhombohedral, a = 8.836 Å, α = 107.70°, U = 563.2 Å^3, Z = 6, D_x = 4.65.

Space group $R\bar{3}$ (C_{3i}^2)

Atomic positions Rhombohedral setting All atoms in 6(*f*)

	x	*y*	*z*
Ge	0.465	0.229	0.056
Zn(1)	0.798	0.562	0.389
Zn(2)	0.132	0.896	0.723
O(1)	0.317	0.005	-0.034
O(2)	0.625	0.312	0.272
O(3)	0.973	0.641	0.636
O(4)	0.579	0.249	-0.078

Interatomic distances (Å)

Errors uncertain but probably about 0.02-0.03 Å

Ge - O(2)	1.75
Ge - O(3)	1.76
Ge - O(1)	1.76
Ge - O(4)	1.79

Angles in degrees subtended at Ge by

	O(3)	O(1)	O(4)
O(2)	114	114	107
O(3)		113	100
O(1)			108

Zn(1) - O(4)	1.90
Zn(1) - O(2)	1.97
Zn(1) - O(3)	2.00
Zn(1) - O(2)'	2.00

Angles in degrees subtended at Zn(1) by

	O(2)	O(3)	O(2)'
O(4)	104	108	104
O(2)		114	114
O(3)			112

Zn(2) - O(4)	1.89
Zn(2) - O(1)	1.96
Zn(2) - O(1)'	1.99
Zn(2) - O(3)	2.00

Angles in degrees subtended at Zn(2) by

	O(1)	O(1)'	O(3)
O(4)	106	100	110
O(1)		113	113
O(1)'			112

Discussion

ZnO_4 and SiO_4 tetrahedra share corners to form columns along the hexagonal *c* axis. The columns are joined by a triply shared corner O atom lying on the 3_1 axis. Six such triple columns again corner share into tubes leaving open channels centered on the $\bar{3}$ axis.

Details of analysis

Crystals prepared by hydrothermal synthesis. Lattice parameters from X-ray rotational diagrams. Intensities estimated visually from Weissenberg photographs of layers with l = 0 to 6 and h = 0. Structure solved from Patterson and electron-density functions and refined by least squares to give R = 0.14 for 250 reflexions of Si-Willemite and 0.13 for 226 reflexions of Ge-Willemite.

SODIUM SAMARIUM GERMANATE

I. [Crystal structure of sodium samarium orthogermanate.] E.A. KHARAKH, A.V. ČIČAGOV and N.V. BELOV, 1970. *Kristallografija*, 15, 1064-1065 [*Soviet Physics - Crystallography*, 15, 924-925].

$NaSmGeO_4$ F.W. = 309.98

Orthorhombic, a = 5.27±2, b = 11.70±3, c = 6.50±2 Å, U = 401 Å^3, D_m = 4.9, Z = 4, D_x = 5.13.

Space group $Pbn2_1$ (C_{2v}^3) Systematic absences and piezoelectric effect

Atomic positions All atoms in $(x,y,z;\bar{x},\bar{y},\frac{1}{2}+z;\frac{1}{2}+x,\frac{1}{2}-y,\frac{1}{2}+z;\frac{1}{2}-x,\frac{1}{2}+y,z)$

	x	*y*	*z*
Sm	0.0013	0.221	0
Ge	0.057	0.094	0.500
Na	0.500	0.002	0.250
O(1)	0.711	0.330	0.200
O(2)	0.217	0.170	0.301
O(3)	0.345	0.453	0.500
O(4)	0.770	0.110	0.500

Interatomic distances (Å)

Errors uncertain but probably greater than 0.03 Å.
These values differ appreciably from those given in I.

Sm - O(3)	2.20
Sm - O(1)	2.32
Sm - O(2)	2.34
Sm - O(2)'	2.36
Sm - O(1)'	2.38
Sm - O(4)	2.44

Ge - O(4)	1.53
Ge - O(3)	1.73
Ge - O(1)	1.77
Ge - O(2)	1.78

Angles in degrees subtended at Ge by

	O(3)	O(1)	O(2)
O(4)	114	113	114
O(3)		110	110
O(1)			94

Na - O(2)	2.49
Na - O(3)	2.50
Na - O(4)	2.50
Na - O(3)'	2.51
Na - O(1)	2.55

Discussion

Structure similar to that of $NaYSiO_4$ (1).

Details of analysis

Intensities estimated visually from Weissenberg photographs of layers with $h = 0$ to 4 taken with MoK_α radiation. Structure determined from Patterson and electron-density functions and refined by Fourier methods to give $R = 0.15$ for 516 observed reflexions [not listed].

1. B.A. MAKSIMOV, V.V. ILJUKHIN and N.V. BELOV, 1966. *Kristallografija*, 11, 681.

SODIUM TITANOGERMANATE

$Na_2TiO(GeO_4)$ F.W. = 246.47

I. [Crystal structure of Na-germanotitanate.] V. JA. VERKHOVSKIJ, E.A. KUZ'MIN, V.V. ILJUKHIN and N.V. BELOV, 1970. *Dokl. Akad. Nauk, SSSR.*, 190, 91-93 [*Soviet Physics - Doklady*, 15, 7-8].

Tetragonal, $a = 6.67$, $c = 5.16$ Å, $U = 230$ Å^3, $Z = 2$, $D_x = 3.56$.

Space group $P4/nmm$ (D_{4h}^7)

Atomic positions (non centrosymmetric setting)

		x	*y*	*z*
2 Ti	in 2(*c*)	1/2	0	0.924
2 Ge	in 2(*a*)	0	0	0
4 Na	in 4(*e*)	1/4	1/4	1/2
8 O(1)	in 8(*i*)	0.213	0	0.808*
2 O(2)	in 2(*c*)	1/2	0	0.257

[*inferred from interatomic distances since part of this number is missing in I.]

Interatomic distances (Å)

[Error not given but probably about 0.02 Å]

Fourfold coordination around Ge (crystallographic symmetry D_{2d})

Ge - O(1) 1.74 (×4)

Environment of Ti (crystallographic symmetry D_{4h})

Ti - O(1)	2.00 (×4)
Ti - O(2)	1.72
Ti - O(2)	3.44

Sixfold coordination around Na (crystallographic symmetry C_{2h})

Na - O(1)	2.32 [2.30]	(×4)
Na - O(2)	2.67	(×2)

Discussion

The titanium atoms are surrounded by an octahedron of oxygen atoms which are linked through shared corners into chains along *c*. The titanium atoms are, however, so far displaced from the centre of the octahedra in the *c* direction that they are effectively only coordinated to five oxygen atoms at the corners of a square pyramid. The chains of TiO_6 octahedra are linked into a three-dimensional network by GeO_4 tetrahedra with which they share corners, while the sodium atoms fit into octahedral holes in this framework.

Details of analysis

Prepared hydrothermally from system Na_2O-GeO_2-H_2O in concentrated NaOH. 300 reflexions measured on layers h = 0 to 4, R = 0 to 1 with MoK_α radiation. Structure solved from Patterson function, R = 0.128. Structure factors not given.

ERBIUM DIGERMANATE

I. [Determination of the crystal structure of erbium pyrogermanate, $Er_2Ge_2O_7$.] JU.I. SMOLIN, 1970. *Kristallografija*, 15, 47-51 [*Soviet Physics - Crystallography*, 15, 36-39].

$Er_2Ge_2O_7$ F.W. = 591.67

Tetragonal, a = 6.778±5, c = 12.34±1 Å, U = 5669 Å^3, D_m = 6.99, Z = 4, D_x = 6.94.

Space group $P4_12_12$ (D_4^4) systematic absences

Atomic positions

	x	*y*	*z*	$B(Å^2)$
Er in 8(*b*)	0.87559±5	0.35274±5	0.13550±5	0.34
Ge in 8(*b*)	0.90037±11	0.15205±13	0.61955±12	0.33
O(1) in 4(*a*)	0.8055±9	0.1945±9	0.75	1.01
O(2) in 8(*b*)	0.0764±9	-0.0331±9	0.6236±9	0.61
O(3) in 8(*b*)	0.0638±8	0.3355±8	0.5751±8	0.43
O(4) in 8(*b*)	0.6833±8	0.1449±10	0.5427±9	0.49

Interatomic distances (Å)

Ge_2O_7 group (crystallographic point symmetry C_2)

Ge - O(2)	1.733±6
Ge - O(4)	1.752±8
Ge - O(3)	1.754±6

Angles in degrees subtended at Ge by

	O(3)	O(1)
O(4)	112.4±4	101.1±4
O(3)		113.7±4

Ge - O(1) - Ge 136.0±4°

Er - O(2)	2.196±6
Er - O(4)	2.217±1
Er - O(4)	2.268±7
Er - O(3)	2.278±6
Er - O(2)	2.372±6
Er - O(3)	2.517±6

Isostructural compounds

	a (Å)	*c* (Å)
$Yb_2Ge_2O_7$	6.745±5	12.26±1
$Ho_2Ge_2O_7$	6.806	12.38
$Y_2Ge_2O_7$	6.806	12.37
$Dy_2Ge_2O_7$	6.828	12.43
$Tb_2Ge_2O_7$	6.856	12.47

Discussion

The erbium atoms lie at the centres of bipyramids which share edges with five adjacent ErO_7 polyhedra.

Details of analysis

Crystals grown from solution in fused Bi_2O_3. Lattice parameters measured from diffractometer [no calibration given]. Isostructural compounds identified from powder photographs. Intensities measured on single-crystal diffractometer with MoK_α radiation and corrected for Lorentz effect, polarization and absorption (crystal = 0.28 mm diameter sphere). Structure solved from Patterson and difference electron-density function and refined by least squares to give R = 0.056 for 870 observed unequivalent reflexions [not listed].

LITHIUM HEPTAGERMANATE

I. Die Kristallstruktur des Lithium Heptagermanate $Li_2[Ge_7O_{15}]$. H. VÖLLENKLE, A. WITTMANN and H. NOWOTNY, 1970. *Mh. Chem.*, 101, 46-56.

$Li_2Ge_7O_{15}$ F.W. = 762.0

Orthorhombic, a = 7.36, b = 16.76, c = 9.69 Å, U = 1193 Å^3, Z = 4, D_x = 4.235.

Space group Pbcn (D_{2h}^{14})

Atomic positions

	x	*y*	*z*	$B(\AA^2)$
4 Ge(1) in 4(*c*)	0	0.4133±2	0.25	0.71±6
8 Ge(2) in 8(*d*)	0.2138±3	0.0582±1	0.2501±3	0.76±5
8 Ge(3) in 8(*d*)	0.2844±4	0.3403±1	0.0328±3	0.66±5
8 Ge(4) in 8(*d*)	0.2872±4	0.3446±2	0.4713±3	0.79±5
4 O(1) in 4(*c*)	0	0.0143±15	0.25	2.11±46
8 O(2) in 8(*d*)	0.2837±23	0.0882±9	0.4097±16	0.71±24
8 O(3) in 8(*d*)	0.1893±23	0.1482±8	0.1518±15	0.45±23
8 O(4) in 8(*d*)	0.1978±23	0.2476±9	0.4418±15	0.71±25
8 O(5) in 8(*d*)	0.1425±23	0.3283±8	0.1704±14	0.35±23
8 O(6) in 8(*d*)	0.5047±34	0.3578±10	0.0936±19	1.99±30
8 O(7) in 8(*d*)	0.1475±23	0.4169±8	0.4083±15	0.57±24
8 O(8) in 8(*d*)	0.1416±21	0.4928±7	0.1676±13	0.00±21
4 Li(1) in 4(*b*)	0	0.5	0	0.99±78
4 Li(2) in 4(*c*)	0	0.2369±42	0.25	2.68±123

Interatomic distances (Å)

Ge(1) - O(8)	1.871±13 (×2)
Ge(1) - O(7)	1.880±15 (×2)
Ge(1) - O(5)	1.930±15 (×2)

Angles in degrees subtended at Ge(1) by

	O(8)'	O(7)'	O(5)'
O(8)	89.2±6	87.2±6	93.0±6
O(8)'		90.2±6	177.7±6
O(7)		176.3±6	92.1±6
O(7)'			90.7±6
O(5)			84.8±6

Ge(2) - O(2)	1.706±16
Ge(2) - O(8)	1.724±15
Ge(2) - O(1)	1.737±11
Ge(2) - O(3)	1.793±14

Angles in degrees subtended at Ge(2) by

	O(8)	O(1)	O(3)
O(2)	114.9±7	113.5±6	105.3±7
O(8)		106.8±8	110.5±6
O(1)			105.4±8

Ge(3) - O(5)	1.706±15
Ge(3) - O(4)	1.722±15
Ge(3) - O(6)	1.750±24
Ge(3) - O(2)	1.764±16

Angles in degrees subtended at Ge(3) by

	O(4)	O(6)	O(2)
O(5)	110.2±7	108.9±8	115.8±7
O(4)		104.2±8	104.9±7
O(6)			112.2±8

Ge(4) - O(6)	1.670±24
Ge(4) - O(7)	1.702±15
Ge(4) - O(3)	1.762±15
Ge(4) - O(4)	1.777±16

Angles in degrees subtended at Ge(4) by

	O(7)	O(3)	O(4)
O(6)	108.9±8	105.9±9	113.6±8
O(7)		111.5±7	111.7±7
O(3)			105.0±7

Li(1) - O(8)	1.933±13 (×2)
Li(1) - O(7)	1.977±15 (×2)
Li(1) - O(2)	2.342±16 (×2)

Angles in degrees subtended at Li(1) by

	O(8)	O(7)'	O(2)'
O(8)	180	82.8±6	84.7±5
O(8)'		97.2±6	95.3±5
O(7)		180	84.5±6
O(7)'			95.5±6
O(2)			180

Li(2) - O(5)	2.01±5 (×2)
Li(2) - O(3)	2.25±5 (×2)
Li(2) - O(4)	2.37±2 (×2)

Angles in degrees subtended at Li(2) by

	O(5)'	O(3)'	O(4)'
O(5)	80.7±27	91.0±6	87.8±15
O(5)'		171.6±27	85.6±14

	O(3)'	O(4)'
O(3)	97.2±27	90.1±12
O(3)'		95.7±12
O(4)		171.3±34

Discussion

GeO_6 octahedra and GeO_4 tetrahedra are linked into rings of 6 polyhedra perpendicular to *a* by corner sharing. The centres of the rings form channels along *a* containing the Li ions.

Details of analysis

Lattice parameters from powder photographs (CrK_α radiation). Intensities measured visually from Weissenberg photographs of layers with $h = 0$ to 6. (CuK_α radiation). Lorentz, polarization and cylindrical absorption corrections ($\mu R = 0.55$, crystal size 0.05×0.05×0.1 mm^3). Structure found from Patterson function and 54 parameters refined by least squares to give $R = 0.072$ for 484 observed reflexions listed but 658 reflexions were unobserved.

BASIC SODIUM TIN GERMANATE

I. The crystal structure of $Na_4Sn_2Ge_4O_{12}(OH)_4$. A.N. CHRISTENSEN, 1970. *Acta Chem. Scand.*, 24, 1287-1293.

$Na_4Sn_2Ge_4O_{12}(OH)_4$ F.W. = 879.72

Triclinic, $a = 5.778$, $b = 11.615$, $c = 5.549$ Å, $\alpha = 107.62°$, $\beta = 75.75°$, $\gamma = 94.75°$, $U = 343.4$ Å^3, $Z = 1$, $D_x = 4.24$.

Space group $P\bar{1}$ (C_i^1)

Atomic positions

	x	*y*	*z*	
2 O(1) in 2(*i*)	0.1681±17	0.0535±9	0.3031±19	
2 O(2) in 2(*i*)	0.6798±15	0.0364±9	0.2522±19	
2 O(3) in 2(*i*)	0.4919±16	0.2098±9	0.0740±21	
2 O(4) in 2(*i*)	0.4415±15	0.2321±8	0.6024±19	
2 O(5) in 2(*i*)	0.0241±17	0.1687±8	0.9668±20	
2 O(6) in 2(*i*)	0.2440±17	0.4028±9	0.0626±18	
2 O(7) in 2(*i*)	0.7365±16	0.4398±9	0.2449±20	
2 O(8) in 2(*i*)	0.9381±18	0.3612±9	0.6925±23	
2 Na(1) in 2(*i*)	0.5830±10	0.4378±5	0.6710±11	
2 Na(2) in 2(*i*)	0.8251±10	0.1509±6	0.6435±12	
2 Ge(1) in 2(*i*)	0.2808±2	0.2501±1	0.9317±3	
2 Ge(2) in 2(*i*)	0.4378±2	0.1251±1	0.3009±3	
1 Sn(1) in 1(*a*)	0	0	0	
1 Sn(2) in 1(*c*)	0	0.5	0	
2 H(1) in 2(*i*)	0.697±40	0.360±23	0.191±49	see discussion
2 H(2) in 2(*i*)	0.316±32	0.271±18	0.658±39	

Anisotropic temperature factors given.

Interatomic distances (Å)

Sn(1) - O(5)	2.013 (×2)
Sn(1) - O(2)	2.016 (×2)
Sn(1) - O(1)	2.055 (×2)

Angles in degrees subtended at Sn(1) by

	O(2)	O(1)
O(5)	90.9	90.9
O(5)'	89.1	89.1
O(2)		90.6
O(2)'		89.4

Sn(2) - O(7)	2.009	(×2)
Sn(2) - O(6)	2.033	(×2)
Sn(2) - O(8)	2.042	(×2)

Angles in degrees subtended at Sn(2) by

	O(6)	O(8)
O(7)	88.9	90.6
O(7)'	91.1	89.4
O(6)		87.8
O(6)'		92.2

Ge(1) - O(5)	1.696
Ge(1) - O(6)	1.710
Ge(1) - O(3)	1.768
Ge(1) - O(4)	1.790

Angles in degrees subtended at Ge(1) by

	O(6)	O(3)	O(4)
O(5)	114.2	112.7	115.2
O(6)		106.3	104.1
O(3)			103.2

Ge(2) - O(1)	1.706
Ge(2) - O(2)	1.727
Ge(2) - O(4)	1.758
Ge(2) - O(3)	1.775

Angles in degrees subtended at Ge(2) by

	O(2)	O(4)	O(3)
O(1)	116.8	111.7	109.5
O(2)		102.3	110.8
O(4)			105.0

Na(1) - O(7)	2.297
Na(1) - O(7)	2.316
Na(1) - O(6)	2.320
Na(1) - O(8)	2.347
Na(1) - O(4)	2.406
Na(1) - O(6)	2.650

Na(2) - O(5)	2.309
Na(2) - O(8)	2.427
Na(2) - O(1)	2.452
Na(2) - O(2)	2.478
Na(2) - O(1)	2.484
Na(2) - O(4)	2.568
Na(2) - O(3)	2.616

σ(distances) = 0.010 Å; σ(angles) = 0.5°

Discussion

GeO_4 tetrahedra are linked through shared corners into chains along *c* which are cross linked by SnO_6 octahedra and Na ions. The hydrogen atoms were assumed to be bonded to the bridging oxygen atoms (O(3),O(4)) of the GeO_3 chain but on refinement one (H(1)) moved close to O(7). [The other (H(2)) is manifestly in an absurd position only 1.6 Å from Ge(1) and probably should be associated with O(8). In any case both hydrogen atoms are only 2.4 Å from Na(1) which is improbably close. It is unlikely that the positions suggested for the hydrogen atoms have any justification].

Details of analysis

Prepared by hydrothermal synthesis. Weissenberg photographs of layers with l = 0 and 1 and precession photographs of layers with h = 0 and k = 0 (CuK_α radiation, λ = 1.5418 Å). Intensities measured on equi-inclination diffractometer with monochromated MoK_α radiation and corrected for Lorentz, polarization and absorption effects (μ = 13.0 mm^{-1}, crystal size 0.17×0.17×0.04). Structure determined by heavy-atom (Sn) method and electron-density function and refined by least squares to give R = 0.054 for 1194 reflexions listed. Inclusion of the H atoms at positions found from difference synthesis did not alter R.

NITRATES AND NITRITES

I. Evidence of additional disorder in the rhombohedral form of potassium nitrite (KNO_2-II). K.O. STRØMME, 1970. *Acta Chem. Scand.*, 24, 1475-1477.

II. On the crystal structure of the high temperature form of silver nitrate ($AgNO_3$-I). *Idem*, 1970. *Ibid.*, 24, 1477-1479.

III. On the crystal structure of lithium nitrate above room temperature. *Idem*, 1970. *Ibid.*, 24, 1479-1481.

Models for the high-temperature disordered phases of KNO_2, $AgNO_3$ and $LiNO_3$ are proposed based on considerations of low-temperature structures and entropy.

AMMONIUM DIHYDROGEN NITRATE

$NH_4NO_3.2HNO_3$

I. A refinement of the crystal structure of $NH_4NO_3.2HNO_3$. F.W.B. EINSTEIN and D.G. TUCK, 1970. *Acta Cryst.*, B26, 1117-1120.

Atomic positions

	x	*y*	*z*	*B*(Å^2)
2 O(11) in 2(*b*)	0.2043±15	½	0	
4 O(12) in 4(*c*)	0.4902±13	0.4468±6	0.1871±21	
4 O(21) in 4(*c*)	0.3222±13	0.2134±6	0.1654±18	
4 O(22) in 4(*c*)	0.1106±12	0.1736±6	0.5147±21	
4 O(23) in 4(*c*)	0.2408±13	0.3318±6	0.4950±18	
2 N(11) in 2(*b*)	0.3933±16	½	0	
4 N(2) in 4(*c*)	0.2248±13	0.2337±6	0.3847±19	
2 NH_4 in 2(*a*)	0.1182±19	0	0	0.0274±21

Anisotropic thermal parameters are given.

Interatomic distances (Å) *and angles*

N(11)-O	1.252*, 1.273*,	O-N(11)-O	120,119°
N(2)-O	1.233*, 1,241*, 1,353*	O-N(2)-O	126, 118, 116
O(12)-O(23)	2.601		

* Bond lengths are corrected for thermal motion assuming that the oxygen atoms ride on the nitrogen atom. Errors ±0.013 on bond lengths and 1° on angles.

Structure

Average shift of the order of 0.002 Å from the values reported by Duke and Llewellyn (1). Nitrate groups are associated into $[H_2N_3O_9]^-$ anions containing a two-fold axis.

Details of analysis

Refinement by least-squares method using the data of Duke and Llewellyn. Final *R* value is 0.094. Space group used was $P22_12_1$, although Duke and Llewellyn indicated the true space group was $P2_1$.

1. *Structure Reports*, 13, 299.

CUPRIC NITRATE

I. The crystal structure of $Cu(NO_3)_2.2{\cdot}5H_2O$, B. MOROSIN, 1970. *Acta Cryst.*, B26, 1203.

$Cu(NO_3)_2.2{\cdot}5H_2O$ F.W. = 232.59

Monoclinic, $a = 16.4539 \pm 4$, $b = 4.9384 \pm 3$, $c = 15.96321 \pm 3$ Å, $\beta = 93.764 \pm 2°$, [$U = 1294.18$ Å^3], $D_m = 2.35$, $Z = 8$, $D_x = 2.38$.

Space group I2/a (C^6_{2h})

Atomic positions

			x	*y*	*z*
8 Cu	in	8(*f*)	0.12613±2	0.01352±6	0.11376±2
8 N(1)	in	8(*f*)	0.1683±1	-0.0243±4	0.2797±1
8 N(2)	in	8(*f*)	0.4016±1	0.0765±4	0.9888±1
8 O(1)	in	8(*f*)	0.1914±1	0.0299±4	0.3534±1
8 O(2)	in	8(*f*)	0.1580±1	0.1685±4	0.2264±1
8 O(3)	in	8(*f*)	0.1557±1	-0.2597±4	0.2556±1
8 O(4)	in	8(*f*)	0.4235±1	-0.0602±4	0.0500±1
8 O(5)	in	8(*f*)	0.3725±1	-0.0160±4	-0.0785±1
8 O(6)	in	8(*f*)	0.4096±1	0.3393±4	0.9941±1
8 O(7)	in	8(*f*)	0.2407±1	-0.0172±4	0.0833±1
8 O(8)	in	8(*f*)	0.0119±1	0.0457±4	0.1399±1
4 O(9)	in	4(*e*)	1/2	0.0458±6	1/4
8 H(1)	in	8(*f*)	0.267±3	0.11±1	0.120±3
8 H(2)	in	8(*f*)	0.265±3	-0.18±1	0.099±3
8 H(3)	in	8(*f*)	-0.025±3	0.05±1	0.091±3
8 H(4)	in	8(*f*)	0.000±3	0.20±1	0.175±3
8 H(5)	in	8(*f*)	0.452±4	-0.05±1	0.247±4

Anisotropic thermal parameters are given

Interatomic distances (Å) *and angles*

Cu-O(8)	1.959	O(8)-Cu-O(6)	91.20°
Cu-O(7)	1.959	O(8)-Cu-O(5)	81.91
Cu-O(6)	1.978	O(8)-Cu-O(3)	90.72
Cu-O(2)	1.992	O(8)-Cu-O(2)	90.99
Cu-O(5)	2.391	O(8)-Cu-O(5)	90.14
Cu-O(3)	2.653	O(6)-Cu-O(5)	102.99
Cu-O(5')	2.675	O(6)-Cu-O(7)	89.18
		O(5')-Cu-O(7)	98.37
		O(3)-Cu-O(7)	89.11
		O(2)-Cu-O(5)	80.40
		O(2)-Cu-O(7)	88.63
		O(5)-Cu-O(7)	89.73
N(1)-O(3)	1.238	O(3)-N(1)-O(1)	122.1
N(1)-O(1)	1.241	O(3)-N(1)-O(2)	118.6
N(1)-O(2)	1.281	O(1)-N(1)-O(2)	119.2
N(2)-O(4)	1.222	O(4)-N(2)-O(5)	124.6
N(2)-O(5)	1.235	O(4)-N(2)-O(6)	118.3
N(2)-O(6)	1.306	O(5)-N(2)-O(6)	117.0

Description of the structure

A more accurate structure than that reported by Dornberger-Schiff and Lierejewicz (1) and Garaj (2).

Details of the analysis

2335 intensities measured with MoK_α radiation and refined by least-squares to a final R value of 0.034.

1. K. DORNBERGER-SCHIFF and J. LEIREJEWICZ, 1958, *Acta Cryst.*, 11, 285.
2. J. GARAJ, 1968. *Acta Chem. Scand.*, 22, 1710.

ZINC NITRATE TETRAHYDRATE

I. Die Kristallstruktur von $Zn(NO_3)_2.4H_2O$. B. RIBAR and W. NOWACKI, 1970. *Z. Kristallogr.*, 131, 175-185.

$Zn(NO_3)_4.4H_2O$ F.W. = 259.44

Monoclinic ($Ni(NO_3)_2.4H_2O$ type, 1) a = 5.321±3, b = 27.36±2, c = 5.730±3 Å, β = 113.67±33°, U = 764.0 Å^3, D_m = 2.22, Z = 4, D_x = 2.25.

Space group $P2_1/n$ (C_{2h}^5)

Atomic positions All atoms in 4(*e*) (x,y,z; ½+x, ½−y, ½+z)

	x	y	z	B(Å^2)
Zn	0.9791±3	0.1258±1	0.2760±3	1.87
O(1)	0.2420±20	0.0747±3	0.2032±21	2.32
O(2)	0.0466±18	0.100±3	0.2787±20	1.95
O(3)	0.4235±17	0.0033±3	0.2250±18	1.77
O(4)	0.0456±20	0.1601±3	0.9763±19	2.14
O(5)	0.1882±21	0.2215±3	0.8282±19	2.27
O(6)	0.0842±21	0.2326±3	0.1471±19	2.23
O(7)	0.3578±18	0.1541±3	0.5603±17	1.71
O(8)	0.9267±22	0.0887±3	0.5683±20	2.58
O(9)	0.6290±17	0.0917±3	0.0002±16	1.39
O(10)	0.7160±18	0.1765±3	0.3180±18	1.65
N(1)	0.2390±20	0.0276±3	0.2389±20	1.39
N(2)	0.1067±21	0.2055±3	0.9863±21	1.51
H(1)	0.420±50	0.120±87	0.700±46	6.10
H(2)	0.290±34	0.180±60	0.660±32	.20
H(3)	0.960±50	0.061±87	0.660±46	6.90
H(4)	0.980±49	0.098±86	0.720±46	5.38
H(5)	0.510±43	0.082±75	0.093±40	2.80
H(6)	0.660±47	0.068±83	0.890±44	4.60
H(7)	0.730±40	0.210±69	0.330±37	1.80
H(8)	0.570±46	0.170±82	0.470±44	3.80

Anisotropic temperature factors also given for non-H atoms.

Interatomic distances (Å) *and angles*

Zn - O(10)	2.053
Zn - O(8)	2.071
Zn - O(4)	2.108
Zn - O(9)	2.116
Zn - O(1)	2.135
Zn - O(7)	2.164

σ = 0.010 Å

Angles in degrees subtended at Zn by

	O(8)	O(4)	O(9)	O(1)	O(7)
O(10)	85.2	98.1	87.0	175.7	98.2
O(8)		176.7	91.4	98.6	87.4
O(4)			88.5	78.1	92.5
O(9)				90.9	174.5
O(1)					84.0

σ = 0.4°

N(1) - O(3)	1.22
N(1) - O(2)	1.24
N(1) - O(1)	1.30

Angles in degrees subtended at N(1) by

	O(2)	O(1)
O(3)	123	118
O(2)		119

N(2) - O(6)	1.22
N(2) - O(5)	1.23
N(2) - O(4)	1.28

Angles in degrees subtended at N(2) by

	O(5)	O(4)
O(6)	120	121
O(5)		118

σ(N-O) = 0.02 Å; σ(O-N-O) = 1°

O(7)-H(2)...O(5)	2.77
O(7)-H(1)...O(9)	2.90
O(8)-H(3)...O(2)	2.82
O(8)-H(4)...O(4)	2.91
O(9)-H(5)...O(1)	2.78
O(9)-H(6)...O(3)	2.86
O(10)-H(7)...O(5)	2.80
O(10)-H(8)...O(7)	2.84

The H atoms are about 1 Å from the water O atoms but their positions is not accurately enough known to make useful comparisons.

Discussion

The structure consists of discrete *cis*-$(NO_3)_2(H_2O)_4Zn$ molecules linked to each other through hydrogen bonds.

Details of analysis

Crystal mounted in Lindemann glass capillaries. Lattice parameters from rotation and Weissenberg photographs [no calibration given]. Intensities measured photometrically from integrated Weissenberg photographs of layers with h = 0 to 2 and l = 0 to 3 and the "usual corrections" made as well as corrections for absorption (μR = 0.8). Structure assumed to be the same as that for $Ni(NO_3)_2.4H_2O$ (1) and 150 parameters refined by least squares to give R = 0.113 for 1402 reflexions listed. The inclusion of the hydrogen atoms did not alter R.

1. P. GALLEZOT, D. WEIGEL and M. PRETTRE, 1967. *Acta Cryst.*, 22, 699.

BASIC ZINC NITRATE

$Zn_5(OH)_8(NO_3)_2.2H_2O$

I. The crystal structure of zinc hydroxide nitrate, $Zn_5(OH)_8(NO_3)_2.2H_2O$. W. STAHLIN and H.R. OSWALD, 1970. *Acta Cryst.*, B26, 860-863.

$Zn_5(OH)_8(NO_3)_2.2H_2O$ F.W. = 622.948

Monoclinic, a = 19.480±5, b = 6.238±1, c = 5.517±1 Å, β = 93.28 2°, U = 669.2 Å^3, D_m = 3.02, Z = 2, D_x = 3.115.

Space group C2/m (C^3_{2h}) from crystal habit and confirmed by structure

Atomic positions

	x	y	z
Zn(1)	0	½	0
Zn(2)	0	0.2518±5	½
Zn(3)	0.0978±1	0	0.0079±4
OH(1)	0.0622±4	0.2512±20	0.1882±13
OH(2)	0.4402±7	½	0.3397±20
OH(3)	0.4492±6	½	0.3091±21
O(1)	0.2863±7	½	0.4974±29
O(2)	0.1845±8	½	0.3090±25
O(3)	0.3032±8	½	0.3014±28
H O	0.3123±7	½	0.0125±25
N	0.2206±9	½	0.4950±35

Anisotropic thermal parameters are given.

Interatomic distances (Å) *and angles*

N-O		1.277±23, 1.209±25, 1.242±23
O-N-O		123±2, 115±2, 122±2°
Zn(1)-OH	(4×)	2.193±12, (2×) 2.020±11
Zn(2)-OH	(2×)	2.160±12, (2×) 2.118±11, (2×) 2.089±11
Zn(3)-OH	(2×)	1.937±12, 1.963±11
$-H_2O$		1.953±11

Discussion of the structure

The structure contains sheets of edge sharing $Zn(OH)_6$ as in structure *C*6 (CdI_2-type) with one-quarter of the zinc atoms removed from the octahedral sites and Zn ions in tetrahedral sites above and below this empty site. Sheets stack along [010] with adjacent sheets shifted by *b*/2 and with NO_3 ions between sheets.

Materials

Single crystals grown from an aqueous solution of $Zn(NO_3)_2$ and urea using powder obtained from ZnO_2 or $Zn(OH)_2$ in aqueous $Zn(NO_3)_2$.

Details of analysis

491 reflexions were measured from multiple-film integrated Weissenberg photographs using a microdensitometer. Cu radiation, absorption corrected. Least-squares refined to *R* value of 0.064 with separate scales for the various layers.

POTASSIUM TETRANITRATOAURATE (III)

I. The crystal structure of anhydrous nitrates and their complexes. Part V. Potassium tetranitratoaurate (III). C.D. GARNER and S.C. WALLWORK, 1970. *J. Chem. Soc. A*, 3092-3095.

$KAu(NO_3)_4$ F.W. = 484.08

Monoclinic, a = 9.21±2, b = 7.14±2, c = 10.04±2 Å, β = 128.6±3°, U = 515.98 Å^3, D_m = 2.97, Z = 2, D_x = 3.11.

Space group $P2_1/c$ (C_{2h}^5) Systematic absences

Atomic positions

		x	*y*	*z*
2 K	in 2(*b*)	0.5	0	0
2 Au	in 2(*a*)	0	0	0
4 N(1)	in 4(*e*)	0.2070±20	0.2471±24	0.1434±17
4 N(2)	in 4(*e*)	0.1646±23	-0.1082±21	0.3417±18
4 O(1)	in 4(*e*)	0.2216±17	0.1070±17	0.0295±15
4 O(2)	in 4(*e*)	0.4353±20	0.3030±22	0.1633±21
4 O(3)	in 4(*e*)	0.2386±24	0.2998±24	0.2115±19
4 O(4)	in 4(*e*)	0.1724±17	-0.1495±17	0.2107±14
4 O(5)	in 4(*e*)	0.2660±27	-0.2067±23	0.4632±17
4 O(6)	in 4(*e*)	0.0658±26	0.0223±18	0.3195±20

Anisotropic temperature factors given in I.

Interatomic distances (Å)

Au-O(4)	1.987±10 (×2)	O(4)-Au-O(1)	89.0±7° (×2)
Au-O(1)	2.025±17 (×2)		91.0±7° (×2)
K-O(3)	2.74±2 (×2)		
K-O(5)	2.86±2 (×2)		
K-O(1)	2.87±2 (×2)		
K-O(2)	2.99±2 (×2)		

N(1) - O(2)	1.14±3
N(1) - O(3)	1.24±3
N(1) - O(1)	1.34±2

Angles in degrees subtended at N(1) by

	O(3)	O(1)
O(2)	129±2	114±2
O(3)		118±2

N(2) - O(5)	1.20±2
N(2) - O(6)	1.22±3
N(2) - O(4)	1.39±3

Angles in degrees subtended at N(2) by

	O(6)	O(4)
O(5)	131±2	112±2
O(6)		118±2

Au-O(1)-N(1)	114.1±15°
Au-O(4)-N(2)	114.4±10°

Discussion

The structure consists of K^+ and $Au(O_3N)_4^-$ ions with the Au atom coordinated by a square plane of O atoms, one from each NO_3 group.

Details of analysis

Crystal sealed in Lindemann glass capillary. Intensities measured on linear diffractometer for layers h = 0 - 9 with Mo$K\alpha$ radiation using an ω scan. Corrected for Lorentz and polarization effects but not absorption (μ = 15.1 mm^{-1}, crystal size 0.33×0.2×0.2 mm^3). Structure solved from Patterson and electron-density functions and refined by least squares to give R = 0.061 for 829 reflexions listed in 1.

1. SUP 20044.

SILVER THIOCYANATE DINITRATE

I. Détermination de la structure cristalline du sel double de thiocyanate et dinitrate d'argent. M. KABOUNY, 1970. *Bull. Soc. Fr. Minér. Crist.*, 93, 433-436.

$Ag_3SCN(NO_3)_2$ F.W. = 505.69

Triclinic, a = 7.6, b = 7.9, c = 6.5 Å, α = 95°, β = 90°, γ = 104°, U = 377 Å^3, Z = 2, D_x = 4.4.

Space group $P\bar{1}$ (C_i^1)

Atomic positions

	x	y	z	B(Å^2)
1 Ag(1) in 1(a)	0	0	0	3.20
1 Ag(2) in 1(b)	0	0	0.5	3.23
2 Ag(3) in 2(i)	0.729±1	0.362±1	0.046±2	2.43
2 Ag(4) in 2(i)	0.357±2	0.679±2	0.475±2	4.15
2 in 2(i)	0.214±4	0.898±4	0.228±6	2.44
2 O(1) in 2(i)	0.190±10	0.362±10	0.422±14	2.99
2 O(2) in 2(i)	0.084±16	0.514±16	0.231±22	8.79

2 O(3) in 2(*i*)	0.932±10	0.227±10	0.293±14	3.10
2 O(4) in 2(*i*)	0.677±10	0.862±10	0.274±15	2.45
2 O(5) in 2(*i*)	0.538±9	0.592±9	0.236±14	2.42
2 O(6) in 2(*i*)	0.797±11	0.688±10	0.084±15	3.35
2 N(1) in 2(*i*)	0.056±11	0.347±12	0.350±17	2.27
2 N(2) in 2(*i*)	0.696±20	0.720±19	0.140±26	6.98
2 N(3) in 2(*i*)	0.514±11	0.183±11	0.257±16	2.04
2 C in 2(*i*)	0.392±11	0.057±12	0.240±16	0.67

Interatomic distances (Å)

Errors about 0.1 Å

Ag(1) - S	2.52	(×2)
Ag(1) - O(3)	2.64	(×2)
Ag(1) - O(6)	2.67	(×2)
Ag(2) - O(3)	2.48	(×2)
Ag(2) - S	2.61	(×2)
Ag(2) - O(4)	2.79	(×2)
Ag(2) - N(1)	2.92	(×2)
Ag(2) - O(1)	2.96	(×2)
Ag(3) - N(3)	2.40	
Ag(3) - O(2)	2.42	
Ag(3) - O(6)	2.49	
Ag(3) - O(3)	2.68	
Ag(3) - S	2.71	
Ag(3) - O(5)	2.80	
Ag(3) - O(5)	2.84	
Ag(3) - O(2)	2.89	
Ag(4) - N(3)	2.08	
Ag(4) - O(5)	2.25	
Ag(4) - O(1)	2.51	
Ag(4) - O(2)	2.62	
Ag(4) - S	2.85	
Ag(4) - O(3)	2.87	
Ag(4) - O(4)	2.89	
Ag(4) - C	2.97	

[other distances are too inaccurate to be useful]

Details of analysis

Intensities estimated visually from Weissenberg photographs taken with Cu radiation. R = 0.12 for 305 reflexions [not listed].

ANTIMONY PHOSPHATE

$SbPO_4$ F.W. = 216.7

I. The crystal structure of $SbPO_4$. B. KINBERGER, 1970. *Acta Chem. Scand.*, 24, 320-328.

Monoclinic, a = 5.0868±6, b = 6.7547±7, c = 4.7247±5 Å, β = 94.66±1°, U = 161.80 Å^3, D_m = 4.3, Z = 2, D_x = 4.45.

Space group $P2_1/m$ (C_{2h}^2) systematic absences and final structure.

Atomic positions

		x	y	z	$B(Å^2)$
2 Sb	in 2(e)	0.1802±2	1/4	0.2053±2	0.43±2
2 P	in 2(e)	0.6110±7	1/4	0.7223±7	0.36±4
2 O(1)	in 2(e)	0.3346±22	1/4	0.8332±21	0.65±12
2 O(2)	in 2(e)	0.5546±24	1/4	0.3933±25	0.85±13
4 O(3)	in 4(f)	0.7692±19	0.0700±15	0.8183±22	1.33±12

Interatomic distances (Å)

Fourfold coordination around P (crystallographic site symmetry C_s)

P - O(1)	1.540
P - O(2)	1.558
P - O(3)	1.507 (×2)

σ = 0.010 Å

Angles in degrees subtended at P by

	O(1)	O(3)'
O(2)	103.9	110.5
O(3)	112.2	107.5

σ = 0.6°

Environment of Sb (crystallographic site symmetry C_s)

Sb - O(1)	1.983
Sb - (2)	2.035
Sb - (3)	2.181 (×2)
Sb - (3)'	2.927 (×2)

σ = 0.010 Å

Angles in degrees subtended at Sb by

	O(1)	O(3)'
O(2)	87.9	85.0
O(3)	84.1	164.8

σ = 0.5°

Discussion

The coordination around Sb can be approximated by an octahedron with four O atoms at four vertices and two adjacent vertices vacant. Such octahedra are linked into chains along b through PO_4 tetrahedra with which they share their axial oxygen atoms. The equatorial oxygen atoms are shared with PO_4 groups in adjacent chains to form a corrugated sheet perpendicular to a^*, adjacent chains in the sheet being arranged back to back. The Sb atoms lie on the outside of the sheets facing O atoms on the adjacent sheets [not facing other Sb atoms as stated in the abstract of I].

Details of analysis

Crystal (0.024×0.073×0.155 mm^3) obtained by treatment of Sb_2O_3 with H_3PO_4 and heating the microcrystalline $SbPO_4$ at 800°C for 8 hours. Lattice parameters from Guinier-Hägg focusing powder camera. Intensities measured with microdensitometer from multiple-film integrating Weissenberg photographs of layers with l = 0 to 9 (MoK_α radiation). Lorentz and polarization corrections (absorption correction not needed). Structure from three-dimensional Patterson, electron-density and difference electron-density functions refined by least squares to give R = 0.074 for 856 reflexions (not listed) and isotropic temperature factors. Anisotropic temperature factors reduced R to 0.068 but the difference was not regarded as meaningful.

MAGNESIUM AMMONIUM PHOSPHATE HEXAHYDRATE

I. Crystal structure of struvite, $MgNH_4PO_4.6H_2O$. A. WHITAKER and J.W. JEFFERY, 1970. *Acta Cryst.*, B26, 1429-1440.

II. X-ray evidence for a single-bonded rotating ammonium ion in struvite. A. WHITAKER and J.W. JEFFERY, 1970. *Acta Cryst.*, B26, 1440-1444.

$MgNH_4PO_4.6H_2O$ F.W. = 245.41

Orthorhombic, a = 6.941±2, b = 6.137±2, c = 11.199±4 Å, [U = 477.03 Å^3], D_m = 1.708, Z = 2, D_x = 1.708.

Space group Pmn2 (C^7_{2v})

Atomic positions

		x	*y*	*z*	*B*(Å^2)
2 P	in 2(*a*)	0	-0.00661±14	-0.00052±8	-
2 O(1)	in 2(*a*)	0	-0.02221±37	0.86279±20	-
2 O(2)	in 2(*a*)	0	0.76229±32	0.05350±21	-
4 O(3)	in 4(*b*)	0.18165±22	0.11371±25	0.04172±15	-
2 Mg	in 2(*a*)	0	0.37648±19	0.37184±11	-
2 O(W1)	in 2(*a*)	0	0.68219±41	0.28610±22	-
2 O(W2)	in 2(*a*)	0	0.08012±42	0.46451±25	-
4 O(W3)	in 4(*b*)	0.21797±24	0.26377±29	0.26205±14	-
4 O(W4)	in 4(*b*)	0.21055±22	0.48534±28	0.48494±16	-
2 NH4	in 2(*a*)	0	0.36849±50	0.73063±25	-
2 H(1)	in 2(*a*)	0	0.3034±75	0.7507±38	4.00
2 H(2)	in 2(*a*)	0	0.7131±73	0.2222±44	4.00
2 H(3)	in 2(*a*)	0	0.7814±78	0.3150±44	4.00
4 H(4)	in 4(*b*)	0.0989±48	0.0233±45	0.4947±27	4.00
4 H(5)	in 4(*b*)	0.1883±59	0.1874±59	0.2059±35	4.00
4 H(6)	in 4(*b*)	0.-242±44	0.8150±65	0.7930±26	4.00
4 H(7)	in 4(*b*)	0.2348±51	0.5868±60	0.4994±31	4.00
4 H(8)	in 4(*b*)	0.1846±54	0.5995±55	0.0026±29	4.00

Anisotropic temperature factors are given.

Interatomic distances (Å) *and angles*

P-O(1)	1.534±2	
-O(2)	1.542±2	
-O(3)	1.536±2	(×2)
Mg-O(W1)	2.108±3	
-O(W2)	2.094±3	
-O(W3)	2.068±2	(×2)
-O(W4)	2.046±2	(×2)

Angles in degrees subtended at P by

	O(1)	O(3')
O(1)	109.5±1	109.7±8
O(2)		108.75±8
O(3)		110.4±1

Discussion of the structure

Structure proposed by Gonzalez and de Lerma (1) and Gonzalez and Garcia-Balaneo (1) for magnesium ammonium arsenate hexahydrate. The ammonium group rotates about the NH_4-O(1) axis.

Materials

Crystals obtained from tin of salmon.

Details of analysis

Lattice parameters and intensities measured with Cu radiation and Weissenberg photographs. Absorption correction imposed on data from spheres. The final is R = 0.028. Hydrogen temperature factors assumed.

1. *Structure Reports*, 19, 440.

LOW LEAD PHOSPHATE

I. Die Struktur der Tieftemperaturform des Bleiphosphates, $Pb_3(PO_4)_2$. U. KEPPLER, 1970. *Z. Kristallogr.*, 132, 228-235.

$Pb_3(PO_4)_2$ F.W. = 811.54

Monoclinic, a = 13.816±35, b = 5.692±15, c = 9.429±24 Å, β = 102.36±5°, U = 724.32 Å^3, D_m = 7.38±5, Z = 4, D_x = 7.43.

Space group C2/c (C_{2h}^6)

Atomic positions

	x	y	z	B(Å^2)
4 Pb(1) in 4(*e*)	0	0.2872±2	0.25	1.01±6
8 Pb(2) in 8(*f*)	0.3184±2	0.3104±2	0.3517±2	0.87±8
8 P in 8(*f*)	0.3981±6	0.2455±6	0.0540±6	0.45±8
8 O(1) in 8(*f*)	0.350±3	0.029±3	0.110±2	0.5±3
8 O(2) in 8(*f*)	0.359±4	0.466±4	0.123±3	0.8±4
8 O(3) in 8(*f*)	0.140±3	0.227±4	0.104±3	0.9±3
8 O(4) in 8(*f*)	0.512±4	0.234±3	0.086±3	1.2±3

Interatomic distances (Å) *and angles*

σ(distances) = 0.03; σ(angles) = 1°

Pb(1) - O(1)	2.60	(×2)
Pb(1) - O(3)	2.63	(×2)
Pb(1) - O(2)	2.76	(×2)
Pb(1) - O(4)	3.00	(×2)
Pb(1) - O(4)	3.14	(×2)
Pb(2) - O(4)	2.34	
Pb(2) - O(3)	2.46	
Pb(2) - O(2)	2.50	
Pb(2) - O(1)	2.74	
Pb(2) - O(2)	2.80	
Pb(2) - O(1)	2.89	
Pb(2) - O(3)	3.05	
Pb(2) - O(1)	3.07	
Pb(2) - O(2)	3.19	

P - O(3)	1.48
P - O(4)	1.54
P - O(1)	1.55
P - O(2)	1.56

Angles in degrees subtended at P by

	O(4)	O(1)	O(2)
O(3)	109	110	105
O(4)		113	113
O(1)			106

Discussion

Low-$Pb_3(PO_4)_2$ is a distorted form of the rhombohedral high-$Pb_3(PO_4)_2$ (isostructural with $Ba_3(PO_4)_2$ (1) with a^* corresponding to the unique high-temperature axis.

Details of analysis

Intensities measured photometrically from precession photographs of layers with h = 0 to 9 using AgK_α radiation and corrected for absorption. Structure determined from Patterson and difference electron-density syntheses and refined [by least squares?] to give R = 0.088 for 637 observed reflexions listed.

1. *Structure Reports*, 11, 388.

TRIAMMONIUM ORTHOPHOSPHATE TRIHYDRATE

I. Die Kristallstruktur von $(NH_4)_3PO_4.3H_2O$, Triammonium-orthophosphat-trihydrate. D. MOOTZ and H. WUNDERLICH, 1970. *Acta Cryst.*, B26, 1826-1835.

$(NH_4)_3PO_4.3H_2O$ F.W. = 203.1

Monoclinic, a = 6.686±3, b = 6.218±2, c = 22.349±7 Å, β = 94.13±4°, U = 926.7 Å^3, D_m = 1.42, Z = 4, D_x = 1.456.

Space group $P2_1/c$ (C^5_{2h})

Atomic positions All atoms in 4(*e*)

	x	*y*	*z*
P	0.20982±5	0.36939±5	0.3419±1
O(1)	0.30374±16	0.44129±17	0.28448±4
O(2)	0.32911±16	0.46467±17	0.39728±4
O(3)	0.21415±16	0.12357±15	0.34679±5
O(4)	-0.00732±15	0.44949±17	0.34040±5
O(5)	-0.12362±19	0.26091±19	0.52996±5
O(6)	0.49355±17	0.36126±15	0.10423±5
O(71)*	-0.27468±35	0.30850±40	0.41628±11
O(72)*	-0.27172±41	0.39647±54	0.42559±12
N(1)	0.12965±21	0.37593±20	0.16797±6
N(2)	0.69120±19	0.37179±19	0.25110±6
N(3)	0.26124±21	0.17547±23	0.49246±6
H(11)	0.240±3	0.392±3	0.1471±10
H(12)	0.180±3	0.404±3	0.2084±9
H(13)	0.005±4	0.470±4	0.1592±10
H(14)	0.074±3	0.243±4	0.1616±10
H(21)	0.783±3	0.404±3	0.2800±10
H(22)	0.580±3	0.412±3	0.2702±10
H(23)	0.716±3	0.474±4	0.2212±10
H(24)	0.698±3	0.226±3	0.2383±9
H(31)	0.148±3	0.225±4	0.5082±9
H(32)	0.303±3	0.268±4	0.4630±11
H(33)	0.354±3	0.167±4	0.5202±10
H(34)	0.239±3	0.035±3	0.4753±9
H(51)	-0.204±4	0.311±5	0.4035±12
H(52)	-0.178±5	0.358±4	0.5553±13
H(61)	0.581±4	0.457±4	0.1186±11
H(62)	0.549±4	0.244±5	0.1080±13
H(71)	-0.184±5	0.392±5	0.3987±14
H(72)	-0.390±5	0.387±5	0.4095±15

Anisotropic thermal parameters were determined for all atoms.
*Half occupied sites.

Interatomic distances (Å) *and angles*

P-O(1)	1.547±1*	O(1)-P-O(2)	109.9±1°
-O(2)	1.551±1*	-O(3)	110.1±1
-O(3)	1.543±1*	-O(4)	109.3±1
-O(4)	1.544±1*	O(2)- -O(3)	108.7±1
		-O(4)	109.1±1
		O(3)- -O(4)	109.8±1

N(1)-H	four	0.92 - 1.06*
N(2)-H	"	0.91 - 0.96*
N(3)-H	"	0.88 - 0.99*
O-H average		0.93*

*Corrected for the effects of thermal motion using the riding approximation.

Structure

The phosphate groups are hydrogen bonded. Each NH_4 forms four hydrogen bonds while two of the water molecules form four, and the remaining two, arising from two half occupied sites, form three hydrogen bonds.

Details of analysis

1583 reflexions measured by diffractometer (CuK_α). Absorption correction applied. Refined by least-squares method to an *R* value of 0.031.

SCHOLZITE

$CaZn_2(PO_4)_2.2H_2O$ F.W. = 396.79

I. Die Kristallstruktur des Minerals Scholzit. K.J. TAXER, 1970. *Naturwissenschaften*, 57, 192.

Monoclinic, *a* = 17.149±3, *b* = 22.236±2, *c* = 6.667±1 Å, β = 90°, *U* = 2542 Å³, *Z* = 12, D_x = 3.11.

Space group P2/c (C_{2h}^4) from refinement of full cell, but the crystals have a sub-cell with pseudosymmetry *Pbcn* (D_{2h}^{14}) and *b'* = *b*/3

Atomic positions (in subgroup *Pbcn*)

	x	*y*	*z*
8 Zn in 8(*d*)	0.279	0.5	0.188
8 P in 8(*d*)	0.350	0.179	0.438
4 Ca in 4(*c*)	0	0.380	1/4
8 O(1) in 8(*d*)	0.424	0.072	0.447
8 O(2) in 8(*d*)	0.277	0.047	0.440
8 O(3) in 8(*d*)	0.155	0.209	0.131
8 O(4) in 8(*d*)	0.051	0.148	0.445
8 O(5) in 8(*d*)	0.345	0.292	0.247

Discussion

The structure consists of ZnO_4 tetrahedra chains along *c* linked by isolated PO_4 tetrahedra. The Ca atoms lie in a distorted octahedral environment of O (4 from four different PO_4 ions and two water molecules). Previous work given in 1.

Details of analysis

Crystals from the "Feldspatgruben" Hagendorf, Ostbayern, Germany. Structure determined from a Patterson function of a subcell with *b'* = *b*/3 and space group *Pbcn*. The weak reflexions could be accounted for by a monoclinic symmetry. The apparent orthorhombic symmetry arises from twinning but most crystals showed diffuse streaks parallel to *a* and did not show extinctions compatible with orthorhombic space group symmetry. *R* = 0.195 for monoclinic structure but dropped to 0.165 when twinning was allowed for. No intensities listed and no details given of how the diffraction patterns were recorded.

1. *Structure Reports*, 20, 311.

FAIRFIELDITE

I. The crystal structure of fairfieldite. L. FANFANI, A. NUNZI and P.F. ZANAZZI, 1970. *Acta Cryst.*, B26, 640-645.

$Ca_2Mn(PO_4)_2.2H_2O$ F.W. = 359.06

Triclinic, *a* = 5.79±1, *b* = 6.57±1, *c* = 5.51±1 Å, α = 102°16'±15', β = 108°40°±15', γ = 90°±18'±15', D_m = 3.08-3.11, *Z* = 1, D_x = 3.095.

Space group $P\bar{1}$ (C_i^1) confirmed by structure resolution

Atomic positions

		x	*y*	*z*	$B(Å^2)$
1 Mn	in 1(*a*)	0	0	0	0.32±3
2 Ca	in 2(*i*)	0.6020±3	0.2337±2	0.3332±3	0.22±3
2 P	in 2(*i*)	0.2463±3	0.2325±2	0.6472±3	0.12±3
2 O(1)	in 2(*i*)	0.0031±11	0.3264±8	0.6157±11	0.48±7
2 O(2)	in 2(*i*)	0.4552±11	0.3995±8	0.6940±10	0.41±7
2 O(3)	in 2(*i*)	0.2461±11	0.0699±8	0.3965±10	0.45±7
2 O(4)	in 2(*i*)	0.3058±10	0.1177±7	0.8785±10	0.43±7
2 O(W)	in 2(*i*)	-0.1543±1	0.3004±8	0.0663±11	0.58±7

Interatomic distances (Å) *and angles*

P-O(1), O(2), O(3), O(4) 1.514±7, 1.550±7, 1.556±6, 1.563±7
O-P-O 104°36' to 112°38'
Mn-O(3) (2×) 2.142±9
-O(4) (2×) 2.269±7
-O(W) (2×) 2.180±7

Ca-O eight in range 2.336-2.724

Discussion of the structure

Structure consists of chains of $MnO_4(H_2O)_2$ octahedra along *c* which are bridged by corner-sharing with PO_4 tetrahedra, a pair of PO_4 lying between each pair of Mn groups. Chains are connected by Ca ions as shown in Fig. 41.

Material

Source was the Foote Mine, North Carolina.

Details of analysis

1089 reflexions from integrated precession photographs taken with Mo radiation were measured with a microdensitometer. Absorption corrections applied. Refined by least-squares methods to *R* = 0.075.

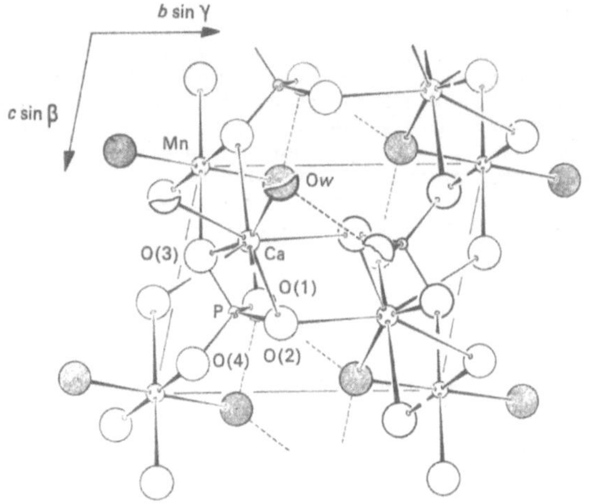

Fig. 41. Crystal structure of fairfieldite projected along the *a* axis. Shaded circles indicate water molecules.

DISODIUM HYDROGEN ORTHOPHOSPHATE (AND ARSENATE) HEPTAHYDRATES

I. The crystal structure of $Na_2HAsO_4.7H_2O$. G. FERRARIS and G. CHIARI, 1970. *Acta Cryst.*, B26, 1574-1583.

II. On the crystal chemistry of salt hydrates. VI. The crystal structure of disodium hydrogen orthoarsenate heptahydrate and of disodium hydrogen orthophosphate heptahydrate. H. BAUR and A.A. KHAN, *Ibid.*, 1584-1596.

$Na_2HAsO_4.7H_2O$ F.W. = 312.01
$Na_2HPO_4.7H_2O$ F.W. = 268.08

Monoclinic ($Na_2HAsO_4.7H_2O$,I) a = 10.6586±17, b = 11.0107±22, c = 9.3940±14 Å, β = 95°26°±1', D_m = 1.874, Z = 4, D_x = 1.843.

($Na_2HAsSO_4.7H_2O$,II) a = 9.401±4, b = 11.025±4, c = 10.674±40 Å, β = 95.50±1°, U = 1101.3 Å^3, D_m = 1.8825 (1), Z = 2, D_x = 1.882.

($Na_2HPO_4.7H_2O$,II) a = 9.258±3, b = 11.007±3, c = 10.437±3 Å, β = 95.61±1°, U = 1058.6 Å^3, D_m = 1.6789 (1).

Space group $P2_1/n$ (C_{2h}^5)

Atomic positions All atoms in (±x,±y,±z;½±x,½±y,½±z)

$Na_2HAsO_4.7H_2O$ (I)

	x	y	z	B(Å^2)
As(1)	0.39609±5	0.26557±5	0.10272±6	
Na(1)	0.9796±2	0.3957±2	0.1356±3	
Na(2)	0.9789±2	0.1018±2	0.1460±3	
O(1)	0.3124±4	0.1778±4	-0.01595±4	
O(2)	0.5521±3	0.2477±3	0.0986±4	
O(3)	0.3520±4	0.2541±3	0.2670±4	
O(4)	0.3619±4	0.4168±4	0.0608±5	
W(1)	0.3643±4	0.0376±4	0.4121±5	
W(2)	0.0950±4	0.2511±4	0.3032±5	
W(3)	0.1460±4	-0.0229±4	0.0710±5	
W(4)	0.5553±4	0.2539±4	0.4816±4	
W(5)	0.6113±4	0.0076±4	0.1845±5	
W(6)	0.6329±4	0.4946±4	0.1991±5	
W(7)	0.8175±4	0.2498±3	0.1931±5	
H(1)	0.862±7	0.075±1	0.476±8	2.0±2
H(2)	0.341±10	0.091±9	0.348±10	4.0±2
H(3)	0.204±10	0.517±9	0.102±10	4.0±2
H(4)	0.176±8	0.253±7	0.312±9	2.0±1
H(5)	0.089±8	0.249±7	0.385±9	2.0±2
H(6)	0.216±10	0.010±10	0.081±10	5.0±2
H(7)	0.338±6	0.418±7	0.372±7	2.0±1
H(8)	0.497±7	0.256±6	0.405±7	1.0±1
H(9)	0.114±12	0.228±11	-0.049±13	8.0±4
H(10)	0.596±10	0.096±10	0.172±11	6.0±3
H(11)	0.883±13	0.467±12	0.380±14	9.0±4
H(12)	0.694±9	0.473±9	0.235±10	4.0±2
H(13)	0.608±10	0.402±11	0.176±11	4.0±3
H(14)	0.802±7	0.253±6	0.280±9	3.0±2
H(15)	0.729±8	0.238±7	0.148±9	3.0±2

Anisotropic thermal parameters refined for the non-hydrogen atoms and hydrogen atom isotropic parameters refined in I.

$Na_2HPO_4.7H_2O$ (II)

	x	*y*	*z*
P	0.1024±2	0.2633±2	0.8955±2
Na(1)	0.1404±4	0.1043±3	0.4756±3
Na(2)	0.6338±4	0.1053±3	0.9771±3
O(1)	0.2564±5	0.2522±5	0.8540±4
O(2)	0.0997±6	0.2492±5	0.0403±5
O(3)	0.9899±6	0.1832±6	0.8190±5
O(W4)	0.5613±6	0.0968±5	0.3612±6
O(W5)	0.4759±6	0.2542±5	0.0575±6
O(W6)	0.8041±6	0.2507±5	0.0951±6
O(W7)	0.1944±6	0.2503±5	0.3098±6
O(W8)	0.9212±7	0.0232±6	0.3518±6
O(W9)	0.3063±6	0.9872±6	0.3680±5
O(W10)	0.4082±6	0.0347±6	0.8618±6
O(W11)	0.1781±6	0.0046±6	0.1085±6

Description of the structure

The two structures are isotypic. Both Na ions lie in octahedral environments of water molecules and form pairs, represented as $Na_2(H_2O)_7^{2+}$ by sharing a common face. These dimers share edges to form chains running parallel to the *b* axis, as seen in Fig. 42. The chains are bonded through XO_4H groups, *X* = As or P, and by hydrogen bonding to adjacent chains.

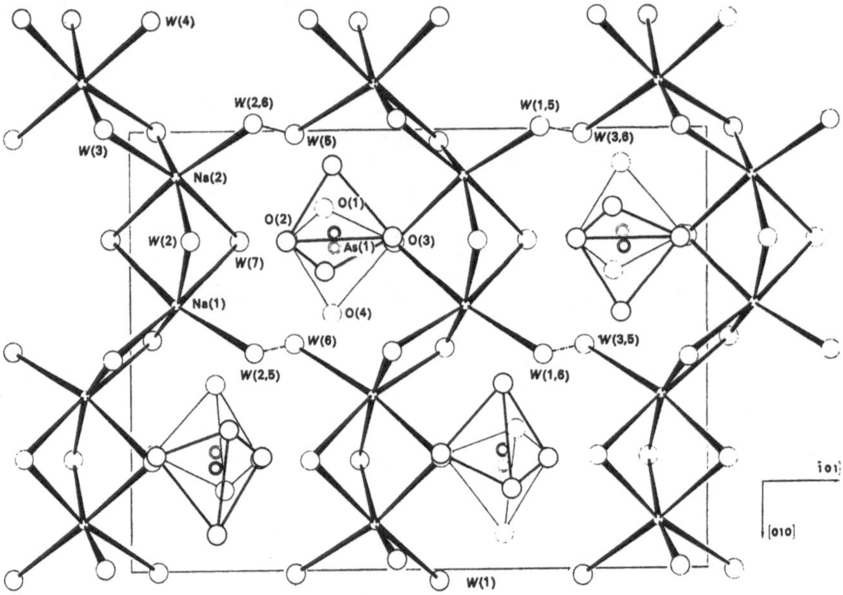

Fig. 42. The $Na_2HPO_4.7H_2O$ structure viewed along [101] showing one Na-polyhedra layer with adjacent AsO_4 tetrahedra; the H_2O-H_2O type hydrogen bridges are shown.

Materials

Grown from solution at slightly elevated temperatures.

Details of analysis

I. 2384 unique reflexions collected with diffractomer (CuK_α). No absorption correction. Refined by least-squares methods to $R = 0.047$.

II. 1177 and 1252 reflexions used in the refinement of the arsenate and phosphate respectively. Collected with diffractometer using AgK_α radiation. Refined by least-squares methods to $R = 0.048$ (arsenate) and $R = 0.081$ (phosphates). Hydrogen atoms could not be located.

MOLYBDENYL PHOSPHATE

$MoOPO_4$ F.W. = 413.82

I. A refinement of the crystal structure of $MoOPO_4$. P. KIERKEGAARD and J.M. LONGO, 1970. *Acta Chem. Scand.*, 24, 427-432.

Tetragonal [form 1 $a = 6.1768 \pm 3$, $c = 4.2932 \pm 3$ Å, $U = 163.8$ Å^3, $D_m = 4.06$, $Z = 2$, $D = 4.19$].

Space group P4/n (C_{4h}^3)

Atomic positions

		x	*y*	*z*
Mo	in 2(*c*)	1/4	1/4	0.80244±13
P	in 2(*b*)	1/4	3/4	1/2
O(1)	in 2(*c*)	1/4	1/4	0.18726±124
O(2)	in 8(*g*)	0.80876±53	0.44256±49	0.29562±72

Anisotropic temperature factors given in I.

Interatomic distances (Å) *and angles*

Fivefold coordination around Mo (crystallographic site symmetry C_4)

Mo-O(1)	1.652±6	O(1)-Mo-O(2)	102.3±1°
Mo-O(2)	1.979±3 (×4)	O(2)-Mo-O(2)	155.4±1°
			87.4±1°

Fourfold coordination around P (crystallographic site symmetry S_4)

P-O(2) 1.522±3 (×4)

O-P-O angles are tetrahedral within experimental error (0.2°)

Details of analysis

Structure of 1 refined. Intensities measured on a single-crystal diffractometer with crystal 0.064×0.064×0.020 mm³ ($MoK\alpha$ radiation). Lorentz, polarization absorption ($\mu = 4.28$ mm^{-1}) and extinction corrections. Twenty one parameters refined by least squares to give $R = 0.045$ for 616 reflexions listed. [A further 153 were omitted because $2 < F_{obs}/F_{calc} < 0.5$ and 302 omitted as unobserved.]

1. P. KIERKEGAARD and M. WESTERLUND, 1964. *Acta Chem. Scand.*, 18, 2217.

DUFRENITE
ROCKBRIDGEITE

I. Crystal chemistry of the basic iron phosphates. P.B. MOORE, 1970. *Amer. Min.*, 55, 135-169.

DUFRENITE

$CaFe_{12}(OH)_{12}(H_2O)_4(PO_4)_8$ F.W. = 1786

Monoclinic, a = 25.84±2, b = 5.126±3, c = 13.78 1 Å, β = 111.20±6°, U = 1702 Å^3; Z = 2, D_x = 3.48 (cf.3).

Space group C2/c (C_{2h}^6)

Atomic positions

	x	y	z	B(Å^2)
4 Fe(1) in 4(a)	0	0	0	1.15±8
4 Fe(2) in 4(c)	0.25	0.25	0	1.25±8
8 Fe(3) in 8(f)	0.1529±1	-0.0150±8	0.1116±3	1.01±6
8 Fe(4) in 8(f)	0.1401±2	-0.2220±8	0.3545±3	1.18±6
8 P(1) in 8(f)	0.2185±3	0.2612±13	0.3312±5	1.20±9
8 P(2) in 8(f)	0.0790±3	0.2808±14	0.3970±5	1.38±10
4 Ca in 4(e)	0	-0.1474±28	0.25	1.78±20
8 O(1) in 8(f)	0.0887±7	0.0660±34	0.3293±13	1.14±26
8 O(2) in 8(f)	0.0769±8	0.5468±39	0.3422±15	1.77±31
8 O(3) in 8(f)	0.0193±8	0.2255±38	0.4030±14	1.85±30
8 O(4) in 8(f)	0.1224±8	0.2896±37	0.5055±14	1.60±29
8 O(5) in 8(f)	0.1727±8	0.2183±37	0.0170±14	1.58±29
8 O(6) in 8(f)	0.2134±8	0.0108±36	0.3878±14	1.47±28
8 O(7) in 8(f)	0.2029±8	-0.5144±38	0.3907±15	2.13±31
8 O(8) in 8(f)	0.1289±9	-0.2574±39	0.2046±15	1.85±33
8 O(9) in 8(f)	0.1766±8	0.2529±36	0.2203±14	1.75±29
8 O(10) in 8(f)	0.2223±8	-0.2034±36	0.1670±14	1.62±28
8 O(11) in 8(f)	0.0769±8	0.1439±38	0.0645±14	2.07±30
8 O(12) in 8(f)	0.0243±9	-0.2853±41	0.1119±16	2.25±34

Interatomic distances (Å) *and angles*

Sixfold coordination around Fe(1) (Crystallographic site symmetry C_i)

Fe(1) - O(3)	1.96	(×2)
Fe(1) - O(11)	2.00	(×2)
Fe(1) - O(12)	2.05	(×2)

Angle in degrees subtended at Fe(1) by

	O(11)	O(11)'	O(12)	O(12)'
O(3)	86	94	91	90
O(11)			93	87

Sixfold coordination around Fe(2) (crystallographic site symmetry C_i)

Fe(2) - O(5)	2.10	(×2)
Fe(2) - O(6)	2.00	(×2)
Fe(2) - O(7)	2.06	(×2)

Angle in degrees subtended at Fe(2) by

	O(6)	O(6)'	O(7)	O(7)'
O(5)	81	100	79	102
O(6)			84	97

Sixfold coordination around Fe(3)

Fe(3) - O(4)	1.98
Fe(3) - O(5)	1.97
Fe(3) - O(8)	2.04
Fe(3) - O(9)	1.96
Fe(3) - O(10)	1.93
Fe(3) - O(11)	2.00

Angle in degrees subtended at Fe(3) by

	O(5)	O(8)	O(9)	O(10)	O(11)
O(4)	93	84	174	90	88
O(5)		177	90	98	90
O(8)			92	84	88
O(9)				94	88
O(10)					172

Sixfold coordination around Fe(4)

Fe(4) - O(1)	1.93
Fe(4) - O(2)	1.97
Fe(4) - O(5)	2.09
Fe(4) - O(6)	2.14
Fe(4) - O(7)	2.13
Fe(4) - O(8)	1.98

Angle in degrees subtended at Fe(4) by

	O(2)	O(5)	O(6)	O(7)	O(8)
O(1)	87	100	96	174	93
O(2)		96	173	98	93
O(5)			77	77	165
O(6)				78	94
O(7)					89

Fourfold coordination around P(1)

P(1) - O(6)	1.53
P(1) - O(7)	1.55
P(1) - O(9)	1.52
P(1) - O(10)	1.53

Angle in degrees subtended at P(1) by

	O(7)	O(9)	O(10)
O(6)	105	110	111
O(7)		109	110
O(9)			112

Fourfold coordination around P(2)

P(2) - O(1)	1.52
P(2) - O(2)	1.55
P(2) - O(3)	1.60
P(2) - O(4)	1.51

Angle in degrees subtended at P(2) by

	O(2)	O(3)	O(4)
O(1)	109	106	114
O(2)		108	110
O(3)			109

Eightfold coordination around Ca (Crystallographic site symmetry C_2)

Ca - O(12)	2.32 (×2)
Ca - O(1)	2.42 (×2)
Ca - O(2)	2.49 (×2)
Ca - O(3)	2.75 (×2)

Possible hydrogen bond

O(8) - O(12) 2.54

σ(distances) = 0.02 Å; σ(angles) = 1°

Discussion (see below)

ROCKBRIDGEITE

$Fe_6(OH)_5(PO_4)_3$ F.W. = 705.1

Orthorhombic, a = 5.172±4, b = 13.783±12, c = 16.805±9 Å, U = 1198 Å^3, Z = 4, D_x = 3.95 (cf. 4).

Space group Cmcm (D_{4h}^{17}) [converted from *Bbmm*] systematic absences (see *Details of analysis*)

Atomic positions

	x	y	z	B(Å^2)
4 Fe(1) in 4(*a*)	0	0	0	0.10±9
8 Fe(2) in 8(*f*)	0	0.0687±5	0.1574±4	1.64±9
8*Fe(3) in 16(*h*)	0.2385±14	0.3214±4	0.1385±3	0.10±8
8 P(1) in 8(*f*)	0.5	0.1420±6	0.0432±5	0.08±11
4 P(2) in 4(*c*)	0	0.4806±10	0.25	0.76±19
8 O(1) in 8(*g*)	0.2508±66	0.0477±19	0.25	2.38±43
16 O(2) in 16(*h*)	0.2615±32	0.0829±10	0.0605±8	0.56±20
4 O(3) in 8(*g*)	0.3876±95	0.3132±31	0.25	0.81±57

*[This figure should be 12 if the occupation numbers are to agree with the formula]

	x	y	z	$B(\AA^2)$
8 O(4) in 8(f)	0	0.3126±22	0.0357±18	1.86±51
8 O(5) in 8(f)	0	0.2175±18	0.1720±14	0.75±37
8 O(6) in 8(f)	0.5	0.4204±18	0.1071±14	0.31±37
8 O(7) in 8(f)	0	0.4171±18	0.1763±15	0.91±37
8 O(8) in 8(f)	0.5	0.2276±16	0.1044±13	0.10±33

Interatomic distances (Å) *and angles*

Sixfold coordination around Fe(1) (crystallographic site symmetry C_{2h})

Fe(1) - O(2)	2.05±2	(×4)
Fe(1) - O(6)	2.11±2	(×2)

Angles in degrees subtended at Fe(1) by

	O(2)	O(6)
O(2)	83±1	83±1

Sixfold coordination around Fe(2) (crystallographic site symmetry C_s)

Fe(2) - O(1)	2.05±2	(×2)
Fe(2) - O(2)	2.13±2	(×2)
Fe(2) - O(5)	2.07±3	
Fe(2) - O(6)	2.22±3	

Angle in degrees subtended at Fe(2) by

	O(1)'	O(2)	O(2)'	O(5)	O(6)
O(1)	79±1	177±1	100±1	93±1	99±1
O(2)			79±1	90±1	78±1
O(5)					164±1

Sixfold coordination around Fe(3)

Fe(3) - O(3)	2.04±2
Fe(3) - O(4)	2.12±2
Fe(3) - O(5)	1.98±2
Fe(3) - O(6)	2.00±2
Fe(3) - O(7)	1.92±2
Fe(3) - O(8)	1.96±2

Angle in degrees subtended at Fe(3) by

	O(4)	O(5)	O(6)	O(7)	O(8)
O(3)	165±1	86±1	91±1	89±1	87±1
O(4)		80±1	103±1	87±1	96±1
O(5)			176±1	90±1	91±1
O(6)				93±1	85±1
O(7)					176±1

Fourfold coordination around P(1) (crystallographic site symmetry C_s)

P(1) - O(2)	1.51±2	(×2)
P(1) - O(4)	1.46±3	
P(1) - O(8)	1.59±2	

Angles in degrees subtended at P(1) by

	O(2)'	O(4)	O(8)
O(2)	109±1	115±1	106±1
O(4)			106±1

Fourfold coordination around P(2) (crystallographic symmetry C_{2v})

P(2) - O(1)	1.59±3	(×2)
P(2) - O(7)	1.52±3	(×2)

Angles in degrees subtended at P(2) by

	O(1)'	O(7)'
O(1)	109±1	110±1
O(7)		109±1

Discussion

Both structures contain centrosymmetric clusters of three face-sharing FeO_6 octahedra (Fe(2) and Fe(4) sharing the O(5), O(6), and O(7) face in dufrenite; Fe(1) and Fe(2) sharing O(2), O(2)' and O(6) face in rockbridgeite). In both structures these clusters are linked into sheets by $Fe(3)O_6$ octahedra and PO_4 tetrahedra, the sheets being parallel to the *bc* plane in dufrenite and the *ab* plane in rockbridgeite. The structures differ in the way these sheets are arranged, being directly connected through a shared octahedral edge and a shared PO_4 tetrahedron in rockbridgeite, but being separated in dufrenite by an additional FeO_6 octahedron and the Ca atom. The structure of the sheet proposed for rockbridgeite differs further in that it is disordered, giving rise to an apparent crystallographic mirror plane perpendicular to *a*.

Details of analysis

Dufrenite:

Crystals from Cornwall, U.K. Analysis (theoretical values given in parentheses) CaO, 1.50 (3.21); FeO, - (8.23); Fe_2O_3, 55.63 (45.73); P_2O_5, 30.26 (32.51); H_2O 10.62 (10.32); Rem, 1.48 (-). Lattice parameters from rotation and Weissenberg photographs. Powder diagram given. Intensities measured on Weissenberg diffractometer for layers k = 0 to 4 and corrected for absorption (crystal size: 0.30×0.30×0.08 mm^3). Structure solved from three-dimensional Patterson and Fourier functions and similarity with beraunite (1). Refined [by least squares] to give R = 0.13 for 720 observed reflexions listed in 2.

Rockbridgeite:

Crystal from Irish Creek, Va., U.S.A. Analysis (theoretical values in parentheses) CaO, 1.12 (-); FeO, 6.14 (11.07); Fe_2O_3, 50.84 (49.20); P_2O_5, 31.76 (32.80); H_2O, 8.53 (6.93); Rem (=MgO, MnO, Al_2O_3), 1.48 (-). Lattice parameters from rotation and Weissenberg photographs. Intensities from visually estimated multiple-film Weissenberg photographs (MoK_α) of layers l = 0 to 5. No absorption correction (crystal size ∿ 0.08 mm). Structure solved from Patterson function and chemical considerations and refined by least squares to give R = 0.12 for 340 non-zero reflexions listed in 2. Refinement in space group *Bbm*2 gave R = 0.14 but was rejected on chemical grounds in favour of the higher symmetry disordered structure.

1. L. FANFANI and P.F. ZANAZZI, 1967. *Acta Cryst.*, 22, 173.
2. NAPS DOCUMENT #00747, ASIS National Auxiliary Publications Service, c/o CCM Information Sciences, Inc. 22 West 34th St., New York, N.Y. 10001.
3. M.E. MROSE, 1955. *Geol. Soc. Amer. Bull.*, 66, 1660.
4. *Structure Reports*, 12, 249.

SILICON PYROPHOSPHATE

I. Die Kristallstruktur von monoklinem Silizumphosphat SiP_2O_7AIII: Eine Phase mit [SiO_6]-Oktaedern. G. BISSERT and F. LIEBAU, 1970. *Acta Cryst.*, B26, 223-240.

SiP_2O_7 F.W. = 202.03

Monoclinic, a = 4.73, b = 6.33, c = 14.71 Å, β = 90.1°, [U = 440.4 Å^3], Z = 4, D_x = 3.046.

Space group $P2_1/c$ (C_{2h}^5)

Atomic parameters All atoms in 4(*e*)

	x	*y*	*z*	*B*(Å^2)
Si	0.2495±4	0.0825±3	0.1351±1	0.30±3
P(1)	0.7457±4	0.8080±3	0.0725±1	0.23±3
P(2)	0.7381±4	0.3937±3	0.1684±1	0.28±3
O(1)	0.7110±13	0.5622±10	0.0881±4	0.61±9
O(2)	0.6607±12	0.8314±10	-0.0248±4	0.58±9
O(3)	0.5449±12	0.9171±9	0.1376±4	0.28±9
O(4)	0.0498±12	0.8643±10	0.0916±4	0.54±9
O(5)	0.9507±13	0.2404±10	0.1311±4	0.68±10
O(6)	0.4515±12	0.2951±10	0.1794±4	0.53±9
O(7)	0.8387±13	0.5084±10	0.2525±4	0.62±9

Interatomic distances (Å)

Si-O	1.732-1.793
P-O(terminal)	1.495-1.515
P-O(bridging)	1.581, 1.596

Interatomic angles in degrees

O-P-O range from 102 to 114°
P-O-P 139
O-Si-O 88-92, or 173 to 178

Description of the structure

The anion, $P_2O_7^{4-}$, has the staggered configuration. These groups use terminal oxygen atoms only to corner-share with SiO_6^{4+} groups. Thus if the terminal oxygen atoms are considered to form distorted octahedra about the bridging oxygen atoms then these octahedra and the SiO_6 octahedra are arranged in a pseudo-rock salt structure with each octahedron of one kind corner-sharing with six octahedra of the other kind. Alternatively Fig. 43 shows the structure packing as bridged tetrahedra and octahedra.

Materials

Crystals were grown in a thermal gradient between 800 and 1000°C in a quartz ampoule.

Details of the analysis

810 reflexions were measured with a densitometer from integrated Weissenberg photographs taken with CuK_α radiation. Least-squares refinement lead to an R value of 0.07. Errors in unit cell parameters were not given.

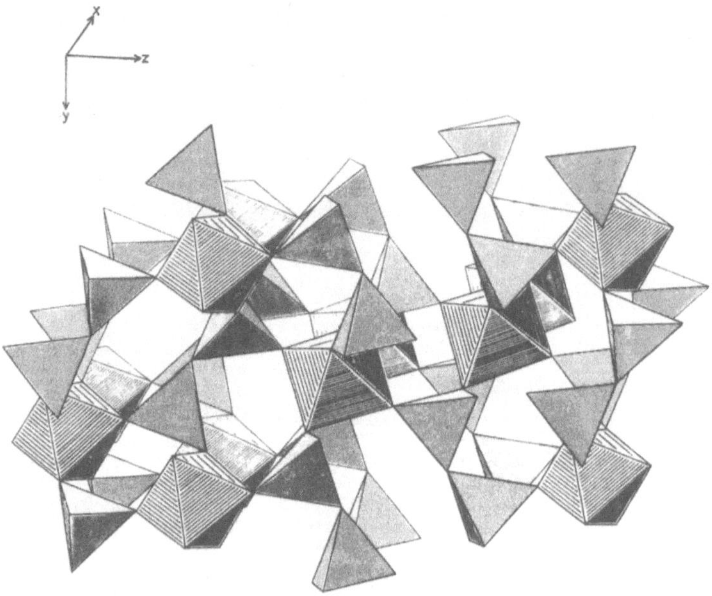

Fig. 43. The packing of the coordination polyhedra in SiP_2O_7 AIII.

α-ZINC PYROPHOSPHATE

I. Crystal structure of α $Zn_2P_2O_7$. B.E. ROBERTSON and C. CALVO, 1970. *J. Solid State Chem.*, 1, 120-133.

$Zn_2P_2O_7$ F.W. = 300.7

Monoclinic, a = 20.068±15, b = 8.259±6, c = 9.099±8 Å, β = 106.35±5°, U = 1447.1 Å^3, D_m = 4.1, Z = 12, D_x = 4.13.

Space group I2/c (C^6_{2h}) Systematic absences and final structure

Atomic positions Non-standard setting. Coordinates as given in International Tables for X-ray Crystallography but (0,0,0;½,½,½) instead of (0,0,0;½,½,0). All atoms in 8(*f*) except O(1A) in 4(*e*).

	x	*y*	*z*	*B*(Å^2)
Zn(A)	0.01302±5	0.31182±22	-0.02382±11	*
P(A)	0.06618±11	0.02024±43	0.20417±23	*
O(1A)	0	0.0886±17	0.25	0.88±13
O(2A)	-0.12407±29	0.0059±11	0.1437±6	0.47±7
O(3CA)	-0.08360±31	0.1574±11	0.3928±8	0.54±8
O(3TA)	0.05120±38	-0.1393±12	0.1197±8	0.95±10
Zn(1)	0.16348±6	0.18296±18	0.00104±14	*
Zn(2)	0.18239±5	0.81089±19	-0.02402±12	*
P(1)	0.10189±11	0.50679±45	0.28423±22	*
P(2)	0.23880±10	0.51842±45	0.20192±22	*
O(1)	0.16450±30	0.5564±12	0.2168±7	0.60±8
O(21)	0.04049±29	0.4939±11	0.1422±6	0.31±6
O(22)	0.28953±28	0.5101±10	0.3612±6	0.51±7
O(3C1)	0.09400±33	0.6513±11	0.3829±8	0.56±9
O(3C2)	0.25367±32	0.6595±11	0.1090±7	0.64±8
O(3T1)	0.11792±39	0.3550±12	0.3758±9	1.11±11
O(3T2)	0.23510±37	0.3576±12	0.1162±9	0.89±10

*Anisotropic temperature factors given.

Interatomic distances (Å) *and angles*

Zn(A) - O(3TA)	1.955±8
Zn(A) - O(3CA)	2.026±7
Zn(A) - O(3C1)	2.058±8
Zn(A) - O(21)	2.059±8
Zn(A) - O(21)	2.092±7

Angles in degrees subtended at Zn(A) by

	O(3CA)	O(3C1)	O(21)	O(21)
O(3TA)	93.9±3	115.0±3	98.8±3	148.5±3
O(3CA)		80.4±3	166.9±4	91.7±3
O(3C1)			91.5±3	96.5±3
O(21)				79.1±3

Environment of P(A)

P(A) - O(3TA)	1.512±10
P(A) - O(3CA)	1.536±9
P(A) - O(2A)	1.541±5
P(A) - O(1A)	1.603±5

Angles in degrees subtended at P(A) by

	O(3CA)	O(2A)	O(1A)
O(3TA)	113.3±5	112.4±5	112.2±5
O(3CA)		109.7±4	103.3±5
O(2A)			105.3±3

P(A) - O(1A)-P(A)' 138.8±10°

Zn(1) - O(3C1)	2.033±8
Zn(1) - O(2A)	2.051±8
Zn(1) - O(22)	2.083±7
Zn(1) - O(3T2)	2.097±8
Zn(1) - O(3CA)	2.103±7
Zn(1) - O(3C2)	2.179±7

Angles in degrees subtended at Zn(1) by

	O(2A)	O(22)	O(3T2)	O(3CA)	O(3C2)
O(3C1)	95.1±3	164.0±4	94.2±3	79.1±3	108.4±3
O(2A)		80.3±3	158.5±3	90.7±3	81.0±3
O(22)			95.5±3	85.5±3	86.2±3
O(3T2)				110.1±3	77.6±3
O(3CA)					169.1±7

Environment of Zn(2)

Zn(2) - O(3T1)	1.929±9
Zn(2) - O(3C2)	2.025±7
Zn(2) - O(2A)	2.032±8
Zn(2) - O(22)	2.056±7
Zn(2) - O(3T2)	2.093±9

Angles in degrees subtended at Zn(2) by

	O(3C2)	O(2A)	O(22)	O(3T2)
O(3T1)	96.4±3	93.8±3	149.6±3	116.3±4
O(3C2)		169.7±5	90.5±3	81.2±3
O(2A)			81.4±3	92.9±3
O(22)				93.9±3

Environment of P(1)

P(1) - O(3T1)	1.489±10
P(1) - O(21)	1.518±5
P(1) - O(3C1)	1.528±9
P(1) - O(1)	1.600±7

Angles in degrees subtended at P(1) by

	O(21)	O(3C1)	O(1)
O(3T1)	114.4±5	111.7±5	111.1±5
O(21)		111.2±4	103.3±4
O(3C1)			104.4±5

Environment of P(2)

P(2) - O(3C2)	1.518±9
P(2) - O(22)	1.520±5
P(2) - O(3T2)	1.531±10
P(2) - O(1)	1.567±7

Angles in degrees subtended at P(2) by

	O(22)	O(3T2)	O(1)
O(3C2)	112.4±4	111.8±5	103.6±5
O(22)		111.7±4	109.0±4
O(3T2)			107.9±5

P(1)-O(1)-P(2) 148.4±5°

Discussion

The structure is derived from the β-$Zn_2P_2O_7$ structure (2) by the displacement of the bridging oxygen atoms of the P_2O_7 groups from their special positions on centres of symmetry to give a bent P-O-P bridge. Adjacent groups in the *c* direction have the oxygen atoms displaced in opposite directions along *b* but, in the *a* direction, three adjacent groups have displacements up and the next three down. The zinc atoms are also displaced, resulting in two having a square pyramidal environment of O atoms and one having close to an octahedral environment. The structure is related to those of α-$Mg_2P_2O_7$ (3) and α-$Cu_2P_2O_7$(1).

Details of analysis

Crystals grown from melt. Lattice parameters found by least squares fit to powder photograph [no calibration or wavelength given]. Intensities measured by microdensitometer from integrating precession photographs of layers with l = 0-2, h = 0 and $l = k-h/3$ taken with a crystal 0.13×0.13×0.35 mm^3, and multiple-film integrating Weissenberg photographs of layers with k = 0-5 taken with a crystal 0.06×0.11×0.06 mm^3. MoK_α radiation used with both crystals. Lorentz, polarization and absorption corrections (μ = 11.0 mm^{-1}). Structure solved by similarity with α-$Cu_2P_2O_7$ and electron-density projections and refined initially in space group *Ic* and finally in *I2/c* by least squares to give R_2 = 0.098 (in *Ic*) and 0.089 (in *I2/c*) for 1841 reflexions listed.

1. B.E. ROBERTSON and C. CALVO, 1967. *Acta Cryst.*, 22, 665.
2. C. CALVO, 1965. *Can. J. Chem.*, 43, 1147.
3. C. CALVO, 1967. *Acta Cryst.*, 23, 289.

LITHIUM DIHYDROGEN PHOSPHITE

I. The crystal structure of lithium hydrogen phosphite, LiH_2PO_3. **E. PHILIPPOT** and O. LINDQVIST, 1970. *Acta Chem. Scand.*, 24, 2803-2810.

LiH_2PO_3 F.W. = 87.93

Orthorhombic, a = 5.169±2, b = 11.024±2, c = 5.060±2 Å, U = 288.3 Å^3, D_m = 2.02, Z = 4, D_x = 2.02.

Space group $P2_1nb$ (C_{2v}^9) Absences and final structure

Atomic positions All atoms in 4(*a*) at (x,y,z;½+x,−y,−z;½+x,½−y,½+z;x,½+y,½−z)

	x	y	z	B(Å^2)
P	0.2460*	0.1534±1	0.1800±3	1.17±7
O(1)	0.2007±15	0.2868±4	0.2316±10	1.76±13
O(2)	0.5398±14	0.1219±5	0.1684±10	1.87±15
O(3)	0.1242±16	0.0702±4	0.3777±10	1.98±13
Li	0.2559±38	0.4214±9	-0.0186±25	1.69±24
H(1)	0.665±14	0.368±4	0.470±11	-1.0±9
H(2)	0.119±38	0.351±9	0.459±31	5.0±34

*Chosen arbitrarily to fix origin.

Interatomic distances (Å) *and angles*

P - H(1)	1.171±6
P - O(3)	1.496±6
P - O(1)	1.512±5
P - O(2)	1.560±7

Angles in degrees subtended at P by

	O(3)	O(1)	O(2)
H(1)	110±3	107±2	106±3
O(3)		114.6±3	107.4±4
O(1)			112.0±3

Li - O(3)	1.914±13
Li - O(1)	1.971±12
Li - O(3)'	1.978±21
Li - O(2)	1.996±16

Angles in degrees subtended at Li by

	O(1)	O(3)	O(2)
O(3)	110.8±7	113.7±7	107.5±8
O(1)		110.2±8	104.4±7
O(3)			109.9±7

O(2)-H(2)	1.18±16
H(2)-O(1)	1.42±15
P-O(2)-H(2)	108±9°
O(2)-H(2)-O(1)	165±13°
O(2)-O(1)	2.567±7

Discussion

$HPO_2(OH)$ ions are connected by H bonds into chains along c. Li atoms are so placed as to bond to four different anions.

Details of analysis

Lattice parameters from 1. Intensities estimated visually from multiple-film equi-inclination Weissenberg photographs of layer with h = 0 to 4 taken with CuK_α radiation. Corrected for Lorentz, polarization, extinction and absorption effects (μ = 6.30 mm^{-1}, crystal size 0.34×0.05×0.08 mm^3). Structure determined from Patterson and electron-density functions and refined by least squares to give R = 0.063 for 285 observed reflexions listed.

1. G. BRUN, *C.R. Acad Sci. Paris*. To be published.

COPPER POTASSIUM POLYPHOSPHATE

I. Structure cristalline du polyphosphate mixte de cuivre-pottassium: $CuK_2(PO_3)_4$. I. TORDJMAN, DUC TRAN QUI and M. LAUGT, 1970. *Bull. Soc. Fr. Minér. Crist.*, 93, 160-165.

$CuK_2(PO_3)_4$ F.W. = 457.6

Monoclinic, *a* = 10.940±2, *b* = 12.250±3, *c* = 7.9037±16 Å, β = 100.4894±3, *U* = 1041.51 Å^3, *Z* = 4, D_x = 2.91.

Space group Cc (C_s^4) Systematic absences [see below]

Atomic positions All atoms in 4(*a*)

	x	*y*	*z*	*B*(Å^2)
Cu	0.0304±17	0.1572±4	0.7507±34	0.014
P(1)	0.4168±27	0.4278±17	0.8262±41	0.100
P(2)	0.1929±22	0.3388±14	0.5753±41	0.298
P(3)	0.3491±20	0.1849±9	0.4267±43	0.100
P(4)	0.1408±25	0.0749±19	0.1953±43	0.433
K(1)	0.1676±14	0.3729±10	0.0883±28	0.346
K(2)	0.3622±18	0.1370±11	0.9058±31	0.782
O(1)	0.1125±30	0.4198±23	0.4739±60	0.324
O(2)	0.1702±32	0.2668±20	0.7257±60	0.300
O(3)	0.3107±31	0.4076±26	0.6611±60	0.300
O(4)	0.2377±35	0.2673±27	0.4460±60	0.652
O(5)	0.4196±39	0.2185±27	0.2900±70	0.354
O(6)	0.4300±35	0.1513±25	0.5734±60	0.689
O(7)	0.2751±46	0.0751±35	0.3236±76	1.996
O(8)	0.1520±60	0.1407±42	0.0460±100	1.041
O(9)	0.1199±43	-0.0395±32	0.1574±73	0.197
O(10)	0.0386±35	0.1266±25	0.2650±68	1.195
O(11)	0.3952±48	0.3604±35	0.9691±81	0.212
O(12)	0.4182±50	0.5559±35	0.8376±85	0.624

Discussion

The structure consists of PO_4 tetrahedra sharing corners to form a chain running along (101) with P-O distances ranging from 1.4 to 1.7 Å. The copper is coordinated to four O atoms at the corners of a square (Cu-O ∿ 2.0) and two further axial O atoms at a distance of 2.5 Å. The K atoms are 8 and 9 coordinate with K-O distances ranging between 2.6 and 3.3 Å. The errors on individual distances are large (0.06 to 0.10 Å).

Details of analysis

Intensities estimated visually from multiple-film Weissenberg photographs of layers *l*=0-5 (Mo radiation). Lorentz, polarization and α_1, α_2 correction made but no correction made for absorption (crystal size 0.04 and 0.06 mm). Structure determined by Patterson and electron-density projections and 81 parameters refined by least squares to give *R* ∿0.08 for 212 reflexions listed. [It is not clear whether structure has a centre of symmetry or not. The heavy ions occupying positions close to those required in *C2/c* and the uncertainty of the light atom positions is so large as to make it difficult to rule out *C2/c* as a possible space group].

MANGANESE AMMONIUM POLYPHOSPHATE

I. Structure cristalline du polyphosphate de manganèse-ammonium. DUC TRAN QUI, I. TORDJMAN, M.-T. ARERBUCH and G. BASSI, 1970. *Bull. Soc. Fr. Minér. Cristallogr.*, 93, 413-420.

$Mn(NH_4)_2(PO_3)_4$ F.W. = 406.90

Monoclinic, a = 11.297, b = 12.993, c = 7.839 Å, β = 101.5°, U = 1127.5 Å^3, Z = 4, D_x = 2.39.

Space group $P2_1/n$ (C_{2h}^5)

Atomic positions All atoms in 4(*e*) at ($x,y,z;\bar{x},\bar{y},\bar{z};\frac{1}{2}-x,\frac{1}{2}+y,\frac{1}{2}-z;\frac{1}{2}+x,\frac{1}{2}-y,\frac{1}{2}+z$)

	x	*y*	*z*	*B*(Å^2)
Mn	0.0164±6	0.1492±4	0.7645±10	0.42±10
P(1)	0.4119±9	0.4063±7	0.8380±17	0.10±16
P(2)	0.2065±10	0.3108±7	0.5922±18	0.18±22
P(3)	0.3412±9	0.1635±7	0.4125±17	0.24±18
P(4)	0.1268±10	0.0683±7	0.1815±17	0.16±16
N(1)	0.1581±24	0.3650±20	0.0865±48	0.32±48
N(2)	0.3638±30	0.1151±22	0.9274±50	0.52±51
O(1)	0.1340±26	0.3975±21	0.5142±38	0.26±38
O(2)	0.1532±25	0.2467±20	0.7233±45	0.92±46
O(3)	0.3434±22	0.3423±23	0.6950±40	3.05±59
O(4)	0.2325±28	0.2452±21	0.4539±41	0.77±42
O(5)	0.4104±26	0.2164±19	0.2779±43	0.77±42
O(6)	0.4124±27	0.1138±19	0.5671±42	0.72±43
O(7)	0.2640±23	0.0802±18	0.2978±38	0.30±34
O(8)	0.1229±25	0.1366±21	0.0377±41	0.73±44
O(9)	0.1006±29	-0.0423±21	0.1430±43	0.93±44
O(10)	0.0554±26	0.1105±21	0.3187±41	0.83±45
O(11)	0.3997±28	0.3588±26	0.0019±45	2.83±59
O(12)	0.3849±31	0.5187±20	0.8140±45	1.31±46

Interatomic distances (Å) *and angles*

σ(distances) = 0.03 Å, σ(angles) = 1°

Mn - O(2)	2.07
Mn - O(5)	2.13
Mn - O(9)	2.14
Mn - O(12)	2.18
Mn - O(11)	2.22
Mn - O(8)	2.24

Angles in degrees subtended at Mn by

	O(5)	O(9)	O(12)	O(11)	O(8)
O(2)	87	168	90	102	85
O(5)		97	167	81	102
O(9)			89	90	83
O(12)				87	90
O(11)					172

P(1) - O(11)	1.46
P(1) - O(3)	1.48
P(1) - O(12)	1.50
P(1) - O(10)	1.67

Angles in degrees subtended at P(1) by

	O(3)	O(12)	O(10)
O(11)	108	118	107
O(3)		113	103
O(12)			107

P(1)-O(3)-P(2) 144°

P(2) - O(1)	1.45
P(2) - O(4)	1.45
P(2) - O(2)	1.54
P(2) - O(3)	1.65

Angles in degrees subtended at P(2) by

	O(4)	O(2)	O(3)
O(1)	109	116	114
O(4)		110	102
O(2)			105

P(2)-O(4)-P(3) 140°

P(3) - O(6)	1.46
P(3) - O(7)	1.56
P(3) - O(5)	1.59
P(3) - O(4)	1.70

Angles in degrees subtended at P(3) by

	O(7)	O(5)	O(4)
O(6)	109	119	114
O(7)		102	102
O(5)			109

P(3)-O(7)-P(4) 138°

P(4) - O(8)	1.43
P(4) - O(9)	1.49
P(4) - O(10)	1.57
P(4) - O(7)	1.64

Angles in degrees subtended at P(4) by

	O(9)	O(10)	O(7)
O(8)	118	113	105
O(9)		112	109
O(10)			98

P(4)-O(10)-P(1) 131°

N(1) - O(6)	2.76
N(1) - O(11)	2.94
N(1) - O(8)	3.01
N(1) - O(7)	3.02
N(1) - O(2)	3.23
N(1) - O(4)	3.23
N(1) - O(9)	3.33
N(1) - O(1)	3.44
N(1) - O(5)	3.48
N(1) - O(6)	3.49

N(2) - O(1)	2.86
N(2) - O(6)	2.98
N(2) - O(1)	3.00
N(2) - O(5)	3.00
N(2) - O(8)	3.03
N(2) - O(2)	3.10
N(2) - O(11)	3.23
N(2) - O(12)	3.30
N(2) - O(7)	3.35
N(2) - O(3)	3.45

Discussion

The structure consists of spirals of linked PO_4 tetrahedra consisting of eight members in a repeating unit.

Details of analysis

Intensities of layers with l = 0 to 6 measured visually from Weissenberg photographs taken with MoK_α radiation. Structure solved by isomorphous replacement (Cd for Mn) and refined by least squares to give R = 0.10 for 378 reflexions listed.

AMMONIUMTETRAMETAPHOSPHATE

$(NH_4)_4P_4O_{12}$ F.W. = 388.1

I. Reinvestigation of the crystal structure of ammoniumtetrametaphosphate $(NH_4)_4P_4O_{12}$. D.A. KOSTER and A.J. WAGNER, 1970. *J. Chem. Soc. A*, 435-441.

Orthorhombic, a = 10.433±2, b = 10.871±3, c = 12.588±3 Å, U = 1428 Å^3, D_m = 1.79, Z = 4, D_x = 1.81.

Space group Cmca (D_{2h}^{18}) from systematic absences and refinement

Atomic positions

	x	y	z
8 P(1) in 8(*f*)	0	0.0339±1	0.1501±1
8 P(2) in 8(*d*)	0.2112±1	0	0
8 O(1) in 8(*f*)	0	-0.1005±2	0.1639±2
8 O(2) in 8(*f*)	0	0.1197±2	0.2416±2
16 O(3) in 16(*g*)	0.2837±1	-0.0916±2	0.0633±1
16 O(4) in 16(*g*)	0.1173±1	0.0770±1	0.0754±1
8 N(1) in 8(*e*)	1/4	0.2724±2	1/4
8 N(2) in 8(*f*)	0	0.3196±2	-0.0431±2
8 H(1) in 8(*f*)	0	0.371±3	-0.103±3
8 H(2) in 8(*f*)	0	0.252±6	-0.059±3
16 H(3) in 16(*g*)	0.070±3	0.339±2	-0.008±2
16 H(4) in 16(*g*)	0.315±2	0.226±2	0.248±2
16 H(5) in 16(*g*)	0.248±3	0.327±2	0.197±2

Anisotropic temperature factors given for all atoms except H for which only isotropic temperature factors are listed.

Interatomic distances (Å) *and angles*

Environment of P(1) (crystallographic symmetry C_s)

P(1) - O(1)	1.472±4
P(1) - O(2)	1.481±4
P(1) - O(4)	1.613±4 (×2)

Angles in degrees subtended at P(1) by

	O(1)	O(2)	O(4)'
O(2)	122.2±2		
O(4)	110.9±1	105.7±2	98.7±2

Environment of P(2) (crystallographic symmetry C_2)

P(2) - O(3)	1.483±3 (×2)
P(2) - O(4)	1.600±4 (×2)

Angles in degrees subtended at P(2) by

	O(3)'	O(4)	O(4')
O(3)	118.7±3	110.2±2	106.2±2
O(4)			104.5±2

P(1)-O(4)-P(2) = 131.2±2°

Environment of N(1) (crystallographic symmetry C_2)

	N-H (Å)	N ... O (Å)	N-H-O(°)
N(1)-H(4) ... O(2) (×2)	0.85	3.09	173
N(1)-H(5) ... O(3) (×2)	0.89	2.80	164

Angles in degrees subtended at N(1) by

	H(4)'	H(5)	H(5)'
H(4)	106	114	114
H(5)			96

Environment of N(2) (crystallographic symmetry C_2)

	N-H(Å)	N...O(Å)	N-H-O(°)
N(2)-H(1) ... O(2)	0.93	2.79	147
N(2)-H(2) ... O(1)	0.77	2.83	156
N(2)-H(3) ... O(3) (×2)	0.87	2.80	171

Angles in degrees subtended at N(2) by

	H(1)	H(2)	H(3)'
H(2)	111		
H(3)	105	111	112

Discussion

The cyclic $P_4O_{12}^{4-}$ ions have crystallographic symmetry C_{2h} and are arranged at the points of a face-centered lattice with *a* as the direction of their two-fold axis. The NH_4^+ ions lie close to four similar lattices displaced from the $P_4O_{12}^{4-}$ lattice by (0, 1/3, 0) (0, 2/3, 0) (1/4, 1/4, 1/4) and (1/4, 1/4, 3/4) respectively. The result is a structure in which layers of $P_4O_{12}^{4-}$ ions alternate along the *b* direction with layers of NH_4^+ ions. This structure was first reported by 1 and refined by 2.

Details of analysis

Prepared from orthophosphoric acid and cupric oxide. Crystals grown from a water-ethanol solution. Lattice parameters from Weissenberg photographs of layers *h* = 0, *k* = 0 and CuK_α radiation ($K_{\alpha 1}$ = 1.54051, $K_{\alpha 2}$ = 1.54433 Å). 1650 reflexions measured using Θ-2θ scan from crystal $(0.2)^3$ mm^3 on Nonius single crystal diffractometer with filtered Mo radiation. Lorentz-polarization, but no absorption corrections made. Refinement of structure of 2 by weighted block-diagonal least squares; weighted *R* for 1429 observed reflexions was 0.055. Structure factors listed.

1. *Structure Reports*, 15, 268.
2. D.W.J. CRUICKSHANK, 1965. *Acta Cryst.*, 17, 675.

COBALT ARSENATE

$Co_3(AsO_4)_2$ F.W. = 454.63

I. Crystallographic studies of cobalt arsenates. I. Crystal structure of $Co_3(AsO_4)_2$. N. KRISHNAMACHARI and C. CALVO, 1970. *Can. J. Chem.*, 48, 881-889.

Monoclinic, *a* = 5.830±4, *b* = 9.675±2, *c* = 10.34±2 Å, β = 93.42±5°, *U* = 582.0 Å^3, Z = 4, D_x = 5.2.

Space group $P2_1/c$ (C_{2h}^5)

Atomic positions

All atoms in 4(*e*)

	x	*y*	*z*
Co(1)	-0.1129±2	0.1352±1	-0.4670±1
Co(2)	-0.3502±2	0.1401±1	-0.2163±1
Co(3)	0.3547±2	0.1235±1	0.0586±1
As(1)	0.3675±2	0.1549±1	0.3809±1
As(2)	-0.1283±2	0.0523±1	0.2032±1
O(11)	0.3835±12	0.2217±7	0.2312±6
O(12)	0.6035±14	0.2389±8	0.4534±6
O(13)	0.1582±13	0.2283±7	0.4636±6
O(14)	-0.3638±12	0.0177±7	0.6204±7
O(21)	0.3723±12	0.4867±7	0.3777±6
O(22)	-0.1365±12	0.2255±6	0.1909±6
O(23)	0.1083±12	-0.0104±8	0.1325±7
O(24)	0.1212±13	0.5061±7	0.1391±6

Anisotropic temperature factors given.

Interatomic distances (Å) and angles

Fourfold coordination around As(1)

As(1) - O(11)	1.685
As(1) - O(12)	1.722
As(1) - O(13)	1.688
As(1) - O(14)	1.670

Angles in degrees subtended at As(1) by

	O(12)	O(13)	O(14)
O(11)	97.5	112.6	112.6
O(12)		99.2	119.0
O(13)			114.6

Fourfold coordination around As(2)

As(2) - O(21)	1.728
As(2) - O(22)	1.681
As(2) - O(23)	1.710
As(2) - O(24)	1.689

Angles in degrees subtended at As(2) by

	O(22)	O(23)	O(24)
O(21)	108.1	108.9	110.0
O(22)		109.9	109.6
O(23)			110.2

Sixfold coordination around Co(1)

Co(1) - O(12)	2.072
Co(1) - O(13)	1.991
Co(1) - O(14)	2.099
Co(1) - O(22)	2.128
Co(1) - O(24)	2.173
Co(1) - O(24)'	2.181

Angles in degrees subtended at Co(1) by

	O(13)	O(14)	O(22)	O(24)	O(24)'
O(12)	105.8	82.6	85.0	88.5	165.4
O(13)		171.6	94.5	86.8	88.8
O(14)			86.2	93.4	82.8
O(22)				173.4	94.5
O(24)					92.0

Sixfold coordination around Co(2)

Co(2) - O(11)	2.096
Co(2) - O(12)	2.138
Co(2) - O(21)	2.292
Co(2) - O(22)	2.075
Co(2) - O(14)	2.061
Co(2) - O(23)	2.042

Angle in degrees subtended at Co(2) by

	O(12)	O(21)	O(22)	O(14)	O(23)
O(11)	74.4	85.4	86.4	99.4	169.8
O(12)		79.4	98.5	170.2	96.3
O(21)			171.8	92.7	88.8
O(22)				88.6	99.4
O(14)					89.2

Sixfold coordination around Co(3)

Co(3) - O(11)	2.021
Co(3) - O(12)	2.287
Co(3) - O(13)	2.048
Co(3) - O(21)	2.161
Co(3) - O(21)'	2.143
Co(3) - O(23)	2.111

Angle in degrees subtended at Co(3) by

	O(12)	O(13)	O(21)	O(21)'	O(23)
O(11)	97.3	96.3	172.4	89.9	89.5
O(12)		73.4	79.0	91.6	172.7
O(13)			89.1	164.4	103.3
O(21)				83.6	94.6
O(21)'					91.1

σ(distances) = 0.007 Å, σ(angles) 0.3°

Discussion

The structure consists of close packed planes of O atoms perpendicular to *b* stacked in an ABAC sequence. The Co atoms lie in distorted octahedral interstices, the As atoms in distorted tetrahedral interstices. Xanthiosite ($Ni_3(AsO_4)_2$) probably is isostructural (1).

Details of analysis

Crystals prepared by melting hydrated $Co_3(AsO_4)$. Lattice parameters from X-ray photographs calibrated with TiO_2 (a = 4.5929, c = 2.9591 Å) Intensities of layers with l = 0 to 3 and k = 0 to 4 measured from integrated precession photographs (MoK_α radiation) and of layers with h = 0 to 4 on a manual Weissenberg diffractometer (ω scan). Corrections for Lorentz, polarization and absorption effects (crystal size 0.1×0.1×0.2 mm^3). Structure solved from Patterson functions and refined by least squares to give a weighted R = 0.082 for about 1400 observed reflexions listed. A further 800 reflexions were unobserved.

1. R.J. DAVIS, M.H. HEY and A.W.G. KINGSBURY, 1965. *Miner. Mag.*, 35, 72.

CADMIUM ORTHOARSENATE

I. Die Kristallstruktur des Cadmiumorthoarsenates $Cd_3(AsO_4)_2$. G. ENGEL and W.E. KLEE, 1970. *Z. Kristallogr.*, 132, 332-339.

$Cd_3(AsO_4)_2$ F.W. = 615.04

Monoclinic, a = 9.285±1, b = 11.936±1, c = 6.599±1 Å, β = 98.45±2, U = 723.4 $Å^3$, Z = 4.

Space group $P2_1/c$ (C^5_{2h})

Atomic positions All atoms in 4(*e*)

	x	*y*	*z*
Cd(1)	0.9475±1	0.1177±1	0.8442±2
Cd(2)	0.7155±1	0.0821±1	0.3334±2
Cd(3)	0.3535±1	0.1909±1	0.1288±2
As(1)	0.0830±2	0.1358±1	0.3934±2
O(1)	0.0810±12	0.0647±10	0.1686±18
O(2)	0.9228±12	0.2024±9	0.4040±18
O(3)	0.2102±12	0.2341±10	0.3731±19
O(4)	0.1252±13	0.0606±10	0.6119±19
As(2)	0.6054±2	0.0887±1	0.8109±1
O(5)	0.4726±13	0.1819±10	0.8450±20
O(6)	0.7064±14	0.1370±11	0.6349±21
O(7)	0.7343±15	0.0774±12	0.0193±22
O(8)	0.5315±12	0.9608±10	0.7461±18

Anisotropic temperature factors given in I for Cd and As, isotropic temperature factors for O.

Interatomic distances (Å) *and angles*

σ(distances) = 0.013; σ(angles) = 0.5°

Cd(1) -	O(1)	2.393
	O(1)'	2.192
	O(2)	2.201
	O(4)	2.506
	O(6)	2.462
	O(7)	2.481

Angles in degrees subtended at Cd(1) by

	O(1)'	O(2)	O(4)	O(6)	O(7)
O(1)	79.3	98.7	100.2	146.4	82.7
O(1)'		163.9	77.9	88.8	74.3
O(2)			118.0	84.2	89.7
O(4)				108.0	151.0
O(6)					63.8

Cd(2) -	O(2)	2.392
	O(4)	2.250
	O(6)	2.108
	O(7)	2.106
	O(8)	2.333

Angles in degrees subtended at Cd(2) by

	O(4)	O(6)	O(7)	O(8)
O(2)	86.1	76.8	91.5	155.8
O(4)		101.6	89.4	118.1
O(6)			163.3	96.1
O(7)				89.7

Cd(3) -	O(3)	2.313
	O(3)'	2.180
	O(5)	2.296
	O(5)'	2.258
	O(8)	2.200

Angles in degrees subtended at Cd(3) by

	O(3)'	O(5)	O(5)'	O(8)
O(3)	95.8	167.1	71.7	102.7
O(3)'		72.7	113.5	146.6
O(5)			107.2	98.7
O(5)'				98.7

As(1) -	O(1)	1.706
	O(2)	1.697
	O(3)	1.684
	O(4)	1.695

Angles in degrees subtended at As(1) by

	O(2)	O(3)	O(4)
O(1)	111.5	101.5	116.6
O(2)		107.8	107.9
O(3)			111.0

As(2) -	O(5)	1.700
	O(6)	1.698
	O(7)	1.691
	O(8)	1.703

Angles in degrees subtended at As(2) by

	O(6)	O(7)	O(8)
O(5)	110.6	112.7	110.3
O(6)		100.8	112.0
O(7)			110.1

Discussion

$Cd_3(AsO_4)$ has the graftonite structure (1).

Details of analysis

Prepared by heating CdO and As_2O_3 treated with HNO_3 to 850°C. Lattice parameters Guinier photographs. Intensities measured with Nb-filtered MoK_α radiation on an automatic four-circle diffractometer, corrected for Lorentz and polarization, but not absorption (crystal size 0.2×0.2×0.6 mm^3). Graftonite structure (1) assumed and refined by least squares to give R = 0.077 for 1844 observed reflexions [not listed].

1. C. CALVO, 1968. *Amer. Min.*, 53, 743.

DIAMMONIUM HYDROGEN ARSENATE

I. The crystal structure of diammonium hydrogen arsenate, $(NH_4)_2HAsO_4$. A.A. KHAN, M.E. STRAUMANIS and W.J. JAMES, 1970. *Acta Cryst.*, B26, 1889-1892.

$(NH_4)_2HAsO_4$ F.W. = 175.92

Monoclinic, a = 11.378±5, b = 6.908±6, c = 8.129±5 Å, β = 113.0±2°, U = 584±1 Å , D_m = 1.989, Z = 4, D_x = 2.00.

Space group $P2_1/c$ (C_{2h}^5)

Atomic positions All atoms in 4(*e*)

	x	y	z	$B(\text{Å}^2)$
As	0.2495±2	0.8910±3	0.4285±1	2.25±5
O(1)	0.2083±7	0.9865±22	0.2142±9	2.93±13
O(2)	0.2599±7	0.0974±16	0.5438±10	2.52±14
O(3)	0.3870±6	0.7682±18	0.4968±8	2.46±13

	x	y	z	B(Å^2)
O(4)	0.1314±6	0.7459±15	0.4293±8	2.16±12
N(1)	0.4489±8	0.1186±19	0.1535±11	2.68±16
N(2)	0.1210±8	0.3774±19	0.2626±11	2.72±16

Interatomic distances (Å) *and angles*

As-O(1)	1.747±8	O(1)-As-O(2)	99.7±6°
As-O(2)	1.684±10	O(1)-As-O(3)	112.0±3
As-O(3)	1.673±7	O(1)-As-O(4)	109.9±4
As-O(4)	1.678±7	O(2)-As-O(3)	113.0±4
		O(2)-As-O(4)	112.7±3
N(1)-H...O(2)	2.781±13	O(3)-As-O(4)	110.0±5
N(1)-H...O(3)	2.807±10		
N(1)-H...O(3)	2.837±11		
N(1)-H...O(3)	2.925±17		
N(1)-H...O(1)	3.093±9		

Preparation

From evaporation of liquid ammonia (in excess) and aqueous As_2O_5.

Discussion of the structure

NH_4^+ and AsO_3OH^{2-} groups are held together by hydrogen bonds through O(1)-H...O(4) with O(2), O(3) and O(4) acting as acceptors to hydrogen bonds from the NH_4^+ ion as shown in Fig. 44.

Details of analysis

619 intensities measured from a cylinder with Cu radiation and corrected for absorption. Refinement by least squares yields an *R* of 0.086.

WEILITE

I. The crystal structure of $CaHAsO_4$ (Weilite). G. FERRARIS and G. CHIARI, 1970. *Acta Cryst.*, B<u>26</u>, 403-410.

$CaHAsO_4$ F.W. = 179.99

Triclinic, *a* = 7.0591±8, *b* = 6.8906±9, *c* = 7.2006±16 Å, α = 97°26'±1', β = 103°33'±1', γ = 87°45'±1', *U* = 337.62 Å^3, D_m = 3.50, *Z* = 4, D_x = 3.541.

Space group $P\bar{1}$ (C_i^1) centrosymmetric except possibility for the hydrogen ions

Atomic positions All atoms in 2(*i*)

	x	y	z
2 As(1)	0.2048±2	0.3776±2	0.7243±2
2 As(2)	0.2978±2	0.9414±2	0.2087±2
2 Ca(1)	0.3024±3	0.4323±3	0.2796±3
2 Ca(2)	0.1724±3	0.8339±3	0.6606±3
2 O(1)	0.3219±11	0.3363±10	0.9531±11
2 O(2)	0.3565±12	0.4976±11	0.6274±10
2 O(3)	0.1361±11	0.1656±10	0.5893±11
2 O(4)	0.0235±11	0.5384±10	0.7410±11
2 O(5)	0.3366±10	0.8260±11	0.0003±11
2 O(6)	0.4650±12	0.1132±12	0.3036±11
2 O(7)	0.0819±11	0.0651±11	0.1601±12
2 O(8)	0.2878±13	0.7757±1	0.3537±12

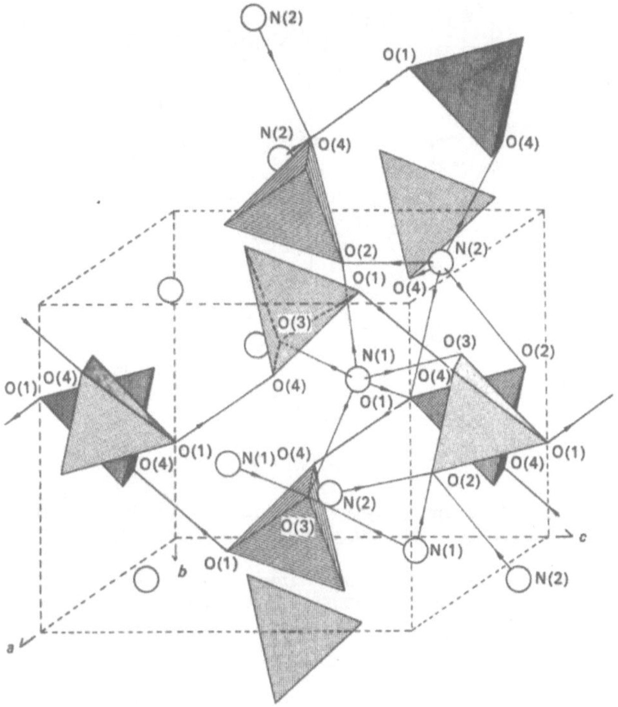

Fig. 44. A three-dimensional view of the structure of $(NH_4)_2HAsO_4$ with tetrahedra and circles representing the AsO_4 groups and the N atoms respectively. Arrows indicate the hydrogen bonds.

Interatomic distances (Å) *and angles*

As(1)-O(1)	1.722±8	O(1)-As(1)-O(2)	109.5±4°	O(2)-As(1)-O(3)	110.6±4°
As(1)-O(2)	1.701±8	O(1)-As(1)-O(3)	109.4±4	O(2)-As(1)-O(4)	104.0±4
As(1)-O(3)	1.661±7	O(1)-As(1)-O(4)	108.3±4	O(3)-As(1)-O(4)	114.9±4
As(1)-O(4)	1.677±7				
As(2)-O(5)	1.676±8	O(5)-As(2)-O(6)	110.5±4	O(6)-As(2)-O(7)	105.2±9
As(2)-O(6')	1.667±8	O(5)-As(2)-O(7)	108.3±4	O(6)-As(2)-O(8)	113.7±4
As(2)-O(7')	1.704±8	O(5)-As(2)-O(8)	108.9±9		
As(2)-O(8)	1.668±8			O(7)-As(2)-O(8)	110.2±9

Structure

Isostructural with monetite ($CaHPO_4$) (2,3). Hydrogen bonding proposed. See Fig. 45.

Materials

Prepared by method of Pierrot (1).

Details of analysis

Structure was refined by least squares with 1475 reflexions measured with an automatic diffractometer and corrected for absorption. Anisotropic thermal parameters varied but adjusted to keep positive-definite. Final *R* value 0.069.

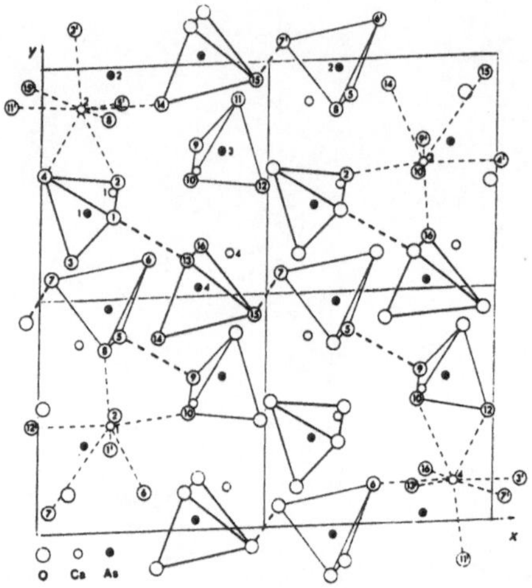

Fig. 45. Packing of AsO_4 groups, hydrogen bonds and Ca coordination in weilite. Four unit cells projected onto the *xy* plane.

1. R. PIERROT, 1964. *Bull. Soc. Franç. Minér. Crist.*, 87, 169.
2. N.A. CURRY, W.A. DEENE and D.W. JONES, 1968. *Bull. Soc. Franç. Minér. Crist.*, 86, 368.
3. *Structure Reports*, 26, 458.

LITHIUM MOLYBDENYL ARSENATE

I. The crystal structure of $LiMoO_2AsO_4$. B. LINNROS, 1970. *Acta Chem. Scand.*, 24, 3711-3722.

$LiMoO_2AsO_4$ F.W. = 273.80

Orthorhombic, a = 9.3792±5, b = 8.9114±4, c = 4.9043±2 Å, U = 409.91 Å^3, D_m = 4.35, Z = 4, D_x = 4.43.

Space group $Pn2_1a$ (C_{2v}^9)

Atomic positions All atoms in 4(a) ($x,y,z;x,\frac{1}{2}+y,z;\frac{1}{2}-x,\frac{1}{2}+y,\frac{1}{2}+z;\frac{1}{2}+x,y,\frac{1}{2}-z$)
Non-standard setting

	x	*y*	*z*	*B*(Å^2)
Mo	0.1606±1	0	0.1107±2	*
As	0.0513±2	0.6802±3	0.4005±3	*
O(1)	0.0144±11	0.8445±11	0.2439±21	1.08±21
O(2)	0.4968±10	0.5359±11	0.2944±16	0.77±17
O(3)	0.2329±11	0.6745±14	0.4474±22	1.30±20
O(4)	0.2482±14	0.8687±13	0.9204±23	1.34±23
O(5)	0.4761±9	0.6717±14	0.7873±21	1.06±19
O(6)	0.2458±12	0.4963±16	0.9035±18	1.56±23
Li	0.387±6	0.848±7	0.592±8	4.65±120

*Anisotropic temperature factors given.

Interatomic distances (Å) and angles

Mo - O(6)	1.683±10
Mo - O(4)	1.707±12
Mo - O(3)	2.014±12
Mo - O(1)	2.056±10
Mo - O(2)	2.165±9
Mo - O(5)	2.176±11

Angles in degrees subtended at Mo by

	O(4)	O(3)	O(1)	O(2)	O(5)
O(6)	101.6±6	95.5±5	93.6±5	165.8±5	88.9±5
O(4)		94.2±5	91.9±5	92.2±5	169.4±5
O(3)			167.8±4	86.5±4	84.7±4
O(1)				82.7±4	87.4±4
O(2)					77.3±3

As - O(2)	1.682±9
As - O(5)	1.687±10
As - O(1)	1.689±10
As - O(3)	1.720±10

Angles in degrees subtended at As by

	O(5)	O(1)	O(3)
O(2)	110.8±5	110.0±5	110.8±5
O(5)		111.5±5	106.9±5
O(1)			106.8±5

Li - O(5)	2.02
Li - O(1)	2.04
Li - O(6)	2.04
Li - O(2)	2.07
Li - O(4)	2.08
Li - O(3)	2.23

σ(Li-O) = 0.06

Angles in degrees subtended at Li by

	O(1)	O(6)	O(2)	O(4)	O(3)
O(5)	97	166	106	88	83
O(1)		90	86	175	96
O(6)			85	85	85
O(2)				93	170
O(4)					84

σ(O-Li-O) = 2°

Discussion

The structure consists of a connected nextwork of MoO_6, AsO_4 and LiO_6 polyhedra. Four O atoms take part in all three coordination spheres, two others form only short bonds to Mo and normal bonds to Li.

Details of analysis

Lattice parameters from Guinier powder photograph taken with monochromatic Cu$K\alpha_1$ radiation, calibrated with KCl (a = 6.29288 Å) as internal standard. Intensities measured with θ-2θ scan on manual diffractometer using Cu$K\alpha$ radiation, corrected for Lorentz, polarization, extinction, and absorption effects (μ = 37.67 mm^{-1}, crystal size = 0.008×0.048×0.072 mm^3). Structure determined from Patterson and electron-density functions and refined by least squares to give R = 0.040 for 293 reflexions listed. The absolute configuration given above is correct.

MAGNESIUM DIARSENATE

$Mg_2As_2O_7$ F.W. = 310.9

I. Refinement of the structure of $Mg_2As_2O_7$. C. CALVO and K. NEELAKANTAN, 1970. *Can. J. Chem.*, 48, 890-894.

Monoclinic, a = 6.567±2, b = 8.524±4, c = 4.739±1 Å, β = 103.8±1°, U = 247.6 Å^3, D_m = 3.982, Z = 2, D_x = 4.004.

Space group C2/m (C^3_{2h}) systematic absences and analogy with thortveitite (1)

Atomic positions

		x	*y*	*z*
4 Mg	in 4(*h*)	0	0.3119±5	1/2
4 As	in 4(*i*)	0.2285±2	0	0.9031±3
2 O(1)	in 2(*a*)	0	0	0
4 O(2)	in 4(*i*)	0.4002±11	0	0.2214±20
8 O(3)	in 8(*j*)	0.2313±11	0.1646±8	0.7163±17

Anisotropic temperature factors given.

Interatomic distances (Å) *and angles*

As_2O_7 ion (crystallographic point symmetry C_{2h})

As - O(1)	1.671 (see discussion)
As - O(2)	1.653
As - O(3)	1.661 (×2)

Angles in degrees subtended at As by

	O(1)	O(3)'
O(2)	102.1	113.2
O(3)	105.8	115.2

O(1) - As - O(1) 180° (by symmetry, but see discussion)

Sixfold coordination around Mg (crystallographic site symmetry C_2)

Mg - O(2)	2.080 (×2)
Mg - O(3)	2.046 (×2)
Mg - O(3)*	2.252 (×2)

Angles in degrees subtended at Mg by

	O(3)	O(3)'	O(2)'	O(3)*	O(3)*'
O(2)	94.1	150.9	79.2	78.8	93.3
O(3)		104.6	150.9	113.3	73.4
O(3)*					169.8

Primed atoms are related to unprimed by the two-fold axis.

σ(distances) = 0.007 Å, σ(angles) = 0.3°

Discussion

A refinement of the structure first reported by 2. Isostructural with thortveitite (1). O(1) shows a large temperature factor perpendicular to As - O(1)- - As suggesting that the true As-O(1) length might be nearer 1.71 Å and that either the As - O(1) - As angle is less than 180° and the O(1) atom is randomly displaced from its ideal position, or that O(1) is undergoing a large thermal motion. It was not possible to test either hypothesis adequately.

Details of analysis

Lattice parameters refined by least squares, powder photograph with MoK_α radiation (λ = 0.70939 Å). Intensities estimated visually from Weissenberg photographs of the layers with l = 0 to 4 and the first 3 layers of the [121] zone, and from precession photographs of layers with k = 0 to 2 and h = 0 to 2. Lorentz and polarization corrections but not absorption (crystal size 0.15×0.15×0.10 mm^3 leads to errors of order 5% in F's). Structure of 2 refined by least squares to give weighted R = 0.107 for 769 reflexions (587 observed) listed in 3.

1. *Strukturbericht*, 2, 523.
2. K. LUKASZEWICZ, 1963. *Bull. Acad. Polon. Sci. Ser. Sci. Chem.*, 11, 361.
3. National Depository for Unpublished Data, National Research Council of Canada, Ottawa, Canada.

COBALT AERUGITE

I. Crystallographic studies of cobalt arsenates II. Crystal structure of $Co_8As_3O_{16}$. N. KRISHNAMACHARI and C. CALVO, 1970. *Can. J. Chem.*, 48, 3124-3131.

$Co_8As_3O_{16}$ F.W. = 952.2

Rhombohedral (Hexagonal setting), a = 6.046±1, c = 28.062±8 Å, U = 888.4 Å^3, Z = 3, D_x = 5.32.

Space group $R\bar{3}m$ (D_{3d}^5) Systematic absences, Laue symmetry, Friedel's law and final structure.

Atomic positions

		x	y	z
7.8±2	Co(1) in 12(*e*)	0	1/2	0
15.6±2	Co(2) in 24(*h*)	-0.1652±5	-x	0.9116±1
3	As(1) in 3(*a*)	0	0	0
6	As(2) in 6(*c*)	0	0	0.8079±1
18	O(1) in 18(*h*)	0.1432±23	-x	-0.0377±4
6	O(2) in 6(*c*)	0	0	0.3743±7
6	O(3) in 6(*c*)	0	0	0.1318±6
18	O(4) in 18(*h*)	0.4860±25	-x	-0.1224±3

Anisotropic temperature factor given.

Interatomic distances (Å) *and angles*

As(1)-O(1)	1.84 (×6)	O(1)-As(1)-O(1)	89.9°
As(2)-O(3)	1.69	O(3)-As(2)-O(4)	108.3°
As(2)-O(4)	1.68 (×2)	O(4)-As(2)-O(4)	110.6°
Co(1)-O(1)	2.16 (×4)	O(1)-Co(1)-O(1)	73.9°
Co(1)-O(2)	3.09 (×2)	O(1)-Co(1)-O(2)	91.2°
Co(2)-O(4)	2.06 (×2)		
Co(2)-O(3)	2.12		
Co(2)-O(1)	2.16 (×2)		
Co(2)-O(2)	2.21		

Angles in degrees subtended at Co(2) by

	O(4)'	O(1)'	O(1)	O(2)
O(4)	105.1	163.9	90.4	88.0
O(3)	93.2	90.4	90.4	178.0
O(1)'	90.4		73.9	88.0

σ(distances) = 0.01 Å, σ(angles) = 0.5°

Discussion

The crystal is composed of a cubic close packed array of oxygen atoms with As(1) and Co(1) occupying the octahedral holes in one layer, Co(2) occupying octahedral holes in the next layer and As(2) occupying tetrahedral holes in the third. Each As(1)O_6 octahedron shares edges with 6 Co(1)O_6 and 6 Co(2)O_6 octahedra. However, the Co sites are only 65% occupied.

Details of analysis

Dark purple crystals prepared by fusing hydrated $Co_3(AsO_4)_2$. Lattice parameters from photographs calibrated with TiO (a = 4.5929, c = 2.9591 Å). Intensities measured on manual Weissenberg diffractometer (ω scan) with [010] along goniometer axis, and photometrically from three zero-layer precession photographs from a crystal 0.01×0.01×0.15 mm³ with MoK_α radiation. [Lorentz and polarization] but no absorption correction (μR≈1). Structure solved from Patterson and electron-density functions and 41 parameters refined by least squares to give R_2 = 0.102 for 613 unequivalent reflexions listed.

SYNADELPHITE

I. Crystal chemistry of the basic manganese arsenates: IV. Mixed arsenic valences in the crystal structure of synadelphite. P.B. MOORE, 1970. *Amer. Min.*, 55, 2023-2037.

$Mn_9(OH)_9(H_2O)_2(AsO_3)(AsO_4)_{12}$ F.W. = 1084.29

Orthorhombic, a = 10.754±11, b = 18.865±17, c = 9.884±14 Å, U = 2005.21 Å³, Z = 4, D_x = 3.58.

Space group Pnma (D_{2h}^{16}) Systematic absences and intensity statistics

Atomic positions

	x	y	z	B(Å²)
4 Mn(1) in 4(c)	0.2941±5	0.75	0.3868±5	0.88±7
8 Mn(2) in 8(d)	0.5613±4	0.6635±2	0.3729±4	0.85±5
8 Mn(3) in 8(d)	0.8269±4	0.5863±2	0.3614±4	0.82±5
8 Mn(4) in 8(d)	0.0887±4	0.5039±2	0.3627±4	0.93±5
8 Mn(5) in 8(d)	0.3564±3	0.4225±2	0.3575±4	1.05±5
4 As(1) in 4(c)	0.9672±3	0.75	0.3618±3	0.12±4
8 As(2) in 8(d)	0.3225±2	0.5811±1	0.5654±2	0.12±3
4 O(1) in 4(c)	0.1151±23	0.75	0.2854±23	0.99±32
4 O(2) in 4(c)	0.4748±27	0.75	0.4832±26	1.33±38
8 O(3) in 8(d)	0.2401±18	0.6535±9	0.5163±18	1.33±28
8 O(4) in 8(d)	0.4018±17	0.6818±9	0.2373±18	1.32±26
8 O(5) in 8(d)	0.4714±16	0.5867±8	0.5176±15	1.10±22
8 O(6) in 8(d)	0.6630±15	0.5927±8	0.2420±15	0.72±20
8 O(7) in 8(d)	0.7158±16	0.6491±8	0.4997±17	0.97±24
8 O(8) in 8(d)	0.7417±18	0.4940±9	0.4987±17	0.88±24
8 O(9) in 8(d)	0.9293±17	0.5115±8	0.2381±16	0.77±23
8 O(10) in 8(d)	0.9757±16	0.5726±8	0.4980±17	1.11±22
8 O(11) in 8(d)	0.1838±16	0.4225±8	0.2349±15	0.83±21
8 O(12) in 8(d)	0.4170±19	0.3306±10	0.2232±18	2.00±29

Interatomic distances (Å) *and angles*

σ(distances) = 0.02 Å; σ(angles) = 0.7°

Mn(1) - O(2)	2.17	
Mn(1) - O(1)	2.17	
Mn(1) - O(4)	2.28	(×2)
Mn(1) - O(3)	2.30	(×2)

Angles in degrees subtended at Mn(1) by

	O(1)	O(4)'	O(3)'
O(2)	178.6	80.1	89.0
O(1)		98.8	91.9
O(4)		68.8	92.4
O(4)'			159.6
O(3)			104.6

Mn(2) - O(7)	2.10	
Mn(2) - O(6)	2.16	
Mn(2) - O(2)	2.17	
Mn(2) - O(4)	2.20	
Mn(2) - O(5)	2.25	
Mn(2) - O(1)	2.33	

Angles in degrees subtended at Mn(2) by

	O(6)	O(2)	O(4)	O(5)	O(1)
O(7)	82.9	97.9	178.2	83.0	107.1
O(6)		169.5	97.3	101.6	84.5
O(2)			81.6	88.9	85.3
O(4)				98.7	71.1
O(5)					168.9

Mn(3) - O(10)	2.11	
Mn(3) - O(6)	2.17	
Mn(3) - O(9)	2.17	
Mn(3) - O(7)	2.17	
Mn(3) - O(4)	2.20	
Mn(3) - O(8)	2.39	

Angles in degrees subtended at Mn(3) by

	O(6)	O(9)	O(7)	O(4)	O(8)
O(10)	176.6	83.9	94.7	96.1	80.7
O(6)		98.5	82.0	90.7	92.2
O(9)			172.4	95.6	92.3
O(7)				92.0	80.1
O(4)					171.1

Mn(4) - O(10)	2.11	
Mn(4) - O(9)	2.12	
Mn(4) - O(6)	2.12	
Mn(4) - O(10)	2.22	
Mn(4) - O(11)	2.24	
Mn(4) - O(6)	2.28	

Angles in degrees subtended at Mn(4) by

	O(9)	O(6)	O(10)	O(11)	O(8)
O(10)	99.2	168.4	80.0	92.8	83.3
O(9)		88.2	82.4	95.1	175.0
O(6)			92.3	95.4	88.7
O(10)				171.9	93.8
O(11)					89.1

Mn(5) - O(9)	2.08	
Mn(5) - O(7)	2.10	
Mn(5) - O(11)	2.21	
Mn(5) - O(5)	2.23	
Mn(5) - O(12)	2.28	
Mn(5) - O(8)	2.37	

Angles in degrees subtended at Mn(5) by

	O(7)	O(11)	O(5)	O(12)	O(8)
O(9)	164.3	93.9	90.0	104.0	84.5
O(7)		93.3	83.4	90.5	81.9
O(11)			175.5	85.5	87.4
O(5)				91.4	95.2
O(12)					169.3

As(1) - O(1)	1.76	
As(1) - O(4)	1.76	(×2)

Angles in degrees subtended at As(1) by

	O(4)	O(4)
O(1)	97.0	97.0
O(4)		93.7

As(2) - O(11)	1.68	(×2)
As(2) - O(8)	1.70	(×2)

Angles in degrees subtended at As(2) by

	O(11)	O(8)	O(3)
O(5)	108.8	109.7	111.6
O(11)		108.0	107.3
O(8)			110.6

Discussion

The structure is based on a double hexagonal close packed array of oxygen atoms (stacking direction *c*) with 1/12 of the O sites vacant. The MnO_6 octahedra share edges to form a V in one layer of formula Mn_9O_{37}. As occurs in two states As(III) with trigonal pyramidal coordination and As(V) with tetrahedral coordination.

Details of analysis

Crystal from Swedish Natural History Museum (NRMS No. 251554). Lattice parameters from powder photographs taken with FeK_α radiation using silicon as standard (a = 5.430 Å). Intensities measured on equi-inclination diffractometer of layers with h = 0 to 12 using monochromated Mo radiation and θ-2θ scan. No absorption correction (crystal size 0.12×0.14×0.18 mm^3). Structure solved by symbolic addition and refined by least squares to give R = 0.102 for 1473 observed reflexions listed in 1.

1. NAPS 00178.

TRANSITION METAL ORTHOVANADATES

I. Structure aux rayons X, neutrons et propriétés magnétiques des orthovanadates de nickel et de cobalt, H. FUESS, E.F. BERTAUT, R. PAUTHENET and A. DURIF, 1970. *Acta Cryst.*, B26, 2036-2046.

$Ni_3V_2O_8$ F.W. = 406.01
$Co_3V_2O_8$ F.W. = 406.67

Orthorhombic $Ni_3V_2O_8$: a = 8.24±1, b = 11.38±1, c = 5.406±8 Å, [U = 553.8 Å^3], Z = 4, [D_x = 4.86].

$Co_3V_2O_8$: a = 8.30±1, b = 11.50±1, c = 6.030±5 Å, [U = 375.6 Å^3], Z = 4, [D_x = 4.54].

Space group Abam (D_{2h}^{18})

Atomic positions

	M = Ni			M = Co		
	x	y	z	x	y	z
4 M(1) in 4(a)	0	0	0	0	0	0
8 M(2) in 8(e) (¼,y,¼)		0.126±4			0.140±3	
8 V in 8(f) (x,y,0)	0.127±7	0.380±3		0.128±10	0.379±5	
8 O(1) in 8(f)	0.233±11	0.240±5		0.252±9	0.250±2	
8 O(2) in 8(f)	0.245±5	-0.004±5		0.248±4	0.005±2	
16 O(3) in 16(g)	0.003±5	0.117±2	0.274±3	0.010±3	0.114±2	0.282±1

Interatomic distances (Å)

	M = Ni	M = Co
M(1)-O(2)	2.00±3 (2×)	2.04±2 (2×)
-O(3)	2.09±2 (4×)	2.12±2 (4×)
M(2)-O(1)	1.99±4 (2×)	2.15±4 (2×)
-O(2)	2.08±3 (2×)	2.01±3 (2×)
-O(3)	2.05±2 (2×)	2.00±3 (2×)
V -O(1)	1.82±4	1.78±3
-O(2)	1.72±3	1.67±3
-O(3)	1.71±2 (2×)	1.72±2 (2×)

Discussion of the structure

Pseudo cubic close packing of oxygen atoms with M in octahedral and V in tetrahedral sites. $Ni_3V_2O_8$ is paramagnetic between 4.2 and 300°K while $Co_3V_2O_8$ has a ferromagnetic transition at 10°K and magnetic space group *Ab'a'm*.

Materials

Powder prepared from stoichiometric quantities heated to 800°C for $Ni_3V_2O_8$ and 1050°C for $Co_3V_2O_8$.

Details of analysis

The structure was determined using powder X-ray and neutron diffraction data. Weighted mean values are reported. Temperature factors are unreported.

BARIUM ORTHOVANADATE

I. The structure of $Ba_3(VO_4)_2$. P. SÜSSE and M.J. BUERGER, 1970. *Z. Kristallogr.*, 131, 161-174.

$Ba_3(VO_4)_2$ F.W. = 641.88

Rhombohedral $Sr_3(PO_4)_2$ type (2) [Hexagonal setting], a = 5.762±8, c = 21.29±3 Å, U = 312.1 Å^3, Z = 3, D_x = 5.20.

Space group $R\bar{3}m$ (D_{3d}^5) from final structure

Atomic positions

	x	y	z
3 Ba(1) in 3(a)	0	0	0
6 Ba(2) in 6(c)	0	0	0.20525±2
6 V in 6(c)	0	0	0.40758±7
6 O(1) in 6(c)	0	0	0.3278±4
18 O(2) in 18(h)	0.1610±4	$-x$	0.5654±5

Anisotropic temperature factors given

Interatomic distances (Å) *and angles*

V-O(1)	1.70±3
V-O(2)	1.71±3 (×3)
V(2)-O-V(2)'	108.8±4°
V(1)-O-V(2)	109.2±3
Ba(1)-O(2)	2.76±1
B(2)-Ba(1)-O(2)'	114.6±2° (×6)
Ba(2)-O(1)	2.61±2
Ba(2)-O(2)'	2.80±2 (×3)
Ba(2)-O(2)"	2.94±2 (×3)
Ba(2)-O(2)"'	3.94±2 (×3)

The errors are probably less than those given.

Discussion

The structure proposed by 1 (see 2) is confirmed and refined. Ba(1) is octahedrally coordinated by O(2) atoms and shares 2 opposite faces with the irregular coordination polyhedron around Ba(2). The VO_4 groups are accurately tetrahedral.

Details of analysis

Lattice parameters from precession photographs [no calibration given]. Intensities measured from spherical crystal with four-circle diffractometer using monochromated MoK_α radiation and corrected for Lorentz, polarization and absorption (crystal radius = 0.18 mm). Structure of 1 refined to give R = 0.048 for 1136 independent reflexion measurements listed [189 unequivalent reflexions].

1. *Structure Reports*, 23, 439.
2. *Ibid.*, 11, 388.

NICKEL URANYL VANADATE TETRAHYDRATE

I. Structure cristalline de l'uranyl-vanadate de nickel tetrahydrate $Ni(UO_2)_2(VO_4)_2.4H_2O$. J. BORENE and F. CESBRON, 1970. *Bull. Soc. Fr. Minér. Cristallogr.*, 93, 426-432.

$Ni(UO_2)_2(VO_4)_24H_2O$ F.W. = 900.70

Orthorhombic, a = 10.60, b = 8.25, c = 15.12 Å, U = 1322 Å^3, Z = 4, D_x = 4.51.

Space group Pnam (D_{2h}^{16})

Atomic positions Atoms at $\pm(x,y,z;\frac{1}{2}+x,\frac{1}{2}-y,\frac{1}{2}-z;\bar{x},\bar{y},\frac{1}{2}+z;\frac{1}{2}-x,\frac{1}{2}+y,z)$

	x	y	z	B(Å^2)
8 U in 8(*d*)	0.1827±1	-0.0150±2	-0.0022±1	*
8 V in 8(*d*)	0.0379±7	0.6461±8	0.0508±5	*
4 Ni in 4(*c*)	0.1895±11	0.0443±14	0.25	*
8 O(1) in 8(*d*)	0.1107±26	0.4436±34	0.0044±21	0.63
8 O(2) in 8(*d*)	0.0250±29	0.6175±38	0.1581±20	1.08
8 O(3) in 8(*d*)	-0.0078±31	0.8463±38	0.0215±20	1.37
8 O(4) in 8(*d*)	0.1992±29	0.7174±37	0.0387±19	0.93
8 O(5) in 8(*d*)	0.1813±28	-0.0719±37	-0.1169±20	0.69
8 O(6) in 8(*d*)	0.1892±26	0.0478±35	0.1153±19	0.27
4 O(7) in 4(*c*)	0.0046±65	-0.0101±80	0.25	3.42
4 O(8) in 4(*c*)	0.2290±61	-0.1972±74	0.25	3.32
4 O(9) in 4(*c*)	0.3726±55	0.1153±74	0.25	2.78
4 O(10) in 4(*c*)	0.1525±47	0.2869±62	0.25	1.75

*Anisotropic temperature factors given.

Interatomic distances (Å) *and angles*

Errors about 0.03 Å, 2° unless specified

U - O(5)	1.80
U - O(6)	1.85
U - O(1)	2.22
U - O(4)	2.30
U - O(3)	2.34
U - O(3)'	2.35
U - O(4)'	2.36
V - O(2)	1.64
V - O(3)	1.78
V - O(4)	1.82
V - O(1)	1.93
V - O(1)'	1.97
Ni - O(7)	2.01
Ni - O(9)	2.03
Ni - O(8)	2.04
Ni - O(6)	2.04 (×2)
Ni - O(10)	2.04

Possible hydrogen bonds

O(7) - O(5)	2.90±6	(×2)
O(8) - O(2)	2.99±6	(×2)
O(9) - O(2)	2.87±6	(×2)
O(10)- O(5)	2.92±5	(×2)

Angles in degrees subtended at V by

	O(3)	O(4)	O(1)	O(1)'
O(2)	111	103	108	105
O(3)		86	92	144
O(4)			148	82
O(1)				81

Angles in degrees subtended at Ni by

	O(9)	O(8)	O(6)'	O(10)
O(7)	176	89	90	92
O(9)		95	90	84
O(8)			91	179
O(6)			178	89

Discussion

The coordination around U is a pentragonal bipyramid with O(5) and O(6) along the axis. The vanadium ions form V_2O_8 groups composed of two VO_5 square pyramids sharing a basal edge.

Details of analysis

Lattice parameters and intensities measured on an automatic diffractometer with CuK_α radiation, corrected for Lorentz, polarization and absorption (μ=110 mm^{-1}, crystal size = 0.16×0.12×0.024 mm^3). Structure solved by Patterson and electron-density functions and refined by least squares to give R = 0.07 for 737 observed reflexions listed.

SULPHURIC ACID DIHYDRATE

I. Hydrogen bond studies. XXXVII. The crystal structure of sulfuric acid dihydrate $(H_3O^+)_2SO_4^{2-}$. I. TAESLER and I. OLOVSSON, 1970. *J. Chem. Phys.*, 51, 4213-4219.

$(H_3O)_2SO_4$ F.W. = 134.1

Monoclinic, a = 13.008±1, b = 7.979±1, c = 14.881±1 Å, β = 101.60±1°, U = 1513.0 $Å^3$, D_m = 1.761, Z = 12, D_x = 1.766.

Space group C2/c (C^6_{2h}) Systematic absences and final structure

Atomic positions

	x	y	z
8 S(1) in 8(*f*)	0.1573±1	0.0878±2	0.0815±1
4 S(2) in 4(*e*)	0.5	0.0961±2	0.75
8 O(1) in 8(*f*)	0.2440±3	0.1878±5	0.1317±3
8 O(2) in 8(*f*)	0.0698±3	0.1939±5	0.0343±3
8 O(3) in 8(*f*)	0.1970±3	0.9848±5	0.0129±3
8 O(4) in 8(*f*)	0.1155±3	0.9722±4	0.1437±2
8 O(5) in 8(*f*)	0.9066±3	0.3027±4	0.2180±2
8 O(6) in 8(*f*)	0.9827±3	0.5134±5	0.3265±3
8 O(7) in 8(*f*)	0.9194±3	0.0154±4	0.1486±2
8 O(8) in 8(*f*)	0.6885±3	0.1822±4	0.1962±2
8 O(9) in 8(*f*)	0.6067±3	0.0134±4	0.4856±2

Anisotropic temperature factors given.

Interatomic distances (Å) *and angles*

S(1) - O(1)	1.459±4
S(1) - O(2)	1.478±4
S(1) - O(3)	1.482±5
S(1) - O(4)	1.485±4

Angles in degrees subtended at S(1) by

	O(2)	O(3)	O(4)
O(1)	111.9±2	108.4±2	111.3±2
O(2)		109.4±2	108.0±2
O(3)			107.9±2

S(2) - O(5)	1.457±4 (×2)
S(2) - O(6)	1.488±4 (×2)

Angles in degrees subtended at S(2) by

	O(5)'	O(6)'
O(5)	112.7±2	108.7±2
O(5)'		109.3±2
O(6)		108.1±2

Hydrogen bonds around O(7)

O(7) - O(5)	2.533±4
O(7) - O(3)	2.570±5
O(7) - O(4)	2.590±5

Angles in degrees subtended at O(7) by

	O(3)	O(4)
O(5)	107.9±2	106.1±2
O(3)		111.8±2

Hydrogen bonds around O(8)

O(8) - O(1)	2.535±5
O(8) - O(4)	2.563±4
O(8) - O(6)	2.566±5

Angles in degrees subtended at O(8) by

	O(4)	O(6)
O(1)	108.9±2	104.9±2
O(4)		100.1±2

Hydrogen bonds around O(9)

O(9) - O(2)	2.520±5
O(9) - O(3)	2.559±6
O(9) - O(6)	2.583±5

Angles in degrees subtended at O(9) by

	O(3)	O(6)
O(2)	109.1±2	98.3±2
O(3)		116.5±2

Discussion

SO_4^{2-} ions are held together by H_3O^+ ions to which they form short O-HO bonds. The hydronium ions are not hydrogen bonded to each other (H_3O-H_3O> 2.9 Å)

Details of analysis

Crystals grown in capillary tubes at -48°C. Lattice parameters measured from oscillation photographs taken at -190°C and calibrated with quartz (a = 4.9131 Å at 25°C). Intensities measured visually from multiple-film equi-inclination Weissenberg photographs of 12 layers along *a* taken at -190°C with unfiltered Cu*K* radiation. They were corrected for Lorentz, polarization and absorption (μ = 5.2 mm^{-1}, radius of crystal = 0.11 mm). Structure solved from Patterson and electron-density syntheses and refined by least squares to give *R* = 0.093 in *Cc* (188 parameters) and 0.102 in *C2/c* (96 parameters) for 1265 reflexions listed. The latter was considered more likely. Addition of H atoms lowered *R* to 0.098.

HYDRAZINIUM SULPHATE

I. Neutron and X-ray diffraction studies of hydrazinium sulphate, $N_2H_6SO_4$.
P.G. JONSSON and W.C. HAMILTON, 1970. *Acta Cryst.*, B26, 536-546.

$N_2H_6SO_4$ F.W. = 130.12

Orthorhombic, *a* = 8.251±5, *b* = 9.159±1, *c* = 5.532±1 Å, *U* = 418.1 Å^3, *Z* = 4, D_x = 2.067.

Space group $P2_12_12_1$ (D_2^4) from extinctions

Atomic positions On each line neutron results are listed first.

	x	*y*	*z*
S	-0.05125±61	0.06917±49	0.21345±84
	-0.05034±3	0.06919±2	0.21259±4
O(1)	0.07963±32	0.08617±28	0.02780±45
	0.07947±10	0.08637±9	0.02723±15
O(2)	0.02541±40	0.07221±28	0.45427±44
	0.02637±13	0.07187±10	0.45495±14
O(3)	-0.16308±36	0.19281±28	0.19338±58
	-0.16332±12	0.19240±11	0.19335±21
O(4)	-0.13238±36	-0.07091±26	0.17444±52
	-0.13161±13	-0.07145±10	0.17401±17
N(1)	0.00953±22	0.33392±15	0.76654±29
	0.00904±13	0.33380±10	0.76683±17
N(2)	-0.15386±22	0.31239±20	0.69043±40
	-0.15340±13	0.31199±13	0.69114±21

H(1)	0.0776±1	0.3574±6	0.6141±11
	0.0630±28	0.3518±25	0.6595±39
H(2)	0.0112±7	0.4235±6	0.8876±9
	0.0122±27	0.4053±28	0.8644±40
H(3)	0.0512±8	0.2398±6	0.8509±12
	0.0480±33	0.2579±25	0.8145±42
H(4)	-0.1591±12	0.2329±10	0.5698±26
	-0.1599±44	0.2459±35	0.6148±71
H(5)	-0.1916±12	0.4026±11	0.6100±24
	-0.1847±34	0.3848±40	0.6280±76
H(6)	-0.2209±13	0.2969±23	0.8281±27
	-0.2139±37	0.2939±64	0.8484±63

Anisotropic thermal parameters are listed.

Interatomic distances (Å) *and angles*

Neutron diffraction results unless otherwise designated.

S - O(1)+	1.491[1.497]±1	O(1)-S - O(2)+	108.3±0.1°
S - O(2)+	1.483[1.494]±1	O(1)-S - O(3)+	109.0±0.1
S - O(3)+	1.468[1.481]±1	O(1)-S - O(4)+	108.7±0.1
S - O(4)+	1.468[1.478]±1	O(2)-S - O(3)+	108.9±0.1
N(1)-N(2)*	1.42±2	O(2)-S - O(4)+	109.9±0.1
N(1)-H(1)	1.036[1.061]±6	O(3)-S - O(4)+	112.0±0.1
N(1)-H(2)	1.059[1.083]±5	N(2)-N(1)-H(1)	107.5±0.4
N(1)-H(3)	1.039[1.069]±5	N(2)-N(1)-H(2)	107.8±0.4
N(2)-H(4)	0.989[1.086]±7	N(2)-N(1)-H(3)	109.3±0.4
N(2)-H(5)	0.989[1.068]±11	N(1)-N(2)-H(4)	110.0±0.6
N(2)-H(6)	0.952[1.106]±13	N(1)-N(2)-H(5)	108.4±0.6
		N(1)-N(2)-H(6)	109.5±0.8
		H(1)-N(1)-H(2)	110.3±0.5
		H(1)-N(1)-H(3)	111.0±0.5
		H(2)-N(1)-H(3)	110.8±0.5
		H(4)-N(2)-H(5)	107.3±1.1
		H(4)-N(2)-H(6)	113.9±1.3
		H(5)-N(2)-H(6)	107.6±1.4

Values corrected for riding motion are in brackets.

+ X-ray results

* Weighted average of X-ray and neutron results

Structure

Essentially the same as that determined previously (1) but with differences as large as 0.20 Å.

Details of analysis

Neutron (λ=1.073Å) (543 reflexions) and X-ray MoK_α (2076 reflexions) data refined separately by least-squares methods to R values of 0.074 and 0.057 respectively. Absolute configuration determined.

1. *Structure Reports*, 15, 251.

ZIRCONIUM SULPHATES

I. The crystal chemistry of zirconium sulphates. IV The structure of the γ-monohydrate, $Zr(SO_4)_2.H_2O$, a layer compound with the zirconium atom in sevenfold coordination. I.J. BEAR and W.G. MUMME, 1970. *Acta Cryst.*, B26, 1125-1131.

II. *Idem*, V. The structure of α-$Zr(SO_4)_2.H_2O$. *Idem*, 1970. *Ibid.*, B26, 1131-1139.

III. *Idem*, VI. The structure of α-$Zr(SO_4)_2$. *Idem*, 1970. *Ibid.*, B26, 1141-1145.

$Zr(SO_4)_2.H_2O$ γ-$Zr(SO_4)_2.H_2O$ F.W. = 301.37

Triclinic, a = 7.89±1, b = 5.21±1, c = 8.96±1 Å, α = 95.2±1°, β = 99.8±1°, γ = 109.2±1°, U = 339 Å^3, D_m = 2.93, Z = 2, D_x = 2.95.

Space group $P\bar{1}$ (C_i^1) P1 not tried

Atomic positions All atoms in 2(*i*)

	x	*y*	*z*	B(Å^2)
Zr	0.1200±3	0.3025±6	0.2719±2	0.60±4
S(1)	0.8424±8	0.6368±19	0.1025±7	0.74±12
S(2)	0.7586±7	0.0971±18	0.4726±7	0.44±11
O(W1)	0.3951±27	0.333±6	0.2402±27	3.2±5
O(2)	0.6714±26	0.653±5	0.1230±26	2.9±5
O(3)	0.6004±23	0.047±5	0.3527±23	2.0±4
O(4)	0.9335±22	0.186±5	0.4191±22	1.8±4
O(5)	0.2554±21	0.146±5	0.4559±20	1.5±4
O(6)	0.2314±24	0.685±5	0.4070±23	2.2±4
O(7)	0.1593±22	0.445±5	0.0588±22	1.9±4
O(8)	0.8848±22	0.419±5	0.1886±23	1.8±4
O(9)	-0.0072±25	-0.106±5	0.1530±24	2.6±4

Interatomic distances (Å) *and angles*

S(1)-O(8)	1.52±2	average O-S(1)-O	109°
O(2)	1.42±3	range	106±1 to 116±1
O(9)	1.39±2		
O(7)	1.49±2		
S(2)-O(3)	1.44±2	average O-S(2)-O	109°
-O(4)	1.48±2	range	106±1 to 113±1
-O(5)	1.46±2		
-O(6)	1.47±2		
Zr-O		six in range	2.01 to 2.19
-O(W1)			2.19±3

Discussion of structure

Structure contains layers of seven-fold coordinated Zr ions with sulphate groups as bridges. Layers are joined by hydrogen bonding through the water molecules as shown in Fig. 46.

Materials

Prepared by evaporation of aqueous solution $Zr(SO_4)_2.4H_2O$ or from 10m sulphuric acid at 120 - 140°C.

Details of analysis

708 reflexions measured from integrated Weissenberg photographs (CuK_α). Crystals sealed in capillaries. Least-squares refinement to a final R value of 0.126.

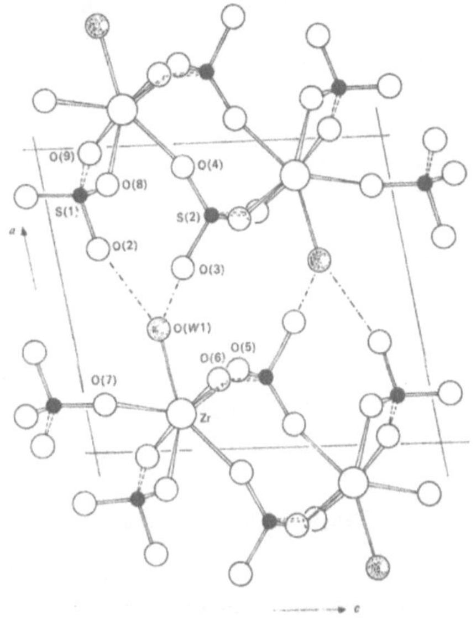

Fig. 46. The structure of γ-$Zr(SO_4)_2.H_2O$ projected on to (010). Large open circles, zirconium atoms; full black circles, sulphur atoms; medium open circles, oxygen atoms; stippled circles, water molecules.

α-$Zr(SO_4)_2.H_2O$.

Monoclinic, a = 7.32±1, b = 8.54±1, c = 11.82±1 Å, β = 106.0±1°, U = 710 $Å^3$, D_m = 2.84, Z = 4, D_x = 2.82.

Space group $P2_1/c$ (C^5_{2h})

Atomic positions All atoms in 4(e)

	x	y	z	$B(Å^2)$
Zr	0.1031±3	0.2539±4	0.1110±2	0.73±3
S(1)	0.2412±9	0.6144±10	0.0234±6	0.87±11
S(2)	0.1591±9	0.4468±9	0.3774±6	0.63±10
O(1)	-0.028±3	0.049±3	0.1394±19	1.7±4
O(W2)	0.375±3	0.162±3	0.2088±18	1.5±4
O(3)	0.139±3	0.318±3	0.2868±18	1.5±4
O(4)	0.433±3	0.666±3	0.0532±18	1.8±4
O(5)	0.359±3	0.522±3	0.3712±20	2.2±5
O(6)	0.220±2	0.373±3	0.4920±15	0.7±3
O(7)	0.235±3	0.456±3	0.0790±15	0.7±3
O(8)	0.106±3	0.720±3	0.0633±17	1.3±4
O(9)	0.155±2	0.894±3	0.3954±16	0.9±3

Interatomic distances (Å) *and angles*

S(1)-O(4)	1.42±2	average O-S-O	109°
-O(8)	1.51±2	range	103±1 to 115±1
-O(7)	1.51±2		
-O(9)	1.47±2		
S(2)-O(3)	1.51±2	average O-S-O	109°
-O(5)	1.38±2	range	105±1 to 115±1
-O(1)	1.47±2		
-O(6)	1.46±2		
Zr-O seven	2.07-2.22		
-O(W2)	2.15±2		

Discussion of the structure

The structure contains layers of $Zr(SO_4)_2.H_2O$ groups, formed of sevenfold coordinate Zr and SO_4 groups, extended in the *bc* plane. Postulated hydrogen bonds couple the layers as shown in Fig. 47.

Material

Equilibration of γ-$Zr(SO_4)_2H_2O$ with 75% H_2SO_4 at 120-150°C.

Details of analysis

807 reflexions measured from Weissenberg photographs using CuK_α radiation. Crystals in capillaries. Least-squares refinements; R = 0.131.

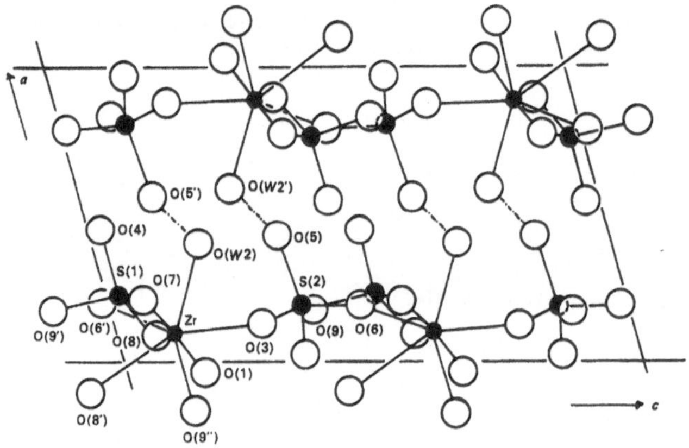

Fig. 47. Crystallographic sites in α-$Zr(SO_4)_2.H_2O$. Small black circles are zirconium atoms; medium black circles, sulphur atoms; large open circles, oxygen atoms and water molecules; dot and dashed lines, postulated hydrogen bonds.

α-$Zr(SO_4)_2$ F.W. = 283.35

Orthorhombic, a = 8.61±1, b = 5.445±10, c = 10.89±1 Å, [U = 510.53 Å^3], D_m = 3.71, Z = 4, D_x = 3.68.

Space group Pnma (D_{2h}^{16}) $Pn2_1a$ not tried.

Atomic positions

	x	*y*	*z*	$B(\text{Å}^2)$
4 Zr in 4(*c*)	0.2547±1	1/4	0.1033±3	0.27±6
4 S(1) in 4(*c*)	0.5009±12	3/4	0.1554±9	0.22±17
4 S(2) in 4(*c*)	0.3543±11	1/4	0.4098±9	0.07±16
4 O(1) in 4(*c*)	0.1042±41	3/4	0.2513±32	1.4±7
4 O(2) in 4(*c*)	0.0720±39	1/4	0.4665±29	1.0±6
4 O(3) in 4(*c*)	0.2402±34	3/4	0.3043±26	0.4±5
4 O(4) in 4(*c*)	0.5140±36	1/4	0.3508±29	1.0±6
8 O(5) in 8(*d*)	0.3929±24	0.532±5	0.1573±19	0.8±4
8 O(6) in 8(*d*)	0.3359±25	0.476±5	0.4801±19	0.5±4

Interatomic distance (Å) *and angles*

S(1)-O(1)	1.35±2	average O-S(1)-O	109°
-O(2)	1.46±2	range	104 to 114
-O(5) (2×)	1.51±2		
S(2)-O(3)	1.51±2	average O-S(2)-O	109°
O(4)	1.52±2	range	105 to 115
O(6)	1.46±2		
Zr-O seven	2.03±2 to 2.16±2		

Discussion of the structure

The structure is formed of sulphate tetrahedra bridging between ZrO_7 polyhedra as shown in Fig. 48. Polyhedra are more regular than in the hydrates.

Materials

Slow evaporation of a solution of $Zr(SO_4)_2.4H_2O$ and H_2SO_4 at 350°C.

Details of analysis

285 reflexions from four layers about the *b* axis were measured from Weissenberg photographs (CuK_α). The crystal was sealed in a capillary. Final *R* value from least-squares refinement is 0.118.

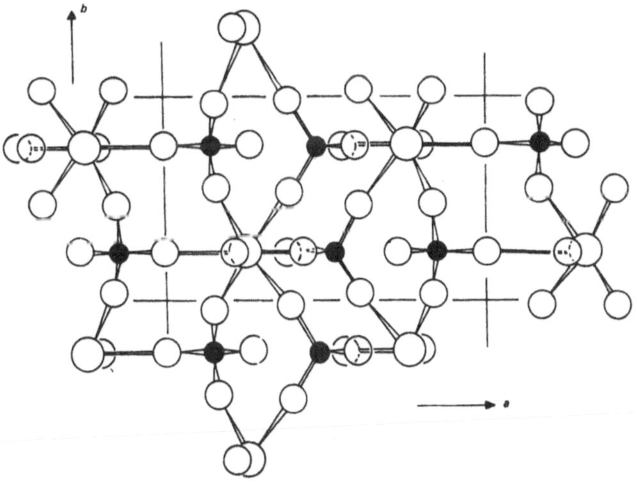

Fig. 48. The structure of α-$Zr(SO_4)_2$ centred at $x = 0$ as viewed along the *c* axis. The large open circles, full black circles and medium open circles represent zirconium, sulphur and oxygen atoms respectively.

α-FERROUS SULPHATE

I. Affinement de la structure de $FeSO_4$-α. D. SAMARAS and J. COING-BOYAT, 1970. *Bull. Soc. Fr. Minér. Crist.*, 93, 190-194.

α-$FeSO_4$ F.W. = 151.9

Orthorhombic ($CrVO_4$ type, 1), a = 5.252, b = 7.978, c = 6.59 Å, U = 276.1 Å^3, Z = 4, D_x = 3.65 (lattice parameters from 2).

Space group Cmcm (D_{2h}^{17})

Atomic positions

	x	y	z
4 Fe in 4(a)	0	0	0
4 S in 4(c)	0	0.3479±15	1/4
8 O(1) in 8(f)	0	0.2481±14	0.0644±30
8 O(2) in 8(g)	0.2036±22	0.4539±20	1/4

Interatomic distances (Å) *and angles*

S-O(1)	1.46±2 (×2)	O(1)-S-O(1)	114±2°
S-O(2)	1.36±2 (×2)	O(1)-S-O(2)	110±1° (×4)
		O(2)-S-O(2)	103±2°
Fe-O(1)	2.02±2 (×2)	O(2)-Fe-O(1)	90 ±1°
Fe-O(2)	2.30±4 (×4)	O(2)-Fe-O(2)	95 ±1°, 85±1°

See details of analysis concerning errors.

Details of analysis

Prepared by heating the heptahydrate to 280°C in a current H_2. Intensities measured on a powder diffractometer with monochromated Co$K\alpha_1$ radiation (λ = 1.7883 Å). Six parameters refined by least squares to give R (intensities) = 0.059 for 25 reflexions listed. No thermal parameters were included. [If the standard errors quoted for the positional parameters are those derived from the least squares refinement they are likely to be rather low in view of the small number of intensities measured.]

1. *Structure Reports*, 9, 181.
2. J. COING-BOYAT, 1966. *Thesis*, Grenoble.

CADMIUM SULPHATE MONOHYDRATE

$CdSO_4.H_2O$ F.W. = 221.46

I. Étude structurale du sulfate de cadmium monohydrate $CdSO_4$,H_2O. J.-M. BREGEAULT and P. HERPIN, 1970. *Bull. Soc. Fr. Minér. Crist.*, 69, 37-42.

Monoclinic, a = 7.60±7, b = 7.54±1, c = 8.18±6 Å, β = 121.86°, U = 394 Å^3, D_m = 3.78, Z = 4, D_x = 3.81.

Space group $P2_1/c$ (C_{2h}^5) From systematic absences

Atomic positions

All atoms in 4(e)

	x	y	z
Cd	0.2187±3	0.2612±9	0.2606±3
S	0.256±1	0.612±3	0.012±1
O(1)	0.138±4	0.490±8	-0.156±4
O(2)	0.105 4	0.769±9	0.002±3

O(3)	0.330±3	0.531±7	0.198±3
O(4)	0.440±3	0.674±9	0.026±3
O(5)	0.271±3	0.123±8	0.035±3

Discussion

CdO_6 octahedra (Cd-O = 2.21 to 2.36) are linked through shared OH_2 groups (Cd-O-Cd = 127°) into chains along *c*. The chains are linked to each other through SO_4 ions (S-O = 1.42 to 1.59 Å) each SO_4 group linking three chains. [The bond lengths cannot be accurate to better than 0.05 Å and the apparent deviations of the SO_4 ion from perfect tetrahedral symmetry is probably not significant. The octahedron of O atoms around Cd, on the other hand, may be somewhat distorted.]

Details of analysis

Multiple-film Weissenberg photographs of layers with h = 0,1 and precession photographs of layers with k = 0-2 (MoK_α radiation) measured with a densitometer. Lorentz and polarization corrections but not absorption (μR = 0.88). Structure solved from Patterson and electron-density projections. Refined by least squares to give R = 0.12 for 520 reflexions listed.

α-VANADYL SULPHATE

I. Structure and magnetic properties of $VOSO_4$. J.M. LONGO and R.J. ARNOTT, 1970. *J. Solid State Chem.*, 1, 394-398.

$VOSO_4$ F.W. = 163.0

Tetragonal, a = 6.261±3, c = 4.101±3 Å, U = 160.8±Å^3, Z = 2, D_x = 3.36 (see also 1).

Space group P4/n (C^3_{4h})

Atomic positions (origin at $\bar{4}$)

	x	y	z
2 V in 2(*c*)	0	½	0.6176±10
2 O(1) in 2(*c*)	0	½	0.220±4
2 S in 2(*b*)	0	0	½
8 O(2) in 8(*g*)	0.538±3	0.683±3	0.285±3

B = 0.4 (Å^2)

Interatomic distances (Å) *and angles*

Sixfold coordination round V (crystallographic site symmetry C_4)

V-O(1)	1.63	
V-O(1)'	2.47	
V-O(2)	2.04	(×4)

Angles in degrees subtended at V by

	O(1)	O(1)	O(2)
O(2)	101	79	88°

Fourfold coordination around S (crystallographic site symmetry S_4)

S-O(2)	1.47	(×4)
O(2)-S-O(2)	106°	(×2)
	111°	(×4)

σ(distances) = 0.02 Å, σ(angles) = 1°

Discussion

The structure which is ismorphous with **$MoOPO_4$** (2) **consists of VO_6 octahedra** linked through an asymmetrically bridging oxygen atom, O(1), into chains along *c*. The four equatorial oxygen atoms are bonded to the SO_4 tetrahedra which link the chains in the *a* and *b* directions. Alternatively, the structure can be considered as a cubic close packed array of oxygen atoms with V in one fifth of the octahedral sites and S in one tenth of the tetrahedral sites. An independent determination is reported in 3.

Details of analysis

α-$VOSO_4$ prepared by dehydration of commercially available hydrate. Integrated intensities were measured from the powder pattern recorded on a Norelco diffractometer with LiF monochromated CuK_α radiation. The 18 reflexions were used for least-squares refinement to give *R* (intensities) = 0.094.

1. J. TUDO, 1965. *Rev. Chim. Minérale*, 2, 53.
2. P. KIERKEGAARD and M. WESTERLUND, 1964. *Acta Chem. Scand.*, 20, 722.
3. G. LUDWIG, 1969. *Z. anorg. Chem.*, 364, 225.

LANARKITE

I. Zur Kristallstruktur von Lanarkit, $Pb_2O(SO_4)$. K. SAHL, 1970. *Z. Kristallogr.*, 132, 99-117.

$Pb_2O(SO_4)$ F.W. = 526.45

Monoclinic, *a* = 13.769±5, *b* = 5.698±3, *c* = 7.079±2 Å, β = 115.93±16°, *U* = 499.48 Å^3, D_m = 7.00±8, *Z* = 4, D_x = 6.99.

*Space group C*2/*m* (C^3_{2h}) Absences, morphology and absence of piezoelectric and pyroelectric effects.

Atomic positions

	x	*y*	*z*	*B*(Å^2)
4 Pb(1) in 4(*i*)	0.1466±1	0	0.1050±2	1.30±3
4 Pb(2) in 4(*i*)	0.5258±1	0	0.2715±2	1.22±3
4 S in 4(*i*)	0.829±1	0	0.344±2	1.2±1
4 O(1) in 4(*i*)	0.111±3	0	0.429±6	2.1±5
8 O(2) in 8(*j*)	0.355±2	0.711±5	0.251±4	2.0±4
4 O(3) in 4(*i*)	0.713±3	0	0.284±6	2.0±5
4 O(4) in 4(*g*)	0	0.246±6	0	1.7±4

Interatomic distances (Å) *and angles*

σ(distances) = 0.03, σ(angles) = 2°

Pb(1)-O(4)	2.30 (×2)
Pb(1)-O(1)	2.55
Pb(1)-O(2)	2.78 (×2)
Pb(1)-O(2)	3.07 (×2)
Pb(1)-O(3)	3.09 (×2)
Pb(2)-O(4)	2.31 (×2)
Pb(2)-O(3)	2.54
Pb(2)-O(2)	2.82 (×2)
Pb(2)-O(1)	3.10 (×2)

S -O(1)	1.45
S -O(3)	1.46
S - O(2)	1.49 (×2)

Angles in degrees subtended at S by

	O(3)	O(2)'
O(1)	110	111
O(3)		108
O(2)		108

Discussion

O(4) is tetrahedrally surrounded by Pb atoms at 2.30 A, each Pb being linked to two O(4) atoms. These, together with an additional O from the SO_4 groups forms with Pb a trigonal prism (O-Pb-O = 77°).

Details of analysis

Lattice parameters from Weissenberg and precession photographs (Mo radiation). Intensities measured photometrically from multiple-film integrated Weissenberg photographs (Mo radiation) of layers k = 0 to 6 and precession photographs (Ag radiation) of layers l = 0 to 4 and h = 0 to 6. Corrections for Lorentz, polarization and absorption (crystal size = 0.12×0.03×0.20 mm^3). Structure determined from Patterson and electron-density functions and refined by least squares to give R = 0.081 for 841 observed reflexions listed.

PARABUTLERITE

I. Structure crystalline de la parabutlerite. J. BORENE, 1970. *Bull. Soc. Fr. Minér. Crist.*, 93, 185-189.

$FeSO_4OH.2H_2O$ F.W. = 207.9

Orthorhombic, a = 20.13, b = 7.38, c = 7.22 Å, U = 1073 Å^3, Z = 8, D_x = 2.47.

Space group Pnma [converted from *Pmnb*] (D_{2h}^{16}). Systematic absences and refinement.

Atomic positions

			x	y	z
8	Fe	in 8(d)	0.1247±1	-0.0001±3	0.7885±2
4	S(1)	in 4(c)	0.1226±2	0.25	0.4236±4
4	S(2)	in 4(c)	0.1269±2	0.75	0.1587±4
8	O(1)	in 8(d)	0.1383±4	0.0877±9	0.5344±8
8	O(2)	in 8(d)	0.1142±5	-0.0884±10	0.0479±9
8	O(W1)	in 8(d)	0.0226±4	0.0105±17	0.7671±12
8	O(W2)	in 8(d)	0.2265±4	-0.0134±14	0.8209±13
4	O(3)	in 4(c)	0.1644±7	0.25	0.2617±13
4	O(4)	in 4(c)	0.0517±7	0.25	0.3833±20
4	O(5)	in 4(c)	0.0824±7	0.75	0.3160±16
4	O(6)	in 4(c)	0.1970±6	0.75	0.2099±25
4	O(H1)	in 4(c)	0.1254±7	0.25	0.8814±13
4	O(H2)	in 4(c)	0.1231±5	0.75	0.6957±12

Anisotropic temperature factors given.

Interatomic distances (Å) *and angles*

S(1) - O(3)	1.440±12
S(1) - O(4)	1.457±15
S(1) - O(1)	1.476±7 (×2)

Angles in degrees subtended at S(1) by

	O(4)	O(1)'
O(3)	114.2±8	108.3±4
O(4)		108.5±4
O(1)		108.7±5

S(2) - O(5)	1.446±13
S(2) - O(6)	1.459±13
S(2) - O(2)	1.460±7 (×2)

Angles in degrees subtended at S(2) by

	O(6)	O(2)'
O(5)	113.6±9	108.8±5
O(6)		108.0±5
O(2)		109.8±5

Fe - O(H2)	1.965±4
Fe - O(1)	1.965±6
Fe - O(H1)	1.966±4
Fe - O(2)	1.995±7
Fe - O(W1)	2.063±8
Fe - O(W2)	2.065±8

Angles in degrees subtended at Fe by

	O(1)	O(H1)	O(2)	O(W1)	O(W2)
O(W2)	89.7±3	179.5±76	90.6±3	89.6±4	90.6±4
O(1)		90.4±4	178.1±22	92.2±3	89.0±4
O(W1)			89.3±4	89.8±5	89.9±5
O(2)				88.7±4	89.0±4
O(W1)					177.7±23

Hydrogen bonds

O(W1) - O(4)	2.669±14
O(W1) - O(5)	2.821±15

Angles in degrees subtended at O(W1) by

	O(4)	O(5)
Fe	124.2±6	141.8±6
O(4)		86.9±4
O(5)		

O(W2) - O(6)	2.608±13
O(W2) - O(3)	2.840±14

Angles in degrees subtended at O(W2) by

	O(6)	O(3)
Fe	121.0±5	141.2±5
O(6)		87.5±4

O(H1) - O(3) 2.856±14

Angles in degrees subtended at O(H1) by

	Fe'	O(3)
Fe	140.1±7	109.3±3
Fe		109.3±3

O(H2) - O(5) 2.861±15

Angles in degrees subtended at O(H2) by

	Fe'	O(5)
Fe	140.1±7	109.4±2
Fe'		109.4±2

Discussion

The structure consists of chains of FeO_6 octahedra linked through shared apices (O(H1), O(H2)) and bridged by SO_4 groups along the *a* direction. Adjacent chains are linked through hydrogen bonds.

Details of analysis

Intensities measured with filtered MoK_α radiation on single crystal diffractometer corrected for Lorentz and polarization but not absorption (μ = 0.32 mm ^{1}). Structure solved from Patterson function and refined by least squares to give *R* = 0.038 for 846 non-zero reflexions [not listed].

CALEDONITE

I. The crystal structure of caledonite. C. GIACOVAZZO, S. MENCHETTI and F. SCORDARI, 1970. *Naturwissenschaften*, 57, 127.

$Cu_2Pb_5(SO_4)_3(CO_3)(OH)_6$ F.W. = 1613.3

Orthorhombic, *a* = 20.10, *b* = 7.15, *c* = 6.56 Å, *U* = 942.8 Å^3, *Z* = 2, D_x = 5.67 (from literature [*sic*]).

Space group $Pmn2_1$ (C_{2v}^7) Piezoelectric effect

Atomic positions

	x	*y*	*z*
2 Pb(1) in 2(*a*)	0	0.082	0.5
4 Pb(2) in 4(*b*)	0.600	0.415	0.040
4 Pb(3) in 4(*b*)	0.839	0.113	0.570
4 Cu in 4(*b*)	0.251	0.490	0.312
2 S(1) in 2(*a*)	0	0.444	0.058
4 S(2) in 4(*b*)	0.337	0.126	0.557
2 O(1) in 2(*a*)	0	0.580	0.230
2 O(2) in 2(*a*)	0	0.559	0.858
4 O(3) in 4(*b*)	0.058	0.313	0.061
4 O(4) in 4(*b*)	0.284	0.281	0.569
4 O(5) in 4(*b*)	0.196	0.053	0.024
4 O(6) in 4(*b*)	0.375	0.108	0.762
4 O(7) in 4(*b*)	0.385	0.172	0.393
4 O(8) in 4(*b*)	0.054	0.252	0.552
2 O(9) in 2(*a*)	0	-0.016	0.634
4 OH(1) in 4(*b*)	0.164	0.378	0.350
4 OH(2) in 4(*b*)	0.159	0.389	0.771
4 OH(3) in 4(*b*)	0.275	0.351	0.066
2 C in 2(*a*)	0	0.161	0.616

Interatomic distances (Å) *and angles*

S(1) - O(1)	1.49
S(1) O(2)	1.55
S(1) - O(3)	1.50 (×2)

Angles in degrees subtended at S(1) by

	O(2)	O(3)'
O(1)	107	114°
O(3)	110	102

S(2) - O(4)	1.54
S(2) - O(5)	1.46
S(2) - O(6)	1.55
S(2) - O(7)	1.48

Angles in degrees subtended at S(2) by

	O(5)	O(6)	O(7)
O(4)	109	111	109°
O(5)		106	112
O(6)			109

C-O(8)	1.33 (×2)
C-O(9)	1.27

O(8)-C-O(8)	109°
O(8)-C-O(9)	121°

Pb(1) - O(3)	2.06 (×2)
Pb(1) - O(9)	2.50
Pb(1) - C	2.58
Pb(1) - S(1)	2.62
Pb(1) - O(7)	3.02 (×2)

Pb(2) - OH(1)	2.32
Pb(2) - OH(2)	2.38
Pb(2) - O(8)	2.56
Pb(2) - OH(3)	2.56
Pb(2) - O(1)	2.86
Pb(2) - O(6)	2.90
Pb(2) - O(2)	2.90
Pb(2) - O(7)	2.91

Pb(3)	- O(8)	2.37
Pb(3)	- OH(2)	2.37
Pb(3)	- OH(1)	2.38
Pb(3)	- O(6)	2.66
Pb(3)	- O(4)	2.75
Pb(3)	- O(7)	3.08
Pb(3)	- O(5)	3.09
Pb(3)	- O(5)	3.12

Cu	- OH(1)	1.94
Cu	- OH(3)	1.96
Cu	- OH(2)	2.02
Cu	- OH(3)	2.09
Cu	- O(4)	2.35
Cu	- O(4)	2.39

Angles in degrees subtended at Cu by

	OH(3)	OH(2)	OH(3)	O(4)	O(4)
OH(1)	96	179	84	84	96
OH(3)		84	177	101	83
OH(2)			96	97	83
OH(3)				81	95
O(4)					176

Possible hydrogen bonds

OH(1)-O(3)	2.89
OH(2)-O(3)	2.84
OH(3)-O(5)	2.67

Errors uncertain but not less than 0.02 Å.

Details of analysis

Integrating Weissenberg photographs measured photometrically. Three-dimensional Patterson synthesis. Final R = 0.07.

ETTRINGITE

I. Crystal structure of ettringite. A.E. MOORE and H.F.W. TAYLOR, 1970. *Acta Cryst.*, B26, 386-393.

$Ca_6[Al(OH)_6]_2(SO_4)_3.26H_2O$ F.W. = 1255.1

Trigonal, a = 11.26, c = 21.48 Å, [U = 2358.5 Å^3], Z = 2, [D_x = 1.77].

Space group $P31c$ (C^4_{3v}) Models with both $P31c$ and $P\bar{3}1c$ symmetry were tried.

Atomic positions

	x	y	z	B(Å^2)
2 Al(1) in 2(a)	0	0	0	1.2±1
2 Al(2) in 2(a)	0	0	0.250±1	1.2±1
6 Ca(1) in 6(c)	0.009±1	0.816±1	0.875±1	1.7±1
6 Ca(2) in 6(c)	0.994±1	0.189±1	0.125±1	1.8±1
6 O(1)H in 6(c)	0.994±1	0.134±3	0.948±1	2.2±5
6 O(2)H in 6(c)	0.996±3	0.865±2	0.057±1	1.0±4
6 O(3)H in 6(c)	0.004±3	0.146±2	0.805±1	1.6±4
6 O(4)H in 6(c)	0.004±3	0.876±2	0.198±1	1.0±4
6 H_2O(5) in 6(c)	0.000±4	0.348±3	0.047±2	3.9±6
6 H_2O(6) in 6(c)	0.010±2	0.663±2	0.958±1	1.8±4
6 H_2O(7) in 6(c)	0.997±3	0.345±3	0.199±1	3.6±6
6 H_2O(8) in 6(c)	0.996±3	0.665±3	0.788±1	3.1±5
6 H_2O(9) in 6(c)	0.263±3	0.405±3	0.618±2	2.9±6
6 H_2O(10) in 6(c)	0.744±4	0.593±4	0.374±2	4.8±9
6 H_2O(11) in 6(c)	0.239±3	0.406±3	0.126±2	4.8±9
6 H_2O(12) in 6(c)	0.768±2	0.598±2	0.870±2	1.7±5

2 O(13)	in 2(*b*)	1/3	2/3	0.420±3	1.9±9
2 O(14)	in 2(*b*)	1/3	2/3	0.814±3	7.8±16
2 O(15)	in 2(*b*)	1/3	2/3	0.070±3	4.2±13
2 O(16)	in 6(*c*)	0.195±3	0.642±3	0.518±2	4.7±8
2 O(17)	in 6(*c*)	0.195±3	0.620±3	0.723±1	4.3±7
2 O(18)	in 6(*c*)	0.192±2	0.585±2	0.482±1	2.1±4
2 O(19)	in 6(*c*)	0.197±5	0.637±5	0.243±3	5.4±15
2 S(1)	in 2(*b*)	1/3	2/3	0.491±1	2.2±5
2 S(2)	in 2(*b*)	1/3	2/3	0.750±2	1.6±2
2 S(3)	in 2(*b*)	1/3	2/3	0.009±1	1.4±4

Interatomic distances (Å)

Al(1)-O(1)	(3×)	1.92
-O(2)	(3×)	1.92
Al(2)-O(3)	(3×)	2.00
-O(4)	(3×)	1.82
S(1)-O(13)		1.51
-O(16)	(3×)	1.56
S(2)-O(14)		1.36
-O(17)	(3×)	1.49
-O(15)		1.31
-O(18)	(3×)	1.48
Ca(1)-OH	(4)	2.36 - 2.52
$-OH_2$	(4)	2.48 - 2.62
Ca(2)-OH	(4)	2.35 - 2.46
$-OH_2$	(4)	2.43 - 2.75

Description of the structure

The structure is based upon columns with composition $[Ca_3[Al(OH)_6].12H_2O]^{3+}$ running paralled to *c* with ordered SO_4^{2-} and fractionally occupied H_2O sites. Postulated hydrogen atom positions provide bonding between SO_4^{2-} and H_2O in the columns.

Materials

Natural specimen from Scawt Hill, N. Ireland.

Details of analysis

Least-squares refinement of combined diffractometer (MoK_α) and film data set (CuK_α) (∿600 reflexions) to an *R* of 0.088. Unit-cell errors were not specified.

POTASSIUM NITRILOTRISULPHONATE DIHYDRATE

I. Crystal structure of potassium nitrilotrisulphonate dihydrate. J.V. TILLACK and C.H.L. KENNARD, 1970. *J. Chem. Soc. Lond. A*, 1637-1640.

$K_3N(SO_3)_3.2H_2O$ F.W. = 407.5

Orthorhombic, *a* = 11.18±1, *b* = 5.776±7, *c* = 18.06±1 Å, *U* = 1166 Å^3, D_m = 2.30, *Z* = 4, D_x = 2.33.

Space group Pnma (D_{2h}^{16}) Systematic absences and statistical test of intensities.

Atomic positions

		x	*y*	*z*	*B*(Å^2)
4 K(1)	in 4(*c*)	0.3856±4	0.25	0.0079±3	1.7±1
4 K(2)	in 4(*c*)	0.1263±4	0.25	0.6746±3	2.0±1
4 K(3)	in 4(*c*)	0.4768±5	0.75	0.4104±3	2.0±1
4 S(1)	in 4(*c*)	0.3668±4	0.25	0.3323±3	1.3±1
4 S(2)	in 4(*c*)	0.1037±4	0.25	0.3583±3	1.3±1

	x	y	z	B(Å^2)
4 S(3) in 4(c)	0.2694±4	0.25	0.4852±2	1.1±1
4 N in 4(c)	0.247±2	0.25	0.392±1	2.3±3
4 O(1) in 4(c)	0.256±2	0.25	0.139±1	4.0±4
4 O(2) in 4(c)	0.372±2	0.25	0.716±1	2.3±4
4 O(11) in 4(c)	0.467±2	0.25	0.3820±9	2.1±3
8 O(12) in 8(d)	0.355±1	0.042±3	0.2920±7	3.7±2
4 O(21) in 4(c)	0.121±1	0.25	0.2776±9	2.3±3
8 O(22) in 8(d)	0.048±1	0.048±2	0.3864±6	2.4±2
4 O(31) in 4(c)	0.150±1	0.25	0.5153±8	1.8±3
8 O(32) in 8(d)	0.332±1	0.042±2	0.5012±6	2.3±2

Interatomic distances (Å) *and angles*

$N(SO_3)_3$ ion (crystallographic site symmetry C_s)

x	N - S(x)	S(x) - O(x1)	S(x) - O(x2)	N - S(x) - O(x1)	N - S(x) - O(x2)
1	1.72	1.44	1.42 (×2)	103°	104°
2	1.72	1.47	1.42 (×2)	103°	107°
3	1.70	1.44	1.42 (×2)	104°	106°

x	O(x1) - S(x) - O(x2)	O(x2) - S(x) - O(x2)'
1	114°	117°
2	114°	111°
3	112°	116°

S(1) - N - S(2)	121°
S(2) - N - S(3)	119°
S(3) - N - S(1)	120°

σ (N - S) = 0.02 Å, σ (S - O) = 0.02 Å, σ (angles) = 1°

K(1) - O	6 distances	2.6 to 3.0
K(2) - O	5 "	2.7 to 3.2
K(3) - O	5 "	2.7 to 3.0

Details of analysis

Lattice parameters from Weissenberg photographs calibrated with Al (a = 4.04934 Å). Intensities estimated visually from Weissenberg photographs (CuK_α radiation) of layers with k = 0 to 5. Corrections for Lorentz and polarization, but not absorption or extinction. Structure determined by statistical analysis of intensities and electron-density syntheses and refined by least squares to give a weighted R = 0.16 (unweighted R = 0.12) for 805 reflexions listed.

NICKEL DITHIONATE HEXAHYDRATE

I. The structure of nickel dithionate hexahydrate. J. CHAN and E. STANLEY, 1970. *Z. Kristallogr.*, 132, 404-410.

$NiS_2O_6.6H_2O$ F.W. = 326.83

Triclinic, a = 6.479±5, b = 8.226±5, c = 9.561±10 Å, α = 138°28±30', β = 116°25±30', γ = 53°10±30', U = 270.3 Å^3, D_m = 1.908, Z = 1, D_x = 2.00.

Space group $P\bar{1}$ (C_i^1)

Atomic positions Ni in 1(*a*) all other atoms in 2(*i*)

	x	*y*	*z*
Ni	0	0	0
S	0.188	0.143	0.496
O(1)	0.316	0.253	0.336
O(2)	0.140	0.342	0.439
O(3)	0.292	0.080	0.712
O(4)	0.319	0.138	0.131
O(5)	0.813	0.149	0.969
O(6)	0.167	0.356	0.713

Discussion

The dithionate ion has S-S = 2.12±2 and S-O = 1.46±2 Å. The Ni atom is octahedrally surrounded (Ni-O = 2.02±2 Å) by six water molecules which form hydrogen bonds of lengths 2.64 to 2.81 Å.

Details of analysis

Intensities measured visually from three zero-layer Weissenberg photographs taken with CuK_α radiation. Structure solved from the Patterson functions and each projection refined separately. R = 0.08 for 240 reflexions listed.

SODIUM SELENATE

I. A note on the structure of Na_2SeO_4. A. KALMAN and D.W.J. CRUICKSHANK, 1970. *Acta Cryst.*, B26, 436.

Na_2SeO_4 F.W. = 188.94

Orthorhombic, a = 6.099±6, b = 12.590±16, c = 10.163±12 Å, [U = 894.4 Å^3], Z = 8, [D_x = 2.806].

Space group Fddd (D_{2h}^{24})

Atomic positions

	x	*y*	*z*
8 Se in 8(*a*)	1/8	1/8	1/8
16 Na in 16(*g*)	1/8	1/8	0.4410±15
32 O in 32(*h*)	-0.32±4	0.0509±13	0.2211±17

Interatomic distances (Å) *and angles*

Se-O (4×)	1.654±21	O-Se-O	109.5±2°, 111.4±9°
Na-O	2.60, 2.31, 2.46 each twice		

Discussion of the structure

Accurate refinement of previously reported structure and data (1).

Details of analysis

Refined by least-squares method to an R of 0.11.

1. I. NARAY-SZABÓ and Gy. ARGAY, 1963. *Acta Chim. Hung.*, 39, 85.

POTASSIUM SELENATE

I. The crystal structure of K_2SeO_4. A. KALMAN, J.S. STEPHENS and D.W.J. CRUICKSHANK, 1970. *Acta Cryst.*, B26, 1451-1454.

K_2SeO_4 F.W. = 221.16

Orthorhombic, a = 6.661±4, b = 10.466±8, c = 6.003±3 Å, U = 481.32 Å^3, D_m = 3.032, Z = 4, D_x = 3.052.

Space group Pnam (D_{2h}^{16})

Atomic positions

	x	y	z
4 Se in 4(c)	0.2242±2	0.4200±1	1/4
4 K(1) in 4(c)	0.1705±5	0.0843±3	1/4
4 K(2) in 4(c)	-0.0057±4	-0.2905±3	1/4
8 O(1) in 8(d)	0.2931±10	0.3471±6	0.0271±10
4 O(2) in 4(c)	0.3024±16	-0.4356±8	1/4
4 O(3) in 4(c)	0.0126±21	0.4251±10	1/4

Anisotropic temperature factors are given.

Interatomic distances (Å) *and angles*

Se-O(1)	1.649±6	(×2)	(thermally corrected)
-O(2)	1.644±9		" "
-O(3)	1.648±16		" "

K(1)-O(1)	3.20	(×2)	3.26(×2)	3.00(×2)
-O(2)	3.23,	3.02(×2).		
-O(3)	2.64,	3.76		

K(2)-O(1)	2.83(×2), 2.73(×2)
-O(2)	2.80, 2.79
-O(3)	2.98, 3.32(×2)

Angles in degrees subtended at Se by

	O(1')	O(2)	O(3)
O(1)	110.5±4	108.4±3	109.8±3°
O(2)			109.8±6

Structure

Isomorphous with βK_2SO_4 (1).

Details of analysis

362 independent reflexions collected with a diffractometer and CuK_α radiation using a cylindrical crystal. Refined by full-matrix least squares (R=0.049). Absorption corrections applied.

1. *Structure Reports*, 22, 447. See also *Strukturbericht*, 2, 86 and *Structure Reports*, 16, 284.

NICKEL SELENATE HEXAHYDRATE

I. Die Kristallstruktur von Nickelselenat-hexahydrat. H. FUESS, 1970. *Z. anorg. Chem.*, 379, 204-212.

$NiSeO_4.6H_2O$ F.W. = 309.76

Tetragonal, $NiSO_4.6H_2O$ type (1), $a = 6.926±1$, $c = 18.416±3$ Å, $U = 883.4$ Å^3, $Z = 4$, $D_x = 2.32$.

Space group $P4_12_12$ (D_4^4)

Atomic positions

	x	*y*	*z*	*B*(Å^2)
4 Ni in 4(*a*)	0.216±1	0.216±1	0	1.5
4 Se in 4(*a*)	0.714±1	0.714±1	0	1.3
8 O(1) in 8(*b*)	0.170±1	0.960±2	0.052±2	2.1
8 O(2) in 8(*b*)	0.466±1	0.245±2	0.059±2	1.6
8 O(3) in 8(*b*)	0.071±1	0.364±1	0.082±2	1.8
8 O(4) in 8(*b*)	0.618±1	0.613±2	0.074±2	2.7
8 O(5) in 8(*b*)	0.948±1	0.672±1	0.007±3	2.1

Interatomic distances (Å) *and angles*

Ni - O(1)	2.04±2 (×2)
Ni - O(2)	2.05±2 (×2)
Ni - O(3)	2.08±2 (×2)

Angles in degrees subtended at Ni by

	O(1)'	O(2)'	O(3)'
O(1)	87	88	90
O(1)'		175	91
O(2)		97	89
O(2)'			91
O(3)			179

σ = 1°

Se - O(5)	1.65±1 (×2)
Se - O(4)	1.67±3 (×2)

Angles in degrees subtended at Se by

	O(5)'	O(4)'
O(5)	110	104
O(5)'		113
O(4)		109

σ = 2°

Possible hydrogen bonds

O(1) - O(5)	2.65±2
O(1) - O(3)	2.87±4

Angles in degrees subtended at O(1) by

	O(5)	O(3)
Ni	127	117
O(5)		115

σ = 1°

O(2) - O(4)	2.77±2
O(2) - O(5)	2.78±3

Angles in degrees subtended at O(2) by

	O(4)	O(5)
Ni	118	97
O(4)		122

σ = 1°

O(3) - O(5)	2.68±4
O(3) - O(4)	2.78±3

Angles in degrees subtended at O(3) by

	O(5)	O(4)	O(1)
Ni	100	112	111
O(5)		132	114
O(4)			89

Details of analysis

Lattice parameters from powder photographs (CuK_α radiation) and single crystal diffractometer for layers with k = 0-4 and corrected for Lorentz, polarization, absorption (crystal a sphere of radius 0.4 mm) and extinction. Structure assumed from $NiSO_4.6H_2O$(1) and refined by Fourier and least squares to give R = 0.08 for 432 reflexions listed.

1. B.H. O'CONNOR and D.H. DALE, 1966. *Acta Cryst.*, 21, 705.

AMMONIUM CHROMATE

I. A re-investigation of the crystal structure of $(NH_4)_2CrO_4$. J.S. STEPHENS and D.W.J. CRUICKSHANK, 1970. *Acta Cryst.*, B 26, 437-439.

$(NH_4)_2CrO_4$ F.W. = 152.08

Monoclinic, a = 12.21±1, b = 6.258±3, c = 7.630±4 Å, β = 115.2±2°, [U = 527.59 Å^3], Z = 4, [D_x = 1.914].

Space group C2/m (C^3_{2h})

Atomic positions

	x	y	z	B(Å^2)
4 Cr in 4(i)	0.17184±3	0	0.21239±5	0.0167±2
4 O(1) in 4(i)	0.3181±2	0	0.3579±3	
4 O(2) in 4(i)	0.0905±2	0	0.3374±4	
8 O(3) in 8(j)	0.1407±1	0.2152±2	0.0767±2	
4 N(1) in 4(i)	0.5171±2	0	0.2516±4	
4 H(1) in 4(i)	0.449±6	0	0.264±8	0.038
4 H(3) in 4(i)	0.582±5	0	0.386±8	0.038
8 H(5) in 8(j)	0.523±3	0.120±6	0.184±5	0.038
4 N(2) in 4(i)	0.8405±20	0	0.2650±3	
4 H(2) in 4(i)	0.906±5	0	0.250±8	0.038
4 H(4) in 4(i)	0.856±5	0	0.390±8	0.038
8 H(6) in 8(j)	0.795±3	0.116±6	0.204±5	0.038

Anisotropic thermal parameters are given for non-hydrogen atoms.

Interatomic distances (Å) and angles*

Cr-O	1.65±4, 1.644±3, 1.641±2 (2×)
	1.669 1.659 1.652
O-Cr-O	108.5, 109.3, 110.2, 110.9° all±0.1°
N-H	0.87±8, 0.99±5, 0.93±4 (2×)

*Corrected for librational motion.

Structure

More accurate than previsouly reported structure (1). Proton N.M.R. line width is 4.8 gauss indicative of substantial NH_4 motional narrowing.

Details of analysis

966 reflexions collected on a diffractometer with Mo radiation. Least-squares refinement to R=0.037. Spherical scattering curve for NH_4^+.

1. B.M. GATEHOUSE and P. LEVERETT, 1969. *J. Chem. Soc.*, (A), 1857.

RUBIDIUM DICHROMATE

I. Crystal structure of triclinic rubidium dichromate, $Rb_2Cr_2O_7$. N. CH. PANAGIOTOPOULOS and I.D. BROWN, 1970. *Can. J. Chem.*, 48, 537-543.

$Rb_2Cr_2O_7$ F.W. = 386.9

Triclinic, a = 13.554±1, b = 7.640±3, c = 7.735±2 Å, α = 93.64±20, β = 98.52±1, γ = 88.80±4°, U = 790.49 Å^3, D_m = 3.12, Z = 4, D_x = 3.25.

Space group $P\bar{1}$ (C_i^1)

Atomic positions

	x	*y*	*z*
Rb(1)	0.13528±11	0.11118±19	0.87628±16
Rb(2)	-0.15303±11	0.22589±19	0.62431±16
Rb(3)	0.34205±11	0.55555±19	0.80287±16
Rb(4)	0.62878±11	0.78601±19	0.70265±16
Cr(1)	0.1107±2	0.2853±3	0.3915±3
Cr(2)	-0.0853±2	0.3531±3	0.1361±3
Cr(3)	0.3883±2	0.0613±3	0.7559±3
Cr(4)	0.5826±2	0.3022±3	0.7157±3
O(B1)	0.0238±7	0.4283±13	0.2736±12
O(11)	0.1994±7	0.4025±13	0.4953±12
O(12)	0.1543±7	0.1463±13	0.2571±12
O(13)	0.0546±7	0.1833±13	0.5225±12
O(21)	-0.1478±7	0.5180±13	0.0506±12
O(22)	-0.1528±7	0.2493±13	0.2534±12
O(23)	-0.0546±7	0.2166±13	-0.0158±12
O(B3)	0.4712±7	0.1857±13	0.6530±12
O(31)	0.2966±7	0.0102±13	0.6072±12
O(32)	0.3496±7	0.1818±13	0.9121±12
O(33)	0.4430±7	-0.1138±13	0.8289±12
O(41)	0.6197±7	0.3717±13	0.5466±12
O(42)	0.6678±7	0.1800±13	0.8150±12
O(43)	0.5604±7	0.4666±13	0.8438±12

Anisotropic temperature factors given

Interatomic distances (Å) *and angles*

Rb(1) - O(21)	2.857±10
Rb(1) - O(13)	2.875±9
Rb(1) - O(22)	2.888±10
Rb(1) - O(23)	2.904±10
Rb(1) - O(12)	2.914±9
Rb(1) - O(32)	2.935±10
Rb(1) - O(23)	3.072±10
Rb(1) - O(31)	3.284±10
Rb(2) - O(22)	2.886±9
Rb(2) - O(23)	2.908±9
Rb(2) - O(31)	2.992±9
Rb(2) - O(13)	3.039±10
Rb(2) - O(12)	3.044±10
Rb(2) - O(42)	3.059±10
Rb(2) - O(11)	3.066±10
Rb(2) - O(B1)	3.192±10
Rb(2) - O(41)	3.238±9

Rb(3) - O(43)	2.868±9
Rb(3) - O(33)	2.880±10
Rb(3) - O(41)	2.916±10
Rb(3) - O(22)	2.930±9
Rb(3) - O(43)	2.998±9
Rb(3) - O(11)	3.024±9
Rb(3) - O(32)	3.024±10
Rb(3) - O(21)	3.092±10
Rb(3) - O(42)	3.491±9
Rb(4) - O(B3)	2.900±9
Rb(4) - O(33)	2.904±10
Rb(4) - O(32)	2.946±9
Rb(4) - O(43)	2.956±10
Rb(4) - O(12)	2.963±10
Rb(4) - O(42)	3.114±10
Rb(4) - O(31)	3.223±10
Rb(4) - O(11)	3.237±10
Rb(4) - O(41)	3.313±10

Cr(1) - O(11)	1.600±9
Cr(1) - O(13)	1.602±10
Cr(1) - O(12)	1.604±10
Cr(1) - O(B1)	1.780±10

Angles in degrees subtended at Cr(1) by

	O(13)	O(12)	O(B1)
O(11)	111.5±5	109.6±5	107.7±5
O(13)		109.5±5	108.8±5
O(12)			109.6±5

Cr(2) - O(23)	1.616±10
Cr(2) - O(21)	1.624±10
Cr(2) - O(22)	1.630±11
Cr(2) - O(B1)	1.772±9

Angles in degrees subtended at Cr(2) by

	O(21)	O(22)	O(B1)
O(23)	110.4±5	108.2±5	109.5±5
O(21)		110.2±5	110.0±5
O(22)			108.4±5

Cr(1)-O(B1)-Cr(2) 123.0±6°

Cr(3) - O(31)	1.600±9
Cr(3) - O(33)	1.609±10
Cr(3) - O(32)	1.615±10
Cr(3) - O(B3)	1.790±10

Angles in degrees subtended at Cr(3) by

	O(33)	O(32)	O(B3)
O(31)	109.8±5	110.1±5	106.0±5
O(33)		110.9±5	110.6±5
O(32)			109.3±5

Cr(4) - O(41)	1.594±10
Cr(4) - O(43)	1.601±10
Cr(4) - O(42)	1.606±9
Cr(4) - O(B3)	1.758±10

Angles in degrees subtended at Cr(4) by

	O(43)	O(42)	O(B3)
O(41)	109.0±5	109.5±5	109.6±5
O(43)		109.6±5	108.3±5
O(42)			110.9±5

Cr(3)-O(B3)-Cr(4) 137.5±5°

Discussion

The structure consists of $Cr_2O_7^{2-}$ and K^+ ions forming sheets perpendicular to a^*. Adjacent sheets are similar but crystallographically distinct and are rotated by 90° with respect to each other.

Details of analysis

Crystals grown from aqueous solution. Lattice parameters measured from Weissenberg photographs calibrated with rutile (a = 4.59369, c = 2.95814 Å). Intensities measured photometrically from integrated Weissenberg photographs of layers k = 0-4 (CuK_α radiation, crystal 0.06×0.20×0.06 mm^3) and precession photographs of layers h = 0-2, l = 0-2 (MoK_α radiation, crystal 0.09×0.03×0.09 mm^3) and corrected for Lorentz and polarization effects, but not absorption ($\mu(CuK_\alpha)$ = 40.4 mm^{-1}), $\mu(MoK_\alpha)$ = 15.8 mm^{-1}). Structure determined from Patterson function and difference synthesis and refined by least squares to give R_2 = 0.058 for 2881 reflexions, some listed in I, all listed in 1.

1. Depository of Unpublished Data, National Research Council of Canada, Ottawa, Canada.

LITHIUM DICHROMATE DIHYDRATE

I. [Crystal structure of lithium bichromate dihydrate $Li_2Cr_2O_7$-$2H_2O$.] I.D. DATT, N.V. RANNEV, T.G. BALICEVA and R.P. OZEROV, 1970. *Kristallografija*, 15, 949-952 [*Soviet Physics-Crystallography*, 15, 826-828].

$Li_2Cr_2O_7.2H_2O$ F.W. = 265.86

Monoclinic, a = 13.30, b = 7.51, c = 7.67 Å, β = 93.5°, U = 765 Å^3, D_m = 2.27, Z = 4, D_x = 2.32.

Space group $C2/c$ (C^6_{2h})

Atomic positions

	x	y	z	B(Å^2)
8 Cr in 8(f)	0.101	0.158	0.357	2.39
8 O(1) in 8(f)	0.060	0.300	0.496	2.62
8 O(2) in 8(f)	0.183	0.050	0.439	2.62
8 O(3) in 8(f)	0.157	0.352	0.223	0.17
4 O(4) in 4(e)	0	0.066	0.25	2.78
8 O(5) in 8(f)	0.140	0.267	0.881	1.87
8 Li in 8(f)	0.078	0.480	0.843	1.35

Interatomic distances (Å) *and angles*

Errors around 0.05 Å

Cr - O(2)	1.52 [1.47]
Cr - O(1)	1.62
Cr - O(4)	1.74 [1.68]
Cr - O(3)	1.84 [1.96]

Angles in degrees subtended at Cr by

	O(1)	O(4)	O(3)
O(2)	111	122	110
O(1)		108	91
O(4)			112

Cr-O(4)-Cr 131°

Discussion

The structure is the same as $(NH_4)_2Cr_2O_7$ (1) with $Li(H_2O)$ substituting for NH_4.

Details of analysis

Lattice parameters from rotation photographs. Intensities measured from Weissenberg photographs of layers with k=0 and l=0 (72 reflexions) and corrected for Lorentz and polarization effects. Structure determined from Patterson and electron-density functions. [This is reported as a preliminary analysis and in general is probably correct. Little value can be placed on the actual coordinates given].

1. *Structure Reports*, 15, 288.

SODIUM DICHROMATE DIHYDRATE

I. [Determination of the crystal structure of sodium bichromate $Na_2Cr_2O_7.2H_2O$] JU.A. KHARITONOV, E.A. KUZ'MIN and N.V. BELOV, 1970. *Kristallografija*, 15, 942-968 [*Soviet Physics-Crystallography*, 15, 820-825].

$Na_2Cr_2O_7.2H_2O$ F.W. = 297.96

Monoclinic, a = 6.21, b = 10.90, c = 12.94 Å, β = 95°, U = 872.6 Å^3, D_m = 2.35, Z = 4, D_x = 2.26.

Space group $P2_1$ (C_i^2)

Atomic positions All atoms in 2(*a*)

	x	*y*	*z*		*x*	*y*	*z*
Cr(1)	0.146	0	0.952	O(6)	0.380	0.439	0.182
Cr(2)	0.052	0.458	0.834	O(7)	0.282	0.591	0.333
Cr(3)	0.343	0.948	0.487	O(8)	0.059	0.447	0.971
Cr(4)	0.346	0.450	0.306	O(9)	0.190	0.346	0.349
Na(1)	0.183	0.771	0.728	O(10)	0.400	0.917	0.626
Na(2)	0.384	0.298	0.029	O(11)	0.084	0.913	0.456
Na(3)	0.149	0.254	0.513	O(12)	0.492	0.366	0.570
Na(4)	0.466	0.769	0.254	O(13)	0.392	0.092	0.468
O(1)	0.385	0.974	0.012	O(14)	0.121	0.859	0.220
O(2)	0.131	0.932	0.840	O(15)	0.355	0.690	0.073
O(3)	0.114	0.152	0.939	O(16)	0.194	0.178	0.698
O(4)	0.011	0.094	0.200	O(17)	0.155	0.643	0.568
O(5)	0.305	0.432	0.806	O(18)	0.475	0.176	0.184

B(Cr) = 2.1 Å^2; B(Na) = 2.6 Å^2; B(O) = 2.5 Å^2
σ(Cr) = 0.001 Å; σ(Na) = 0.002 Å; σ(O) = 0.004 Å.

Interatomic distances (Å) *and angles*

σ(Cr-O) = 0.005 Å, σ(Na-O) = 0.006 Å [The true error is probably rather larger]

Cr(1) - O(2)	1.623
Cr(1) - O(1)	1.639
Cr(1) - O(3)	1.675
Cr(1) - O(8)	1.780

Angles in degrees subtended at Cr(1) by

	O(1)	O(3)	O(8)
O(2)	108.9	111.4	111.3°
O(1)		108.3	109.9
O(3)			107.0

Cr(2) - O(4)	1.586
Cr(2) - O(14)	1.636
Cr(2) - O(5)	1.667
Cr(2) - O(8)	1.774

Cr(1)-O(8)-Cr(2) 125.5°

Angles in degrees subtended at Cr(2) by

	O(14)	O(5)	O(8)
O(4)	111.2	108.2	108.9°
O(14)		113.0	109.9
O(5)			105.3

Cr(3) - O(12)	1.589
Cr(3) - O(13)	1.622
Cr(3) - O(11)	1.668
Cr(3) - O(10)	1.834

Angles in degrees subtended at Cr(3) by

	O(13)	O(11)	O(10)
O(12)	109.7	114.0	105.9
O(13)		112.0	107.7
O(11)			107.1

Cr(4) - O(9)	1.621	Angles in degrees subtended at Cr(4) by			
Cr(4) - O(7)	1.633		O(7)	O(6)	O(10)
Cr(4) - O(6)	1.641				
Cr(4) - O(10)	1.775	O(9)	114.9	114.6	102.7
		O(7)		109.4	107.9
Cr(3)-O(10)-Cr(4)	122.4°	O(6)			106.6

Na(1) - O(2)	2.316	Na(2) - O(1)	2.480	Na(3) - O(12)	2.509
Na(1) - O(17)	2.490	Na(2) - O(6)	2.508	Na(3) - O(16)	2.526
Na(1) - O(4)	2.497	Na(2) - O(3)	2.522	Na(4) - O(14)	2.363
Na(1) - O(10)	2.530	Na(2) - O(8)	2.648	Na(4) - O(16)	2.366
Na(1) - O(18)	2.541	Na(3) - O(11)	2.315	Na(4) - O(5)	2.444
Na(1) - O(9)	2.574	Na(3) - O(9)	2.381	Na(4) - O(12)	2.504
Na(2) - O(18)	2.432	Na(3) - O(17)	2.404	Na(4) - O(7)	2.514
Na(2) - O(15)	2.475	Na(3) - O(13)	2.426	Na(4) - O(15)	2.534

Details of analysis

Lattice parameters from rotation [and Weissenberg?] photographs. Intensities estimated visually from equi-inclination Weissenberg photographs of layers with $h = 0$ to 4 and $k = 0$ to 2 taken with MoK_α radiation corrected for Lorentz and polarization but not absorption effects. Structure from Patterson function refined by electron-density and least squares to give $R(F^2) = 0.083$ for 900 observed reflexions [not listed].

DIALKALIMETAL HYDROXY BIS CHROMATE(VI) CHROMIUM(III)

I. The crystal structure of $Na_2Cr_3O_8OH$ and $K_2Cr_3O_8OH$. O. JONSSON, 1970. *Acta Chem. Scand.*, 24, 3627-3666.

DIPOTASSIUM HYDROXY BIS CHROMATE(VI) CHROMIUM(III)

$K_2CrOH(CrO_4)_2$ F.W. = 378.18

Monoclinic, $a = 9.267\pm1$, $b = 6.077\pm1$, $c = 7.813\pm1$ Å, $\beta = 114.08\pm1$, $U = 401.7$ Å^3, $D_m = 3.13$, $Z = 2$, $D_x = 3.12$.

Space group $P2_1/m$ (C^2_{2h})

Atomic positions

	x	*y*	*z*	*B*(Å^2)
2 K(1) in 2(*e*)	0.0867±4	0.75	0.1778±5	1.86±7
2 K(2) in 2(*e*)	0.2445±4	0.75	0.7521±5	1.95±8
2Cr(1) in 2(*d*)	0.5	0	0.5	0.85±6
2Cr(2) in 2(*e*)	0.1736±3	0.25	0.4578±4	1.20±6
2Cr(3) in 2(*e*)	0.3345±3	0.25	0.0682±3	1.11±6
4 O(1) in 4(*f*)	0.2864±8	0.0273±16	0.5035±10	1.76±14
2 O(2) in 2(*e*)	0.4460±14	0.25	0.3183±18	2.25±24
2 O(3) in 2(*e*)	0.4311±12	0.75	0.3237±15	1.24±18
2 O(4) in 2(*e*)	0.0524±14	0.25	0.2405±18	2.37±24
2 O(5) in 2(*e*)	0.0730±14	0.25	0.5832±17	2.35±24
4 O(6) in 4(*f*)	0.2229±9	0.0389±17	0.0106±11	2.25±16
2 O(7) in 2(*e*)	0.4484±15	0.25	0.9583±19	2.69±25

Interatomic distances (Å) *and angles*

K(1) - O(6)	2.780±11 (×2)	K(2) - O(1)	2.716±10 (×2)
K(1) - O(5)	2.817±16	K(2) - O(6)	2.744±10 (×2)
K(1) - O(3)	2.919±11	K(2) - O(4)	2.776±15
K(1) - O(6)	2.932±8 (×2)	K(2) - O(7)	2.822±11
K(1) - O(1)	2.983±8 (×2)	K(2) - O(5)	3.038±10
K(1) - O(4)	2.984±13	K(2) - O(2)	3.133±15
K(1) - O(4)	3.115±4 (×2)	K(2) - O(5)	3.433±5 (×2)

Cr(1) - O(3)	1.974±7	(×2)
Cr(1) - O(1)	1.998±8	(×2)
Cr(1) - O(2)	1.999±8	(×2)

Angles in degrees subtended at Cr(1) by

	O(1)	O(2)
O(3)	91.9±4	100.0±4
O(3)'	88.1±4	80.0±4
O(1)		89.1±5
O(1)'		90.9±5

Cr(2) - O(5)	1.605±16	
Cr(2) - O(4)	1.609±12	
Cr(2) - O(1)	1.658±9	(×2)

Angles in degrees subtended at Cr(2) by

	O(4)	O(1)'
O(5)	108.4±7	109.9±4°
O(4)		109.6±4
O(1)		109.5±4

Cr(3) - O(6)	1.593±9	(×2)
Cr(2) - O(7)	1.610±17	
Cr(3) - O(2)	1.800±12	

Angles in degrees subtended at Cr(3) by

	O(6)'	O(7)	O(2)
O(6)	107.3±5	109.9±4	109.0±4°
O(7)			111.6±6

DISODIUM HYDROXY BIS CHROMATE(VI) CHROMIUM(III)

$Na_2CrOH(CrO_4)_2$ F.W. = 345.36

Monoclinic, a = 8.518±1, b = 5.998±1, c = 7.536±1 Å, β = 111.12±1°, U = 359.2 Å^3, D_m = 3.19, Z = 2, D_x = 3.19.

Space group $P2_1/m$ (C_{2h}^2)

Atomic positions

	x	y	z	B(Å^2)
2 Na(1) in 2(*e*)	0.0930±12	0.75	0.1666±14	2.64±19
2 Na(2) in 2(*e*)	0.2413±11	0.75	0.7658±13	2.01±17
2 Cr(1) in 2(*d*)	0.5	0	0.5	0.54±7
2 Cr(2) in 2(*e*)	0.1668±4	0.25	0.4948±5	0.84±7
2 Cr(3) in 2(*e*)	0.3237±4	0.25	0.0646±5	0.78±7
4 O(1) in 4(*f*)	0.2879±11	0.0250±13	0.5557±13	1.41±17
2 O(2) in 2(*e*)	0.4250±16	0.25	0.3175±20	1.41±26
2 O(3) in 2(*e*)	0.4047±15	0.75	0.3222±18	0.77±23
2 O(4) in 2(*e*)	0.0592±20	0.25	0.2663±23	2.05±30
2 O(5) in 2(*e*)	0.0422±18	0.25	0.6087±21	1.77±28
4 O(6) in 4(*f*)	0.2075±12	0.0329±14	0.9968±14	1.80±19
2 O(7) in 2(*e*)	0.4681±21	0.25	0.9720±27	3.04±38

Interatomic distances (Å) *and angles*

Na(1) - O(5)	2.366±22	
Na(1) - O(3)	2.486±15	
Na(1) - O(6)	2.523±14	(×2)
Na(1) - O(6)	2.737±12	(×2)
Na(2) - O(1)	2.416±12	(×2)
Na(2) - O(4)	2.483±20	
Na(2) - O(6)	2.521±13	(×2)
Na(2) - O(7)	2.551±17	
Na(2) - O(5)	2.981±15	

Cr(1) - O(1)	2.003±10	(×2)
Cr(1) - O(2)	1.979±9	(×2)
Cr(1) - O(3)	1.980±7	(×2)

Angles in degrees subtended at Cr(1) by

	O(3)	O(1)
O(2)	98.8±4	90.0±5°
O(2)'	81.2±4	90.0±5
O(3)		89.0±5
O(3)'		91.0±5

Cr(2) - O(5)	1.587±18	
Cr(2) - O(4)	1.633±15	
Cr(2) - O(1)	1.660±8	(×2)

Angles in degrees subtended at Cr(2) by

	O(4)	O(1)'
O(5)	109.8±8	108.7±5°
O(4)		110.4±5
O(1)		108.8±4

Cr(3) - O(6)	1.604±9	(×2)
Cr(3) - O(7)	1.618±22	
Cr(3) - O(2)	1.787±14	

Angles in degrees subtended at Cr(3) by

	O(6)'	O(7)	O(2)
O(6)	108.6±5	109.9±5	110.2±5°
O(7)			108.1±8

Discussion

The two compounds are isostructural and contain the groups $Cr^{III}(OH)_2(Cr^{IV}O_4)_4$ linked into chains through a shared octahedron edge and the (CrO_4) groups. The chains are linked through alkali metal atoms.

Details of analysis

Lattice parameters measured with Guinier-Hägg focusing powder camera using KCl (a=6.2919 Å) as standard. Intensities measured visually from multiple-film Weissenberg photographs taken with CuK_α radiation and corrected for Lorentz, polarization and absorption effects ($\mu(K_2Cr(OH)(CrO_4)_2)$ = 44.6 mm^{-1}, crystal size 0.015×0.02×0.05 mm^3). Structure solved by Patterson methods and refined by least squares to give R = 0.08 for $K_2Cr_3O_8OH$ for 564 observed reflexions and R = 0.098 for $Na_2Cr_3O_8OH$ for 483 observed reflexions, both sets listed.

POTASSIUM TETRACHROMATE

I. [Crystal structure of potassium tetrachromate $K_2Cr_4O_{13}$.] V.P. GOLOVAČEV, E.A. KUZ'MIN, JU. A. KHARITONOV and N.V. BELOV, 1970. *Dokl. Akad. Nauk SSSR*, 132, 1272-1274 [*Soviet Physics-Doklady*, 15, 530-532].

$K_2Cr_4O_{13}$ F.W. = 494.18

Monoclinic, a = 8.71, b = 7.55, c = 9.37 Å, β = 93°, U = 615 Å^3, Z = 2, D_x = 2.66.

Space Group Pc (C_s^2)

Atomic positions All atoms in 2(a)

	x	y	z	
Cr(1)	0.441	0.439	0.335	
Cr(2)	0.434	0.105	0.102	
Cr(3)	0.759	0.930	0.829	[*sic* probably 0.929]
Cr(4)	0.095	0.429	0.383	
K(1)	0.796	0.405	0.104	
K(2)	0.079	0.892	0.246	
O(1)	0.906	0.064	0.802	[*sic* probably 0.902]
O(2)	0.588	0.069	0.001	
O(3)	0.282	0.103	0.997	
O(4)	0.578	0.563	0.345	
O(5)	0.959	0.562	0.332	

O(6)	0.263	0.568	0.323
O(7)	0.070	0.244	0.268
O(8)	0.119	0.633	0.050
O(9)	0.447	0.349	0.166
O(10)	0.448	0.311	0.466
O(11)	0.425	0.970	0.238
O(12)	0.750	0.128	0.309
O(13)	0.767	0.779	0.093

Interatomic distances (Å)

Cr(1) - O(4)	1.52	K(1) - O(5)	2.77
Cr(1) - O(10)	1.56	K(1) - O(13)	2.84
Cr(1) - O(9)	1.73	K(1) - O(12)	2.88
Cr(1) - O(6)	1.83	K(1) - O(5)	2.99
		K(1) - O(4)	3.01
Cr(2) - O(3)	1.61	K(1) - O(7)	3.02
Cr(2) - O(11)	1.64	K(1) - O(9)	3.15
Cr(2) - O(2)	1.70	K(1) - O(2)	3.24
Cr(2) - O(9)	1.94	K(1) - O(4)	3.25
Cr(4) - O(5)	1.61	K(2) - O(7)	2.67
Cr(4) - O(8)	1.64	K(2) - O(8)	2.72
Cr(4) - O(7)	1.77	K(2) - O(5)	2.84
Cr(4) - O(6)	1.91	K(2) - O(3)	2.87
		K(2) - O(6)	2.99
		K(2) - O(11)	3.08
		K(2) - O(13)	3.12

[Bonds to Cr(3) and O(1) are omitted as there is a manifest error in their coordinates. Cr(3) bonds to O(1), O(2), O(12) and O(13). O(1) probably bonds to Cr(3) and K(1)].

[Errors uncertain but are around 0.10 Å]

Discussion

Linear Cr_4O_{13} ions (bridging Cr-O-Cr angles 110-144°) are sandwiched between layers of K atoms in the *bc* plane.

Details of analysis

Intensities measured visually from rotating crystal photographs of layers with k = 0 to 6 and h = 0 and 1. Structure determined from Patterson function and refined by least squares to give R = 0.108 for 550 non-zero reflexions [not listed].

HYDRAZINIUM FLUOROBORATE AND HYDRAZINIUM PERCHLORATE

I. The crystal structures of the isostructural compounds hydrazinium fluoroborate and hydrazinium perchlorate. J.W. CONANT and R.B. ROOF, JR., 1970. *Acta Cryst.*, B26, 1928-1932.

$NH_2NH_3ClO_4$ F.W. = 132.49
$NH_2NH_3BF_4$ F.W. = 119.85

Monoclinic, (HPC): a = 14.412±7, b = 5.389±5, c = 12.797±3 Å, β = 113.09±5°, [U = 993.89 Å^3], Z = 8, D_x = 1.910.

(HFB): a = 14.006±8, b = 5.316±4, c = 12.387±5 Å, β = 112.87±5°, [U = 922.27 Å^3], Z = 8, D_x = 1.859.

Space group C2/c (C^6_{2h}) based on distribution of the intensities.

Atomic positions All atoms in 8(*f*)

a) Hydrazinium perchlorate

	x	*y*	*z*
Cl	0.1067±1	0.0772±4	0.4231±2
O(1)	0.1515±6	0.0678±14	0.3415±6
O(2)	0.1783±6	0.0348±17	0.5347±6
O(3)	0.0663±7	0.3211±15	0.4218±8
O(4)	0.0293±6	-0.1031±18	0.3987±7
N(1)	0.1556±5	0.3982±13	0.1522±6
N(2)	0.1599±5	0.5904±13	0.2344±5
H(1)	0.091±3	0.304±11	0.132±5
H(2)	0.157±4	0.477±3	0.082±2
H(3)	0.214±3	0.281±10	0.184±5
H(4)	0.099±3	0.701±8	0.202±2
H(5)	0.161±5	0.512±3	0.306±2

Anisotropic thermal parameters are given.

b) Hydrazinium fluoroborate

	x	*y*	*z*
B	0.107±2	0.072±6	0.424±2
F(1)	0.147±1	0.076±2	0.337±1
F(2)	0.186±1	0.025±3	0.524±1
F(3)	0.065±1	0.303±3	0.425±1
F(4)	0.035±1	-0.108±3	0.400±1
N(1)	0.155±1	0.391±3	0.152±1
N(2)	0.158±1	0.590±3	0.230±1
N(1)	0.086±5	0.306±23	0.126±9
H(2)	0.168±9	0.452±5	0.081±5
H(3)	0.212±7	0.270±20	0.198±9
H(4)	0.106±7	0.724±12	0.192±4
H(5)	0.150±10	0.529±5	0.302±5

Anisotropic thermal parameters are given.

Interatomic distances (Å) *and angles*

	HPC	HFB
Bonds	*X*=Cl, *Y*=O	*X*=B, *Y*=F
N(1)-N(2)	1.461±9	1.425±19
X - *Y*(1)	1.428±6	1.379±26
- *Y*(2)	1.415±7	1.325±29
- *Y*(3)	1.435±7	1.360±31
- *Y*(4)	1.418±7	1.339±30
H(1)-*Y*(3)	2.10±4	1.96±6
H(2)-*Y*(3)	2.23±4	2.32±7
H(3)-N(2)	1.99±2	1.94±3
H(4)-*Y*(4)	2.08±1	2.05±4
Angles		
Y(1)-*X* - *Y*(2)	111.87±0.01°	106.46±0.09°
- *Y*(3)	108.79±0.01	106.94±0.08
- *Y*(4)	110.96±0.01	109.50±0.08
Y(2)-*X* - *Y*(3)	106.90±0.01	111.87±0.09
- *Y*(4)	108.10±0.01	111.16±0.08
Y(3)-*X* - *Y*(4)	110.14±0.01	110.70±0.09
N(1)-H(1)-*Y*(3)	147.13±0.28	153.63±1.02
N(1)-H(2)-*Y*(3)	144.61±0.20	130.48±0.86
N(1)-H(3)-N(2)	170.87±0.10	164.81±0.46
N(2)-H(4)-*Y*(4)	167.57±0.01	160.22±0.51

Description of the structure

The two compounds are isostructural. The anion consists of a tetrahedrally coordinated ClO_4 or BF_4 group and the cation is a bi-tetrahedron joined together at a corner and with a nitrogen atom at each centre. The six corners are occupied by five hydrogen atoms and a lone pair of electrons.

These structures consist of sheets parallel to the *ab* plane which are formed from alternate rows of anions and cations. One hydrogen atom provides bonding between rows within the sheet and one apparently is not involved in any hydrogen bonding. The remaining three provide bonding between the cation and anions and cations on neighbouring sheets.

Materials

Hydrazinium fluoroborate was produced from neutralizing a solution of fluoroboric acid with hydrazine, whereas perchloric acid was used to obtain hydrazinium perchlorate. The former compound was recrystallized from 2-propanol and the latter from ethanol.

Details of analysis

233 reflexions for the fluoroborate and 649 reflexions for the perchlorate were measured using a diffractometer and MoK_α radiation with balanced filters. Absorption and extinction parameters were applied. Structure refined by least-squares method to *R* values of 0.106 and 0.076 for the fluoroborate and perchlorate respectively, compared to a statistical agreement between equivalent reflexions of 0.038 and 0.023.

COBALT COMPOUND

I. Die Kristall und Molekülstruktur von $[Co_3(NH_3)_8(OH)_2(NO_2)_2(CN)_2](ClO_4)_3$: $NaClO_4.2H_2O$. J. WEISS, H. SIEBERT and K. WIEGHARDT, 1970. *Acta Cryst.*, B26, 1709-1712.

$Co_3(NH_3)_8(OH)_2(NO_2)_2(CN)_2(ClO_4)_3.NaClO_4.2H_2O$ F.W. = 739.70

Monoclinic, *a* = 14.114, *b* = 7.864, *c* = 14.843 Å, β = 112.3°, [*U* = 1647.454 Å^3], *Z* = 2, D_x = 2.07.

Space group, *A*2 (C_2^3) from Patterson function

Atomic positions

	x	*y*	*z*	*B*(Å^2)
2 Co(1) in 2(*a*)	0	0	0	2.2±1
4 Co(2) in 4(*c*)	0.1720±3	0.2334±8	0.1631±3	2.2±1
4 C in 4(*c*)	0.0979±28	0.821±5	0.027±3	4.2±8
4 N(1) in 4(*c*)	0.0688±19	0.410±4	0.141±2	3.2±6
4 N(2) in 4(*c*)	0.2399±21	0.295±4	0.307±2	4.0±6
4 N(3) in 4(*c*)	0.2663±16	0.388±3	0.136±2	2.8±4
4 N(4) in 4(*c*)	0.2685±21	0.037±4	0.177±2	3.5±7
4 N(5) in 4(*c*)	0.1501±19	0.708±4	0.041±2	4.0±6
4 N(6) in 4(*c*)	0.0153±17	0.003±4	0.137±2	2.9±5
4 O(1) in 4(*c*)	0.0816±15	0.091±3	0.198±1	3.2±4
4 O(2) in 4(*c*)	0.1011±15	0.177±3	0.026±1	2.5±4
4 O(3) in 4(*c*)	0.0424±18	0.422±4	0.336±2	5.0±5
4 Cl(1) in 4(*c*)	0.2369±6	0.762±1	0.3655±6	3.5±2
4 O(4) in 4(*c*)	0.1253±19	0.627±4	0.437±2	5.5±6
4 O(5) in 4(*c*)	0.3114±25	0.876±5	0.3811±2	8.3±8
4 O(6) in 4(*c*)	0.2038±27	0.696±6	0.2691±26	9.8±1.0
4 O(7) in 4(*c*)	0.1612±28	0.885±6	0.3759±25	10.6±1.0
4 Cl(2) in 4(*c*)	0.4572±6	0.739±2	0.1288±6	4.4±2
4 O(8) in 4(*c*)	0.4017±18	0.220±4	0.6943±17	6.0±6

O(9)	in 4(*c*)	0.4042±25	0.830±5	0.0471±23	7.9±8
O(10)	in 4(*c*)	0.4835±32	0.084±6	0.3718±30	11.8±1.2
O(11)	in 4(*c*)	0.4560±32	0.343±3	0.3712±29	11.3±1.2
O(12)	in 4(*c*)	0.4530±18	0.205±4	0.1004±18	5.5±6
2 Na	in 2(*b*)	1/2	0.401±3	0	4.9±5

Interatomic distances (Å) and angles

Co(1)-C (2×)	1.90±4	O(2)-Co(1)-N(6)	90.1°	N(2)-Co(2)-O(1)	84.4°
-O(2) (2×)	1.92±2	O(2)-Co(1)-C	94.0	N(2)-Co(2)-N(3)	91.9
-N(6) (2×)	1.97±3	O(2)-Co(1)-O(2)'	87.5	N(3)-Co(2)-O(2)	92.8
Co(2)-N(1)	1.95±3	N(6)-Co(1)-O(2)'	89.0	N(3)-Co(2)-N(4)	90.6
-N(2)	2.02±3	N(6)-Co(1)-C'	90.6	N(4)-Co(2)-O(1)	90.9
-N(3)	1.96±3	C Co(1)-C'	84.4	N(4)-Co(2)-O(2)	89.6
-N(4)	2.02±3	N(6)-Co(1)-C	90.2	O(1)-Co(2)-O(2)	90.9
-O(1)	1.93±2	N(1)-Co(2)-O(2)	86.1	Co(1)-N(6)-O(1)	120.9
-O(2)	1.94±2	N(1)-Co(2)-O(1)	86.2	O(1)-N(6)-O(3)	118.3
		N(1)-Co(2)-N(3)	92.4	Co(1)-N(6)-O(3)	120.6
		N(1)-Co(2)-N(2)	95.1	Co(2)-O(1)-N(6)	122.1
		N(2)-Co(2)-N(4)	89.0	Co(1)-O(2)-Co(2)	115.1

N-O	1.24±3, 1.21±4		
N-C	1.13±4		
Cl(1)-O (4)	1.34-1.49	O-Cl(1)-O	95.3-117.9°
Cl(2)-O (4)	1.36-1.48	O-Cl(2)-O	95.5-123.1
Na-O(9) (2×)	2.32		
-O(12) (2×)	2.41		

Discussion of the structure

The cation is shown in Fig. 49. It lies on two-fold axes and is formed from three octahedrally coordinated cobalt ions bridged in pairs by one OH and one NO_2. The central Co has two CN^- ligands in the *cis* position.

Details of analysis

905 reflexions were visually estimated from Weissenberg photographs taken with MoK_α radiation. Refined by least-squares methods to R = 0.086.

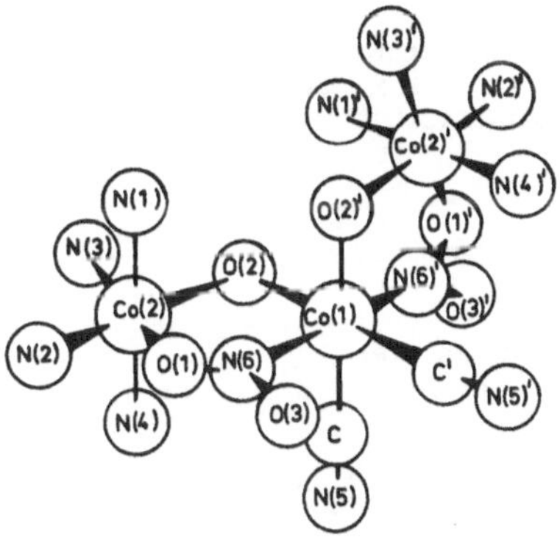

Fig. 49. The cation consisting of $[Co_3(NH_3)_8(OH)_2(NO_2)_2(CN)_2]^{3+}$.

SODIUM IODATE

I. Refinement of the structure of $NaIO_4$. A. KALMAN and D.W.J. CRUICHSHANK, 1970. *Acta Cryst.*, B26, 1782-1785.

$NaIO_4$ F.W. = 213.89

Tetragonal, a = 5.337±1, c = 11.947±2 Å, U = 340.29 Å^3, D_m = 4.12, Z = 2, D_x = 4.174.

Space group $I4_1/a$ (C_{2h}^6)

Atomic parameters

2 Na in 4(b)
4 I in 4(a)
16 O in 16(f) with x = −0.2341±1, y = 0.1501±13, z = 0.0806±6

Anisotropic thermal parameters are given.

Interatomic distances (Å) *and angles*

I-O	1.775±7 (corrected for libration)	O-I-O (2×)	114.1±4°
		O-I-O (4×)	107.3±4

Na-O (4×)	2.54
-O (4×)	2.60

Discussion of the structure

Seheelite-type. A more accurate determination of the structure than previous work (1,2).

Details of analysis

927 reflexions visually measured from Weissenberg photographs (MoK_α radiation). Independent scales for layer lines. Refined by least-squares method to an R value of 0.089.

MAGNESIUM TRIHYDROGEN HEXAOXOIODATE

I. Crystal and molecular structure of hexaquomagnesium trihydrogen-hexaoxoiodate(VII). F. BIGOLI, A.M.M. LANFREDI, A.T. and M.T. CAMELLINI, 1970. *Acta Cryst.*, B26, 1075-1079.

$Mg(OH_2)_6H_3IO_6$ F.W. = 358.35

Monoclinic, a = 5.180±6, b = 9.886±13, c = 10.025±7 Å, β = 116.90±15°, U = 485.4 Å^3, D_m = 2.45, Z = 2, D_x = 2.45.

Space group Pc (C_s^2) All atoms in 2(a)

Atomic positions

	x	y	z
I	0	0.2206±2	1/4
Mg	-0.029±4	0.287±2	0.752±2
O(1)	-0.103±10	0.105±2	0.078±4
O(2)	-0.204±9	0.106±3	0.301±5
O(3)	0.109±9	0.332±1	0.408±4
O(4)	0.207±10	0.330±3	0.176±5
O(5)	-0.345±8	0.325±3	0.134±4
O(6)	0.346±7	0.109±3	0.354±4
O(7)	0.253±10	0.154±2	0.917±5

O(8)	-0.341±10	0.148±3	0.740±6
O(9)	-0.329±9	0.399±3	0.587±4
O(10)	0.298±9	0.417±3	0.772±5
O(11)	-0.085±9	0.409±2	0.902±4
O(12)	0.063±8	0.167±3	0.613±5

Interatomic distances (Å)

I-O(1)	2.01±3	
I-O(2)	1.78±5	
I-O(3)	1.87±3	
I-O(4)	1.93±5	
I-O(5)	1.95±4	
I-O(6)	1.97±4	
Mg-O	2.06-2.15	(six)

Structure

$Mg(OH_6^{2+}$ octahedra and $H_3IO_6^{2-}$ groups held together by a network of hydrogen bonds.

Materials

Crystals were grown from aqueous solution with equivalent amounts of $MgCO_3$ and HIO_4.

Details of analysis

Positive piezoelectric response. 831 reflexions measured with a microdensitometer from integrated Weissenberg photographs taken with Cu radiation were corrected for absorption, but not anomalous dispersion. Refined by difference synthesis with anisotropic thermal parameters to an *R* of 0.088.

CADMIUM TRIHYDROGENHEXAOXOIODATE

I. Crystal and molecular structure of cadmium trihydrogenhexaoxoiodate (VII) trihydriate. A. BRAIBANTI, A. TIRIPICCHIO, F. BIGOLI and M.A. PELLINGHELLI, 1970. *Acta Cryst.*, B26, 1069-1074.

$Cd[H_3IO_6].3H_2O$ F.W. = 392.39

Monoclinic, *a* = 5.957±7, *b* = 14.380±11, *c* = 9.715±7 Å, β = 120.8±2, *U* = 714.8 Å^3, D_m = 3.61, *Z* = 4, D_x = 3.65.

Space group $P2_1/c$ (C_{2h}^5) from extinctions

Atomic positions All atoms in 4(*e*)

	x	*y*	*z*
I	0.2197±5	0.2411±1	0.0510±3
Cd	0.2426±7	0.0804±4	0.3236±4
O(1)	0.041±10	0.3214±9	0.125±6
O(2)	0.053±12	0.1378±19	0.073±8
O(3)	0.481±5	0.2202±9	0.2753±22
O(4)	0.416±9	0.1576±10	-0.009±5
O(5)	0.425±9	0.3388±14	0.052±6
O(6)	-0.029±6	0.2650±13	-0.163±3
$H_2O(1)$	0.288±8	-0.0356±8	0.530±4
$H_2O(2)$	0.282±14	-0.048±4	0.204±9
$H_2O(3)$	0.200±13	0.026±2	0.213±8

Interatomic distances (Å)

I-O 6 between 1.86 - 1.94
Cd-O 8 between 2.23 - 2.79

Description of the structure

Anions probably consist of $IO_3(OH)_3$ with cadmium ions bridging between them.

Details of analysis

1449 reflexions were measured with a microdensitometer from integrated photograhs and refined by differential synthesis to a final *R* value of 0.109.

POTASSIUM PENTACYANONITROSYLVANADATE(I)

I. The crystal structure of potassium pentacyanonitrosylvanadate(I). $K_3V(CN)_5NO.2H_2O$. S. JAGNER and N.-G. VANNERBERG, 1970. *Acta Chem. Scand.*, 24, 1988-2002.

$K_3V(CN)_5NO.2H_2O$ F.W. = 361.84

Orthorhombic, a = 15.498±3, b = 7.145±1, c = 11.666±2 Å, U = 1291.8 Å^3, D_m = 1.83, Z = 4, D_x = 2.14.

Space group Pccn (D_{2h}^{10}) Systematic absences

Atomic positions

	x	*y*	*z*	*B*(Å^2)
4 K(1) in 4(*d*)	0.25	0.75	0.3804±4	*
8 K(2) in 8(*e*)	-0.0012±2	0.1047±4	0.3454±2	*
4 V in 4(*c*)	0.25	0.25	0.1555±2	*
8 O(1) in 8(*e*)	-0.0002±4	0.2317±13	0.0696±7	*
8 C(1) in 8(*e*)	0.3331±7	0.1237±17	0.2859±9	*
8 C(3) in 8(*e*)	0.1654±8	0.0155±22	0.1784±9	*
8 N(1) in 8(*e*)	0.3814±6	0.0480±13	0.3453±9	*
8 N(3) in 8(*e*)	0.1232±8	0.8880±15	0.1924±10	*
4 O(2) in 8(*e*)	0.1485±15	0.0316±37	0.4888±19	3.41
4 C(2) in 8(*e*)	0.3153±20	0.3747±62	0.5465±24	1.76
4 N(2) in 8(*e*)	0.3386±18	0.4597±46	0.4701±24	3.34
4 N(4) in 8(*e*)	0.1963±21	0.1340±67	0.5547±30	3.11

*Anisotropic temperature factors given.

Interatomic distances (Å) *and angles*

K(1) - N(2)	2.70
K(1) - O(2)	2.85
K(1) - N(1)	2.98 (×2)
K(1) - N(3)	3.11 (×2)
K(2) - N(3)	2.80
K(2) - O(1)	2.84
K(2) - O(1)	2.86
K(2) - N(2)	2.95
K(2) - N(3)	3.05
K(2) - N(1)	3.07
K(2) - N(1)	3.10

V - N(4)	1.66		Angles in degrees subtended at V by				
V - C(2)	1.85		C(2)'	C(3)	C(3)'	C(1)	C(1)'
V - C(3)	2.14 (×2)						
V - C(1)	2.19 (×2)	N(4)	91.6	90.3	99.9	172.9	89.5
		C(2)'		97.3	92.5	87.5	175.4
		C(3)			165.7	82.9	87.2
		C(3)'				87.2	82.9
		C(1)					91.9

C(1) - N(1)	1.15	V - C(1) - N(1)	175°
C(2) - N(2)	1.14	V - C(2) - N(2)	165°
C(3) - N(3)	1.13	V - C(3) - N(3)	177°
N(4) - O(2)	1.29	V - N(4) - O(2)	171°

O(1)-H...N(1)	2.89
O(1)-H...O(2)	3.01

σ(distances between ordered atoms) = 0.02 Å
σ(distances involving C(2), N(2), N(4), O(2)) = 0.04 Å
σ(angles) = 2°

Discussion

The crystals contain discrete $V(CN)_5NO$ ions bonded by K ions and H_2O molecules. The anion is disordered with respect to the interchange of C(2)N(2) and N(4)O(2) around the two-fold axis along *c*.

Details of analysis

Lattice parameters measured from focusing Guinier camera calibrated with $Pb(NO_3)_2$ (a = 7.8564 Å). Intensities estimated visually from multiple-film equi-inclination Weissenberg photographs of layers k = 0 to 4 and l = 0 to 2 taken with CuK_α radiation. Corrections made for Lorentz, polarization and absorption effects (μ = 15.2 mm^{-1}, crystal size = 0.03×0.16×0.07 mm^3). The structure was solved by Patterson and electron-density functions and 81 parameters refined by least squares to give R = 0.078 for 481 independent reflexions listed. The composition was confirmed by chemical analysis.

DI-μ-HYDROXOTETRAKIS-μ-HYDROXOBIS- (TETRAAMMINECHROMIUM (III)) BIS(DIAMMINECHROMIUM (III)) BROMINE DIHYDRATE

$[Cr_4(OH)_6(NH_3)_{12}]Br_6.2H_2O$. F.W. = 1030.0

I. The crystal structure of di-μ-hydroxotetrakis-μ-hydroxobis-(tetraminechromium(III)) bis (diamminechromium (III)) bromide dihydrate. E. BAND and T. NARASIMHAYYA, 1970. *Acta. Chem. Scand.*, 24, 275-284.

Orthorhombic, a = 22.58±1, b = 13.93±1, c = 10.00±1 Å, U = 3145 Å^3, D_m = 2.17, Z = 4, D_x = 2.17.

Space group Pbca (D_{2h}^{15}) from systematic absences

Atomic positions

All atoms in 8(*c*)

	x	*y*	*z*	*B*(Å^2)
Cr(1)	0.0302±1	0.0368±3	0.8808±3	1.51±6
Cr(2)	0.1304±1	0.9674±4	0.1437±3	1.71±7
O(1)	0.0063±6	0.0853±14	0.0605±14	1.66±28
O(2)	0.0483±6	0.9328±15	0.1943±15	1.93±29
O(3)	0.1059±6	0.0036±13	0.9369±13	1.37±27
N(1)	0.0590±9	0.9860±20	0.6932±22	2.90±45
N(2)	0.0622±8	0.1733±19	0.8278±19	2.14±37

N(3)	0.1592±10	0.9358±22	0.3415±23	3.43±49
N(4)	0.2175±9	0.9994±20	0.0858±23	3.08±47
N(5)	0.1164±9	0.1099±21	0.1930±23	2.76±42
N(6)	0.3530±10	0.1754±21	0.5917±23	3.28±49
Br(1)	0.1593±1	0.1601±3	0.5400±3	3.39±7
Br(2)	0.4700±1	0.1890±3	0.9647±2	3.05±6
Br(3)	0.3048±1	0.1342±30	0.2754±3	3.47±7
O(W)	0.2196±10	0.1554±20	0.8452±24	5.31±56

Interatomic distances (Å) *and angles*

$[Cr_4(OH)_6(NH_3)_{12}]^{6+}$ ion (crystallographic site symmetry C_i)

Environment of Cr(1)

Cr(1) - O(1)	1.97
Cr(1) - O(1)'	1.99
Cr(1) - O(2)	1.97
Cr(1) - O(3)	1.95
Cr(1) - N(2)	2.11
Cr(1) - N(1)	2.10

Angles in degrees subtended at Cr(1) by

	O(1)'	O(2)	O(3)	N(2)	N(1)
O(1)	84	85	92	174	96
O(1)'		92	86	91	178
O(2)			177	92	91
O(3)				92	92
N(2)					88

Environment of Cr(2)

Cr(2) - O(2)	1.98
Cr(2) - O(3)	1.94
Cr(2) - N(3)	2.12
Cr(2) - N(4)	2.09
Cr(2) - N(5)	2.09
Cr(2) - N(6)	2.06

Angles in degrees subtended at Cr(2) by

	O(3)	N(3)	N(4)	N(5)	N(6)
O(2)	92	89	178	89	92
O(3)		177	87	94	85
N(3)			91	88	92
N(4)				88	91
N(5)					178

Cr(1) - Cr(1)' 2.93

Cr(1) - O(1) - Cr(1)' 96°
Cr(1) - O(2) - Cr(2)' 134°
Cr(1) - O(3) - Cr(2) 135°

Primed atoms are related to umprimed by inversion through the ionic symmetry centre.

σ(Cr-O) = 0.02; σ(Cr-N) - 0.02 Å; σ(angles) = 1°

Possible hydrogen bonds

O(1) - Br(2)	3.26	N(1) - Br(2)	3.40
O(2) - Br(2)	3.22	N(4) - Br(3)	3.32
O(3) - Br(3)	3.36	N(5) - Br(2)	3.48
O(4) - N(3)	3.01	N(5) - Br(3)	3.40
O(4) - N(4)	3.24	N(5) - Br(3)	3.39
O(4) - Br(1)	3.34		

σ(O-Br) - 0.01; σ(N-Br) = 0.02 Å

Details of analysis

Lattice parameters from Guinier photographs (Cu radiation) calibrated with Si. Intensities measured visually from multiple-film Weissenberg, oscillation and precession photographs (Cu radiation, crystal size 0.12×0.24×0.19 mm^3). No absorption correction. Structure solved using three-dimensional Patterson and Fourier syntheses and refined by least squares to give R = 0.09 for 1300 reflexions listed.

POTASSIUM HEXACYANOMANGANESE (III)

I. The OD-structure of $K_3Mn(CN)_6$. N.-G. VANNERBERG, 1970. *Acta Chem. Scand.*, 24, 2335-2348.

$K_3Mn(CN)_6$ F.W. = 328.35

The crystals show stacking disorder between the two possible maximum degrees of ordering structures MDO1 and MDO2.

MDO1

Orthorhombic, a = 13.529±2, b = 10.502±1, c = 8.436±1 Å, U = 1198.8 Å^3, Z = 4, D_x = 1.82.

Space group Pnca (D_{2h}^{14})

MDO2

Monoclinic, a = 7.087±1, b = 10.502±1, c = 8.436±1 Å, β = 107.31±1°, U = 599.4 Å^3, Z = 2, D_x = 1.82.

Space group $P2_1/c$ (C_{2h}^5)

Atomic positions in $P2_1/c$ (C_{2h}^5)

	x	*y*	*z*
2 Mn in 2(*a*)	0	0	0
2 K(1) in 2(*c*)	0	0	0.5
4 K(2) in 4(*e*)	0.50266±4	0.22894±3	0.62556±4
4 C(1) in 4(*e*)	0.8752±15	0.1756±14	0.9689±19
4 C(2) in 4(*e*)	0.1919±19	0.0550±15	0.2144±20
4 C(3) in 4(*e*)	0.1879±20	0.0559±14	0.8809±20
4 N(1) in 4(*e*)	0.7972±15	0.2714±11	0.9486±18
4 N(2) in 4(*e*)	0.2959±28	0.0899±19	0.3364±22
4 N(3) in 4(*e*)	0.3020±17	0.0822±15	0.8129±21

Interatomic distances (Å) *and angles*

Mn - C(3)	1.981±17 (×2)
Mn - C(2)	1.998±14 (×2)
Mn - C(1)	2.028±14 (×2)

Angles in degrees subtended at Mn(1) by

	C(2)	C(1)
C(3)	89.7±6	89.9±6
C(3)'	90.3±6	90.1±6
C(2)		90.4±6
C(2)'		89.6±6

C(1) - N(1)	1.136±18	Mn - C(1) - N(1)	176.9±11°
C(2) - N(2)	1.134±21	Mn - C(2) - N(2)	177.3±15
C(3) - N(3)	1.156±24	Mn - C(3) - N(3)	176.4±14

K - N(1)	2.765±11 (×2)	K(2) - N(2)	2.847±18
K - N(2)	2.986±22 (×2)	K(2) - N(3)	2.869±17
K - N(3)	2.989±14 (×2)	K(2) - N(1)	2.905±15
K - C(3)	3.146±15 (×2)	K(2) - N(1)	2.932±12
K - C(2)	3.153±18 (×2)		

Discussion

The MDO2 structure is composed of a nearly cubic close packed array of 3K + 1 $[Mn(CN)_6]$ groups the anion being surrounded by 12 K ions and the K ions with 4 $[Mn(CN)_6]$ ions. The actual structure is ordered in the *bc* plane but adjacent planes stack with the *a* translation making an angle of either 107° or 77° with *c*. The MDO2 structure results when all such translations are parallel, the MDO1 structure when they alternate.

Details of analysis

Lattice parameters measured from Guinier powder photograph calibrated with KCl [value assumed for a_0 not given] using CuK_α, radiation (λ = 1.54050 Å). Intensities measured with Weissenberg camera and equi-inclination diffractometer (crystal size 0.035×0.036×0.14 mm^3 elongated along *c*, the diffractometer axis). Corrections for Lorentz, polarization and absorption effects. Structure solved from Patterson and electron-density functions of superposition structure. Final structure refined in both the MDO1 and MDO2 forms to give *R* for the latter = 0.072 for 898 reflexions listed.

COBALT CHLORIDE HEXAHYDRATE

$CoCl_2.6H_2O$ F.W. = 237.9

I. Neutron-diffraction study of $CoCl_2.6H_2O$. R. KLEINBERG, 1970. *J. Chem. Phys.*, 53, 2660-2662.

Monoclinic (isomorphous with $NiCl_2.6H_2O$,1), *a* = 10.34, *b* = 7.06, *c* = 6.67 Å, β = 122°20', *Z* = 2.

Space group $C2/m$ (C^3_{2h})

Discussion

Heavy-atom positions from the X-ray study (2) and the hydrogen positions found in $NiCl_2.6H_2O$ gave *R* = 0.12 for 28 reflexions.

Details of analysis

Crystals grown from aqueous solution. Lattice parameter from 22 neutron diffraction nuclear reflexions.

1. R. KLEINBERG, 1967. *J. Appl. Phys.*, 38, 1453.
2. *Structure Reports*, 24, 280.

TRIBROMOTRIINDIUMTETRACOBALTPENTADECACARBONYL

I. The crystal structure of tribromotriindiumtetracobaltpentadecacarbonyl. P.D. CRADWICK and D. HALL, 1970. *J. Organometal. Chem.*, 22, 203-211.

$Br_3In_3Co_4(CO)_{15}$ F.W. = 1237.95

Monoclinic, *a* = 14.25±1, *b* = 17.62±2, *c* = 16.87±2 Å, β = 133.2±5°, D_m = 2.69, *Z* = 2, D_x = 2.64.

Space group $P2_1/c$ (C^5_{2h})

Atomic positions are given together with anisotropic temperature factors (In, Br, Co) and isotropic temperature factors (C,O) in I. The structure is described in I in terms of a *B* face centered cell in which β = 99.3°

Discussion

The molecule consists of an alternating In_3Br_3 ring in chair conformation. Each In atom is bonded to a $Co(CO)_4$ group. In addition a $Co(CO)_3$ group caps the ring by forming bonds to all three In atoms. Four of the In-Br bonds do not differ significantly from 2.727 Å but two others are at 2.800±6 and 2.696±5 Å. Four of the In-Co bonds do not differ significantly from 2.58 Å but two others are at 2.532±5 and 2.637±6 Å. The Co-C distances are all 1.79±5 Å and C-O distances 1.15±5 Å. There is no apparent explanation for the deviant In-Br and In-Co bonds. The CO groups of the $Co(CO)_3$ moiety lie directly over the Br atoms and two of the $Co(CO)_4$ groups have one carbonyl directed parallel to the axis of the molecule away from the capping Co.

Details of analysis

Intensities of reflexions from layers with *l* = 0 to 11 (*B* centered cell) were measured using an ω scan on a crystal monochromated equi-inclination diffractometer. The structure was solved from the Patterson and electron-density functions and was refined by least squares to give *R* = 0.072 for 1920 observed reflexions listed.

TETRAKIS ANTIMONY COBALT TRICARBONYL

I. Organametallic pnicogen complexes. V. Preparation, structure, and bonding of the tetrameric antimony-cobalt cluster system, $Co_4(CO)_{12}Sb_4$: the first known (main group element)-(metal carbonyl) cubane-type structure. A.S. FOUST and L.F. DAHL, 1970. *J. Amer. Chem. Soc.*, 92, 7337-7341.

$Co_4(CO)_{12}Sb_4$ F.W. = 1058.85

Tetragonal, a = 14.561±3, c = 10.507±4 Å, U = 2227.7 Å^3, Z = 4, D_x = 3.15.

Space group $I4_1/amd$ (D_{2h}^{19})

Atomic positions origin at 2/*m*

		x	*y*	*z*
16 Sb	in 16(*h*)	0	0.3582±1	-0.0183±1
16 Co	in 16(*h*)	0.1416±2	0.25	0.0126±3
16 C(1)	in 16(*h*)	0.1249±16	0.25	0.1822±27
16 O(1)	in 16(*h*)	0.1200±120	0.25	0.2921±17
32 C(2)	in 32(*i*)	0.2204±12	0.3440±14	-0.0053±14
32 O(2)	in 32(*i*)	0.2736±9	0.4020±9	-0.0202±12

Anisotropic temperature factors given in I.

Interatomic distances (Å) *and angles*

Co - C(2)	1.796±19 (×2)
Co - C(1)	1.799±28
Co - Sb	2.613±3
Co - Sb	2.615±2 (×2)
C(1) - O(1)	1.16±4
C(2) - O(2)	1.16±3
Co - Sb - Co	103.5±1° (×2)
	104.1±1
Sb - Co - Sb	74.4±1 (×2)
	74.1±1

C(1) - Co - C(2)	101±1°
C(2) - Co - C(2)	99±1
Sb - Co - C(1)	91±1 (×2)
	162±1
Sb - Co - C(2)	91±1 (×2)
	162±1 (×2)
	92±1 (×2)
Co - C(1) - O(1)	176±2
Co - C(2) - O(2)	177±1

Discussion

The molecule contains 4Sb and 4Co atoms arranged at the alternate corners of a cube with each Co additionally bonded to three carbonyl groups. Within experimental error the tetramer has T_d symmetry.

Details of analysis

Preparation described in I. Lattice parameters and intensities measured on automatic diffractometer using θ-2θ scan and filtered MoK_α radiation. Correction for Lorentz and polarization effects but not absorption (transmission coefficients range from 0.36 to 0.40). Structure solved by symbolic addition and electron-density synthesis and refined by least squares to give R = 0.052 for 282 independent reflexions deposited in document NAPS-01148.

TRIDEUTERIUM COBALT HEXACYANIDE

I. Neutron diffraction study of $D_3Co(CN)_6$. H.-U. GÜDEL, A. LUDI, P. FISCHER and W. HÄLG, 1970. *J. Chem. Phys.*, 53, 1917-1923.

$D_3Co(CN)_6$ F.W. = 221.0

Trigonal, a = 6.431±4, c = 5.695±4 Å, U = 204.0 Å^3, Z = 1, D_x = 1.80.

Space group $P\bar{3}1m$ (D^1_{3d})

Atomic positions

	x	y	z	B(Å^2)
3 D in 3(g)	1/2	0	1/2	4.11±57
1 Co in 1(a)	0	0	0	1.22±148
6 C in 6(k)	0.2381±12	0	0.1924±20	1.14±14
6 N in 6(k)	0.3760±11	0	0.3202±18	1.94±12

Interatomic distances (Å) *and angles*

Sixfold coordination around Co (Crystallographic site symmetry D_{3d})

Co-C	1.883±9 (×6)
C-Co-C	90.5 ±4° (×6)
	89.5 ±4° (×6)
C-N	1.147±13
Co-C-N	176±2°
N-D	1.30±1 (but see discussion)
N-N	2.596±13
C-N-D	167±1°

Discussion

The structure of $H_3Co(CN)_6$ determined by X-rays (1) agrees with the present refinement. The N-D-N hydrogen bond is particularly short but it was not possible to tell if it is symmetrical (as the model implies) or asymmetric and disordered.

Details of analysis

Crystals grown from D_2O solution. Neutron diffraction intensities measured with Al monochromated neutrons (λ = 1.102-1.104 Å) on a powder diffractometer. Structure refined by least squares to give R_2(intensities) = 0.072 for 55 intensities listed.

1. H.-U. GÜDEL, A. LUDI and H. BÜRKI, 1968. *Helv. Chem. Acta*, 51, 1383.

MANGANESE(II)HEXACYANOCOBALT(II)HYDRATE

I. The structural chemistry of prussian blue analogs. A single-crystal study of manganese(II)hexacyanocobaltate(III), $Mn_3[Co(CN)_6]_2.xH_2O$. A. LUDI, H.-U. GÜDEL and M. RÜEGG, 1970. *Inorg. Chem.*, 9, 2224-2227.

$Mn_3[Co(CN)_6]_2.13H_2O$ F.W. = 829.10

Cubic, a = 10.421±5 Å, U = 1131.7 Å^3, D_m = 1.65, Z = 1 1/3, D_x = 1.62.

Space group Fm3m (O_h^5)

Atomic positions

	x	y	z	B(Å^2)
4 Mn in 4(*a*)	0	0	0	3.2±2
2 2/3 Co in 4(*b*)	1/2	1/2	1/2	2.3±2
16 C in 24(*e*)	0.321±2	0	0	3.2±4
16 N in 24(*e*)	0.212±2	0	0	4.0±4
8 O(1) in 192(*l*)	0.223±6	0.010±24	0.053±7	5.1±20
8 O(2) in 192(*l*)	0.305±7	0.274±17	0.259±14	5.9±20

Interatomic distances (Å)

Co-C	1.865±2	(×6)
C-N	1.136±3	
Mn-N	2.209±2	(×4)
Mn-H_2O(1)	2.391±6	(×2)
O(1)-O(2)	2.67±2	
	2.72±3	
	2.80±3	
	2.90±3	

Discussion

The structure proposed by 1 is incorrect. Mn atoms lie on a face centered lattice with $Co(CN)_6$ lying on a second lattice half a unit cell away and oriented so that Co-C-N-Mn bridges are formed. One third of the $Co(CN)_6$ ions are missing and the N atoms are replaced by H_2O. Further H_2O molecules lie at the centres of the small cubes defined by the metal atoms. The amount of water present is variable and it is possible that additional water molecules are found on the vacant Co sites.

Details of analysis

Crystals grown by slow interdiffusion of aqueous solutions of $MnCl_2$ and $H_3Co(CN)_6$. Lattice parameter from Weissenberg photograph calibrated with Si (a_o = 5.4308). Intensities measured photometrically from integrated precession photographs of layers with l = 0-4 and h=k taken with MoK_α radiation from a crystal 0.45×0.34×0.26 were corrected for Lorentz and polarization effects but not absorption (μ = 2.2 mm^{-1}). Structure refined by least squares to give R_1= 0.092

for 80 observed reflexions listed. Refinement of the model of 1 without water gave R = 0.105 and with water gave R = 0.22.

1. *Strukturbericht*, 4, 190.

μ-CHLORO-μ-AMIDO-OCTAAMMINEDICOBALT TETRACHLORIDE TETRAHYDRATE

I. The crystal structure of a dimeric cobalt compound containing a chloro bridge. R. BARRO, R.E. MARSH and W.P. SCHAEFER, 1970. *Inorg. Chem.*, 9. 2131-2137.

$(NH_3)_4CoClNH_2Co(NH_3)_4Cl_4.4H_2O$ F.W. = 519.5

Triclinic, a = 7.812±1, b = 6.666±2, c = 20.577±4 Å, α = 107.02±2, β = 90.52±2, γ = 107.14±2°, U = 973.76 Å^3, D_m = 1.77 2, Z = 2, D_x = 1.76.

Space group P$\bar{1}$ (C_i^1)

Atomic positions All atoms in 2(i). A second molecule occurs at $x, y, z+\frac{1}{2}$

	x	y	z
4 Co(1)	0.0072±1	0.0087±1	0.3276±1
2 Cl(1)	0.0340±4	0.2782±5	0.2747±2
4 Cl(2)	0.6299±2	0.3693±3	0.2833±1
4 Cl(3)	0.3115±2	0.1390±2	0.0139±1
4 N(1)	0.2674±6	0.0671±11	0.3360±3
4 N(2)	-0.2527±6	-0.0460±9	0.3264±3
4 N(3)	-0.0177±8	-0.2154±9	0.3729±4
4 N(4)	0.0437±8	0.2459±8	0.4128±3
2 N(5)	0.0250±12	0.1688±17	0.2635±4
4 O(1)	0.7031±8	0.3394±10	0.4525±3
4 O(2)	0.6731±7	0.3649±9	0.1223±3
4 H(1)	0.317±10	0.054±11	0.376±4
4 H(2)	0.312±7	0.155±9	0.327±3
4 H(3)	0.300±9	-0.058±11	0.310±3
4 H(4)	-0.273±6	-0.053±7	0.358±3
4 H(5)	-0.304±11	0.041±13	0.320±4
4 H(6)	-0.314±8	-0.184±10	0.295±3
4 H(7)	0.071±10	-0.246±10	0.377±3
4 H(8)	-0.050±8	-0.183±9	0.403±3
4 H(9)	-0.089±10	-0.355±13	0.346±4
4 H(10)	0.124±11	0.370±13	0.410±4
4 H(11)	0.096±11	0.236±13	0.448±4
4 H(12)	-0.050±8	0.278±9	0.429±3
4 H(13)	0.662±12	0.456±14	0.454±11
4 H(14)	0.648±14	0.284±17	0.476±5
4 H(15)	0.636±11	0.395±13	0.161±5
4 H(16)	0.590±11	0.303±12	0.097±4

Anistropic temperature factors given for heavy atoms, isotropic temperature factors for H.

Interatomic distances (Å) *and angles*

The $(NH_3)_4Co(NH_2)ClCo(NH_3)_4$ ion (pseudo symmetry C_i)

Co(1) - N(5)	1.869±8 (×½)
Co(1) - N(5)'	1.907±11 (×½)
Co(1) - N(4)	1.938±5
Co(1) - N(3)	1.946±8
Co(1) - N(1)	1.950±5
Co(1) - N(2)	1.953±5
Co(1) - Cl(1)	2.320±4 (×½)
Co(1) - Cl(1)'	2.337±3 (×½)

Angles in degrees subtended at Co(1) by

	N(4)	N(3)	N(1)	N(2)	Cl(1)	Cl(1)'
N(5)	166.9±4	99.6±4	93.9±3	89.8±3	81.0±4	13.5±4
N(5)	100.5±3	165.9±3	91.0±4	92.2±3	14.7±3	79.8±3
N(4)		93.5±3	86.9±2	89.9±2	85.9±2	179.4±16
N(3)			89.2±3	88.4±3	179.1±15	86.1±2
N(1)				175.8±6	91.4±2	92.6±2
N(2)					91.0±2	90.7±2

Co(1)-Cl(1)-Co(1)'	85.5±1°
Co(1)-N(5)-Co(1)'	113.7±5°

N(1)-H(1)	0.94±9
N(1)-H(2)	0.67±6
N(1)-H(3)	0.96±7

Angles in degrees subtended at N(1) by

	H(2)	H(3)	Co(1)
H(1)	116±6	89±6	116±5
H(2)		109±6	113±6
H(3)			112±4

N(2)-H(4)	0.68±6
N(2)-H(5)	0.83±10
N(2)-H(6)	0.94±7

Angles in degrees subtended at N(2) by

	H(5)	H(6)	Co(1)
H(4)	104±7	107±6	106±4
H(5)		106±6	123±5
H(6)			110±4

N(3)-H(7)	0.79±8
N(3)-H(8)	0.67±6
N(3)-H(9)	0.93±7

Angles in degrees subtended at N(3) by

	H(8)	H(9)	Co(1)
H(7)	112±7	97±7	114±5
H(8)		112±6	109±6
H(9)			113±6

N(4)-H(10)	0.90±8
N(4)-H(11)	0.86±9
N(4)-H(12)	0.86±6

Angles in degrees subtended at N(4) by

	H(11)	H(12)	Co(1)
H(10)	98±8	108±7	111±5
H(11)		102±8	118±5
H(12)			118±3

O(1)-H(13)	0.92±12
O(1)-H(14)	0.76±11
H(13)-O(1)-H(14)	105±14°
O(2)-H(15)	0.84±10
O(2)-H(16)	0.76±8
H(15)-O(2)-H(16)	106±8°

Discussion

The $(NH_3)_4Co(NH_2)ClCo(NH_3)_4$ ion consists of two CoN_5Cl octahedra bridged through shared N and Cl atoms. Adjacent molecules along *c* are disordered with respect to the arrangement of the bridging groups giving rise to a pseudo centre of symmetry in the middle of the ion. The Cl^- and H_2O molecules are held between the ions by a series of long hydrogen bonds.

Details of analysis

Lattice parameters from least-squares analysis of 21 reflexions measured on a diffractometer with $FeK_{\alpha 1}$ and $FeK_{\alpha 2}$ radiation (λ = 1.93597 and 1.93991 Å resp.). Intensities measured on G.E. single-crystal diffractometer with a spheroidal crystal of average diameter 0.22 mm using filtered Fe radiation. Lorentz, polarization and absorption corrections applied (μ = 16.28 mm^{-1}). Structure solved for pseudocell with $c' = c/2$ (the reflexions with l odd were very weak) using Patterson and Fourier syntheses. 159 parameters refined by least squares to give R_1 = 0.040 for 1018 observed reflexions with l even listed in I. The superstructure could not be determined.

THIOSULPHATE PENTAAMMINECOBALT(III)CHLORIDE MONOHYDRATE

I. Crystal and molecular structure of thiosulphatopenta-amminecobalt(III) chloride monohydrate. S. BAGGIO, 1970. *J. Chem. Soc.(London)* A, 2384-2387.

$Co(NH_3)_5(S_2O_3)Cl.H_2O$ F.W. = 309.5

Orthorhombic, a = 13.12±3, b = 7.97±2, c = 10.29±2 Å, U = 1089.0 [1075.99] Å^3, D_m = 1.87±2, Z = 4, D_x = 1.91 [1.80].

Space group Pnma (D_{2h}^{16}) Systematic absences

Atomic positions

	x	y	z
4 Co in 4(*c*)	0.1616±2	0.25	0.2458±2
4 Cl in 4(*c*)	0.1034±4	0.75	0.0147±5
4 S(1) in 4(*c*)	0.1985±3	0.25	0.0287±4
4 S(2) in 4(*c*)	0.3537±3	0.25	0.0001±4
4 O(1) in 4(*c*)	0.3629±11	0.25	-0.1394±11
8 O(2) in 8(*d*)	0.3972±6	0.4005±11	0.0594±9
4 O(3) in 4(*c*)	0.4203±23	0.75	0.2328±34
4 N(1) in 4(*c*)	0.1295±13	0.25	0.4325±14
8 N(2) in 8(*d*)	0.2609±8	0.4271±11	0.2822±10
8 N(3) in 8(*d*)	0.0578±8	0.0770±15	0.2095±11

Interatomic distances (Å) *and angles*

(The distances listed in I for the $Co(NH_3)_5(S_2O_3)$ ion are all about 0.003 Å shorter than those given below)

Co - N(2)	1.959±10 (×2)
Co - N(1)	1.967±15
Co - N(3)	1.976±11 (×2)
Co - S(1)	2.286±5

Angles in degrees subtended at Co by

	N(2)'	N(1)	N(3)	N(3)'	S(1)
N(2)	92.2±5	87.5±5	178.1±46	89.6±6	92.6±3
N(1)			92.1±5	92.1±5	179.9±21
N(3)				88.5±6	87.8±3

S(2) - O(1)	1.441±12
S(2) - O(2)	1.462±9 (×2)
S(2) - S(1)	2.062±5
Co-S(1)-S(2)	110.5±2°

Angles in degrees subtended at S(2) by

	O(2)	S(1)
O(1)	112.5±5	103.0±6
O(2)	110.2±6	109.1±4

Cl - N(2)	3.302±11 (×2)
Cl - N(3)	3.342±13 (×2)
Cl - N(3)	3.423±12 (×2)

Discussion

The $Co(NH_3)_5S_2O_3$ ion consists of an octahedral arrangement of ligands around Co, the thiosulphate group being bonded through its terminal S atom. The Cl ion and water molecule lie between them. All possible hydrogen bonds (except N(2)-O(2)) are larger than 3 Å.

Details of analysis

Lattice parameters from zero-layer Weissenberg photographs ($\lambda(MoK_\alpha)$ = 0.7107 Å). Intensities measured with a microdensitometer from multiple-film equi-inclination integrated precession photographs of layers with k = 0-9 and integrated precession photographs of layers with l = 0,1 and h = 0 taken with MoK_α radiation from a crystal 0.08×0.7×0.08 mm^3. Lorentz and polarization corrections but absorption correction not necessary. Structure solved from three-dimensional Patterson function and refined by least squares to give R_2 = 0.101 for 738 observed reflexions listed in <u>1</u>.

1. Supplementary Publication Scheme SUP 20016, National Lending Laboratory, Boston Spa, UK.

μ-AMIDO-μ-NITRO-BIS TETRAAMMINECOBALT(III) TETRACHLORIDE TETRAHYDRATE

I. The structure of μ-amido-μ-nitro-bis[tetraamminecobalt(III)] tetrachloride tetrahydrate. U. THEWALT and R.E. MARSH, 1970. *Inorg. Chem.*, 9, 1604-1610.

$[(NH_3)_4Co(NH_2)(NO_2)Co(NH_3)_4]Cl_4.4H_2O$ F.W. = 530.0

Monoclinic, a = 10.328±2, b = 15.103±2, c = 7.756±1 Å, β = 124.10±2°, U = 1001.8 Å^3, D_m = 1.76±1, Z = 2, D_x = 1.757.

Space group $P2_1/m$ (C^2_{2h}) Systematic absences and statistical tests.

Atomic positions

	x	*y*	*z*
2 Co(1) in 2(*e*)	0.2253±1	0.25	0.2995±2
2 Co(2) in 2(*e*)	-0.0932±1	0.25	0.2971±2
4 Cl(1) in 4(*f*)	0.0502±2	0.5838±1	0.2043±2
4 Cl(2) in 4(*f*)	0.5039±2	0.3947±1	0.1828±2
4 N(1) in 4(*f*)	0.2336±5	0.3794±3	0.3003±6
2 N(2) in 2(*e*)	0.4562±7	0.25	0.4664±9
2 N(3) in 2(*e*)	0.2114±7	0.25	0.0370±9
4 N(4) in 4(*f*)	-0.0921±5	0.3800±3	0.3054±7
2 N(5) in 2(*e*)	-0.1795±7	0.25	0.4672±10
2 N(6) in 2(*e*)	-0.3018±7	0.25	0.0457±9
2 N(7) in 2(*e*)	-0.0001±7	0.25	0.1409±9
2 N(8) in 2(*e*)	0.2324±7	0.25	0.5516±9
2 O(1) in 2(*e*)	0.1064±5	0.25	0.5528±7
2 O(2) in 2(*e*)	0.3542±6	0.25	0.7262±7
4 O(3) in 4(*f*)	0.3661±5	0.5929±3	0.2235±7
4 O(4) in 4(*f*)	0.2581±5	0.4457±3	0.6770±6

Anistropic temperature factors and hydrogen atom positions also given.

Interatomic distances (Å) *and angles*

Co(1) - N(8)	1.915±8
Co(1) - N(7)	1.930±6
Co(1) - N(1)	1.956±4 (×2)
Co(1) - N(3)	1.959±8
Co(1) - N(2)	1.976±6

Angles in degrees subtended at Co(1) by

	N(7)	N(1)'	N(3)	N(2)
N(8)	89.6±3	91.2±2	178.3±7	89.4±3
N(7)		92.1±1	88.8±3	179.0±24
N(1)		175.2±3	88.9±2	87.9±1
N(3)				92.2±3

Co(2) - O(1)	1.897+4
Co(2) - N(7)	1.923±9
Co(2) - N(6)	1.932±5
Co(2) - N(5)	1.963±9
Co(2) - N(4)	1.964±4 (x2)

Angles in degrees subtended at Co(2) by

	N(7)	N(6)	N(5)	N(4)'
O(1)	91.4±2	176.7±9	86.2±2	89.0±1
N(7)		91.9±3	177.6±4	91.4±2
N(6)			90.5±3	90.9±1
N(5)				88.5±2
N(4)				176.6±4

N(8) - O(2)	1.227±6
N(8) - O(1)	1.307±10

Co(1)-N(8)-O(2)	123.7±7°
Co(1)-N(8)-O(1)	122.6±4
O(1)-N(8)-O(2)	113.7±7
N(8)-O(1)-Co(2)	119.7±4

Bonding of water molecules (σ = 0.005 Å, 0.2°)

O(3)...H-N(6)	2.979
O(3)...H-N(2)	3.136
O(3)-H...Cl(1)	2.185
O(3)-H...Cl(2)	3.404

Angles in degrees subtended at O(3) by

	N(2)	Cl(1)	Cl(2)
N(6)	74.9	100.5	125.2
N(2)		102.3	129.9
Cl(1)			115.4

O(4)...H-N(1)	2.962
O(4)...H-N(4)	3.269
O(4)-H...Cl(2)	3.169
O(4)-H...Cl(2)	3.356

Angles in degrees subtended at O(4) by

	N(4)	Cl(2)	Cl(2)
N(1)	124.8	102.5	131.7
N(4)		71.9	102.3
Cl(2)			80.8

[There are an additional 4 atoms within 3.3 Å of O(4)]

Hydrogen bonds from N atoms

N(1)-H...O(4)	2.962±7	
N(1)-H...Cl(2)	3.396±6	
N(1)-H...Cl(1)	3.370±4	
N(2)-H...O(3)	3.136±5	(×2)
N(3)-H...Cl(1)	3.382±4	(×2)
N(4)-H...O(4)	3.186±7	
N(5)-H...Cl(1)	3.280±4	(×2)
N(6)-H...O(3)	2.979±6	(×2)

Discussion

The cation, $(NH_3)_4Co(NH_2)(NO_2)Co(NH_3)_4$, lies in a mirror plane with the Co atoms bridged in the plane by -N(7)- of the amide group and -O(1)-N(8)- of the NO_2 group.

Details of analysis

Lattice parameters from Straumanis Weissenberg photographs using CuK_α radiation ($\lambda(\alpha_1)$ = 1.54051, $\lambda(\alpha_2)$ = 1.54433 Å). Intensities measured with filtered Fe radiation on four-circle diffractometer using θ-2θ scan (crystal 0.06×0.04×0.05 mm^3). Lorentz and polarization but no absorption correction (μ = 13.8 mm^{-1}). Structure determined from Patterson function and electron-density maps and refined by least squares to give R = 0.048 for 1071 reflexions listed. Hydrogen atom parameters estimated from difference electron-density maps and not refined.

HEXAAMMINECOBALT(III) TETRACHLOROZINCATE (II) CHLORIDE

$[Co(NH_3)_6][ZnCl_4]Cl$ F.W. = 403.8

I. The crystal structure of hexaamminecobalt (III) tetrachlorozincate (II) chloride, $[Co(NH_3)_6][ZnCl_4]Cl$. D.W. MEEK and J.A. IBERS, 1970. *Inorg. Chem.*, 9, 465.

Orthorhombic, a = 21.219±9, b = 7.653±4, c = 8.106±3 Å, U = 1316 Å^3, D_m = 2.02±1, Z = 4, D_x = 2.037.

Space group Pnma (D_{2h}^{16}) from systematic absences and final structure

Atomic positions

		x	*y*	*z*
4 Zn	in 4(*c*)	0.09188±3	¼	0.13355±8
4 Co	in 4(*c*)	0.35909±3	¼	0.11604±8
4 Cl(1)	in 4(*c*)	0.05777±6	¼	0.41035±18
8 Cl(2)	in 8(*d*)	0.15259±5	0.00606±13	0.12585±11
4 Cl(3)	in 4(*c*)	0.00975±6	¼	0.03664±19
4 Cl(4)	in 4(*c*)	0.19895±6	¼	0.23741±18
4 N(1)	in 4(*c*)	0.2997 ±2	¼	0.3025 ±6
8 N(2)	in 8(*d*)	0.4101 ±1	0.0672 ±4	0.2248 ±4
8 N(3)	in 8(*d*)	0.3081 ±1	0.0697 ±4	0.0097 ±4
4 N(4)	in 4(*c*)	0.4166 ±2	¼	-0.0753 ±6
4 H(11)	in 4(*c*)	0.2621	¼	0.2694
8 H(12)	in 8(*d*)	0.3053	0.1602	0.3630
8 H(21)	in 8(*d*)	0.3903	-0.0301	0.2280
8 H(22)	in 8(*d*)	0.4447	0.0510	0.1766
8 H(23)	in 8(*d*)	0.4181	0.0950	0.3255
8 H(31)	in 8(*d*)	0.2735	0.1098	-0.0234
8 H(32)	in 8(*d*)	0.3276	0.0275	-0.0745
8 H(33)	in 8(*d*)	0.3011	-0.0148	0.0750
8 H(41)	in 8(*d*)	0.4105	0.3398	-0.1362
4 H(42)	in 4(*c*)	0.4547	¼	-0.0459

Interatomic distances (Å) *and angles*

$Co(NH_3)_6$ ion (crystallographic point symmetry C_s)

Co - N(1)	1.968±5	
Co - N(2)	1.977±3	(×2)
Co - N(2)	1.953±3	(×2)
Co - N(4)	1.974±5	

Angles in degrees subtended at Co by

	N(1)	N(2)	N(2)'	N(3)'
N(2)	90.4		90.1	179.55
N(3)	89.1	90.0		89.9
N(4)	178.4	90.7		89.7

$\sigma < 0.2°$

$ZnCl_4$ ion (crystallographic point symmetry C_s)

Zn - Cl(1)	2.358±2	
Zn - Cl(2)	2.269±1	(×2)
Zn - Cl(3)	2.223±2	

Angles in degrees subtended at Zn by

	Cl(1)	Cl(2)'
Cl(2)	101.56	110.72
Cl(3)	110.49	115.35

$\sigma = 0.05°$

Hydrogen bonds

N(2) - H(1)...Cl(4)	3.368
N(3) - H(1)...Cl(4)	3.359
N(3) - H(2)...Cl(2)	3.273
N(3) - H(3)...Cl(4)	3.196
N(4) - H(2)...Cl(1)	3.280

Discussion

The $Co(NH_3)_6$ ion is an almost regular octahedron but the deviations are possibly significant. The deviations from I_d symmetry of the $ZnCl_4$ ion on the other hand are definitely significant but are not large enough for the ion to be formulated as $ZnCl_5$ since the Zn-Cl(4) distance (3.769 Å) is too long to be considered as a covalent bond. The Cl(4) ion is surrounded trigonally by six N - H ...Cl hydrogen bonds.

Details of analysis

Prepared from $Co(NH_3)_6Cl_3$ and $ZnCl_2$ in aqueous HCl solution. Lattice parameters and intensities measured on a four-circle diffractometer (MoK_α radiation, λ = 0.70930 Å) the former obtained by least-squares refinement of 10 accurately located reflexions. Crystal 0.21×0.10×0.48 mm^3 mounted with the long axis (*b*) along goniometer head axis, θ-2θ scan to measure 1637 reflexions with 2θ < 56°. Lorentz, polarization and absorption corrections. Structure determined from three-dimensional Patterson and Fourier synthesis and refined by least squares to give weighted *R* = 0.046. Although hydrogen atoms were included, their positions were assumed and not refined. 1329 Reflexions listed.

TRIFLUOROSILYLTETRACARBONYLCOBALT

$Co(SiF_3)(CO)_4$ F.W. = 256.1

I. The crystal and molecular structure of trifluorosilyltetracarbonylcobalt. K. EMERSON, P.R. IRELAND and W.T. ROBINSON, 1970. *Inorg. Chem.*, 9, 436-440.

Orthorhombic, *a* = 8.224±4, *b* = 4.418±10, *c* = 18.591±7 Å, *U* = 675.5 Å^3, *Z* = 8, D_x = 1.95.

Space group Cmca (D_{2h}^{18}) from systematic absences and statistical analysis of intensities

Atomic positions

		x	*y*	*z*
8 Co	in 8(*f*)	0	0.1662±2	0.1308±1
8 Si	in 8(*f*)	0	-0.0281±6	0.1410±2
8 F(1)	in 8(*f*)	0	-0.0932±9	0.0708±6
16 F(2)	in 16(*g*)	0.1454±12	-0.0779±7	0.1820±6
8 C(1)	in 8(*f*)	0	0.1656±15	0.2272±10
8 O(1)	in 8(*f*)	0	0.1639±12	0.2871±7
16 C(2)	in 16(*g*)	0.1850±16	0.1482±9	0.0820±6
16 O(2)	in 16(*g*)	0.3022±12	0.1375±8	0.0512±5
8 C(3)	in 8(*f*)	0	0.3237±18	0.1253±11
8 O(3)	in 8(*f*)	0	0.4222±12	0.1184±8

Individual anisotropic temperature factors given.

Interatomic distances (Å) *and angles*

Fivefold coordination around Co
(crystallographic point symmetry C_s)

Co - Si	2.226±5	
Co - C(1)	1.78 ±2	
Co - C(2)	1.78 ±1	(×2)
Co - C(3)	1.80 ±2	

Angles in degrees subtended at Co by

	Si	C(1)	C(2)
C(1)	84.9		
C(2)	85.9	120.5	117.3
C(3)	178.4	93.5	94.9

σ = 0.5°

Fourfold coordination around Si
(crystallographic point symmetry C_s)

Si - Co	2.226±5	
Si - F(1)	1.50 ±1	
Si - F(2)	1.53 ±1	(×2)

Angles in degrees subtended at Si by

	Co	F(2)
F(1)	114.9	104.3
F(2)	114.4	103.2

σ = 0.5°

	C - O	Co - C - O
C(1) - O(1)	1.11±2	179±2°
C(2) - O(2)	1.13±1	180±1
C(3) - O(3)	1.13±2	177±2

Discussion

The crystal is composed of discrete molecules of $F_3SiCo(CO)_4$ having symmetry not differing significantly from C_{3v}. One carbonyl group (C(3)) is directed along the Si - Co axis of the molecule and the other three are staggered with respect to the three fluorine atoms. The molecule is librating about its axis with a large amplitude which suggests that the Si-F distances quoted above are likely to be as much as 0.1 Å too short. The Co-C and C-O distances are also probably a little short but not by more than 0.02 Å. The Co atom lies just a little above the centre of a trigonal bipyramid formed by its five ligands, the Si atom just a little above the centre of the tetrahedron formed by its ligands, the displacement in both cases being towards the axial carbonyl group.

Details of analysis

Prepared by the method of 1. Lattice parameters from least-squares refinement of measurements from precession photographs with CoK_α radiation. Intensities measured photometrically from integrated Weissenberg photographs of five layers with the crystal mounted about [110]. Lorentz and polarization corrections, films scaled by means of common reflexions. No correction was possible for absorption although some reflexions might have been significantly affected. Structure determined by symbolic addition and refined by least-squares analysis. Final weighted R = 0.089 for 375 observed reflexions listed.

1. A.J. CHALK and J.F. HARROD, 1965. *J. Amer. Chem. Soc.* 87, 1133.

HYDRIDOTETRAKIS(TRIFLUOROPHOSPHINE)COBALT(I)

I. The structure of hydridotetrakis(trifluorophosphine)cobalt(I), $CoH(PF_3)_4$. B.A. FRENZ and J.A. IBERS, 1970. *Inorg. Chem.*, 9, 2403-2408.

$CoH(PF_3)_4$ F.W. = 411.8

Monoclinic, a = 13.02±1, b = 7.63±1, c = 13.08±1 Å, β = 122.4±1°, U = 1097 Å^3, Z = 4, D_x = 2.49.

Space group C2/c (C^6_{2h}) Systematic absences and final structure

Atomic positions

	x	y	z
4 Co in 4(*e*)	0	0.1477±3	0.25
8 P(1) in 8(*f*)	-0.1304±3	0.0095±4	0.1029±3
8 P(2) in 8(*f*)	-0.0816±4	0.3175±4	0.3072±3
8 F(11) in 8(*f*)	-0.2681±7	0.0442±12	0.0477±9
8 F(12) in 8(*f*)	-0.1393±9	0.0210±13	-0.0182±6
8 F(13) in 8(*f*)	-0.1349±8	-0.1947±9	0.1102±8
8 F(21) in 8(*f*)	-0.0857±9	0.5172±12	0.2876±10
8 F(22) in 8(*f*)	-0.2188±9	0.2900±14	0.2581±10
8 F(23) in 8(*f*)	-0.0341±10	0.3238±12	0.4440±7

Anisotropic temperature factors given.

Interatomic distances (Å) *and angles*

Co - P(1)	2.049±3 (×2)
Co - P(2)	2.055±5 (×2)

Angles in degrees subtended at Co by

	P(1)'	P(2)'
P(1)	118.0±2	109.6±2
P(1)'		108.2±2
P(2)		101.8±2

P(1) - F(12)	1.528±10
P(1) - F(11)	1.556±9
P(1) - F(13)	1.564±7

Angles in degrees subtended at P(1) by

	F(11)	F(13)
F(12)	94.7±6	97.8±6
F(11)		97.2±5

P(2) - F(21)	1.541±10
P(2) - F(23)	1.549±10
P(2) - F(22)	1.556±12

Angles in degrees subtended at P(2) by

	F(23)	F(22)
F(21)	96.8±6	97.8±6
F(23)		98.1±8

Discussion

The Co atom lies on a two-fold axis and is surrounded by a distorted tetrahedron of PF_3 groups with the F atoms lying in the planes defined by the Co atom and the tetrahedron edges. The H atom was not found but could lie below P(1) on the two-fold axis or more likely on a tetrahedral face, in which case the molecules would have to be disordered.

Details of analysis

Crystals grown at -125±6°C in 0.3 mm quartz capillary tubes. Lattice parameters from precession photographs with MoK_α radiation (λ = 0.7107 Å). Intensities estimated visually from precession photographs with l = 0, k = h-4, h-2, h, h+2, h+4, h-2l, h+l, h+2l, h+2l+2 and h+4l/3 taken with MoK_α radiation corrected for Lorentz and polarization but not absorption or extinction (μ = 2.36 mm^{-1}, crystal size 0.6×0.3 mm(diameter)). Structure determined from Patterson function and difference-Fourier synthesis and 78 parameters refined by least squares to give R_2 = 0.109 for 664 inequivalent observed reflexions listed.

TETRASODIUM CUPRAMMONIUM BIS COPPERDITHIOSULPHATE HYDRATE

$Na_4Cu(NH_3)_4[Cu(S_2O_3)_2]_2 \cdot H_2O$

I. The electronic properties and crystal structure of $Na_4Cu(NH_3)_4[Cu(S_2O_3)_2]_2 \cdot L$ (L = H_2O or NH_3). B.J. HATHAWAY and F. STEPHENS, 1970. *J. Chem. Soc. Lond. A*, 884-888.

The structure proposed by 1 has been further refined independently of 2 to give R = 0.076.

By synthesizing the two isomorphous crystals with respectively water and ammonia of crystallization, I argue that the crystal studied by 1 was the hydrate and not the ammoniate. In either case the H_2O or NH_3 is to be found at (0,0,½).

1. A. FERRARI, A. BRIABANTI and A. TIRIPICCHIO, 1966. *Acta Cryst.*, 21, 605.
2. B. MOROSIN and A.C. LARSON, 1969. *Acta Cryst.*, B25, 1417.

TETRAAMMINECOPPER (II) DIHALOCUPRATES (I)

$Cu(NH_3)_4(CuI_2)_2$ F.W. = 766.4

I. The structures of the tetraamminecopper (II) dihalocuprates (I). I. The structure of $Cu(NH_3)_4(CuI_2)_2$. J.A. BAGLIO, H.A. WEAKLIEM, F. DEMELIO and P.A. VAUGHAN, 1970. *J. Inorg. Nucl. Chem.*, 32, 735-801.

Triclinic, a = 7.386±2, b = 9.985±2, c = 6.931±2 Å, α = 102.58±2°, β = 81.85±2°, γ = 133.52±2°, U = 341.6 Å^3, D_m = 3.81, Z = 1, D_x = 3.73.

Space group $P\bar{1}$ (C_i^1)

Atomic positions

	x	y	z
1 Cu(1) in 1(*a*)	0	0	0
2 Cu(2) in 2(*i*)	0.0019±18	0.4966±17	0.2457±17
2 I(1) in 2(*i*)	0.9100±7	0.6692±7	0.1194±6
2 I(2) in 2(*i*)	0.3697±7	0.7335±6	0.5345±6
2 N(1) in 2(*i*)	0.0842±89	0.1760±71	0.2839±74
2 N(2) in 2(*i*)	0.6390±99	0.8918±83	0.0759±86

Anisotropic temperature factors given.

Interatomic distances (Å) *and angles*

Six-fold coordination around Cu(1) (= Cu(II)) (crystallographic site symmetry C_i)

Cu(1) - N(1)	2.16±7
Cu(1) - N(2)	2.11±7
Cu(1) - I(1)	3.171±6

Angles in degrees subtended at Cu(1) by

	N(2)	I(1)
N(1)	81	102
N(2)		95

σ = 2°

Four-fold coordination around Cu(2) (=Cu(I))

Cu(2) - I(1)	2.575±15
Cu(2) - I(1)'	2.959±15
Cu(2) - I(2)	2.760±15
Cu(2) - I(2)'	2.606±15

Angles in degrees subtended at Cu(2) by

	I(1)'	I(2)	I(2)'
I(1)	103.8	110.3	107.9
I(1)'		118.9	117.2
I(2)			98.5

σ = 0.5°

Cu(1) - Cu(2)	4.870±14
Cu(2) - Cu(2)'	3.506±20
Cu(2) - I(1) - Cu(2)'	76.2±5°
Cu(2) - I(2) - Cu(2)"	81.6±5

Details of analysis

Crystals (0.3×0.3×0.6 mm^3)coated with petroleum jelly. Intensities estimated visually from multiple-film equi-inclination Weissenberg photographs (CuK_α radiation). Lorentz and polarization corrections but not absorption or extinction. Structure determined from Patterson projections and three-dimensional difference electron density and refined by least squares to give R = 0.13 for 900 reflexions listed.

II. The structures of the tetraamminecopper (II) dihalocuprates (I) II. The structures of $Cu(NH_3)_4(CuBr_2)_2$ and $Cu(NH_3)_4(CuCl_2)_2 \cdot H_2O$. J.A. BAGLIO and P.A. VAUGHAN, 1970. *J. Inorg. Nucl. Chem.*, 32, 803-810.

$Cu(NH_3)_4(CuBr_2)_2$ F.W. = 578.37

Tetragonal, a = 10.37±2, c = 5.713±8 Å, U = 615.0 Å^3, Z = 2, D_x = 3.12.

Space group I4/mmm (D_{4h}^{17}) systematic absences and final structure.

Atomic positions

		x	y	z
2 Cu(1)	in 2(*a*)	0	0	0
4 Cu(2)	in 4(*d*)	0	½	¼
8 Br	in 8(*j*)	0.1984±3	½	0
8 N	in 8(*h*)	0.1362±19	0.1362±19	0

Anisotropic temperature factors given.

$Cu(NH_3)_4(CuCl_2)_2 \cdot H_2O$ F.W. = 418.60

Tetragonal, a = 10.19±2, c = 5.468±8 Å, U = 567.8 Å^3, Z = 2, D_x = 2.44.

Space group I4/mmm(D_{4h}^{17}) systematic absences and final structure.

Atomic positions

		x	y	z
2 Cu(1)	in 2(*a*)	0	0	0
4 Cu(2)	in 4(*d*)	0	½	¼
8 Cl	in 8(*j*)	0.1887±6	½	0
8 N	in 8(*h*)	0.1405±14	0.1405±14	0
2 O	in 2(*b*)	0	0	½

Anisotropic temperature factors given.

Interatomic distances (Å) *and angles*

	$Cu(NH_3)_4(CuB_2)_2$	$Cu(NH_3)_4(CuCl_2)_2 \cdot H_2O$
Square planar coordination around Cu(1) (= Cu(II)) (crystallographic site symmetry D_{4h})		
Cu(1) - N (×4)	2.00±3	2.02±2
Cu(1) - X (×4)	4.234±13	4.186±7
Cu(1) - O (×2)		2.734±4
Four-fold coordination around Cu(2) (=Cu(I)) (crystallographic site symmetry D_{2d})		
Cu(2) - X	2.503±4	2.359±6
X - Cu(2) - X	110.4±2°	109.2±1°
	109.0±1	109.6±1
Cu(2) - X - Cu(2)	69.6±1	70.8±1

Discussion

The three crystals have similar structures. In all three there are chains of edge sharing $Cu^{I}I_4$ tetrahedra running along the *c* axis. Between the chains (and in the bromide and chloride perpendicular to them) lie the planar $Cu^{II}(NH_3)_4$ groups. In the chloride two water molecular complete a distorted octahedral environment around Cu^{II}. In the iodide a similar effect is achieved by a distortion of the structure so that two iodine atoms from the chain are brought close to Cu^{II}. However in the bromide Cu^{II} has no close neighbours other than the four NH_3 groups.

Details of analysis

Lattice parameters from least-squares refinement of measurements from Weissenberg photographs. Intensities measured from multiple-film Weissenberg photographs (CuK_α radiation, crystal 0.2 mm wide, 0.4 mm long). Twelve layers recorded around (110) and four around (010) for Br complex and eight layers recorded around (010) for Cl complex. Corrections for Lorentz and polarization effects but not for absorption or extinction. Structure from Patterson and Fourier sections refined by least squares to give R = 0.104 for 158 reflexions listed (Br complex) and R = 0.097 for 145 reflexions listed (Cl complex).

AMMONIUM μ-DIHYDROXY HEXACARBONATODIZIRCONIUM(IV)

I. [Crystal structure of $(NH_4)_6[Zr_2(OH)_2(CO_3)_6].3H_2O$.] JU.E. GORBUNOVA, G.N. NOVICKAJA and V.G. KUZNECOV, 1970. *Ž. Strukt. Khim.*, 11, 560-562 [*J. Struct. Chem.*, 11, 523-524].

$(NH_4)_6[Zr_2(OH)_2(CO_3)_6].3H_2O$ F.W. = 738.78

Monoclinic, a = 18.06±5, b = 11.58±3, c = 6.99±2 Å, β = 114±1°, U = 1335 Å^3, Z = 2, D_x = 1.83.

Space group $P2_1/a$ (C_{2h}^5)

Atomic positions All atoms in 4(e) at ±(x,y,z;½+x,½−y,z)

	x	y	z
Zr	0.059	0.115	0.475
O(1)	0.471	0.439	0.592
O(2)	0.290	0.046	0.821
O(3)	0.168	0.077	0.801
O(4)	0.179	0.059	0.524
O(5)	0.048	0.274	0.664
O(6)	0.126	0.272	0.460
O(7)	0.134	0.422	0.698
O(9)	0.069	0.117	0.183
O(9)	0.435	0.321	0.872
O(10)	0.446	0.317	0.191
O(11)	0.473	0.060	0.205
O(12)	0.318	0.193	0.176
C(1)	0.217	0.054	0.713
C(2)	0.100	0.312	0.574
C(3)	0.000	0.148	0.069
N(1)	0.159	0.303	0.134
N(2)	0.300	0.462	0.839
N(3)	0.393	0.101	0.640

Discussion

Apparently isostructural with $K_6[Zr_2(OH)_2(CO_3)_6]6H_2O$ (1). The interatomic distances are not sufficiently accurate to be meaningful. $\overline{Zr\text{-}Zr}$ = 3.49 Å.

Details of analysis

Intensities estimated visually from RKOP and KFOR photographs of layers with k = 0 and l = 0 to 3 using Mo and Cu radiation. Lorentz and polarization corrections. Structure from Patterson and electron-density functions, one carbon atom not found (position assumed). R = 0.19 for 560 observed reflexions.

1. JU.E. GORBUNOVA, V.G. KUZNECOV and E.S. KOVALEVA, 1968. *Ž. Strukt. Khim.*, 9, 918.

AMMONIUM OCTACHLORODIMOLYBDATE CHLORIDE HYDRATE

$(NH_4)_5Mo_2Cl_9H_2O$ F.W. = 619.10

I. Stoichiometric and structural characterization of the compound $(NH_4)_5Mo_2Cl_9$-H_2O. J.V. BRENCIC and F.A. COTTON, 1970, 1970. *Inorg. Chem.*, 9, 346-351.

Orthorhombic, a = 10.528±4, b = 13.219±5, c = 13.129±5 Å, U = 1827 Å^3, D_m = 2.22±4, Z = 4, D_x = 2.25.

Space group $Cmc2_1$ (C_{2v}^{12}) from systematic absences and refined structure

Atomic positions

	x	y	z
4 Mo(1) in 4(*a*)	0	0.0798±3	0
4 Mo(2) in 4(*a*)	0	0.1830±2	0.1265±4
4 Cl(1) in 4(*a*)	0	0.9055±6	0.0712±9
8 Cl(2) in 8(*b*)	0.2279±7	0.0496±5	0.9687±8
4 Cl(3) in 4(*a*)	0	0.1894±8	0.8529±10
4 Cl(4) in 4(*a*)	0	0.3562±6	0.0521±10
8 Cl(5) in 8(*b*)	0.2292±6	0.2099±5	0.1636±8
4 Cl(6) in 4(*a*)	0	0.0762±8	0.2821±13
4 Cl(7) in 4(*a*)	0	0.3448±11	0.3414±12
4 O(1) in 4(*a*)	0	0.5518±28	0.3032±33
4 N(1) in 4(*a*)	0	0.6309±25	0.1081±23
8 N(2) in 8(*b*)	0.2514±23	0.4508±17	0.2181±23
8 N(3) in 8(*b*)	0.2274±22	0.1957±18	0.4222±26

Anisotropic temperature factors given for all atoms.

Interatomic distances (Å) *and angles*

$Mo_2Cl_8^{4-}$ ion (crystallographic symmetry C_{2h})

Mo(1) - Mo(2)	2.150±5
Mo(1) - Cl(1)	2.486±9
Mo(1) - Cl(2)	2.467±7 (×2)
Mo(1) - Cl(3)	2.415±13

Angles in degrees subtended at Mo(1) by

	Cl(1)	Cl(2)'	Cl(3)
Mo(2)	107.3±2	103.4±2	103.7±3
Cl(1)		85.0±2	149.0±4
Cl(2)		153.2±4	87.9±3

Mo(2) - Cl(4)	2.490±10
Mo(2) - Cl(5)	2.487±7 (×2)
Mo(2) - Cl(6)	2.483±16

Angles in degrees subtended at Mo(2) by

	Cl(4)	Cl(5)'	Cl(6)
Mo(1)	106.3±3	104.0±2	106.0±4
Cl(4)		86.8±2	147.7±5
Cl(5)		151.9±2	85.4±3

Environment of Cl(7) (crystallographic symmetry C_2)

Cl(7) - O(1)	2.78
Cl(7) - N(3)	3.27 (×2)
Cl(7) - N(2)	3.40 (×2)
Cl(7) - N(1)	3.51
Cl(7) - Mo(2)	3.54

Environment of O(1) (crystallographic symmetry C_s)

O(1) - Cl(7)	2.78
O(1) - N(1)	2.76
O(1) - N(2)	3.16 (×2)

Environment of N(1) (crystallographic symmetry C_s)

N(1) - O(1)	2.76
N(1) - Cl(1)	3.66
N(1) - Cl(3)	3.99
N(1) - Cl(4)	3.70

Environment of N(2)

N(2) - O(1)	3.16
N(2) - Cl(4)	3.64
N(2) - Cl(5)	3.27
N(2) - Cl(7)	3.40

Environment of N(3)

N(3) - Cl(1)	3.36
N(3) - Cl(2)	3.29
N(3) - Cl(5)	3.40
N(3) - Cl(6)	3.40
N(3) - Cl(7)	3.27

Discussion

The $Mo_2Cl_8^{6-}$ ion has effectively D_{4h} symmetry with each Mo atom bonded to the other Mo atom and four Cl atoms, arranged at the apices of a square pyramid. A fifth chlorine atom lies 3.53 Å away from each molybdenum atom along the axis of the ion. The remainder of the structure consists of free Cl^- ions, NH_4^+ ions and water molecules, the latter presumably hydrogen bonded to the free Cl.

Details of analysis

Prepared by the treatment of dimolybdenum tetracetate with NH_4Cl in concentrated HCl. Lattice parameters and intensities of 534 independent reflexions were measured from a crystal, 0.02×0.02×0.06 mm^3, elongated along *c*, mounted on a GE-XRD5 diffractometer using CuK_α radiation. Lorentz and polarization corrections made, but not absorption. Structure solved from three-dimensional Patterson function and difference Fourier; refined by weighted least squares to give weighted R = 0.092. Structure factors given.

SODIUM HEXAMOLYBDOCHROMATE(III) OCTAHYDRATE

I. The crystal structure of sodium hexamolybdochromate(III) octahydrate. A. PERLOFF, 1970. *Inorg. Chem.*, 9, 2228-2239.

$Na_3(CrMo_6O_{24}H_6).8H_2O$ F.W. = 1236.5

Triclinic, *a* = 10.908±4, *b* = 10.9807±4, *c* = 6.4679±2 Å, α = 107.59±2, β = 84.438±2, γ = 112.465±3°, *U* = 682.28 Å^3, D_m = 2.95, *Z* = 1, D_x = 2.99.

Space group $P\bar{1}$ (C_i^1) Absence of piezoelectric effect and structure

Atomic positions

		x	*y*	*z*
1 Cr	in 1(*a*)	0	0	0
2 Mo(1)	in 2(*i*)	0.27552±3	-0.06622±3	-0.00253±5
2 Mo(2)	in 2(*i*)	0.28333±3	0.18775±3	-0.18424±5
2 Mo(3)	in 2(*i*)	0.00896±3	0.25164±3	-0.18745±5
1 Na(1)	in 1(*f*)	0.5	0	0.5
2 Na(2)	in 2(*i*)	0.4368±2	-0.4761±2	0.2798±3
2 O(1)	in 2(*i*)	0.1032±2	-0.0218±3	0.2137±4
2 O(2)	in 2(*i*)	0.1378±2	-0.0316±3	-0.2052±4
2 O(3)	in 2(*i*)	0.1096±2	0.1956±2	0.0359±4
2 O(4)	in 2(*i*)	0.1257±3	-0.2381±3	-0.0254±4

2	O(5) in 2(*i*)	0.3346±3	0.1285±3	0.0352±4
2	O(6) in 2(*i*)	0.1442±3	0.1840±3	-0.3545±4
2	O(7) in 2(*i*)	0.0745±3	-0.2336±3	0.4139±5
2	O(8) in 2(*i*)	0.3595±3	-0.0690±3	0.2036±6
2	O(9) in 2(*i*)	0.3535±3	-0.1189±3	-0.2347±5
2	O(10) in 2(*i*)	0.3737±3	0.1349±4	-0.3964±5
2	O(11) in 2(*i*)	0.3685±3	0.3616±3	-0.0824±5
2	O(12) in 2(*i*)	0.1016±3	0.4217±3	-0.0920±5
2	O(13) in 2(*i*)	0.2303±3	0.3739±3	0.3982±5
2	O(14) in 2(*i*)	0.1422±7	-0.4243±5	-0.4236±7
2	O(15) in 2(*i*)	0.3199±5	-0.3612±5	0.1805±14
2	O(16) in 2(*i*)	0.4737±4	-0.3616±4	-0.3354±6

Anisotropic temperature factors given.

Interatomic distances (Å) *and bond angles*

σ(Cr-O) = 0.002; σ(Mo-O) = 0.003; σ(Na-O) = 0.004; σ(O-O) = 0.005 Å; σ(angles) = 0.1°

Cr - O(2)	1.968±2	(×2)
Cr - O(3)	1.971±2	(×2)
Cr - O(1)	1.986±3	(×2)

Angles in degrees subtended at Cr by

	O(3)'	O(1)'
O(2)	96.4±1	94.9±1
O(2)'	83.6±1	85.1±1
O(3)		84.8±1
O(3)'		95.2±1

Mo(1) - O(8)	1.703±4
Mo(1) - O(9)	1.708±3
Mo(1) - O(5)	1.926±3
Mo(1) - O(4)	1.943±3
Mo(1) - O(2)	2.289±1
Mo(1) - O(1)	2.346±2

Angles in degrees subtended at Mo(1) by

	O(9)	O(5)	O(4)	O(2)	O(1)
O(8)	105.9±2	101.3±1	98.0±1	164.8±1	95.3±1
O(9)		101.7±1	98.3±1	89.0±1	157.4±1
O(5)			147.0±1	72.0±1	81.2±1
O(4)				82.5±1	70.5±1
O(2)					70.5±1

Mo(2) - O(10)	1.699±3
Mo(2) - O(11)	1.716±3
Mo(2) - O(5)	1.934±3
Mo(2) - O(6)	1.941±3
Mo(2) - O(3)	2.270±2
Mo(2) - O(2)	2.294±3

Angles in degrees subtended at Mo(2) by

	O(11)	O(5)	O(6)	O(3)	O(2)
O(10)	105.2±1	100.7±2	96.9±1	161.2±1	93.7±1
O(11)		98. ±1	101.2±1	92.0±1	160.3±1
O(5)			149.1±1	83.8±1	71.8±1
O(6)				72.0±1	82.0±1
O(3)					70.2±1

Mo(3) - O(12)	1.694
Mo(3) - O(7)	1.720±3
Mo(3) - O(4)	1.908±3
Mo(3) - O(6)	1.985±3
Mo(3) - O(3)	2.243±3
Mo(3) - O(1)	2.306±3

Angles in degrees subtended at Mo(3) by

	O(7)	O(4)	O(6)	O(3)	O(1)
O(12)	106.0±2	99.1±1	100.2±1	92.8±1	162.7±1
O(7)		102.5±1	94.5±1	158.7±1	90.7±1
O(4)			149.6±1	84.0±1	72.0±1
O(6)				71.8±1	82.8±1
O(3)					71.8±1

Na(1) - O(1)	2.307±4	(×2)
Na(1) - O(8)	2.324±4	(×2)
Na(1) - O(9)	2.539±3	(×2)

Angles in degrees subtended at Na(1) by

	O(8)'	O(9)
O(10)	75.5±1	82.2±1
O(10)'	104.5±1	97.8±1
O(8)		76.3±1
O(8)'		103.7±1

Na(2) - O(15)	2.347±8
Na(2) - O(16)	2.419±4
Na(2) - O(13)	2.443±3
Na(2) - O(11)	2.453±3
Na(2) - O(16)	2.458±6
Na(2) - O(11)	2.475±3

Angles in degrees subtended at Na(2) by

	O(16)	O(13)	O(11)	O(16)	O(11)
O(15)	101.6±2	89.8±2	84.2±2	168.4±2	89.6±2
O(16)		83.2±1	166.8±1	87.1±2	109.9±1
O(13)			85.0±1	83.5±1	167.6±1
O(11)				85.7±1	82.6±1
O(16)					94.9±1

Possible hydrogen bonds

O(1)-H...O(6)	2.960		
O(2)-H...O(7)	2.706		
O(3)-H...O(13)	2.629		
O(13)-H...O(14)	2.652	O(14)-O(13)-O(6)	109.0°
O(13)-H...O(6)	2.831		
O(14)-H...O(4)	2.803	O(4)-O(14)-O(14)'	103.2°
O(14)-H...O(14)'	3.004		
O(15)-H...O(12)	2.897	O(8)-O(15)-O(12)	116.5°
O(15)-H...O(8)	3.028		
O(16)-H...O(5)	2.896		

Discussion

Columns of the disc like $(Cr(OH)_6Mo_6O_{18})^{3-}$ ions are stacked along the *c* axis with the Na ions and the water molecules in the channels between them.

Details of analysis

Lattice parameters from least-squares refinement of 19 high-angle reflexions measured on diffractometer with CuK_α radiation (λ = 1.54051 Å). Intensities measured using stationary crystal-stationary counter diffractometer (MoK_α radiation) with a crystal 0.20×0.22×0.29 mm^3 mounted in capillary tube. They were corrected for Lorentz and polarization effects, but not absorption (μ = 3.21 mm^{-1}) Structure solved from Patterson and electron-density functions and about 200 parameters refined by least squares to give R_1 = 0.033 for 3683 observed reflexions listed. Hydrogen-atom positions on anion proposed from charge neutrality principle and consideration of possible hydrogen bonds.

AMMONIUM FLUORIDE OXOPEROXOTETRAFLUOROMOLYBDATE(VI)

I. Studies on peroxomolybdates. VII The crystal structure of (tri)ammonium fluoride oxoperoxotetrafluoromolybdate(VI), $(NH_4)_3F[MoO(O_2)F_4]$. I. LARKING and R. STOMBERG, 1970. *Acta Chem. Scand.*, 24, 2043-2054.

$(NH_4)_3F[MoO(O_2)F_4]$ F.W. = 292.95

Orthorhombic, *a* = 16.521±4, *b* = 15.956±4, *c* = 6.793±2 Å, *U* = 1790.7 Å^3, D_m = 2.09, Z = 8, D_x = 2.17.

Space group Pbca (D_{2h}^{15}) Systematic absences

Atomic positions All atoms in 8(*c*)

	x	*y*	*z*
Mo	0.1290±1	0.0833±1	0.2041±3
O(1)	0.1749±12	0.3948±11	0.4912±25
O(2)	0.3594±15	0.4646±16	0.2260±39
O(3)	0.4282±14	0.4846±12	0.1264±31
F(1)	0.0727±8	0.0740±7	0.4695±18
F(2)	0.1411±7	0.2003±9	0.3050±20
F(3)	0.2256±9	0.0702±9	0.3555±26
F(4)	0.0229±7	0.1342±9	0.1195±18
F(5)	0.3670±7	0.2941±11	0.4770±19
N(1)	0.2624±11	0.3043±13	0.1597±30
N(2)	0.3863±11	0.0630±14	0.1936±33
N(3)	0.0049±11	0.3123±11	0.2296±26

Anisotropic temperature factors given.

Interatomic distances (Å) *and bond angles*

Mo - O(1)	1.67±2
Mo - O(2)	1.91±3
Mo - F(3)	1.91±2
Mo - O(3)	1.91±2
Mo - F(2)	2.00±2
Mo - F(4)	2.02±1
Mo - F(1)	2.03±1
O(2) - O(3)	1.36±3

Angles in degrees subtended at Mo by

	O(2)	F(3)	O(3)	F(2)	F(4)	F(1)
O(1)	103.3±10	96.3±8	99.1±9	93.2±7	93.7±7	172.1±7
O(2)		76.4±9	41.7±10	152.8±9	120.6±8	84.5±8
F(3)			118.2±8	80.4±6	157.6±6	84.1±6
O(3)				156.0±7	79.8±8	87.6±7
F(2)					79.0±5	79.0±5
F(4)						83.4±5

N(1) - F(5)	2.64±2
N(1) - F(5)	2.77±2
N(1) - F(2)	2.78±2
N(1) - F(3)	2.94±3
N(1) - O(1)	3.04±3
N(1) - O(2)	3.05±3
N(1) - F(2)	3.14±2
N(2) - F(5)	2.73±3
N(2) - F(1)	2.75±2
N(2) - F(4)	2.83±2
N(2) - F(3)	2.88±2
N(2) - O(3)	3.11±3
N(2) - O(2)	3.24±3
N(2) - F(1)	3.28±2
N(3) - F(5)	2.69±2
N(3) - F(1)	2.77±2
N(3) - F(4)	2.80±2
N(3) - F(2)	2.92±2
N(3) - F(4)	2.95±2
N(3) - O(3)	3.18±3

Discussion

The complex anion has a pentagonal bipyramidal configuration with the peroxo group and three fluorine atoms lying in the equatorial plane. The ion has approximate C_s symmetry.

Details of analysis

Lattice parameters measured from Guinier photograph calibrated with lead nitrate (a = 7.8566 Å) using CuK_α radiation. Intensities estimated visually from multiple-film equi-inclination Weissenberg photographs of layers with k = 0 and l = 0 to 6 taken with CuK_α radiation, corrected for Lorentz and polarization effects, but not absorption (crystal size 0.06×0.06×0.2 mm^3). Structure determined from Patterson and electron-density functions and refined by least squares to give R = 0.129 for 1079 reflexions listed.

AMMONIUM OXYPENTABROMOMOLYBDENUM(V)

I. [The x-ray crystal structure of $(NH_4)_2MoOBr_5$]. L.O. ATOVMJAN, O.A. DIJAČENKO and E.B. LOBKOVSKIJ, 1970. *Ž. Strukt. Khim.*, 11, 469-471 [*J. Struct. Chem.*, 11, 429-431].

$(NH_4)_2MoOBr_5$ F.W. = 543.50

Orthorhombic, *a* = 14.28±6, *b* = 10.49±4, *c* = 7.32±3 Å, *U* = 1097 Å^3, *Z* = 4, D_x = 3.29.

Space group Pnma (D_{2h}^{16})

Atomic positions

		x	*y*	*z*	*B*(Å^2)
4 Mo	in 4(*c*)	0.0948±1	0.25	0.1816±5	0.53±5
4 Br(1)	in 4(*c*)	0.9999±2	0.25	0.4767±6	1.42±6
4 Br(2)	in 4(*c*)	0.2490±2	0.25	0.4225±7	1.33±6
4 Br(3)	in 4(*c*)	0.2237±2	0.25	0.9416±7	1.70±6
8 Br(4)	in 8(*d*)	0.1073±2	0.0071±2	0.1943±5	1.20±5
4 O	in 4(*c*)	0.999±1	0.25	0.0096±3	1.1±3
8 N	in 8(*d*)	0.146±1	0.006±1	0.671±3	1.4±3

Interatomic distances (Å) *and bond angles*

Mo - O	1.86±2	
Mo - Br(3)	2.55±1	
Mo - Br(1)	2.55±1	
Mo - Br(4)	2.56±1	(×2)
Mo - Br(2)	2.83±1	

Angles in degrees subtended at Mo by

	Br(3)	Br(1)	Br(4)'	Br(2)
O	94.0±15	100.3±15	94.4±15*	176.1±20
Br(3)		165.8±15	88.5±15	82.3±15
Br(1)			90.4±15*	83.5±15
Br(4)			170.9±34	85.6±15

[*These two angles have been interchanged in I]

Discussion

The structure is closely related to that of K_2PtCl_6 (1) but with distortions arising from the lack of ideal octahedral symmetry in the $MoOBr_5$ ion.

Details of analysis

Intensities estimated visually from Weissenberg photographs of layers with *l* = 0 to 4, *h* = 0 and *k* = 0 taken with unfiltered Cu radiation. Correction for secondary extinction but not absorption. Structure determined from Patterson function and refined by least squares to give *R* = 0.12 for 592 observed reflexions [not listed].

1. *Stukturbericht*, 1, 429.

OCTACYANOWOLFRAMIC ACID

I. The square antiprismatic configuration for the octacyanotungstate (IV) ion. The crystal structure of $H_4W(CN)_8.6H_2O$. S.S. BASSON, L.D.C. BOK and J.G. LEIPOLDT, 1970. *Acta Cryst.*, B26, 1209-1216.

$H_4W(CN)_8.6H_2O$ F.W. = 504.1

Orthorhombic, *a* = 14.30, *b* = 9.881, *c* = 6.276 Å, *U* = 885.5 Å^3, D_m = 1.904, *Z* = 2, D_x = 1.889.

Space group Pmma (D_{2h}^{5}) other possibilities were not tested.

Atomic positions

	x	y	z	$B(\text{Å}^2)$
2 W in 2(f)	¼	¼	0.0437±13	2.82±5
4 C(1) in 4(k)	¼	0.31±2	0.23±3	3.3±6
4 C(2) in 4(j)	0.13±2	½	0.24±2	3.1±6
8 C(3) in 8(l)	0.16±1	0.37±2	0.86±2	3.8±5
4 N(1) in 4(k)	¼	0.21±2	0.32±2	3.7±6
4 N(2) in 4(j)	0.06±2	½	0.36±2	3.7±6
8 N(3) in 8(l)	0.11±1	0.30±2	0.76±2	4.8±5
4 O(1) in 4(i)	0.07±3	0	0.16±3	9.5±1.1
2 O(2) in 2(e)	¼	0	0.94±4	9.7±1.7
2 O(3) in 2(e)	¼	0	0.55±5	11.0±2.0
4 O(4) in 4(h)	0	0.17±2	½	6.4± .7

Mean interatomic distances (Å)

W-C 2.2
C-N 1.1

Mean bond angles subtend at C by W and N 177°.

Discussion of the structure

The structure has approximate square antiprisms of CN ligands about each W atom which form linear columns normal to (001). Structure is shown in projection in Fig. 50. The anion has *mm* symmetry.

Material

Silver salts from $K_4W(CN)_8.2H_2O$ and $K_4Mo(CN)_8.2H_2O$ were dissolved in HCl and the acids precipitated.

Details of analysis

The intensities of 837 refleixons from integrated multiple-film equi-inclination Weissenberg photographs taken with Cu radiation were estimated visually. No absorption corrections. Final *R* at 0.138 for isotropic thermal parameters. $W(CN)_8$ has an approximate square antiprism configuration. Water molecules appear to be disordered. Crystals blacken with exposure to X-rays. Unit cell and bond length errors not reported.

SODIUM OCTACYANOWOLFRAMATE

I. A second configuration for the octacyanide group. The crystal structure of $Na_3W(CN)_8.4H_2O$. L.D.C. BOK, J.G. LEIPOLDT and S.S. BASSON, 1970. *Acta Cryst.*, B26, 684.

$Na_3W(CN)_8.4H_2O$ F.W. = 533.02

Monoclinic, a = 6.126, b = 16.13, c = 17.42 Å, β = 94°45', U = 1715.4 Å^3, D_m = 2.05, Z = 4, D_x = 2.06.

Space group $P2_1/c$ (C_{2h}^5)

Atomic positions All atoms in 4(e)

	x	y	z	$B(\text{Å}^2)$
W	0.0424±2	0.2263±1	0.1036±1	3.47±4
C(1)	0.252±4	0.182±1	0.199±1	3.6±5
C(2)	0.210±5	0.123±2	0.057±2	4.5±6
C(3)	0.222±5	0.274±2	0.014±2	3.7±6
C(4)	0.211±4	0.335±2	0.159±2	3.7±6
C(5)	-0.126±4	0.269±1	0.197±1	2.6±5
C(6)	-0.145±5	0.331±2	0.059±2	4.1±6

4 C(7)	-0.157±5	0.185±2	0.006±2	3.9±6
4 C(8)	-0.123±7	0.118±3	0.143±2	4.5±11
4 N(1)	0.364±4	0.163±1	0.251±1	4.4±5
4 N(2)	0.312±4	0.068±2	0.035±1	4.4±5
4 N(3)	0.311±4	0.296±1	-0.042±1	5.0±6
4 N(4)	0.306±4	0.392±2	0.173±2	4.7±6
4 N(5)	-0.208±4	0.289±2	0.255±1	5.3±6
4 N(6)	-0.247±4	0.386±2	0.040±2	5.4±7
4 N(7)	-0.260±5	0.160±2	-0.048±2	4.3±5
4 N(8)	-0.217±4	0.061±1	0.165±1	4.3±5
4 Na(1)	0.481±2	0.454±6	0.4070±6	4.1±2
4 Na(2)	-0.454±2	0.299±7	0.2920±6	4.6±3
4 Na(3)	-0.433±2	0.2351±6	0.3573±6	4.2±2
4 O(1)	0.272±5	0.041±2	0.388±2	9.4±8
4 O(2)	0.274±4	0.348±1	0.340±1	6.0±5
4 O(3)	-0.284±4	0.463±1	0.296±1	5.6±5
4 O(4)	-0.194±5	0.111±2	0.368±2	8.3±8

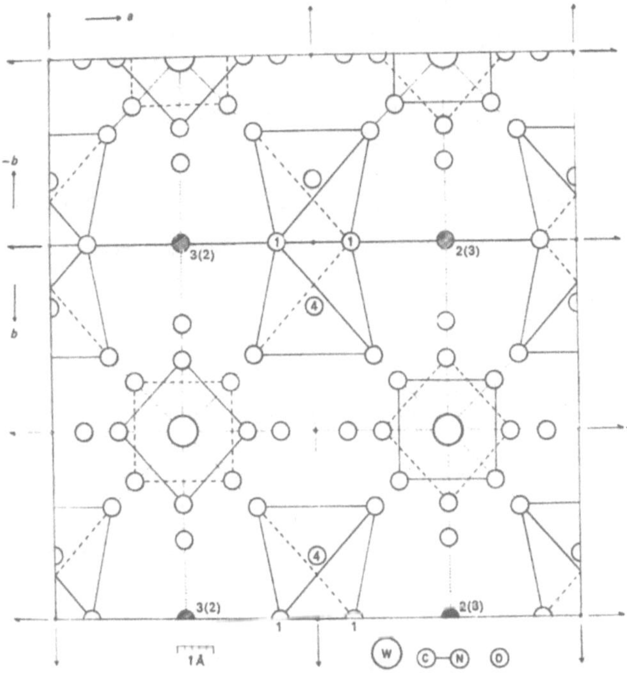

Fig. 50. Projection of the structure along *c*. The polyhedra around W and water molecules. Dotted lines indicate H-bonding between O(3) and N(1) atoms. Shaded circles refer to superposition of two oxygen atoms.

Interatomic distances (Å) *and bond angles*

8 W-C	2.12 - 2.16	4 Na(1) - N	
8 C-N	1.11 - 1.20	4 Na(2) - N	2.41 - 2.79
		4 Na(3) - N	
		2 Na(1) - O	
		2 Na(2) - O	2.37 - 2.55
		2 Na(3) - O	

W-C-N bond angle 174 - 178°

Structure

The anion has antiprism configuration. Sodium ions are bonded to four nitrogen and two oxygen atoms.

Preparation

$Na_3W(CN)_8.4H_2O$ prepared by the procedure of Baadsgaard and Treadwell(1) and crystal grown from aqueous solution in the dark.

Details of analysis

Crystals covered with Canada balsam. About 2600 reflexions visually estimated from integrated Weissenberg photographs taken about *a* axis with Cu radiation. Absorption corrections. Least-squares refinement gave R = 0.132.

1. H. BAADSGAARD and W.D. TREADWELL, 1955, *Helv. Chim. Acta*, 38, 1669.

AMMONIUM OCTACHLORODITECHNICATE DIHYDRATE

I. Characterization of the trinegative octachloroditechnicate ion. W.K. BRATTON and F.A. COTTON, 1970. *Inorg. Chem.*, 9, 789-793.

$(NH_4)_3Tc_2Cl_82H_2O$ F.W. = 571.6

Trigonal, a = 13.04±2, c = 8.40±1 Å, U = 1237 Å^3, D_m = 2.4, Z = 3, D_x = 2.30.

Space group $P3_121$ (D_3^4) or $P3_221$ (D_3^6) from systematic absences

Atomic positions (in space group $P3_221$)

	x	y	z	B(Å^2)
6 Tc in 6(*c*)	0.6336±5	0.5520±4	0.0639±7	3.7±2
6 Cl(1) in 6(*c*)	0.7402±16	0.6837±14	0.2682±22	5.0±4
6 Cl(2) in 6(*c*)	0.4940±13	0.4384±14	0.2642±23	4.6±4
6 Cl(3) in 6(*c*)	0.5695±13	0.3645±13	-0.0595±18	3.9±3
6 Cl(4) in 6(*c*)	0.8132±15	0.6174±14	-0.0683±25	4.5±4
6 N(1) in 6(*c*)	0.483±5	0.673±4	0.416±7	5. ±1
3 N(2) in 3(*a*)	-0.299±7	0	0.6667	7. ±2
3 O(1) in 6(*c*)	0.180±8	0.231±7	-0.275±11	4. ±2
3 O(2) in 3(*b*)	0	0.053±7	-0.167±10[sic]	13. ±3

Interatomic distances (Å) *and bond angles*

Tc_2Cl_8 ion (crystallographic point symmetry C_2)

Tc - Tc'	2.13
Tc - Cl(1)	2.33
Tc - Cl(2)	2.38
Tc - Cl(3)	2.39
Tc - Cl(4)	2.33

Angles in degrees subtended at Tc by

	Cl(1)	Cl(2)	Cl(3)	Cl(4)
Tc	105.4	104.5	106.6	103.8
Cl(1)		85.8	147.9	87.8
Cl(2)			84.4	151.8
Cl(3)				86.6

σ(Tc-Tc) = 0.01, σ(Tc-Cl) = 0.02, σ(angles) = 0.7°

Nine-fold coordination around N(1)

N(1) - Cl(1)	3.52	N(1) - Cl(4)	3.42	N(1) - Cl(2)"	3.40
N(1) - Cl(2)	3.38	N(1) - Cl(1)'	3.52	N(1) - Cl(3)'	3.30
N(1) - Cl(3)	3.33	N(1) - Cl(2)'	3.42	N(1) - Cl(3)"	3.41

σ(N - Cl) = 0.06

Eight-fold coordination around N(2) (crystallographic site symmetry C_2)

N(2) - Cl(1)	3.33 (×2)	N(2) - Cl(2)	3.23 (×2)
N(2) - Cl(2)	3.17 (×2)	N(2) - O(1)	2.7 (×2)
	σ(N - Cl) = 0.06	σ(N - O) = 0.1	
O(1) - O(2)	2.5±1		
O(1) - Cl(4)	3.1±1		
	all other O-Cl distances are longer		

Discussion

The $Tc_2Cl_8{}^{3-}$ ions occur in the eclipsed configuration. The arrangement of NH_4^+ and H_2O involves a hydrogen bonding scheme of some complexity because of the partial occupancy of the H_2O sites.

Details of analysis

Lattice parameters from oscillation, Weissenberg and precession photographs. Intensities estimated visually from multiple-film equi-inclination Weissenberg photographs taken with crystal 0.15 mm in diameter and 0.40 mm long. Corrections for Lorentz, polarization and absorption (μ = 27 mm^{-1}) effects. Structure from three-dimensional Patterson and Fourier functions, refined by least squares (omitting 120 unobserved reflexions) with fixed occupation numbers for O(1) and O(2) to give a weighted R = 0.13 for 292 observed reflexions listed. Anomolous dispersion effects were not large enough to allow the space group to be chosen unequivocally.

TRI-μ-CHLORO-PENTACARBONYL (TRICHLOROSTANNYL) DIRUTHENIUM

$(SnCl_3)Ru_2Cl_3(CO)_5$ F.W. = 674

I. Crystal and molecular structure of tri-μ-chloro-pentacarbonyl (trichlorostannyl) diruthenium. M. ELDER and D. HALL, 1970. *J. Chem. Soc. Lond.*, A 1970, 245-248.

Monoclinic, a = 10.414±8, b = 11.164±14, c = 14.674±9 Å, β = 96.7±1°, U = 1694 Å^3, D_m = 2.6, Z = 4, D_x = 2.64.

Space group $P2_1/n$ (C_{2h}^5)

Atomic positions

All atoms in 4(*e*) (x,y,z; $\bar{x},\bar{y},\bar{z}$; ½-x,½+y,½-z; ½+x,½-y,½+z)

	x	y	z
Sn	0.0440±2	0.2726±3	0.2483±2
Ru(1)	0.1908±3	0.2214±2	0.1252±2
Ru(2)	0.4935±3	0.2575±4	0.1479±2
Cl(1)	0.3605±10	0.1345±12	0.2347±7
Cl(2)	0.3520±9	0.1886±12	0.0182±6
Cl(3)	0.3205±9	0.4022±11	0.1603±7
Cl(4)	0.1352±12	0.3451±14	0.3911±7
Cl(5)	-0.1145±11	0.4157±13	0.1994±9
Cl(6)	-0.0932±11	0.1230±13	0.2954±9
C(1)	0.070±4	0.293±5	0.034±3
O(1)	0.003±4	0.339±4	-0.021±3
C(2)	0.109±5	0.076±6	0.109±3
O(2)	0.059±3	-0.016±4	0.094±2
C(3)	0.613±4	0.148±4	0.127±3
O(3)	0.685±3	0.063±3	0.115±2
C(4)	0.592±4	0.322±4	0.260±3
O(4)	0.644±3	0.335±3	0.328±2

C(5)	0.576±4	0.366±4	0.077±2
O(5)	0.635±3	0.424±3	0.033±2

Anisotropic temperature factors given.

Interatomic distances (Å) *and bond angles*

Sixfold coordination around Ru(2)

Ru(2) - C(3)	1.80
Ru(2) - C(4)	1.97
Ru(2) - C(5)	1.87
Ru(2) - Cl(1)	2.42
Ru(2) - Cl(2)	2.39
Ru(2) - Cl(3)	2.44

Angles in degrees subtended at Ru(2) by

	C(4)	C(5)	Cl(1)	Cl(2)	Cl(3)
C(3)	95	88	99	91	174
C(4)		90	92	173	91
C(5)			172	92	90
Cl(1)				84.2	83.1
Cl(2)					83.1

Sixfold coordination around Ru(1)

Ru(1) - C(1)	1.90
Ru(1) - C(2)	1.83
Ru(1) - Sn	2.565
Ru(1) - Cl(1)	2.44
Ru(1) - Cl(2)	2.46
Ru(1) - Cl(3)	2.45

Angles in degrees subtended at Ru(1) by

	C(2)	Sn	Cl(1)	Cl(2)	Cl(3)
C(1)	91	90	175	95	96
C(2)		89	91	98	173
Sn			94.0	172.3	91.6
Cl(1)				82.3	81.3
Cl(2)					81.6

Fourfold coordination around Sn

Sn - Ru(1)	2.565
Sn - Cl(4)	2.34
Sn - Cl(5)	2.35
Sn - Cl(6)	2.35

Angles in degrees subtended at Sn by

	Cl(4)	Cl(5)	Cl(6)
Ru(1)	119.7	113.0	118.7
Cl(4)		103.7	100.4
Cl(5)			98.2

Carbonyl groups

	C - O	Ru - C - O
C(1) - O(1)	1.13	176°
C(2) - O(2)	1.17	176
C(3) - O(3)	1.23	173
C(4) - O(4)	1.09	166
C(5) - O(5)	1.14	174

Ru(1) - Cl(1) - Ru(2)	81.0°
Ru(1) - Cl(2) - Ru(2)	81.2
Ru(1) - Cl(3) - Ru(2)	80.4
Ru - Ru	3.157

Standard errors in distances and angles

Ru - Sn	0.004	Sn - Ru - Cl	0.3°
Ru - Cl	0.01	Sn - Ru - C	1.0
Ru - C	0.05	C - Ru - C	2.0
Sn - Cl	0.01	Cl - Ru - C	1.0
C - O	0.06	Cl - Ru - Cl	0.4
Ru - Ru	0.004	Ru - Sn - Cl	0.4
		Ru - C - O	4.0

Discussion

The molecule consists of two octahedrally coordinated Ru atoms bridged by three Cl atoms. The carbonyl and $SnCl_3$ groups complete the coordination around ruthenium.

Details of analysis

Lattice parameters from precession photographs and diffractometer ($\lambda(MoK_{\alpha 1})$ = 0.70926 Å). Intensities of layers with k = 0 to 8 measured on diffractometer (PAILRED) with crystal monochromated MoK_α radiation. Crystal in needle form 0.7 mm long (b axis) × 0.062 mm diameter. Lorentz and polarization corrections, absorption correction not necessary. Structure from Patterson, Fourier and difference Fourier functions refined by least squares to give weighted R = 0.085 for 886 non-zero reflexions. Structure factors not listed but are available from first named author at University of Alberta, Edmonton, Canada.

CAESIUM TRANS-TETRACHLOROAQUOCARBONYLRUTHENATE(II)

I. Crystal structure of cesium trans-tetrachloroaquocarbonylruthenate(II). J.A. STANKO and S. CHAIPAYUNGPUNDHU, 1970. *J. Amer. Chem. Soc.*, 92, 5580-5585.

$Cs_2[Cl_4(H_2O)RuCO]$ F.W.= 554.70

Orthorhombic, a = 8.096±3, b = 17.19±1, c = 7.437±2 Å, U = 1035 Å^3, Z = 4, D_x = 3.54.

Space group Amam (D_{4h}^{17})

Atomic positions Atoms at ±(x,y,z; $-x,-y,z$; ½$-x,y,z$; ½$+x,-y,z$)

		x	y	z
4 Cs(1)	in 4(c)	0.25	0.4781±1	0
4 Cs(2)	in 4(c)	0.25	0.7551±1	0
4 Ru	in 4(c)	0.25	0.1219±1	0
16 Cl	in 16(h)	0.4616±2	0.1162±1	0.2246±2
4 C	in 4(c)	0.25	0.2268±9	0
4 O(1)	in 4(c)	0.25	0.2902±9	0
4 O(2)	in 4(c)	0.25	-0.0087±6	0

Anisotropic temperature factors given.

Interatomic distances (Å) *and bond angles*

Ru-C	1.804±15	C-O(2)	1.09±2
Ru-O(1)	2.245±11	Ru-C-O(2)	180° (by symmetry)
Ru-Cl	2.395±2 (×4)		
Cl-Ru-C	92.3±1°		
Cl-Ru-O(1)	87.7±1		
Cl-Ru-Cl	91.4±1, 88.5±1		
Cs(1)-O(1)	3.23±1		
Cs(1)-Cl	3.504±2 (×4)		
Cs(1)-Cl'	3.573±2 (×4)		
Cs(1)-Cl(2)	3.725±1		
Cs(2)-Cl	3.582±2 (×4)		
Cs(2)-Cl'	3.625±2 (×4)		
Cs(2)-C	3.750±2 (×2)		
Cs(2)-O(1)	3.767±2 (×2)		

Discussion

The $[Cl_4(CO)(H_2O)Ru]^{2-}$ ion is *trans* with a rather short Ru-C bond.

Details of analysis

Lattice parameters and intensities measured on automatic four-circle diffractometer ($\lambda(CuK_\alpha)$ = 1.5418 Å, intensities measured with Zr-filtered Mo radiation). Reflexions in range $2\theta<60°$ measured with θ-2θ scan. Structure solved from Patterson function was refined by least squares to give R_2 = 0.046 for 712 reflexions with $I>2\sigma(I)$. The final structure showed slight disorder of the CO and H_2O groups.

POTASSIUM TRICHLOROAMMINEPLATINATE (II) MONOHYDRATE

$K(PtCl_3NH_3)H_2O$ F.W. = 375.60

I. The crystal structure of potassium trichloroammineplatinate (II) monohydrate. Y.P. JEANNIN and D.R. RUSSELL, 1970. *Inorg. Chem.*, 9, 778-783.

Orthorhombic, a = 4.572±2, b = 21.025±9, c = 8.125±4 Å, U = 681.0 Å^3, D_m = 3.181±3, Z = 4, D_x = 3.19.

Space group $Cmc2_1$ (C_{2v}^{12}) systematic absences (but see below under *Details of analysis*)

Atomic positions

All atoms in 4(a) (0,y,z)

	y	z
Pt	0.34338±3	0.25 (arbitrary choice)
Cl(1)	0.4536±2	0.2506±28
Cl(2)	0.3373±9	0.5311±39
Cl(3)	0.3425±6	-0.0306±21
N	0.2467±9	0.2277±63
K	0.0395±3	0.4640±9
O	0.0756±11	0.1374±31

Anisotropic temperature factors given as well as atomic positions and temperature factors for the refinement in space group *Cmcm*.

Interatomic distances (Å) *and bond angles*

$PtCl_3NH_3$ ion (crystallographic site symmetry C_s)

Pt - Cl(1)	2.317±7
Pt - Cl(2)	2.288±27
Pt - Cl(3)	2.280±14
Pt - N	2.042±19

Angles in degrees subtended at Pt by

	Cl(2)	Cl(3)
Cl(1)	93.1±6	90.6±5
N	91.9±9	84.4±7

Corrections for thermal motion add between 0.004 and 0.028 to these. The results of the refinement in *Cmcm* do not differ significantly from those given above.

Eight-fold coordination around K (crystallographic site symmetry C_s)

K - Cl(1)	3.39 (×2)
K - Cl(2)	3.37 (×2)
K - Cl(3)	3.26 (×2)
K - O	2.80
K - O'	2.76

Four-fold coordination around water (crystallographic site symmetry C_s)

O - K	2.80	K - O - K'	104°
O - K'	2.76	Cl - O - Cl'	97°
O...Cl(2)	3.05 (×2)		
N...Cl(2)	3.30		
N...Cl(3)	3.55 (×2)		

Discussion

The structure is very similar to that of $K(PtCl_3NH_3)$ (1) with the K shifted slightly to allow room for H_2O. Another model in space group *Cmcm* with the K and H_2O randomly arranged is also discussed.

Details of analysis

Lattice parameters from precession photographs (MoK_α) calibrated with NaCl (a = 5.6399 Å). Intensities measured with diffractometer (XRD-5) from a crystal 0.08×0.08×0.16 mm^3 were corrected for Lorentz, polarization and absorption effects. Structure solved by trial and error and electron-density syntheses. Refinement performed in two of the space groups allowed by systematic absences and these gave $R(Cmcm)$ = 0.068 and $R(Cm2_1)$ = 0.069 for 830 reflexions listed. The latter refinement was favoured on the basis of possible anomalous dispersion effects, the Hamilton R factor ratio test and chemical likelihood, but a disordered structure in *Cmcm* is possible even though it would result in some K atoms coming to within 2.8 Å of each other.

1. *Structure Reports*, 15, 159.

POTASSIUM TETRACHLOROAQUONITRIDE OSMIUM HYDRATE

I. [The structure of crystals of $K[OsNCl_4H_2O]H_2O$]. L.O. ATOVMJAN, and V.V. TKAČEV, 1970. *Ž. Strukt. Khim.*, 11, 933-934 [*J. Struct. Chem.*, 11, 868-869].

$K[OsNCl_4H_2O]H_2O$ F.W. = 421.15

Orthorhombic, a = 9.049±3, b = 13.171±4, c = 7.325±2 Å, U = 873.0 Å^3, Z = 4, D_x = 3.20.

Space group $Cmc2_1$ (C_{2v}^{12})

Atomic positions

		x	y	z	B(Å^2)
4 Os	in 4(*a*)	0	0.1491±1	0	1.90
8 Cl(1)	in 8(*b*)	0.183±1	0.0334±6	0.030±1	2.01
8 Cl(2)	in 8(*b*)	0.183±1	0.2786±7	0.068±1	1.82
4 K	in 4(*a*)	0.5	0.141±2	0.924±3	6.66
4 O	in 4(*a*)	0	0.147±2	0.341±4	2.14
4 O	in 4(*a*)	0.5	0.073±3	0.270±6	6.78
4 N	in 4(*a*)	0	0.144±4	-0.237±6	2.19

Interatomic distances (Å) *and bond angles*

Os - N	1.75±4	
Os - Cl(2)	2.34±1	(×2)
Os - Cl(1)	2.35±1	(×2)
Os - O	2.50±3	

Angles in degrees subtended at Os by

	Cl(2)	Cl(1)	O
N	106.8±12	90.7±12	169.8±18
Cl(2)	90.2±3	87.4±3	80.2±4
Cl(2)		162.3±3	80.2±4
Cl(1)		89.5±3	82.1±5
Cl(1)			

K - O	2.69±5	
K - O	2.86±4	
K - O	3.04±5	
K - Cl(2)	3.27±2	(×2)
K - Cl(1)	3.29±2	(×2)
K - Cl(2)	3.55±2	(×2)

Discussion

The crystals were reported (1) to have the formula $K_2O_5NCl_5$ (first modification). The H_2O is *trans* to N.

Details of analysis

All measurements made on crystal 0.25×0.30×0.45 mm^3. Lattice parameters refined from diffractometer measurements. Intensities measured on a KFOR with unfiltered Mo radiation. Structure solved from Patterson and difference synthesis and refined by least squares to give R = 0.105 for 467 observed reflexions [not listed].

1. H. DUFET, 1902. *Bull. Soc. Minér.*, 25, 141;
L.O. AFOVMJAN and G.B. BOKIJ, 1960. *Ž. Strukt. Khim.*, 1, 501.

TRIPLOIDITE

$Mn_{1.5}Fe_{0.5}PO_4(OH)$ F.W. = 222.31

I. The crystal structure of triploidite and its relation to the structures of other minerals of the triplite-triploidite group. L. WALDROP, 1970. *Z. Kristallogr.*, 131, 1-20.

Monoclinic, a = 12.366±1, b = 13.276±2, c = 9.943±1 Å, β = 108.230±4°, U = 1550.4 Å^3, D_m = 3.697, Z = 16, D_x = 3.80.

Space group $P2_1/a$ (C^5_{2h})

(substructure has $b_s = \frac{1}{2}b$ and space group $I2/a$)

Atomic positions All atoms in 4(*e*) $x,y,z;\ \bar{x},\bar{y},\bar{z};\ \frac{1}{2}+x,\frac{1}{2}-y,z;\ \frac{1}{2}-x,\frac{1}{2}+y,\bar{z}$.

	x	y	z	B(Å^2)
M(1)	0.1872±2	0.4784±2	0.1882±3	0.40±5
M(2)	0.1969±2	0.9968±2	0.2102±3	0.80±5
M(3)	0.3046±2	0.7518±2	0.2958±3	0.74±5
M(4)	0.3183±2	0.2696±2	0.3086±3	0.50±5
M(5)	0.0972±2	0.0710±2	0.4660±3	0.21±5
M(6)	0.0835±2	0.5748±2	0.4447±3	0.26±5
M(7)	0.3939±2	0.6746±2	0.0343±3	0.21±4
M(8)	0.4230±2	0.1780±2	0.0448±3	0.54±4
P(1)	0.0782±4	0.8219±4	0.3791±5	0.42±8
P(2)	0.0746±4	0.3268±4	0.3824±5	0.38±8
P(3)	0.4233±4	0.4224±4	0.1144±5	0.34±8

	x	*y*	*z*	*B*(Å^2)
P(4)	0.4242±4	0.9280±4	0.1200±5	0.39±8
O(1)	0.0476±7	0.4131±8	0.4659±9	0.38±16
O(2)	0.0599±7	0.9054±8	0.4757±10	0.57±17
O(3)	0.4300±6	0.8412±7	0.0177±9	0.64±15
O(4)	0.4571±6	0.3375±8	0.0279±9	0.70±15
O(5)	0.0254±7	0.0472±8	0.2385±10	0.81±17
O(6)	0.0450±7	0.5536±8	0.2261±10	0.67±16
O(7)	0.4706±8	0.7025±9	0.2639±10	0.16±19
O(8)	0.4684±8	0.2014±9	0.2640±11	0.59±20
O(9)	0.1718±7	0.8491±7	0.3119±9	0.29±16
O(10)	0.1694±7	0.3593±7	0.3195±9	0.34±17
O(11)	0.3254±7	0.3840±6	0.1674±8	0.63±15
O(12)	0.3402±7	0.9053±6	0.2023±8	0.47±14
O(13)	0.1191±9	0.7280±10	0.4740±12	0.78±22
O(14)	0.1154±9	0.2333±10	0.4768±12	0.68±22
O(15)	0.3878±9	0.0215±9	0.0284±11	0.66±20
O(16)	0.3808±9	0.5120±9	0.0183±11	0.68±21
O(17)*	0.2565±5	0.0313±5	0.4354±7	0.31±13
O(18)*	0.2053±6	0.1612±5	0.1911±7	0.61±13
O(19)*	0.2410±5	0.7205±5	0.0693±7	0.53±13
O(20)*	0.3024±6	0.5823±5	0.3117±7	0.69±14

*OH-oxygen atoms.

Interatomic distances (Å) *and angles*

Fivefold coordination around M(1)

M(1)-O(6)	2.156
M(1)-O(10)	2.106
M(1)-O(11)	2.181
M(1)-O(15)	2.140
M(1)-O(20)	2.084

σ = 0.010 Å

Angles in degrees subtended at M(1) by

	O(10)	O(11)	O(15)	O(20)
O(6)	89.4	171.6	86.1	92.0
O(10)		82.2	136.7	108.4
O(11)			99.4	91.4
O(15)				114.7

σ = 0.4°

Sixfold coordination around M(2)

M(2)-O(5)	2.323
M(2)-O(9)	2.272
M(2)-O(12)	2.169
M(2)-O(16)	2.182
M(2)-O(17)	2.176
M(2)-O(18)	2.197

σ = 0.010 Å

Angles in degrees subtended at M(2) by

	O(9)	O(12)	O(16)	O(17)	O(18)
O(5)	86.4	162.2	88.7	81.0	78.1
O(9)		76.1	117.8	76.9	155.9
O(12)			96.5	97.9	119.6
O(16)				161.7	80.5
O(17)					82.5

σ = 0.4°

Sixfold coordination around M(3)

M(3)-O(7)	2.269
M(3)-O(9)	2.134
M(3)-O(12)	2.339
M(3)-O(14)	2.179
M(3)-O(19)	2.181
M(3)-O(20)	2.257

σ = 0.010 Å

Angles in degrees subtended at M(3) by

	O(9)	O(12)	O(14)	O(19)	O(20)
O(7)	159.4	84.5	88.6	81.4	75.7
O(9)		75.4	95.8	99.0	124.9
O(12)			115.4	78.9	154.8
O(14)				161.8	80.0
O(19)					82.8

σ = 0.4°

Fivefold coordination around M(4)

M(4)-O(8)	2.232
M(4)-O(10)	2.223
M(4)-O(11)	2.089
M(4)-O(13)	2.128
M(4)-O(18)	2.090

σ = 0.010 Å

Angles in degrees subtended at M(4) by

	O(10)	O(11)	O(13)	O(18)
O(8)	168.6	87.1	92.0	93.5
O(10)		81.6	97.6	88.7
O(11)			142.5	106.7
O(13)				110.9

σ = 0.4°

Sixfold coordination around M(5)

M(5)-O(2)'	2.217
M(5)-O(2)	2.254
M(5)-O(5)	2.179
M(5)-O(14)	2.166
M(5)-O(17)	2.151
M(5)-O(20)	2.175

σ = 0.010 Å

Angles in degrees subtended at M(5) by

	O(2)	O(5)	O(14)	O(17)	O(20)
O(2)'	85.3	100.7	86.1	171.2	89.6
O(2)		83.1	170.9	88.6	94.8
O(5)			101.4	84.9	169.2
O(14)				99.5	82.1
O(17)					84.5

σ = 0.4°

Fivefold coordination around M(6)

M(6)-O(1)	2.215
M(6)-O(1)'	2.084
M(6)-O(6)	2.094
M(6)-O(13)	2.083
M(6)-O(17)	2.048

σ = 0.010 Å

Angles in degrees subtended at M(6) by

	O(1)	O(6)	O(13)	O(17)
O(1)	80.2	88.9	166.7	81.8
O(1)'		119.9	91.4	120.8
O(6)			104.2	115.5
O(13)				94.0

σ = 0.4°

Sixfold coordination around M(7)

M(7)-O(3)	2.273
M(7)-O(4)	2.125
M(7)-O(7)	2.212
M(7)-O(16)	2.166
M(7)-O(18)	2.200
M(7)-O(19)	2.116

σ = 0.010 Å

Angles in degrees subtended at M(7) by

	O(4)	O(7)	O(16)	O(18)	O(19)
O(3)	81.1	83.2	167.5	93.7	86.4
O(4)		100.5	87.4	87.4	165.9
O(7)			104.0	170.8	84.2
O(16)				80.8	104.4
O(18)					87.0

σ = 0.4°

Fivefold coordination around M(8)

M(8)-O(3)	2.113
M(8)-O(4)	2.176
M(8)-O(8)	2.096
M(8)-O(15)	2.119
M(8)-O(19)	2.069

σ = 0.010 Å

Angles in degrees subtended at M(8) by

	O(4)	O(8)	O(15)	O(19)
O(3)	83.7	110.4	91.9	131.1
O(4)		86.6	171.4	82.7
O(8)			101.9	115.3
O(16)				94.7

σ = 0.4°

Fourfold coordination around P(1)

P(1)-O(2)	1.529
P(1)-O(7)	1.495
P(1)-O(9)	1.549
P(1)-O(13)	1.550

σ = 0.011 Å

Angles in degrees subtended at P(1) by

	O(7)	O(9)	O(13)
O(2)	111.6	112.3	106.5
O(7)		109.1	109.5
O(9)			107.7

σ = 0.6°

Fourfold coordination around P(2)

P(2)-O(1)	1.512
P(2)-O(8)	1.511
P(2)-O(10)	1.553
P(2)-O(14)	1.542

σ = 0.011 Å

Angles in degrees subtended at P(2) by

	O(8)	O(10)	O(14)
O(1)	109.2	109.6	111.2
O(8)		109.7	108.3
O(10)			108.9

σ = 0.6°

Fourfold coordination around P(3)

P(3)-O(4)	1.552
P(3)-O(5)	1.518
P(3)-O(11)	1.550
P(3)-O(16)	1.513

σ = 0.011 Å

Angles in degrees subtended at P(3) by

	O(5)	O(11)	O(16)
O(4)	110.1	108.9	108.8
O(5)		110.5	109.9
O(11)			108.6

σ = 0.6°

Fourfold coordination around P(4)

P(4)-O(3)	
P(4)-O(6)	
P(4)-O(12)	
P(4)-O(15)	

σ = 0.011 Å

Angles in degrees subtended at P(4) by

	O(6)	O(12)	O(15)
O(3)	109.3	112.3	106.2
O(6)		109.6	108.6
O(12)			110.8

σ = 0.6°

Discussion

The structure is very similar to that of triplete (1) differing in small shifts and an ordering of the OH groups in triploidite which replace the disordered F atoms in triplite. One result is the occurrence of smaller coordination numbers around the metal atoms. This results in the F atoms in triplite being tetrahedrally coordinated by metal atoms while in triploidite the OH ions are only triangularly coordinated by metal atoms. It was not possible to determine whether the Fe and Mn atoms were ordered on the M sites or not.

Details of analysis

Triploidite crystals from Branchville, Connecticut, U.S.A. corresponding to $Mn_{1.5}Fe_{0.5}PO_4(OH)$.

Lattice parameters from back-reflexion Weissenberg photographs refined by least squares. Intensities measured on equi-inclination diffractometer using ω scan and FeK_α radiation. Lorentz, polarization and absorption (crystal size = 0.18×0.18×0.13 mm^3) corrections. Structure solved by similarity with triplite(1) and about 130 parameters refined by least squares to give R_2 = 0.055 for 769 observed reflexions (listed). Anisotropic thermal motion not indicated in final difference electron density.

1. L. WALDROP, 1969. *Z. Kristallogr.*, 130, 1.

RANSOMITE

I. The crystal structure of ransomite. M.M. WOOD, 1970. *Amer. Min.*, 55, 729-734.

$CuFe_2(SO_4)_4.6H_2O$ F.W. = 667.6

Monoclinic, *a* = 4.811±2, *b* = 16.217±4, *c* = 10.403±2 Å, β = 93.02°±3, *U* = 810.5 Å^3, *Z* = 2, D_x = 2.73.

Space group $P2_1/c$ (C_{2h}^5)

Atomic positions

			x	*y*	*z*	*B*(Å^2)
2	Cu	in 2(*a*)	0	0	0	1.33±8
4	Fe	in 4(*e*)	0.7377±6	0.6272±1	0.0223±2	0.81±6
4	S(1)	in 4(*e*)	0.7906±9	0.2869±2	0.1399±3	0.82±7
4	S(2)	in 4(*e*)	0.2502±8	0.4298±2	0.1704±3	0.60±7
4	O(1)	in 4(*e*)	0.0309±26	0.2144±6	0.5549±10	1.57±19
4	O(2)	in 4(*e*)	0.5880±26	0.1470±6	0.5927±9	1.62±17
4	O(3)	in 4(*e*)	0.6687±30	0.2922±7	0.6322±7	2.33±22
4	O(4)	in 4(*e*)	0.8927±28	0.1906±7	0.7713±11	2.05±21
4	O(5)	in 4(*e*)	0.1427±25	0.4554±6	0.0999±9	1.24±16
4	O(6)	in 4(*e*)	0.0705±26	0.5999±6	0.1405±9	1.25±17
4	O(7)	in 4(*e*)	0.5340±25	0.5474±5	0.1267±9	1.17±16
4	O(8)	in 4(*e*)	0.2717±26	0.5128±6	0.3098±10	1.58±18
4	O(W1)	in 4(*e*)	0.7150±26	0.0661±6	0.9072±9	1.28±17
4	O(W2)	in 4(*e*)	0.3508±26	0.2814±6	0.8570±9	1.18±17
4	O(W3)	in 4(*e*)	0.1937±31	0.1065±7	0.0483±12	2.40±22

Interatomic distances (Å) *and angles*

Cu - O(W1)	1.955±11	(×2)
Cu - O(W3)	2.014±12	(×2)
Cu - O(8)	2.437±11	(×2)

Angles in degrees subtended at Cu by

	O(W3)	O(8)
O(W1)	87.5±5	92.2±4
O(W1)	92.5±5	87.8±4
O(W3)		90.8±5
O(W3)		89.2±5

Fe - O(2)	1.948±11
Fe - O(5)	1.954±10
Fe - O(7)	1.982±10
Fe - O(1)	1.995±11
Fe - O(W2)	2.003±10
Fe - O(6)	2.016±11

Angles in degrees subtended at Fe by

	O(5)	O(7)	O(1)	O(12)	O(6)
O(2)	88.2±4	92.2±5	94.8±5	94.1±4	176.8±12
O(5)		94.6±4	92.0±4	174.1±5	89.7±4
O(7)			170.5±4	90.7±4	85.6±4
O(1)				82.5±4	87.6±5
O(W2)					88.2±4

S(1) - O(3)	1.411±12
S(1) - O(4)	1.474±12
S(1) - O(1)	1.493±13
S(1) - O(2)	1.513±11

Angles in degrees subtended at S(1) by

	O(4)	O(1)	O(2)
O(3)	113.3±6	106.6±6	111.5±7
O(4)		108.8±7	107.6±6
O(1)			108.9±6

S(2) - O(6)	1.452±11
S(2) - O(8)	1.474±11
S(2) - O(7)	1.490±12
S(2) - O(5)	1.490±10

Angles in degrees subtended at S(2) by

	O(8)	O(7)	O(5)
O(6)	111.5±6	109.3±6	109.9±6
O(8)		108.5±6	109.7±6
O(7)			107.8±6

O(W1)-Cu	1.96±1
O(W1)-O(4)	2.63±2
O(W1)-O(8)	2.64±2

Angles in degrees subtended at O(W1) by

	O(4)	O(8)
Cu	116.5±6	117.1±5
O(4)		116.3±5

O(W2)-Fe	2.00±1
O(W2)-O(4)	2.76±2
O(W2)-O(3)	2.87±2

Angles in degrees subtended at O(W2) by

	O(4)	O(3)
Fe	113.9±6	127.3±5
O(4)		102.8±4

O(W3)-Cu	2.01±1
O(W3)-O(3)	2.91±2
Cu-O(W3)-O(3)	154.8±7°

Discussion

FeO_6 octahedra are linked into chains along *a*, two SO_4 groups linking each pair of octahedra. The chains are joined into double chains by the axial O of one FeO_6 group being shared with an $S(2)O_4$ group of an adjacent chain. Planar $Cu(H_2O)_4$ groups join the double chains into sheets perpendicular to *b* through the formation of long axial Cu-O bonds by the remaining oxygen on the $S(2)O_4$ group.

Details of analysis

Crystals from United Verde mine, Jerome, Arizona, U.S.A. Intensities measured from Weissenberg photographs with filtered CuK_α radiation. [Lorentz, polarization] and absorption corrections (crystal 0.06×0.10×1.00 mm³). Structure determined by Patterson and electron-density functions refined by least squares to give R = 0.12 for 1186 observed reflexions listed in 1.

1. NAPS document No. 0947, National Auxiliary Publications Service of ASIS c/o CCM Information Corporation, 909 Third Ave., New York, N.Y. 10022, U.S.A.

COQUIMBITE

I. Crystal structures and mineral chemistry of hydrated ferric sulfates. I. The crystal structure of coquimbite. J.H. FANG and P.D. ROBINSON, 1970. *Amer. Min.*, 55, 1534-1540.

$Fe_{1.5}Al_{0.5}(SO_4)_3.9H_2O$ F.W. = 547.56

Trigonal, a = 10.922±9, c = 17.084±14 Å, U = 1764.92 Å³, Z = 4, D_x = 2.06.

Space group $P\bar{3}1c$ (D^2_{3d})

Atomic positions

	x	y	z	B(Å²)
2 Al* in 2(*a*)	0	0	0	1.07±14
2 Fe(1) in 2(*c*)	1/3	2/3	1/4	1.71±10
4 Fe(2) in 4(*f*)	2/3	1/3	1/4	2.19±9
12 S in 12(*i*)	0.2444±2	0.4146±2	0.1232±1	0.49±7
12 O(1) in 12(*i*)	0.3187±7	0.3451±7	0.0909±4	1.42±14

		x	*y*	*z*	$B(\text{Å}^2)$
12	O(2) in 12(*i*)	0.1081±7	0.3106±7	0.1548±4	1.03±14
12	O(3) in 12(*i*)	0.2197±7	0.4946±7	0.0597±4	1.21±14
12	O(4) in 12(*i*)	0.3349±7	0.5158±6	0.1838±3	0.74±13
12	O(5) in 12(*i*)	0.1645±7	0.0698±7	0.0622±4	1.18±14
12	O(6) in 12(*i*)	0.4485±7	0.1153±7	0.2101±4	1.71±16
12	O(7) in 12(*i*)	0.5720±8	0.1616±8	0.0720±4	1.70±15

*includes 10% Fe

Interatomic distances (Å) *and angles*

Al-O(5)	1.889±8 (×6)	O(5)-Al-O(5)	91.4±2° (×6)
			88.6±2° (×6)
Fe(1)-O(4)	2.006±4 (×6)	O(4)-Fe-O(4)	91.3±3° (×6)
			89.3±2° (×3)
			88.1±2° (×3)
			179.2±4° (×3)
Fe(2)-O(3)	1.968±7 (×3)	O(3)-Fe-O(3)	93.5±3° (×3)
Fe(2)-O(7)	2.014±7 (×3)	O(3)-Fe-O(7)	93.1±3° (×3)
			173.2±4° (×3)
			84.7±2° (×3)
		O(7)-Fe-O(7)	88.8±3° (×3)

S - O(2)	1.452±6
S - O(1)	1.469±9
S - O(4)	1.476±5
S - O(3)	1.500±8

Angles in degrees subtended at S by

	O(1)	O(4)	O(3)
O(2)	110.6±4	111.5±4	108.1±4
O(1)		108.3±4	109.7±4
O(4)			108.6±4

Environment of water molecules

O(5)-Al	1.889±7
O(5)-O(1)	2.656±9
O(5)-O(2)	2.694±11

Angles in degrees subtended at O(5) by

	O(1)	O(5)
Al	119.8±4	119.9±3
O(1)		117.9±3

O(6)-O(7)	2.639±10
O(6)-O(2)	2.741±7
O(6)-O(6)	2.743±9
O(6)-O(6)	2.748±9

Angles in degrees subtended at O(6) by

	O(2)	O(6)	O(6)
O(7)	92.4±3	120.0±3	94.7±4
O(2)		121.2±4	129.5±4
O(6)			97.5±3

O(7)-Fe(2)	2.014±7
O(7)-O(6)	2.639±10
O(7)-O(1)	2.646±12

Angles in degrees subtended at O(7) by

	O(6)	O(1)
Fe(2)	132.9±5	125.4±4
O(6)		94.8±3

Discussion

The structure consists of $Al(OH_2)_6$ ions and $(H_2O)_3Fe(SO_4)_3Fe(SO_4)_3Fe(H_2O)_3$ clusters with FeO_6 octahedra and SO_4 tetrahedra sharing corners. The remaining water atoms lie in channels between the ions.

Details of analysis

Crystals from Tierra Amarilla, Chile from U.S. National Museum, Smithsonian catalogue no. 12548. Lattice parameters from precession photographs. Intensities measured on automatic diffractometer with Mn filtered Fe radiation and corrected for Lorentz, polarization and absorption effects (crystal a sphere of diameter 0.35 mm). Structure determined from Patterson and electron-density functions and refined by least squares to give R = 0.061 for 420 observed reflexions listed in 1.

1. NAPS document 01111.

ROEMERITE

$Fe(H_2O)_6[Fe(H_2O)_4(SO_4)_2]_2$

I. The crystal structure of roemerite. L. FANFANI, A. NUNZI and P.F. ZANAZZI, 1970. *Amer. Min.*, 55, 78-89.

Triclinic, a = 6.463±8, b = 15.309±18, c = 6.341±8 Å, α = 90° 32±10', β = 101° 5 ±10', γ = 85° 44±10', U = 308 Å^3, D_m = 2.18, Z = 1, D_x = 2.173.

Space group $P\bar{1}$ (C_i^1)

Atomic positions

All atoms 2(i) except Fe(1) in 1(a)

	x	y	z
Fe(1)	0	0	0
Fe(2)	0.33850±11	0.32040±5	0.59415±12
S(1)	0.59285±20	0.16794±8	0.36415±21
S(2)	0.75980±18	0.38682±8	0.90730±21
O(1)	0.43090±68	0.20817±27	0.48397±73
O(2)	0.58301±76	0.21947±32	0.16464±73
O(3)	0.80677±60	0.16991±27	0.49380±70
O(4)	0.53956±72	0.07819±30	0.32383±91
O(5)	0.62142±65	0.33029±29	0.75584±75
O(6)	0.64356±68	0.42242±27	0.06814±72
O(7)	0.82510±67	0.45660±27	0.78307±78
O(8)	0.94487±60	0.33021±27	0.00778±65
O(9)	0.25089±89	0.04938±49	-0.12976±10
O(10)	0.11209±77	0.04923±35	0.31091±81
O(11)	0.17529±102	-0.11732±35	0.09135±79
O(12)	0.25760±62	0.25577±27	0.84407±64
O(13)	0.03495±59	0.30719±26	0.44316±63
O(14)	0.21791±65	0.43565±26	0.69570±73
O(15)	0.37811±65	0.38822±26	0.33431±69
H(1)	0.337±18	0.018±7	-0.175±19
H(2)	0.276±19	0.117±7	-0.134±19
H(3)	0.233±19	0.063±7	0.350±19
H(4)	0.065±19	0.095±8	0.358±20
H(5)	0.170±19	-0.141±7	0.205±20
H(6)	0.225±19	-0.147±8	0.014±20
H(7)	0.343±19	0.231±7	0.935±19
H(8)	0.160±19	0.277±7	0.908±20
H(9)	-0.040±19	0.259±7	0.440±20
H(10)	0.011±19	0.302±7	0.311±19
H(11)	0.300±19	0.477±7	0.789±19
H(12)	0.055±19	0.431±8	0.709±20
H(13)	0.337±18	0.440±7	0.315±20
H(14)	0.460±19	0.389±8	0.237±20

Anisotropic temperature factors listed.

Interatomic distances (Å) *and angles*

Sixfold coordination around Fe(1) = Fe^{2+} (Crystallographic site symmetry C_i)

Fe(1)-O(9)	2.141	(×2)
Fe(1)-O(10)	2.118	(×2)
Fe(1)-O(11)	2.076	(×2)

σ (Fe-O) = 0.007 Å

Angles in degrees subtended at Fe(1) by

	O(10)	O(10)'	O(11)	O(11)'
O(9)	93.2	86.8	92.2	87.8
O(10)			88.8	91.2

σ (O-Fe-O) - 0.3°

Sixfold coordination around Fe(2) = Fe^{3+}

Fe(2)-O(1)	1.945*
Fe(2)-O(5)	1.933*
Fe(2)-O(12)	2.047
Fe(2)-O(13)	2.035
Fe(2)-O(14)	2.025
Fe(2)-O(15)	2.023

σ (Fe-O) = 0.005 Å

Angles in degrees subtended at Fe(2) by

	O(5)	O(12)	O(13)	O(14)	O(15)
O(1)	89.5	89.0	91.2	175.3	93.8
O(5)		91.0	176.0	95.2	95.8
O(12)			85.1	90.6	172.6
O(13)				84.1	88.0
O(14)					86.0

σ (O-Fe-O) = 0.2°

Fourfold coordination around S(1)

S(1)-O(1)	1.501
S(1)-O(2)	1.486
S(1)-O(3)	1.469
S(1)-O(4)	1.450

σ (S-O) = 0.005 Å

Angles in degrees subtended at S(1) by

	O(2)	O(3)	O(4)
O(1)	108.7	110.8	105.4
O(2)		108.2	113.3
O(3)			110.2

σ (O-S-O) = 0.3°

Fourfold coordination around S(2)

S(2)-O(5)	1.497*
S(2)-O(6)	1.458
S(2)-O(7)	1.466
S(2)-O(8)	1.467

σ (S-O) - 0.005 Å

Angles in degrees subtended at S(2) by

	O(6)	O(7)	O(8)
O(5)	109.2	108.5	106.5
O(6)		111.5	110.9
O(7)			110.5

σ (O-S-O) = 0.3°

*Oxygen atoms bonding to both S and Fe atoms

Water Molecules

A		B	C	ABC	AB	AC	BC
O(4)	....	H(1)	-O(9)	161°	1.95	2.71	0.80
		H(2)	-O(9)	-	-	-	1.06
H(1)	-	O(9)	-H(2)	114°			
O(4)	....	H(3)	-O(10)	156°	2.05	2.82	0.82
O(3)	....	H(4)	-O(10)	149°	2.26	2.99	0.82
H(3)	-	O(10)	-H(4)	91°			
O(3)	....	H(5)	-O(11)	165°	1.94	2.74	0.82
O(2)	....	H(6)	-O(11)	169°	2.09	2.84	0.76
H(5)	-	O(11)	-H(6)	114°			

O(2)	....H(7) -O(12)	156°	1.91	2.66	0.79
O(8)	....H(8) -O(12)	172°	1.78	2.63	0.86
H(7)	- O(12)-H(8)	105°			
O(3)	....H(9) -O(13)	168°	1.82	2.64	0.84
O(8)	....H(10)-O(13)	167°	1.93	2.74	0.82
H(9)	- O(13)-H(10)	89°			
O(6)	....H(11)-O(14)	158°	1.80	2.73	0.98
O(7)	....H(12)-O(14)	159°	1.66	2.70	1.08
H(11)	- O(14)-H(12)	118°			
O(7)	....H(13)-O(15)	163°	1.87	2.67	0.82
O(6)	....H(14)-O(15)	163°	1.85	2.71	0.89
H(13)	- O(15)-H(14)	93°			

σ(O-H) = 0.12, σ(O-O) = 0.007 Å, σ(O-H-O) = 10°

Discussion

The crystal is composed of octahedral $Fe(H_2O)_6^{2+}$ ions hydrogen bonded to $Fe(H_2O)_4(SO_4)_2^-$ ions in which the iron is octahedrally coordinated to four water molecules and two sulphate groups.

Details of analysis

Roemerite from Dexter Mine, Utah. Analysis (theoretical values for assumed stoichiometry are given in brackets) FeO, 9.73 (8.94); Fe_2O_3, 21.39 (19.86); SO_3, 40.02 (39.83); H_2O, 28.71 (31.37). Lattice parameters from precession photographs agree with those of 1. Intensities measured with microdensitometer from integrated precession photographs of layers with h=0-3 and l=0-3 (MoK_α radiation). Lorentz, polarization and absorption corrections (μR = 0.8 for assumed spherical crystal). Structure from three-dimensional Patterson and Fourier functions refined by least squares to give R = 0.061 for 2320 observed reflexions listed in 2.

1. P.R. VAN LOAN and E.W. NUFFIELD, 1959. *Canad. Min.*, 6, 348.
2. *NAPs Document #00710* from ASIS National Auxiliary Publications Service c/o CCM Information Services, Inc. 22 West 34th St., New York, NY 10001.

LOEWEITE

I. Crystal structure and mineral chemistry of double-salt hydrates: II. The crystal structure of loeweite. J.H. FANG and P.D. ROBINSON, 1970. *Amer. Min.*, 55, 378-386.

$Na_{12}Mg_7(SO_4)_{13}15(H_2O)$ F.W. = 1976.9

Rhombohedral, a = 11.769±9 Å, α = 106.5±2°, U = 1375.67 Å^3, D_m = 2.36±1, Z = 3, D_x = 7.14.

Space group $R\bar{3}$ (C_{3i}^2)

Atomic positions

	x	y	z	B(Å^2)
Na(1)	0.9103±3	0.5064±3	0.3569±3	1.54±6
Na(2)	0.3121±3	0.4552±3	0.1928±3	1.79±7
Mg(1)	0.6890±2	0.2068±2	0.1214±2	0.55±5
Mg(2)	0.5	0.5	0.5	0.49±0
S(1)	0.8328±2	0.4537±2	0.0535±2	0.49±3
S(2)	0.5647±2	0.4112±2	0.2335±2	0.49±4
S(3)	0.1195	0.1195	0.1195	1.04±8

	x	y	z	B(Å^2)
O(1)	0.6210±5	0.0255±5	0.1152±5	1.28±0
O(2)	0.7861±6	0.3817±7	0.1269±7	1.51±1
O(3)	0.8511±6	0.0657±6	0.4220±6	1.61±1
O(4)	0.5255±6	0.0269±6	0.2721±6	2.27±2
O(5)	0.6942±5	0.5151±5	0.3116±5	1.08±0
O(6)	0.4874±5	0.4095±5	0.3135±5	0.90±9
O(7)	0.4977±5	0.4319±5	0.1189±5	0.96±9
O(8)	0.5734±5	0.2841±5	0.1942±5	0.89±9
O(9)	0	0	0	3.75± 1
O(W1)	0.8120±5	0.2764±5	0.3161±5	1.22±10
O(W2)	0.8350±5	0.1634±6	0.0701±7	2.57±13
O(W3)	0.1620±11	0.2367±11	0.0762±10	6.96± 9

Interatomic distances (Å) *and angles*

S(1)-O(1)	1.476±6	S(2)-O(5)	1.473±6	S(3)-O(9)	1.601±2
S(1)-O(2)	1.480±6	S(2)-O(6)	1.485±6	S(3)-O(W3)	1.601±13 (×3)
S(1)-O(3)	1.463±6	S(2)-O(7)	1.481±6		
S(1)-O(4)	1.455±6	S(2)-O(8)	1.478±6		

Mg(2)-O(W1)	2.096±6	Mg(2)-O(6)	2.090±5 (×6)
Mg(2)-O(1)	2.026±6		
Mg(2)-O(2)	2.023±6		
Mg(2)-O(3)	2.007±6		
Mg(2)-O(8)	2.082±6		
Mg(2)-O(W2)	2.119±8		

Na(1)-O(5)	2.485±6	Na(2)-O(5)	2.580±6
Na(1)-O(W1)	2.462±6	Na(2)-O(6)	2.441±6
Na(1)-O(7)	2.404±6	Na(2)-O(6)	2.464±6
Na(1)-O(7)	2.616±6	Na(2)-O(7)	2.624±6
Na(1)-O(2)	2.428±6	Na(2)-O(4)	2.310±8
Na(1)-O(3)	2.866±6	Na(2)-O(8)	2.473±6
Na(1)-O(4)	2.352±6	Na(2)-O(W3)	2.378±12

Angles subtended at S(1) by

	O(2)	O(3)	O(4)
O(1)	108.3	110.9	111.1
O(2)		106.5	110.1
O(3)			109.8

Angles subtended at S(2) by

	O(6)	O(7)	O(8)
O(5)	109.0	111.0	110.5
O(6)		108.8	106.8
O(7)			110.5

O(9)-S(3)-O(W3)	108.9
O(W3)-S(3)-O(W3)	110.0

Discussion

The composition $6NaSO_4.7MgSO_4.15H_2O$ has been confirmed by least-squares refinement in the space group $R\bar{3}$. This implies that one of the three independant sulphate groups is disordered over two equivalent sites.

Details of analysis

Three-dimensional intensities collected by counter methods from synthetic material were used in the crystal structure determination, solved by the reiterative application of Sayre's equation and refined to an *R* factor of 6.2%.

MANGANOSTIBITE

I. Manganostibite; a novel cubic close-packed structure type. P.B. MOORE, 1970. *Amer. Min.*, 55, 1489-1499.

Mn_7SbAsO_{12} F.W. = 773.25

Orthorhombic, a = 8.727±5, b = 18.847±6, c = 6.062±4 Å, U = 997.06 Å^3, D_m = 4.95, Z = 4, D_x = 5.14.

Space group Ibmm (D_{2h}^{28})

Atomic positions general position $(0,0,0;\frac{1}{2},\frac{1}{2},\frac{1}{2})\pm(x,y,z;\bar{x},y,\bar{z};\bar{x},y,\frac{1}{2}+z;x,y,\frac{1}{2}-z)$

		x	y	z	B(Å^2)
4 Sb	in 4(c)	0.25	0.25	0.25	0.32±3
4 As	in 4(e)	0.3678±6	0	0.75	0.50±6
8 Mn(1)	in 8(i)	0.3817±5	0.1746±2	0.75	0.29±5
4 Mn(2)	in 4(a)	0	0	0	0.46±7
8 Mn(3)	in 8(f)	0	0.1575±2	0	0.50±6
8 Mn(4)	in 8(i)	0.2794±5	0.0821±3	0.25	0.66±6
8 O(1)	in 8(i)	0.0228±24	0.0755±10	0.25	0.57±26
8 O(2)	in 8(i)	0.0236±22	0.2400±10	0.25	0.47±24
8 O(3)	in 8(i)	-0.0265±23	0.4280±9	0.25	0.45±24
8 O(4)	in 8(h)	0.2589±26	0	-0.0193±31	0.60±23
16 O(5)	in 16(j)	0.2538±19	0.1728±6	0.0268±19	0.30±14

Interatomic distances (Å) *and angles*

Sb-O(2)	1.986±19	(×2)
Sb-O(5)	1.987±11	(×4)

Angles in degrees subtended at Sb by

	O(5)	O(5)"
O(2)	87.0±6	87.0±6
O(2)'	93.0±6	93.0±6
O(5)		85.8±5
O(5)'		94.2±5

As-O(3)	1.641±18	(×2)
As-O(4)	1.691±20	(×2)

Angles in degrees subtended at As by

	O(3)'	O(4)'
O(3)	111.6±10	108.4±5
O(4)		111.6±11

Mn(1)-O(5)	2.016±13	(×2)
Mn(1)-O(2)	2.031±19	
Mn(1)-O(3)	2.093±18	

Angles in degrees subtended at Mn(1) by

	O(5)'	O(2)	O(3)
O(5)	112.7±6	110.6±4	101.3±4
O(2)			119.9±8

Mn(2)-O(1)	2.088±13	(×3)
Mn(2)-O(4)	2.264±23	

Angles in degrees subtended at Mn(2) by

	O(1)"	O(4)
O(1)	85.9±5	93.3±7
O(1)'	94.1±5	86.7±7

Mn(3)-O(1)	2.174±14	(×2)
Mn(3)-O(2)	2.181±14	(×2)
Mn(3)-O(5)	2.241±16	(×2)

Angles in degrees subtended at Mn(3) by

	O(1)'	O(2)'	O(5)'
O(1)	89.4±5	90.8±5	87.2±6
O(1)'		179.8±6	103.5±6
O(2)		89.1±5	93.0±6
O(2)'			76.4±6
O(5)			165.2±7

Mn(4)-O(3)	2.166±20	
Mn(4)-O(5)	2.192±12	(×2)
Mn(4)-O(1)	2.244±21	
Mn(4)-O(4)	2.256±14	(×2)

Angles in degrees subtended at Mn(4) by

	O(5)'	O(1)	O(4)'
O(3)	99.8±6	171.8±12	91.1±7
O(5)	76.3±5	86.6±6	94.6±4
O(5)'		86.6±6	166.8±7
O(1)			83.3±7
O(4)			92.7±5

Discussion

Manganostibite is a cubic close-packed O structure related to spinel but with a different ordering of the cations. In particular a corner sharing tetrahedra grouping of composition Mn_2AsO_{10} occurs.

Details of analysis

Specimen number 28939 (Swedish Natural History Museum) from Brattfors Mine, Nordmarks Odalfält, Värmland, Sweden, contains Mg, Fe and Zn substituted for Mn and Si substituted for As. Lattice parameters from powder photograph calibrated with Si (a=5.4301 Å). Intensities measured on an equi-inclination diffractometer for layers with h = 0 to 12 using monochromated MoK_α radiation and corrected for [Lorentz, polarization and] absorption (μ=15.5 mm^{-1}; crystal a flat bead of volume 0.0039 mm^3). Structure solved from Patterson function and by trial and error and refined to give R = 0.11 for 657 observed reflexions listed in 1.

1. NAPS 01051.

THOREAULITE

I. The crystal structure of $SnTa_2O_7$, Thoreaulite, an example of tin in five-fold coordination. W.G. MUMME, 1970. *Amer. Min.*, 55, 367-377.

$SnTa_2O_7$ F.W. = 592.6

Monoclinic, a = 17.140±5, b = 4.865±2, c = 5.548±2 Å, β = 91.00±5°, U = 469.6 [462.6] Å^3, D_m = 7.6-7.9, Z = 4, D_x = 7.5 [8.49].

Space group C2/c (C^6_{2h})

Atomic positions

	x	y	z	B(Å^2)
4 Sn(1) in 4(e)	0	0.2300±58	0.25	1.2±2
8 Ta(1) in 8(f)	0.1698±4	0.2403±38	0.6675±11	0.9±1
4 O(1) in 4(e)	0	0.392±50	0.75	3.6±24
8 O(2) in 8(f)	0.075±9	0.097±29	0.599±25	2.6±22
8 O(3) in 8(f)	0.124±10	0.567±35	0.427±27	3.6±24
8 O(4) in 8(f)	0.217±8	0.077±30	0.342±23	3.4±19

Interatomic distances (Å) *and angles*

Environment of Sn(1) (Crystallographic site symmetry C_2)

Sn(1)-O(1)	1.84±24	
Sn(1)-O(2)	2.22±14	(×2)
Sn(1)-O(2)'	2.40±14	(×2)

Angles in degrees subtended at Sn(1) by

	O(2)	O(2)'
O(1)	136±4	106±3
O(2)	88±6	79±5
O(2)		79±5
O(2)'		149±5

Ta(1)-O(2)	1.80±15
Ta(1)-O(3)	1.90±16
Ta(1)-O(4)	1.99±14
Ta(1)-O(4)'	2.14±14
Ta(1)-O(4)"	2.14±13
Ta(1)-O(3)'	2.21±16

Angles in degrees subtended at Ta(1) by

	O(3)	O(4)	O(4)'	O(4)"	O(3)'
O(2)	88	99	166	92	81
O(3)		101	102	171	87
O(4)			88	88	172
O(4)'				77	90
O(4)"					84

σ(O-Ta-O) = 6°

Discussion

The structure consists of double layers of edge sharing TaO_6 octahedra (composition TaO_3) linked by 5 coordinated Sn. A preliminary account of this structure has been published by 1 who gave a different Sn coordination.

Details of analysis

Thoreaulite from Monono, Belgium Congo (American Museum of Natural History specimen no. 21932). Lattice parameters from least-squares analysis of measurements from Guinier focussing powder photograph calibrated with KCl [assumed lattice constant for KCl not given].

Intensities estimated visually from multiple-film integrated Weissenberg photographs (l=0, k = 0-5) of crystals 0.02×0.02×0.1 mm^3 using MoK_α radiation. No correction for absorption. Structure solved by Patterson and Fourier methods and refined by least squares to give R = 0.15. Lack of further refinement attributed to anisotropic thermal vibrations and replacement of Sn and Ta by other elements.

1. N.V. MAKSIMOVA and V.V. ILJUKHIN, 1967. *Soviet Physics-Crystallography*, 12, 105.

LANDAUITE

I. [Crystal structure of Landauite]. L.E. NIKOLAEVA and N.V. BELOV, 1970. *Dokl. Akad. Nauk SSSR*, 190, 1098-1110 [*Soviet Physics-Doklady*, 15, 91-93].

$Ti_2(Ti_{0.87}Zn_{0.39}Fe_{0.48}Mn_{0.15})O_7$

Monoclinic, a = 5.225±5, b = 8.955±7, c = 9.783±7 Å, β = 107°, U = 437.7 Å^3, D_m = 4.42, Z = 4, D_x = 4.60 [assuming all Ti].

Space group A2/a (C^6_{2h})

Atomic positions Atoms in (0,0,0;0,½,½)±(x,y,z;½+$x,\bar{y},z$)

	x	y	z	B(Å^2)
8 Ti(1) in 8(f)	0.000	0.100	0.141	1.36
8 Ti(2) in 8(f)	0.25	0.375	0.143	1.58
8 O(1) in 8(f)	0.000	0.482	0.182	2.35
4 O(2) in 4(e)	0.25	0.000	0	2.64
8 O(3) in 8(f)	0.000	0.231	0.000	2.85
8 O(4) in 8(f)	0.25	0.725	0.235	1.89

Substitution occurs in site Ti(1)

Interatomic distances (Å)

Ti(1)-O(3)	1.81
Ti(1)-O(2)	1.83
Ti(1)-O(4)	1.87
Ti(1)-O(1)	1.88
Ti(1)-O(2)	2.34
Ti(1)-O(4)	2.39
Ti(2)-O(1)	1.75
Ti(2)-O(1)	1.79
Ti(2)-O(4)	1.80
Ti(2)-O(3)	2.06
Ti(2)-O(3)	2.53

Errors uncertain but probably 0.02 Å or greater.

Discussion

The bi- and tri-valent ions are in the site Ti(1), the Ti^{4+} ions in Ti(2) which shows a distorted tetragonal pyramidal coordination [the long Ti-O bond is apparently not the axial bond but one of the equatorial bonds]. These pyramids form pairs by sharing one equatorial edge.

Details of analysis

Intensities estimated visually from Weissenberg photographs of layers with h=0 to 4, k=0 to 1 corrected for Lorentz and polarization effects. Structure solved from Patterson function and refined to give R = 0.16 for 547 observed reflexions [not listed].

STRONTIOGINORITE

I. Crystal structure of strontioginorite, $(Sr,Ca)_2B_{14}O_{20}(OH)_6.5H_2O$. J.A. KONNERT, J.R. CLARK and C.L. CHRIST, 1970. *Amer. Min.*, 55, 1911-1931.

$(Sr,Ca)_2B_{14}O_{20}(OH)_6.5H_2O$ F.W. = 791.14

Monoclinic, a = 12.817±8, b = 14.448±8, c = 12.783±8 Å, β = 101.42±8; U = 2320.29 Å^3, D_m = 2.25, Z = 4, D_x = 2.26.

Space group $P2_1/a$ (C^5_{2h})

Atomic positions All atoms in 4(e) ±(x,y,z;½+x,½-y,z)

	x	y	z	B(Å^2)
Sr	0.6178±1	0.1711±1	0.7364±1	*
Ca	0.1264±1	0.2187±1	0.2463±1	*
B(1)	0.3903±8	0.2010±7	-0.1236±8	0.7±2
B(2)	0.4334±8	0.3031±7	0.0504±8	1.0±2
B(3)	0.2622±8	0.1961±7	0.0125±8	0.8±2
B(4)	0.5752±8	0.2551±7	-0.0426±8	0.7±2
B(5)	0.3410±8	0.2573±7	0.1920±8	0.8±2
B(6)	0.3046±9	0.0571±7	-0.0796±8	1.1±2
O(1)	0.3667±4	0.2306±4	-0.0165±4	0.6±1
O(2)	0.4324±5	0.2890±4	0.1635±5	1.0±1
O(3)	0.2627±5	0.2156±4	0.1239±5	1.0±1
O(4)	0.2574±5	0.0967±4	-0.0046±5	1.2±1
O(5)	0.3645±5	0.1038±4	-0.1369±5	1.1±1
O(6)	0.5032±5	0.2121±4	-0.1213±5	1.0±1
O(7)	0.5443±5	0.2942±4	0.0439±5	1.2±1
OH(8)	0.2923±5	-0.0375±5	-0.1003±5	2.1±1
OH(9)	0.3876±5	0.3922±4	0.0146±5	1.7±1
O(10)	0.1774±5	0.2449±4	-0.0557±5	1.2±1

	x	y	z	$B(Å^2)$
O(11)	0.3221±5	0.2682±4	0.2933±5	0.9±1
B(7)	0.3902±8	0.3189±7	0.3781±8	0.7±2
B(8)	0.4223±8	0.2021±7	0.5384±8	0.9±2
B(9)	0.2549±8	0.3110±7	0.5049±8	0.8±2
B(10)	0.5704±8	0.2546±7	0.4567±8	0.8±2
B(11)	0.3364±8	0.2530±7	0.6868±7	0.5±2
B(12)	0.2922±8	0.4571±7	0.4152±8	0.8±2
O(12)	0.3628±4	0.2839±4	0.4816±4	0.5±1
O(13)	0.4122±5	0.2040±4	0.6507±5	0.9±1
O(14)	0.2589±5	0.3007±4	0.6183±4	0.9±1
O(15)	0.2349±5	0.4071±4	0.4757±4	0.9±1
O(16)	0.3680±5	0.4175±4	0.3694±5	1.0±1
O(17)	0.5030±5	0.3049±4	0.3809±4	0.8±1
O(18)	0.5358±5	0.2136±4	0.5399±5	0.9±1
OH(19)	0.2707±5	0.5488±5	0.3972±5	1.9±1
O(20)	0.3810±5	0.1160±4	0.4898±5	1.0±1
O(25)	0.1729±5	0.2518±4	0.4419±4	0.9±1
O(26)	0.3285±5	0.2568±4	0.7919±5	1.0±1
B(13)	0.4099±8	0.0668±7	0.4092±8	0.9±2
B(14)	0.5247±8	0.0623±7	0.2736±8	0.9±2
OH(21)	0.3730±5	-0.0219±4	0.3899±5	1.5±1
O(22)	0.4742±5	0.1041±4	0.3465±5	1.3±1
OH(23)	0.5046±5	-0.0264±5	0.2415±5	1.8±1
OH(24)	0.5952±5	0.1170±4	0.2329±5	1.5±1
$H_2O(25)$	0.1096±6	0.1489±5	0.7207±6	2.3±1
$H_2O(26)$	0.1537±6	0.0614±5	0.2715±6	2.6±1
$H_2O(27)$	0.2367±5	0.4502±5	0.8493±5	2.1±1
$H_2O(28)$	0.0167±6	0.0384±5	0.8661±6	2.8±1
$H_2O(29)$	0.0509±6	-0.0246±5	0.4008±6	2.8±1

*Anisotropic temperature factors given.

Interatomic distances (Å) *and angles*

Sr	-O(27)	2.571±6
Sr	-O(18)	2.597±6
Sr	-O(14)	2.608±6
Sr	-O(25)	2.609±7
Sr	-O(6)	2.623±7
Sr	-OH(23)	2.663±7
Sr	-O(13)	2.690±6
Sr	-OH(21)	2.710±6
Sr	-O(26)	2.852±6
Sr	-O(10)	2.884±6
Ca	-O(26)	2.312±7
Ca	-OH(24)	2.408±6
Ca	-O(25)	2.499±5
Ca	-O(2)	2.506±6
Ca	-O(11)	2.564±6
Ca	-O(3)	2.567±7
Ca	-O(17)	2.581±6
Ca	-O(7)	2.600±6

B(1)- O(5)	1.44±1
B(1)- O(26)	1.45±1
B(1)- O(6)	1.45±1
B(1)- O(1)	1.52±1

Angles in degrees subtended at B(1) by

	O(26)	O(6)	O(1)
O(5)	111.9±7	108.3±7	107.3±7
O(26)		110.4±8	109.7±7
O(6)			109.0±7

B(2)- O(7)	1.45±1
B(2)-OH(9)	1.45±1
B(2)- O(2)	1.46±1
B(2)- O(1)	1.51±1

Angles in degrees subtended at B(2) by

	OH(9)	O(2)	O(1)
O(7)	113.7±8	104.2±7	111.1±8
OH(9)		110.7±8	106.8±6
O(2)			110.4±7

B(3)- O(10)	1.44±1
B(3)- O(3)	1.45±1
B(3)- O(4)	1.45±1
B(3)- O(1)	1.54±1

Angles in degrees subtended at B(3) by

	O(3)	O(4)	O(1)
O(10)	110.9±8	112.9±7	106.6±7
O(3)		109.4±7	109.8±7
O(4)			107.2±7

B(4)- O(10)	1.35±1
B(4)- O(7)	1.37±1
B(4)- O(6)	1.37±1

Angles in degrees subtended at B(4) by

	O(7)	O(6)
O(10)	122.8±8	115.9±9
O(7)		121.3±9

B(5)- O(3)	1.33±1
B(5)- O(11)	1.37±1
B(5)- O(2)	1.37±1

Angles in degrees subtended at B(5) by

	O(11)	O(2)
O(3)	114.6±9	123.3±9
O(11)		122.1±8

B(6)- O(5)	1.34±1
B(6)- O(4)	1.36±1
B(6)-OH(8)	1.40±1

Angles in degrees subtended at B(6) by

	O(4)	OH(8)
O(5)	123.8±8	116.4±9
O(4)		119.8±9

B(7)- O(11)	1.45±1
B(7)- O(16)	1.45±1
B(7)- O(17)	1.45±1
B(7)- O(12)	1.52±1

Angles in degrees subtended at B(7) by

	O(16)	O(17)	O(12)
O(11)	111.0±7	113.4±8	105.9±7
O(16)		108.5±7	108.5±8
O(17)			109.4±6

B(8)- O(20)	1.44±1
B(8)- O(18)	1.46±1
B(8)- O(13)	1.47±1
B(8)- O(12)	1.52±1

Angles in degrees subtended at B(8) by

	O(18)	O(13)	O(12)
O(20)	112.5±8	110.0±8	111.1±7
O(18)		105.4±7	108.9±7
O(13)			108.7±7

B(9)- O(15)	1.45±1
B(9)- O(14)	1.45±1
B(9)- O(25)	1.47±1
B(9)- O(12)	1.52±1

Angles in degrees subtended at B(9) by

	O(14)	O(25)	O(12)
O(15)	109.0±7	110.3±7	108.5±7
O(14)		111.7±8	108.4±7
O(25)			108.9±7

B(10)- O(18)	1.37±1
B(10)- O(25)	1.37±1
B(10)- O(17)	1.37±1

Angles in degrees subtended at B(10) by

	O(25)	O(17)
O(18)	123.2±8	121.5±8
O(25)		115.2±8

B(11)- O(13)	1.36±1
B(11)- O(26)	1.37±1
B(11)- O(14)	1.37±1

Angles in degrees subtended at B(11) by

	O(26)	O(14)
O(13)	123.6±8	121.5±8
O(26)		114.9±8

B(12)- O(16)	1.36±1
B(12)-OH(19)	1.36±1
B(12)- O(15)	1.37±1

Angles in degrees subtended at B(12) by

	OH(19)	O(15)
O(16)	118.4±9	122.0±8
OH(19)		119.6±9

B(13)- O(20)	1.36±1
B(13)- O(22)	1.37±1
B(13)-OH(21)	1.37±1

Angles in degrees subtended at B(13) by

	O(22)	OH(21)
O(20)	121.5±8	119.2±9
O(22)		119.3±9

B(14)-OH(23)	1.36±1
B(14)- O(22)	1.38±1
B(14)-OH(24)	1.38±1

Angles in degrees subtended at B(14) by

	O(22)	OH(24)
OH(23)	122.3±9	121.9±9
O(22)		115.9±8

Ring angles in degrees

B(1)-O(1)-B(2)	121.0±7	B(7)-O(12)-B(8)	119.8±7
B(2)-O(7)-B(4)	121.9±7	B(8)-O(13)-B(10)	121.0±7
B(4)-O(6)-B(1)	126.1±7	B(10)-O(17)-B(7)	125.1±7
B(1)-O(1)-B(3)	117.6±6	B(7)-O(12)-B(9)	117.4±6
B(3)-O(4)-B(6)	120.8±7	B(9)-O(15)-B(12)	124.3±7
B(6)-O(5)-B(1)	124.4±7	B(12)-O(16)-B(7)	121.8±7
B(2)-O(1)-B(3)	120.7±7	B(8)-O(12)-B(9)	119.9±7
B(3)-O(3)-B(5)	125.8±7	B(9)-O(14)-B(11)	124.6±7
B(5)-O(2)-B(2)	119.0±7	B(11)-O(13)-B(8)	123.4±7

Hydrogen bonds

OH(8)-OH(24)	2.69±1	OH(19)-H_2O(25)	2.76±1
OH(8)-OH(9)	2.92±1	OH(19)- O(20)	2.82±1
OH(9)-H_2O(27)	2.70±1	OH(21)- O(15)	2.62±1
OH(9)-H_2O(28)	2.75±1	OH(21)- O(23)	2.78±1
OH(9)-OH(8)	2.92±1	OH(23)- O(5)	2.60±1
OH(9)-H_2O(28)	2.93±1	OH(23)- OH(21)	2.78±1
		OH(24)- OH(8)	2.69±1
		H_2O(25)-H_2O(28)	2.88±1
		H_2O(25)-H_2O(29)	2.93±1
		H_2O(26)-H_2O(29)	2.62±1
		H_2O(26)-H_2O(27)	2.79±1
		H_2O(26)-H_2O(28)	2.90±1

Angles in degrees subtended at OH(8) by

	OH(24)	OH(9)
B(6)	118.9±6	109.7±6
OH(24)		130.4±3

Angles in degrees subtended at OH(9) by

	$H_2O(27)$	$H_2O(28)$
B(2)	134.8±5	113.5±5
$H_2O(27)$		111.6±3

Angles in degrees subtended at OH(19) by

	$H_2O(25)$	O(20)
B(12)	119.0±6	112.7±6
$H_2O(25)$		127.9±3

Angles in degrees subtended at OH(21) by

	O(15)	Sr	OH(23)
B(13)	117.1±6	142.8±6	84.8±6
O(15)		99.8±2	155.5±3
Sr			58.0±2

Angles in degrees subtended at OH(23) by

	O(5)	Sr	OH(21)
B(14)	117.4±6	142.3±6	82.9±6
O(5)		99.9±3	154.6±4
Sr			59.7±2

Angles in degrees subtended at OH(24) by

	Ca	OH(8)
B(14)	130.6±6	117.8±6
Ca		111.3±3

Angles in degrees subtended at $H_2O(25)$ by

	OH(19)	$H_2O(28)$	$H_2O(29)$
Sr	124.1±3	121.3±3	131.6±3
OH(19)		114.0±3	78.4±3
$H_2O(28)$			70.5±3

Angles in degrees subtended at $H_2O(26)$ by

	$H_2O(29)$	$H_2O(27)$	$H_2O(28)$
Ca	118.2±4	124.6±4	109.2±3
$H_2O(29)$		116.5±3	74.6±3
$H_2O(27)$			77.3±3

Angles in degrees subtended at $H_2O(27)$ by

	OH(9)	$H_2O(26)$	O(4)
Sr	118.8±3	112.9±3	144.7±3
OH(9)		104.5±3	76.9±2
$H_2O(26)$			90.6±3

Angles in degrees subtended at $H_2O(28)$ by

	$H_2O(25)$	$H_2O(26)$	OH(9)
OH(9)	126.0±4	98.9±3	98.7±3
$H_2O(25)$		104.0±3	125.9±3
$H_2O(26)$			96.2±3

Angles in degrees subtended at $H_2O(29)$ by

	$H_2O(25)$	O(16)
$H_2O(26)$	110.1±3	99.2±3
$H_2O(25)$		77.3±3

Discussion

Strontioginorite contains polymerised borate polyions each containing two units of three condensed B_3O_3 rings. The polyions form (010) sheets containing also the Sr and Ca atoms but the sheets are held together only by hydrogen bonds and Sr atoms giving rise to a characteristic platy habit.

Details of analysis

Sample described by 1. Milliprobe analysis confirms Sr: Ca ratio of 1:1. Lattice parameters from calibrated precession photographs [calibration not given]. Intensities measured on automatic diffractometer using 2θ scan and CuK_α radiation and corrected for Lorentz, polarization absorption (μ=6.71 mm^{-1}, crystal size 0.27×0.20×0.06 mm^3). Structure solved by symbolic addition and electron-density functions and refined by least squares to give R_2 = 0.070 for 2350 observed reflexions listed in 2.

1. O. BRAITSCH, 1959. *Contrib. Miner. Petrology*, 6, 366.
2. NAPS 01112.

PROSOPITE

I. [Structure of prosopite, $Ca[Al_2F_4(OH)_4]$, and some crystallochemical features of fluoroaluminates]. Z.V. PUDOVKINA and JU.A. PJATENKO, 1970. *Dokl. Akad. Nauk SSSR*, 190, 665-667 [*Doklady Akad. Nauk SSSR Earth Sciences*, 190, 131-133].

$CaAl_2F_4(OH)_4$ F.W. = 238.04

Monoclinic, a = 6.76±3, b = 11.12±4, c = 7.32±3Å, β = 95.0±3°, U = 538 Å^3, Z = 4, D_x = 2.88.

Space group C2/c (C^6_{2h})

Atomic positions

		x	y	z
4 Ca	in 4(e)	0	0.461	¼
4 Al(1)	in 4(c)	¼	¼	0
4 Al(2)	in 4(e)	0	0.143	¼
8 F(1)	in 8(f)	0.077	0.382	0.965
8 F(2)	in 8(f)	0.182	0.021	0.285
8 O(1)	in 8(f)	0.201	0.264	0.257
8 O(2)	in 8(f)	0.023	0.150	0.997

Interatomic distances (Å)

Al(1)-F	1.88 (×2)	Ca-F	2.29 (×2)
Al(1)-O	1.89 (×2)	Ca-F	2.36 (×2)
Al(1)-O	1.93 (×2)	Ca-F	2.39 (×2)
Al(2)-F	1.83 (×2)	Ca-O	2.58 (×2)
Al(2)-O	1.87 (×2)		
Al(2)-O	1.91 (×2)		

Discussion

The AlO_4F_2 octahedra share O-O edges to form chains running along [$\bar{1}$01]. The chains are linked by Ca atoms.

Details of analysis

Lattice parameters from oscillation photographs. Intensities estimated visually from layer line photographs (h=0, l=0,1) taken with a spherical crystal (μR= 0.23). Structure solved from Patterson function and refined by electron-density projections calculated assuming space group *Cc* (in order to eliminate overlapping peaks [*sic*]). $R \sim 0.11$ for 259 observed zero-layer reflexions [not listed]. Interchange of O and F atoms increased R by 0.02.

β-EUCRYPTITE

$LiAlSiO_4$ F.W. = 125.90

I. Structure and superstructure of β-eucryptite. V. TSCHERRY and H. SCHULZ, 1970. *Naturwissenschaften*, 57, 194.

Hexagonal, a = 10.4918±3, c = 11.1749±7 Å, U = 1231 Å^3, Z = 12, D_x = 2.03.

Space group $P6_422$ (D_6^5)

Atomic positions

	x	y	z
3 Li(1) in 3(c)	½	0	0
3 Li(2) in 3(b)	0	0	½
6 Li(3) in 6(f)	½	0	0.333
6 Si(1) in 6(g)	0.249	0	0
6 Si(2) in 6(i)	0.249	2x	0
6 Al(1) in 6(h)	0.252	0	½
6 Al(2) in 6(j)	0.252	2x	½
12 O(1) in 12(k)	0.202	0.087	0.091
12 O(2) in 12(k)	0.397	0.107	0.930
12 O(3) in 12(k)	0.403	0.097	0.407
12 O(4) in 12(k)	0.593	0.203	0.250

B(Li) = 1.6, B(Si) = 0.4, B(Al) = 0.3, B(O) = 0.8 Å^2
σ = 0.001 Å.

Discussion

β-eucryptite has the stuffed high quartz structure with Si and Al alternating in layers along the c axis. The Li atoms in the channel at x = 0, y = 0 lie at the same level as the Al atoms while those lying in the remaining three channels are at the same level as the Si atoms.

The average bond distances are: Si - O = 1.604, Al - O = 1.738, and Li - O = 2.03 Å (all four-coordinate). Previous structural work is given in 1.

Details of analysis

The crystal used had composition $Li_{0.95}Al_{0.98}Si_{1.03}O_4$ and showed sharp *a* and *c* reflexions. Intensities measured on a single-crystal diffractometer (MoK_α radiation, θ - 2θ scan). No extinction correction. Structure refined first in the high-quartz subcell ($a' = a/2$, $b' = b/2$) and then in the supercell to give $R = 0.050$ for the 497 non-zero subcell reflexions and 0.11 for the 376 non-zero supercell reflexions (no intensities given).

1. *Structure Reports*, 11, 474.

HIGH-QUARTZ LITHIUM ALUMINUM SILICATE

I. The crystal structure of $Li_2Al_2Si_3O_{10}$ (high quartz solid solution). C.-T. Li, 1970. *Z. Kristallogr.*, 132, 118-128.

$Li_2Al_2Si_3O_{10}$

Hexagonal, $a = 5.238\pm1$, $c = 5.472\pm1$ Å, $U = 130.02$ Å^3, $D_m = 2.39$, $Z = 0.6$, $D_x = 2.39$.

Space group $P6_222$ (D_6^4)

Atomic positions

	x	*y*	*z*
(Si,Al)	0	0	0
O	0.2041 5	0.4082 5	0.5
Li	0	0	0

Interatomic distances (Å) *and angles*

Al,Si-O	1.649±1 (×4)
O-Al,Si-O	104.0±3 111.7±2 112.8±1°
Li-O	2.064±4 (×4)
O-Li-O	78.0±1, 126.70±2, 127.6±1°
Si,Al-O-Si,Al	150.8±3°

Discussion

The structure is based on the high-quartz framework. However, in contrast to more silica-rich material, there is no reconstructive change to the keatite structure at elevated temperatures. This behaviour has been rationalised in terms of the lack of repulsive forces due to the long T - Li distances, the lower concentration of Li and the relatively low strain energy of the framework linkages.

Details of analysis

Synthetic crystals of composition $1.0Li_2O.Al_2O_3.3.04SiO_2$ grown from a glass melt were used in the collection of three-dimensional data by counter methods. The structure was refined to an *R* factor of 4.1%.

NEPHELINE

I. Refinement of the nepheline structure at several temperatures. N. FOREMAN and D.R. PEACOR, 1970. *Z. Kristallogr.*, 132, 45-70.

II. Least squares refinement of the structure of a plutonic nepheline. W.A. DOLLASE, 1970. *Z. Kristallogr.*, 132, 27-44.

$(Na,Ca,K)_8Al_8Si_8O_{32}$

Hexagonal, a = 9.993±2, c = 8.374±3 Å, U = 724 Å^3, Z = 1 (I)
a = 10.007±7, c = 8.385±5 Å, U = 727 Å^3, Z = 1 (II)

Space group $P6_3$ (C_6^6)

Atomic positions

I reports positions at 8°C, 214°C, 399°C, 608°C and 908°C
II reports positions at room temperature (given below)

	x	y	z
K	0	0	1/4
Na	0.9965±4	0.4427±4	1/4
T(1)	1/3	2/3	0.4445±3
T(2)	1/3	2/3	0.0555±3
T(3)	0.0939±1	1/3	0.5638±2
T(4)	0.0930±1	0.3333±1	0.9362±2
O(1) (1/3 occupied)	0.2914±19	0.6315±23	1/4
O(2)	0.0279±7	0.3179±8	3/4
O(3)	0.1721±10	0.5248±9	0.9844±18
O(4)	0.1638±10	0.5086±9	0.4954±18
O(5)	0.2258±4	0.2860±8	0.5609±5
O(6)	0.2258±4	0.2689±8	0.9391±5

Interatomic distances (Å) *and angles*

K - O(6)	2.964±5
K - O(2)	3.052±4
K - O(5)	3.057±5
Na - O(5)	2.513±5
Na - O(4)	2.524±13
Na - O(2)	2.527±10
Na - O(6)	2.554±5
Na - O(1)	2.591±15
Na - O(3)	2.699±13
Na - O(3)	2.714±14
Na - O(4)	2.856±14
Na - O(1)	3.125±7

T(1)-O(1)	1.677±6
T(1)-O(4)	1.697±10 (×3)

Angles in degrees subtended at T(1) by

	O(4)	O(4)'	O(4)"
O(1)	112.0±6	109.8±6	91.1±9
O(4)		113.9±5	113.9±5

T(2)-O(3)	1.639±10 (×3)
T(2)-O(1)	1.677±5

Angles in degrees subtended at T(2) by

	O(3)"	O(1)
O(3)	107.6±5	97.9±8
O(3)'	107.6±5	118.0±5
O(3)"		117.0±6

T(3)-O(5)	1.611±7
T(3)-O(6)	1.630±4
T(3)-O(4)	1.633±9
T(3)-O(2)	1.672±3

Angles in degrees subtended at T(3) by

	O(6)	O(4)	O(2)
O(5)	109.9±3	109.6±4	109.4±3
O(6)		104.5±5	112.0±2
O(4)			111.4±6

T(4)-O(2)	1.669±3
T(4)-O(3)	1.716±9
T(4)-O(5)	1.719±4
T(4)-O(6)	1.744±7

Angles in degrees subtended at T(4) by

	O(3)	O(5)	O(6)
O(2)	105.3±5	109.0±2	108.3±3
O(3)		109.0±4	113.4±4
O(5)			111.5±3

Discussion

I. It was concluded that plutonic nepheline has virtually complete Al-Si ordering. The satellites to the 001 reflexions are probably due to the ordering of the K site vacancies which results in formation of domains; this provides a rationalization for the regular non-stoichiometric composition observed in nephelines of this type. The thermal expansion of the *a* (2.3×10^{-4}Å/°C) and *c* (1.06×10^{-4}Å/°C) axes can be explained by a rotation of the individual Si and Al tetrahedra.

II. A non-centric model ($P6_3$) was used for the refinement. The constraints were applied such that the majority of the structure obeyed the space group $P6_3/m$. The positionally disordered nature of the O(1) atom has been confirmed and explained as being the result of the substitution of Al for Si on the *T*(1) site.

Details of analysis

I. Natural material from Bancroft, Ontario, Canada of composition $Na_{5.64}Ca_{0.36}K_{1.32}\square_{0.68}Al_{7.93}Si_{8.07}O_{32}$ was used to gather three-dimensional counter collected intensity data at a sequence of temperatures. The structures were refined to the following *R* factors: 5.3%(8°C), 5.6%(214°C), 6.5%(399°C), 7.7%(608°C), 10.7%(908°C) and 4.8%(8°C).

II. Natural material from Larvik, Norway of composition $K_{1.40}Na_{5.47}Ca_{0.30}Al_{7.47}Si_{8.53}O_{32}$ was used in the collection of three-dimensional counter data. The structure refined to an *R* factor of 8.0%.

SODA MELILITE

I. The crystal structure of synthetic soda melilite. S.J. LOUISNATHAN, 1970. *Z. Kristallogr.*, 131, 314-321.

$CaNaAlSi_2O_7$ F.W. = 258.21

Tetragonal, *a* = 7.6344±6, *c* = 5.0513±6 Å, *U* = 294.41 Å^3, *Z* = 2, D_x = 2.91.

Space group $P\bar{4}2_1m$ (D^3_{2d}) Systematic absences

Atomic positions

		x	*y*	*z*
2 Ca+2 Na	in 4(*e*)	0.3399±2	0.1601	0.5134±3
2 Al	in 2(*a*)	0	0	0
4 Si	in 4(*e*)	0.1416±2	0.3584	0.9531±3
2 O(1)	in 2(*c*)	0.5	0	0.1687±12
4 O(2)	in 4(*e*)	0.1428±6	0.3572	0.2652±9
8 O(3)	in 8(*f*)	0.0852±4	0.1747±5	0.8120±6

Anisotropic temperature factors given.

Interatomic distances (Å) and angles

Ca,Na – O(1)	2.454±4	
Ca,Na – O(3)	2.463±3	(×2)
Ca,Na – O(2)	2.470±4	
Ca,Na – O(2)	2.572±2	(×2)
Ca,Na – O(3)	2.793±3	(×2)

Al – O(3)	1.762±3	(×4)	O(3)–Al–O(3)'	114.8±1 (×2)
			O(3)–Al–O(3)"	106.9±1 (×4)

Si – O(2)	1.577±5	
Si – O(3)	1.631±4	(×2)
Si – O(1)	1.648±2	

Angles in degrees subtended at Si by

	O(3)'	O(1)
O(2)	115.7±2	112.4±2
O(3)	105.3±2	103.2±2

Si–O(1)–Si 136.2±4°

Discussion

Soda melilite has the melilite structure with the Si and Al atoms ordered on tetrahedral sites and the Ca and Na atoms disordered on sites with eightfold coordination.

Details of analysis

Sample prepared at 1150°C and 10 Kbar pressure. Lattice parameters from powder patterns using Si as internal standard [no value given for a_o]. Intensities measured on manual Weissenberg diffractometer for layers k = 0 to 6 using MoK_α radiation. Corrected for Lorentz, polarization transmission and absorption (μ = 1.64 mm^{-1}, crystal size 0.06×0.1×0.2 mm^3). Structure of melilite assumed and refined by least squares to give R_2 = 0.051 for 491 reflexions listed.

STILBITE

I. Crystal structure of stilbite. M. SLAUGHTER, 1970. *Amer. Min.*, 55, 387–397.

$Na_{2.7}Ca_{3.4}Mg_{0.2}(Al_{9.9}Si_{26.4}O_{72.0})24.5H_2O$

Monoclinic, a = 13.69±1, b = 18.25±1, c = 11.31±2 Å, β = 128.2±2°, U = 2220.61 Å^3, Z = 8.

Space group C2/m (C^3_{2h})

Atomic positions

	x	y	z	B(Å^2)
Si(1)	0.9828±3	0.1960±3	0.2421±3	0.88
Si(2)	0.2646±3	0.3093±4	0.2610±3	0.77
Si(3)	0.1892±3	0.0889±3	0.4841±3	0.90
Si(4)	0.1110±3	0.3168±3	0.5004±3	0.75
Si(5)	0	0.2623±5	0	1.27
O(1)	0.9635±9	0.2078±9	0.0843±8	1.48
O(2)	0.1158±7	0.3154±11	0.1190±10	1.33
O(3)	0.0499±11	0.2673±10	0.3508±10	2.05
O(4)	0.0614±12	0.1191±9	0.3223±7	1.80
O(5)	0.2938±12	0.2307±9	0.3456±8	1.66
O(6)	0.3036±11	0.3787±11	0.3755±12	1.78
O(7)	0.3410±11	0.3165±10	0.1934±9	1.55
O(8)	0.3153±13	0.1126±12	0.5012±9	1.88
O(9)	0.1863±12	0	0.4872±11	1.96
O(10)	0	0.3509±10	0.5	1.50

	x	y	z	$B(\text{Å}^2)$
Ca	0.2806±6	0	0.0959±5	2.30
Na(1) (×0.25)	0.4861±40	0.0721±3	0.9535±57	3.30
Na(2) (×0.04)	0.1260±41	0.055±28	0.725±31	3.30
Na(3) (×0.25)	0.035±44	0.035±53	0.542±60	3.30
X (×0.04)	0.475±45	0.056±29	0.579±72	3.30
O(W1)	0.3337±29	0.1241±13	0.1038±23	4.00
O(W2)	0.1479±16	0.0793±24	0.1168±28	4.30
O(W3)	0.1007±38	0	0.8385±29	3.90
O(W4)	0.3668±46	0	0.3554±24	4.10
O(W5)	0.5130±25	0	0.2472±46	4.10
O(W6)	0.3281±37	0.0209±56	0.8960±28	4.10

Interatomic distances (Å) *and angles*

Environment of Si(1)

Si(1)-O(3)	1.63±2
Si(1)-O(1)	1.65±1
Si(1)-O(4)	1.66±14
Si(1)-O(7)	1.68±2

Angles in degrees subtended at Si(1) by

	O(1)	O(4)	O(7)
O(3)	110.5±8	113.0±27	110.8±9
O(1)		108.6±11	106.5±5
O(4)			107.1±23

Environment of Si(2)

Si(2)-O(5)	1.63±2
Si(2)-O(7)	1.64±2
Si(2)-O(2)	1.64±1
Si(2)-O(6)	1.65±2

Angles in degrees subtended at Si(2) by

	O(7)	O(2)	O(6)
O(5)	110.4±9	109.8±9	111.9±7
O(7)		107.6±6	108.4±9
O(2)			108.6±9

Environment of Si(3)

Si(3)-O(9)	1.62±1
Si(3)-O(6)	1.64±2
Si(3)-O(4)	1.66±5
Si(3)-O(8)	1.67±2

Angles in degrees subtended at Si(3) by

	O(6)	O(4)	O(8)
O(9)	109.1±9	109.4±55	107.1±10
O(6)		109.8±30	110.8±7
O(4)			110.6±24

Environment of Si(4)

Si(4)-O(3)	1.62±1
Si(4)-O(5)	1.64±1
Si(4)-O(10)	1.64±1
Si(4)-O(8)	1.64±2

Angles in degrees subtended at Si(4) by

	O(5)	O(10)	O(8)
O(3)	111.9±8	109.2±7	110.5±8
O(5)		109.3±7	109.7±7
O(10)			106.1±1

Environment of Si(5)

Si(5)-O(2)	1.62±1	(×2)
Si(5)-O(1)	1.66±1	(×2)

Angles in degrees subtended at Si(5) by

	O(2)	O(1)
O(2)	106.6±10	111.7±6
O(2)		110.4±7
O(1)		106.1±9

Environment of Ca

Ca - O(W1)	2.36±3	(×2)
Ca - O(W3)	2.38±2	
Ca - O(W4)	2.39±3	
Ca - O(W2)	2.45±4	(×2)
Ca - O(W5)	2.52±3	
Ca - O(W6)	2.74±5	(×2)

Partially occupied cation sites

Na(1)-O(2)	2.601	Na(2)-O(4)	2.56	Na(3)-O(6)	2.38
Na(1)-O(2)	2.328	Na(2)-O(6)	2.25	Na(3)-O(9)	2.58
Na(1)-O(W5)	2.630	Na(2)-O(7)	2.46	Na(3)-O(W3)	2.95
Na(1)-O(W6)	2.500	Na(2)-O(W6)	2.59		

X-O(8)	2.06
X-O(10)	2.04
X-O(W4)	2.24
X-O(W5)	2.13

Discussion

Stilbite differs from other related zeolites in that it has a relatively large framework cavity occupied by Ca ions which are surrounded by water molecules rather than being bonded directly to the framework. Al has a slight preference for the Si(1) cation site.

Details of analysis

The structure of material from Nova Scotia, Canada of composition $(Na_{6\cdot08}K_{0\cdot02}Ca_{2\cdot02}Mg_{0\cdot01})(Al_{10\cdot38}Si_{25\cdot62}O_{72})20\cdot38H_2O$ was refined using three-dimensional counter-collected intensities to an *R* factor of 11%.

CANCRINITE

I. Chemistry of soil minerals. Part VII. Synthesis, properties, and crystal structure of salt-filled cancrinites. R.M. BARRER, J.F. COLE and H. VILLIGER, 1970. *J. Chem. Soc. Lond.* A 1523-1531.

BASIC CANCRINITE

$6(NaAlSiO_4).1\cdot08(SiO_2).0\cdot57(Na_2O).4\cdot05(H_2O)$

Trigonal, a = 12.72±2, c = 5.19±1 Å, U = 728 Å^3, D_m = 2.32±2, Z = 1, D_x = 2.34.

Space group $P3$ (C_3^1)

Atomic positions

Based on pseudo space group $P6_3$ (C_6^6)

	Position based on $P6_3$	x	y	z	B(Å^2)
6 Al	in 6(*c*)	0.0765±48	0.4147±28	0.750	0.2±5
6 Si	in 6(*c*)	0.3285±51	0.4110±36	0.740±18	1.7±5
6 O(1)	in 6(*c*)	0.2070±60	0.4009±27	0.642±11	1.5±5
6 O(2)	in 6(*c*)	0.1228±32	0.5698±73	0.703±16	4.3±8
6 O(3)	in 6(*c*)	0.0287±54	0.3638±48	0.051±16	2.1±8
6 O(4)	in 6(*c*)	0.3168±46	0.3558±47	0.034±12	1.4±9
0.98±10 O(5)	in 2(*a*)	0	0	0.426±53	5.3±21
2.64±18 O(6)	in 6(*c*)	0.1184±73	0.1061±80	0.452±28	1.7±14

	Position based on $P6_3$	x	y	z	$B(\text{Å}^2)$
2.16±18 O(7)	in 6(*c*)	0.063±37	0.0809±88	0.608±30	6.9±27
1.82±14 O(8)	in 2(*b*)	0.3333	0.6667	0.176±26	7.4±21
4.20±12 Na(1)	in 6(*c*)	0.1410±50	0.2930±24	0.259±13	1.6±6
1.50±12 Na(2)	in 6(*c*)	0.115±15	0.1933±94	0.228±52	3.7±21
1.44±12 Na(3)	in 2(*b*)	0.3333	0.6667	0.628±22	7.9±24

NITRATE CANCRINITE

$6(NaAlSiO_4).1{\cdot}86(NaNO_3).1{\cdot}44(H_2O)$

Trigonal, a = 12.67±2, c = 5.19±1 Å, U = 722 Å^3, D_m = 2.32±2, Z = 1, D_x = 2.38.

Space group P3 (C_3^1)

Atomic positions

Based on the pseudo-space group $P6_3$

	Position based on $P6_3$	x	y	z	$B(\text{Å}^2)$
6 Al	in 6(*c*)	0.0733±50	0.4097±45	0.7500	3.9±8
6 Si	in 6(*c*)	0.3332±52	0.4139±24	0.752±17	1.2±5
6 O(1)	in 6(*c*)	0.2170±65	0.4030±46	0.644±15	5.1±13
6 O(2)	in 6(*c*)	0.1231±52	0.5646±68	0.689±15	4.1±13
6 O(3)	in 6(*c*)	0.0340±44	0.3164±40	0.014±18	0.9±13
6 O(4)	in 6(*c*)	0.3279±59	0.3442±53	0.028±17	2.3±12
2.52±24 O(5)	in 6(*c*)	0.112±9	0.031±10	0.290±33	2.8±21
1.62±34 O(6)	in 2(*b*)	0.3333	0.6667	0.437±30	3.7±32
0.88±30 O(7)	in 2(*b*)	0.3333	0.6667	0.168±79	6.4±62
5.22±18 Na(1)	in 6(*c*)	0.1231±48	0.2676±27	0.268±13	1.6±7
1.80±18 Na(2)	in 2(*a*)	0	0	0.407±23	6.7±27
1.14±14 Na(3)	in 2(*b*)	0.3333	0.6667	0.860±26	1.0±20

Interatomic distances (Å)

	Basic cancrinite	Nitrate cancrinite
Al-O(1)	1.86	1.89
Al-O(2)	1.77	1.78
Al-O(3)	1.69	1.54
Al-O(4)	1.77	1.64
	σ(Al-O) 0.10	σ(Al-O) 0.12
Si-O(1)	1.57	1.53
Si-O(2)	1.63	1.65
Si-O(3)	1.59	1.83
Si-O(4)	1.66	1.62
	σ(Si-O) 0.10	σ(Si-O) 0.10

Basic cancrinite		Nitrate cancrinite	
Na(1)-O(1)	2.32±6	Na(1)-O(1)	2.47±7
Na(1)-O(3)	2.31±8	Na(1)-O(3)	2.26±11
Na(1)-O(3)	3.02±8	Na(1)-O(3)	2.83±10
Na(1)-O(4)	2.29±8	Na(1)-O(4)	2.64±10
Na(1)-O(4)	2.85±8	Na(1)-O(4)	2.72±11
Na(1)-O(6)	2.46±13	Na(1)-O(5)	2.23±17
Na(1)-O(7)	2.71±45		

Na(2)-O(4)	2.57±19	Na(2)-O(5)	2.37±17 (×3)
Na(2)-O(5)	2.38±22		
Na(2)-O(5)	2.66±24	Na(3)-O(2)	2.48±7 (×3)
Na(2)-O(7)	2.15±49	Na(3)-O(7)	2.12±20
Na(3)-O(1)	2.94±3 (×3)		
Na(3)-O(2)	2.36±4 (×3)		
Na(3)-O(8)	2.35±14		

CHROMATE CANCRINITE

Trigonal, a = 12.72±2, c = 5.19±1 Å, U = 728 Å^3.

Space group $P3$ (C_3^1)

MOLYBDATE CANCRINITE

Trigonal, a = 12.75±2, c = 5.19±1 Å, U = 731 Å^3.

Discussion

The main framework in both basic and nitrate cancrinite is similar to that found by 1. A slightly different arrangement of Na atom sites is suggested for basic cancrinite than that given by 1. The Si atoms and some interstitial water atoms could not be located. Similar difficulty was experienced in locating the nitrate groups in nitrate cancrinite although several models were tested. Some nitrate occurs in the small cavities but O(5) belongs to a nitrate group in the main channel alternating along c with Na ions.

The chromate and molybdate ions in the chromate and molybdate cancrinites appear to be in the small cages, although it was not possible to refine these structures.

Details of analysis

Basic cancrinite grown from gel with excess aqueous NaOH at 390°. The other samples were prepared from kaolinite with excess NaOH and the included salt. The nitrate prepared at 190° [C], the chromate and molybdate at 80° [C]. About 66 powder lines measured on diffractometer for each material. Structure refined in pseudo space group $P6_3$ because of the smaller number of parameters (17) needed. The structure of 1 initially assumed and was refined by Fourier and, using intensities, least-squares methods. R = 0.07 for basic cancrinite, 0.12 for nitrate cancrinite (reflexions listed) and 0.35 for chromate and molybdate cancrinite (no reflexions or coordinates listed).

1. O. JARCHOW, 1965. *Z. Kristallogr.*, 122, 407.

GEDRITE

I. Gedrites crystal structures and intra crystalline cation distributions. J.J. PAPIKE, 1970. *Amer. Min.*, 55, 1945-1972.

Orthorhombic, a = 18.531±4, b = 17.741±4, c = 5.249±5 Å, U = 1725.8±14 Å^3, D_x = 3.18 (Mason Mountain Sample).

a = 18.601±4, b = 17.839±3, c = 5.284±2 Å, U = 1753.2±6 Å^3, D_x = 3.29 (New Hampshire Sample).

Space group $Pnma$ (D_{2h}^{16}) No deviations from this symmetry were noted.

Atomic coordinates and interatomic distances for both samples are given in I.

Discussion

Polymorphism among *Pnma*, $P2_1/m$ and $C2/m.Fe^{2+}$-Mg amphiboles can be rationalized in terms of a partial O-rotation of the tetrahedral chains (A and B) which lead to an octahedral coordination of both the A and M(4) sites. Octahedral Al is in the M(2) site, Fe^{2+} and Mg are distributed over all M sites such that the Fe^{2+} concentration in M(4) is greater than in M(1), M(2) and M(3).

Details of analysis

The crystal structures of two gedrites of composition $(Ca_{0.03}Na_{0.47})(Mg_{4.52}Fe^{2+}_{1.14}Al_{1.21}Mn_{0.02}Ti_{0.06})(Si_{6.25}Al_{1.75})O_{22}(OH)_2$ from Mason Mountain, N. Carolina and $(Ca_{0.04}Na_{0.54}K_{0.01})(Mg_{3.01}Fe^{2+}_{2.35}Fe^{3+}_{0.14}Al_{1.37}Mn_{0.03}Ti_{0.02}Li_{0.02}Cr_{0.02})(Si_{5.95}Al_{2.05})O_{22}(OH)_2$ from S. New Hamshire were refined using three-dimensional counter collected intensity data to *R* factors of 7.6% and 7.2% respectively.

VERMICULITE

I. The crystal structure of a clay-organic complex of 6-amino hexanoic acid and vermiculite. F. KANAMARU and V. VAND, 1970. *Amer. Min.*, 55, 1550-1561.

$(Mg,Al,Fe)_3(Al,Si)_4O_{10}(OH)_2Na_x,(H_2O)_y,(HO_2C(CH_2)_5NH_3)_z$

Monoclinic, a = 5.33±1, b = 9.18±2, c = 17.45±4 Å, β = 97°, Z = 2.

Space group Probably *C*2

Atomic positions given in I

Discussion

Due to the poor quality of the crystals only a rudimentary refinement could be obtained. However, zigzag chains of amino acid could be identified in the α-configuration lying sub-parallel to the *b* axis and tilted to the silicate chain by 50°.

Details of analysis

The material was prepared by soaking a fragment of vermiculite (Kenya) in a 1N-NaCl solution at room temperature for 2-3 weeks and then 4 weeks in 1N 6-amino hexanoic acid. Two-dimensional intensity data were estimated from Weissenberg and oscillation photographs.

HEMIHEDRITE

I. The crystal structure of hemihedrite. W.J. MCLEAN and J.W. ANTHONY, 1970. *Amer. Min.*, 55, 1103-1114.

$ZnF_2[Pb_5(CrO_4)_3SiO_4]_2$ F.W. = 3053.48

Triclinic, a = 9.497±3, b = 11.443±5, c = 10.841±4 Å, α = 120.50±4, β = 92.10±4, γ = 55.84±3°, U = 787.38 Å³, D_m = 6.42, Z = 1, D_x = 6.43.

Space group $P\bar{1}$ (C_i^1)

Atomic positions

	x	y	z	B(Å²)
Pb(1)	0.2591±1	0.1116±1	0.2594±1	*
Pb(2)	0.2616±1	0.0862±1	0.6568±1	*

	x	*y*	*z*	$B(\text{Å}^2)$
Pb(3)	0.9306±1	0.2442±1	0.0306±1	*
Pb(4)	0.7307±1	0.4165±1	0.7513±1	*
Pb(5)	0.3171±1	0.4530±1	0.5316±1	*
Zn	0	0.5	0	*
Cr(1)	0.9591±4	0.0752±4	0.3555±4	*
Cr(2)	0.5642±4	0.1758±4	0.1535±4	*
Cr(3)	0.4470±4	0.3246±4	0.8369±4	*
Si	0.0229±6	0.4516±6	0.6594±5	*
F	0.1425±25	0.2488±24	0.9318±21	0.87±32
O(1)	0.7535±27	0.2193±27	0.4835±23	2.12±36
O(2)	0.1079±35	0.0778±34	0.4436±29	2.38±49
O(3)	0.9940±26	0.1184±25	0.7361±22	1.92±33
O(4)	0.9677±28	0.1120±27	0.2294±23	2.36±37
O(5)	0.5101±26	0.1407±25	0.2712±21	1.54±33
O(6)	0.4318±24	0.1992±23	0.0573±20	2.16±30
O(7)	0.7772±27	0.0082±26	0.0274±22	2.06±34
O(8)	0.5405±24	0.3585±23	0.2686±20	1.76±29
O(9)	0.5990±24	0.2969±24	0.9141±21	1.87±31
O(10)	0.4559±31	0.3985±29	0.7442±26	2.45±41
O(11)	0.2385±27	0.4815±26	0.9767±23	0.97±35
O(12)	0.4794±21	0.1360±20	0.7183±18	1.54±25
O(13)	0.2089±25	0.3053±24	0.5120±21	1.60±32
O(14)	0.0421±25	0.3859±24	0.7677±21	1.20±32
O(15)	0.9931±20	0.3717±20	0.2568±17	0.92±25
O(16)	0.8475±27	0.4788±26	0.6134±23	1.18±35

*Anisotropic temperature factors given.

Interatomic distances (Å)

Environment of Pb(1)

Pb(1)-O(13)	2.32±2
Pb(1)-O(5)	2.58±3
Pb(1)-O(15)	2.62±2
Pb(1)-O(12)	2.65±2
Pb(1)-O(4)	2.77±3
Pb(1)-O(7)	2.79±2
Pb(1)-O(2)	2.80±4
Pb(1)-O(6)	3.11±2
Pb(1)-O(9)	3.38±3

Environment of Pb(2)

Pb(2)-O(14)	2.35±2
Pb(2)-F	2.42±2
Pb(2)-O(12)	2.43±2
Pb(2)-O(3)	2.48±3
Pb(2)-O(2)	2.68±3
Pb(2)-O(5)	2.77±2
Pb(2)-O(1)	3.12±3
Pb(2)-O(13)	3.40±3

Environment of Pb(3)

Pb(3)-F	2.35±3
Pb(3)-O(7)	2.39±2
Pb(3)-O(15)	2.39±2
Pb(3)-O(11)	2.65±3
Pb(3)-O(3)	2.67±2
Pb(3)-O(4)	3.10±2
Pb(3)-O(4)	3.12±3
Pb(3)-O(10)	3.13±2
Pb(3)-O(9)	3.20±2

Environment of Pb(4)

Pb(4)-O(16)	2.46±4
Pb(4)-O(8)	2.53±2
Pb(4)-O(1)	2.53±2
Pb(4)-O(11)	2.64±2
Pb(4)-O(10)	2.73±4
Pb(4)-O(14)	2.79±3
Pb(4)-O(3)	2.79±2
Pb(4)-O(6)	3.06±2

Environment of Pb(5)

Pb(5)-O(16)	2.28±3
Pb(5)-O(13)	2.35±3
Pb(5)-O(5)	2.62±2
Pb(5)-O(15)	2.68±2
Pb(5)-O(10)	2.78±3
Pb(5)-O(8)	2.92±3
Pb(5)-O(8)	3.08±2

Environment of Zn

Zn	-F	2.05±2	(×2)
Zn	-O(14)	2.09±2	(×2)
Zn	-O(11)	2.17±3	(×2)

Angles in degrees subtended at Zn by

	O(14)	O(11)
F	83±1	94±1
F'	97±1	86±1
O(14)		88±1
O(14)'		92±1

Environment of Cr(1)

Cr(1)-O(4)	1.64±3
Cr(1)-O(1)	1.65±2
Cr(1)-O(3)	1.67±3
Cr(1)-O(2)	1.68±3

Angles in degrees subtended at Cr(1) by

	O(1)	O(3)	O(2)
O(4)	112±1	108±1	110±2
O(1)		106±1	110±1
O(3)			111±1

Environment of Cr(2)

Cr(2)-O(6)	1.60±2
Cr(2)-O(8)	1.68±2
Cr(2)-O(7)	1.68±2
Cr(2)-O(5)	1.69±3

Angles in degrees subtended at Cr(2) by

	O(8)	O(7)	O(5)
O(6)	112±1	107±1	110±1
O(8)		111±1	105±1
O(7)			112±1

Environment of Cr(3)

Cr(3)-O(9)	1.60±2
Cr(3)-O(10)	1.64±4
Cr(3)-O(12)	1.70±2
Cr(3)-O(11)	1.70±2

Angles in degrees subtended at Cr(3) by

	O(10)	O(12)	O(11)
O(9)	108±2	111±1	109±1
O(10)		111±1	108±1
O(12)			110±1

Environment of Si

Si	-O(16)	1.62±3
Si	-O(13)	1.63±2
Si	-O(14)	1.64±3
Si	-O(15)	1.65±2

Angles in degrees subtended at Si by

	O(13)	O(14)	O(15)
O(16)	113±1	105±2	113±1
O(13)		109±1	103±1
O(14)			113±1

Discussion

The structure is pseudosymmetric approximating space group $P2_1/m$ very closely and can be considered as a low symmetry member of the tsumebite structural series. The chemical formula determined during the refinement was $ZnF_2[Pb_5(CrO_4)_3SiO_4]_{2 \cdot 1}$.

Details of analysis

Three-dimensional intensity data were collected by counter methods from natural material from Arizona. The structure was solved by the heavy-atom method and refined to an R factor of 4.1%.

SODIUM ZIRCONIUM OXYORTHOSILICATE

$Na_2ZrO(SiO_4)$ F.W. = 245.28

I. [The crystal structure of Na, Zr-oxyorthosilicate $Na_2ZrSiO_5 = Na_2O.Zr(SiO_4)$]. E.N. TREUŠNIKOV, V.V. ILJUKHIN and N.V. BELOV, 1970. *Dokl. Akad. Nauk SSSR*, 190, 334-337 [*Soviet Physics-Doklady*, 15, 9-11].

Monoclinic, a = 13.92, b = 5.46, c = 13.70 Å, β = 120°, U = 1041 Å^3, Z = 8, D_x = 3.12.

Space group $P2_1/c$ (C_{2h}^5)

Atomic positions All atoms in 4(*e*)

	x	*y*	*z*		*x*	*y*	*z*
Zr(1)	0.0647	0.7608	0.3719	O(2)	0.289	0.254	0.150
Zr(2)	0.4349	0.2612	0.3068	O(3)	0.408	0.789	0.041
Si(1)	0.1583	0.2277	0.0688	O(4)	0.211	0.714	0.362
Si(2)	0.3413	0.7711	0.4088	O(5)	0.091	0.303	0.129
Na(1)	0.314	0.741	0.145	O(6)	0.378	0.596	0.338
Na(2)	0.079	0.733	0.148	O(7)	0.134	0.937	0.027
Na(3)	0.184	0.264	0.330	O(8)	0.363	0.064	0.392
Na(4)	0.420	0.231	0.067	O(9)	0.007	0.474	0.271
O(1)	0.127	0.102	0.461	O(10)	0.492	0.966	0.265

Interatomic distances (Å) [standard errors not given but must be at least 0.02 Å]

Six-fold coordination around Zr(1)

Zr(1) - O(1)	2.16	
Zr(1) - O(5)'	2.09	(2.17)
Zr(1) - O(4)	2.14	(2.12)
Zr(1) - O(7)	2.16	(2.14)
Zr(1) - O(9)	2.08	(2.08)
Zr(1) - O(9)'	1.99	(1.97)

Six-fold coordination around Zr(2)

Zr(2) - O(2)	2.09
Zr(2) - O(3)'	2.14
Zr(2) - O(6)	2.12
Zr(2) - O(8)	2.17
Zr(2) - O(10)'	2.06
Zr(2) - O(10)	2.00

Four-fold coordination around Si(1)

Si(1) - O(1)'	1.61
Si(1) - O(2)	1.58
Si(1) - O(5)	1.60
Si(1) - O(7)	1.66

Four-fold coordination around Si(2)

Si(2) - O(3)'	1.63
Si(2) - O(4)	1.60
Si(2) - O(6)	1.62
Si(2) - O(8)	1.67

Environment of Na(1)

Na(1) - O(2)	2.68
Na(1) - O(2)'	2.83
Na(1) - O(3)	2.39
Na(1) - O(6)	2.50
Na(1) - O(7)	2.42
Na(1) - O(10)	2.49
Na(1) - O(10)'	2.75

Environment of Na(2)

Na(2) - O(1)'	2.55
Na(2) - O(4)	2.60
Na(2) - O(5)	2.38
Na(2) - O(5)'	3.13
Na(2) - O(7)	2.44
Na(2) - O(9)'	2.39
Na(2) - O(9)	2.75

Environment of Na(3)

Na(3) - O(1)	2.47
Na(3) - O(4)	2.49
Na(3) - O(5)	2.44
Na(3) - O(8)	2.42
Na(3) - O(9)	2.42
Na(3) - O(9)'	2.76
Na(3) - O(4)'	3.03

Environment of Na(4)

Na(4) - O(2)	2.58
Na(4) - O(3)'	2.43
Na(4) - O(6)'	2.51
Na(4) - O(8)'	2.42
Na(4) - O(10)	2.40
Na(4) - O(10)'	2.82
Na(4) - O(3)	3.06

Discussion

Like $Na_2TiO(GeO_4)$, this structure can be considered as being composed of a framework of ZrO_6 octahedra and SiO_4 tetrahedra linked through shared corners, with the sodium atoms occupying holes in the framework. As with the TiO_6 octahedra in $Na_2TiO(GeO_4)$, the ZrO_6 octahedra are linked through a shared corner into chains running along the *c* axis but these chains are arranged in a hexagonal array rather than the square array found in the titanogermanate.

Details of analysis

Intensities for layers with h = 0-1 and k = 0-4 estimated visually from photographs. Structure determined from Patterson function and refined by least squares to give R = 0.13 [intensities not listed].

TRICALCIUM DICHLOROSILICATE

I. [Crystal structure of $Ca_3[SiO_4]Cl_2$, a metastable phase in cement clinker. E.N. TREUŠNIKOV, V.V. ILJUKHIN and N.V. BELOV, 1970. *Dokl. Akad. Nauk, SSSR*, 193, 1048-1051.

$Ca_3(SiO_4)Cl_2$ F.W. = 355.73

Monoclinic, a = 9.79, b = 6.76, c = 10.96 Å, β = 105.4°, U = 699 Å^3, Z = 4, D_x = 2.68.

Space group $P2_1/c$ (C^5_{2h})

Atomic positions All atoms in 4(*e*)

	x	*y*	*z*
Ca(1)	0.1955	0.7428	0.0453
Ca(2)	0.0789	0.4565	0.2715
Ca(3)	0.4286	0.2230	0.1118
Cl(1)	0.5078	0.7899	0.1240
Cl(2)	0.2833	0.8233	0.3174
Si	0.1209	0.2493	0.0306
O(1)	0.235	0.416	0.109
O(2)*	0.022	0.069	0.003
O(3)	0.017	0.199	0.121
O(4)	0.017	0.189	0.393

*The distances calculated with these coordinates do not agree with those quoted in I. There is obviously some error in the coordinates.

Interatomic distances (Å)

Ca(1) - O(1)	2.32	Ca(3) - O(2)*	2.30
Ca(1) - O(4)	2.38	Ca(3) - O(1)	2.29
Ca(1) - O(3)	2.41	Ca(3) -Cl(2)	2.80
Ca(1) - O(2)*	2.29	Ca(3) -Cl(1)	2.82
Ca(1) -Cl(2)	2.88	Ca(3) -Cl(1)	2.83
Ca(1) -Cl(2)	2.93	Ca(3) -Cl(1)	3.03
Ca(1) -Cl(1)	2.96		
Ca(2) - O(3)	2.36	Si - O(1)	1.66
Ca(2) - O(3)	2.36	Si - O(2)*	1.65
Ca(2) - O(4)	2.39	Si - O(3)	1.63
Ca(2) - O(4)	2.41	Si - O(4)	1.63
Ca(2) - O(1)	2.65		
Ca(2) - O(2)*	2.56		

[Errors uncertain but probably around 0.02 Å]

*See note above.

Details of analysis

Intensities estimated visually from photographs of the k = 0 to 5 and h = 0 to 1 layers taken with MoK_{α} radiation. Structure determined from Patterson function and refined by least squares to give R = 0.11 for 1400 observed reflexions [not listed].

LEUCOPHOENICITE

I. Edge-sharing silicate tetrahedra in the crystal structure of leucophoenicite. P.B. MOORE, 1970. *Amer. Min.*, 55, 1146-1166.

$Mn_7(SiO_4)_3(OH)_2$ F.W. = 722.91

Monoclinic, a = 10.842±19, b = 4.826±6, c = 11.324±9 Å, β = 103.93±9°, U = 575.09 Å^3, Z = 2, D_x = 4.17.

Space group $P2_1/a$ (C^5_{2h})

Atomic positions Non-standard setting

	x	y	z	B(Å^2)
2 Mn(1)	0	0	0	0.73±3
4 Mn(2)	0.3149±2	0.0150±5	0.1396±2	0.80±3
4 Mn(3)	0.3308±2	0.4942±5	0.4110±2	0.69±3
4 Mn(4)	0.0781±2	-0.0105±5	0.2967±2	0.80±3
2 Si(1)	0.0246±6	0.4144±13	0.4381±5	0.47±8
4 Si(2)	0.1287±3	0.5731±6	0.1439±3	0.32±4
4 O(1)	0.4907±7	-0.2135±17	0.1458±7	0.48±11
4 O(2)	0.3345±8	0.2137±18	-0.0265±7	0.73±12
4 O(3)	0.2289±8	-0.2879±19	0.2614±7	0.76±12
4 O(4) (=½OH+½O)	0.4207±9	0.2348±20	0.3058±8	1.26±14
4 O(5) (=½OH+½O)	0.1736±8	0.2626±20	0.4391±8	0.90±12
4 O(6)	0.1290±8	0.2390±19	0.1450±8	0.79±12
4 O(7)	0.5254±9	0.7708±20	0.4379±9	1.19±14

Interatomic distances (Å) *and angles*

Environment of Mn(1)

Mn(1)-O(1)	2.174±8 (×2)
Mn(1)-O(6)	2.209±8 (×2)
Mn(1)-O(2)	2.229±9 (×2)

Angles in degrees subtended at Mn(1) by

	O(6)	O(2)
O(1)	84.5±3	108.3±2
O(1)'	95.5±3	71.8±3
O(6)		97.1±4
O(6)'		82.9±4

Environment of Mn(2)

Mn(2)-O(2)	2.167±9
Mn(2)-O(1)	2.191±8
Mn(2)-O(4)	2.224±9
Mn(2)-O(6)	2.304±9
Mn(2)-O(2)	2.321±8
Mn(2)-O(3)	2.352±9

Angles in degrees subtended at Mn(2) by

	O(1)	O(4)	O(6)	O(2)	O(3)
O(2)	89.1±3	112.6±4	94.9±3	90.0±3	156.9±3
O(1)		86.8±4	176.0±39	101.0±4	97.7±4
O(4)			91.8±4	156.3±5	89.9±3
O(6)				78.9±4	78.6±4
O(2)					67.1±3

Environment of Mn(3)

Mn(3)-O(3)	2.073±8
Mn(3)-O(4)	2.121±10
Mn(3)-O(5)	2.128±10
Mn(3)-O(5)	2.145±10
Mn(3)-O(7)	2.391±9
Mn(3)-O(7)	2.456±10

Angles in degrees subtended at Mn(3) by

	O(4)	O(5)	O(5)	O(7)	O(7)
O(3)	93.9±4	95.7±4	103.4±4	169.9±9	95.7±4
O(4)		105.6±4	156.4±4	77.1±4	83.8±4
O(5)			91.6±4	91.2±4	164.7±11
O(5)				83.8±3	75.9±4
O(7)					78.9±4

Environment of Mn(4)

Mn(4)-O(5)	2.147±9
Mn(4)-O(4)	2.188±10
Mn(4)-O(1)	2.194±8
Mn(4)-O(7)	2.214±11
Mn(4)-O(3)	2.225±9
Mn(4)-O(6)	2.272±10

Angles in degrees subtended at Mn(4) by

	O(4)	O(1)	O(7)	O(3)	O(6)
O(5)	80.4±4	176.4±14	88.2±4	105.1±4	94.6±4
O(4)		102.1±4	88.3±4	172.6±18	92.8±4
O(1)			94.6±3	72.2±3	82.6±3
O(7)				96.8±4	177.1±11
O(3)					81.9±3

Environment of Si(1)

Si(1)-O(7)	1.519±11
Si(1)-O(7)	1.765±12
Si(1)-O(5)	1.774±11
Si(1)-O(4)	1.795±10

Angles in degrees subtended at Si(1) by

	O(7)	O(5)	O(4)
O(7)	113.4±6	114.0±6	113.7±5
O(7)		106.7±5	104.7±6
O(5)			103.2±6

Environment of Si(2)

Si(2)-O(6)	1.612±10
Si(2)-O(2)	1.626±9
Si(2)-O(3)	1.645±8
Si(2)-O(1)	1.650±8

Angles in degrees subtended at Si(2) by

	O(2)	O(3)	O(1)
O(6)	115.0±5	113.7±4	114.3±5
O(2)		104.2±5	104.0±5
O(3)			104.4±5

Discussion

The structure is based on the hexagonal close packing of oxygen anions and has the peculiarity of a half-occupied symmetry-related pair of edge sharing silica tetrahedra (Si(1)).

Details of analysis

Natural crystals of composition $Mn_7(SiO_4)_2[(SiO_4)(OH)_2]$ were used to collect three-dimensional counter data. The final *R* factor for the refinement of the structure which was determined by Patterson synthesis was 7%.

ALLEGHANYITE

I. The crystal structure of alleghanyite, $Mn_5[(OH)_2(SiO_4)_2]$. P.J. RENTZEPERIS, 1970. *Z. Kristallogr.*, 132, 1-8.

$Mn_5(OH)_2(SiO_4)_2$ F.W. = 490.85

Monoclinic, a = 8.2747±8, b = 4.8503±5, c = 10.7198±10 Å, β = 108.64±1°, U = 407.67 Å^3, D_m = 4.02, Z = 2, D_x = 3.99.

Space group $P2_1/c$ (C_{2h}^5) From absences

Atomic positions

	x	*y*	*z*	$B(Å^2)$
4 Mn(1) in 4(*e*)	0.3110±2	0.0087±4	0.1719±2	2.29±5
4 Mn(2) in 4(*e*)	0.0772±2	0.0137±4	0.3812±2	2.48±5
2 Mn(3) in 2(*d*)	0.5	0	0.5	2.34±5
4 Si in 4(*e*)	0.2976±4	0.5773±6	0.3569±3	2.02±6
4 O(1) in 4(*e*)	0.2945±8	0.7123±16	0.4952±7	2.26±13
4 O(2) in 4(*e*)	0.1347±9	0.7185±16	0.2469±7	2.21±13
4 O(3) in 4(*e*)	0.4632±8	0.7143±16	0.3303±7	2.10±13
4 O(4) in 4(*e*)	0.2976±8	0.2430±16	0.3537±7	1.94±13
4 O(5) in 4(*e*)	0.0973±10	0.2530±17	0.0566±7	2.54±14

Anisotropic temperature factors also given.

Interatomic distances (Å) *and angles*

Mn(1)-O(3)	2.125±7
Mn(1)-O(1)	2.142±8
Mn(1)-O(5)	2.159±7
Mn(1)-O(3)'	2.267±7
Mn(1)-O(4)	2.289±8
Mn(1)-O(2)	2.346±8

Angles in degrees subtended at Mn(1) by

	O(1)	O(5)	O(3)'	O(4)	O(2)
O(3)	91.9±3	107.4±3	91.4±3	93.4±3	158.5±2
O(1)		87.7±3	102.1±3	173.9±2	97.6±3
O(5)			158.7±3	87.8±3	92.3±3
O(3)'				80.9±3	67.9±2
O(4)					78.5±3

Mn(2)-O(2)	2.100±7
Mn(2)-O(5)	2.155±8
Mn(2)-O(5)'	2.178±9
Mn(2)-O(2)'	2.188±8
Mn(2)-O(4)	2.235±8
Mn(2)-O(1)	2.337±7

Angles in degrees subtended at Mn(2) by

	O(5)	O(5)'	O(2)'	O(4)	O(1)
O(2)	97.9±3	88.7±3	101.0±3	103.0±3	168.6±3
O(5)		84.9±3	161.0±3	90.9±3	92.3±3
O(5)'			97.3±3	168.0±3	87.2±3
O(2)'				83.1±3	89.0±2
O(4)					81.7±3

Mn(3)-O(1)	2.188±7 (×2)
Mn(3)-O(3)	2.229±7 (×2)
Mn(3)-O(4)	2.230±6 (×2)

Angles in degrees subtended at Mn(3) by

	O(3)	O(4)
O(1)	71.4±3	94.8±2
O(1)'	108.6±3	85.2±2
O(3)		97.0±3
O(3)'		83.0±3

Si -O(4)	1.622±8
Si -O(3)	1.628±8
Si -O(1)	1.628±8
Si -O(2)	1.631±7

Angles in degrees subtended at Si by

	O(3)	O(1)	O(2)
O(4)	113.5±4	114.9±4	114.2±4
O(3)		104.7±4	104.4±4
O(1)			103.9±4

O(5)-H...O(5) 2.973±11 (Possibly hydrogen bond)

Angles in degrees subtended at O(5) by

	Mn(1)	Mn(2)'	O(5)
Mn(2)	127.5±4	95.1±3	101.3±3
Mn(1)		130.2±4	91.0±3
Mn(2)'			106.2±3

Discussion

The basic structure is similar to that of chondrodite (1) but alleghanyite shows a number important structural differences.

Details of analysis

Crystal from Bald Knob, Alleghany Co., N. Carolina, U.S.A. Lattice parameters from Weissenberg and precession photographs (no calibration given). Intensities measured on a manual equi-inclination diffractometer for layers with k = 0 to 5 using filtered CuK_α radiation. Cylindrical absorption correction applied (μ= 68.1 mm^{-1}, crystal size 0.36×0.14×0.12 mm^3). The structure of chondrodite (1) assumed and refined by least squares to give R = 0.075 with anisotropic temperature factors for 746 reflexions listed.

1. *Strukturbericht*, 1, 401.

CHONDRODITE

I. The crystal structures of the humite minerals. II. Chondrodite. G.V. GIBBS, P.H. RIBBE and C.P. ANDERSON, 1970. *Amer. Min.*, 55, 1182-1194.

$Mg_5(SiO_4)_2F_{1.3}(OH)_{0.7}$

Monoclinic, a = 4.7284±3, b = 10.2539±3, c = 7.8404±2 Å, α = 109.059±2°, U = 359.30 Å^3, D_m = 3.18 0, Z = 2.

Space group $P2_1/b$ (C^5_{2h})

Atomic positions Unique axis a non-standard

	x	y	z	B(Å^2)
Mg(1)	0.5	0	0.5	0.49±2
Mg(2)	0.0091	0.1731	0.3055	0.45±2
Mg(3)	0.4915	0.8867	0.0791	0.43±2
Si	0.0768	0.1441	0.7038	0.20±1
O(1)	0.7787	0.0009	0.2937	0.36±3
O(2)	0.7280	0.2404	0.1252	0.40±3
O(3)	0.2255	0.1682	0.5275	0.35±3
O(4)	0.2649	0.8546	0.2943	0.42±3
F,OH	0.2656	0.0582	0.1018	0.50±2

Interatomic distances (Å) *and angles*

Environment of Mg(1)

Mg(1)-O(1)	2.089 (×2)
Mg(1)-O(3)	2.113 (×2)
Mg(1)-O(4)	2.119 (×2)

Angles in degrees subtended at Mg(1) by

	O(3)	O(4)
O(1)	104.9 0	85.5 0
O(1)'	75.1 0	94.5 9
O(3)		95.9 0
O(3)'		84.1 0

Environment of Mg(2)

Mg(2)-O(3)	2.034
Mg(2)-F	2.045
Mg(2)-O(1)	2.051
Mg(2)-O(4)	2.173
Mg(2)-O(3)	2.184
Mg(2)-O(2)	2.209

Angles in degrees subtended at Mg(2) by

	F	O(1)	O(4)	O(3)	O(2)
O(3)	101.6	92.5	92.7	91.0	161.9
F		90.7	87.7	163.5	94.5
O(1)			174 7	99.4	95.5
O(4)				81.1	79.6
O(3)					71.7

Environment of Mg(3)

Mg(3)-O(2)	1.995
Mg(3)-F	2.015
Mg(3)-F	2.043
Mg(3)-O(4)	2.113
Mg(3)-O(2)	2.120
Mg(3)-O(1)	2.182

Angles in degrees subtended at Mg(3) by

	F	F	O(4)	O(2)	O(1)
O(2)	96.8	89.5	98.4	98.6	171.4
F		85.9	91.1	164.2	91.6
F			171.8	97.9	89.1
O(4)				89.0	83.4
O(2)					73.2

Si - O(1)	1.642
Si - O(2)	1.636
Si - O(3)	1.640
Si - O(4)	1.617

Angles in degrees subtended at Si by

	O(2)	O(3)	O(4)
O(1)	103.0	102.7	114.3
O(2)		103.6	115.3
O(3)			115.6

Standard errors 0.002 Å

Discussion

It is postulated that the small amount of Fe is ordered into the M(1) octahedron, whereas the larger M(2) and smaller M(3) octahedra contain only Mg. This distribution gives more equal temperature factors on refinement and is consistent with crystal-field theory hypotheses.

Details of analysis

Crystals of composition $Mg_{3.95}Fe_{0.05}(SiO_4)_2Mg_{1.0}F_{1.3}(OH)_{0.7}$ from Hangleby, Sibbo, Finland were used to collect three-dimensional intensity data. The structure refined to a weighted *R* factor of 3.9%.

CEMENT PHASE Y

I. [Crystal structure of cement phase Y = $Ca_6[Si_2O_7][SiO_4](OH)_2$]. R.M. GANIEV, V.V. ILJUKHIN and N.V. BELOV, 1970. *Dokl. Akad. Nauk SSSR*, 190, 831-834 [*Soviet Physics-Doklady*, 15, 85-87].

$Ca_6(Si_2O_7)(SiO_4)(OH)_2$ F.W. = 534.74

Triclinic, a = 6.85, b = 6.95, c = 12.90 Å, α = 90.75±25, β = 97.33±33, γ = 98.25±25°, U = 603 Å^3, Z = 2, D_x = 2.94.

Space group $P\bar{1}$ (C_i^1)

Atomic positions All atoms in 2(*i*)

	x	*y*	*z*
Ca(1)	0.082	0.194	0.103
Ca(2)	0.192	0.206	0.489
Ca(3)	0.315	0.191	0.876
Ca(4)	0.560	0.034	0.321
Ca(5)	0.894	0.468	0.291
Ca(6)	0.595	0.477	0.703
Si(1)	0.618	0.291	0.098
Si(2)	0.809	0.285	0.885
Si(3)	0.715	0.299	0.509
O(1)	0.614	0.490	0.166
O(2)	0.394	0.138	0.053
O(3)	0.615	0.184	0.806
O(4)	0.747	0.153	0.174
O(5)	0.988	0.149	0.910

	x	y	z
O(6)	0.229	0.165	0.300
O(7)	0.850	0.130	0.486
O(8)	0.192	0.210	0.685
O(9)	0.530	0.215	0.565
O(10)	0.743	0.336	0.999
O(11)	0.643	0.368	0.393
O(12)	0.836	0.482	0.581
O(13)	0.112	0.510	0.170

Interatomic distances (Å)

[Errors uncertain but about 0.05 Å]

Ca(1) - O(13)	2.32	Ca(4) - O(3)	2.31
Ca(1) - O(5)	2.36	Ca(4) - O(9)	2.35
Ca(1) - O(2)	2.39	Ca(4) - O(11)	2.45
Ca(1) - O(5)	2.50	Ca(4) - O(4)	2.51
Ca(1) - O(4)	2.56	Ca(4) - O(6)	2.55
Ca(1) - O(6)	2.63	Ca(4) - O(8)	2.57
Ca(1) - O(10)	2.84	Ca(4) - O(7)	2.72
Ca(2) - O(7)	2.32	Ca(5) - O(13)	2.29
Ca(2) - O(7)	2.34	Ca(5) - O(12)	2.30
Ca(2) - O(12)	2.38	Ca(5) - O(11)	2.33
Ca(2) - O(9)	2.39	Ca(5) - O(1)	2.37
Ca(2) - O(6)	2.50	Ca(5) - O(8)	2.42
Ca(2) - O(8)	2.53	Ca(5) - O(4)	2.64
Ca(3) - O(1)	2.29	Ca(6) - O(11)	2.32
Ca(3) - O(5)	2.32	Ca(6) - O(1)	2.38
Ca(3) - O(2)	2.32	Ca(6) - O(13)	2.41
Ca(3) - O(3)	2.35	Ca(6) - O(12)	2.42
Ca(3) - O(4)	2.43	Ca(6) - O(3)	2.46
Ca(3) - O(8)	2.51	Ca(6) - O(9)	2.48
		Ca(6) - O(6)	2.61
Si(1) - O(1)	1.63		
Si(1) - O(10)	1.64		
Si(1) - O(4)	1.67		
Si(1) - O(2)	1.76		
Si(2) - O(3)	1.63		
Si(2) - O(10)	1.64		
Si(2) - O(13)	1.65		
Si(2) - O(5)	1.66		
Si(3) - O(9)	1.58		
Si(3) - O(11)	1.62		
Si(3) - O(12)	1.62		
Si(3) - O(7)	1.64		

Discussion

The crystal contains discrete Si_2O_7 and SiO_4 groups linked by Ca ions.

Details of analysis

Structure determined from Patterson function calculated with reflexions from the layers h = 0-4 and k = 0 and refined by Fourier methods to give R = 0.17 for 1800 reflexions [not listed].

TILLEYITE

I. Crystal structure of tilleyite: Refinement and coordination. S.J. LOUISNATHAN and J.V. SMITH, 1970. *Z. Kristallogr.*, 132, 288-306.

$Ca_5Si_2O_7(CO_3)_2$ F.W. = 488.58

Monoclinic, *a* = 15.108±3, *b* = 10.241±1, *c* = 7.579±1 Å, β = 105.17±1°, *U* = 1128.1 Å^3, *Z* = 4, D_x = 2.88.

Space group $P2_1/a$ (C^5_{2h})

Atomic positions All atoms in 4(*e*) ±(*x*,*y*,*z*;½-*x*,½+*y*,-*z*)

	x	*y*	*z*
Ca(1)	0.0033±1	0.9938±2	0.7501±3
Ca(2)	0.1769±1	0.2135±2	0.0908±2
Ca(3)	0.1813±1	0.2064±2	0.6087±2
Ca(4)	0.1239±1	0.5895±2	0.0449±2
Ca(5)	0.1359±1	0.5787±2	0.5599±2
Si(1)	0.2032±1	0.9209±2	0.1381±3
Si(2)	0.2023±1	0.9142±2	0.5689±3
C (1)	0.0301±6	0.3058±9	0.3402±11
C (2)	0.0248±6	0.3018±9	0.8157±11
O (1)	0.0752±5	0.2084±8	0.2983±10
O (2)	0.0713±5	0.2052±8	0.7867±10
O (3)	0.0705±4	0.3772±7	0.4807±9
O (4)	0.0547±4	0.3766±7	0.9563±9
O (5)	0.0519±5	0.6642±7	0.7549±10
O (6)	0.0584±5	0.6826±7	0.2834±10
O (7)	0.2253±4	0.7744±7	0.0878±8
O (8)	0.2272±5	0.7665±7	0.6290±8
O (9)	0.1008±5	0.9760±8	0.0528±9
O (10)	0.1018±4	0.9711±7	0.5524±9
O (11)	0.2263±4	0.5308±6	0.8839±8
O (12)	0.2214±4	0.5219±6	0.3202±8
O (13)	0.2253±4	0.9161±7	0.3639±8

Isotropic temperature factors given in I.

Interatomic distances (Å) *and angles*

σ(Si-O) = σ(Ca-O) = 0.007; σ(C-O) = 0.01

C(1)-O(1)	1.29	C(2)-O(2)	1.27
C(1)-O(3)	1.30	C(2)-O(4)	1.29
C(1)-O(5)	1.30	C(2)-O(6)	1.29
Si(1)-O(7)	1.605	Si(2)-O(8)	1.596
Si(1)-O(9)	1.614	Si(2)-O(10)	1.601
Si(1)-O(11)	1.588	Si(2)-O(12)	1.656
Si(1)-O(13)	1.656	Si(2)-O(13)	1.678
Ca(1)-O(1)	2.367	Ca(2)-O(1)	2.470
Ca(1)-O(2)	2.381	Ca(2)-O(2)	2.436
Ca(1)-O(9)	2.457	Ca(2)-O(4)	2.499
Ca(1)-O(9')	2.387	Ca(2)-O(7)	2.335
Ca(1)-O(10)	2.383	Ca(2)-O(8)	2.322
Ca(1)-O(10')	2.450	Ca(2)-O(9)	2.673
		Ca(2)-O(11)	2.352

Ca(3)-O(2)	2.399	Ca(4)-O(4)	2.435
Ca(3)-O(1)	2.446	Ca(4)-O(5)	2.313
Ca(3)-O(3)	2.440	Ca(4)-O(6)	2.468
Ca(3)-O(8)	2.616	Ca(4)-O(7)	2.403
Ca(3)-O(7)	2.463	Ca(4)-O(11)	2.287
Ca(3)-O(10)	2.676	Ca(4)-O(12)	2.322
Ca(3)-O(12)	2.368		
Ca(5)-O(3)	2.299		
Ca(5)-O(6)	2.369		
Ca(5)-O(5)	2.354		
Ca(5)-O(8)	2.344		
Ca(5)-O(12)	2.557		
Ca(5)-O(11)	2.527		
O(7) -Si(1)-O(9)	118.1°	O(8) -Si(2)-O(10)	120.8°
O(7) -Si(1)-O(11)	116.8	O(8) -Si(2)-O(12)	114.2
O(7) -Si(1)-O(13)	102.9	O(8) -Si(2)-O(13)	100.6
O(9) -Si(1)-O(11)	108.0	O(10)-Si(2)-O(12)	108.3
O(9) -Si(1)-O(13)	109.5	O(10)-Si(2)-O(13)	110.2
O(11)-Si(1)-O(13)	99.7	O(12)-Si(2)-O(13)	100.3

Discussion

The structure is the same as that described in 1 and 2. [The bridging Si-O distances are anomolously long.]

Details of analysis

Structure determined by 1. Lattice parameters measured on powder diffractometer (Si calibration, a = 5.43062 Å). Sample from Crestmore, California, U.S.A. Intensities measured with MoK_α radiation on equi-inclination automatic diffractometer for layers with k = 0 to 13. Corrections made for Lorentz, polarization and absorption. Structure of 1 refined by least squares to give R_1 = 0.11 and R_2 = 0.06 for 3450 reflexions listed. The layer with k = 3 was omitted as having been improperly measured. 139 reflexions (including 17% of k = 0 layer were omitted as they had $|\Delta F| \sim \frac{1}{2}|F_o|$. [Including them would raise R_1 to about 0.12, a rather large value for diffractometer measured intensities and much larger than would be expected from the agreement between equivalent reflexions.]

1. *Structure Reports*, 17, 551.
2. N.V. BELOV, 1963. *Crystal Chemistry of Large-cation Silicates*, Consultants Bureau, New York.

SYNTHETIC FEBEFSITE

I. [Crystal structure of $Fe_3BeSi_3O_9(F,OH)_2$, an example of a woolastonite-like silicon-oxygen chain based on Fe.] V.V. BAKAKIN and L.P. SOLOV'EVA, 1970. *Kristallografija*, 15, 1144-1151 [*Soviet Physics-Crystallography*, 15, 999-1005].

$Fe_3BeSi_3O_9(OH,F)_2$ F.W. = 438.80 [nominal]

Orthorhombic, a = 9.49±1, b = 12.23±4, c = 7.10±3 Å, U = 824 Å^3, D_m = 3.568, Z = 4, D_x = 3.51.

Space group $Pna2_1$ (C_{2v}^9)

Atomic positions All atoms in 4(*a*)

	x	*y*	*z*	*B*(Å^2)
Fe(1)	0.4924±4	0.4995±4	0.5	0.86±4
Fe(2)	0.5015±6	0.2766±3	0.228±1	0.60±4
Fe(3)	0.4959±6	0.2747±4	0.762±1	1.13±6
Si(1)	0.2281±10	0.4279±7	0.228±2	1.06±11
Si(2)	0.2242±10	0.4254±7	0.789±2	0.82±10
Si(3)	0.2056±7	0.6133±6	0.513±3	0.73±8
Be	0.2185±26	0.2580±20	0.517±9	1.00±37
O(1)	0.1822±13	0.4331±10	0.011±7	0.77±19
O(2)	0.1716±18	0.3131±13	0.318±3	0.94±28
O(3)	0.1722±18	0.5412±13	0.318±3	1.00±28
O(4)	0.3972±17	0.4249±13	0.251±3	1.04±24
O(5)	0.1569±17	0.3196±13	0.690±3	0.82±26
O(6)	0.1600±19	0.5355±14	0.689±4	1.28±31
O(7)	0.3944±16	0.4245±13	0.763±3	0.95±24
O(8)	0.3736±13	0.6389±10	0.517±5	0.74±20
O(9)	0.1089±13	0.7180±10	0.500±9	0.87±20
O(10)	0.1195±12	0.1415±10	0.523±4	0.48±18
O(11)	0.3810±12	0.2251±10	0.513±5	0.56±17

Interatomic distances (Å) *and angles*

Fe(1)-O(8)	2.05±1
Fe(1)-O(10)	2.11±1
Fe(1)-O(4)	2.18±2
Fe(1)-O(7)	2.20±2
Fe(1)-O(4)	2.26±2
Fe(1)-O(7)	2.28±2

Angles in degrees subtended at Fe(1) by

	O(10)	O(4)	O(7)	O(4)	O(7)
O(8)	172.0±13	99.7±9	87.8±9	82.4±9	93.6±9
O(10)		87.6±8	97.2±8	90.5±8	80.8±8
O(4)			76.1±7	176.9±7	109.0±7
O(7)				101.7±7	174.3±7
O(4)					73.0±7

Fe(2)-O(2)	2.05±2
Fe(2)-O(9)	2.06±5
Fe(2)-O(4)	2.07±2
Fe(2)-O(8)	2.17±3
Fe(2)-O(11)	2.41±3
Fe(2)-O(10)	2.58±2

Angles in degrees subtended at Fe(2) by

	O(9)	O(4)	O(8)	O(11)	O(10)
O(2)	117.3±7	144.0±8	92.3±7	88.4±7	67.2±6
O(9)		97.1±7	84.3±13	109.2±12	174.9±7
O(4)			84.2±6	86.3±6	78.7±6
O(8)				164.4±8	98.0±8
O(11)					68.0±7

Fe(3)-O(5)	1.98±2
Fe(3)-O(7)	2.07±2
Fe(3)-O(9)	2.08±5
Fe(3)-O(11)	2.16±3
Fe(3)-O(10)	2.30±2
Fe(3)-O(8)	2.44±3

Angles in degrees subtended at Fe(3) by

	O(7)	O(9)	O(11)	O(10)	O(8)
O(5)	151.0±8	112.6±7	90.8±7	71.1±6	93.0±7
O(7)		94.0±7	90.9±7	81.1±6	81.4±6
O(9)			109.2±12	172.1±11	77.5±13
O(11)				77.3±8	170.2±7
O(10)					95.5±8

Si(1)-O(1)	1.60±5
Si(1)-O(4)	1.61±2
Si(1)-O(3)	1.62±2
Si(1)-O(2)	1.63±2

Angles in degrees subtended at Si(1) by

	O(4)	O(3)	O(2)
O(1)	112±1	105±1	109±1
O(4)		108±1	106±1
O(3)			118±1

Si(2)-O(5)	1.60±2
Si(2)-O(7)	1.63±2
Si(2)-O(1)	1.63±5
Si(2)-O(6)	1.64±2

Angles in degrees subtended at Si(2) by

	O(7)	O(1)	O(6)
O(5)	110±1	112±1	109±1
O(7)		111±1	109±1
O(1)			106±1

Si(3)-O(9)	1.58±2
Si(3)-O(8)	1.62±1
Si(3)-O(6)	1.63±3
Si(3)-O(3)	1.67±3

Angles in degrees subtended at Si(3) by

	O(8)	O(6)	O(3)
O(9)	114±1	111±2	106±2
O(8)		111±2	108±2
O(6)			106±1

Be - O(5)	1.56±6
Be - O(11)	1.59±3
Be - O(2)	1.63±6
Be - O(10)	1.71±3

Angles in degrees subtended at Be by

	O(11)	O(2)	O(10)
O(5)	120±4	112±2	100±2
O(11)		111±3	109±2
O(2)			102±3

Discussion

The structure consists of chains running along *c*. Each chain is in the form of a sandwich with the centre part composed of a ribbon of edge sharing FeO_6 octahedra, three octahedra wide. The sides of the chain are composed of 3 SiO_4 and 1 BeO_4 tetrahedra corner sharing in four-member rings which are joined to each other by further corner sharing (O(1)).

Details of anlaysis

Crystals of composition $(Fe^{2+}_{2.72}Fe^{3+}_{0.13}Mn_{0.08}Al_{0.08})BeSi_{2.97}O_{9.10}F_{1.18}(OH)_{0.72}$ grown hydrothermally. Intensities measured visually from equi-inclination Weissenberg photographs of layers with l = 0 to 6 taken with Mo radiation. Crystal ground to sphere of diameter 0.35 mm. Structure solved by Patterson and electron-density functions and refined by least squares to give R = 0.075 for 1000 non-zero reflexions [not listed]. Because of the pseudo centre of symmetry, simultaneous refinement of all parameters was not possible and strongly correlated parameters were refined in different cycles.

PIGEONITE

I. Refinement of the structure of pigeonite. N. MORIMOTO and N. GÜVEN, 1970. *Amer. Min.*, 55, 1195-1209.

$(Mg_{0.39}Fe_{0.52}Ca_{0.09})SiO_3$

Monoclinic, a = 9.706±2, b = 8.950±1, c = 5.246±1 Å, β = 108.590±10°, U = 431.94 Å^3, D_m = 3.30, Z = 8, D_x = 3.56.

Space group $P2_1/c$ (C^5_{2h})

Atomic positions

	x	*y*	*z*
M(1)	0.2508±4	0.6548±2	0.2328±8
M(2)	0.2564±3	0.0183±1	0.2308±6
Si(A)	0.0427±3	0.3398±5	0.2797±6
Si(B)	0.5504±3	0.8367±5	0.2372±6
O(1A)	0.8659±8	0.3404±13	0.1715±15
O(2A)	0.1220±9	0.4970±10	0.3306±16
O(3A)	0.1037±5	0.2633±6	0.5779±13

	x	y	z
O(1B)	0.3743±8	0.8342±13	0.1344±15
O(2B)	0.6290±8	0.9877±11	0.3765±17
O(3B)	0.6053±5	0.7087±6	0.4773±12

Interatomic distances (Å) *and angles*

Environment of *M*(1) = 0.720 Mg + 0.280 Fe

M(1)-O(2A)	2.057±10
M(1)-O(1A)	2.062±8
M(1)-O(1B)	2.068±8
M(1)-O(3B)	2.087±10
M(1)-O(1A)	2.158±11
M(1)-O(1B)	2.163±11

Angles in degrees subtended at *M*(1) by

	O(1A)	O(1B)	O(3B)	O(1A)	O(1B)
O(2A)	95.3±4	88.8±4	90.8±4	93.7±4	175.4±4
O(1A)		175.6±5	84.5±4	93.8±4	84.5±4
O(1B)			96.9±4	84.4±4	91.3±4
O(3B)				175.3±4	93.7±4
O(1A)					81.7±4

Environment of *M*(2) = 0.060 Mg + 0.760 Fe + 0.18 Ca

M(2)-O(3B)	2.006±8
M(2)-O(2A)	2.093±8
M(2)-O(1A)	2.144±11
M(2)-O(1B)	2.156±11
M(2)-O(3A)	2.430±5
M(2)-O(3B)	2.614±6

Angles in degrees subtended at *M*(2) by

	O(2A)	O(1A)	O(1B)	O(3A)	O(3B)
O(3B)	173.0±4	84.4±3	93.3±3	116.2±3	104.8±3
O(2A)		88.6±3	85.6±3	66.5±3	82.1±3
O(1A)			82.2±4	110.5±3	168.9±3
O(1B)				148.3±3	91.1±3
O(3A)					71.2±2

Environment of Si(A)

Si(A)-O(2A)	1.585±10
Si(A)-O(1A)	1.627±8
Si(A)-O(3A)	1.636±7
Si(A)-O(3A)	1.650±7

Angles in degrees subtended at Si(A) by

	O(1A)	O(3A)	O(3A)
O(2A)	117.2±6	101.7±4	110.5±4
O(1A)		110.8±4	109.4±4
O(3A)			106.6±4

Environment of Si(B)

Si(B)-O(3B)	1.608±10
Si(B)-O(1B)	1.621±8
Si(B)-O(3B)	1.660±7
Si(B)-O(3B)	1.665±7

Angles in degrees subtended at Si(B) by

	O(1B)	O(3B)	O(3B)
O(3B)	117.4±5	104.1±4	110.4±4
O(1B)		107.0±4	107.7±4
O(3B)			110.1±4

Discussion

The structure has been interpreted as an average of the clino-ferrosilite and diopside structures. Ca and Fe are ordered into the *M*(2) site; excess Fe and Mg occupy the *M*(1) site.

Details of analysis

The structural refinement was made on a crystal from the Isle of Mull, Scotland of composition $(Mg_{0.39}Fe_{0.52}Ca_{0.09})SiO_3$ using three-dimensional counter collected intensities. The final *R* factor was 3.6%.

NORDITE

I. The crystal structure of nordite and its relationship to melilite and datolite-gadolinite. V.V. BAKAKIN, N.V. BELOV, S.V. BORISOV and L.P. SOLOVYEVA, 1970. *Amer. Min.*, 55, 1167-1181.

$(La,Ce,Ca)_4(Sr,Ca)_4Na_2(Na,Mn)(Zn,Mg,Fe,Mn)_4Si_6O_{17}$

Orthorhombic, a = 14.27±3, b = 5.16±1, c = 19.45±15 Å, U = 1432.17 Å^3, D_m = 3.48, Z = 4,

Space group Pcca (D_{2h}^8)

Atomic positions

	x	y	z	B(Å^2)
La	0.25	0	0.3194	0.39±3
Sr	0.25	0	0.0217±2	0.06±5
Ca	0.25	0.5	0.1687±3	0.91±7
Si(1)	0.0980±7	0.4542±23	0.0623±4	0.61±14
Si(2)	0.1019±7	0.5433±23	0.2747±4	0.46±13
Si(3)	0.1116±5	0.5516±17	-0.0819±4	0.31±11
Na(1)	0	0	0	2.05±36
Na(2)	0.0696±10	0.0125±81	0.1693±8	2.10±26
O(1)	0.9955±16	0.3389±53	0.0811±14	0.84±10
O(2)	0.1701±18	0.2995±71	0.1101±12	0.70±10
O(3)	0.1172±18	0.3537±59	-0.0131±12	1.49±46
O(4)	0.1019±16	0.7640±68	0.0672±12	1.64±45
O(5)	0	0.6366±96	0.25	1.28±58
O(6)	0.1783±18	0.7090±69	0.2321±12	0.83±33
O(7)	0.1221±19	0.6503±58	0.3532±13	1.00±37
O(8)	0.1178±18	0.2392±58	0.2714±12	1.26±40
O(9)	0.1857±15	0.2226±58	0.4173±14	1.11±33

Interatomic distances (Å) *and angles*

Environment of La

La	-O(9)	2.41±3	(×2)
La	-O(8)	2.44±3	(×2)
La	-O(6)	2.49±3	(×2)
La	-O(7)	2.65±3	(×2)

Environment of Sr

Sr	-O(9)	2.51±3	(×2)
Sr	-O(2)	2.58±3	(×2)
Sr	-O(4)	2.60±3	(×2)
Sr	-O(3)	2.72±3	(×2)

Environment of Ca

Ca	-O(2)	1.92±3	(×2)
Ca	-O(6)	1.93±3	(×2)

Angles in degrees subtended at Ca by

	O(2)	O(6)'
O(2)	107±1	113±1
O(2)'		111±1
O(6)		101±1

Environment of Si(1)

Si(1)-O(3)	1.58±3
Si(1)-O(2)	1.60±3
Si(1)-O(4)	1.60±4
Si(1)-O(1)	1.62±3

Angles in degrees subtended at Si(1) by

	O(2)	O(4)	O(1)
O(3)	105±2	112±2	104±2
O(2)		116±2	106±2
O(4)			113±2

Environment of Si(2)

Si(2)-O(8)	1.59±3
Si(2)-O(5)	1.61±2
Si(2)-O(6)	1.62±3
Si(2)-O(7)	1.65±3

Angles in degrees subtended at Si(2) by

	O(5)	O(6)	O(7)
O(8)	114±2	114±2	110±1
O(5)		106±2	110±6
O(6)			100±2

Environment of Si(3)

Si(3)-O(9)	1.58±3
Si(3)-O(1)	1.63±2
Si(3)-O(7)	1.64±3
Si(3)-O(3)	1.68±3

Angles in degrees subtended at Si(3) by

	O(1)	O(7)	O(3)
O(9)	112±2	114±1	115±1
O(1)		108±1	104±1
O(7)			103±2

Environment of Na(1)

Na(1)-O(4)	2.30±3 (×2)
Na(1)-O(1)	2.36±3 (×2)
Na(1)-O(3)	2.49±3 (×2)

Environment of Na(2)

Na(2)-O(2)	2.36±4
Na(2)-O(8)	2.41±4
Na(2)-O(4)	2.41±4
Na(2)-O(6)	2.52±4
Na(2)-O(1)	2.63±4
Na(2)-O(5)	2.69±5

Discussion

The structure of nordite was found to be a derivative of the structures of melilite and gadolinite.

Details of analysis

Metamict material from Lovozerski Massif (Kola Pennisula) with chemical composition $(La_{2.1}Ce_{1.7}Ca_{0.2})(Sr_{2.1}Ca_{1.9})Na_8(Na_{3.2}Mn_{0.8})(Zn_{1.5}Mg_{1.3}Fe^{2+}_{0.7}Mn_{0.5})Si_8Si_8(Si_{7.4}Fe_{0.6})O_{68}$ was annealed at 800°C for 10 hours. Intensities were visually estimated from three-dimensional Weissenberg data and the structure was refined to an *R* factor of 12.2%.

EPIDIDYMITE

I. The crystal structure of epididymite. P.D. ROBINSON and J.H. FANG, 1970. *Amer. Min.*, 55, 1541-1549.

$HNaBeSi_3O_8$

Orthorhombic, a = 12.74±1, b = 13.63±1, c = 7.33±1 Å, U = 1273 Å^3, Z = 8, D_x = 2.55.

Space group Pnma (D_{2h}^{16}) Statistical analysis of intensities

Atomic positions

		x	*y*	*z*	$B(\text{Å}^2)$
8 Na	in 8(*d*)	0.1030±3	0.0698±3	0.5014±6	2.2±1
8 Be	in 8(*d*)	0.4951±8	0.0002±9	0.3433±16	0.5±2
8 Si(1)	in 8(*d*)	0.1589±2	0.1380±2	0.0155±3	0.3±1
8 Si(2)	in 8(*d*)	0.3429±2	0.1366±2	0.7314±3	0.3±1
8 Si(3)	in 8(*d*)	0.3317±2	0.1378±2	0.3054±3	0.4±1
4 O(1)	in 4(*c*)	0.1207±7	0.25	0.0005±11	1.1±2
8 O(2)	in 8(*d*)	0.2174±4	0.1234±4	0.2098±8	0.9±1
8 O(3)	in 8(*d*)	0.0603±4	0.0664±5	0.0067±7	0.7±1
8 O(4)	in 8(*d*)	0.2373±4	0.1142±4	0.8498±8	1.0±1
8 O(5)	in 8(*d*)	0.4345±4	0.0645±4	0.7907±8	0.8±1
4 O(6)	in 4(*c*)	0.3784±6	0.25	0.7656±11	0.8±1
8 O(7)	in 8(*d*)	0.3048±5	0.1239±4	0.5204±8	0.7±1
4 O(8)	in 4(*c*)	0.3709±7	0.25	0.2674±11	0.8±1
8 O(9)	in 8(*d*)	0.4151±4	0.0629±5	0.2256±8	0.9±1
4 O(10)	in 4(*c*)	0.0577±8	0.25	0.4890±14	2.3±2

Interatomic distances (Å) *and angles*

Be - O(9)	1.586±13
Be - O(5)	1.596±13
Be - O(3)	1.647±13
Be - O(3)'	1.661±13

Angles in degrees subtended at Be by

	O(5)	O(3)	O(3)'
O(9)	108.9±7	113.1±8	114.4±7
O(5)		115.4±7	112.4±8
O(3)			91.9±6

Si(1)-O(3)	1.593±6
Si(1)-O(4)	1.606±6
Si(1)-O(1)	1.606±4
Si(1)-O(2)	1.620±6

Angles in degrees subtended at Si(1) by

	O(4)	O(1)	O(2)
O(3)	109.7±3	109.9±4	108.9±3
O(4)		109.2±4	110.7±3
O(1)			108.5±4

Si(2)-O(5)	1.587±6
Si(2)-O(6)	1.630±4
Si(2)-O(7)	1.630±6
Si(2)-O(4)	1.631±6

Angles in degrees subtended at Si(2) by

	O(6)	O(7)	O(4)
O(5)	109.9±4	114.4±3	110.2±3
O(6)		109.2±4	109.0±4
O(7)			103.8±3

Si(3)-O(9)	1.586±6
Si(3)-O(7)	1.624±6
Si(3)-O(2)	1.629±6
Si(3)-O(8)	1.633±4

Angles in degrees subtended at Si(3) by

	O(7)	O(2)	O(8)
O(9)	115.1±3	111.3±3	109.6±4
O(7)		102.4±3	109.8±4
O(2)			108.3±4

Na - O(5)	2.442±7
Na - O(9)	2.455±8
Na - O(10)	2.525±5
Na - O(5)'	2.635±7
Na - O(7)	2.680±7
Na - O(2)	2.689±7
Na - O(9)'	2.918±7

Possible hydrogen bonds

O(10)-H...O(6)	2.91
O(10)-H...O(8)	3.03
O(6)-O(10)-O(8)	76°

Discussion

The structure differs from that proposed by 1. It consists of double chains of corner sharing SiO_4 tetrahedra of composition $(Si_6O_{15})_\infty$ running along *c*. The chains are joined by corner sharing with Be_2O_6 groups formed by two edge sharing BeO_4 tetrahedra. The structure is completed by the Na atom and water molecule.

Details of analysis

Crystals from Navsaruk, Greenland obtained from U.S. National Museum (Smithsonian R84350). Lattice parameters from precession photographs and diffractometer. Intensities measured on diffractometer with CuK_α radiation (crystal size 0.25 mm). Structure solved from Sayre's equation and difference electron-density function and refined by least squares to give $R = 0.084$ for 738 reflexions listed in 2. O(10) was identified as a water atom using the valence analysis of 3.

1. *Structure Reports*, 24, 493.
2. NAPS 01111.
3. G. DONNAY and R. ALLMANN, 1970. *Amer. Min.*, 55, 1003.

PHENAXITE

I. [The crystal structure of phenaxite $FeNaK(Si_4O_{10})$ $(KNaFe[Si_4O_{10}])$. V.P. GOLOVAČEV, JU.N. DROZDOV, E.A. KUZ'MIN and N.V. BELOV, 1970. *Dokl. Akad. Nauk SSSR*, 194, 818-820 [*Soviet Physics-Doklady*, 15, 902-904].

$FeNaKSi_4O_{10}$ F.W. = 390.28

Triclinic, $a = 6.98$, $b = 8.24$, $c = 9.98$ Å, $\alpha = 114.20$, $\beta = 80.22$, $\gamma = 115.60°$, $U = 472$ Å^3, $D_m = 2.74$, $Z = 2$, $D_x = 2.74$.

Space group $P\bar{1}$ (C_i^1)

Atomic positions

	x	*y*	*z*
K	0.140	0.809	0.010
Na	0.525	0.270	0.410
Fe	0.049	0.299	0.407
Si(1)	0.669	0.913	0.213
Si(2)	0.230	0.631	0.270
Si(3)	0.380	0.304	0.131
Si(4)	0.803	0.632	0.273
O(1)	0.630	0.856	0.041
O(2)	0.294	0.457	0.133
O(3)	0.632	0.424	0.167
O(4)	0.431	0.838	0.286
O(5)	0.791	0.798	0.230
O(6)	0.790	0.139	0.280
O(7)	0.033	0.629	0.205
O(8)	0.175	0.589	0.418
O(9)	0.760	0.655	0.436
O(10)	0.251	0.192	0.237

Interatomic distances (Å)

Si(1) - O(6)	1.56	K - O(7)	2.73
Si(1) - O(5)	1.59	K - O(1)	2.74
Si(1) - O(1)	1.62	K - O(3)	2.87
Si(1) - O(4)	1.66	K - O(10)	2.88
Si(2) - O(8)	1.60	K - O(5)	2.99
Si(2) - O(7)	1.61	K - O(6)	3.03
Si(2) - O(4)	1.64	K - O(2)	3.03
Si(2) - O(2)	1.69	K - O(7)	3.14
Si(3) - O(10)	1.59	K - O(1)	3.33
Si(3) - O(2)	1.61	Na - O(9)	2.44
Si(3) - O(3)	1.62	Na - O(6)	2.46
Si(3) - O(1)	1.67	Na - O(8)	2.48
Si(4) - O(9)	1.55	Na - O(10)	2.49
Si(4) - O(5)	1.63	Na - O(9)	2.79
Si(4) - O(3)	1.63	Fe - O(6)	1.99
Si(4) - O(7)	1.64	Fe - O(9)	2.04
		Fe - O(8)	2.12
		Fe - O(10)	2.16
		Fe - O(8)	2.31

[Errors in distances are uncertain but are around 0.05 Å]

Discussion

The structure contains double chains running along *a*; each chain is composed of linked Si_4O_{12} rings.

Details of analysis

Lattice parameters refined by high-angle measurements from Weissenberg photographs with CuK_α radiation. Intensities measured visually from Weissenberg photographs of layers with h = 0-7 and k = 0 to 5 (Mo radiation). Corrections made for spot shape but not absorption (crystal size 1×0.2×0.2 mm^3). Structure solved by Patterson and electron-density methods and refined by least squares to give R = 0.15. [No intensities given].

HOWLITE

I. Crystal structure of the calcium silicoborate, howlite. J.J. FINNEY, I. KUMBASAR, J.A. KONNERT and J.R. CLARK, 1970. *Amer. Min.*, 55, 716-728.

$Ca_2SiB_5O_9(OH)_5$ F.W. = 391.3

Monoclinic, a = 12.78±3, b = 9.33±2, c = 8.60±2 Å, β = 104.83±10°, U = 991.28 Å^3, D_m = 2.50, Z = 4, D_x = 2.62.

Space group $P2_1/c$ (C^5_{2h})

Atomic positions

	x	y	z	B(Å^2)
Ca(1)	0.1171±5	0.1838±8	0.1493±8	0.8±1
Ca(2)	0.4118±5	0.8216±8	0.5228±8	0.7±1
Si	0.1411±8	0.5549±11	0.0115±11	0.7±2
B(1)	0.039±3	0.471±5	0.257±4	1.2±6
B(2)	0.323±3	0.384±5	0.033±4	1.0±6
B(3)	0.492±3	0.423±5	0.256±5	1.4±7
B(4)	0.343±3	0.261±5	0.284±4	1.1±6
B(5)	0.157±3	0.650±5	0.476±5	1.3±7
O(1)	0.081±2	0.512±3	-0.174±3	0.9±4
O(2)	0.127±2	0.723±3	0.047±3	1.4±5
O(3)	0.102±2	0.450±2	0.137±3	0.8±4
O(4)	0.269±2	0.525±3	0.030±3	0.9±4

	x	*y*	*z*	*B*(Å^2)
O(5)	0.061±2	0.597±3	0.355±3	0.5±4
O(6)	0.438±2	0.407±3	0.078±3	1.1±4
O(7)	0.292±2	0.171±2	0.364±3	0.8±4
O(8)	0.287±2	0.282±3	0.130±3	1.0±4
O(9)	0.438±2	0.333±2	0.352±2	0.8±4
OH(10)	0.497±2	0.570±3	0.295±3	1.1±4
OH(11)	0.601±2	0.356±3	0.272±3	0.9±4
OH(12)	0.246±2	0.685±3	0.398±3	1.0±4
OH(13)	0.197±2	0.542±3	0.599±3	0.8±4
OH(14)	0.055±2	0.338±2	0.351±2	0.5±4

Interatomic distances (Å) *and angles*

Environment of Ca(1)

Ca(1) - O(8)	2.41±3
Ca(1) - O(5)	2.41±3
Ca(1) -OH(13)	2.43±3
Ca(1) - O(3)	2.49±2
Ca(1) -OH(14)	2.49±2
Ca(1) - O(1)	2.49±3
Ca(1) - O(7)	2.52±2
Ca(1) -OH(14)	2.53±2

Environment of Ca(2)

Ca(2) - O(4)	2.34±3
Ca(2) - O(9)	2.42±2
Ca(2) - O(6)	2.44±3
Ca(2) -OH(11)	2.46±3
Ca(2) -OH(12)	2.47±2
Ca(2) -OH(11)	2.52±3
Ca(2) -OH(10)	2.52±2
Ca(2) - O(6)	2.58±3

Environment of Si

Si - O(2)	1.62±3
Si - O(3)	1.63±3
Si - O(4)	1.63±3
Si - O(1)	1.63±2

Angles subtended at Si' by

	O(3)	O(4)	O(1)
O(2)	113±1	108±1	112±1
O(3)		108±1	111±1
O(4)			105±1

Environment of B(1)

B(1) - O(5)	1.43±5
B(1) - OH(14)	1.47±4
B(1) - O(3)	1.48±5
B(1) - O(1)	1.53±4

Angles in degrees subtended at B(1) by

	OH(14)	O(3)	O(1)
O(5)	113±2	117±3	102±3
OH(14)		104±3	109±3
O(3)			111±2

Environment of B(2)

B(2) - O(8)	1.42±5
B(2) - O(6)	1.44±4
B(2) - O(4)	1.48±5
B(2) - O(7)	1.50±4

Angles in degrees subtended at B(2) by

	O(6)	O(4)	O(7)
O(8)	114±3	113±3	107±3
O(6)		108±3	108±3
O(4)			106±2

Environment of B(3)

B(3) -OH(10)	1.41±5
B(3) - O(9)	1.47±5
B(3) -OH(11)	1.50±5
B(3) - O(6)	1.52±5

Angles in degrees subtended at B(3) by

	O(9)	OH(11)	O(6)
OH(10)	115±3	113±3	108±3
O(9)		106±3	111±3
OH(11)			103±3

Environment of B(4)

B(4) - O(8)	1.35±4
B(4) - O(7)	1.36±5
B(4) - O(9)	1.38±4

Angles in degrees subtended at B(4) by

	O(7)	O(9)
O(8)	113±3	122±3
O(7)		125±3

Environment of B(5)

B(5) - O(2)	1.43±5
B(5) -OH(13)	1.46±5
B(5) - O(5)	1.48±4
B(5) -OH(12)	1.50±5

Angles in degrees subtended at B(5) by

	OH(13)	O(5)	OH(12)
O(2)	110±3	108±3	109±3
OH(13)		111±3	108±3
O(5)			110±3

Hydrogen bonds

OH(14)-O(2)	2.90	OH(11)-OH(13)	2.70

Discussion

The structure results from the linkage of three major units: colemanite like chains, silicoborate spirals and single and double chains of Ca coordinated polyhedra. Of the three hydroxyl ions only three appear to be involved in hydrogen bonding on the basis of bond length and charge balance criteria.

Details of analysis

From the Sterling Borax Mine, Tick Canyon, California. Intensity data visually estimated from Weissenberg photographs were used in the three-dimensional refinement of the structure. The structure was solved using the heavy-atom method and refined to an *R* factor of 10%.

POTASSIUM TRITHIOCARBONATE MONOHYDRATE

I. The crystal structure of potassium trithiocarbonate monohydrate, $K_2CS_3.H_2O$, E. PHILIPPOT and O. LINDQVIST, 1970. *Acta Cryst.*, B26, 877-881.

$K_2CS_3.H_2O$ F.W. = 204.42

Monoclinic, a = 6.759±5, b = 17.066±4, c = 6.418±2 Å, β = 95.42±5°, U = 737.0 Å^3, D_m = 1.82, Z = 4, D_x = 1.83.

Space group $P2_1/n$ (C^5_{2h})

Atomic coordinates

	x	y	z	B(Å^2)
K(1)	0.5867±3	0.3006±1	0.5661±2	3.50±7
K(2)	0.7872±2	0.5399±1	0.7710±3	3.21±7
S(1)	0.4921±3	0.3176±1	0.0639±3	3.26±7
S(2)	0.3855±3	0.1353±1	0.4366±3	3.08±7
S(3)	0.7033±3	0.4541±1	0.2657±3	3.51±7
O	0.7936±10	0.6433±4	0.4439±9	4.69±20
C	0.6940±10	0.3783±4	0.0913±9	2.48±18

Interatomic distances (Å) *and angles*

C-S(1)	1.709±7
C-S(2)	1.720±7
C-S(3)	1.708±6
K(1)-O	2.738±7
6 K(1)-S	3.207-3.404
K(2)-O	2.745±6
7 K(2)-S	3.206-3.590

Angles subtended at C by S are within 1σ (0.4°) of 120°

Discussion of the structure

CS_3^{2-} ion has D_{3h} symmetry within experimental error. The crystal packing is shown in Fig. 51.

Details of analysis

1396 reflexions were visually estimated from Weissenberg photographs using Cu radiation and a crystal sealed in a capillary. Absorption corrections were applied. The structure was refined by least squares, using anisotropic thermal parameters, to an *R* value of 0.067.

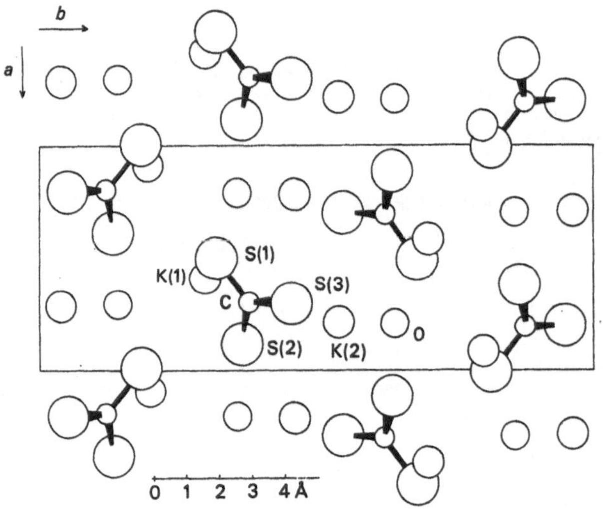

Fig. 51. The structure of $K_2CS_3.H_2O$ projected along the *c* axis.

SUBJECT INDEX

This index contains the names of substances printed at the heads of the reports, and the names of substances referred to therein. The numerical prefixes *mono* (not di, tri, bis...), *o*, *m*, *p*, D, L, etc. are disregarded in fixing the alphabetical order, although sometimes double entries are made.

FORMULA INDEX

The *symbols* in each formula are arranged alphabetically, and the formulae are then arranged alphabetically. Within groups of formulae containing the same elements the order is governed by the subscripts of the first, second, ... symbols in turn.

Solvents of crystallization are not included in the alphabetical formula, but placed separately at the end. Inorganic compounds containing water possibly or certainly as OH groups may be entered twice, once with the water alphabetized with the rest of the formula, and again with the water separated at the end.

Intermetallic compounds and alloys are also included in the index. When solid solutions are indexed, it is sometimes necessary to ignore second or third elements in parenthesis. Thus for example $(Al,Si)_2Mo$ would follow AlMo in the index and not AlSi.

AUTHOR INDEX

Modified letters [å, à, č, è, é, š, ž, *etc.*] are placed in, or immediately following the positions where the unmodified letters would occur in the normal English alphabetical order. Names beginning with Mc or Mac are collected at the beginning of the M's. Names beginning with a separated *de*, *van* or *von* are listed according to the main component, thus de Vries occurs with the V's.